# The Fundamentals of Engineering Drawing and Graphic Technology

# THE FUNDAMENTALS OF ENGINEERING DRAWING AND GRAPHIC TECHNOLOGY

**FOURTH EDITION**

**Thomas E. French**
Late Professor of Engineering Drawing
The Ohio State University

**Charles J. Vierck**
Visiting Professor to the Graphics Division of the
Department of Mechanical Engineering
University of Florida

**McGraw-Hill Book Company**

New York • St. Louis • San Francisco • Auckland • Bogotá • Düsseldorf • Johannesburg
London • Madrid • Mexico • Montreal • New Delhi • Panama • Paris • São Paulo
Singapore • Sydney • Tokyo • Toronto

## The Fundamentals of Engineering Drawing and Graphic Technology

1 2 3 4 5 6 7 8 9 0   DODO   7 8 3 2 1 0 9 8

This book was set in News Gothic by York Graphic
Services, Inc. The editors were Michael Elia and
J. W. Maisel; the designer was Wladislaw Finne;
the production supervisor was John F. Harte.
R. R. Donnelley & Sons Company
was printer and binder.


**Library of Congress Cataloging in Publication Data**

French, Thomas Ewing,       date
    The fundamentals of engineering drawing and
graphic technology.

    First-2d ed. published under title: Fundamentals of
engineering drawing.
    Bibliography:   p.
    Includes index.
    1.   Mechanical drawing.   I.   Vierck, Charles J.,
joint author.   II.   Title.
T353.F837   1978       604'.2'4       78-1093
ISBN 0-07-022171-5

# CONTENTS

Preface                                                                ix

## 1. INTRODUCTION                                                      3
This chapter outlines the graphic language in theory and practice

## 2. GRAPHIC INSTRUMENTS AND THEIR USE                                 29
Accurate representation of shape, relationship, and size is
accomplished through the use of instruments

## 3. GRAPHIC GEOMETRY                                                  71
Accurate, concise representation requires a knowledge of
geometric constructions

## 4. LETTERING                                                         113
Complete graphic depiction and specification requires word
supplements

## 5. ORTHOGRAPHIC DRAWING AND SKETCHING                                131
Orthographic drawing and sketching is the basic graphic form of
representation for design and construction drawings.

## 6. AUXILIARY VIEWS                                                   203
Auxiliary views are special projections used to clarify and complete
orthographic descriptions of shape

## 7. SECTIONAL VIEWS AND CONVENTIONS                                   235
These special views and practices are specific aids to complete
and accurate orthographic representation

**8. PICTORIAL DRAWING AND SKETCHING**     **265**

Pictorial methods are used either as a basic form of shape description or as a supplement to orthographic depiction

**9. METHODS USED IN MANUFACTURE**     **325**

Concise and complete specification of size requires a knowledge of manufacturing procedures

**10. DIMENSIONS, NOTES, LIMITS, AND PRECISION**     **341**

These are the elements used to describe size, which along with a description of shape comprise a complete graphic representation

**11. CHARTS, GRAPHS, AND DIAGRAMS**     **405**

These are graphic representations of data and are fundamental to all of science and engineering

**12. SURFACE INTERSECTIONS**     **427**

Geometric components of an object or assembly meet in intersections which must be shown in order to complete the graphic description

**13. DEVELOPED VIEWS**     **447**

These show a representation of the shape and size of thin material used to make an object by folding, rolling, or forming

**14. COMPUTER GRAPHICS AND NUMERICAL CONTROL**     **465**

A discussion of the computer and its importance in graphics

**15. DRAWINGS FOR ENGINEERING DESIGN AND CONSTRUCTION**     **483**

This chapter discusses professional practices employed in the making of design, detail, assembly, production, construction, and other drawings

**Problem Drawings in Metric Dimensions**                536

**Glossary**                                              A1

**Appendix A**   **The Hand Calculator**                 A9

**Appendix B**   **Mathematical Tables**                 A22

**Appendix C**   **Standard Parts, Sizes, Symbols,
                 and Abbreviations**                      A47

**Appendix D**   **Supply Catalogues and
                 Technical Manuals**                      A105

**Bibliography of Allied Subjects**                       A130

Index                                                     1

# PREFACE

By any known method of appraisal it is evident that the last seven or eight decades have produced more changes, principally scientific, than any other period in history. For example, consider the development of the automobile, the airplane, radio (AM and FM), television, advancements in optics and chemistry, the telephone, audio equipment and techniques, motion picture excellence in both black-and-white and color, new methods of reproduction and copying, the computer, the portable hand calculator, rocketry and space science including human travel to the moon and the recent placing of scientific instruments for measuring physical phenomena on Mars, and also, new procedures in the humanities, ecology, conservation, and many other areas.

Engineering drawing and graphic technology is the principal method of communication in engineering and science and, as such, vitally concerns itself with all new developments.

The graphics of engineering design and construction may very well be the most important course of all studies for an engineering or technical career. The indisputable reason why graphics is so extremely important is that it is the *language of the designer, technician, and engineer,* used to communicate designs and construction details to others. An engineer, no matter how knowledgeable concerning the highly complex technical and scientific aspects of the profession, would be completely ineffectual without a command of graphics, simply because all efforts to transmit designs to others would fail miserably. All the technical people working under the direction of an engineer must also have the same command of the language. The language of graphics is written in the form of drawings which represent the shape, size, and specifications of physical objects. The language is read by interpreting drawings so that physical objects can be constructed exactly as originally conceived by the designer.

Recently two factors have appeared which are of important interest to communication in design and construction, (a) the possible adoption of the metric system in the United States and (b) the development of computers and calculators to facilitate problem solving. The metric system, because of its fundamental 10x construction, greatly facilitates computation. Therefore the emphasis and coverage of this text has been altered to conform. A discussion of the metric system is given in Chapter 1. Many problems have been dimensioned in the metric system. Also, descriptions and a discussion of computer methods are given in a new chapter. Hand calculators are described in the Appendix.

With the adoption of the metric system, standards and specifications must, of course, be changed. These changes are presently being studied, and some are in the process of adoption. However, it will be some time before complete standards are available.

This textbook represents many years of study not only in teaching, engineering experience, and writing, but also in painstaking attention to book design, principles, and usage. A carefully considered plan to make this book more readable and readily usable has been employed.

A book of this scope and completeness requires cooperation between the author and all other persons involved in bringing the work to fruition. The frank, honest, and sometimes extensive discussion with associates is highly valued, along with the collaboration of Professor Richard I. Hang and Elizabeth M. Vierck.

**Charles J. Vierck**

# INTRODUCTION

# 1

This Chapter Outlines the Graphic Language in Theory and Practice

This Chapter Outlines the Graphic Language Basic Concepts—Writing and Reading • Essentials—Lines and Lettering • Methods of Expression • Methods of Shape • Description— Orthographic and Pictorial • Auxiliaries, Sections, Intersections, Developments • Methods of Size Description • Measurement and Evaluation • Metric and English Systems • Charts and Graphs • The Fundamentals of Design • Basic Machine Elements—Screw Threads, Fasteners, Keys, Springs • Welding and Riveting • Jigs, Fixtures, Gears, Cams • Piping; Drawings of Electrical Systems, Structures, Maps and Topography • Drawings for Engineering Design and Construction

## 1. The Graphic Language— Theory and Practice.

In beginning the study of graphics, you are embarking upon a rewarding educational experience and one that will be of real value in your future career. When you have become proficient in it, you will have at your command a method of communication used in all branches of technical industry, a language unequaled for accurate description of physical objects.

The importance of this graphic language can be seen by comparing it with word languages. All who attend elementary and high school study the language of their country and learn to read, write, and speak it with some degree of skill. In high school and college most students study a foreign language. These word languages are highly developed systems of communication. Nevertheless, any word language is inadequate for describing the size, shape, and relationship of physical objects. Study the photograph in Fig. 1 and try to describe it verbally so that someone who has not seen it can form an accurate and complete mental picture. It is almost impossible to do this. Furthermore, in trying to describe the picture, you may want to use pencil and paper to sketch all or a part in an attempt to make the word description more complete, meaningful, and accurate, or tend to use your hands, gesturing to aid in explaining shape and relationship. From this we can see that a word language is often without resources for accurate and rapid communication of shape and size and the relationships of components.

Engineering is applied science, and communication of physical facts *must* be complete and accurate. Quantitative relationships are expressed mathematically. The written word completes many descriptions. But whenever machines and structures are designed, described, and built, graphic representation is necessary. Although the works of artists (or photography and other methods of reproduction) would provide pictorial representation, they cannot serve as engineering descriptions. Shaded pictorial drawings and photographs are used for special purposes, but the great bulk of engineering drawings are made in line only, with separate views arranged in a logical system of projection. To these views, dimensions and special notes giving operations and other directions for manufacture are added. This is the language of graphics which can be defined as *the graphic representation of physical objects and relationships.*

It is accomplished by describing *shape* and *size.*

As the foundation upon which all designing and subsequent manufacture are based, engineering graphics is one of the most important single branches of study in a technical school. Every engineering student must know how to make and how to read drawings. The subject is essential in all types of engineering practice, and should be understood by all connected with, or interested in, technical industry. All designs and directions for manufacture are prepared by draftsmen, professional writers of the language, but even one who may never make drawings must be able to read and understand them or be professionally illiterate. Thorough training in engineering graphics is particularly important for the engineer because he is responsible for and specifies the drawings required in his work and must therefore be able to interpret every detail for correctness and completeness.

Our object is to study the language of graphics

**FIG. 1.** Try to describe in words the shape, the relative size, and the position of the objects in this picture.

so that we can write it, expressing ourselves clearly to one familiar with it, and read it readily when written by another. To do this, we must know its basic theory and composition, and be familiar with its accepted conventions and abbreviations. Since its principles are essentially the same throughout the world, a person who has been trained in the practices of one nation can readily adapt himself to the practices of another.

This language is entirely graphic and written, and is interpreted by acquiring a visual knowledge of the object represented. A student's success with it will be indicated not alone by his skill in execution, but also by his ability to interpret lines and symbols and to visualize clearly in space.

In the remainder of this chapter we shall introduce briefly the various aspects of graphics that will be discussed at length later. It is hoped that this preview will serve as a broad perspective against which the student will see each topic, as it is studied, in relation to the whole. Since our subject is a graphic language, illustrations are helpful in presenting even this introductory material; figures are used both to clarify the text and to carry the presentation forward.

## 2. Essentials of Graphic Writing: Lines and Lettering.

Drawings are made up of lines that represent the surfaces, edges, and contours of objects. Symbols, dimensional sizes, and word notes are added to these lines, collectively making a complete description. Proficiency in the methods of drawing straight lines, circles, and curves, either freehand or with instruments, and the ability to letter word statements are fundamental to writing the graphic language. Furthermore, lines are connected according to the geometry of the object represented, making it necessary to know the geometry of plane and solid figures and to understand how to combine circles, straight lines, and curves to represent separate views of many geometric combinations.

## 3. Methods of Expression.

There are two fundamental methods of writing the graphic language: freehand and with instruments.

**THE USE OF GRAPHIC INSTRUMENTS.** Facility in the use of instruments makes for speed and accuracy.

**LETTERING.** The standard lettering for engineering drawings in known as "commercial Gothic." Both vertical and inclined styles are used.

**GRAPHIC GEOMETRY.** A knowledge of, and facility in, the construction of lines and geometric figures promotes efficiency.

Freehand drawing is done by sketching the lines with no instruments other than pencils and erasers. It is an excellent method during the learning process because of its speed and because at this stage the study of projection is more important than exactness of delineation. Freehand drawings are much used commercially for preliminary designing and for some finished work. Instrument drawing is the standard method of expression. Most drawings are made "to scale," with instruments used to draw straight lines,

**FREEHAND DRAWINGS.** The freehand method is fine for early study because it provides training in technique, form, and proportion. It is used commercially for economy.

**INSTRUMENT DRAWINGS.** Because of the necessity of drawing "to scale," most drawings are made with instruments.

circles, and curves concisely and accurately. Training in both freehand and instrument work is necessary for the engineer so that he will develop competence in writing the graphic language and the ability to judge work done under his direction.

## 4. Methods of Shape Description.

Delineation of the *shape* of a part, assembly, or structure is the primary element of graphic communication. Since there are many purposes for which drawings are made, the engineer must select, from the different methods of describing shape, the one best suited to the situation at hand. Shape is described by projection, that is, by the process of causing an image to be formed by rays of sight taken in a particular direction from an object to a picture plane.

Following projective theory, two methods of *representation* are used: orthographic views and pictorial views.

For the great bulk of engineering work, the orthographic system is used, and this method, with its variations and the necessary symbols and abbreviations, constitutes an important part of this book. In the orthographic system, separate views arranged according to the projective theory are made to show clearly all details of the object represented. The figures that follow illustrate the fundamental types of orthographic drawings and orthographic views.

"Pictorial representation" designates the methods of projection resulting in a view that shows the object approximately as it would be seen by the eye. Pictorial representation is often used for presentation drawings; text, operation, and maintenance book illustrations; and some working drawings.

There are three main divisions of pictorial projection: axonometric, oblique, and perspective. Theoretically, axonometric projection is projection in which only one plane is used, the object being turned so that three faces show. The main axonometric positions are isometric, dimetric, and trimetric.

Oblique projection is a pictorial method used principally for objects with circular or curved features only on one face or on parallel faces; and for such objects the oblique is easy to draw and dimension.

Perspective projection gives a result identical with what the eye or a single-lens camera would record.

## 5. Special Practices.

These are used to simplify and clarify.

**ONE-VIEW DRAWINGS.** These are used whenever views in more than one direction are unnecessary, for example, for parts made of thin material.

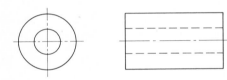

**TWO-VIEW DRAWINGS.** Parts such as cylinders require only two views. More would duplicate the two already drawn.

**THREE-VIEW DRAWINGS.** Most objects are made up of combined geometric solids. Three views are required to represent their shape.

**ISOMETRIC DRAWING.** This method is based on turning the object so that three mutually perpendicular edges are equally foreshortened.

**DIMETRIC DRAWING.** This method is based on turning the object so that two mutually perpendicular edges are equally foreshortened.

**TRIMETRIC DRAWING.** This method is based on turning the object so that three mutually perpendicular edges are all unequally foreshortened.

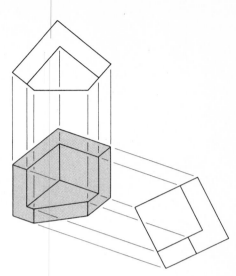

**AXONOMETRIC PROJECTION FROM OR-THOGRAPHIC VIEWS.** Pictorials—sometric, dimetric, or trimetric—may be obtained by projection from orthographic views. The views are located by a geometric method.

**OBLIQUE DRAWING.** This pictorial method is useful for portraying cylindrical parts. Projectors are oblique to the picture plane. Cavalier and cabinet drawings are specific forms.

**OBLIQUE PROJECTION FROM ORTHOGRA-PHIC VIEWS.** In this pictorial method views are arranged so that, by projection, an oblique drawing results.

**PERSPECTIVE.** In this method of pictorial projection angular and parallel forms are both used extensively.

(a)

(b)

**AUXILIARIES: EDGE AND NORMAL VIEWS.** Auxiliaries are used to show the normal view (true size and shape) of (a) an inclined surface (at an angle to two of the planes of projection) or (b) of a skew surface (inclined to all three planes of projection).

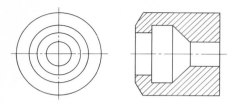

**SECTIONAL VIEWS.** These are used to clarify the representation of objects with complicated internal detail.

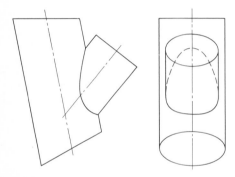

**INTERSECTIONS.** Geometric surfaces or solids often are combined so that additional projection is needed to determine the line of intersection between the parts.

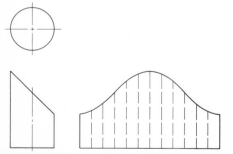

**DEVELOPMENTS.** These are the projections of geometric surfaces into a flat pattern.

## 6. Methods of Manufacture.

To delineate a drawing properly and give accurate size description and specifications for manufacture, a knowledge of processing methods is essential.

## 7. Methods of Size Description.

After delineation of shape, *size* is the second element of graphic communication, completing the representation of the object. Size is shown by "dimensions," which state linear distances, diameters, radii, and other necessary magnitudes.

The dimensions put on the drawing are *not necessarily* those used in making the drawing but are those required for the proper functioning of the part after assembly, selected so as to be readily usable by the workers who are to make the piece. Before dimensioning the drawing, study the machine and understand its functional requirements; then put yourself in the place of the patternmaker, diemaker, machinist, etc., and mentally construct the object to discover which dimensions would best give the information.

## 8. Methods of Measurement and Evaluation.

There is much more to the description of size than learning the standards and rules for the placement of dimensions on drawings.

First of all, the system of measurement and evaluation must be known and understood. History records many different evaluation systems. These are well known and documented. Emerging from all recorded methods two systems have withstood the test of time—the English system (now usually considered the usual system in the United States) and the metric system. The metric system has been adopted and used by almost all countries except the United States, but recently much pressure has been put upon our government to adopt the metric system as the standard. This has come about because of economy and also because the metric system is far superior from the standpoints of accuracy and usability. Further, because the metric system is based on multiples of ten, it lends itself easily to computer (Chap. 20) and calculator (Appendix) calculations, in addition to eliminating difficult equivalent computations.

The metric system is really much easier to learn and use, but after using the usual American system for a long period of time, new values and conceptions must be learned.

The modernized metric system is known as the *SI system* and is approved by the National Bureau of Standards.

Approvals and definitions given by the National Bureau of Standards (NBS Special Publication 330, 1972 edition) are as follows.

### I.2 The three classes of SI units
SI units are divided into three classes:

> base units,
> derived units,
> supplementary units.

From the scientific point of view division of SI units into these three classes is to a certain extent arbitrary, because it is not essential to the physics of the subject.

Nevertheless the General Conference,[1] considering the advantages of a single, practical, worldwide system for international relations, for teaching and for scientific work, decided to base the International System on a choice of seven well-defined units which by convention are regarded as dimensionally independent: the metre, the kilogram, the second, the ampere, the kelvin, the mole, and the candela (see II.1). These SI units are called *base units.*[2]

The second class of SI units contains *derived units*, i.e., units that can be formed by combining base units according to the algebraic relations linking the corresponding quantities. Several of these algebraic expressions in terms of base units can be replaced by special names and symbols which can themselves be used to form other derived units (see II.2).

Although it might be thought that SI units can only be base units or derived units, the 11th CGPM (1960) admitted a third class of SI units, called *supplementary units*, for which it declined to state whether they were base units or derived units (see II.3).

The SI units of these three classes form a coherent set in the sense normally attributed to the expression "coherent system of units".

The decimal multiples and sub-multiples of SI units formed by means of SI prefixes must be given their full name *multiples and sub-multiples of SI units* when it is desired to make a distinction between them and the coherent set of SI units.

## II. SI UNITS

### II.1 Base units

#### 1. Definitions
*a) Unit of length.*—The 11th CGPM (1960) replaced the definition of the metre based on the

[1] General Conference of Weights and Measures, hereinafter called CGPM.
[2] *Translators' note.* The spellings "metre" and "kilogram" are used in this USA/UK translation in the hope of securing worldwide uniformity in the English spelling of the names of the units of the International System.

international prototype of platinum-iridium, in force since 1889 and amplified in 1927, by the following definition:

*The metre is the length equal to* 1 650 763.73 *wavelengths in vacuum of the radiation corresponding to the transition between the levels* $2p_{10}$ *and* $5d_5$ *of the krypton-86 atom.* (11th CGPM (1960), Resolution 6)

The old international prototype of the metre which was legalized by the 1st CGPM in 1889 is still kept at the International Bureau of Weights and Measures under the conditions specified in 1889.

*b) Unit of mass.*—The 1st CGPM (1889) legalized the international prototype of the kilogram and declared: *this prototype shall henceforth be considered to be the unit of mass.*

With the object of removing the ambiguity which still occurred in the common use of the word "weight", the 3rd CGPM (1901) declared: *the kilogram is the unit of mass* [and not of weight or of force]; *it is equal to the mass of the international prototype of the kilogram.*

This international prototype made of platinum-iridium is kept at the BIPM[3] under conditions specified by the 1st CGPM in 1889.

*c) Unit of time.*—Originally the unit of time, the second, was defined as the fraction 1/86 400 of the mean solar day. The exact definition of "mean solar day" was left to astronomers, but their measurements have shown that on account of irregularities in the rotation of the Earth the mean solar day does not guarantee the desired accuracy. In order to define the unit of time more precisely the 11th CGPM (1960) adopted a definition given by the International Astronomical Union which was based on the tropical year. Experimental work had however already shown that an atomic standard of time-interval, based on a transition between two energy levels of an atom or a molecule, could be realized and reproduced much more accurately. Considering that a very precise definition of the unit of time of the International System, the second, is indispensa-

ble for the needs of advanced metrology, the 13th CGPM (1967) decided to replace the definition of the second by the following:

*The second is the duration of* 9 192 631 770 *periods of the radiation corresponding to the transition between the two hyperfine levels of the ground state of the cesium-133 atom.* (13th CGPM (1967), Resolution 1)

*d) Unit of electric current.*—Electric units, called "international", for current and resistance, had been introduced by the International Electrical Congress held in Chicago in 1893, and the definitions of the "international" ampere and the "international" ohm were confirmed by the International Conference of London in 1908.

Although it was already obvious on the occasion of the 8th CGPM (1933) that there was a unanimous desire to replace those "international" units by so-called "absolute" units, the official decision to abolish them was only taken by the 9th CGPM (1948), which adopted for the unit of electric current, the ampere, the following definition:

*The ampere is that constant current which, if maintained in two straight parallel conductors of infinite length, of negligible circular cross section, and placed* 1 *metre apart in vacuum, would produce between these conductors a force equal to* $2 \times 10^{-7}$ *newton per metre of length.* (CIPM (1946), Resolution 2 approved by the 9th CGPM, 1948)

The expression "MKS unit of force" which occurs in the original text has been replaced here by "newton" adopted by the 9th CGPM (1948, Resolution 7).

*e) Unit of thermodynamic temperature.*—The definition of the unit of thermodynamic temperature was given in substance by the 10th CGPM (1954, Resolution 3) which selected the triple point of water as fundamental fixed point and assigned to it the temperature 273.16°K by definition. The 13th CGPM (1967, Resolution 3) adopted the name *kelvin* (symbol K) instead of "degree Kelvin" (symbol °K) and in its Resolution 4 defined the unit of thermodynamic temperature as follows:

*The kelvin, unit of thermodynamic temperature, is the fraction* 1/273.16 *of the thermodynamic tempera-*

[3] The International Bureau of Weights and Measures.

*ture of the triple point of water.* (13th CGPM (1967), Resolution 4)

The 13th CGPM (1967, Resolution 3) also decided that the unit kelvin and its symbol K should be used to express an interval or a difference of temperature.

*Note.*—In addition to the thermodynamic temperature (symbol $T$), expressed in kelvins, use is also made of Celsius temperature (symbol $t$) defined by the equation

$$t = T - T_0$$

where $T_0 = 273.15$ K by definition. The Celsius temperature is in general expressed in degrees Celsius (symbol °C). The unit "degree Celsius" is thus equal to the unit "kelvin" and an interval or a difference of Celsius temperature may also be expressed in degrees Celsius.

*f*) *Unit of amount of substance.*—Since the discovery of the fundamental laws of chemistry, units of amount of substance called, for instance, "gram-atom" and "gram-molecule", have been used to specify amounts of chemical elements or compounds. These units had a direct connection with "atomic weights" and "molecular weights", which were in fact relative masses. "Atomic weights" were originally referred to the atomic weight of oxygen (by general agreement taken as 16). But whereas physicists separated isotopes in the mass spectrograph and attributed the value 16 to one of the isotopes of oxygen, chemists attributed that same value to the (slightly variable) mixture of isotopes 16, 17, and 18, which was for them the naturally occurring element oxygen. Finally an agreement between the International Union of Pure and Applied Physics (IUPAP) and the International Union of Pure and Applied Chemistry (IUPAC) brought this duality to an end in 1959/60. Physicists and chemists have ever since agreed to assign the value 12 to the isotope 12 of carbon. The unified scale thus obtained gives values of "relative atomic mass".

It remained to define the unit of amount of substance by fixing the corresponding mass of carbon 12; by international agreement, this mass has been fixed at 0.012 kg, and the unit of the quantity, "amount of substance",[3] has been given the name *mole* (symbol mol).

Following proposals of IUPAP, IUPAC, and ISO, the CIPM gave in 1967, and confirmed in 1969, the following definition of the mole, adopted by the 14th CGPM (1971, Resolution 3):

*The mole is the amount of substance of a system which contains as many elementary entities as there are atoms in 0.012 kilogram of carbon 12.*

Note. *When the mole is used, the elementary entities must be specified and may be atoms, molecules, ions, electrons, other particles, or specified groups of such particles.*

Note that this definition specifies at the same time the nature of the quantity whose unit is the mole.[4]

*g*) *Unit of luminous intensity.*—The units of luminous intensity based on flame or incandescent filament standards in use in various countries were replaced in 1948 by the "new candle". This decision had been prepared by the International Commission on Illumination (CIE) and by the CIPM before 1937, and was promulgated by the CIPM at its meeting in 1946 in virtue of powers conferred on it in 1933 by the 8th CGPM. The 9th CGPM (1948) ratified the decision of the CIPM and gave a new international name, *candela* (symbol cd), to the unit of luminous intensity. The text of the definition of the candela, as amended in 1967, is as follows.

*The candela is the luminous intensity, in the perpendicular direction, of a surface of 1/600 000 square metre of a blackbody at the temperature of freezing platinum under a pressure of 101 325 newtons per square metre.* (13th CGPM (1967), Resolution 5)

---

[4]The name of this quantity, adopted by IUPAP, IUPAC, and ISO is in French "quantité de matière" and in English "amount of substance"; (the German and Russian translations are "Stoffmenge" and "количество вещества"). The French name recalls "quantitas materiae" by which in the past the quantity now called mass used to be known; we must forget this old meaning, for mass and amount of substance are entirely different quantities.

## 2. *Symbols*

The base units of the International System are collected in table 1 with their names and their symbols (10th CGPM (1954), Resolution 6; 11th CGPM (1960), Resolution 12; 13th CGPM (1967), Resolution 3; 14th CGPM (1971), Resolution 3).

The general principle governing the writing of unit symbols had already been adopted by the 9th CGPM (1958), Resolution 7, according to which:

*Roman* [upright] *type, in general lower case, is used for symbols of units; if however the symbols are derived from proper names, capital roman type is used* [for the first letter]. *These symbols are not followed by a full stop* [period].

Unit symbols do not change in the plural.

## II.2 Derived units

### 1. *Expressions*

Derived units are expressed algebraically in terms of base units by means of the mathematical symbols of multiplication and division. Several derived units have been given special names and symbols which may themselves be used to express other derived units in a simpler way than in terms of the base units.

Derived units may therefore be classified under three headings. Some of them are given in tables 2, 3, and 4.

*Note* a—The values of certain so-called dimensionless quantities, as for example refractive index, relative permeability or relative permittivity, are ex-

### Table 1. SI Base Units

| Quantity | Name | Symbol |
|---|---|---|
| length | metre | m |
| mass | kilogram | kg |
| time | second | s |
| electric current | ampere | A |
| thermodynamic temperature* | kelvin | K |
| amount of substance | mole | mol |
| luminous intensity | candela | cd |

*Celsius temperature is in general expressed in degrees Celsius (symbol °C) (see Note, p. 13).

### Table 2. Examples of SI Derived Units Expressed in Terms of Base Units

| Quantity | SI unit | |
|---|---|---|
| | Name | Symbol |
| area | square metre | $m^2$ |
| volume | cubic metre | $m^3$ |
| speed, velocity | metre per second | m/s |
| acceleration | metre per second squared | $m/s^2$ |
| wave number | 1 per metre | $m^{-1}$ |
| density, mass density | kilogram per cubic metre | $kg/m^3$ |
| concentration (of amount of substance) | mole per cubic metre | $mol/m^3$ |
| activity (radioactive) | 1 per second | $s^{-1}$ |
| specific volume | cubic metre per kilogram | $m^3/kg$ |
| luminance | candela per square metre | $cd/m^2$ |

pressed by pure numbers. In this case the corresponding SI unit is the ratio of the same two SI units and may be expressed by the number 1.

*Note* b—Although a derived unit can be expressed in several equivalent ways by using names of base units and special names of derived units, CIPM sees no objection to the preferential use of certain combinations or of certain special names in order to distinguish more easily between quantities of the same dimension; for example, the hertz is often used in preference to the reciprocal second for the frequency of a periodic phenomenon, and the newton-metre in preference to the joule for the moment of a force, although, rigorously, $1 \text{ Hz} = 1 \text{s}^{-1}$, and $1 \text{ N} \cdot \text{m} = 1 \text{ J}$.

### 2. *Recommendations*

The International Organization for Standardization (ISO) has issued additional recommendations with the aim of securing uniformity in the use of units, in particular those of the International System (see the series of Recommendations R 31 and Recommendation R 1000 of Technical Committee

## Table 3.  SI Derived Units With Special Names

| Quantity | SI unit | | | |
|---|---|---|---|---|
| | Name | Symbol | Expression in terms of other units | Expression in terms of SI base units |
| frequency | hertz | Hz | | $s^{-1}$ |
| force | newton | N | | $m \cdot kg \cdot s^{-2}$ |
| pressure | pascal | Pa | $N/m^2$ | $m^{-1} \cdot kg \cdot s^{-2}$ |
| energy, work, quantity of heat | joule | J | $N \cdot m$ | $m^2 \cdot kg \cdot s^{-2}$ |
| power, radiant flux | watt | W | $J/s$ | $m^2 \cdot kg \cdot s^{-3}$ |
| quantity of electricity, electric charge | coulomb | C | $A \cdot s$ | $s \cdot A$ |
| electric potential, potential difference, electromotive force | volt | V | $W/A$ | $m^2 \cdot kg \cdot s^{-3} \cdot A^{-1}$ |
| capacitance | farad | F | $C/V$ | $m^{-2} \cdot kg^{-1} \cdot s^4 \cdot A^2$ |
| electric resistance | ohm | $\Omega$ | $V/A$ | $m^2 \cdot kg \cdot s^{-3} \cdot A^{-2}$ |
| conductance | siemens | S | $A/V$ | $m^{-2} \cdot kg^{-1} \cdot s^3 \cdot A^2$ |
| magnetic flux | weber | Wb | $V \cdot s$ | $m^2 \cdot kg \cdot s^{-2} \cdot A^{-1}$ |
| magnetic flux density | tesla | T | $Wb/m^2$ | $kg \cdot s^{-2} \cdot A^{-1}$ |
| inductance | henry | H | $Wb/A$ | $m^2 \cdot kg \cdot s^{-2} \cdot A^{-2}$ |
| luminous flux | lumen | lm | | $cd \cdot sr$ [a] |
| illuminance | lux | lx | | $m^{-2} \cdot cd \cdot sr$ [a] |

[a] In this expression the steradian (sr) is treated as a base unit.

## Table 4.  Examples of SI Derived Units Expressed by Means of Special Names

| Quantity | SI unit | | |
|---|---|---|---|
| | Name | Symbol | Expression in terms of SI base units |
| dynamic viscosity | pascal second | Pa·s | $m^{-1} \cdot kg \cdot s^{-1}$ |
| moment of force | metre newton | N·m | $m^2 \cdot kg \cdot s^{-2}$ |
| surface tension | newton per metre | N/m | $kg \cdot s^{-2}$ |
| heat flux density, irradiance | watt per square metre | W/m² | $kg \cdot s^{-2}$ |
| heat capacity, entropy | joule per kelvin | J/K | $m^2 \cdot kg \cdot s^{-2} \cdot K^{-1}$ |
| specific heat capacity, specific entropy | joule per kilogram kelvin | J/(kg·K) | $m^2 \cdot s^{-2} \cdot K^{-1}$ |
| specific energy | joule per kilogram | J/kg | $m^2 \cdot s^{-2}$ |
| thermal conductivity | watt per metre kelvin | W/(m·K) | $m \cdot kg \cdot s^{-3} \cdot K^{-1}$ |
| energy density | joule per cubic metre | J/m³ | $m^{-1} \cdot kg \cdot s^{-2}$ |
| electric field strength | volt per metre | V/m | $m \cdot kg \cdot s^{-3} \cdot A^{-1}$ |
| electric charge density | coulomb per cubic metre | C/m³ | $m^{-3} \cdot s \cdot A$ |
| electric flux density | coulomb per square metre | C/m³ | $m^{-2} \cdot s \cdot A$ |
| permittivity | farad per metre | F/m | $m^{-3} \cdot kg^{-1} \cdot s^4 \cdot A^2$ |
| current density | ampere per square metre | A/m² | |
| magnetic field strength | ampere per metre | A/m | |
| permeability | henry per metre | H/m | $m \cdot kg \cdot s^{-2} \cdot A^{-2}$ |
| molar energy | joule per mole | J/mol | $m^2 \cdot kg \cdot s^{-2} \cdot mol^{-1}$ |
| molar entropy, molar heat capacity | joule per mole kelvin | J/(mol·K) | $m^2 \cdot kg \cdot s^{-2} \cdot K^{-1} \cdot mol^{-1}$ |

ISO/TC 12 "Quantities, units, symbols, conversion factors and conversion tables").

According to these recommendations:

*a)* The product of two or more units is preferably indicated by a dot. The dot may be dispensed with when there is no risk of confusion with another unit symbol

*for example:* N · m or N m *but not:* mN

*b)* A solidus (oblique stroke, /), a horizontal line, or negative powers may be used to express a derived unit formed from two others by division

*for example:* m/s, $\dfrac{m}{s}$ or m · s$^{-1}$

*c)* The solidus must not be repeated on the same line unless ambiguity is avoided by parentheses. In complicated cases negative powers or parentheses should be used

*for example:* m/s$^2$ or m · s$^{-2}$ *but not:* m/s/s
m · kg/(s$^3$ · A) or m · kg · s$^{-3}$ · A$^{-1}$    m · kg/s$^3$/A

## II.3 Supplementary units

The General Conference has not yet classified certain units of the International System under either base units or derived units. These SI units are assigned to the third class called "supplementary units", and may be regarded either as base units or as derived units.

For the time being this class contains only two, purely geometrical, units: the SI unit of plane angle, the *radian,* and the SI unit of solid angle, the *steradian* (11th CGPM (1960), Resolution 12).

*The radian is the plane angle between two radii of a circle which cut off on the circumference an arc equal in length to the radius.*

## Table 5.  SI Supplementary Units

| Quantity | SI unit | |
|---|---|---|
| | Name | Symbol |
| plane angle | radian | rad |
| solid angle | steradian | sr |

## Table 6.  Examples of SI Derived Units Formed by Using Supplementary Units

| Quantity | SI unit | |
|---|---|---|
| | Name | Symbol |
| angular velocity | radian per second | rad/s |
| angular acceleration | radian per second squared | rad/s$^2$ |
| radiant intensity | watt per steradian | W/sr |
| radiance | watt per square metre steradian | W·m$^{-2}$·sr$^{-1}$ |

*The steradian is the solid angle which, having its vertex in the center of a sphere, cuts off an area of the surface of the sphere equal to that of a square with sides of length equal to the radius of the sphere.*

Supplementary units may be used to form derived units. Examples are given in table 6.

# III.  DECIMAL MULTIPLES AND SUB-MULTIPLES OF SI UNITS

## III.1 SI Prefixes

The 11th CGPM (1960, Resolution 12) adopted a first series of names and symbols of prefixes to form decimal multiples and sub-multiples of SI units. Prefixes for $10^{-15}$ and $10^{-18}$ were added by the 12th CGPM (1964, Resolution 8).

## III.2 Recommendations

ISO recommends the following rules for the use of SI prefixes:

*a)* Prefix symbols are printed in roman (upright) type without spacing between the prefix symbol and the unit symbol.

*b)* An exponent attached to a symbol containing a prefix indicates that the multiple or sub-multiple of the unit is raised to the power expressed by the exponent,

*for example:* $1 \text{ cm}^3 = 10^{-6} \text{ m}^3$
$1 \text{ cm}^{-1} = 10^2 \text{ m}^{-1}$

**Table 7. SI Prefixes**

| Factor | Prefix | Symbol | Factor | Prefix | Symbol |
|---|---|---|---|---|---|
| $10^{12}$ | tera | T | $10^{-1}$ | deci | d |
| $10^{9}$ | giga | G | $10^{-2}$ | centi | c |
| $10^{6}$ | mega | M | $10^{-3}$ | milli | m |
| $10^{3}$ | kilo | k | $10^{-6}$ | micro | $\mu$ |
| $10^{2}$ | hecto | h | $10^{-9}$ | nano | n |
| $10^{1}$ | deka | da | $10^{-12}$ | pico | p |
| | | | $10^{-15}$ | femto | f |
| | | | $10^{-18}$ | atto | a |

*c*) Compound prefixes, formed by the juxtaposition of two or more SI prefixes, are not to be used.

*for example:* 1 nm   *but not:* 1 m$\mu$m

### III.3 The kilogram

Among the base units of the International System, the unit of mass is the only one whose name, for historical reasons, contains a prefix. Names of decimal multiples and sub-multiples of the unit of mass are formed by attaching prefixes to the word "gram" (CIPM (1967), Recommendation 2).

## IV. UNITS OUTSIDE THE INTERNATIONAL SYSTEM

### IV.1 Units used with the International System

The CIPM (1969) recognized that users of SI will wish to employ with it certain units not part of it, but which are important and are widely used. These units are given in table 8. The combination of units of this table with SI units to form compound units should, however, be authorized only in limited cases; in particular, the kilowatt-hour should eventually be abandoned.

Visual comparisons of the English and metric systems are given in the *graphic* comparison figures on pages 18 and 19. Comparison of mathematical values are given in the appendix.

**Table 8. Units in Use with the International System**

| Name | Symbol | Value in SI unit |
|---|---|---|
| minute | min | 1 min = 60 s |
| hour[a] | h | 1 h = 60 min = 3 600 s |
| day | d | 1 d = 24 h = 86 400 s |
| degree | ° | $1° = (\pi/180)$ rad |
| minute | ′ | $1′ = (1/60)° = (\pi/10\ 800)$ rad |
| second | ″ | $1″ = (1/60)′ = (\pi/648\ 000)$ rad |
| litre[a] | l | $1\ l = 1\ dm^3 = 10^{-3}\ m^3$ |
| tonne[a] | t | $1\ t = 10^3$ kg |

(a) The symbol of this unit is included in Resolution 7 of the 9th CGPM (1948). The litre is defined in Resolution 6 of the 12th CGPM (1964).

### 9. Charts, Graphs, and Diagrams.

These are graphic plots for the analysis of engineering data or the presentation of statistics. Special standards designate preferred construction, lines, symbols, and lettering.

Charts, graphs, and diagrams fall roughly into two classes: (1) those used for purely technical purposes and (2) those used in advertising or in presenting information in a way that will have popular appeal. The engineer is concerned mainly with those of the first class, but he should be acquainted also with the preparation of those of the second and understand their potential influence. The aim here is to give a short study of the types of charts, graphs, and diagrams with which engineers and those in allied professions should be familiar.

## LENGTH

| U.S. | | METRIC |
|---|---|---|
| 1 inch | = 25.4 | millimetres |
| 1 foot | = 30.48 | centimetres |
| 1 yard | = 0.914 | metre |

| METRIC | | U.S. |
|---|---|---|
| 1 millimetre | = 0.039 | inch |
| 1 centimetre | = 0.394 | inch |
| 1 metre | = 39.37 | inches |

METRIC (cm, metres)

2 cm  3 cm  4 cm  6 cm 8 cm 10 cm  15 cm 20 cm  30 cm 40 cm  60 cm 80  1 m  1.5 m 2 m  3 m  4 m  60 cm 80  1 m  1.5 m 2 m  3 m  4 m  6 m  8 m 10 m

ENGLISH (in., ft., yds.)

1″  1½″  2″  3″  4″ 5″ 6″  8″ 10″ 12″  1½′  2′  3′  4′ 5′ 6′  8′ 10′  1 yd.  1½  2 yds.  3 yds. 4 yds.  6 yds. 8  10 yds.

## THICKNESS

1 mil = .025 millimetre

1 millimetre = 39.37 mils

MM

.0025 .003  .004 .005 .006  .008 .010  .015  .020 .025 .030  .04  .05 .06  .08  .10  .15  .20 .25 .30  .4  .5  .6  .8 1.0  1.5  2.0 2.5 3.0  4  5  6

MILS

.10  .15  .20 .25 .30  .4  .5 .6  .8  1.0  1.5  2.0 2.5 3.0  4  5  6  8  10  15  20 25 30  40 50 60  80 100  150  200 250

## AREA

1 sq. foot = 929.03 sq. centimetres

1 sq. centimetre = 0.155 sq. inch

CM²

400  500  600  800  1000  1500  2000 2500 3000  4000  5000 6000  8000  10000  15000  20000  25000 30000  40000  50000 60000  80000 100000

SQ. FT.

.5  .6  .7 .8 .9 1.0  1.5  2.0  2.5  3.0 3.5 4.0 4.5 5.0  6  7  8  9 10  15  20  25  30 35 40 45 50  60  70 80 90 100

## VOLUME

1 gallon = 3.785 litres

1 litre = 0.264 gallon

LITRES

.3  .4  .5  .6  .8 1.0  1.5  2.0  3  4  5  6  8  10  15  20  30  40 50 60  80  100  150  200  300  400

GALLONS

.10  .15  .20  .30  .4  .5 .6  .8 1.0  1.5  2.0  3  4  5  6  8  10  15  20  30  40 50 60  80 100

## VOLUME

1 cu. inch = 16.387 cu. centimetres

1 cu. centimetre = 0.061 cu. inch

CC

15  20  25 30  40  50 60  80  100  150  200 250 300  400 500 600  800 1000  1500  2000 2500 3000 4000 5000 6000 8000  10000  15000  20000  30000 40000

CU. IN.

1.0  1.5  2.0 2.5 3.0  4  5  6  8  10  15  20 25 30  40 50 60  80 100  150  200 250 300  400 500 600  800 1000  1500  2000

## TEMPERATURE

$$°Fahrenheit = \frac{(°Celcius) 9}{5} + 32$$

$$°Celcius = \frac{(°Fahrenheit - 32) 5}{9}$$

°C

-40  -20  0  20  40  60  80  100  120  140  160  180  200  220  240  260  280  300  320

°F

-40 -20  0  20 40 60  80 100 120 140 160 180 200 220 240 260 280 300 320 340 360 380 400 420 440 460 480 500 520 540 560 580 600

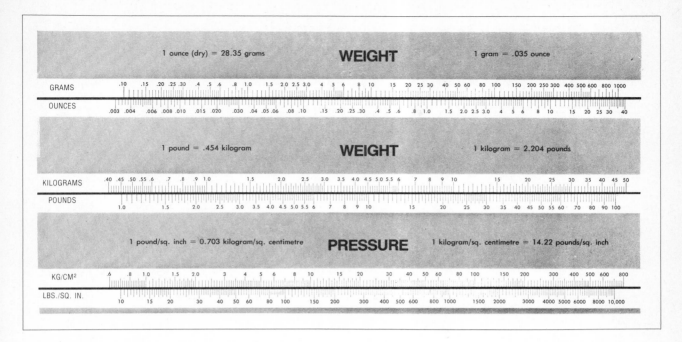

## 10. The Fundamentals of Design.

The chapter on design discusses materials, procedures, and the aspects of good mechanical design and explains how a designer thinks and works.

The word "design" has many meanings. A digest of various dictionary definitions is: to plan, conceive, invent, and to designate so as to transmit the plan to others. *Design* has many purely artistic connotations. For example, the design of fabrics, clothing, furniture, etc. In engineering, design has come to mean that broad category of invention leading to the production of useful devices.

*Design*, from the latin "designare" (to mark out), is the process of developing plans, schemes, directions, and specifications for something new. Thus it is within context to speak or Hitler's designs for world conquest; the design (conception) of a book, play, or motion picture; or the design of fabric, clothing, furniture, appliances, or toher completely physical objects. Design is distinguished from production and craftsmanship; design is the creative original plan, and the production and craftsmanship are a part of the execution of the plan.

Design means creation in the purest sense. Specifically, design does not go beyond creation, and it logically follows that the execution of a design, that is, the carrying out of the plan by presentation, action, production, manufacture, craftsmanship, and use are not design at all but are simply and positively the products of the design. Also, when examining a finished product, it is proper to speak of the *design* of it, and by this term of reference we mean the original plan or scheme and not the product itself.

## 11. Basic Machine Elements.

Many machine elements occur repeatedly in all kinds of engineering work. Familiarity with these elements is necessary so that dimensioning and specifications on the drawings will be correct. The material that follows introduces basic machine elements and shop processes, illustrating the principles of graphics laid out in the previous sections.

**DIMENSIONING ORTHOGRAPHIC DRAW-INGS.** Dimensions showing the magnitude and relative position of each portion of the object are placed on the view where each dimension is most meaningful.

**DIMENSIONING PICTORIAL DRAWINGS.** The descriptions of magnitudes and positions are shown on the pictorial by dimensions placed so as to be easily readable.

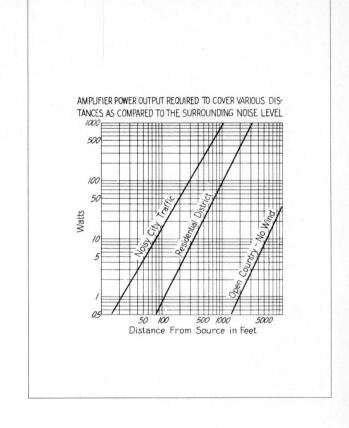

The earliest records of the screw are found in the writings of Archimedes (278 to 212 B.C.). Although specimens of ancient Greek and Roman screws are so rare as to indicate that they were seldom used, there are many from the later Middle Ages; and it is known that crude lathes and dies were used to cut threads in the latter period. Most early screws were made by hand, by forging the head, cutting the slot with a saw, and fashioning the screw with a file. In America in colonial times wood screws were blunt on the ends; the gimlet point did not appear until 1846. Iron screws were made for each threaded hole. There was no interchanging of parts, and nuts had to be tied to their own bolts. In England, Sir Joseph Whitworth made the first attempt to set up a uniform standard in 1841.

**DESIGN DRAWING.**  Made by Leonardo da Vinci about 1500.

## 12. Special Fields and Practices.

The methods of projection, the choice of representational type (shape description), and the accompanying size description are more or less uniform in all fields because these are the elements of which all drawings are composed. However, for most special fields certain symbols and notation, drawing, and dimensioning practices have been standardized, and a knowledge of these is necessary so that drawings will be readily and efficiently understood.

The special practices or fields given in this book include: various portions of mechanisms and systems; welding; electrical drawing; structural and topographic practices; and commercial practices and economies.

A "production," or "working," drawing is any drawing used to give information for the manufacture or construction of a machine or the erection of a structure. Complete knowledge for the production of a machine or structure is given by a *set* of working drawings conveying all the facts fully and explicitly so that further instructions are not required.

## 13. Drawings for Engineering Design and Construction.

There are many types of drawings used in a wide diversity of fields. Typical are drawings for presentation, proposal, and design; drawings for manufacture, construction, and assembly; drawings giving specialized information for a particular field; and many others.

When a new machine or structure is designed, the first drawings are usually in the form of freehand sketches on which the original ideas, scheming, and inventing are worked out. These drawings are accompanied or followed by calculations to prove the

**FASTENERS.** Bolts and screws in a wide variety of forms are used for fastenings. Familiarity with the standard sizes available and their drawing and specification is fundamental.

**KEYS.** Keys are used to prevent cylindrical elements such as gears and pulleys from rotation on their shafts.

suitability of the design. Working from the sketches and calculations, the design department produces a *design* drawing. This is a preliminary pencil drawing on which more details of the design are worked out. It is accurately made with instruments, full size if possible, and shows the shape and position of the various parts. Little attempt is made to show all the

SHARP V  AMERICAN NATIONAL (UNIFIED)  SQUARE  ACME

**SCREW THREADS.** Screw threads are used on fasteners, on devices for making adjustments, and for the transmission of power and motion. The screw thread is the most extensively used machine element.

Compression Spring  Torsion Spring  Extension Spring

**SPRINGS.** A spring is an elastic body that stores energy when deflected. Basic forms are illustrated.

intricate detail; only the essential dimensions, such as basic calculated sizes, are given. On the drawing, or separately as a set of written notes, will be the designer's general specifications for materials, heat-treatments, finishes, clearances, or interferences, etc., and any other information needed by the draftsman in making up the individual drawings of the separate parts.

**REPRESENTATION OF WELDS.** Welds are represented by symbols standardized by the American Welding Society (AWS) and the American Standards Association (ASA).

**RIVETS.** Rivets are permanent fasteners. They are cylinders of metal with a head on one end. When placed in position, the opposite head is formed by impact.

**JIGS AND FIXTURES.** In modern high-quantity production these devices are used for holding parts during machining. Drawings are made to conform to tool-engineering standards.

**GEARS.** Special practices include calculation of detailed sizes and their tabulation on drawings.

### 14. Terminology.

During any course of study in graphics the student must become thoroughly familiar with the terminology of design and construction. The glossary of technical, structural, architectural, and welding terms in the Appendix should be consulted as needed. Form the habit of never passing over a word without understanding its meaning completely.

### 15. Appendix Tables.

Many details of graphics depend upon specifications for drills, bolts and screws, fits, limits, pipe sizes, symbols, and other standards. It is important to become well acquainted with this material, which is given in the Appendix.

Note: All fittings 2" M.I. unless otherwise noted
All valves 2" Iron body.

**PIPING.** Piping drawings are made to scale or symbolically. Symbols for pipe, valves, fittings, and other details are standardized by the ANSI.

**DRAWINGS OF ELECTRICAL SYSTEMS.** Most electrical drawings are made symbolically, using ANSI standards.

## 16. Computer Graphics.

This is the classification given to the science of obtaining a readout of one kind or another and using the readout *mathematically* to plot information or *directly* to draw a relationship, control a machine or portion of a machine. Readouts are also used for numerical control (see Chap. 20).

**CAMS.**  Special practices include full-size layouts and also calculation and dimensioning of points on cam surfaces.

**MAPS AND TOPOGRAPHY.**  Maps and topographic drawings are made up chiefly of symbols.

**STRUCTURES.** Architectural standards of the American Institute of Architects (AIA) and structural standards of the American Institute of Steel Construction (AISC) prevail in the building industry.

**SPECIALIZED PRACTICES.** These include simplified drawing practices, special uses of reproduction processes, photography, and modelmaking. The figure above is a photo drawing.

**A DETAIL DRAWING.** This is typical of a drawing used to manufacture a single part.

# GRAPHIC INSTRUMENTS AND THEIR USE

Accurate Representation of Shape, Relationship, and Size is Accomplished through the Use of Instruments

This Chapter Discusses:
Selection of Instruments • Boards, Paper, Tape, Thumbtacks • Pencils, Erasers, Pens, Ink • T square, Triangles, Scales, Curves • Case Instruments • The Use of T Square, Triangles, Scales, Dividers, Compasses, Ruling Pens, Curves, Erasers • The Alphabet of Lines • Tangents • Problems

## 1. Graphic Instruments— Why They Are Needed.

To record information on paper (or another surface), instruments and equipment are required. Even for drawings made freehand, pencils, erasers, and sometimes coordinate paper or other special items are used.

The lines made on drawings are *straight* or *curved* (including circles and arcs). They are made with drawing instruments, which are the necessary tools for laying down lines on a drawing in an accurate and efficient manner. In order to position the lines, a measuring device, a scale, is needed.

The various instruments will be described in detail later, but the opening of this chapter will serve as an introduction. To draw straight lines, the T square, with its straight blade and perpendicular head, or a triangle is used to support the stroke of the pencil. To draw circles, a compass is needed. In addition to the compass, the draftsman needs dividers for spacing distances and a small bow compass for drawing small circles. To draw curved lines other than circles, a French curve is required. A scale is used for making measurements. To complete drawings in ink, the draftsman will require pens for inking and an ink bottle in holder.

## 2. Selection of Instruments.

In selecting instruments and materials for drawing, secure the *best* you can afford. For one who expects to do work of professional grade, it is a mistake to buy inferior instruments. Sometimes a beginner is tempted to get cheap instruments for learning, expecting to buy better ones later. With reasonable care a set of good instruments will last a lifetime, whereas poor ones will be an annoyance from the start and worthless after short use. Since poor instruments can be difficult to distinguish from good ones, it is well to seek trustworthy advice before buying. Instruments have been greatly improved in recent years.

## 3. Drawing Boards.

The drawing surface may be the table top itself or a separate board. In either case the working surface should be made of well-seasoned clear white pine

or basswood, cleated to prevent warping. The working edge must be straight and should be tested with a steel straightedge. Some boards and table tops are supplied with a hardwood edge or a steel insert on the working edge, thus ensuring a better wearing surface. Figure 1 illustrates a drawing board which, because of its design and the materials of which it is made, is rigid, sturdy, and light.

## 4. Drawing Paper.

Drawing paper is made in a variety of qualities and may be had in sheets or rolls. White drawing papers that will not turn yellow with age or exposure are used for finished drawings, maps, charts, and drawings for photographic reproduction. For pencil layouts and working drawings, cream or buff detail papers are preferred as they are easier on the eyes and do not show soil so quickly as white papers. In general, paper should have sufficient grain, or "tooth," to take the pencil, be agreeable to the eye, and have a hard surface not easily grooved by the pencil and good erasing qualities. Formerly imported papers were considered the best, but American mills are now making practically all the paper used in this country. Cheap manila papers should be avoided.

**FIG. 1.** Drawing board. This one is light and rigid. *Courtesy of Keuffel & Esser Co.*

## 5. Tracing Paper.

Tracing papers are thin papers, *natural* or *transparentized*, on which drawings are traced, in pencil or ink, and from which blueprints or similar contact prints can be made. In most drafting rooms original drawings are penciled on tracing papers, and blueprints are made directly from these drawings, a practice increasingly successful because of improvements both in papers and in printing. Tracing papers vary widely in color, thickness, surface, etc., and the grade of pencil and the technique must be adjusted to suit the paper; but with the proper combination, good prints can be obtained from such drawings.

## 6. Tracing Cloth.

Finely woven cloth coated with a special starch or plastic is used for making drawings in pencil or ink. The standard tracing cloth is used for inked tracings and specially made pencil cloth for pencil drawings or tracings. Cloth is more permanent than paper as it will stand more handling. Tracing and duplicating processes are described later.

## 7. Drafting Tape.

The paper to be used is usually attached to the drawing board by means of Scotch drafting tape, with a short piece stuck across each corner or with tape along the entire edge of the paper. Drafting tape is not the same as masking tape (made by the same company); the latter has a heavier coating of adhesive and does not come off the drawing paper so cleanly as the former.

## 8. Thumbtacks.

The best thumbtacks are made with thin heads and steel points screwed into them. Cheaper ones are made by stamping. Use tacks with tapering pins of small diameter and avoid flat-headed (often colored) map pins, as the heads are too thick and the pins rather large.

## 9. Pencils.

The basic instrument is the graphite lead pencil, made in various hardnesses. Each manufacturer has special methods of processing designed to make the lead strong and yet give a smooth clear line. Figure 2 shows five varieties of pencils. At the left are two

STANDARD
PENCILS

SEMIAUTOMATIC
PENCILS

(a)          (b)          (c)          (d)          (e)

**FIG. 2.** Drafting pencils. The semiautomatics are most convenient to use.

ordinary pencils, with the lead set in wood. (*a*) is of American and (*b*) of foreign manufacture. Both are fine but have the disadvantage that in use, the wood must be cut away to expose the lead (a time-consuming job), and the pencil becomes shorter until the last portion must be discarded. Semiautomatic pencils, (*c*) to (*e*), with a chuck to clamp and hold the lead, are more convenient. (*c*) has a plastic handle and changeable tip (for indicating the grade of lead). (*d*) has an aluminum handle and indexing tip for indication of grade. (*e*) is the rather elegant Alteneder pencil with rosewood handle, hardened-steel chuck ring, and eraser. Drawing pencils are graded by numbers and letters from 6B, very soft and black, through 5B, 4B, 3B, 2B, B, and HB to F, the medium grade; then H, 2H, 3H, 4H, 5H, 6H, 7H, and 8H to 9H, the hardest. The soft (B) grades are used primarily for sketching and rendered drawings and the hard (H) grades for instrument drawings.

## 10. Pencil Pointer.

After the wood of the ordinary pencil is cut away with a pocketknife or mechanical sharpener, the lead must be formed to a long, conic point. A lance-tooth steel file, Fig. 3A, about 6 in. long is splendid for the purpose. Some prefer the standard sandpaper pencil-pointer pad, Fig. 3B.

## 11. Erasers.

The Ruby pencil eraser, Fig. 3C, large size with beveled ends, is the standard. This eraser not only removes pencil lines effectively but is better for ink than the so-called ink eraser, as it removes ink without seriously damaging the surface of paper or cloth. A good metal erasing shield, Fig. 3D, aids in getting clean erasures.

Artgum or a *soft*-rubber eraser, Fig. 3E, is useful for cleaning paper and cloth of finger marks and smears that spoil the appearance of the completed drawing.

**FIG. 3.** Accessories. These are used for fastening paper, sharpening pencils, and erasing.

## 12. Penholders and Pens.

The penholder should have a grip of medium size, small enough to enter the mouth of a drawing-ink bottle easily yet not so small as to cramp the fingers while in use. A size slightly larger than the diameter of a pencil is good.

An assortment of pens for lettering, grading from coarse to fine, may be chosen from those listed in Chap. 4.

A pen wiper of lintless cloth or thin chamois skin should always be at hand for both lettering and ruling pens.

## 13. Drawing Ink.

Drawing ink is finely ground carbon in suspension, with natural or synthetic gum added to make the mixture waterproof. Nonwaterproof ink flows more freely but smudges easily. Drawing ink diluted with distilled water or Chinese ink in stick form rubbed up with water on a slate slab is used in making wash drawings and for very fine line work.

Bottleholders prevent the bottle from upsetting and ruining the drawing table or floor. They are made in various patterns; one is illustrated in Fig. 4. As a temporary substitute, the lower half of the paper container in which the ink is sold may be fastened to the table with a thumbtack, or a strip of paper or cloth with a hole for the neck of the bottle may be tacked down over the bottle. Several companies are now supplying ink in small-necked plastic bottles or tubes, Fig. 4, which are squeezed to supply the ink, a drop at a time. These containers prevent spillage and protect the ink from deterioration through evaporation or contamination.

## 14. The T Square.

The fixed-head T square, Fig. 5, is used for all ordinary work. It should be of hardwood, and the blade should be perfectly straight. The transparent-edged blade is much the best. A draftsman will have several fixed-head squares of different lengths and will find an adjustable-head square of occasional use.

## 15. Triangles.

Triangles, Fig. 5, are made of transparent celluloid (fiberloid) or other plastic material. Through internal strains they sometimes lose their accuracy. Triangles should be kept flat to prevent warping. For ordinary work, a 6- or 8-in. 45° and a 10-in. 30–60° are good sizes.

DROPPER BOTTLE IN BOTTLEHOLDER

PLASTIC "SQUEEZE" DROPPER TUBE

**FIG. 4.** Standard bottle in holder; Leroy pen-filler cartridge. The cartridge is particularly convenient.

T SQUARE

45° TRIANGLE

30–60° TRIANGLE

**FIG. 5.** T square and triangles. These are standard equipment for drawing straight lines.

## 16. Scales.

Scales, Figs. 6 to 9, are made in a variety of graduations to meet the requirements of many different kinds of work. For convenience, scales are classified according to their most-common uses.

*Mechanical Engineer's Scales.* These are divided and numbered so that fractions of inches represent inches. The most common ranges are ⅛, ¼, ½, and 1 in. to the *inch*. These scales are known as the *size* scales because the designated reduction also represents the ratio of size, as, for example, *one-eighth* size. A full- and half-size scale is illustrated in Fig. 6. Mechanical engineer's scales are almost always "full divided"; that is, the smallest divisions run throughout the entire length. They are often graduated with the marked divisions numbered from right to left, as well as from left to right, as shown in Fig. 6. Mechanical engineer's scales are used mostly for drawings of machine parts and small structures where the drawing size is never less than one-eighth the size of the actual object.

*Civil Engineer's Scales.* These are divided into decimals with 10, 20, 30, 40, 50, 60, and 80 divisions to the inch (Fig. 7). Such a scale is usually full divided and is sometimes numbered both from left to right and right to left. Civil engineer's scales are most used for plotting and drawing maps, although they are convenient for any work where divisions of the inch in tenths is required.

*Architect's Scales.* Divided into proportional feet and inches, these scales have divisions indicating ⅛, ¼, ⅜, ½, ¾, 1½, and 3 in. to the *foot* (Fig. 8).

**FIG. 6.** A mechanical engineer's full- and half-size scale. Divisions are in inches to sixteenths of an inch. (The half-size scale is on the back of the full-size scale.)

**FIG. 7.** A civil engineer's scale. Divisions are 10, 20, 30, 40, 50, and 60 parts to the inch.

**FIG. 8.** An architect's (or mechanical engineer's) scale. Reduction in size is based on proportion of a foot.

They are usually "open divided"; that is, the units are shown along the entire length, but only the end units are subdivided into inches and fractions. These scales are much used by all engineers—mechanical, industrial, chemical, etc.—for both machine and structural drawings and are sometimes called *mechanical engineer's* scales.

*Metric Scales.* Metric scales are supplied in all of the styles and sizes and materials of other scales. They are usually full-divided and are numbered in meters, centimeters, or millimeters depending upon the reduction in size. Typical size reductions are: 1:1, 1:2, 1:5, 1:10, 1:20, 1:25, 1:33.3, 1:50, 1:75, 1:80, 1:100, 1:150. Also available are *metric equivalent* scales for direct conversion of English scales. These are: metric to English measure, English to metric measure, and *dual* scales giving both metric and English values. The best scales are engine-divided. See Fig. 9.

Scales are made with various cross-sectional shapes, as shown in Fig. 10. The triangular form, (*a*) and (*b*), has long been favored because it carries six scales as a unit and is very stiff. However, many prefer the flat types as being easier to hold flat to a board and having a particular working scale more readily available. The "opposite-bevel" scale, (*c*) and (*d*), is easier to pick up than the "flat-bevel" scale,

(*f*); moreover, it shows only one graduation at a time. The "double-bevel" scale, (*e*), in the shorter lengths is convenient as a pocket scale, but it can be had in lengths up to 24 in.

Practically all drafting scales of good quality were formerly made of boxwood, either plain or with white edges of celluloid. Although metal scales have been available for more than 30 years, they were seldom used until about 1935, when drafting machines equipped with metal scales made rather important gains in popularity. Extruded medium-hard aluminum alloys are the preferred metals. The Second World War brought a period of considerable experimentation with various types of plastic for all kinds of drafting scales. Up to the present time the white-edge scale has retained a large measure of its popularity. Another type of scale is now getting considerable attention, a metal scale with a white plastic coating that carries the graduations. Both magnesium and aluminum have been used successfully. These scales have all the reading advantages of a white-edge boxwood scale together with the stability afforded by metal.

## 17. Curves.

Curved rulers, called "irregular curves" or "French curves," are used for curved lines other than circle

FIG. 9. Examples of metric scales.

TRIANGULAR    OPPOSITE BEVEL    DOUBLE BEVEL    FLAT BEVEL

*(a)*    *(b)*    *(c)*    *(d)*    *(e)*    *(f)*

**FIG. 10.** Scale types. Flat scales are the most effective.

arcs. The patterns for these curves are laid out in parts of ellipses and spirals or other mathematical curves in various combinations. For the student, one ellipse curve of the general shape of Fig. 11*a* or *d* and one spiral, either a logarithmic spiral, (*b*), or one similar to the one used in Fig. 61, is sufficient. (*c*) is a useful small curve.

## 18. The "Case" Instruments.

We have so far, with the exception of curves, considered only the instruments (and materials) needed for drawing straight lines. A major portion of any drawing is likely to be circles and circle arcs, and the so-called "case" instruments are used for these. The basic instruments are shown in Fig. 12. At the left is a divider of the "hairspring" type (with a screw for fine adjustment), used for laying off or transferring measurements. Next is the large compass with lengthening bar and pen attachment. The three "bow" instruments are for smaller work. They are almost always made without the conversion feature (pencil to pen). The ruling pen, at the extreme right,

is used for inking straight lines. A set of instruments of the type shown in Fig. 12 is known as a three-bow set.

Until recently the three-bow set was considered the standard, in design and number of pieces, for all ordinary drafting work. However, the trend now is for more rigid construction and fewer pieces. The 6-in. compass in Fig. 13 embraces practically a whole set in one instrument. Used as shown, the instrument is a pencil compass; with pencil replaced by pen, it is used to ink circles; with steel point installed in place of pencil point, it becomes a divider; the pen point, placed in the handle provided, makes a ruling pen; and there is a small metal container for steel points and lead. Similar instruments of other manufacture are shown in Figs. 14 and 15. The instrument in the upper right of Fig. 15 is a "quick-change" design (with vernier adjustment), a great convenience when changing from a very small setting to a large one.

Figure 14*b* shows a quick-action bow employing special nylon nuts in which the center screw oper-

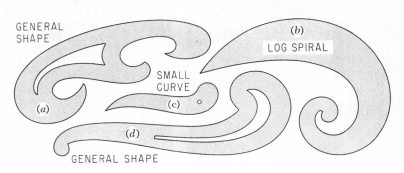

**FIG. 11.** Irregular curves. These are used for drawing curves where the radius of curvature is not constant.

DIVIDER    COMPASS, LENGTHENING BAR, PEN ATTACHMENT    BOW PEN, PENCIL AND DIVIDER    PEN

**FIG. 12.** The basic three-bow set. Large compass and dividers; bow pen, pencil, and dividers; ruling pen.

PENCIL COMPASS

PEN HANDLE

PEN LEADS, POINTS

**FIG. 13.** Large bow. It performs triple service as dividers, pen, and pencil compass. Pen leg in handle makes ruling pen.

The standard 6-in. compass will open to only approximately 5 in. The lengthening bar of the older-type compass (Fig. 12) will extend the radius to about 8 in. Lengthening bars for the newer spring-bow instruments are of the beam type. The instrument that is shown in Fig. 17 has a center point on the beam, thus employing the pencil and pen of the instrument itself, while the compass that is shown in Fig. 18 uses the center point of the instrument and separate pen and pencil attachments on the beam. A standard metal beam compass for drawing circles up to 16-in. radius is shown in Fig. 19.

Various ruling pens are shown in Fig. 20. Those shown in (a) to (e) are standard types. Note that (d) uses the compass-pen leg in a handle. (f) is a contour pen, (g) is a border pen for wide lines, (h) and (i) are "railroad" pens for double lines, and (j) and (k) are border pens with large ink capacity.

All manufacturers supply instruments made up in sets, with a leather, metal, or plastic case. Figure 21 shows a standard three-bow combination. Figure 22 shows a large-bow set with beam compass.

ates. The design allows the legs to be moved in or out to change the setting, after which the center screw is used for fine adjustment.

Even though it is possible to find an instrument that serves practically all purposes, it is convenient to have several instruments, thus saving the time required to convert from pencil leg to ink leg or divider points. For this reason the newer more rigid instruments are also made as separate pieces in different sizes. Figure 16 shows the 6-in. dividers and large 6-in. bow compass; three 3-in. bow compasses, two with pencil and one with ink leg, which can also be used as dividers by changing attachments; and pen points, which are used on the 6-in. compass or in a handle (not shown) to make a ruling pen.

### 19. Lettering Devices.
The Braddock-Rowe triangle and the Ames lettering instrument (Figs. 2a and b) are convenient devices used in drawing guide lines for lettering.

### 20. Checklist of Instruments and Materials.
Set of drawing instruments, including: 6-in. compass with fixed needle-point leg, removable pencil and pen legs, and lengthening bar; 6-in. hairspring dividers; $3\frac{1}{2}$-in. bow pencil, bow pen, and bow dividers; two ruling pens; box of leads. Or large-bow set containing $6\frac{1}{2}$-in. bow compass; $4\frac{1}{2}$-in. bow compass; $6\frac{1}{2}$-in. friction dividers and pen attachment for compass; $5\frac{1}{2}$-in. ruling pen; beam compass with extension beam; box for extra leads and points

DIVIDER
OR
COMPASS

PEN

(a)

(b)

**FIG. 14.** (a) A very rugged bow. Serves as dividers, or pen or pencil compass. (b) A quick-action bow instrument, with pencil and pen legs and extension bar.

BOW PENCIL     FAST-ACTION BOW

LARGE PENCIL COMPASS
WITH PEN AND
DIVIDER LEGS

**FIG. 15.** Three different designs of large bow instruments. All are rigid and efficient.

DIVIDER      LARGE BOW

SMALL BOWS AND ATTACHMENTS

**FIG. 16.** A large- and small-bow set. Combines the advantages of both standard three-bow and large-bow designs.

BEAM

PENCIL            ADJUSTABLE
                   POINT

**FIG. 17.** Extender for large bow. This type extends the range of its center point, with the pen or pencil on the bow.

Drawing board
T square
45° and 30–60° triangles
Three mechanical engineer's scales, flat pattern, or the equivalent triangular scale
Lettering instrument or triangle
French curves
Drawing pencils, 6H, 4H, 2H, H, and F
Pocketknife or pencil sharpener
Pencil pointer (file or sandpaper)
Pencil eraser (Ruby)
Artgum or cleaning rubber
Penholder, pens for lettering, and penwiper
Bottle of drawing ink and bottleholder
Scotch drafting tape or thumbtacks
Drawing paper to suit
Tracing paper and cloth
Dustcloth or brush

To these may be added:

Civil engineer's scale

POINT

BEAM

PEN    PENCIL

**FIG. 18.** Extender for large bow. This type uses the center point of the bow and has its own pen and pencil attachments.

BEAM

CONNECTOR

PEN

POINT    PENCIL

EXTENSION BEAM

**FIG. 19.** A beam compass. This compass is adjustable for radii of 1 to 16 inches.

STANDARD PENS

CONTOUR PEN

BORDER PEN

RAILROAD PENS

BORDER PENS

(a)  (b)  (c)  (d)  (e)   (f)   (g)   (h)   (i)   (j)   (k)

**FIG. 20.** Various ruling pens. Standard pens are all purpose instruments. Special pens (*f*) to (*k*) are convenient and efficient for contours, double lines, and heavy lines.

Protractor
Erasing shield
Slide rule
Six-foot steel tape
Clipboard or sketchbook
Hard Arkansas oilstone
Piece of soapstone
Cleaning powder or pad

The student should mark all his instruments and materials plainly with his initials or name as soon as they have been purchased and approved.

## 21. Additional Instruments.

The instruments and materials described in this chapter are all that are needed for ordinary practice and are, with the exception of such supplies as paper, pencils, ink, and erasers, what a draftsman is as a rule expected to take with him into a drafting room.

There are many other special instruments and devices that are not necessary in ordinary work but with which the draftsman should be familiar, as they may be convenient in special cases and are often found as a part of drafting-room equipment. Consult manufacturers' catalogues. An example is shown in Fig. 23.

## 22. The Use of Instruments.

In beginning to use drawing instruments, it is important to learn to handle them correctly. Carefully read the instructions and observe strictly all details of technique.

Facility will come with continued practice, but it is essential to adhere to good form from the outset. Bad form in drawing can be traced in every instance

**FIG. 21.** A three-bow set, in case. It contains large dividers, large compass with extension bars and pen attachment, two ruling pens, bow pen, bow pencil, bow dividers, and bone center.

**FIG. 22.** A large-bow set, in case. It contains large- and small-bow compasses with attachments, dividers, ruling pen, and beam compass.

to the formation of bad habits in the early stages of learning. Once formed, these habits are difficult to overcome.

It is best to make a few drawings solely to become familiar with the handling and feel of the instruments so that later, in working a drawing problem, you will not lose time because of faulty manipulation. Practice accurate penciling first, and do not attempt inking until you have become really proficient in penciling. With practice, the correct, skillful use of drawing instruments will become a subconscious habit.

For competence in drawing, *accuracy* and *speed* are essential, and in commercial work neither is worth much without the other. It is well to learn as a beginner that a *good* drawing can be made as quickly as a *poor* one. Erasing is expensive and most of it can be avoided. The draftsman of course erases occasionally, and a student must learn to make

corrections, but in beginning to use instruments, strive for sheets without blemish or inaccuracy.

### 23. Preparation for Drawing.

The drawing table should be set so that the light comes from the left, and it should be adjusted to a convent height, that is, 36 to 40 in., for use while sitting on a standard drafting stool or while standing. There is more freedom in drawing standing, especially when working on large drawings. The board, for use in this manner, should be inclined at a slope of about 1 to 8. Since it is more tiring to draw standing, many modern drafting rooms use tables so made that the board can be used in an almost vertical position and can be raised or lowered so that the draftsman can use a lower stool with swivel seat and backrest, thus working with comfort and even greater freedom than when an almost horizontal board is used.

The instruments should be placed within easy reach, on the table or on a special tray or stand which is located beside the table. The table, the

**FIG. 23** A "technical pen." These are supplied in single pens or in sets. A set of twelve pens gives line widths of about $\frac{1}{200}$ in. to about $\frac{3}{32}$ in.

board, and the instruments should be wiped with a dustcloth before starting to draw.

## 24. The Pencil and Its Use.

The grade of pencil must be selected carefully, with reference to the surface of the paper as well as to the line quality desired. For a pencil layout on detail paper of good texture, a pencil as hard as 5H or 6H may be used; for finished pencil drawings on the same paper, 2H, 3H, or 4H pencils give the blacker line needed. For finished pencil drawings or tracings on vellum, softer pencils, H to 3H, are employed to get printable lines. The F pencil is used for technical sketching, and the H is popular for lettering. In each case the pencil must be hard enough not to blur or smudge but not so hard as to cut grooves in the paper under reasonable pressure.

*To sharpen a pencil,* cut away the wood from the unlettered end with a pen-knife or mechanical sharpener, as shown in Fig. 24a, and then sharpen the lead to make a long, conic point, as at (b), by twirling the pencil as the lead is rubbed with long even strokes against the sandpaper pad or file or placed in a special lead sharpener.

A flat or wedge point will not wear away in use so fast as a conic point, and on that account some prefer it for straight-line work. The long, wedge point illustrated at (c) is made by first sharpening, as at (a), then making the two long cuts on opposite sides, as shown, then flattening the lead on the sandpaper pad or file, and finishing by touching the corners to make the wedge point narrower than the diameter of the lead.

Have the sandpaper pad within easy reach, and *keep the pencils sharp.* Some hang the pad or file on a cord attached to the drawing table. The professional draftsman sharpens his pencil every few minutes. After sharpening the lead, wipe off excess graphite dust before using the pencil. Form the habit of sharpening the lead as often as you might dip a writing pen into the inkwell. Most commercial and many college drafting rooms are equipped with Dexter or other pencil sharpeners to save time.

Not only must pencil lines be clean and sharp,

(a) REMOVE WOOD

(b) MAKE POINT ON SANDPAPER PAD OR FILE

(c) WEDGE POINT FOR STRAIGHT LINES

**FIG. 24.** Sharpening the pencil. Wood is removed first, then the point is made with sandpaper or a file or special lead sharpener. For semiautomatic pencils, adjust the lead to the proper length and make the point with sandpaper or a file.

but for pencil drawings and tracings to be blue-printed, it is absolutely necessary that all the lines of each kind be uniform, firm, and opaque. This means a careful choice of pencils and the proper use of them. The attempt to make a dark line with too hard a pencil results in cutting deep grooves in the paper. Hold the pencil firmly, yet with as much ease and freedom as possible.

Keep an even constant pressure on the pencil, and when using a conic point, rotate the pencil as the line is drawn so as to keep both the line and pencil sharp. Use a draftsman's brush or soft cloth occasionally to dust off excess graphite from the drawing.

Too much emphasis cannot be given to the importance of clean, careful, accurate penciling. Never entertain the thought that poor penciling can be corrected in tracing.

### 25. Placing the Paper.

Since the T-square blade is more rigid near the head than toward the outer end, the paper, if much smaller than the size of the board, should be placed close to the left edge of the board (within an inch or so) with its lower edge several inches from the bottom of the board. With the T square against the left edge of the board, square the top of the paper; hold it in this position, slipping the T square down from the edge, and put a thumbtack in each upper corner, pushing it in up to the head so that the head aids in holding the paper. Then move the T square down over the paper to smooth out possible wrinkles, and put thumbtacks in the other two corners. Drafting tape may be used instead of thumbtacks.

### 26. Use of the T Square.

The T square and the triangles have straight edges and are used for drawing straight lines. Horizontal lines are drawn with the T square, which is used with its head against the left edge of the drawing board and manipulated as follows: Holding the head of the tool, as shown in Fig. 25a, slide it along the edge of the board to a spot very near the position desired. Then, for closer adjustment, change your hold either to that shown at (b), in which the thumb remains on top of the T-square head and the other

(a)

SLIDE TO POSITION

(b)

ADJUST CAREFULLY

(c)

HOLD IN POSITION
AS LINE IS DRAWN

**FIG. 25.** Manipulating the T square. The head must be firmly against the straight left edge of the board.

fingers press against the underside of the board, or, as is more usual, to that shown at (c), in which the fingers remain on the T square and the thumb is placed on the board.

**FIG. 26.** Drawing a horizontal line. Hold the T square with the left hand; draw the line from left to right; incline the pencil in the direction of stroke, so that the pencil "slides" over the paper.

Figure 26 shows the position of the hand and pencil for drawing horizontal lines. Note that the pencil is inclined in the direction the line is drawn, that is, toward the right, and also slightly away from the body so that the pencil point is as close as possible to the T-square blade.

In drawing lines, take great care to keep them accurately parallel to the guiding edge of the T square. The pencil should be held lightly, but close against the edge, and the angle should not vary during the progress of the line. Horizontal lines should always be drawn from left to right. A T-square blade can be tested for straightness by drawing a sharp line through two points and then turning the square over and with the same edge drawing another line through the points, as shown in Fig. 27.

## 27. Use of the Triangles.

Vertical lines are drawn with the triangle, which is set against the T square with the perpendicular edge nearest the head of the square and thus toward the light (Fig. 28). These lines are always drawn upward, from bottom to top.

**FIG. 27.** To test a T square. Turn the T square upside down; draw the line; turn it right side up and align it with the original line; draw the second line and compare it with the first.

**FIG. 28.**  Drawing a vertical line. With the T square and triangle in position, draw the line from bottom to top—always away from the body.

In drawing vertical lines, the T square is held in position against the left edge of the board by the thumb and little finger of the left hand while the other fingers of this hand adjust and hold the triangle. You can be sure that the T square is in contact with the board when you hear the little double click as the two come together, and slight pressure of the thumb and little finger toward the right will maintain the position. As the line is drawn, pressure of all the fingers against the board will hold the T square and triangle firmly in position.

As in using the T square, care must be taken to keep the line accurately parallel to the guiding edge. Note the position of the pencil in Fig. 28.

In both penciling and inking, the triangles must always be used in contact with a guiding straight-edge. To ensure accuracy, never work to the extreme corner of a triangle; to avoid having to do so, keep the T square below the lower end of the line to be drawn.

With the T square against the edge of the board, lines at 45° are drawn with the standard 45° triangle, and lines at 30° and 60° with the 30–60° triangle, as shown in Fig. 29. With vertical and horizontal lines included, lines at increments of 45° are drawn with the 45° triangle as at (*b*), and lines at 30° increments with the 30–60° triangle as at (*a*). The two triangles are used in combination for angles of 15, 75, 105°, etc. (Fig. 30). Thus any multiple of 15° is drawn directly; and a circle is divided with the 45° triangle into 8 parts, with the 30–60° triangle into 12 parts, and with both into 24 parts.

*To Draw One Line Parallel to Another (Fig. 31).* Adjust to the given line a triangle held against a straightedge, hold the guiding edge in position, and slide the triangle on it to the required position.

*To Draw a Perpendicular to Any Line (Fig. 32).* Place a triangle with one edge against the T square (or another triangle), and move the two until the hypotenuse of the triangle is coincident with the line, as at position *a*; hold the T square in position and turn the triangle, as shown, until its other side is against the T square; the hypotenuse will then be perpendicular to the original line. Move the triangle to the required position. A quicker method is to set the triangle with its hypotenuse against the guiding edge, fit one side to the line, slide the triangle to the required point and draw the perpendicular, as shown at position *b*.

Never attempt to draw a perpendicular to a line with only one triangle by placing one leg of the triangle along the line.

Through internal strains, triangles sometimes lose their accuracy. They may be tested by drawing a perpendicular and then reversing the triangle, as shown in Fig. 33.

### 28. The Left-handed Draftsman.
If you are left-handed, reverse the T square and

**FIG. 29.** To draw angles of 30°, 45°, 60°. Multiples of 30° are drawn with the 30–60° triangle, multiples of 45° with the 45° triangle.

**FIG. 30.** To draw angles of 15° and 75°. Angles in increments of 15° are obtained with the two triangles in combination.

**FIG. 31.** To draw parallel lines. With the T square as a base, the triangle is aligned and then moved to the required position.

**FIG. 32.** To test a triangle for right angle, draw a line with the triangle in each position; twice the error is produced.

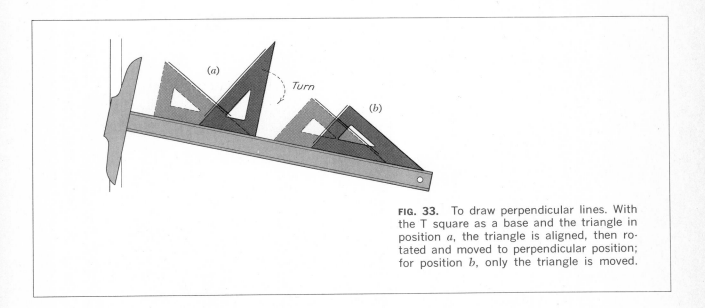

**FIG. 33.** To draw perpendicular lines. With the T square as a base and the triangle in position *a*, the triangle is aligned, then rotated and moved to perpendicular position; for position *b*, only the triangle is moved.

triangles left for right as compared with the regular right-handed position. Use the head of the T square along the right edge of the board, and draw horizontal lines from right to left. Place the triangle with its vertical edge to the right, and draw vertical lines from bottom to top. The drawing table should be placed with the light coming from the right.

## 29. Use of the Scale.

Scale technique is governed largely by the requirements of accuracy and speed. Before a line can be drawn, its relative position must be found by scaling, and the speed with which scale measurement can be made will greatly affect the total drawing time.

Precise layouts and developments, made to scale

**FIG. 34.** Making a measurement. Place the scale in position; the distance is marked on paper by short, light *lines*.

for the workmen, must be very accurately drawn, at the expense of speed; conversely, drawings with figured dimensions need not be quite so carefully scaled, and better speed may be attained.

To make a measurement, place the scale on the drawing where the distance is to be laid off, align the scale in the direction of the measurement, and make a *light* short dash with a sharp pencil at the proper graduation mark (Fig. 34). In layout work where extreme accuracy is required, a "pricker," or needle point set in a wood handle, may be substituted for the pencil, and a *small* hole pricked into the paper in place of the pencil mark. It is best to start with the "zero" of the scale when setting off lengths or when measuring distances. In using an open-divided scale, inches (or fractions) are accounted for in one direction from the zero graduation while feet (or units) are recorded in the opposite direction.

Measurements should not be made on a drawing by taking distances off the scale with dividers, as this method is time-consuming and no more accurate than the regular methods.

To avoid cumulative errors, successive measurements on the same line should, if possible, be made without shifting the scale. In representing objects that are larger than can be drawn to their natural

## SCALES

### MECHANICAL ENGINEER'S

| | |
|---|---|
| $1'' = 1''$ (full size) | $\frac{1}{2}'' = 1''$ ($\frac{1}{2}$ size) |
| $\frac{1}{4}'' = 1''$ ($\frac{1}{4}$ size) | $\frac{1}{8}'' = 1''$ ($\frac{1}{8}$ size) |

### ARCHITECT'S OR MECHANICAL ENGINEER'S

| | | |
|---|---|---|
| $12'' = 1'\text{-}0''$ (full size) | $1'' = 1'\text{-}0''$ ($\frac{1}{12}$ size) | $\frac{1}{4}'' = 1'\text{-}0''$ ($\frac{1}{48}$ size) |
| $6'' = 1'\text{-}0''$ ($\frac{1}{2}$ size) | $\frac{3}{4}'' = 1'\text{-}0''$ ($\frac{1}{16}$ size) | $\frac{3}{16}'' = 1'\text{-}0''$ ($\frac{1}{64}$ size) |
| $3'' = 1'\text{-}0''$ ($\frac{1}{4}$ size) | $\frac{1}{2}'' = 1'\text{-}0''$ ($\frac{1}{24}$ size) | $\frac{1}{8}'' = 1'\text{-}0''$ ($\frac{1}{96}$ size) |
| $1\frac{1}{2}'' = 1'\text{-}0''$ ($\frac{1}{8}$ size) | $\frac{3}{8}'' = 1'\text{-}0''$ ($\frac{1}{32}$ size) | $\frac{3}{32}'' = 1'\text{-}0''$ ($\frac{1}{128}$ size) |

### CIVIL ENGINEER'S

10, 20, 30, 40, 50, 60, or 80 divisions to the inch representing feet, 10 ft, 100 ft, rods, miles, or any other necessary unit

### METRIC
Reductions of
$1:1$, $1:2$, $1:5$, $1:10$, $1:20$, $1:25$, $1:33.3$, $1:50$, $1:75$, $1:80$, $1:100$, $1:150$

**FIG. 35.** Standard scales. Special scales are available. See manufacturers' catalogues.

or full size, it is necessary to reduce the size of the drawing in some regular proportion, and for this purpose one of the standard mechanical engineer's, civil engineer's, or architect's scales is used. Standard scales are given in Fig. 35.

The first reduction is to *half size,* or to the scale of 6″ = 1′-0″. In other words, ½ in. on the drawing represents a distance of 1 in. on the object. Stated in terms used for the architect's scales, a distance of 6 in. on the drawing represents 1 ft on the object. This scale is used even if the object is only slightly larger than could be drawn full size. If this reduction is not sufficient, the drawing is made to *quarter size,* or to the scale of 3″ = 1′-0″. If the quarter-size scale is too large, the next reduction is *eighth size,* or 1½″ = 1′-0″, the smallest proportion usually supplied on standard mechanical engineer's scales; but the architect's scales are used down to ³⁄₃₂″ = 1′-0″, as shown by the listings in Fig. 35.

In stating the scale used on a drawing, the information should be given in accordance with the scale used to make the drawing. If a standard mechanical engineer's scale is employed, the statement may read that the scale is (1) *full size,* (2) *half size,* (3) *quarter size,* or (4) *eighth size.* These scales may also be given as (1) 1″ = 1″, (2) ½″ = 1″, (3) ¼″ = 1″, or (4) ⅛″ = 1″. If a standard architect's scale is used, the statement is given in terms of inches to the foot. Examples are (1) 3″ = 1′-0″, (2) 1½″ = 1′-0″, or (3) 1″ = 1′-0″. In stating the scale, the first figure always refers to the drawing and the second to the object. Thus 3″ = 1′-0″ means that 3 in. on the *drawing* represents 1 ft on the *object.*

Drawings to odd proportions, such as 9″ = 1′-0″, 4″ = 1′-0″, 5″ = 1′-0″, are used only in rare cases when drawings are made for reduction and the conditions of size demand a special scale.

The terms "scale" and "size" have different meanings. The scale ¼″ = 1′-0″ is the usual one for ordinary house plans and is often called by architects the "quarter scale." This term should not be confused with term "quarter size," as the former means ¼ in. to 1 ft and the latter ¼ in. to 1 in.

The size of a circle is generally stated by giving its diameter, while to draw it the radius is necessary. Two scales are usually supplied together on the same

body, for example, half and quarter size. Therefore, in drawing to half size, it is often convenient to lay off the amount of the diameter with the quarter-size scale and use this distance as the radius.

Small pieces are often made "double size," and very small mechanisms, such as watch parts, are drawn to greatly enlarged sizes: 10 to 1, 20 to 1, 40 to 1, and 50 to 1, using special enlarging scales.

For plotting and map drawing, the civil engineer's scales of decimal parts, with 10, 20, 30, 40, 50, 60, and 80 divisions to the inch, are used. These scales are not used for machine or structural work but in certain aircraft drawings.

The important thing in drawing to scale is to think and speak of each dimension in its full size and not in the reduced (or enlarged) size it happens to be on the paper. This practice prevents confusion between *actual* and *represented* size.

## 30. Reading the Scale.
Reading the standard mechanical engineer's scales is rather simple, because the scale is plainly marked in inches, and the smaller graduations are easily recognized as the regular divisions of the inch into ½, ¼, ⅛, and ¹⁄₁₆. Thus the scales for half size, quarter size, and eighth size are employed in exactly the same manner as a full-size scale.

The architect's scales, being open-divided and to stated reductions, such as 3″ = 1′-0″, may require some study by the beginner in order to prevent confusion and mistakes. As an example, consider the scale of 3″ = 1′-0″. This is the first reduction scale of the usual triangular scale; on it the distance of 3 in. is divided into 12 equal parts, and each of these is subdivided into eighths. This distance should be thought of not as 3 in. but as a foot divided into inches and eighths of an inch. Notice that the divisions start with the *zero* on the inside, the inches of the divided foot running to the *left* and the open divisions of feet to the *right,* so that dimensions given in feet and in inches may be read directly, as 1′-0½″ (Fig. 36). On the other end will be found the scale of 1½″ = 1′ 0″, or eighth size, with the distance of 1½ in. divided on the right of the zero into 12 parts and subdivided into quarter inches, with the foot divisions to the left of the zero

**FIG. 36.** Reading an open-divided scale. Feet are given on one side of the zero, inches and fractions on the other.

coinciding with the marks of the 3-in. scale. Note again that in reading a distance in feet and inches, for example, the 2'-7⅛'' distance in Fig. 36, feet are determined to the left of the zero and inches to the right of it. The other scales, such as $\frac{3}{4}'' = 1'\text{-}0''$ and $\frac{1}{4}'' = 1'\text{-}0''$, are divided in a similar way, the only difference being in the value of the smallest graduations. The scale of $\frac{3}{32}'' = 1'\text{-}0''$, for example, can be read only to the nearest 2 in.

### 31. "Laying out" the Sheet.
The paper is usually cut somewhat larger than the desired size of the drawing and trimmed to size after the work is finished. Suppose the finished size is to be 11 by 17 in. with a ½-in. border inside. Lay the scale down on the paper close to the lower edge and measure 17 in., marking the distance with the pencil; at the same time mark ½ in. inside at each end for the border line. Use a short dash forming a continuation of the division line on the scale in

**FIG. 37.** Handling the dividers. The instrument is opened and adjusted with one hand.

laying off a dimension. Do not bore a hole with the pencil. Near the left edge mark 11- and ½-in. border-line points. Through these four marks on the left edge, draw horizontal lines with the T square; and through the points on the lower edge, draw vertical lines, using the triangle against the T square.

### 32. Use of Dividers.
Dividers are used for transferring measurements and for dividing lines into any number of equal parts. Facility in their use is essential, and quick and absolute control of their manipulation must be gained. The instrument should be opened with one hand by pinching the chamfer with the thumb and second finger. This will throw it into correct position with thumb and forefinger outside the legs and the second and third fingers inside, with the head resting just above the second joint of the forefinger (Fig. 37). It is thus under perfect control, with the thumb and forefinger to close it and the other two to open it. Practice this motion until you can adjust the dividers to the smallest fraction. In coming down to small divisions, the second and third fingers must be gradually slipped out from between the legs as they are closed down upon them. Notice that the little finger is not used in manipulating the dividers.

### 33. To Divide a Line by Trial.
In bisecting a line, the dividers are opened at a guess to roughly half the length. This distance is stepped off on the line, holding the instrument by the handle with the thumb and forefinger. If the division is short, the leg should be thrown out to half the remainder (estimated by eye), without removing the

other leg from the paper, and the line spaced again with this new setting (Fig. 38). If the result does not come out exactly, the operation can be repeated. With a little experience, a line can be divided rapidly in this way. Similarly, a line, either straight or curved, can be divided into any number of equal parts, say, five, by estimating the first division, stepping this lightly along the line, with the dividers held vertically by the handle, turning the instrument first in one direction and then in the other. If the last division falls short, one-fifth of the remainder should be added by opening the dividers, keeping one point on the paper. If the last division is over, one-fifth of the excess should be taken off and the line re-spaced. If it is found difficult to make this small adjustment accurately with the fingers, the hair-spring may be used. You will find the bow spacers more convenient than the dividers for small or numerous divisions. Avoid pricking unsightly holes in the paper. The position of a small prick point may be preserved, if necessary, by drawing a small circle around it with the pencil.

## 34. Use of the Compasses.

The compasses have the same general shape as the dividers and are manipulated in a similar way. First of all, the needle should be permanently adjusted. Insert the pen in place of the pencil leg, turn the needle with the shoulder point out, and set it a trifle longer than the pen, as in Fig. 39; replace the pencil leg, sharpen the lead to a long bevel, as in Fig. 40, and adjust it to the needle point. All this is done so that the needle point will be in perfect position for using the pen; the pencil, which must be sharpened frequently, can be adjusted each time to mate in length with the needle point.

FIG. 39. Adjusting the needle point of a large compass. The point is adjusted to the pen; the pen is then replaced by the pencil leg and the pencil adjusted to the point.

FIG. 40. Adjusting the pencil lead. The length is adjusted so that the instrument will be vertically centered.

FIG. 38. Bisecting a line. Half is estimated; then the dividers are readjusted by estimating half the original error.

FIG. 41. Setting the compass to radius size. Speed and accuracy are obtained by adjusting directly on the scale.

FIG. 42. Guiding the needle point. For accuracy of placement, guide with the little finger.

to perhaps 3 in. in diameter can be drawn with the legs of the compass straight, but for larger sizes, both the needlepoint leg and the pencil or pen leg

FIG. 43. Starting a circle. The compass is inclined in the direction of the stroke.

FIG. 44. Completing a circle. The stroke is completed by twisting the knurled handle in the fingers.

*To Draw a Circle.* Set the compass on the scale, as shown in Fig. 41, and adjust it to the radius needed; then place the needle point at the center on the drawing, guiding it with the left hand (Fig. 42). Raise the fingers to the handle and draw the circle in one sweep, rolling the handle with the thumb and forefinger, inclining the compass slightly in the direction of the line (Fig. 43).

The position of the fingers after the rotation is shown in Fig. 44. The pencil line can be brightened, if necessary, by making additional turns. Circles up

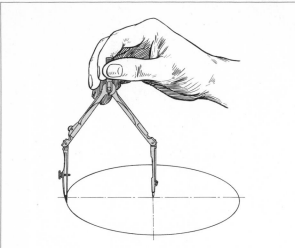

FIG. 45. Drawing a large circle. Knuckle joints are bent to make the legs perpendicular to the paper.

FIG. 47. Adjusting a bow instrument. One hand only is needed. Center-wheel bows are held similarly.

should be bent at the knuckle joints so as to be perpendicular to the paper (Fig. 45). The 6-in. compass may be used in this way for circles up to perhaps 10 in. in diameter; larger circles are made by using the lengthening bar, as illustrated in Fig. 46, or the beam compass (Fig. 19). In drawing concentric circles, the *smallest* should always be drawn first, before the center hole has become worn.

The bow instruments are used for small circles, particularly when a number are to be made of the same diameter. To avoid wear (on side-wheel instruments), the pressure of the spring against the nut can be relieved in changing the setting by holding the points in the left hand and spinning the nut in or out with the finger. Small adjustments should be made with one hand with the needle point in position on the paper (Fig. 47).

When several concentric circles are drawn, time may be saved by marking off the several radii on the paper from the scale and then setting the compass to each mark as the circles are made. In some cases it may be advantageous to measure and mark

FIG. 46. Use of the lengthening bar. The joints must be bent to bring the legs perpendicular. Usually two hands are used because the handle is off center.

the radius on the paper instead of setting the compass directly on the scale. This method must be used whenever the radius is greater than the length of the scale.

When *extreme accuracy* is required, the compass is set, a light circle is drawn on the paper, and the diameter is checked with the scale; if the size is not satisfactory, the compass is adjusted and the operation is repeated until the size needed is obtained.

### 35. The Ruling Pen.

The ruling pen is for inking straight lines and noncircular curves. Several types are illustrated in Fig. 20. The important feature is the shape of the blades; they should have a well-designed ink space between them, and their points should be rounded (actually elliptical in form) equally, as in Fig. 48. If pointed, as in Fig. 49, the ink will arch up as shown and will be provokingly hard to start. If rounded to a blunt point, as in Fig. 50, the ink will flow too freely, forming blobs and overruns at the ends of the lines. Pens in constant use become dull and worn, as illustrated in Fig. 51. It is easy to tell whether a pen is dull by looking for the reflection of light that travels from the side and over the end of the point when the pen is turned in the hand. If the reflection can be seen all the way, the pen is too dull. A pen in poor condition is an abomination, but a well-sharpened one is a delight to use. Every draftsman should be able to keep his pens in fine condition.

High-grade pens usually come from the makers well sharpened. Cheaper ones often need sharpening before they can be used.

### 36. To Sharpen a Pen.

The best stone for the purpose is a hard Arkansas knife piece. It is well to soak a new stone in oil for several days before using. The ordinary carpenter's oilstone is too coarse for drawing instruments.

The nibs must first be brought to the correct shape, as in Fig. 48. Screw the nibs together until they touch and, holding the pen as in drawing a line, draw it back and forth on the stone, starting the stroke with the handle at 30° or less with the stone and swinging it up past the perpendicular as the

**FIG. 48.** Correct shape of pen nibs. A nicely uniform elliptical shape is best.

**FIG. 49.** Incorrect shape of pen nibs. This point is much too sharp. Ink will not flow well.

**FIG. 50.** Incorrect shape of pen nibs. This point is too flat. Ink will blob at the beginning and end of the line.

**FIG. 51.** Shape of worn pen nibs. This point needs sharpening to the shape of Fig. 48.

**FIG. 52.** Sharpening a pen. After bringing the point to the shape of Fig. 48. work down the sides by a rocking motion to conform to blade contour.

line across the stone progresses. This will bring the nibs to exactly the same shape and length, leaving them very dull. Then open them slightly, and sharpen each blade in turn, on the outside only, until the bright spot on the end has just disappeared. Hold the pen, as in Fig. 52, at a small angle with the stone and rub it back and forth with a slight oscillating or rocking motion to conform to the shape of the blade. A stone 3 or 4 in. long held in the left hand with the thumb and fingers gives better control than one laid on the table. Silicon carbide cloth or paper can be substituted for the stone, and for a fine job, crocus cloth may be used for finishing. A pocket magnifying glass may be helpful in examining the points. The blades should not be sharp enough to cut the paper when tested by drawing a line across it without ink. If oversharpened, the blades should again be brought to touch and a line swung very lightly across the stone as in the first operation. When tested with ink, the pen should be capable of drawing clean sharp lines down to the finest hairline. If these finest lines are ragged or broken, the pen is not perfectly sharpened. It should not be necessary to touch the inside of the blades unless a burr has been formed, which might occur if the metal is very soft, the stone too coarse, or the pressure too heavy. To remove such a burr, or wire edge, draw a strip of detail paper between the nibs, or open the pen wide and lay the entire inner surface of the blade flat on the stone and move it with a very light touch.

### 37. Use of the Ruling Pen.

The ruling pen is always used in connection with a guiding edge—T square, triangle, or curve. The T square and triangle should be held in the same positions as for penciling.

To fill the pen, take it to the bottle and touch the quill filler between the nibs. Be careful not to get any ink on the outside of the blades. If the newer plastic squeeze bottle is used, place the small spout against the sides of the nibs and carefully squeeze a drop of ink *between* the nibs. Not more than $\frac{3}{16}$ to $\frac{1}{4}$ in. of ink should be put in; otherwise the weight of the ink will cause it to drop out in a blot. The pen should be held in the fingertips, as illustrated in Fig. 53, with the thumb and second finger against the sides of the nibs and the handle resting on the forefinger. Observe this hold carefully, as the tendency will be to bend the second finger to the position used when a pencil or writing pen is held. The position illustrated aids in keeping the pen at the proper angle and the nibs aligned with the ruling edge.

The pen should be held against the straightedge or guide with the blades parallel to it, the screw on the outside and the handle inclined slightly to the right and always kept in a plane passing through the line and perpendicular to the paper. The pen is thus directed by the upper edge of the guide, as

**FIG. 53.** Correct position of ruling pen. The pen is inclined in the direction of the stroke but held perpendicular to the paper, as in Fig. 54.

**FIG. 54.** Correct pen position. Even though the pen is inclined in the direction of the stroke, both nibs must touch the paper equally.

**FIG. 55.** Inking over pencil line. *Center* ink line over original layout line.

illustrated in actual size in Fig. 54. If the pen point is thrown out from the perpendicular, it will run on one blade and make a line that is ragged on one side. If the pen is turned in from the perpendicular, the ink is likely to run under the edge of the guide and cause a blot.

A line is drawn with a steady, even arm movement, the tips of the third and fourth fingers resting on, and sliding along, the straightedge, keeping the angle of inclination constant. Just before the end of the line is reached, the two guiding fingers on the straightedge should be stopped and, without stopping the motion of the pen, the line finished with a finger movement. Short lines are drawn with this finger movement alone. When the end of the line is reached, the pen is lifted quickly and the straightedge moved away from the line. The pressure on the paper should be light but sufficient to give a clean-cut line, and it will vary with the kind of paper and the sharpness of the pen. The pressure against the T square, however, should be only enough to guide the direction.

If the ink refuses to flow, it may be because it has dried in the extreme point of the pen. If pinching the blades slightly or touching the pen on the finger does not start it, the pen should immediately be wiped out and fresh ink supplied. Pens must be wiped clean after using.

In inking on either paper or cloth, the full lines will be much wider than the pencil lines. You must be careful to have the center of the ink line cover the pencil line, as illustrated in Fig. 55.

Instructions in regard to the ruling pen apply also to the compass. The compass should be slightly inclined in the direction of the line and both nibs of the pen kept on the paper, bending the knuckle joints, if necessary, to effect this.

It is a universal rule in inking that *circles and circle arcs must be inked first.* It is much easier to connect a straight line to a curve than a curve to a straight line.

### 38. Tangents.

It should be noted particularly that two lines are tangent to each other when the center lines of the lines are tangent and not simply when the lines touch each other; thus at the point of tangency, the width will be equal to the width of a single line (Fig. 56). Before inking tangent lines, the point of tangency should be marked in pencil. For an arc tangent to a straight line, this point will be on a line

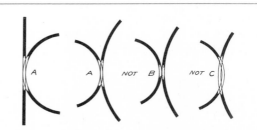

**FIG. 56.** Correct and incorrect tangents. Lines must be the width of *one* line at tangent point.

through the center of the arc and perpendicular to the straight line, and for two circle arcs it will be on the line joining their centers, as described in Secs. 5 to 16, Chap. 3.

### 39. The "Alphabet of Lines."

As the basis of drawing is the line, a set of conventional symbols covering all the lines needed for different purposes may properly be called an alphabet of lines. Figures 57 and 58 show the alphabet of lines adopted by the ANSI as applied to the following:

1. Drawings made directly or traced in pencil on tracing paper or pencil cloth, from which blueprints or other reproductions are to be made (Fig. 57).

2. Tracings in ink on tracing cloth or tracing paper and inked drawings on white paper for display or photoreproductions (Fig. 58).

The ANSI recommends three widths of lines for finished drawings: *thick* for visible outlines and cutting-plane and short-break lines; *medium* for hidden outlines; and *thin* for section, center, extension, dimension, long-break, adjacent-part, alternate-position, and repeat lines. The actual

**FIG. 57.** The alphabet of lines for pencil drawings.

**FIG. 58.** The alphabet of lines for inked drawings.

widths of the three weights of lines, on average drawings, should be about as in Figs. 57 and 58. A convenient line gage is given in Fig. 59. If applied to Fig. 58, this gage would show the heavy lines in ink to be between $\frac{1}{30}$ and $\frac{1}{40}$ in., the medium lines $\frac{1}{60}$ in., and the fine lines $\frac{1}{100}$ in. in width. To use the line gage, draw a line about $1\frac{1}{2}$ in. long in pencil or ink on a piece of the drawing paper and apply it alongside the gage. By this method a good comparison can be made. Note that the standard lines for pencil drawings are some-

what thinner than for inked drawings, the thick line being about $\frac{1}{60}$ in., medium about $\frac{1}{80}$ in., and thin between $\frac{1}{100}$ and $\frac{1}{150}$ in. Study Figs. 57 and 58 carefully and try to make your drawings conform to these standard lines. Professional appearance depends to a great extent upon the line weights used. Line widths for layout drawings are *thin* throughout because the watchword here is accuracy. Layout drawings are often traced in pencil or ink to the weights given in Figs. 57 and 58 and are then *finished* drawings.

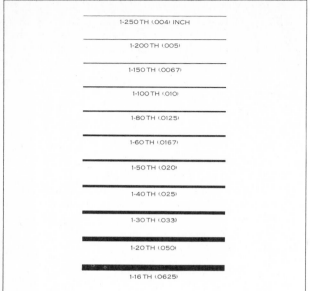

**FIG. 59.** Line gage. Draw a line on the paper to be used and apply it here to determine the width.

**FIG. 60.** Faulty ink lines. The difficulty is indicated in each case.

### 40. Line Practice.

After reading the preceding paragraphs, take a blank sheet of paper and practice making straight lines and circles in all the forms—full, dashed, etc.—shown in Figs. 57 and 58. Include starting and stopping lines, with special attention to tangents and corners.

*In pencil,* try to get all the lines uniform in width and color for each type. Circle arcs and straight lines should match exactly at tangent points.

*In ink,* proceed as for pencil practice and pay particular attention to the weight of lines and to the spacing of dashed lines and center lines.

If the inked lines appear imperfect in any way, ascertain the reason immediately. It may be the fault of the pen, the ink, the paper, or the draftsman; the probabilities are greatly in favor of the last if the faults resemble those in Fig. 60, which illustrates the characteristic appearance of several kinds of poor line. The correction in each case will suggest itself.

### 41. Use of the French Curve.

The French curve is a guiding edge for noncircular curves. When sufficient points have been determined, it is best to sketch in the line lightly in pencil, freehand and without losing the points, until it is clean, smooth, continuous, and satisfactory to the eye. Then apply the curve to it, selecting a part that will fit a portion of the line most nearly and seeing to it, particularly, that the curve is so placed that the direction in which its curvature increases is the direction in which the curvature of the line increases (Fig. 61). In drawing the part of the line matched by the curve, *always* stop a little short of the distance in which the guide and the line seem to coincide. After drawing this portion, shift the curve to find another place that will coincide with the continuation of the line. In shifting the curve, take care to preserve smoothness and continuity and to avoid breaks or cusps. Do this by seeing that in its successive positions the curve is always adjusted so that it coincides for a short distance with the part of the line already drawn. Thus at each junction the tangents will coincide.

If the curved line is symmetrical about an axis, marks locating this axis, after it has been matched accurately on one side, may be made in pencil on the curve and the curve then reversed. In such a case take exceptional care to avoid a "hump" at the joint.

It is often better to stop a line short of the axis on each side and close the gap afterward with another setting of the curve.

When using the curve in inking, the pen should be held perpendicular and the blades kept parallel to the edge. The inking of curves is excellent practice.

Sometimes, particularly at sharp turns, a combination of circle arcs and curves may be used: In inking a long, narrow ellipse, for example, the sharp curves may be inked by selecting a center on the major diameter by trial, drawing as much arc as will practically coincide with the ends of the ellipse, and then finishing the ellipse with the curve. The experienced draftsman will sometimes ink a curve that cannot be matched accurately by varying the distance of the pen point from the ruling edge as the line progresses.

### 42. Erasing.

The erasing of pencil lines and ink lines is a necessary technique to learn. When changing some detail, a designer, working freely but lightly, uses a soft pencil eraser so as not to damage the finish of the paper. Heavier lines are best removed with a Ruby pencil eraser. If the paper has been grooved by the line, it may be rubbed over with a burnisher or even with the back of the thumbnail. In erasing an ink

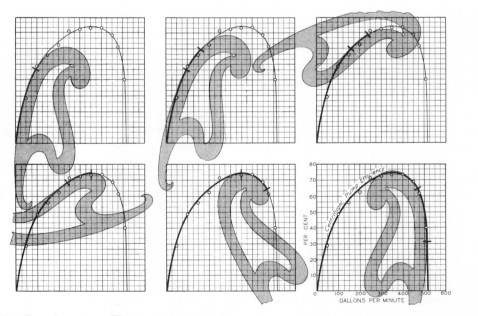

**FIG. 61.** Use of the French curve. The changing curvature of line and curve must match.

line, hold the paper down firmly and rub lightly and patiently, with a Ruby pencil eraser, first along the line and then across it, until the ink is removed. A triangle slipped under the paper or cloth gives a good backing surface.

When an erasure is made close to other lines, select an opening of the best shape on the erasing shield (Fig. 3d) and rub through it, holding the shield down firmly, first seeing that both of its sides are clean. Wipe the eraser crumbs off the paper with a dustcloth or brush. Never scratch out a line or blot with a knife or razor blade, and use so-called ink erasers sparingly, if at all. A skilled draftsman sometimes uses a sharp blade to trim a thickened spot or overrunning end on a line.

For extensive erasing, an electric erasing machine is a great convenience. Several successful models are on the market.

**43. Cautions in the Use of Instruments.** To complete this discussion of instruments, here are a few points worth noting:

**Never** use the scale as a ruler for drawing lines.

**Never** draw horizontal lines with the lower edge of the T square.

**Never** use the lower edge of the T square as a horizontal base for the triangles.

**Never** cut paper with a knife and the edge of the T square as a guide.

**Never** use the T square as a hammer.

**Never** put either end of a pencil into the mouth.

**Never** work with a dull pencil.

**Never** sharpen a pencil over the drawing board.

**Never** jab the dividers into the drawing board.

**Never** oil the joints of compasses.

**Never** use the dividers as reamers, pincers, or picks.

**Never** use a blotter on inked lines.

**Never** screw the pen adjustment past the contact point of the nibs.

**Never** leave the ink bottle uncorked.

**Never** hold the pen over the drawing while filling.

**Never** put into the drawing-ink bottle a writing pen that has been used in ordinary writing ink.

**Never** try to use the same thumbtack holes in either paper or board when putting paper down a second time.

**Never** scrub a drawing all over with an eraser after finishing. It takes the life out of the lines.

**Never** begin work without wiping off the table and instruments.

**Never** put instruments away without cleaning them. This applies with particular force to pens.

**Never** put bow instruments away without opening to relieve the spring.

**Never** work on a table cluttered with unneeded instruments or equipment.

**Never** fold a drawing or tracing.

**44. Exercises in the Use of Instruments.** The following problems can be used as progressive exercises for practice in using the instruments. Do them as finished pencil drawings or in pencil layout to be inked. Line work should conform to that given in the alphabet of lines (Figs. 57 and 58).

The problems in Chap. 3 afford excellent additional practice in accurate penciling.

## PROBLEMS

### Group 1. Straight Lines.

**1.** An exercise for the T square, triangle, and scale. Through the center of the space draw a horizontal and a vertical line. Measuring on these lines as diameters, lay off a 4-in. square. Along the lower side and upper half of the left side measure $\frac{1}{2}$-in. spaces with the scale. Draw all horizontal lines with the T square and all vertical lines with the T square and triangle.

**2.** An interlacement. For T square, triangle, and dividers. Draw a 4-in. square. Divide the left side and lower side into seven equal parts with dividers. Draw horizontal and vertical lines across the square through these points. Erase the parts not needed.

**3.** A street-paving intersection. For 45° triangle and scale. An exercise in starting and stopping short lines. Draw a 4-in. square. Draw its diagonals with 45° triangle. With the scale, lay off $\frac{1}{2}$-in. spaces along the diagonals from their intersection. With 45° triangle, complete the figure, finishing one quarter at a time.

**4.** A square pattern. For 45° triangle, dividers, and scale. Draw a 4-in. square and divide its sides into three equal parts with dividers. With 45° triangle, draw diagonal lines connecting these points. Measure $\frac{3}{8}$ in. on each side of these lines, and finish the pattern as shown in the drawing at the bottom of the page.

**5.** An acoustic pattern. For 45° triangle, T square, and scale. Draw two intersecting 45° diagonals 4 in. long, to form a field. With the scale lay off $\frac{1}{2}$-in. spaces from their intersection. Add the narrow border $\frac{3}{16}$ in. wide. Add a second border $\frac{1}{2}$ in. wide. The length of the border blocks is projected from the corners of the field blocks.

**6.** Five cards. Visible and hidden lines. Five cards $1\frac{3}{4}$ by 3 in. are arranged with the bottom card in the center, the other four overlapping each other and placed so that their outside edges form a 4-in. square. Hidden lines indicate edges covered.

**7.** A Maltese cross. For T square, spacers, and 45° and 30–60° triangles. Draw a 4-in. square and a $1\frac{3}{8}$-in. square. From the corners of the inner square, draw lines to the outer square at 15° and 75°, with the two triangles in combination. Mark points with spacers $\frac{1}{4}$ in. inside each line of this outside cross, and complete the figure with triangles in combination.

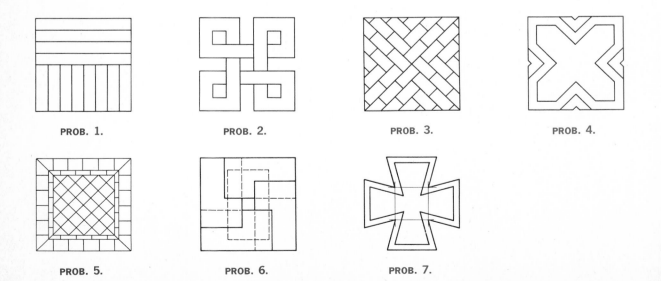

PROB. 1.          PROB. 2.          PROB. 3.          PROB. 4.

PROB. 5.          PROB. 6.          PROB. 7.

## Group 2. Straight Lines and Circles.

**8.** Insignia. For T square, triangles, scale, and compasses. Draw the 45° diagonals and the vertical and horizontal center lines of a 4-in. square. With compass, draw a $\frac{3}{4}$-in.-diameter construction circle, a $2\frac{3}{4}$-in. circle, and a $3\frac{1}{4}$-in. circle. Complete the design by adding a square and pointed star as shown.

**9.** A six-point star. For compass and 30–60° triangle. Draw a 4-in. construction circle and inscribe the six-point star with the T square and 30–60° triangle. Accomplish this with 4 successive changes of position of the triangle.

**10.** A stamping. For T square, 30–60° triangle, and compasses. In a 4-in. circle draw six diameters 30° apart. Draw a 3-in. construction circle to locate the centers of

$\frac{5}{16}$-in.-radius circle arcs. Complete the stamping with perpendiculars to the six diameters as shown.

**11.** Insignia. The device shown is a white star with a red center on a blue background. Draw a 4-in. circle and a $1\frac{1}{4}$-in. circle. Divide the large circle into five equal parts with the dividers and construct the star by connecting alternate points as shown. Red is indicated by vertical lines and blue by horizontal lines. Space these by eye approximately $\frac{1}{16}$-in. apart.

**12.** A 24-point star. For T square and triangles in combination. In a 4-in. circle draw 12 diameters 15° apart, using T square and triangles singly and in combination. With same combinations, finish the figure as shown.

PROB. 8.

PROB. 9.

PROB. 10.

PROB. 11.

PROB. 12.

## Group 3. Circles and Tangents.

**13.** Concentric circles. For compass (legs straight) and scale. Draw a horizontal line through the center of a space. On it mark off radii for eight concentric circles $\frac{1}{4}$ in. apart. In drawing concentric circles, always draw the smallest first.

**14.** A four-centered spiral. For accurate tangents. Draw a $\frac{1}{8}$-in. square and extend its sides as shown. With the upper-right corner as center, draw quadrants with $\frac{1}{8}$- and $\frac{1}{4}$-in. radii. Continue with quadrants from each corner in order until four turns have been drawn.

**15.** A loop ornament. For bow compass. Draw a 2-in.

square, about center of space. Divide $AE$ into four $\frac{1}{4}$-in. spaces with scale. With bow pencil and centers $A$, $B$, $C$, and $D$, draw four semicircles with $\frac{1}{4}$-in. radius, and so on. Complete the figure by drawing the horizontal and vertical tangents as shown.

**16.** A rectilinear chart. For French curve. Draw a 4-in. field with $\frac{1}{2}$-in. coordinate divisions. Plot points at the intersections shown, and through them sketch a smooth curve very lightly in pencil. Finish by marking each point with a $\frac{1}{16}$-in. circle and drawing a smooth line with the French curve.

PROB. 13.

PROB. 14.

PROB. 15.

PROB. 16.

## Group 4. Scales.

**17.** Scale practice.

(*a*) Measure lines *A* to *G* to the following scales: *A*, full size; *B*, $\frac{1}{2}$ size; *C*, 3″ = 1′-0″; *D*, 1″ = 1′-0″; *E*, $\frac{3}{4}$″ = 1′-0″; *F*, $\frac{1}{4}$″ = 1′-0″; *G*, $\frac{3}{16}$″ = 1′-0″.

(*b*) Lay off distances on lines *H* to *N* as follows: *H*, $3\frac{3}{16}$″, full size; *I*, 7″, $\frac{1}{2}$ size; *J*, 2′-6″, $1\frac{1}{2}$″ = 1′-0″; *K*, 7′-$5\frac{1}{2}$″, $\frac{1}{2}$″ = 1′-0″; *L*, 10′-11″, $\frac{3}{8}$″ = 1′-0″; *M*,

28′-4″, $\frac{1}{8}$″ = 1′-0″; *N*, 40′-10″, $\frac{3}{32}$″ = 1′-0″.

(*c*) For engineer's scale. Lay off distances on lines *H* to *N* as follows: *H*, 3.2″, full size; *I*, 27′-0″, 1″ = 10′-0″; *J*, 66′-0″, 1″ = 20′-0″; *K*, 105′-0″, 1″ = 30′-0″; *L*, 156′-0″, 1″ = 40′-0″; *M*, 183′-0″, 1″ = 50′-0″; *N*, 214′-0″, 1″ = 60′-0″.

PROB. 17.

## Group 5. Combinations.

**18.** A telephone dial plate. Draw double size.

**19.** A film-reel stamping. Draw to scale of 6″ = 1′-0″.

**20.** Box cover. Make a one-view drawing for rectangular stamping 3 by 4 in., corners rounded with $\frac{1}{2}$-in. radius. Four holes, one in each corner, $\frac{3}{16}$-in. diameter, 3 and 2 in. center to center, for fasteners. Rectangular hole in center, $\frac{3}{8}$ by 1 in., with 1-in. side parallel to 4-in. side. Two slots $\frac{1}{4}$ in. wide, 2 in. long with semicircular ends, midway between center and 4-in. edges, with 2-in. side parallel to 4-in. side and centered between 3-in. edges.

**21.** Spacer. Make a one-view drawing for circular stamping 4 in. OD (outside diameter), 2 in. ID (inside diameter). Six $\frac{1}{4}$-in.-diameter holes equally spaced on 3-in.-diameter circle, with two holes on vertical center line. Two semicircular notches 180° apart made with $\frac{3}{8}$-in. radius centered at intersections of horizontal center line and 4-in.-OD circle.

**22.** Blank for wheel. Make a one-view drawing for stamping 5 in. OD; center hole $\frac{1}{2}$ in. in diameter; eight spokes $\frac{3}{8}$ in. wide connecting $1\frac{1}{2}$-in.-diameter center portion with $\frac{1}{2}$-in. rim. Eight $\frac{1}{4}$-in.-diameter holes with centers at intersection of center lines of spokes and $4\frac{1}{2}$-in. circle; $\frac{1}{8}$-in. fillets throughout to break sharp corners.

**PROB. 18.**   A telephone dial plate.

**PROB. 19.**   A film-reel stamping.

**23.** Cover plate. Make a one-view drawing for rectangular stamping 3 by 4 in., corners beveled ½ in. each way. Four holes, one in each corner, ¼-in. diameter, 3 and 2 in. center to center, for fasteners. Rectangular hole in center, ½ by 1 in. with 1-in. side parallel to 4-in. side. Two holes, ¾-in. diameter, located midway between the slot and short side of rectangle on center line through slot.

**24.** Drawing of fixture base. Full size. Drill sizes specify the diameter (see Glossary).

**25.** Drawing of gage plate. Scale, twice size. Drill sizes specify the diameter (see Glossary).

**26.** Drawing of milling fixture plate. Scale, twice size. Drill size specifies diameter (see Glossary).

**27.** Drawing of dial shaft. Use decimal scale and draw 10 times size.

**28.** Drawing of inner toggle for temperature control. Use decimal scale and draw 10 times size.

**PROB. 25.** Gage plate.

**PROB. 24.** Fixture base.

**PROB. 26.** Milling fixture plate.

**PROB. 27.** Dial shaft.

**PROB. 28.** Inner toggle for temperature control.

**29.** Drawing of mounting surface—O control. Scale, twice size.
**30.** Drawing of mounting leg—O control. Scale, full size.
**31.** Drawing of cooling fin and tube support. Full size.
**32.** Drawing of cone, sphere, and cylinder combinations. Scale, half size.
**33.** Drawing of cone-and-ball check. Scale, four times size.

**34.** Drawing of bell crank. Stamped steel. Scale, full size.
**35.** Drawing of control plate (aircraft hydraulic system). Stamped aluminum. Scale, full size.
**36.** Drawing of torque disk (aircraft brake). Stamped steel. Scale, half size.
**37.** Drawing of brake shoe. Stamped steel with molded asbestos composition wear surface. Scale, full size.

**PROB. 29.** Mounting surface: O control.

**PROB. 30.** Mounting leg: O control.

**PROB. 31.** Cooling fin and tube support.

**PROB. 32.** Cone, sphere, and cylinder combinations.

**PROB. 33.**  Cone and ball check.

**PROB. 34.**  Bell crank.

**PROB. 35.**  Control plate.

**PROB. 36.**  Torque disk.

**PROB. 37.**  Brake shoe.

# GRAPHIC GEOMETRY

Accurate, Concise Representation Requires a
Knowledge of Geometric Constructions

This Chapter Presents the Geometry of:
Straight Lines, Parallels, Perpendiculars •
Tangents, Tangent Points • Tangent Lines,
Circles, Curves • Bisections, Trisections,
Divisions • Angles • Geometric Shapes • The
Ellipse • Noncircular Curves • The
Parabola • The Hyperbola • Cycloids, Involute,
Spirals • Helices • Tangents to Curves •
Problems

## 1. Graphic Geometry—
## Comparison and Classification.

Strict interpretation of constructional geometry allows use of only the compass and an instrument for drawing straight lines, and with these the geometer, following mathematical theory, accomplishes his solutions. In graphics the principles of geometry are employed constantly, but instruments are not limited to the basic two as T square, triangles, scales, curves, etc., are used to make constructions with speed and accuracy. Since there is continual application of geometric principles, the methods given in this chapter should be mastered thoroughly. It is assumed that students using this book understand the elements of plane geometry and will be able to apply their knowledge.

The constructions given here afford excellent practice in the use of instruments. Remember that the results you obtain will be only as accurate as your skill makes them. Take care in measuring and in drawing so that your work will be accurate and professional in appearance.

For easy reference, the various geometric figures are given in Fig. 101 at the end of this chapter.

This chapter is divided, for convenience and logical arrangement, into four parts: Line Relationships

**FIG. 2.** To draw a line through two points (*alternate method*). Carefully align the triangle or T square with the points, and draw the required line.

**FIG. 3.** To draw a line parallel to another. Align a triangle with the given line *AB* using a base as shown; move it to position through the given point *P* and draw the required line.

**FIG. 4.** To draw a line parallel to and a given distance from a line. Space the distance with circle arc *R;* then align a triangle on a base as shown, move to position tangent to the circle arc, and draw the required line.

**FIG. 1.** To draw a line through two points. Use the pencil as a pivot and align the triangle or T square with the second point.

and Connections, which represent the bulk of the geometry needed in everyday work; Geometry of Straight-line Figures; Geometry of Curved Lines; and Constructions for Lofting and Large Layouts.

# LINE RELATIONSHIPS AND CONNECTIONS

## 2. To Draw Straight Lines (Figs. 1 and 2).

Straight lines are drawn by using the straight edge of the T square or one of the triangles. For short lines a triangle is more convenient. Observe the directions for technique in Secs. 2, 3, and 4.

To draw a straight line through two points (Fig. 1), place the point of the pencil at $Q$ and bring the triangle ( or T square) against the point of the pencil. Then, using this point as a pivot, swing the triangle until its edge is in alignment with point $P$, and draw the line.

To draw a straight line through two points, *alternate method* (Fig. 2), align the triangle or T square with points $P$ and $Q$, and draw the line.

## 3. To Draw Parallel Lines (Figs. 3 and 4).

Parallel lines may be required in any position. Parallel horizontals or verticals are most common. Horizontals are drawn with T square alone, verticals with T square and triangle. The general cases (odd angles) are shown in Figs. 3 and 4.

To draw a straight line through a point, parallel to another line (Fig. 3), adjust a triangle to the given line $AB$, with a second triangle as a base. Slide the aligned triangle to its position at point $P$ and draw the required line.

To draw a straight line at a given distance from and parallel to another line (Fig. 4), draw an arc with the given distance $R$ as radius and any point on the given line $AB$ as center. Then adjust a triangle to line $AB$, with a second triangle as a base. Slide the aligned triangle to position tangent to the circle arc and draw the required line.

**FIG. 5.**   To draw a line perpendicular to another (when the *given line* is *horizontal*). Place the triangle on the T square and draw the required line.

**FIG. 6.**   To draw a line perpendicular to another (general position). Align a triangle with the given line as shown; slide it (on another triangle as a base) to position of perpendicular and draw the required line.

## 4. To Draw Perpendicular Lines (Figs. 5 and 6).

Perpendiculars occur frequently as horizontal-to-vertical (Fig. 5) but also often in other positions (Fig. 6). Note that the construction of a perpendicular utilizes the 90° angle of a triangle.

To erect a perpendicular to a given straight line (*when the given line is horizontal, Fig. 5*), place a triangle on the T square as shown and draw the required perpendicular.

To erect a perpendicular to a given straight line (*general position, Fig. 6*), set a triangle with its hypotenuse against a guiding edge and adjust one side to the given line. Then slide the triangle so that the second side is in the position of the perpendicular and draw the required line.

## 5. Tangents (Figs. 7 to 20).

A *tangent* to a curve is a line, either straight or curved, that passes through two points on the curve infinitely close together. One of the most frequent geometric operations in drafting is the drawing of tangents to circle arcs and the drawing of circle arcs tangent to straight lines or other circles. These should be constructed accurately, and on pencil drawings that are to be inked or traced the points of tangency should be located by short cross marks to show the stopping points for the ink lines. The method of finding these points is indicated in the following constructions. Note in all the following tangent constructions that the location of tangent points is based on one of these geometric facts: (1) the tangent point of a straight line and circle will lie at the intersection of a perpendicular to the straight line that passes through the circle center, and (2) the tangent point of two circles will lie on the circumferences of both circles and on a straight line connecting the circle centers. See Fig. 7.

## 6. Tangent Points (Fig. 7).

To find the point of tangency for line *AB* and a circle with center *D*, draw *DC* perpendicular to line *AB*. Point *C* is the tangent point.

To find the tangent point for two circles (Fig. 7) with centers at *D* and *E*, draw *DE*, joining the centers. Point *P* is the tangent point.

## 7. To Draw a Circle of Given Size Tangent to a Line and Passing through a Point (Fig. 8).

Draw a line *AB*, the given radius distance *R* away from and parallel to the given line. Using the given point *S* as center, cut line *AB* at *O* with the given radius. *O* is the center of the circle. Note that there are two possible positions for the circle.

FIG. 7. Tangent points. The tangent point of a straight line and circle is on the perpendicular from the circle center. The tangent point of two circles is on a line connecting their centers.

FIG. 8. A circle tangent to a line and passing through a point. The circle center must be equidistant from the line and the point.

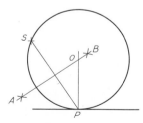

FIG. 9. A circle tangent to a line at a point and passing through a second point. The circle center is at the intersection of the bisector of the chord connecting the points and the perpendicular from the point on the line tangent.

## 8. To Draw a Circle Tangent to a Line at a Point and Passing through a Second Point (Fig. 9).

Connect the two points $P$ and $S$ and draw the perpendicular bisector $AB$ (see Sec. 38). Draw a perpendicular to the given line at $P$. The point where this perpendicular intersects the line $AB$ is the center $O$ of the required circle.

## 9. To Draw a Tangent to a Circle at a Point on the Circle (Fig. 10).

Given the arc $ABC$, draw a tangent at the point $C$. Arrange a triangle in combination with the T square (or another triangle) so that its hypotenuse passes through center $O$ and point $C$. Holding the T square firmly in place, turn the triangle about its square corner and move it until the hypotenuse passes through $C$. The required tangent then lies along the hypotenuse.

## 10. To Draw a Tangent to a Circle from a Point Outside (Fig. 11).

Given the arc $ACB$ and point $P$, arrange a triangle in combination with another triangle (or T square) so that one side passes through point $P$ and is tangent to the circle arc. Then slide the triangle until the right-angle side passes through the center of the circle and mark lightly the tangent point $C$. Bring the triangle back to its original position and draw the tangent line.

## 11. To Draw a Circle Arc of Given Radius Tangent to Two Lines at Right Angles to Each Other (Fig. 12).

Draw an arc of radius $R$, with center at corner $A$, cutting the lines $AB$ and $AC$ at $T$ and $T_1$. Then with $T$ and $T_1$ as centers and with the same radius $R$, draw arcs intersecting at $O$, the center of the required arc.

**FIG. 10.** A tangent at a point on a circle. The tangent line must be perpendicular to the line from the point to the center of the circle.

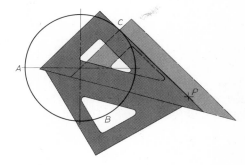

**FIG. 11.** A tangent to a circle from a point outside. The line is drawn by alignment through the point and tangent to the circle; the tangent point is then located on the perpendicular from the circle center.

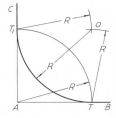

**FIG. 12.** An arc tangent at right-angle corner. The arc center must be equidistant from both lines.

## 12. To Draw an Arc of Given Radius Tangent to Two Straight Lines (Fig. 13).

Given the lines $AB$ and $CD$, set the compass to radius $R$, and at any convenient point on the given lines draw the arcs $R$ and $R_1$. With the method of Fig. 4, draw parallels to the given lines through the limits of the arcs. These parallels are the loci of the centers of all circles of radius $R$ tangent to lines $AB$ and $CD$, and their intersection at point $O$ will be the center of the required arc. Find the tangent points by erecting perpendiculars, as in Fig. 11, to the given lines through the center $O$. The figure above shows the method for an obtuse angle, below for an acute angle.

## 13. To Draw a Tangent to Two Circles (Fig. 14, Open Belt).

Arrange a triangle in combination with a T square or triangle so that one side is in the tangent position. Move to positions 2 and 3, marking lightly the tangent points $T_1$ and $T_2$. Return to the original position and draw the tangent line. Repeat for the other side.

## 14. To Draw a Tangent to Two Circles (Fig. 15, Crossed Belt).

Arrange a triangle in combination with a T square or triangle so that one side is in the tangent position. Move to positions 2 and 3, marking lightly the tangent points $T_1$ and $T_2$. Return to the original position and draw the tangent line. Repeat for the other side.

## 15. To Draw a Circle of Radius $R$ Tangent to a Given Circle and a Straight Line (Fig. 16).

Let $AB$ be the given line and $R_1$ the radius of the given circle. Draw a line $CD$ parallel to $AB$ at a distance $R$ from it. With $O$ as center and radius $R + R_1$, swing an arc intersecting $CD$ at $X$, the desired center. The tangent point for $AB$ will be on a perpendicular to $AB$ from $X$; the tangent point for the two circles will be on a line joining their centers $X$ and $O$.

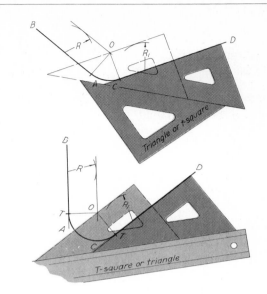

FIG. 13. An arc tangent to two straight lines. The arc center must be equidistant from both straight lines.

FIG. 14. Tangents to two circles (open belt). Tangent lines are drawn by alignment with both circles. Tangent points lie on perpendicular lines from circle centers.

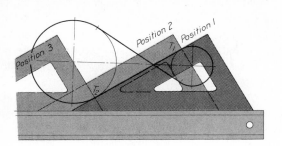

**FIG. 15.** Tangents to two circles (crossed belt). Tangent lines are drawn by alignment with both circles. Tangent points lie on perpendicular lines from circle centers.

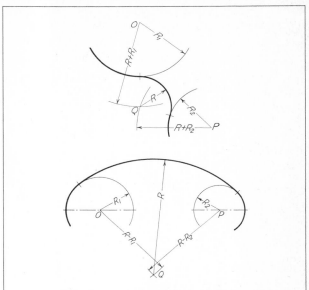

**FIG. 17.** An arc tangent to two circles. The arc center must be equidistant from both circles.

**FIG. 16.** An arc tangent to a straight line and a circle. The arc center must be equidistant from the line and the circle.

## 17. To Draw a Reverse, or Ogee, Curve (Fig. 18).

Given two parallel lines $AB$ and $CD$, join $B$ and $C$ by a straight line. Erect perpendiculars at $B$ and $C$. Any arcs tangent to lines $AB$ and $CD$ at $B$ and $C$ must have their centers on these perpendiculars. On

## 16. To Draw a Circle of Radius $R$ Tangent to Two Given Circles.

*First Case* (*Fig. 17, Top*). The centers of the given circles are outside the required circle. Let $R_1$ and $R_2$ be the radii of the given circles and $O$ and $P$ their centers. With $O$ as center and radius $R + R_1$, describe an arc. With $P$ as center and radius $R + R_2$, swing another arc intersecting the first arc at $Q$, which is the center sought. Mark the tangent points in line with $OQ$ and $QP$.

*Second Case* (*Fig. 17, Bottom*). The centers of the given circles are inside the required circle. With $O$ and $P$ as centers and radii $R - R_1$ and $R - R_2$, describe arcs intersecting at the required center $Q$.

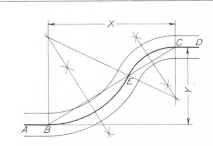

**FIG. 18.** An ogee curve. It is made of circle arcs tangent to each other and to straight lines.

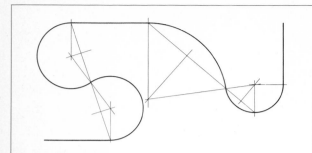

**FIG. 19.** Ogee applications. Note that the circle arcs are tangent on a line connecting the centers.

the line *BC* assume point *E*, the point through which it is desired that the curve shall pass. Bisect *BE* and *EC* by perpendiculars. Any arc to pass through *B* and *E* must have its center somewhere on the perpendicular from the middle point. The intersection, therefore, of these perpendicular bisectors with the first two perpendiculars will be the centers for arcs *BE* and *EC*. This line might be the center line for a curved road or pipe. The construction may be checked by drawing the line of centers, which *must* pass through *E*. Figure 19 illustrates the principle of reverse-curve construction in various combinations.

### 18. To Draw a Reverse Curve Tangent to Two Lines and to a Third Secant Line at a Given Point (Fig. 20).

Given two lines *AB* and *CD* cut by the line *EF* at points *E* and *F*, draw a perpendicular *JH* to *EF* through a given point *P* on *EF*. With *E* as center and radius *EP*, intersect *CD* at *G*. Draw a perpendicular from *G* intersecting *JH* at *H*. With *F* as center and radius *FP*, intersect *AB* at *K*. Draw a perpendicular to *AB* from *K* intersecting *JH* at *J*. *H* and *J* will be the centers for arcs tangent to the three lines.

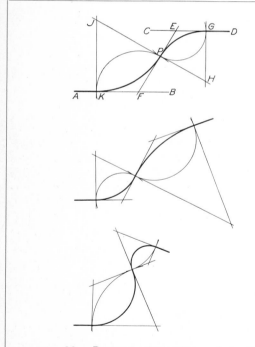

**FIG. 20.** Reverse curve tangent to three lines. Note that the arc centers must lie on a line perpendicular to the tangent line at the point of tangency.

## GEOMETRY OF STRAIGHT-LINE FIGURES

### 19. To Bisect a Line (Fig. 21).

*With Compass.* From the two ends of the line, swing arcs of the same radius, greater than one-half the length of the line, and draw a line through the arc intersections. This line bisects the given line and is also the *perpendicular* bisector. Many geometric problems depend upon this construction.

*With T Square and Triangles (Fig. 22).* At two points, *A* and *B*, on the line, draw lines *AC* and *BC* at equal angles with *AB*. A perpendicular *CD* to *AB* then cuts *AB* at *D*, the midpoint, and *CD* is the perpendicular bisector of *AB*.

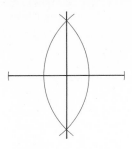

**FIG. 21.** To bisect a line (with compass). The intersections of equal arcs locate two points on the perpendicular bisector.

**FIG. 22.** To bisect a line (with T square and triangle). Equal angles locate one point on the perpendicular bisector.

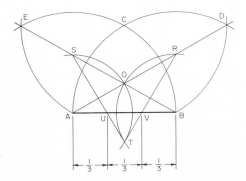

**FIG. 23.** To trisect a line (with compass). Construction is based on the geometry of an equilateral triangle.

## 20. To Trisect a Line.

Trisection of a line is not nearly so often needed as bisection, but will occasionally be required.

*With Compasses and Straightedge (Fig. 23).* On given line *AB* and with radius *AB*, draw two arcs of somewhat more than quarter circles, using *A* and *B* as centers. These arcs will intersect at *C*. Using the same radius *AB* and with *C* as center cut the first arcs at *D* and *E*. Then draw *DA* and *EB*, which intersect at *O*. Using *OA* (or *OB*) as radius and *A* and *B* as centers, cut *AD* and *BE* at *R* and *S*, extend these arcs to intersect at *T*. Then draw *RT* and *ST*, which will intersect *AB* at third points *U* and *V*. You will recognize, upon analysis, that this construction is based on the geometry of an equilateral triangle inscribed in a circle, where the diameter of the circle is equal to one side of the triangle.

*With T Square and Triangles (Fig. 24).* The above construction can be accomplished with less detail by using a 30-60° triangle to obtain the needed angles. From *A* and *B* draw lines at 30° to intersect at *C*. Then at 60° to *AB* draw lines from *C* that intersect *AB* at *U* and *V*, the third points.

## 21. To Divide a Line into 2, 3, 4, 6, 8, 9, 12, or 16 Parts (Figs. 25 and 26).

Divisions into successive halves, 2, 4, 8, etc., are the most common. The principle of Fig. 22 can easily be applied to accomplish the division. As shown in Fig. 25, equal angles from *A* and *B* locate *C*, and the perpendicular from *C* to *AB* then gives the mid-point. Successive operations will give 4, 8, 16, etc., parts.

The principles of bisection (Fig. 22) and trisection (Fig. 24) can be combined to get 6, 9, or 12 parts. As indicated in Fig. 26, lines at 30° to *AB* from *A* and *B* locate *C*, and lines at 60° to *AB* from *C* locate *D* and *E*, the third points. Then 30° lines from *A* and *D* will bisect *AD* at *F*, giving sixth divisions; or trisecting by the principle of Fig. 24 as from *D* and *E*, giving *G* and *H*, will produce ninth points; or as from *E* and *B*, the half point located at *J*, and again the half points of *EJ* and *BJ* located at *K* and *L*, give twelfth divisions of *AB*.

**FIG. 24.** To trisect a line (with T square and triangle). Third points are located by 30° and 60° angles.

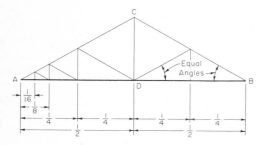

**FIG. 25.** To divide a line into 2, 4, 8, or 16 parts. Equal angles locate a point for bisection.

**FIG. 26.** To divide a line into 3, 6, 9, or 12 parts. The 30-60° triangle is used to obtain third points.

## 22. To Divide a Line into Any Number of Parts.

*First Method (Fig. 27).* To divide a line $AB$ into, say, five equal parts, draw any line $BC$ of indefinite length. On it measure, or step off, five divisions of convenient length. Connect the last point with $A$, and using two triangles as shown in Fig. 3, draw lines through the points parallel to $CA$ intersecting $AB$.

*Second Method (Figs. 28 and 29).* Draw a perpendicular $AC$ from $A$. Then place a scale so that five convenient equal divisions are included between $B$ and the perpendicular, as in Fig. 28. With a triangle and T square draw perpendiculars through the points marked, dividing the line $AB$ as required. Figure 29 illustrates an application in laying off stair risers. This method can be used for dividing a line into any series of proportional parts.

## 23. To Lay Out a Given Angle.

*Tangent Method (Fig. 30).* The trigonometric tangent of an angle of a triangle is the ratio of the length of the side opposite the angle divided by the length of the adjacent side. Thus, tan $A = Y/X$, or $X$ tan $A = Y$. To lay out a given angle, obtain the value of the tangent from a table of natural tangents (see the Appendix), assume any convenient distance $X$, and multiply $X$ by the tangent to get distance $Y$. Note that the angle between the sides $X$ and $Y$ must be a right angle.

## 24. To Lay Out a Given Angle.

*Chordal Method (Fig. 31).* If the length of a chord is known for an arc of given radius and included angle, the angle can be accurately laid out. Given an angle in degrees, to lay out the angle, obtain the chord length for a 1-in. circle arc from the table in the Appendix. Select any convenient arc length $R$ and multiply the chord length for a 1-in. arc by this distance, thus obtaining the chord length $C$ for the radius distance selected. Lay out the chord length on the arc with compass or dividers and complete the sides of the angle.

The chord length for an angle can be had from a sine table by taking the sine of one-half the given angle and multiplying by two.

## 25. To Lay Out an Angle of 45° (Fig. 32).

Angles of 45° occur often, and are normally drawn with the 45° triangle. However, for large constructions or when great accuracy is needed, the method of equal legs is valuable (tangent 45° =

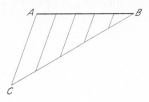

**FIG. 27.** To divide a line. Equal divisions on any line *BC* are transferred to *AB* by parallels to *AC*.

**FIG. 28.** To divide a line. Scale divisions are transferred to given line *AB*.

**FIG. 29.** To divide a line. Scale divisions divide (in this case) the vertical space.

**FIG. 30.** Angle by tangent. The proportion of *Y* to *X* is obtained from a table of natural tangents.

**FIG. 31.** Angle by chord. The proportion of *C* to *R* is obtained from a table of chords.

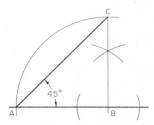

**FIG. 32.** To lay out a 45° angle. The two legs, *AB* and *BC*, are equal and perpendicular.

**FIG. 33.** To construct angles in multiples of 30°. Chords equal to the radius will divide a circle into six equal parts.

1.0). With any distance *AB* on the given line, with center *B* and radius *AB* draw more than a quarter circle. Erect a perpendicular at *B* (to *AB*) which intersects the arc at point *C*. Then *CA* makes 45° with *AB*.

## 26. To Lay Out Angles of 30°, 60°, 90°, 120°, etc. (Fig. 33).

As with 45° angles, angles in multiples of 30° may also be needed for large constructions or when great accuracy is required. On a given line, with any convenient radius *AB*, swing an arc with *A* as center.

With the same radius and *B* as center cut the original arc at *C*. The included angle *CAB* is 60° and by bisection (equal arcs from *C* and *B* to locate *E*) 30° is obtained at *CAE* and *EAB*. For 120°, *AB* laid off again from *C* to *D* gives *DAB* as 120°, and by bisecting *DAC*, 90° is given at *FAB*.

### 27. To Bisect an Angle (Fig. 34).

Given angle *SOR*, using any convenient radius shorter than *OR* or *OS*, swing an arc with *O* as center, locating *A* and *B*. Then with the same radius or another radius longer than one-half the distance from *A* to *B*, swing arcs with *A* and *B* as centers, locating *C*. The bisector is *CO*. With repeated bisection, the angle can be divided into 4, 8, 16, 32, 64, etc., parts.

### 28. To Divide an Angle (Fig. 35).

Sometimes an angle must be divided into some number of parts, say 3, 5, 6, 7, or 9, that cannot be obtained by bisection. It is possible to trisect an angle by employing a movable scale, but this method is more complicated and not as accurate as dividing an arc with the dividers and is of no use for odd parts such as 5 and 7.

To divide an angle into, for example, 5 parts (Fig. 35), draw any arc across the angle as shown, and use the bow dividers to divide the arc into equal parts. Lines through these points to the apex are the required divisions.

### 29. To Construct a Triangle Having Given the Three Sides (Fig. 36).

Given the lengths *A*, *B*, and *C*. Draw one side *A* in the desired position. With its ends as centers and radii *B* and *C*, draw two intersecting arcs as shown. This construction is used extensively in developments by triangulation.

### 30. To Locate the Geometric Center of an Equilateral Triangle (Fig. 37).

With *A* and *B* as centers and radius *AB*, draw arcs as shown. These will intersect at *C*, the third corner.

FIG. 34.   To bisect an angle. Equal arcs from *A* and *B* locate *C* on the bisector.

FIG. 35.   To divide an angle into equal parts. Equal divisions (chords) are stepped off on an arc of the angle.

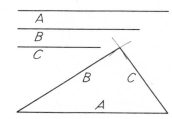

FIG. 36.   To construct a triangle. The legs are laid off with a compass.

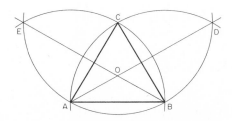

FIG. 37.   To locate the geometric center of an equilateral triangle. Perpendicular bisectors of the sides intersect at the center.

With center *C* and radius *AB* cut the original arcs at *D* and *E*. Then *AD* and *BE* intersect at *O*, the geometric center.

### 31. To Transfer a Polygon to a New Base.

*By Triangulation (Fig. 38).* Given polygon *ABCDEF* and a new position of base *A'B'*, consider each point as the vertex of a triangle whose base is *AB*. With centers *A'* and *B'* and radii *AC* and *BC*, describe intersecting arcs locating the point *C'*. Similarly, with radii *AD* and *BD* locate *D'*. Connect *B'C'* and *C'D'* and continue the operation always using *A* and *B* as centers.

*Box or Offset Method (Fig. 39).* Enclose the polygon in a rectangular "box." Draw the box on the new base and locate the points *ABCEF* on this box. Then set point *D* by rectangular coordinates as shown.

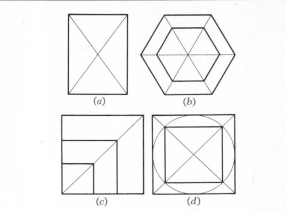

**FIG. 40.** Uses of the diagonal. The diagonals will locate the center of any regular or symmetrical geometric shape that has an even number of sides.

**FIG. 38.** To transfer a polygon: by triangulation. All corners are located by triangles having a common base.

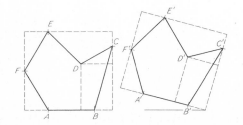

**FIG. 39.** To transfer a polygon: by "boxing." Each corner is located by its position on a rectangle.

### 32. Uses of the Diagonal.

The diagonal is used in many ways to simplify construction and save drafting time. Figure 40 illustrates the diagonal used at (*a*) for locating the center of a rectangle, at (*b*) for enlarging or reducing a geometric shape, at (*c*) for producing similar figures having the same base, and at (*d*) for drawing inscribed or circumscribed figures.

### 33. To Construct a Regular Hexagon, Given the Distance across Corners.

*First Method (Fig. 41).* Draw a circle with *AB* as a diameter. With the same radius and *A* and *B* as centers, draw arcs intersecting the circle and connect the points.

*Second Method (without Compass).* Draw lines with the 30-60° triangle in the order shown in Fig. 42.

### Given the Distance across Flats.

The distance across flats is the diameter of the inscribed circle. Draw this circle, and with the 30-60° triangle draw tangents to it as in Fig. 43.

**FIG. 41.** A hexagon. Arcs equal to the radius locate all points. (Corner distance *AB* is known.)

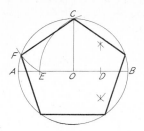

**FIG. 44.** To inscribe a regular pentagon in a circle. Arc construction locates one side; others are stepped off.

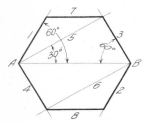

**FIG. 42.** A hexagon. The 30-60° triangle gives construction for all points. (Corner distance *AB* is known.)

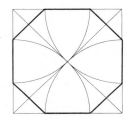

**FIG. 45.** To inscribe a regular octagon in a square. Arcs from the corners of the enclosing square locate all points.

**FIG. 43.** A hexagon. Tangents drawn with the 30-60° triangle locate all points. (Across-flats distance is known.)

**FIG. 46.** To construct a regular polygon. Given one side, a polygon of any number of sides can be drawn by this method.

## 34. To Inscribe a Regular Pentagon in a Circle (Fig. 44).

Draw a diameter *AB* and a radius *OC* perpendicular to it. Bisect *OB*. With this point *D* as center and radius *DC*, draw arc *CE*. With center *C* and radius *CE*, draw arc *EF*. *CF* is a side of the pentagon. Step off this distance around the circle with dividers.

## 35. To Inscribe a Regular Octagon in a Square (Fig. 45).

Draw the diagonals of the square. With the corners of the square as centers and a radius of half the diagonal, draw arcs intersecting the sides and connect these points.

## 36. To Construct a Regular
## Polygon, Given One Side (Fig. 46).

Let the polygon have seven sides. With the side *AB* as radius and *A* as center, draw a semicircle and divide it into seven equal parts with dividers. Through the second division from the left draw radial line *A-2*. Through points, *3, 4, 5,* and *6* extend radial lines as shown. With *AB* as radius and *B* as center, cut line *A-6* at *C*. With *C* as center and the same radius, cut *A-5* at *D*, and so on at *E* and *F*. Connect the points *or*, after *A-2* is found, draw the circumscribing circle. Another method is to guess at *CF* and divide the circle by trial; this is more common than the geometric method because the intersections of arcs are sometimes difficult.

# GEOMETRY OF CURVED LINES

## 37. Definitions.

A *curved line* is generated by a point moving in a constantly changing direction, according to some mathematical or graphic law. Curved lines are classified as single-curved or double-curved.

A *single-curved line* is a curved line having all points of the line in a plane. Single-curved lines are often called "plane curves."

A *double-curved line* is a curved line having no four consecutive points in the same plane. Double-curved lines are also known as "space curves."

We have defined (Sec. 5) a *tangent* to a curved line as a line, either straight or curved, that passes through two points on the curve that are infinitely close together. Note that this definition of tangency places the tangent in the plane of the curve for single-curved lines and in an instantaneous plane of the curve for double-curved lines.

A *normal* is a line (or plane) perpendicular to a tangent line of a curve at the point of tangency. A normal to a single-curved line will be a line in the plane of the curve and perpendicular to a straight line connecting two consecutive points on the curve.

## 38. Arc and Circle Centers
## (Figs. 47 to 50).

The center of any arc must lie on the perpendicular bisector of any and all chords of the arc. Thus, in

Application of circles and arcs

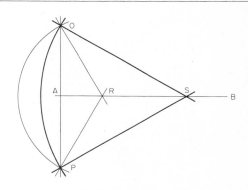

**FIG. 47.** Arc centers. The center of any arc lies on the perpendicular bisector of any chord of the arc.

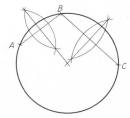

**FIG. 48.** Circle through three points. Perpendicular bisectors of chords locate the center.

Fig. 47, through two points $O$ and $P$, an infinite number of arcs may be drawn but all will have centers such as $R$ and $S$ which lie on the perpendicular bisector $AB$ of chord $OP$. This construction is used in Fig. 9 where the circle is required to pass through points $S$ and $P$.

Using the above principle, only one circle can be drawn through three points, as illustrated in Fig. 48. The center must lie on the perpendicular bisectors of both chords, $AB$ and $BC$. Incidentally, the center will also lie on the perpendicular bisector of chord $AC$. This construction can be used as a check on accuracy.

Any diameter of a circle and a third point on the circumference, connected to form a triangle, will produce two chords of the circle that are perpen-

dicular to each other. Therefore, the center of a circle may be found, as in Fig. 49, by selecting any two points such as $A$ and $B$, drawing the perpendicular chords $AC$ and $BD$ and then the two diameters $CB$ and $DA$, which cross at center $O$. Note in Fig. 50 that all chords connecting the diameter $AB$—$AD$ and $DB$, $AC$ and $CB$, $AE$ and $EB$, $AF$ and $FB$—form a right angle between the chords.

### 39. To Inscribe a Circle in an Equilateral Triangle (Fig. 51).
With center $A$ and radius $AB$, draw an arc as shown. With $B$ and $C$ as centers and radius $AB$, cut this arc at $D$ and $E$. Then $EC$ and $DB$ intersect at $O$, the geometric center, and the radius of the subscribing circle is $OG$ and/or $OF$.

### 40. To Inscribe an Equilateral Triangle in a Circle (Fig. 52).
Using the radius of the circle and starting at $A$, the known point of orientation, step off distances $AB$, $BC$, $CD$, and $DE$. The equilateral triangle is $ACE$. Note that six chords equal in length to the radius will give six equally spaced points on the circumference.

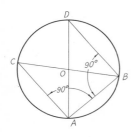

FIG. 49. To locate the center of a circle. Perpendicular chords to chord $AB$ produce diameters $BC$ and $AD$.

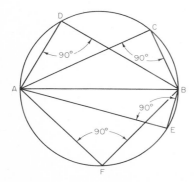

FIG. 50. Chords connecting a diameter. Chords from a point, connected to a diameter, form a right angle between the chords.

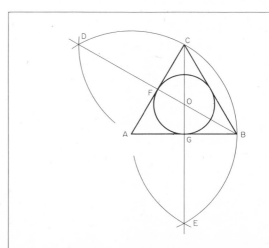

FIG. 51. To inscribe a circle in an equilateral triangle. Construction (to get perpendicular bisectors of two sides) locates the circle center.

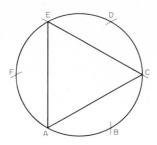

**FIG. 52.** To inscribe an equilateral triangle in a circle. By stepping off the radius twice, corners are located. Compare with Fig. 41.

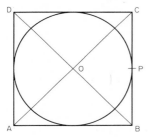

**FIG. 53.** To inscribe a circle in a square. Diagonals locate the circle center.

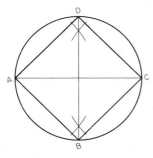

**FIG. 54.** To inscribe a square in a circle. The perpendicular bisector of one diameter locates the two corners to be found.

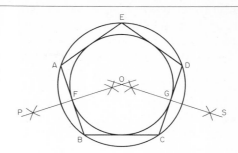

**FIG. 55.** To draw a circle on a regular polygon. Perpendicular bisectors of two sides locate the center.

**FIG. 56.** To approximate the length of an arc. See footnote 1 for accuracy.

### 42. To Inscribe a Square in a Circle (Fig. 54).
From $A$, the known point of orientation, draw diameter $AC$ and then erect $BD$, the perpendicular bisector of $AC$. The square is $ABCD$.

### 43. To Scribe a Circle in or on a Regular Polygon (Fig. 55).
Draw perpendicular bisectors of any two sides, for example $PO$ and $SO$ of $AB$ and $CD$, giving $O$, the center. The inner-circle radius is $OF$ or $OG$ and the external-circle radius is $OA$, $OB$, etc.

### 44. To Lay off on a Straight Line the Approximate Length of a Circle Arc (Fig. 56).
Given the arc $AB$. At $A$ draw the tangent $AD$ and the chord produced, $BA$. Lay off $AC$ equal to half the chord $AB$. With center $C$ and radius $CB$, draw an arc intersecting $AD$ at $D$; then $AD$ will be equal in length to arc $AB$ (very nearly).[1] If the given arc

[1] In this (Professor Rankine's) solution, the error varies as the fourth power of the subtended angle. For 60° the line will be $1/900$ part short, while at 30° it will be only $1/14,400$ part short.

### 41. To Inscribe a Circle in a Square (Fig. 53).
Given the circle $ABCD$, draw diagonals $AC$ and $BD$, which intersect at point $O$, the center. $OP$ is the radius of the circle.

is between 45° and 90°, a closer approximation will result by making $AC$ equal to the chord of half the arc instead of half the chord of the arc.

The usual way of rectifying an arc is to set the dividers to a space small enough to be practically equal in length to a corresponding part of the arc. Starting at $B$, step along the arc to the point nearest $A$ and without lifting the dividers step off the same number of spaces on the tangent, as shown in Fig. 57.

### 45. To Lay Off on a Given Circle the Approximate Length of a Straight Line (Fig. 58).

Given the line $AB$ tangent to the circle at $A$. Lay off $AC$ equal to one-fourth $AB$. With $C$ as center and radius $CB$, draw an arc intersecting the circle $D$. The line $AD$ is equal in length to $AB$ (very nearly).[1] If arc $AD$ is greater than 60°, solve for one-half $AB$.

### 46. Plane Curves: the Conic Sections.

In cutting a right-circular cone (a cone of revolution) by planes at different angles, we obtain four curves called "conic sections" (Fig. 59). These are the *circle*, cut by a plane perpendicular to the axis; the *ellipse*, cut by a plane making a greater angle with the axis than do the elements; the *parabola*, cut by a plane making the same angle with the axis as do the elements; the *hyperbola*, cut by a plane making a smaller angle than do the elements.

### 47. The Ellipse: Major and Minor Diameters (Fig. 60).

An ellipse is the plane curve generated by a point moving so that the sum of its distances from two fixed points ($F_1$ and $F_2$), called "focuses," is a constant equal to the major axis,[2] or *major diameter, AB.*

The minor axis, or *minor diameter, DE,* is the line through the center perpendicular to the major diam-

[2]"Major axis" and "minor axis" are the traditional terms. However, because of the confusion between these and the mathematical $X$ and $Y$ axes and the central axis (axis of rotation), the terms "major diameter" and "minor diameter" will be used in this discussion.

**FIG. 57.** To approximate the length of an arc. The sum of short chords closely approaches the arc length.

**FIG. 58.** To lay off, on an arc, a specified distance. See footnote 1 for accuracy.

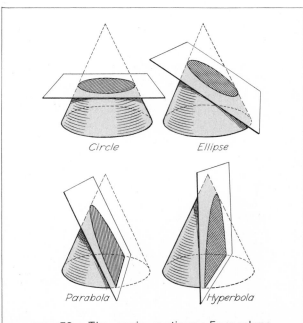

Circle    Ellipse

Parabola    Hyperbola

**FIG. 59.** The conic sections. Four plane curves are produced by "cutting" a cone.

Application of the ellipse (intersection)

*Method (Fig. 62).* The conjugate diameters are *CN* and *JG*. With center *O* and radius *OJ*, draw a semicircle intersecting the ellipse at *P*. The major and minor diameters will be parallel to the chords *GP* and *JP*, respectively.

eter. The focuses may be determined by cutting the major diameter with an arc having its center at an end of the minor diameter and a radius equal to one-half the major diameter.

Aside from the circle, the ellipse is met with in practice much more often than any of the other conics, so it is important to be able to construct it readily. Several methods are given for its construction, both as a true ellipse and as an approximate curve made by circle arcs. In the great majority of cases when the curve is required, its major and minor diameters are known.

The mathematical equation of an ellipse with the origin of rectilinear coordinates at the center of the ellipse is $x^2/a^2 + y^2/b^2 = 1$, where $a$ is the intercept on the $X$ axis and $b$ is the intercept on the $Y$ axis.

## 48. The Ellipse: Conjugate Diameters (Fig. 61).

Any line through the center of an ellipse may serve as *one* of a pair of conjugate diameters. Each of a pair of conjugate diameters is always parallel to the tangents to the curve at the extremities of the other. For example, *AB* and *CD* are a pair of conjugate diameters. *AB* is parallel to the tangents *MN* and *PQ*; and *CD* is parallel to the tangents *MP* and *NQ*. Also, each of a pair of conjugate diameters bisects all the chords parallel to the other. A given ellipse may have an unlimited number of pairs of conjugate diameters.

*To Determine the Major and Minor Diameters, Given the Ellipse and a Pair of Conjugate Diameters. First*

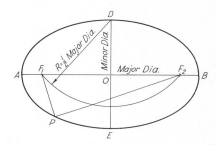

**FIG. 60.** The ellipse: major and minor diameters. These are perpendicular to each other.

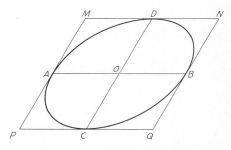

**FIG. 61.** The ellipse: conjugate diameters. Each is parallel to the tangent at the end of the other.

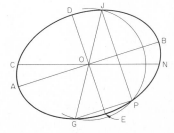

**FIG. 62.** Determination of major and minor diameters from conjugate diameters. Curve is given.

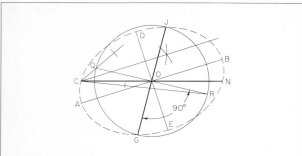

**FIG. 63.** Determination of major and minor diameters from conjugate diameters. Curve is not given.

*Second Method: When the Curve Is Not Given (Fig. 63).* The conjugate diameters *CN* and *JG* are given. With center *O* and radius *OJ*, describe a circle and draw the diameter *QR* at right angles to *JG*. Bisect the angle *QCR*. The major diameter will be parallel to this bisector and equal in length to $CR + CQ$; that of the minor diameter will be $CR - CQ$.

### 49. Ellipse Construction: Pin-and-string Method.

This well-known method, sometimes called the "gardener's ellipse," is often used for large work and is based on the definition of the ellipse. Drive pins at the points *D*, $F_1$, and $F_2$ (Fig. 60), and tie an inelastic thread or cord tightly around the three pins. If the pin *D* is removed and a marking point moved in the loop, keeping the cord taut, it will describe a true ellipse.

### 50. Ellipse Construction: Trammel Method for Major and Minor Diameters.

*First Method (Fig. 64).* On the straight edge of a strip of paper, thin cardboard, or sheet of celluloid, mark the distance *ao* equal to one-half the major diameter and *do* equal to one-half the minor diameter. If the strip is moved, keeping *a* on the minor diameter and *d* on the major diameter, *o* will give points on the ellipse. This method is convenient as no construction is required, but for accurate results great care must be taken to keep the points *a* and *d* exactly on the major and minor diameters.

*Second Method (Fig. 65).* On a strip—as used in the first method—mark the distance *do* equal to one-half the minor diameter and *oa* equal to one-half the major diameter. If this strip is moved, keeping *a* on the minor diameter and *d* on the major diameter, *o* will give points on the ellipse. This arrangement is preferred where the ratio between the major and minor diameters is small.

### 51. Ellipse Construction: Triangle Trammel for Conjugate Diameters (Fig. 66).

The conjugate diameters *AB* and *DE* are given. Erect the perpendicular *AC* to the diameter *ED* and lay off distance *AC* from *E* to locate point *G*. Erect the perpendicular *EF* to the diameter *AB*. Transfer to a piece of paper, thin cardboard, or sheet of celluloid, and cut out the triangle *EFG*. If this triangle is moved, keeping *f* on *AB* and *g* on *ED*, *e* will give points on the ellipse. Extreme care must be taken to keep points *f* and *g* on the conjugate diameters.

### 52. Ellipse Construction: Concentric-circle Method for Major and Minor Diameters (Fig. 67).

This is perhaps the most accurate method for determining points on the curve. On the two principal diameters, which intersect at *O*, describe circles. From a number of points on the outer circle, as *P* and *Q*, draw radii *OP*, *OQ*, etc., intersecting the inner circle at *P'*, *Q'*, etc. From *P* and *Q* draw lines parallel to *OD*, and from *P'* and *Q'* draw lines parallel to *OB*. The intersection of the lines through *P* and *P'* gives one point on the ellipse, the intersection of the lines through *Q* and *Q'* another point, and so on. For accuracy, the points should be taken closer together toward the major diameter. The process may be repeated in each of the four quadrants and the curve sketched in lightly freehand; or one quadrant only may be constructed and repeated in the remaining three by marking the French curve.

**FIG. 64.** An ellipse by trammel (first method). Points on the curve are plotted.

**FIG. 65.** An ellipse by trammel (second method). Points on the curve are plotted.

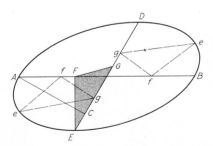

**FIG. 66.** An ellipse by triangle trammel (conjugate diameters). Points are plotted at the apex of the triangle.

## 53. Ellipse Construction: Circle Method for Conjugate Diameters (Fig. 68).

The conjugate diameters *AB* and *DE* are given. On the conjugate diameter *AB*, describe a circle; then from a number of points, as *P*, *Q*, and *S*, draw perpendiculars as *PP'*, *QO*, and *SS'* to the diameter *AB*. From *S* and *P*, etc., draw lines parallel to *QD*, and from *S'* and *P'* draw lines parallel to *OD*. The intersection of the lines through *P* and *P'* gives one

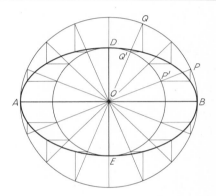

**FIG. 67.** An ellipse by concentric-circle method. Major- and minor-diameter circles and construction give points on the curve.

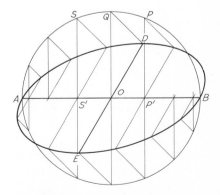

**FIG. 68.** An ellipse by circle method (conjugate diameters). After first construction, parallels plot points on the curve.

point on the ellipse, the intersection of the lines through S and S' another point, and so on.

## 54. Ellipse Construction: Parallelogram Method (Figs. 69 and 70).

This method can be used either with the major and minor diameters or with any pair of conjugate diameters. On the given diameters construct a parallelogram. Divide AO into any number of equal parts and AG into the same number of equal parts, numbering points from A. Through these points draw lines from D and E, as shown. Their intersections will be points on the curve.

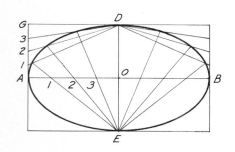

**FIG. 69.** An ellipse by parallelogram method. Points are plotted by construction through equal divisions of AO and AG.

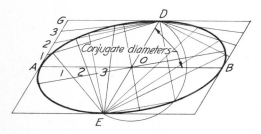

**FIG. 70.** An ellipse by parallelogram method (conjugate diameters). The constructive method of Fig. 69 is used, but here applied to conjugate diameters.

## 55. To Draw a Tangent to an Ellipse.

*At a Given Point on the Curve* (Fig. 71). Draw lines from the given point P to the focuses. The line bisecting the exterior angle of these focal radii is the required tangent.

*Parallel to a Given Line* (Fig. 71). Draw $F_1E$ perpendicular to the given line GH. With $F_2$ as center and radius AB, draw an arc cutting $F_1E$ at K. The line $F_1K$ cuts the ellipse at the required point of tangency T, and the required tangent passes through T parallel to GH.

*From a Point Outside* (Fig. 72). Find the focuses $F_1$ and $F_2$. With the given point P and radius $PF_2$, draw the arc $RF_2Q$. With $F_1$ as center and radius AB, strike an arc cutting this arc at Q and R. Connect $QF_1$ and $RF_1$. The intersections of these lines with the ellipse at $T_1$ and $T_2$ will be the tangent points of tangents to the ellipse from P.

*Concentric-circle Method* (Fig. 73). When the ellipse has been constructed by the concentric-circle method (Fig. 67), a tangent at any point H can be drawn by dropping a perpendicular to AB from the point to the outer circle at K and drawing the auxiliary tangent KL to the outer circle, cutting the major diameter at L. From L draw the required tangent LH.

## 56. To Draw a Normal to an Ellipse (Fig. 73).

From point P on the curve, project a parallel to the minor diameter to intersect the major diameter circle at Q. Draw OQ extended to intersect (at N) an arc with center at O and radius AO + OE. NP is the required normal.

*Or,* normals may be drawn perpendicular to the tangents of Figs. 71 and 72.

## 57. To Draw a Four-centered Approximate Ellipse (Fig. 74).

Join A and D. Lay off DF equal to AO − DO. This is done graphically as indicated on the figure by swinging A around to A' with O as center where now DO from OA' is DA', the required distance. With D as center, an arc from A' to the diagonal AD locates F. Bisect AF by a perpendicular crossing AO

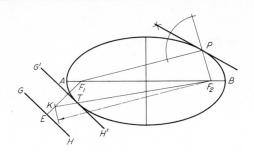

**FIG. 71.** Tangents to an ellipse at a point and parallel to a given line. Focuses are used in construction.

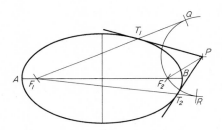

**FIG. 72.** Tangents to an ellipse from an outside point. Tangent points $T_1$ and $T_2$ are accurately located.

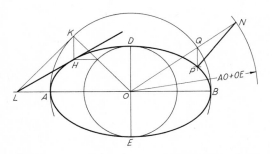

**FIG. 73.** Tangent and normal to an ellipse. The concentric-circle method is used.

Another mehtod is shown in Fig. 75. This should be used only when the minor diameter is at least two-thirds the length of the major diameter.

### 58. To Draw an Eight-centered Approximate Ellipse (Fig. 76).

When a closer approximation is desired, the eight-centered ellipse, the upper half of which is known in masonry as the "five-centered arch," may be constructed. Draw the rectangle $AFDO$. Draw the diagonal $AD$ and a line from $F$ perpendicular to it, intersecting the extension of the minor diameter at $H$. Lay off $OK$ equal to $OD$, and, on $AK$ as a diameter, draw a semicircle intersecting the extension of the minor diameter at $L$. Make $OM$ equal to $LD$. With center $H$ and radius $HM$, draw the arc $MN$. From $A$, along $AB$, lay off $AQ$ equal to $OL$. With $P$ as center and radius $PQ$, draw an arc intersecting $MN$ at $N$; then $P$, $N$, and $H$ are centers for one-quarter of the eight-centered approximate ellipse.

It should be noted that an ellipse changes its radius of curvature at every successive point and that these approximations are therefore not ellipses, but simply curves of the same general shape and, incidentally, not nearly so pleasing in appearance.

### 59. To Draw Any Noncircular Curve (Fig. 77).

This may be approximated by drawing tangent circle arcs: Select a center by trial, draw as much of an arc as will practically coincide with the curve, and then, changing the center and radius, draw the next portion, remembering always that *if arcs are to be tangent, their centers must lie on the common normal at the point of tangency.*

Curves are sometimes inked in this way in preference to using irregular curves.

### 60. The Parabola.

The parabola is a plane curve generated by a point so moving that its distance from a fixed point, called the "focus," is always equal to its distance from a straight line, called the "directrix." Among its practical applications are searchlights, parabolic reflec-

at $G$ and intersecting $DE$ produced (if necessary) at $H$. Make $OG'$ equal to $OG$, and $OH'$ equal to $OH$. Then $G$, $G'$, $H$, and $H'$ will be centers for four tangent circle arcs forming a curve *approximating* the shape of an ellipse.

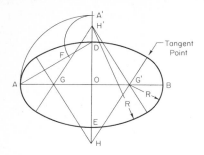

**FIG. 74.** A four-centered approximate ellipse. Four tangent circle arcs give elliptical shape.

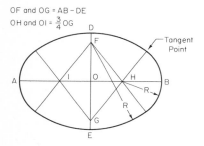

OF and OG = AB − DE

OH and OI = $\frac{3}{4}$ OG

**FIG. 75.** A four-centered approximate ellipse. This method works best when major and minor diameters are nearly equal.

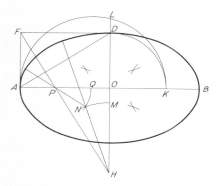

**FIG. 76.** An eight-centered approximate ellipse. This gives a much better approximation than the four-center methods but requires more construction.

Application of the parabola

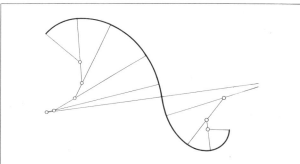

**FIG. 77.** A curve constructed with circle arcs. Note that lines through pairs of centers locate the tangent points of arcs.

tors, some loud-speakers, road sections, and certain bridge arches.

The mathematical equation for a parabola with the origin of rectilinear coordinates at the intercept of the curve with the $X$ axis and the focus on the axis is $y^2 = 2\,px$, where $p$ is twice the distance from the origin to the focus.

To draw a parabola when the focus $F$ and the directrix $AB$ are given (Fig. 78), draw the axis through $F$ perpendicular to $AB$. Through any point $D$ on the axis, draw a line parallel to $AB$. With the

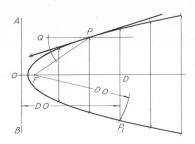

**FIG. 78.** A parabola. Points on the curve are equidistant from the focus *F* and directrix *AB*.

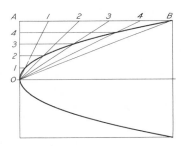

**FIG. 79.** A parabola by parallelogram method. Points are plotted by lines through equal-numbered divisions of *OA* and *AB*.

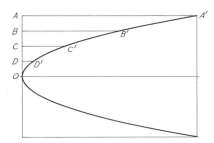

**FIG. 80.** A parabola by offset method. Offsets are proportionate to squares of divisions of *OA*.

distance *DO* as radius and *F* as center, draw an arc intersecting the line, thus locating a point *P* on the curve. Repeat the operation as many times as needed.

*To Draw a Tangent at Any Point P.* Draw *PQ* parallel to the axis and bisect the angle *FPQ*.

## 61. Parabola Construction: Parallelogram Method.

Usually when a parabola is required, the dimensions of the enclosing rectangle, that is, the width and depth of the parabola (or span and rise), are given, as in Fig. 79. Divide *OA* and *AB* into the same number of equal parts. From the divisions on *AB*, draw lines converging at *O*. From the divisions on *OA*, draw lines parallel to the axis. The intersections of these with the lines from the corresponding divisions on *AB* will be points on the curve.

## 62. Parabola Construction: Offset Method (Fig. 80).

Given the enclosing rectangle, the parabola can be plotted by computing the offsets from the line *OA*. These offsets vary in length as the square of their distances from *O*. Thus if *OA* is divided into four parts, *DD'* will be $\frac{1}{16}$ of *AA'*; *CC'*, since it is twice as far from *O* as *DD'*, will be $\frac{4}{16}$ of *AA'*; and *BB'*, $\frac{9}{16}$. If *OA* had been divided into five parts, the relations would be $\frac{1}{25}$, $\frac{4}{25}$, $\frac{9}{25}$, and $\frac{16}{25}$, the denominator in each case being the square of the number of divisions. This method is the one generally used by civil engineers in drawing parabolic arches.

## 63. Parabola Construction: Parabolic Envelope (Fig. 81).

This method of drawing a pleasing curve is often used in machine design. Divide *OA* and *OB* into the

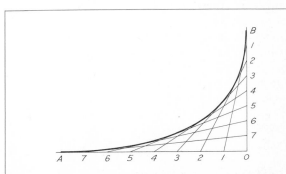

**FIG. 81.** A parabola by envelope method. The curve is drawn tangent to straight-line construction.

Application of the hyperbola (intersection)

gular, hyperbola referred to its asymptotes. With it, the law $PV = c$, connecting the varying pressure and volume of a portion of steam or gas, can be graphically presented.

*To Draw an Equilateral Hyperbola (Fig. 83).* Let $OA$ and $OB$ be the asymptotes of the curve and $P$ any point on it (this might be the point of cutoff on an indicator diagram). Draw $PC$ and $PD$. Mark any points 1, 2, 3, etc., on $PC$, and through these points draw a system of lines parallel to $OA$ and a second system through the same points converg-

same number of equal parts. Number the divisions from $O$ and $B$ and connect the corresponding numbers. The tangent curve will be a portion of a parabola—but a parabola whose axis is not parallel to either coordinate.

### 64. The Hyperbola.
The hyperbola is a plane curve generated by a point moving so that the difference of its distances from two fixed points, called the "focuses," is a constant. (Compare this definition with that of the ellipse.)

The mathematical equation for a hyperbola with the center at the origin of rectilinear coordinates and the focuses on the $X$ axis is $x^2/a^2 - y^2/b^2 = 1$, where $a$ is the distance from the center to the $X$ intercept and $b$ is the corresponding $Y$ value of the asymptotes, lines that the tangents to the curve meet at infinity.

*To Draw a Hyperbola When the Focuses $F_1$ and $F_2$ and the Transverse Axis AB (Constant Difference) Are Given (Fig. 82).* With $F_1$ and $F_2$ as centers and any radius greater than $F_1B$, as $F_1P$, draw arcs. With the same centers and any radius $F_1P - AB$, strike arcs intersecting these arcs, giving points on the curve.

*To Draw a Tangent at Any Point P.* Bisect the angle $F_1PF_2$.

### 65. Equilateral Hyperbola.
The case of the hyperbola of commonest practical interest to the engineer is the equilateral, or rectan-

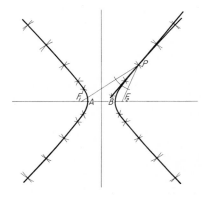

FIG. 82. A hyperbola. From any point, the difference in distance to the focuses is a constant.

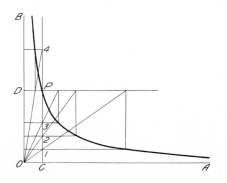

FIG. 83. An equilateral hyperbola. The asymptotes are perpendicular to each other.

ing at $O$. From the intersections of the lines of the second system with $PD$ extended, draw perpendiculars to $OA$. The intersections of these perpendiculars with the corresponding lines of the first system give points on the curve.

## 66. Cycloid Curves.

A cycloid is the curve generated by the motion of a point on the circumference of a circle rolled in a plane along a straight line. If the circle is rolled on the outside of another circle, the curve generated is called an "epicycloid"; if rolled on the inside, it is called a "hypocycloid." These curves are used in drawing the cycloid system of gear teeth.

The mathematical equation for a cycloid (parametric form) is $x = r\theta - r \sin \theta$, $y = r - r \cos \theta$, where $r$ is the radius of a moving point and $\theta$ is the turned angle, in radians, about the center from zero position.

*To Draw a Cycloid (Fig. 84).* Divide the rolling circle into a convenient number of parts (say, eight), and, using these divisions, lay off on the tangent $AB$ the rectified length of the circumference. Draw through $C$ the line of centers $CD$, and project the division points up to this line by perpendiculars to $AB$. Using these points as centers, draw circles representing different positions of the rolling circle, and project, in order, the division points of the original circle across to these circles. The intersections thus determined will be points on the curve. The epicycloid and hypocycloid are drawn similarly, as shown in Fig. 85.

Formation of a cycloid

Application of the involute

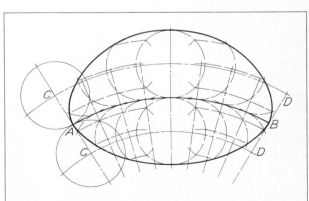

**FIG. 85.** An epicycloid and hypocycloid. Both are formed by a circle rolling on another circle.

**FIG. 84.** A cycloid. A point on the circumference of a rolling wheel describes this curve.

## 67. The Involute.

An involute is the spiral curve traced by a point on a taut cord unwinding from around a polygon or circle. Thus the involute of any polygon can be drawn by extending its sides, as in Fig. 86, and, with the corners of the polygon as successive centers, drawing arcs terminating on the extended sides.

The equation of the involute of a circle (parametric form) is $x = r \ (\sin \theta - \theta \cos \theta)$, $y = r \ (\cos \theta + \theta \sin \theta)$, where $r$ is the radius of the circle and $\theta$ is the turned angle for the tangent point.

In drawing a spiral in design, as for example of bent ironwork, the easiest way is to draw it as the involute of a square.

A circle may be conceived of as a polygon of an infinite number of sides. Thus to draw the involute of a circle (Fig. 87), divide it into a convenient number of parts, draw tangents at these points, lay off on these tangents the rectified lengths of the arcs from the point of tangency to the starting point, and connect the points by a smooth curve. The involute of the circle is the basis for the involute system of spur gearing.

## 68. The Spiral of Archimedes.

The spiral of Archimedes is the plane curve generated by a point moving uniformly along a straight line while the line revolves about a fixed point with uniform angular velocity.

*To Draw a Spiral of Archimedes that Makes One Turn in a Given Circle (Fig. 88).* Divide the circle into a number of equal parts, drawing the radii and numbering them. Divide the radius $O\text{-}8$ into the same number of equal parts, numbering from the center. With $O$ as center, draw concentric arcs intersecting the radii of corresponding numbers, and draw a smooth curve through these intersections. The Archimedean spiral is the curve of the heart cam used for converting uniform rotary motion into uniform reciprocal motion.

The mathematical equation is $p = a\theta$ (polar form).

## 69. Other Plane Curves.

The foregoing paragraphs discuss common plane curves that occur frequently in scientific work. Other curves, whenever needed, can be found in any good

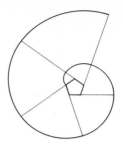

**FIG. 86.** An involute of a pentagon. A point on a cord unwound from the pentagon describes this curve.

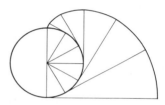

**FIG. 87.** An involute of a circle. As a taut cord is unwound from the circle, it forms a series of tangents to the circle.

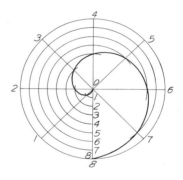

**FIG. 88.** The spiral of Archimedes. This curve increases uniformly in distance from the center as rotation (angular velocity) is constant.

textbook of analytic geometry and calculus. Typical of the curves that may be met with are the catenary, cardioid, sine curve, cosine curve, logarithmic spiral, reciprocal (hyperbolic) spiral, parabolic spiral, logarithmic curve, exponential curve, and curves of velocity and acceleration.

## 70. Double-curved Lines.

The scarcity of geometric double-curved lines, by comparison with the numerous single-curved lines, is surprising. There are only two double-curved lines, the cylindrical and the conic helix, much used in engineering work. Double-curved lines will, however, often occur as the lines of intersection between two curved solids or surfaces. These lines are not geometric but are double-curved lines of general form.

## 71. The Helix.

The helix is a space curve generated by a point moving uniformly along a straight line while the line revolves uniformly about another line as an axis. If the moving line is parallel to the axis, it will generate a cylinder. The word "helix" alone always means a cylindrical helix. If the moving line intersects the axis at an angle less than 90°, it will generate a cone, and the curve made by the point moving on it will be a "conic helix." The distance parallel to the axis through which the point advances in one revolution is called the "lead." When the angle becomes 90°, the helix degenerates into the Archimedean spiral.

## 72. To Draw a Cylindrical Helix (Fig. 89).

Draw the two views of the cylinder (see Chap. 5), and then measure the lead along one of the contour elements. Divide this lead into a number of equal

Application of the helix

parts (say, 12) and the circle of the front view into the same number. Number the divisions on the top view starting at point 1 and the divisions on the front view starting at the front view of point 1. When the generating point has moved one-twelfth of the distance around the cylinder, it has also advanced one-twelfth of the lead; when halfway around the cylinder, it will have advanced one-half the lead. Thus points on the top view of the helix can be found

**FIG. 89.** The cylindrical helix and its development. Any point moves at a constant rate both around and along the axis.

by projecting the front views of the elements, which are points on the circular front view of the helix, to intersect lines drawn across from the corresponding divisions of the lead. If the cylinder is developed, the helix will appear on the development as a straight line inclined to the base at an angle, called the "helix angle," whose tangent is $L/\pi D$, where $L$ is the lead and $D$ the diameter.

### 73. To Draw a Conic Helix (Fig. 90).

First make two views of the right-circular cone (see Chap. 5) on which the helix will be generated. Then lay out uniform angular divisions in the view showing the end view of the axis (in Fig. 90, the top view) and divide the lead into the same number of parts. Points can now be plotted on the curve. Each plotted point will lie on a circle cut from the cone by a plane dividing the lead and will also lie on the angular-division line. Thus, for example, to plot point 14, draw the circle diameter obtained from the front

**FIG. 90.** The conic helix. A point traverses the surface of the cone while moving at a constant rate both around and along the axis.

view, as shown in the top view. Point 14 in the top view then lies at the intersection of this circle and radial-division line 14. Then locate the front view by projection from the top view to plane 14 in the front view.

## CONSTRUCTIONS FOR LOFTING AND LARGE LAYOUTS

### 74.

There are cases when the regular drafting instruments are impractical, either because of size limitations or when extreme accuracy is necessary. The geometric methods for common cases of parallelism, perpendicularity, and tangency are given in the following paragraphs.

### 75. To Draw a Line through a Point and Parallel to a Given Line (Fig. 91).

With the given point $P$ as center and a radius of sufficient length, draw an arc $CE$ intersecting the given line $AB$ at $C$. With $C$ as center and the same radius, draw the arc $PD$. With $C$ as center and radius $DP$, draw an arc intersecting $CE$ at $E$. Then $EP$ is the required line.

### 76. To Draw a Line Parallel to Another at a Given Distance.

*For Straight Lines (Fig. 92).*   With the given distance as radius and two points on the given line as centers (as far apart as convenient), draw two arcs. A line tangent to these arcs will be the required line.

*For Curved Lines (Fig. 93).*   Draw a series of arcs with centers along the line. Draw tangents to these arcs with a French curve (see Fig. 60, Chap. 2).

### 77. To Erect a Perpendicular from a Point to a Given Straight Line (Fig. 94).

With point $P$ as center and any convenient radius $R_1$, draw a circle arc intersecting the given line at $A$ and $B$. With any convenient radius $R_2$ and with

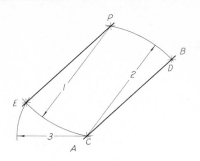

**FIG. 91.** Parallel lines. Two points determine the parallel.

**FIG. 92.** Parallel lines. The parallel is drawn tangent to a pair of arcs.

**FIG. 93.** Curved parallel lines. The curves are drawn tangent to circle arcs.

centers at $A$ and $B$, draw intersecting arcs locating $Q$. The required perpendicular is $PQ$, with $S$ the intersection of the perpendicular and the given line.

### 78. To Erect a Perpendicular from a Point on a Given Straight Line.

*First Method* (Fig. 95). With point $P$ on the line as center and any convenient radius $R_1$, draw circle arcs to locate points $A$ and $B$ equidistant from $P$. With any convenient radius $R_2$ longer than $R_1$ and with centers at $A$ and $B$, draw intersecting arcs locating $Q$. $PQ$ is the required perpendicular.

*Second Method* (Fig. 96). With any convenient center $C$ and radius $CP$, draw somewhat more than a semicircle from the intersection of the circle arc with the given line at $A$. Draw $AC$ extended to meet the circle arc at $Q$. $PQ$ is the required perpendicular.

### 79. To Draw a Tangent to a Circle at a Point on the Circle (Fig. 97).

Given the arc $ABC$ and to draw a tangent at point $C$, draw the extended diameter $CD$ and locate point $M$. Then with any convenient radius $R_1$, locate point $P$ equidistant from $C$ and $M$. With $P$ as center and the same radius $R_1$, draw somewhat more than a semicircle and draw the line $MPQ$. The line $QC$ is a tangent to the circle at point $C$.

### 80. To Draw a Tangent to a Circle from a Point Outside (Fig. 98).

Connect the point $P$ with the center of the circle $O$. Then draw the perpendicular bisector of $OP$, and with the intersection at $D$ as center, draw a semicircle. Its intersection with the given circle is the point of tangency. Draw the tangent line from $P$.

### 81. To Draw a Tangent to Two Circles.

*First Case: Open Belt* (Fig. 99). At center $O$ draw a circle with radius $R_1 - R_2$. From $P$ draw a tangent to this circle by the method of Fig. 98. Extend $OT$ to $T_1$, and draw $PT_2$ parallel to $OT_1$. Join $T_1$ and $T_2$, giving the required tangent.

*Second Case: Crossed Belt* (Fig. 100). Draw $OA$ and $O_1B$ perpendicular to $OO_1$. From $P$, where $AB$ crosses $OO_1$, locate tangents as in Fig. 98.

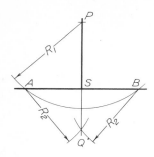

**FIG. 94.** A perpendicular from a point outside. Equal radii determine the perpendicular.

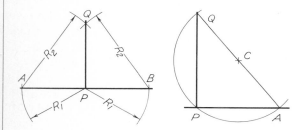

**FIG. 95 (LEFT).** A perpendicular from a point on a line. Equal radii determine a point on the perpendicular.

**FIG. 96 (RIGHT).** A perpendicular from a point on a line. Geometrically, perpendicular chords of a circle produce this solution.

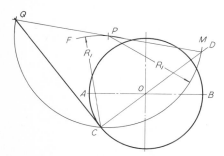

**FIG. 97.** To draw a tangent at a point on a circle. This solution sets up two right-angle chords of a circle. One chord passes through the center of the given circle.

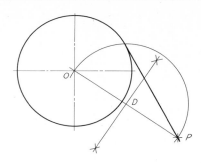

**FIG. 98.** To draw a tangent from an outside point. Here, as in Fig. 97, perpendicular chords determine the solution.

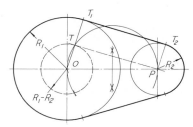

**FIG. 99.** Tangent lines (open belt). The difference in radius of the two circles is the geometric basis. Then perpendicular chords are used.

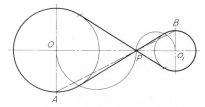

**FIG. 100.** Tangent lines (crossed belt). A duplicate of Fig. 99, except for position of the tangent lines.

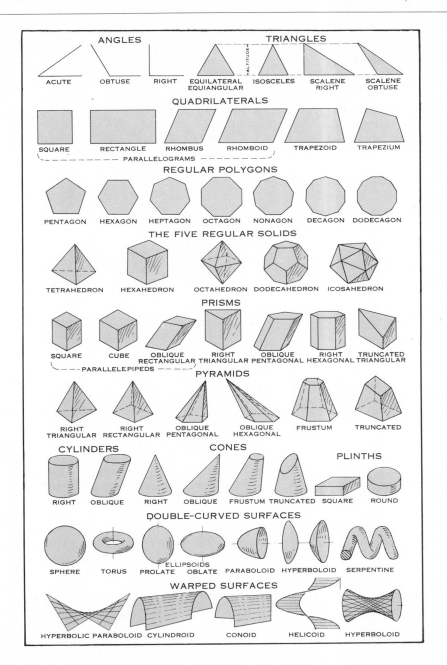

**FIG. 101.** Geometric shapes. These plane figures, solids, and surfaces should be studied and remembered.

# PROBLEMS

To be of value both as drawing exercises and as solutions, geometric problems must be worked accurately. Keep your pencil sharp, and use comparatively light lines.

## Group 1. Lines and Plane Figures.

**1.** Near the center of the working space, draw a horizontal line $4\frac{1}{2}$ in. long. Divide it into seven equal parts by the method of Fig. 27.

**2.** Draw a vertical line 1 in. from the left edge of the space and $3\frac{7}{8}$ in. long. Divide it into parts proportional to 1, 3, 5, and 7.

**3.** Construct a polygon as shown in the problem illustration, drawing the horizontal line $AK$ (of indefinite length) $\frac{3}{8}$ in. above the bottom of the space. From $A$ draw and measure $AB$. Proceed in the same way for the remaining sides. The angles can be obtained by proper combinations of the two triangles (see Figs. 28 and 29, Chap. 2).

**4.** Draw a line $AK$ making an angle of 15° with the horizontal. With this line as base, transfer the polygon of Prob. 3.

## Group 2. Problems in Accurate Joining of Tangent Lines.

**8.** Draw the offset swivel plate.

**PROB. 8.** Offset swivel plate.

**9.** Draw two lines $AB$ and $AC$ making an included angle of 30°. Locate point $P$, 4 in. from $A$ and $\frac{1}{2}$ in. from line $AB$. Draw two circle arcs centered at point $P$, one tangent

Locate a point by two intersecting lines; indicate the length of a line by short dashes across it.

**5.** Draw a regular hexagon having a distance across corners of 4 in.

**6.** Draw a regular hexagon, distance across flats $3\frac{3}{8}$ in.

**7.** Draw a regular dodecagon, distance across flats $3\frac{3}{8}$ in.

**PROB. 3.** Irregular polygon.

to line $AB$, the other to $AC$. Then draw two lines tangent to the opposite sides of the arcs and passing through point $A$. Locate all tangent points by construction.

**10.** Construct an ogee curve joining two parallel lines $AB$ and $CD$ as in Fig. 18, making $X = 4$ in., $Y = 2\frac{1}{2}$ in., and $BE = 3$ in. Consider this as the center line for a rod $1\frac{1}{4}$ in. in diameter, and draw the rod.

**11.** Make contour view of the bracket. In the upper ogee curve, the radii $R_1$ and $R_2$ are equal. In the lower one, $R_3$ is twice $R_4$.

**PROB. 11.** Bracket.

**12.** Draw an arc of a circle having a radius of $3^{13}/_{16}$ in., with its center $\frac{1}{2}$ in. from the top of the space and $1\frac{1}{4}$ in. from the left edge. Find the length of an arc of 60° by construction; compute the length arithmetically, and check the result.

**13.** Front view of washer. Draw half size.

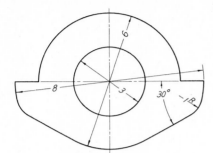

**PROB. 13.** Washer.

**14.** Front view of shim. Draw full size.

**PROB. 14.** Shim.

**15.** Front view of rod guide. Draw full size.

**PROB. 15.** Rod guide.

**16.** Front view of a star knob. Radius of circumscribing circle, $2\frac{3}{8}$ in. Diameter of hub, $2\frac{1}{2}$ in. Diameter of hole, $\frac{3}{4}$ in. Radius at points, $\frac{3}{8}$ in. Radius of fillets, $\frac{3}{8}$ in. Mark tangent points in pencil.

**PROB. 16.** Star knob.

**17.** Front view of sprocket. OD, $4\frac{3}{4}$ in. Pitch diameter, 4 in. Root diameter, $3\frac{1}{4}$ in. Bore, $1\frac{1}{4}$ in. Thickness of tooth at the pitch line is $\frac{9}{16}$ in. Splines, $\frac{1}{4}$ in. wide by $\frac{1}{8}$ in. deep. Mark tangent points in pencil. See Glossary for terms.

**PROB. 17.** Sprocket.

**18.**   Front view of a fan. Draw full size.

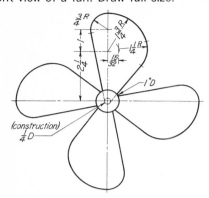

**PROB. 18.**   Fan.

**19.**   Front view of a level plate. Draw full size.

**PROB. 19.**   Level plate.

**20.**   Front view of an eyelet. Draw full size.

**PROB. 20.**   Eyelet.

**21.**   Front view of a stamping. Draw full size.

**PROB. 21.**   Stamping.

**22.** Front view of spline lock. Draw full size.

**PROB. 22.** Spline lock.

**23.** Front view of gage cover plate. Draw full size.

*Scale Detail*

**PROB. 23.** Gage cover plate.

**24.** Drawing of heater tube. Draw twice size.

**PROB. 24.** Heater tube.

**25.** Drawing of exhaust-port contour. Draw twice size.

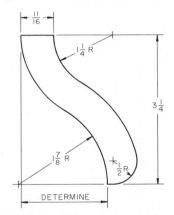

**PROB. 25.** Exhaust-port contour.

**26.** Drawing of toggle spring for leaf switch. Use decimal scale and draw 20 times size.

**PROB. 26.** Toggle spring for leaf switch.

**27.** Drawing of pulley shaft. Draw full size.

**PROB. 27.** Pulley shaft.

**28.** Drawing of cam for type A control. Plot cam contour from the polar coordinates given. Use decimal scale and draw five times size.

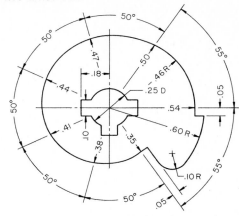

**PROB. 28.** Cam for type A control.

## Group 3. Plane Curves.

**29.** The conjugate diameters of an ellipse measure 3 and 4 in., the angle between them being 60°. Construct the major and minor diameters of this ellipse and draw the ellipse.

**30.** Using the pin-and-string method, draw an ellipse having a major diameter of 6 in. and a minor diameter of $4\frac{1}{4}$ in.

**31.** Using the trammel method, draw an ellipse having a major diameter of $4\frac{1}{2}$ in. and a minor diameter of 3 in.

**32.** Using the trammel method, draw an ellipse having a major diameter of $4\frac{1}{2}$ in. and a minor diameter of 4 in.

**33.** Using the concentric-circle method, draw an ellipse having a major diameter of $4\frac{5}{8}$ in. and a minor diameter of $1\frac{1}{2}$ in.

**34.** Draw an ellipse on a major diameter of 4 in. One point on the ellipse is $1\frac{1}{2}$ in. to the left of the minor diameter and $\frac{7}{8}$ in. above the major diameter.

**35.** Draw an ellipse having a minor diameter of $2\frac{3}{16}$ in. and a distance of $3\frac{1}{4}$ in. between focuses. Draw a tangent at a point $1\frac{3}{8}$ in. to the right of the minor diameter.

**36.** Draw an ellipse whose major diameter is 4 in. A tangent to the ellipse intersects the minor diameter $1\frac{3}{4}$ in. from the center, at an angle of 60°.

**37.** Draw a five-centered arch with a span of 5 in. and a rise of 2 in.

**38.** Draw an ellipse having conjugate diameters of $4\frac{3}{4}$ in. and $2\frac{3}{4}$ in., making an angle of 75° with each other. Determine the major and minor diameters.

**39.** Draw the major and minor diameters for an ellipse having a pair of conjugate diameters 60° apart, one horizontal and $6\frac{1}{4}$ in. long, the other $3\frac{1}{4}$ in. long.

**40.** Using the circle method, draw an ellipse having a pair of conjugate diameters, one making 15° with the horizontal and 6 in. long, the other making 60° with the first diameter and $2\frac{1}{2}$ in. long.

**41.** Using the triangle-trammel method, draw an ellipse having a pair of conjugate diameters 60° apart, one horizontal and 6 in. long, the other 4 in. long.

**42.** Using the triangle-trammel method, draw an ellipse having a pair of conjugate diameters 45° apart, one making 15° with the horizontal and 6 in. long, the other 3 in. long.

**43.** Draw a parabola, axis vertical, in a rectangle 4 by 2 in.

**44.** Draw a parabolic arch, with 6-in. span and a 2½-in. rise, by the offset method, dividing the half span into eight equal parts.

**45.** Draw an equilateral hyperbola passing through a point *P*, ½ in. from *OB* and 2½ in. from *OA*. (Reference letters correspond to Fig. 83.)

**46.** Draw two turns of the involute of a pentagon whose circumscribed circle is ½ in. in diameter.

**47.** Draw one-half of the involute of a circle 3¼ in. in diameter whose center is 1 in. from the left edge of the space. Compute the length of the last tangent and compare with the measured length.

**48.** Draw a spiral of Archimedes making one turn in a circle 4 in. in diameter.

**49.** Draw the cycloid formed by a rolling circle 2 in. in diameter. Use 12 divisions.

**50.** Draw the epicycloid formed by a 2-in.-diameter circle rolling on a 15-in.-diameter directing circle. Use 12 divisions.

**51.** Draw the hypocycloid formed by a 2-in.-diameter circle rolling inside a 15-in.-diameter directing circle. Use 12 divisions.

**52.** Front view of cam. Draw full size.

**PROB. 52.** Front view of cam.

**53.** Front view of fan base. Draw full size.

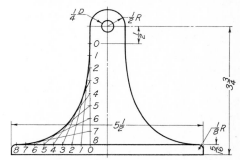

**PROB. 53.** Front view of fan base.

**54.** Front view of trip lever. Draw half size.

**PROB. 54.** Front view of trip lever.

## Group 4. Problems for Calculator and Mathematical Tables.

**55.** Find the altitude of an equilateral triangle, each side of which is 4 in.

**56.** Find the value of one-third of an angle of 231°4′2″.

**57.** A tank is 31 in. in diameter and 4 ft 7½ in. long. Find the capacity in liters.

**58.** A 4-in.-diameter cold-rolled steel bar is 96 cm long. What is the weight in pounds? In grams?

**59.** A circle of 12⅝-in. diameter used as a target for optical devices is to be compared in area with a lens of 80 mm diameter. What percentage is the lens area to the target area?

**60.** A guided missile is fired, hits a target 841 miles away, and during flight rises to a height of 91 miles. What are these distances in kilometers?

**61.** Find one side of a square whose equivalent area is a circle of 6⅜-in. diameter. Answer in centimeters.

**62.** An ellipse has a major axis of 14 in. and a minor axis of 8½ in. How far from the geometric center is each focus? What is the equivalent distance in centimeters?

**63.** A certain pressure is recorded by a column of water 2 ft 4½ in. high. What is the equivalent pressure in inches of mercury?

**64.** Find the area of a tract of land designated by a traverse that is an equilateral triangle each side of which is 228 ft 4 in. long. Answer in square miles. Also calculate the area in square kilometers.

**65.** A square is inscribed in a circle of 4-in. diameter. What is the area of the square?

**66.** An equilateral triangle is inscribed in a circle of 9⅜-in. diameter. What is the sum of the sides of the triangle? What is the circumference of the circle? What is the area of the triangle and of the circle?

**67.** A round copper bar 2 in. in diameter and 4 ft long is to be balanced in weight by a cube of aluminum. What is the size of the cube?

**68.** If a 2-in. round bar of steel 2 ft 6 in. long is to be replaced by a bar of Bakelite, what will be the difference in weight?

**69.** A circular opening 6 in. in diameter is to be replaced by openings 0.70 cm in diameter. How many openings must be used to give an equivalent area?

**70.** A 3⅛-in. cube of aluminum is to be replaced by steel. What is the percentage difference in weight?

**71.** A vertical water tank 4 ft 0 in. square holds 980 gallons. What is the pressure, per square inch, on the bottom of the water tank?

# LETTERING

# 4

Complete Graphic Depiction and Specification
Requires Word Supplements

Discussed and Illustrated in this Chapter:
Single-stroke Lettering • Proportions • Guide
Lines • Lettering in Pencil • Vertical Capital
Letters • Order of Strokes • Vertical Lower-case
Letters • Inclined Letters • Composition •
Titles • Pen-and-ink Lettering • Supplemental
Letter Styles • Problems

## 1. Lettering—Word Supplements on Drawings.

Graphic representation of the shape of a part, machine, or structure gives one aspect of the information needed for its construction. To this must be added, to complete the description, figured dimensions, notes on material and finish, and a descriptive title—all lettered, freehand, in a style that is perfectly legible, uniform, and capable of rapid execution. As far as the appearance of a drawing is concerned, the lettering is the most important part. But the usefulness of a drawing, too, can be ruined by lettering done ignorantly or carelessly, because illegible figures are apt to cause mistakes in the work.

In a broad sense, lettering is a branch of design. Students of lettering fall into two general classes: those who will use letters and words to convey information on drawings, and those who will use lettering in applied design, for example, art students, artists, and craftsmen. The first group is concerned mainly with legibility and speed, the second with beauty of form and composition. In our study of engineering graphics we are concerned only with the problems of the first group. The engineering student takes up lettering as an early part of his work in graphics and continues its practice throughout his course, becoming more and more skillful and proficient.

## 2. Single-stroke Lettering.

By far the greatest amount of lettering on drawings is done in a rapid single-stroke letter, either vertical or inclined, and every engineer must have absolute command of these styles. The ability to letter well can be acquired only by continued and careful practice, but it can be acquired by anyone with normal muscular control of his fingers who will practice faithfully and intelligently and take the trouble to observe carefully the shapes of the letters, the sequence of strokes in making them, and the rules for their composition. It is not a matter of artistic talent or even of dexterity in handwriting. Many persons who write poorly letter very well.

The term "single-stroke," or "one-stroke," does not mean that the entire letter is made without lifting the pencil or pen but that the width of the stroke of the pencil or pen is the width of the stem of the letter.

## 3. General Proportions.

There is no standard for the proportions of letters, but there are certain fundamental points in design and certain characteristics of individual letters that must be learned by study and observation before composition into words and sentences should be attempted. Not only do the widths of letters in any alphabet vary, from $I$, the narrowest, to $W$, the widest, but different alphabets vary as a whole. Styles narrow in their proportion of width to height are called "COMPRESSED," or "CONDENSED," and are used when space is limited. Styles wider than the normal are called **"EXTENDED."**

The proportion of the thickness of stem to the height varies widely, ranging from $\frac{1}{3}$ to $\frac{1}{20}$. Letters with heavy stems are called **"BOLDFACE,"** or **"BLACKFACE,"** those with thin stems "LIGHT-FACE."

## 4. The Rule of Stability.

In the construction of letters, the well-known optical illusion in which a horizontal line drawn across the middle of a rectangle appears to be below the middle must be provided for. In order to give the appearance of stability, such letters as $B$, $E$, $K$, $S$, $X$, and $Z$ and the figures $3$ and $8$ must be drawn smaller at the top than at the bottom. To see the effect of this illusion, turn a printed page upside down and notice the appearance of the letters mentioned.

## 5. Guide Lines.

Always draw light guide lines for both tops and bottoms of letters, using a sharp pencil. Figure 1 shows a method of laying off a number of equally spaced lines. Draw the first base line; then set the bow spacers to the distance wanted between base lines and step off the required number of base lines. Above the last line mark the desired height of the letters. With the same setting, step down from this upper point, thus obtaining points for the top of each line of letters.

The Braddock-Rowe triangle (Fig. 2a) and the Ames

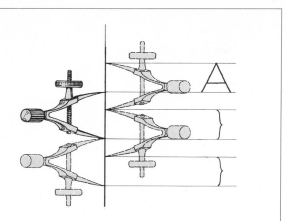

**FIG. 1.** To space guide lines. Bow dividers are spaced for the distance between base lines; this distance stepped off from capital and base lines locates successive lines of lettering.

lettering instrument (Fig. 2b) are convenient devices for spacing lines of letters. In using these instruments, a sharp pencil is inserted in the proper row of countersunk holes, and the instrument, guided by a T-square blade, is drawn back and forth by the pencil, as indicated by Fig. 3. The holes are grouped for capitals and lower case, the numbers indicating the height of capitals in thirty-seconds of an inch; thus no. 6 spacing of the instrument means that the capitals will be $\frac{6}{32}$, or $\frac{3}{16}$, in. high.

## 6. Lettering in Pencil.

Good technique is as essential in lettering as in drawing. The quality of the lettering is important whether it appears on finished work to be reproduced by one of the printing processes or as part of a pencil drawing to be inked. In the first case, the penciling must be clean, firm, and opaque; in the second, it may be lighter. The lettering pencil should be selected carefully by trial on the paper. In one instance, the same grade may be chosen as

**FIG. 2.** (a) Braddock-Rowe triangle. Numbered sets of holes locate capital, waist (see Fig. 18), and base lines of successive lines of lettering. (b) Ames lettering instrument. A center disk, adjusted to letter-height number, gives equal spaces or 2:3 and 3:5 waist-line proportion for successive lines of lettering.

that used for the drawing; in another, a grade or two softer may be preferred. Sharpen the pencil to a long, conic point, and then round the lead slightly on the end so that it is not so sharp as a point used for drawing.

The first requirement in lettering is the correct holding of the pencil or pen. Figure 4 shows the pencil held comfortably with the thumb, forefinger, and second finger on alternate flat sides and the third and fourth fingers on the paper. Vertical, slanting, and curved strokes are drawn with a steady, even, *finger* movement; horizontal strokes are made similarly but with some pivoting of the hand at the wrist (Fig. 5). Exert pressure that is firm and uniform but not so heavy as to cut grooves in the paper. To keep the point symmetrical, form the habit of rotating the pencil after every few strokes.

**FIG. 4.** Vertical strokes. These are made entirely by finger movement.

**FIG. 5.** Horizontal strokes. These are made by pivoting the whole hand at the wrist; fingers move slightly to keep the stroke perfectly horizontal.

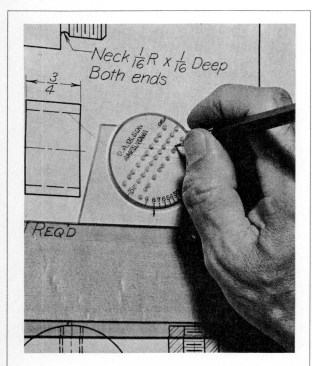

**FIG. 3.** Using the Ames lettering instrument. Lines are drawn as the pencil moves the instrument along the T square (or triangle).

## 7. Single-stroke Vertical Capitals.

The vertical, single-stroke, commercial Gothic letter is a standard for titles, reference letters, etc. As for the proportion of width to height, the general rule is that the smaller the letters are, the more extended they should be in width. A low extended letter is more legible than a high compressed one and at the same time makes a better appearance.

For proficiency in lettering it is essential to learn the form and peculiarity of each of the letters. Although lettering must be based on a careful regard for the fundamental letter forms, this is not to say that it will be without character. Individuality in lettering is often nearly as marked as in handwriting.

## 8. Order of Strokes.

In the following figures an alphabet of slightly extended vertical capitals has been arranged in family groups. Study the shape of each letter, with the order and direction of the strokes forming it, and practice it until its form and construction are perfectly familiar. Practice it first in pencil to large size, perhaps $\frac{3}{8}$ in. high, then to smaller size, and finally directly in ink.

To bring out the proportions of widths to heights and the subtleties in the shapes of the letters, they are shown against a square background with its sides divided into sixths. Several of the letters in this alphabet, such as *A* and *T*, fill the square; that is, they are as wide as they are high. Others, such as *H* and *D*, are approximately five spaces wide, or their width is five-sixths of their height. *These proportions must be learned visually* so well that letters of various heights can be drawn in correct proportion without hesitation.

*The I-H-T Group (Fig. 6).* The letter *I* is the foundation stroke. You may find it difficult to keep the stems vertical. If so, draw direction lines lightly an inch or so apart to aid the eye. The *H* is nearly square (five-sixths wide), and in accordance with the rule of stability, the cross-bar is just above the center. The top of the *T* is drawn first to the full width of the square, and the stem is started accurately as its middle point.

*The L-E-F Group (Fig. 7).* The *L* is made in two strokes. The first two strokes of the *E* are the same as for the *L*; the third, or upper, stroke is slightly shorter than the lower; and the last stroke is two-thirds as long as the lower and just above the middle. *F* has the same proportions as *E*.

*The N-Z-X-Y Group (Fig. 8).* The parallel sides of *N* are generally drawn first, but some prefer to

FIG. 6. The *I-H-T* group. Note the direction of fundamental horizontal and vertical strokes.

FIG. 7. The *L-E-F* group. Note the successive order of strokes.

FIG. 8. The *N-Z-X-Y* group. Note that *Z* and *X* are smaller at the top than at the bottom, in accordance with the rule of stability.

FIG. 9. The *V-A-K* group. The horizontal of *A* is one-third from the bottom; the second and third strokes of *K* are perpendicular to each other.

make the strokes in consecutive order. *Z* and *X* are both started inside the width of the square on top and run to full width at the bottom. This throws the crossing point of the *X* slightly above the center. The junction of the *Y* strokes is at the center.

*The V-A-K Group (Fig. 9).* *V* is the same width as *A*, the full breadth of the square. The *A* bridge is one-third up from the bottom. The second stroke of *K* strikes the stem one-third up from the bottom; the third stroke branches from it in a direction starting from the top of the stem.

*The M-W Group (Fig. 10).* These are the widest letters. *M* may be made in consecutive strokes or by drawing the two vertical strokes first, as with the *N. W* is formed of two narrow *V*'s, each two-thirds of the square in width. Note that with all the pointed letters the width at the point is the width of the stroke.

*The O-Q-C-G Group (Fig. 11).* In this extended alphabet the letters of the *O* family are made as full circles. The *O* is made in two strokes, the left side a longer arc than the right, as the right side is harder to draw. Make the kern of the *Q* straight. A large-size *C* and *G* can be made more accurately with an extra stroke at the top, whereas in smaller letters the curve is made in one stroke (Fig. 19). Note that the bar on the *G* is halfway up and does not extend past the vertical stroke.

*The D-U-J Group (Fig. 12).* The top and bottom strokes of *D* must be horizontal. Failure to observe this is a common fault with beginners. In large letters *U* is formed by two parallel strokes, to which the bottom stroke is added; in smaller letters, it may be made in two strokes curved to meet at the bottom. *J* has the same construction as *U*, with the first stroke omitted.

*The P-R-B Group (Fig. 13).* With *P, R,* and *B,* the number of strokes depends upon the size of the letter. For large letters the horizontal lines are started and the curves added, but for smaller letters only one stroke for each lobe is needed. The middle lines of *P* and *R* are on the center line; that of *B* observes the rule of stability.

*The S-8-3 Group (Fig. 14).* The *S, 8,* and *3* are closely related in form, and the rule of stability must be observed carefully. For a large *S,* three strokes are used; for a smaller one, two strokes; and for a very small size, one stroke only is best. The *8* may be made on the *S* construction in three strokes, or in "head and body" in four strokes. A perfect *3* can be finished into an *8.*

*The 0-6-9 Group (Fig. 15).* The cipher is an ellipse five-sixths the width of the letter *O.* The backbones of the *6* and *9* have the same curve as the cipher, and the lobes are slightly less than two-thirds the height of the figure.

FIG. 10. The *M-W* group. *M* is one-twelfth wider than it is high; *W* is one-third wider than it is high.

FIG. 11. The *O-Q-C-G* group. All are based on the circle.

FIG. 12. The *D-U-J* group. These are made with combinations of straight and curved strokes.

FIG. 13. The *P-R-B* group. Note the rule of stability with regard to *R* and *B.*

*The 2-5-7-& Group (Fig. 16).* The secret in making the *2* lies in getting the reverse curve to cross the center of the space. The bottom of *2* and the tops of *5* and *7* should be horizontal straight lines. The second stroke of *7* terminates directly below the middle of the top stroke. Its stiffness is relieved by curving it slightly at the lower end. The ampersand (&) is made in three strokes for large letters and two for smaller ones and must be carefully balanced.

*The Fraction Group (Fig. 17).* Fractions are always made with horizontal bar. Integers are the same

**FIG. 14.** The *S-8-3* group. A perfect *S* and *3* can be completed to a perfect *8.*

**FIG. 15.** The *0-6-9* group. The width is five-sixths of the height.

**FIG. 16.** The *2-5-7-&* group. Note the rule of stability. The width is five-sixths of the height.

**FIG. 17.** Fractions. The total height of a fraction is twice that of the integer.

**FIG. 18.** Basic forms for lower-case letters. For standard letters, the waist-line height is two-thirds of capital height; capital line and drop line are therefore one-third above and one-third below the body of the letter.

### 9. Vertical Lower-case Letters.

The single-stroke, vertical lower-case letter is not commonly used on machine drawings but is used extensively in map drawing. It is the standard letter for hypsography in government topographic drawing. The bodies are made two-thirds the height of the capitals with the ascenders extending to the capital line and the descenders dropping the same distance below. The basic form of the letter is the combination of a circle and a straight line (Fig. 18). The alphabet, with some alternate shapes, is shown in Fig. 19, which also gives the capitals in alphabetic order.

### 10. Single-stroke Inclined Capitals.

Many draftsmen use the inclined, or slant, letter in preference to the upright. The order and direction of strokes are the same as in the vertical form.

After ruling the guide lines, draw slanting "direction lines" across the lettering area to aid the eye in keeping the slope uniform. These lines may be drawn with a special lettering triangle of about $67\frac{1}{2}°$, or the slope of 2 to 5 may be fixed on the paper by marking two units on a horizontal line and five on a vertical line and using T square and triangle as shown in Fig. 20. The Braddock-Rowe triangle and the Ames instrument (Fig. 2*a* and *b*) both provide for the drawing of slope lines. The form taken by the rounded letters when inclined is illustrated in Fig. 21, which shows that curves are sharp in all upper right-hand and lower left-hand corners and flattened in the other two corners. Take particular care with the letters that have sloping sides, such

height as capitals. The total fraction height is best made twice the height of the integer. The numerator and denominator will be about three-fourths the height of the integer. Be careful to leave a clear space above and below the horizontal bar. Guide lines for fractions are easily obtained with lettering instruments by using the set of uniformly spaced holes or by drawing the integer height above and below the center, the position of the horizontal bar.

**ABBCcDEFGHI**

**JJKLMNOPPQR**

**SsssTUuVWXYZ**

**1234567890**

**aₒᵣabcdefggₒᵣghijklm**

**nopqrstuvwxyₒᵣyz**

**FIG. 19.** Single-stroke vertical capitals and lower case. Note the alternate strokes for small-size capitals and the alternate shapes for lower-case *a*, *g*, and *y*.

**FIG. 20.** Slope guide lines. The standard slope angle of $67\frac{1}{2}°$ (see Fig. 2a and b) can be approximated by a slope of 2 to 5.

*O O GDRS*

**FIG. 21.** Form of curved-stroke inclined capitals. The basic shape is elliptical.

**FIG. 22.** Form of straight-stroke inclined capitals. The "center line" must be a slope line.

**FIG. 23.** Single-stroke inclined capitals and lower case. Note the alternate strokes for small-size capitals and the alternate shapes for the lower case *a*, *g*, and *y*.

as *A*, *V*, and *W*. The sloping sides of these letters must be drawn so that they appear to balance about a slope guide line passing through each vertex, as in Fig. 22. The alphabet is given in Fig. 23. Study the shape of each letter carefully.

Professional appearance in lettering is due to three things: (1) keeping to a uniform slope, (2) having the letters full and well shaped, and (3) keeping them close together. The beginner invariably cramps the individual letters and spaces them too far apart.

## 11. Single-stroke, Inclined Lower-case Letters.

The inclined lower-case letters (Fig. 23) have bodies two-thirds the height of the capitals with the as-

cenders extending to the capital line and the descenders dropping the same distance below the base line. Among older engineers, particularly civil engineers, this letter is generally known as the "Reinhardt letter," in honor of Charles W. Reinhardt, who first systematized its construction. It is legible and effective and, after its swing has been mastered, can be made rapidly. The lower-case letter is suitable for notes and statements on drawings because it is much more easily read than all capitals, since we read words by the word shapes, and it can be done much faster.

All the letters of the Reinhardt alphabet are based on two elements, the straight line and the ellipse, and have no unnecessary hooks or appendages. They may be divided into four groups, as shown in

Figs. 24 to 27. The dots of *i* and *j* and the top of the *t* are on the "*t* line," halfway between the waist line and the capital line. The loop letters are made with an ellipse whose long axis is inclined about 45° in combination with a straight line. In lettering rapidly, the ellipse tends to assume a pumpkin-seed form; guard against this.

The *c*, *e*, and *o* are based on an ellipse of the shape of the capitals but not inclined quite so much as the loop-letter ellipse. In rapid small work, the *o* is often made in one stroke, as are the *e*, *v*, and *w*. The *s* is similar to the capital but, except in letters more than ⅛ in. high, is made in one stroke. In the hook-letter group, note particularly the shape of the hook.

The single-stroke letter may, if necessary, be much compressed and still be clear and legible (Fig. 28). It is also used sometimes in extended form.

## 12. For Left-handers Only.

The order and direction of strokes in the preceding alphabets have been designed for right-handed persons. The principal reason that left-handers sometimes find lettering difficult is that whereas the right-hander progresses away from the body, the left-hander progresses toward the body; consequently his pencil and hand partially hide the work he has done, making it harder to join strokes and

**FIG. 24.** The straight-line inclined lower-case letters. Note that the center lines of the letters follow the slope angle.

**FIG. 25.** The loop letters. Note the graceful combination of elliptical body, ascenders, and descenders.

**FIG. 26.** The ellipse letters. Their formation is basically elliptical.

**FIG. 27.** The hook letters. They are combinations of ellipses and straight lines.

COMPRESSED LETTERS ARE USED
when space is limited. Either vertical
or inclined styles may be compressed.

EXTENDED LETTERS OF A
given height are more legible

**FIG. 28.** Compressed and extended letters. Normal width has *O* the same in width as in height and the other letters in proportion. Wider than high for *O* is extended, narrower is compressed.

to preserve uniformity. Also, in the case of inclined lettering, the slope direction, instead of running toward his eye, runs off into space to the left of his body, making this style so much harder for him that he is strongly advised to *use vertical letters exclusively*.

For the natural left-hander, whose writing position is the same as a right-hander except reversed left for right, a change in the sequence of strokes of some of the letters will obviate part of the difficulty caused by interference with the line of sight. Figure 29 gives an analyzed alphabet with an alternate for some letters. In *E* the top bar is made before the bottom bar, and *M* is drawn from left to right to avoid having strokes hidden by the pencil or pen. Horizontal portions of curves are easier to make from right to left; hence the starting points for *O*, *Q*, *C*, *G*, and *U* differ from the standard right-hand stroking. *S* is the perfect letter for the left-hander and is best made in a single smooth stroke. The figures *6* and *9* are difficult and require extra practice. In the lower-case letters *a*, *d*, *g*, and *q*, it is better

to draw the straight line before the curve even though it makes spacing a little harder.

The hook-wrist left-handed writer, who pushes his strokes from top to bottom, finds vertical lettering more difficult than does the natural left-hander. In Fig. 29, where alternate strokes are given for some of the letters, the hook-wrist writer will probably find the second stroking easier than the first. Some prefer to reverse *all* the strokes, drawing vertical strokes from bottom to top and horizontal strokes from right to left.

By way of encouragement it may be said that many left-handed draftsmen letter beautifully.

### 13. Composition.

Composition in lettering has to do with the selection, arrangement, and spacing of appropriate styles and sizes of letters. On engineering drawings the selection of the style is practically limited to vertical or inclined single-stroke lettering, so composition here means arrangement into pleasing and legible form. After the shapes and strokes of the individual letters

**FIG. 30.** Background areas. Equal areas between letters produce spacing that is visually uniform.

have been learned, the entire practice should be on composition into words and sentences, since proper spacing of letters and words does more for the appearance of a block of lettering than the forms of the letters themselves. Letters in words are not spaced at a uniform distance from each other but are arranged so that the areas of white space (the irregular backgrounds between the letters) are approximately equal, making the spacing *appear* approximately uniform. Figure 30 illustrates these background shapes. Each letter is spaced with reference to its shape and the shape of the letter preceding it. Thus adjacent letters with straight sides would be spaced farther apart than those with curved sides. Sometimes combinations such as *LT* or *AV* may even overlap. Definite rules for spacing are not successful; it is a matter for the draftsman's judgment and sense of design. Figure 31 illustrates word composition. The sizes of letters to use in any particular case can be determined better by sketching them lightly than by judging from the guide lines alone. A finished line of letters always looks larger than the guide lines indicate. Avoid the use of a

coarse pen for small sizes and one that makes thin wiry lines for large sizes. When capitals and small capitals are used, the height of the small capitals should be about four-fifths that of the capitals.

In spacing words, a good principle is to leave the space that would be taken by an assumed letter *I* connecting the two words into one, as in Fig. 32. The space would never be more than the height of the letters.

The clear distance between lines may vary from $\frac{1}{2}$ to $1\frac{1}{2}$ times the height of the letter but for the sake of appearance should not be exactly the same as the letter height. The instruments in Figs. 2 and 3 provide spacing that is two-thirds of the letter height. Paragraphs should always be indented.

**14. Titles.**
The most important problem in lettering composition is the design of titles. Every drawing has a descriptive title that is either all hand-lettered or filled in on a printed form. It gives necessary information concerning the drawing, and the information that is needed will vary with the different kinds of drawings.

The usual form of lettered title is the *symmetrical title*, which is balanced or "justified" on a vertical center line and designed with an elliptical or oval outline. Sometimes the wording necessitates a pyramid or inverted-pyramid ("bag") form. Figure 33 illustrates several shapes in which titles can be composed. The lower right-hand corner of the sheet is, from long custom and because of convenience in filing, the usual location for the title, and in laying

**FIG. 31.** Word composition. Careful spacing of letters and words and the proper emphasis of size and weight are important.

WORDSISPACEDIBYISKETCHINGIANIIIBETWEEN

WORDS SPACED BY SKETCHING AN I BETWEEN

**FIG. 32.** Word spacing. Space words so that they read naturally and do not run together (too close) or appear as separate units (too far apart).

out a drawing, this corner is reserved for it. The space allowed depends on the size and purpose of the drawing. On an 11- by 17-in. working drawing, the title may be about 3 in. long.

### 15. To Draw a Title.

When the wording has been determined, write out the arrangement on a separate piece of paper as in Fig. 34 (or, better, typewrite it). Count the letters, including the word spaces, and make a mark across the middle letter or space of each line. The lines must be displayed for prominence according to their relative importance as judged from the point of view of the persons who will use the drawing. Titles are usually made in all capitals. Draw the base line for the most important line of the title and mark on it the approximate length desired. To get the letter

height, divide this length by the number of letters in the line, and draw the capital line. Start at the center line, and sketch lightly the last half of the line, drawing only enough of the letters to show the space each will occupy. Lay off the length of this right half on the other side, and sketch that half, working forward or backward. When this line is satisfactory in size and spacing, draw the remainder in the same way. Study the effect, shift letters or lines if necessary, and complete in pencil. Use punctuation marks only for abbreviations.

*Scratch-paper Methods.* Sketch each line of the title separately on a piece of scratch paper, using guide lines of determined height. Find the middle point of each of these lines, fold the paper along the base line of the letters, fit the middle point to

**FIG. 33.** Shapes in symmetrical composition. Design for clarity and emphasis.

**FIG. 34.** Title composition. Sketch lightly; then when satisfactory, complete.

LEONARDT 516 F: 506 F
HUNT 512: ESTERBROOK 968
Esterbrook 1000          Spencerian No.1
Gillott 404: Gillott 303    For very fine lines Gillott 170 and 290
                            or Esterbrook 356 and 355

**FIG. 35.** Pen strokes, full size. These are used principally for average-size lettering on working drawings.

the center line on the drawing, and draw the final letters directly below the sketches. *Or* draw the letters along the edge of the scratch paper, using the upper or lower edge as one of the guide lines. *Or* letter the title on scratch paper, cut it apart and adjust until satisfactory, and then trace it.

### 16. Notes and Filled-in Titles.

By far the principal use of lettering on engineering drawings occurs on working drawings, where the dimensions, explanatory notes, record of drawing changes, and title supplement the graphic description of the object.

Styles vary, but the trend is toward the use of capital letters exclusively. Most draftsmen are able to letter in capital letters with greater readability and fewer mistakes than in lower case. However, for some structural drawings and others where space may be at a premium, lower case is used.

On any working drawing the accuracy of the information conveyed is paramount. For this reason it is important to make the shapes of the letters precisely correct with no personal flourishes, embellishments, or peculiarities. For example, a poorly made *3* may read as an *8* or vice versa. A poorly made *5* may be misconstrued as a *6*. Strive for a smooth, professional appearance with, above all, perfect readability. The problems of Group 5 at the end of this chapter give practice in the lettering of notes.

### 17. Lettering Pens.

There are many steel writing pens that are adaptable to or made especially for lettering. The size of the strokes of a few popular ones is shown in full size in Fig. 35. Several special pens made in sets of graded sizes have been designed for single-stroke lettering; among them are those illustrated in Fig. 36. These are particularly useful for large work. The ink-holding reservoir of the Henry tank pen (Fig. 37) assists materially in maintaining uniform weight of line. A similar device can be made by bending a

**FIG. 36.** Barch-Payzant, Speedball, Edco, and Leroy pens. These are used principally for displays, large titles, and number blocks.

brass strip from a paper fastener, a piece of annealed watch spring, or—perhaps best—a strip cut from a piece of shim brass into the shape shown in Fig. 38 and inserting it in the penholder so that the curved end just touches the pen nib.

To remove the oil film, always wet a new pen and wipe it thoroughly before using. A lettering pen well broken in by use is worth much more than a new one. It should be kept with care and never lent. A pen that has been dipped into writing ink should never be put into drawing ink. When in use, a pen should be wiped clean frequently with a cloth pen-wiper.

## 18. Using the Pen.

Use a penholder with cork grip (the small size) and set the pen in it firmly. Many prefer to ink a pen with the quill filler, touching the quill to the under-side of the pen point, rather than dipping the pen in the ink bottle. If the pen is dipped, the surplus ink should be shaken back into the bottle or the pen touched against the neck of the bottle as it is withdrawn. Lettering with too much ink on the pen gives the results shown in Fig. 39.

In lettering, the penholder is held in the fingers firmly but without pinching, in the position shown in Fig. 40. The strokes of the letters are made with a steady, even motion and a slight, uniform pressure on the paper that will not spread the nibs of the pen.

## 19. Other Lettering Styles.

This chapter has been devoted entirely to the commercial Gothic letter because this style predominates in all types of commercial drawings. However, other letter styles are used extensively on architectural drawings and on maps, in design and for display, and for decorative letters on commercial products. Some traditional letter styles and their uses are given in the Appendix.

FIG. 37. Henry tank pen. The reservoir is designed to prevent spreading of the nibs.

FIG. 38. Ink holder. This helps to avoid heavy flow of ink when the pen is full.

# EHMNWTZ

FIG. 39. Too much ink. Fill the pen sparingly and often.

FIG. 40. Holding the pen. Note that the thumb, forefinger, and second finger make a three-point support.

# PROBLEMS

## Group 1. Single-stroke Vertical Capitals.

**1.** Large letters in pencil for careful study of the shapes of the individual letters. Starting $\frac{9}{16}$ in. from the top border, draw guide lines for five lines of $\frac{3}{8}$-in. letters. Draw each of the straight-line letters *I*, *H*, *T*, *L*, *E*, *F*, *N*, *Z*, *Y*, *V*, *A*, *M*, *W*, and *X*, four times in pencil only, studying carefully Figs. 6 to 10. The illustration is a full-size reproduction of a corner of this exercise.

**2.** Same as Prob. 1 for the curved-line letters *O*, *Q*, *C*, *G*, *D*, *U*, *J*, *B*, *P*, *R*, and *S*. Study Figs. 11 to 14.

**3.** Same as Prob. 1 for the figures *3*, *8*, *6*, *9*, *2*, *5*, $\frac{1}{2}$, $\frac{3}{4}$, $\frac{5}{8}$, $\frac{7}{16}$, and $\frac{9}{32}$. Study Figs. 14 to 17.

**4.** Composition. Same layout as for Prob. 1. Read Sec. 13 on composition; then letter the following five lines in pencil: (*a*) WORD COMPOSITION, (*b*) TOPOGRAPHIC SURVEY, (*c*) TOOLS AND EQUIPMENT, (*d*) BRONZE BUSHING, (*e*) JACK-RAFTER DETAIL.

**5.** Quarter-inch vertical letters in pencil and ink. Starting $\frac{1}{4}$ in. from the top, draw guide lines for nine lines of $\frac{1}{4}$-in. letters. In the group order given, draw each letter four times in pencil and then four times directly in ink, as shown in the illustration.

**6.** Composition. Make a three-line design of the quotation from Benjamin Lamme on the Lamme medals: "THE ENGINEER VIEWS HOPEFULLY THE HITHERTO UN-ATTAINABLE."

**7.** One-eighth-inch vertical letters. Starting $\frac{1}{4}$ in. from the top, draw guide lines for 18 lines of $\frac{1}{8}$-in. letters. Make each letter and numeral eight times directly in ink. Fill the remaining lines with a portion of Sec. 13 on composition.

**8.** Composition. Letter the following definition: "Engineering is the art and science of directing and controlling the forces and utilizing the materials of nature for the benefit of man. All engineering involves the organization of human effort to attain these ends. It also involves an appraisal of the social and economic benefits of these activities."

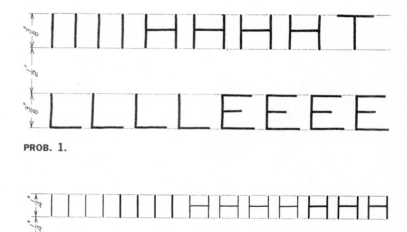

PROB. 1.

PROB. 5.

## Group 2. Single-stroke Inclined Capitals.

**9 to 16.** Same spacing and specifications as for Probs. 1 to 8, but for inclined letters. Study Sec. 10 and Figs. 20 to 23.

## Group 3. Single-stroke Inclined Lower Case.

**17.** Large letters in pencil for use with $\frac{3}{8}$-in. capitals. The bodies are $\frac{1}{4}$ in., the ascenders $\frac{1}{8}$ in. above, and the descenders $\frac{1}{8}$ in. below. Starting $\frac{3}{8}$ in. from the top, draw guide lines for seven lines of letters. This can be done quickly by spacing $\frac{1}{8}$ in. uniformly down the sheet and bracketing capital and base lines. Make each letter of the alphabet four times in pencil only. Study Figs. 23 to 27.

**18.** Lower case for $\frac{3}{16}$-in. capitals. Starting $\frac{1}{2}$ in. from the top, draw capital, waist, and base lines for 13 lines of letters (Braddock or Ames no. 6 spacing). Make each letter six times in pencil and then six times in ink.

**19.** Composition. Same spacing as Prob. 18. Letter the opening paragraph of this chapter.

**20.** Letter Sec. 4.

## Group 4. Titles.

**21.** Design a title for the assembly drawing of a rear axle, drawn to the scale of 6 in. = 1 ft, as made by the Chevrolet Motor Co., Detroit. The number of the drawing is C82746. Space allowed is 3 by 5 in.

**22.** Design a title for the front elevation of a powerhouse, drawn to $\frac{1}{4}$-in. scale by Burton Grant, Architect, for the Citizens Power and Light Company of Punxsutawney, Pennsylvania.

## Group 5. Notes.

**23 to 25.** Vertical capitals. Copy each note in no. 6 (Ames or Braddock) spacing; then in no. 4 spacing.

**26 to 28.** Inclined capitals. Copy in no. 5 or no. 4 spacing.

**29 to 31.** Vertical lower case. Copy in no. 5 spacing.

**32 to 34.** Inclined lower case. Copy in no. 4 spacing.

**35.** For extra practice, select any of Probs. 23 to 34 and letter in a chosen style and size.

PAINT WITH METALLIC SEALER AND TWO COATS LACQUER AS PER CLIENT COLOR ORDER. **PROB. 23.**

THIS PRINT IS AMERICAN THIRD-ANGLE PROJECTION **PROB. 24.**

TO BE REMOVED AFTER MACHINING AND BEFORE ASSEMBLY. **PROB. 25.**

CUTOFF BURR MUST NOT PROJECT BEYOND THIS SURFACE. **PROB. 26.**

ALTERNATE MATERIAL: 1ST. RED BRASS 85 % CU. 2ND. COMM. BRASS 90 % CU. 3RD. COMM. BRASS 95 % CU. **PROB. 27.**

THIS HOLE IN PIECE NO. 821 ONLY. REMOVE BURR ON UPPER SIDE. **PROB. 28.**

Flatten ear on this side to make piece No. 51367. See detail. **PROB. 29.**

Deburring slot must be centered on 0.625 hole within ±0.001. **PROB. 30.**

Pivot point for high pressure bellows. **PROB. 31.**

Material: #19 Ga (0.024) C R Steel. Temper to Rockwell B-40 to 65. **PROB. 32.**

This length varies from 0.245 to 0.627. See table Ⓐ below. **PROB. 33.**

Extrude to 0.082 ± .002 Dia. 3-56 Class 2 tap. **PROB. 34.**

# ORTHOGRAPHIC DRAWING AND SKETCHING

# 5

Orthographic Drawing and Sketching Is the Basic Graphic Form of Representation for Design and Construction Drawings

This Chapter Presents:
Theory • Orthographic Views • The Six Principal Views • Combinations of Views • "Alternate-position" Views • The Three Space Dimensions • Relationship of Planes, View Directions, and Space Dimensions • Classification of Surfaces and Lines • Representation of Lines • Hidden Features • Center Lines • Precedence of Lines • Exercises in Projection • Selection of Views • Projection of Views • Order of Drawing • Freehand Drawing • Freehand Techniques • Reading Orthographic Drawings • Reading Lines, Areas • Learning to Read by Sketching and Modeling • Exercises • Problems

# THEORY

## 1. Methods of Projection—Classification.

The previous chapters have been preparatory to the real subject of graphics as a language. Typically, engineers design and develop machines and structures and direct their construction. Furthermore, to design and then *communicate* every detail to manufacturing groups, descriptions must be prepared that show every aspect of the *shape* and *size* of each part and of the complete machine or structure. Because of this necessity, graphics is the fundamental method of communication. Only as a supplement, for notes and specifications, is the word language used.

In this chapter and Chap. 8 we are concerned with the methods of describing shape. Chapter 10 discusses size description.

Shape is described by projection, that is, by the process of causing an image to be formed by rays of sight taken in a particular direction from an object to a picture plane.[1] Methods of projection vary according to the direction in which the rays of sight are taken to the plane. When the rays are perpendicular to the plane, the projective method is *orthographic*. If the rays are at an angle to the plane, the projective method is called *oblique*. Rays taken to a particular station point result in *perspective projection*. By the methods of perspective the object is represented as it would appear to the eye.

Projective theory is the basis of background information necessary for shape representation. In graphics, two fundamental methods of *shape representation* are used:

(1). *Orthographic views*, consisting of a set of two or more separate views of an object taken from different directions, generally at right angles to each other and arranged relative to each other in a definite way. Each of the views shows the shape of the object for a particular view direction and collectively the views describe the object completely. Orthographic projection *only* is used.

[1] Only in *projective geometry*, a highly theoretical graphic subject, are surfaces other than a plane used.

(2). *Pictorial views*, in which the object is oriented behind and projected upon a single plane. Either orthographic, oblique, or perspective projection is used.

Since orthographic views provide a means of describing the *exact shape* of any material object, they are used for the great bulk of engineering work.

## 2. Theory of Orthographic Projection.

Let us suppose that a transparent plane has been set up between an object and the station point of an observer's eye (Fig. 1). The intersection of this plane with the rays formed by lines of sight from the eye to all points of the object would give a picture that is practically the same as the image formed in the eye of the observer. This is perspective projection.

If the observer would then walk backward from the station point until he reached a theoretically *infinite* distance, the rays formed by lines of sight from his eye to the object would grow longer and finally become infinite in length, parallel to each other, and perpendicular to the picture plane. The image so formed on the picture plane is what is known as "orthographic projection." See Fig. 2.

## 3. Definition.

Basically, orthographic[2] projection could be defined as any single projection made by dropping perpendiculars to a plane. However, it has been accepted through long usage to mean the combination of two or more such views, hence the following definition: *Orthographic projection is the method of representing the exact shape of an object by dropping perpendiculars from two or more sides of the object to planes, generally at right angles to each other; collectively, the views on these planes describe the object completely.* (The term "orthogonal"[3] is sometimes used for this system of drawing.)

## 4. Orthographic Views.

The rays from the picture plane to infinity may be discarded and the picture, or "view," thought of as

[2] Literally, "right writing."
[3] Meaning right-angled.

**FIG. 1.** Perspective projection. The rays of projection converge at a station point from which the object is observed. Rays intersect a picture plane and produce a projection of the object.

**FIG. 2.** Orthographic projection. The station point is at infinity, making the rays parallel to each other. The rays are perpendicular to the picture plane.

being found by extending perpendiculars to the plane from all points of the object, as in Fig. 3. This picture, or projection on a frontal plane, shows the shape of the object when viewed from the front, but it does not tell the shape or distance from front to rear. Accordingly, more than one projection is required to describe the object.

In addition to the frontal plane, imagine another transparent plane placed horizontally above the object, as in Fig. 4. The projection on this plane, found by extending perpendiculars to it from the object, will give the appearance of the object as if viewed from directly above and will show the distance from front to rear. If this horizontal plane is now rotated into coincidence with the frontal plane, as in Fig. 5, the two views of the object will be in the same plane, as if on a sheet of paper. Now imagine a third plane, perpendicular to the first two (Fig. 6). This plane is called a "profile plane," and a third view can be projected on it. This view shows the shape of the object when viewed from the side and the

**FIG. 3.** The frontal plane of projection. This produces the front view of the object.

**FIG. 4.** The frontal and horizontal planes of projection. Projection on the horizontal plane produces the top view of the object. Frontal and horizontal planes are perpendicular to each other.

**FIG. 5.** The horizontal plane rotated into the same plane as the frontal plane. This makes it possible to draw two views of the object on a plane, the drawing paper.

distance from bottom to top and front to rear. The horizontal and profile planes are shown rotated into the same plane as the frontal plane (again thought of as the plane of the drawing paper) in Fig. 7. Thus related in the same plane, they give correctly the three-dimensional shape of the object.

In orthographic projection the picture planes are called "planes of projection"; and the perpendiculars, "projecting lines" or "projectors."

In looking at these theoretical projections, or views, do not think of the views as flat surfaces on the transparent planes, but try to imagine that you are looking *through* the transparent planes at the object itself.

## 5. The Six Principal Views.

Considering the matter further, we find that the object can be entirely surrounded by a set of six planes, each at right angles to the four adjacent to it, as in Fig. 8. On these planes, views can be obtained of the object as it is seen from the top, front, right side, left side, bottom, and rear.

Think now of the six sides, or planes, of the box as being opened up, as in Fig. 9, into one plane, the plane of the paper. The front is already in the plane of the paper, and the other sides are, as it were, hinged and rotated into position as shown. The projection on the frontal plane is the *front view, vertical projection,* or *front elevation;* that on the horizontal plane, the *top view, horizontal projection,* or *plan;* that on the side, or "profile," plane, the

**FIG. 6.** The three planes of projection: frontal, horizontal, and profile. Each is perpendicular to the other two.

**FIG. 7.** The horizontal and profile planes rotated into the same plane as the frontal plane. This makes it possible to draw three views of the object on a plane, the drawing paper.

**FIG. 8.** "The transparent box." This encloses the object with another frontal plane behind, another horizontal plane below, and another profile plane to the left of the object.

**FIG. 9.** The transparent box as it opens and all planes rotate to the plane of the frontal plane. Note that horizontal and profile planes are hinged to the frontal plane and that the "rear-view" plane is hinged to the left profile plane.

*side view, profile projection, side elevation,* or sometimes *end view* or *end elevation.* By reversing the direction of sight, a *bottom view* is obtained instead of a *top view,* or a *rear view* instead of a *front view.* In comparatively rare cases a bottom view or rear view or both may be required to show some detail

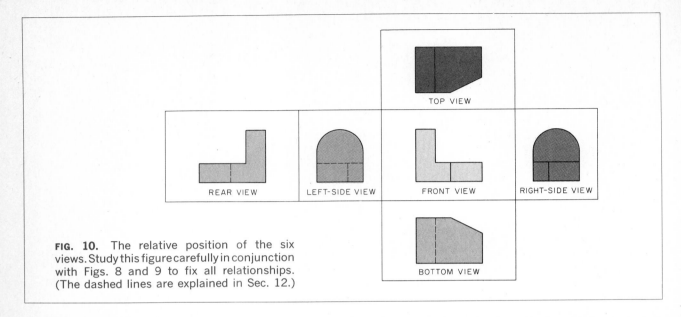

**FIG. 10.** The relative position of the six views. Study this figure carefully in conjunction with Figs. 8 and 9 to fix all relationships. (The dashed lines are explained in Sec. 12.)

of shape or construction. Figure 10 shows the relative position of the six views as set by the ANSI. In actual work there is rarely an occasion when all six principal views are needed on one drawing, but no matter how many are required, their positions relative to one another are given in Fig. 10 (except as noted in Sec. 7). All these views are principal views. Each of the six views shows two of the three dimensions of height, width, and depth.

**FIG. 11.** Top, front, and right-side views. This is the most common combination. Note that the top view is directly above and in projection (alignment) with the front view; and that the right-side view is to the right of and in projection with the front view. Observe also that *two* (and remember which *two*) space dimensions of height, width, and depth are represented in each view.

**FIG. 12.** Top, front, and left-side views. Note that the left-side view is drawn to the left of and in projection with the front view. The left-side view is preferred only when, because of the shape of the object, representation is clearer with the left-side view than with the right-side view.

## 6. Combination of Views.

The most usual combination selected from the six possible views consists of the *top, front,* and *right-side* views, as shown in Fig. 11, which, in this case, best describes the shape of the given block. Sometimes the left-side view helps to describe an object more clearly than the right-side view. Figure 12 shows the arrangement of *top, front,* and *left-side* views for the same block. In this case the right-side view would be preferred, as it shows no hidden edges (see Sec. 12 on hidden features). Note that the *side view of the front face of the object is adjacent to the front view* and that the side view of any point will be the same distance from the front surface as

**FIG. 15.** The transparent box opening for alternate-position views. Note that frontal and profile planes are hinged to the horizontal plane.

**FIG. 13.** Front, bottom, and right-side views. The bottom view is used instead of the top view only when its use gives clearer representation.

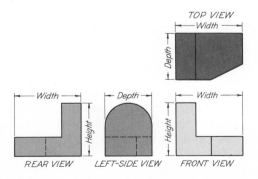

**FIG. 14.** Top, front, left-side, and rear views. The rear view is added *only* when some detail on the rear of the object is important and representation can be improved by its use.

is its distance from the front surface on the top view. The combination of *front, right-side,* and *bottom* views is shown in Fig. 13 and of *front, top, left-side,* and *rear* views in Fig. 14.

## 7. "Alternate-position" Views.

The top of the enclosing transparent box may be thought of as in a fixed position with the front, rear, and sides hinged, as in Fig. 15, thus bringing the sides in line with the top view and the rear view above the top view, Fig. 16. This alternate-position arrangement is of occasional use to save space on the paper in drawing a broad, flat object (Fig. 17). The alternate position for the rear view may be used if this arrangement makes the drawing easier to read (Fig. 18).

## 8. The Three Space Dimensions.

As all material objects, from single pieces to complicated structures, have distinct limits and are measurable by three space dimensions,[4] it is desirable

---

[4]*Space dimensions* and *dimensions of the object* should not be confused. The primary function of orthographic projection is to show the shape of the object. Size is not established until the figured dimensions and/or the scale are placed on the drawing. Space dimensions are *only* the measure of three-dimensional space.

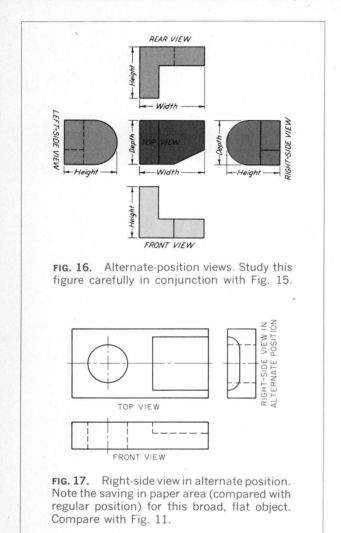

**FIG. 16.** Alternate-position views. Study this figure carefully in conjunction with Fig. 15.

**FIG. 17.** Right-side view in alternate position. Note the saving in paper area (compared with regular position) for this broad, flat object. Compare with Fig. 11.

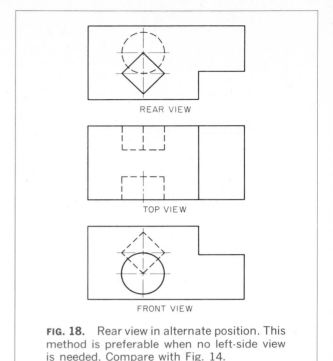

**FIG. 18.** Rear view in alternate position. This method is preferable when no left-side view is needed. Compare with Fig. 14.

for drawing purposes to define these dimensions and to fix their direction.

*Height* is the difference in elevation between any two points, measured as the perpendicular distance between a pair of horizontal planes that contain the points, as shown in Fig. 19. Edges of the object may or may not correspond with the height dimensions. Edge *AB* corresponds with the height dimension, while edge *CD* does not, but the space heights of *A* and *C* are the same, as are *B* and *D*. Height is

always measured in a vertical direction and has no relationship whatever to the shape of the object.

*Width* is the positional distance left to right between any two points measured as the perpendicular distance between a pair of profile planes containing the points. In Fig. 20 the relative width between points *E* and *G* on the left and *H* and *F* on the right of an object is shown by the dimension marked "width." The object edge *EF* is parallel to the width direction and corresponds with the width dimension, but edge *GH* slopes downward from *G* to *H*, so this actual edge of the object is longer than the width separating points *G* and *H*.

*Depth*[5] is the positional distance front to rear between any two points measured as the perpendicular distance between two frontal planes containing the points. Figure 21 shows two frontal

[5] As in the civil engineering sense.

**FIG. 19.** Definition of height. This is the difference in elevation between two points.

**FIG. 21.** Definition of depth. This is the difference from front to rear between two points.

**FIG. 20.** Definition of width. This is the difference from left to right between two points.

**FIG. 22.** Location of points in space. Height, width, and depth must be designated.

planes, one at the front of the object containing points $J$ and $L$, the other at the rear containing points $K$ and $M$. The relative depth separating the front and rear of the object is the perpendicular distance between the planes as shown.

Any point can be located in space by giving its height, width, and depth relative to some other known point. Figure 22 shows a cube with four identified corners $A$, $B$, $C$, and $D$. Assuming that the plane containing points $A$ and $B$ is the front of the object, height, width, and depth would be as marked. Assuming also that point $A$ is fixed in space, point $B$ could be located from point $A$ by giving the width dimension, including the statement that height and depth measurements are zero. $C$ could be located from $A$ by giving width, height, and zero depth. $D$ could be located from $A$ by giving width, height, and depth measurements.

## 9. The Relationship of Planes, View Directions, and Space Dimensions.

As explained in Secs. 4 and 5, the object to be drawn may be thought of as surrounded by transparent planes upon which the actual views are projected. The three space dimensions—height, width, and depth—and the planes of projection are unchangeably oriented and connected with each other and with the view directions (Fig. 23). Each of the planes of projection is perpendicular, respectively, to its own view direction. Thus the frontal plane is perpendicular to the front-view direction, the horizontal plane is perpendicular to the top-view direction, and the profile plane is perpendicular to the side-view direction. The two space measurements for a view are parallel to the plane of that view and perpendicular to the view direction. Therefore height and width are parallel to the frontal plane and perpen-

**FIG. 23.** The relationship between space directions and the planes of projection. View directions are perpendicular to their planes of projection. Height is parallel to frontal and profile planes, width is parallel to frontal and horizontal planes, and depth is parallel to horizontal and profile planes.

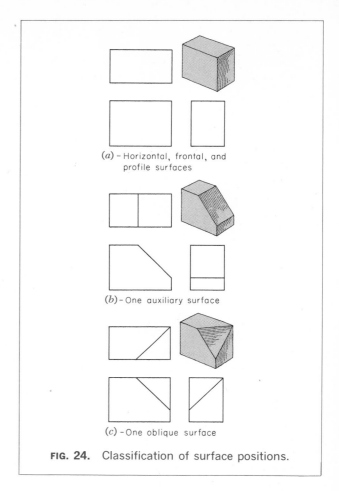

(a) – Horizontal, frontal, and profile surfaces

(b) – One auxiliary surface

(c) – One oblique surface

**FIG. 24.** Classification of surface positions.

dicular to the front-view direction; width and depth are parallel to the horizontal plane and perpendicular to the top-view direction; height and depth are parallel to the profile plane and perpendicular to the side-view direction. Note that the three planes of projection are *mutually* perpendicular, as are the three space measurements and the three view directions. Carefully study the views in Figs. 11 to 17 and note the space dimensions marked on each figure.

## 10. Classification of Surfaces and Lines.

Any object, depending upon its shape and space position, may or may not have some surfaces parallel or perpendicular to the planes of projection.

Surfaces are classified according to their space relationship with the planes of projection (Fig. 24). *Horizontal, frontal,* and *profile* surfaces are shown at (a). When a surface is inclined to two of the planes of projection (but perpendicular to the third), as at (b), the surface is said to be *auxiliary* or *inclined*. If the surface is at an angle to all three planes, as at (c), the term *oblique* or *skew* is used.

The edges (represented by lines) bounding a surface may, because of the shape or position of the object, also be in a simple position or inclined to the planes of projection. A line in, or parallel to, a plane of projection takes its name from the plane. Thus a *horizontal line* is a line in a horizontal plane, a *frontal line* is a line in a frontal plane, and a *profile line* is a line in a profile plane. When a line is parallel to two planes, the line takes the name of both planes, as *horizontal-frontal, horizontal-profile,* or *frontal-profile.* A line not parallel to any plane of projection is called an *oblique* or *skew line.* Figure 25 shows various positions of lines.

An edge appears in true length when it is parallel to the plane of projection, as a point when it is perpendicular to the plane, and shorter than true length when it is inclined to the plane. Similarly, a

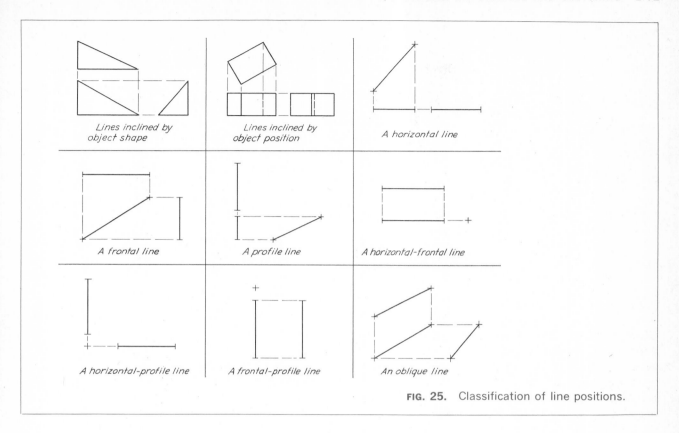

**FIG. 25.** Classification of line positions.

surface appears in true shape when it is parallel to the plane of projection, as a line when it is perpendicular to the plane, and foreshortened when it is inclined to the plane. As an example, Fig. 24a shows an object with its faces parallel to the planes of projection; top, front, and right-side surfaces are shown in true shape; and the object edges appear either in true length or as points. The inclined surface of the object at (b) does not show in true shape in any of the views but appears as an edge in the front view. The front and rear edges of the inclined surface are in true length in the front view and foreshortened in the top and side views. The top and bottom edges of the inclined surface appear in true length in top and side views and as points in the front view. The oblique (skew) surface of the object at (c) does not show in true shape in any of the views, but each of the bounding edges shows

in true length in one view and is foreshortened in the other two views.

## 11. Representation of Lines.

Although uniform in appearance, the lines on a drawing may indicate three different types of directional change on the object. An *edge* view is a line showing the edge of a receding surface that is perpendicular to the plane of projection. An *intersection* is a line formed by the meeting of two surfaces when either one surface is parallel and one at an angle or both are at an angle to the plane of projection. A *surface limit* is a line that indicates the reversal of direction of a curved surface (or the series of points of reversal on a warped surface). Figure 26 illustrates the different line meanings, and these are further explained in Sec. 45.

**FIG. 26.** What a line indicates.

server's eye. The edges, intersections, and surface limits of these hidden parts are indicated by a discontinuous line called a *dashed line*.[6] See the alphabet of lines (Figs. 56 and 57). In Fig. 27 the drilled hole[7] that is visible in the right-side view is hidden in the top and front views, and therefore it is indicated in these views by a dashed line showing the hole and the shape as left by the drill point. The milled slot (see Glossary) is visible in the front and side views but is hidden in the top view.

The beginner must pay particular attention to the execution of these dashed lines. If carelessly drawn, they ruin the appearance of a drawing and make it harder to read. Dashed lines are drawn lighter than full lines, of short dashes uniform in length with the space between them very short, about one-fourth the length of the dash. It is important that they start and stop correctly. A dashed line always starts with a dash except when the dash would form a continu-

## 12. Hidden Features.

To describe an object completely, a drawing should contain lines representing all the edges, intersections, and surface limits of the object. *In any view there will be some parts of the object that cannot be seen from the position of the observer, as they will be covered by portions of the object closer to the ob-*

[6]The line indicating hidden features has been traditionally known as a "dotted" line. However, in this treatise the term "dashed" line will be used because it accurately describes the appearance of the line. The term "hidden line," also sometimes used, is completely inaccurate because there are no *lines* on the object, and the line indicating hidden features *is* visible on the drawing.

[7]See Glossary and Index.

**FIG. 27.** Dashed-line technique. Note especially that a dashed line *begins with a space when it continues in the same direction as a full line.*

ation of a full line; in that case a space is left, as shown in Fig. 27. Dashes always meet at corners. An arc must start with a dash at the tangent point except when the dash would form a continuation of a straight or curved full line. The number of dashes used in a tangent arc should be carefully judged to maintain a uniform appearance (Fig. 28). Study carefully all dashed lines in Figs. 27 and 29.

### 13. Center Lines.

In general, the first lines drawn in the layout of an engineering drawing are the center lines, which are the axes of symmetry for all symmetrical views or portions of views: (1) Every part with an axis, such as a cylinder or a cone, will have the axis drawn as a center line before the part is drawn. (2) Every circle will have its center at the intersection of two mutually perpendicular center lines.

The standard symbol for center lines on finished drawings is a fine line made up of alternate long and short dashes, as shown in the alphabet of lines (Figs. 56 and 57). Center lines are always extended slightly beyond the outline of the view or portion of the view to which they apply. They form the skeleton

**FIG. 29.** Dashed lines and arcs. Study this figure in conjunction with Figs. 27 and 28.

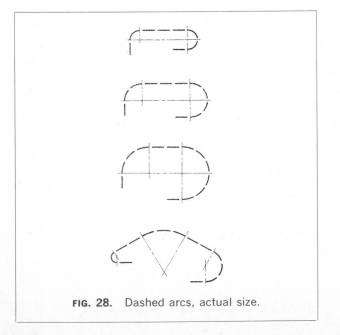

**FIG. 28.** Dashed arcs, actual size.

construction of the drawing; the important measurements are made and dimensions given to and from these lines. Study the center lines in Probs. 90 to 110.

### 14. Precedence of Lines.

In any view there is likely to be a coincidence of lines. Hidden portions of the object may project to coincide with visible portions. Center lines may occur where there is a visible or hidden outline of some part of the object.

Since the physical features of the object must be represented, full and dashed lines take precedence over all other lines. Since the visible outline is more prominent by space position, full lines take precedence over dashed lines. A full line could cover a dashed line, but a dashed line could not cover a full line. It is evident also that a dashed line could not occur as one of the boundary lines of a view.

When a center line and cutting-plane (explained in Chap. 7) line coincide, the one that is more important for the readability of the drawing takes precedence over the other.

Break lines (explained in Chap. 7) should be placed so that they do not spoil the readability of the over-all view.

Dimension and extension lines must always be placed so as not to coincide with other lines of the drawing.

The following list gives the order of precedence of lines:
1. Full line
2. Dashed line
3. Center line or cutting-plane line
4. Break lines
5. Dimension and extension lines
6. Crosshatch lines

Note the coincident lines in Fig. 30.

### 15. Exercises in Projection.
The principal task in learning orthographic projection is to become thoroughly familiar with the theory and then to practice this theory by translating from a picture of the object to the orthographic views. Figures 31 and 32 contain a variety of objects shown by a pictorial sketch and translated into orthographic

**FIG. 30.** Coincident-line study. Coincident lines are caused by the existence of features of identical size or position, one behind the other.

views. Study the objects and note (1) how the object is oriented in space, (2) why the orthographic views given were chosen, (3) the projection of visible features, (4) the projection of hidden features, and (5) center lines.

## WRITING THE GRAPHIC LANGUAGE

### 16. Objectives.
The major objective of a student of the graphic language is to learn orthographic drawing. In addition to an understanding of the theory of orthographic projection, several aspects of drawing are necessary preliminaries to its study. These have been discussed in the preceding chapters and include skill and facility in the use of instruments (Chap. 2), a knowledge of applied geometry (Chap. 3), and fluency in lettering (Chap. 4). In writing the graphic language, a topic we will now take up, always pay careful attention to accuracy and neatness.

### 17. Object Orientation.
An object can, of course, be drawn in any of several possible positions. *The simplest position should be used,* with the object oriented so that the principal faces are perpendicular to the sight directions for the views and parallel to the planes of projection, as shown in Fig. 33. Any other position of the object, with its faces at some angle to the planes of projection, would complicate the drawing, foreshorten the object faces, and make the drawing difficult to make and to read.

### 18. Selection of Views.
In practical work it is important to choose the combination of views that will describe the shape of an object in the best and most economical way. Often only two views are necessary. For example, a cylindrical shape, if on a vertical axis, would require only a front and top view; if on a horizontal axis, only a front and side view. Conic and pyramidal shapes

**FIG. 31.** Projection studies. Study each picture and the accompanying orthographic views and note the projection of all features.

**FIG. 32.** Projection studies. Study each picture and the accompanying orthographic views and note the projection of all features.

can also be described in two views. Figure 34 illustrates two-view drawings. Some shapes will need more than the three regular views for adequate description.

Objects can be thought of as being made up of combinations of simple geometric solids, principally

cylinders and rectangular prisms, and the views necessary to describe any object would be determined by the directions from which it would have to be viewed to see the characteristic contour shapes of these parts. Figure 35, for example, is made up of several prisms and cylinders. If each of these

**FIG. 33.** Object orientation. Use the simplest position. It will give the clearest possible representation and be the easiest to draw.

**FIG. 34.** Two-view drawings. These are sufficient for any object having a third view identical with or similar to one of the two views given.

**FIG. 35.** Geometric shapes combined. Even the most complicated objects can be analyzed as combined geometric shapes.

simple shapes is described and its relation to the others is shown, the object will be fully represented. In the majority of cases the three regular views—top, front, and side—are sufficient to do this.

Sometimes two views are proposed as sufficient for an object on the assumption that the contour in the third direction is of the shape that would naturally be expected. In Fig. 36, for example, the figure at (a) would be assumed to have a uniform cross section and be a square prism. But the two views *might* be the top and front views of a wedge, as shown in three views at (b). Two views of an object, as drawn at (c), do not describe the piece at all. The object at (a) might be assumed to be square in section, but it could as easily be round, triangular, quarter-round, or of another shape, which

should have been indicated by a side view. Sketch several different front views for each top view, Fig. 37a to c.

With the object preferably in its functioning position and *with its principal surfaces parallel to the planes of projection*, visualize the object, mentally

**FIG. 36.** A study of views. Two views do not describe a rectangular object (*a*), but a wedge shape (*b*) or quarter round (*c*) is described by two views.

**FIG. 37.** Top views given. One view does not describe an object unless some additional explanation is given. Sketch several front views for (*a*), (*b*), and (*c*).

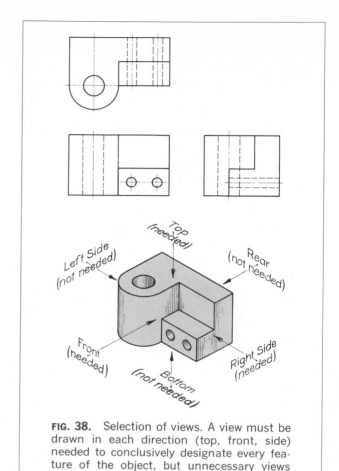

**FIG. 38.** Selection of views. A view must be drawn in each direction (top, front, side) needed to conclusively designate every feature of the object, but unnecessary views must *not* be drawn.

picturing the orthographic views one at a time to decide on the best combination. In Fig. 38, the arrows show the direction of observation for the six principal views of an object, and indicate the mental process of the person making the selection. He notes that the front view would show the two horizontal holes as well as the width and height of the piece, that a top view is needed to show the contour of the vertical cylinder, and that the cutout corner calls for a side view to show its shape. He notes further that the right-side view would show this cut in full lines, while the left-side view would give it in

dashed lines. He observes also that neither a bottom view nor a rear view would be of any value in describing this object. Thus he has correctly chosen the front, top, and right-side views as the best combination for describing this piece. As a rule, the side view containing the fewer dashed lines is preferred. If the side views do not differ in this respect, the right-side view is preferred in standard practice.

In inventive and design work, any simple object should be visualized mentally and the view selected without a picture sketch. In complicated work, a pictorial or orthographic sketch may be used to advantage, but it should not be necessary, in any case, to sketch all possible views in order to make a selection.

Study the drawings in Fig. 39 and determine why each view was chosen.

## 19. Drawing Sizes.

Standard sizes for sheets of drawing paper, based on multiples of $8\frac{1}{2}$ by 11 in. and 9 by 12 in., are specified for drawings by the ANSI. Trimmed sizes of drawing paper and cloth, with suitable border and title dimensions in the final chapter.

## 20. Spacing the Views.

View spacing is necessary so that the drawing will be balanced within the space provided. A little preliminary measuring is necessary to locate the views. The following example describes the procedure: Suppose the piece illustrated in Fig. 40 is to be drawn full size on an 11- by 17-in sheet. With an end-title strip, the working space inside the border will be $10\frac{1}{2}$ by 15 in. The front view will require $7\frac{11}{16}$ in., and the side view $2\frac{1}{4}$ in. This leaves $5\frac{1}{16}$ in. to be distributed between the views and at the ends.

This preliminary planning need not be to exact

**FIG. 39.** Selection-of-view study. Determine in each case (*a* to *k*) why the views shown are the best choice.

**FIG. 40.** Spacing the views on the paper. This is done graphically. Study the text carefully while referring to this figure and go through the steps by laying out the given object on a standard 11- × 17-in. sheet.

dimensions, that is, small fractional values, such as $^{15}/_{64}$ in. or $^{31}/_{32}$ in., can be adjusted to $\frac{1}{4}$ and 1 in., respectively, to speed up the planning. In this case the $7^{11}/_{16}$-in. dimension can be adjusted to $7\frac{3}{4}$ in.

Locate the views graphically and quickly by measuring with your scale along the bottom border line. Starting at the lower right corner, lay off first $2\frac{1}{4}$ in. and then $7\frac{3}{4}$ in. The distance between views can now be decided upon. It is chosen by eye to separate the views without crowding, yet placing them sufficiently close together so that the drawing will read easily (in this case $1\frac{1}{2}$ in.). Measure the distance; half the remaining distance to the left corner is the starting point of the front view. For the vertical location: the front view is 4 in. high, and the top

view $2\frac{1}{4}$ in. deep. Starting at the upper-left corner, lay off first $2\frac{1}{4}$ in. and then 4 in.; judge the distance between views (in this case 1 in.), and lay it off; then a point marked at less than half the remaining space will locate the front view, allowing more space at the bottom than at the top for appearance.

Block out lightly the spaces for the views, and study the over-all arrangement, because changes can easily be made at this stage. If it is satisfactory, select reference lines in each view from which the space measurements of height, width, and depth that appear in the view can be measured. The reference line may be an edge or a center line through some dominant feature, as indicated on Fig. 40 by the center lines in the top and side views and the medium-weight lines in all the views. The directions for height, width, and depth measurements for the views are also shown.

## 21. Projecting the Views.

After laying out the views locate and draw the various features of the object. In doing this, carry the views *along together*, that is, *do not* attempt to complete one view before proceeding to another. Draw first the most characteristic view of a feature and then project it and draw it in the other views before going on to a second feature. As an example, the vertical hole of Fig. 41 should be drawn first in the top view, and then the dashed lines representing the limiting elements or portions should be projected and drawn in the front and side views.

In some cases, one view cannot be completed before a feature has been located and drawn in another view. Study the pictorial drawing in Fig. 41, and note from the orthographic views that the horizontal slot must be drawn on the front view before the edge *AB* on the slanting surface can be found in the top view.

Projections (horizontal) between the front and side views are made by employing the T square to draw the required horizontal line (or to locate a required point), as in Fig. 42.

Projections (vertical) between the front and top views are made by using the T square and a triangle as in Fig. 43.

Projections between the top and side views cannot be projected directly but must be measured and transferred or found by special construction. In carrying the top and side views along together, it is usual to transfer the depth measurement from one to the other with dividers, as in Fig. 44*a*, or with a scale, as at (*b*). Another method, used for an irregular figure, is to "miter" the points around, using a 45° line drawn through the point of intersection of the top and side views of the front face, extended as shown in Fig. 45. The method of Fig. 45, however, requires more time and care than the methods of Fig. 44 and is, therefore, not recommended.

**FIG. 41.** Projection of lines. Carry all views along together. The greatest mistake possible is to try to complete one view before starting another.

**FIG. 42.** Making a horizontal projection. This is the simplest operation in drawing. The T square provides *all* horizontal lines.

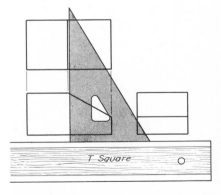

**FIG. 43.** Making a vertical projection. The 90° angle of a triangle with one leg on the horizontal T square produces the vertical.

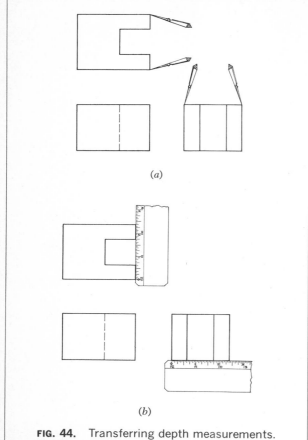

(a)

(b)

**FIG. 44.** Transferring depth measurements. Depth cannot be projected. Transfer the necessary distances with dividers, as at (a), or with scale, as at (b).

**FIG. 45.** Projecting depth measurements. A "miter line" at 45°, with horizontal and vertical projectors, transfers the depth from top to side view (or vice versa).

(a)

(b)

**FIG. 46.** Projections of surfaces bounded by linear edges. Depth measurements cannot be projected directly with T square and triangle. They must be transferred.

## 22. Projections of Surfaces Bounded by Linear Edges.

In drawing projections of inclined surfaces, in some cases the corners of the bounding edges may be used, and in other cases the bounding edges themselves may be projected. In illustration of these methods, Fig. 46a shows a vertical hexagonal hole that is laid out from specifications in the top view. Then the front view is drawn by projecting from the six corners of the hexagon and drawing the four dashed lines to complete the front view. To get the side view, a horizontal projection is made from each corner on the front view to the side view, thus locating the height of the points needed on the side view. Then measurements $D_1$, $D_2$, and $D_3$ taken from the top view and transferred to the side view locate all six corners in the side view. The view is completed by connecting these corners and drawing the three vertical dashed lines. The object in Fig. 46b shows a horizontal slot running out on an inclined surface. In the front view the true width and height of this slot is laid out from specifications. The projection to the side view is a simple horizontal projection for the dashed line, indicating the top surface of the slot. To get the top view, the width is projected from the front view and then the position of the runout line is measured (distance $D_1$) and transferred to the top view.

In summary, it may be stated that, if a line appears at some angle on a view, its two ends must be projected; if a line appears parallel to its path of projection, the complete line can be projected.

## 23. Projections of an Elliptical Boundary.

The intersection of a cylindrical hole (or cylinder) with a slanting (inclined or skew) surface, as shown in Fig. 47, will be an ellipse, and some projections of this elliptical edge will appear as another ellipse. The projection can be made as shown in Fig. 47a by assuming a number of points on the circular view and projecting them to the edge view (front) and then to an adjacent view (side). Thus points 1 to 4 are located in the top view and projected to the front view, and the projectors are then drawn to the side view. Measurements of depth taken from the top view (as $D_1$) will locate the points in the side view. Draw a smooth curve through the points, using a French curve.

For an ellipse on an inclined surface, the projection can also be made by establishing the major and minor diameters of the ellipse, as shown in Fig. 47b. A pair of diameters positioned so as to give the largest and smallest extent of the curve will give the required major and minor diameters. Thus, $AB$ will

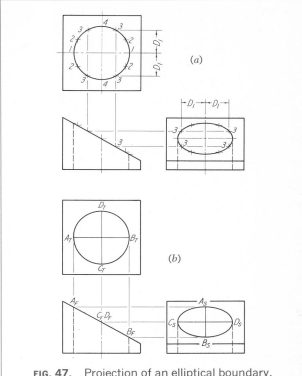

FIG. 47. Projection of an elliptical boundary. Points on the curve are projected to determine the curve (a), or the curve is determined by major and minor diameters (b).

project to the side view as the smaller, or minor, diameter $A_S B_S$, and $CD$ will project as the larger, or major, diameter $C_S D_S$. The ellipse can then be drawn by one of the methods of Secs. 50, 52, and 54 in Chap. 3.

If the surface intersected by the cylinder is skew, as shown in Fig. 48, a pair of perpendicular diameters located in the circular view will give a pair of conjugate diameters in an adjacent view. Therefore, $A_T B_T$ and $C_T D_T$ projected to the front view will give conjugate diameters, which can be employed as explained in Secs. 51, 53, and 54 in Chap. 3 to draw the required ellipse.

In projecting the axes, they can be extended to

**FIG. 48.** Projection of an elliptical boundary by employing conjugate diameters. Refer to Sec. 53 and Fig. 68, both in Chap. 3.

## 24. Projections of a Curved Boundary.

Any nongeometric curve (or a geometric curve not having established axes) must be projected by locating points on the curve. If the surface is in an inclined position, as in Fig. 49a, points may be assumed on the curve laid out from data (assumed in this case to be the top view) and projected first to the edge view (front) and then to an adjacent view (side). Measurements, such as 1, 2, etc., from the top view transferred to the side view complete the projection. A smooth curve is then drawn through the points.

If the surface is skew, as in Fig. 49b, elements of the skew surface, such as 1'-1, 2'-2, etc., located in an adjacent view (by drawing the elements parallel to some known line of the skew surface, such as $AB$) make it possible to project points on the curve 1, 2, etc., to the adjacent view, as shown.

## 25. Projections by Identifying Corners.

In projecting orthographic views or in comparing the views with a picture, it is helpful in some cases to letter (or number) the corners of the object and, with these identifying marks, to letter the corresponding

the straight-line boundary of the skew surface. Thus the line 1–2 located in the front view and intersected by projection of $A_TB_T$ from the top view locates $A_FB_F$. Similarly, lines 3–4 and 5–6 at the ends of the axis $CD$ locate $C_FD_F$.

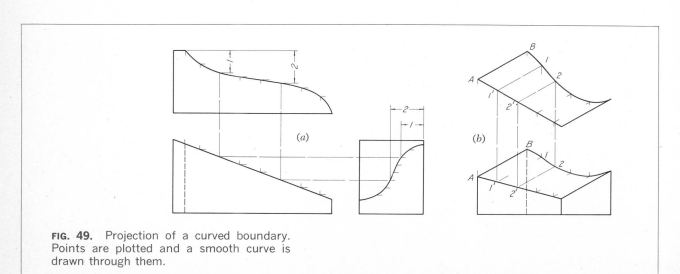

**FIG. 49.** Projection of a curved boundary. Points are plotted and a smooth curve is drawn through them.

points on each of the views, as in Fig. 50. Hidden points directly behind visible points are lettered to the right of the letter of the visible point, and in this figure, they have been further differentiated by the use of "phantom," or dotted, letters. Study Fig. 51, and number or letter the corners of the three views to correspond with the pictorial view.

## 26. Order of Drawing.

The order of working is important, as speed and accuracy depend largely upon the methods used in laying down lines. Avoid duplications of the same measurement and keep to a minimum changing from one instrument to another. Naturally, *all* measurements cannot be made with the scale at one time or *all* circles and arcs drawn without laying down the compass, but as much work as possible should be done with one instrument before shifting to another. An orderly placement of working tools on the drawing table will save time when changing instruments. The usual order of working is shown in Fig. 52.

1. Decide what combination of views will best describe the object. A freehand sketch will aid in choosing the views and in planning the general arrangement of the sheet.

2. Decide what scale to use, and by calculation or measurement find a suitable standard sheet size;

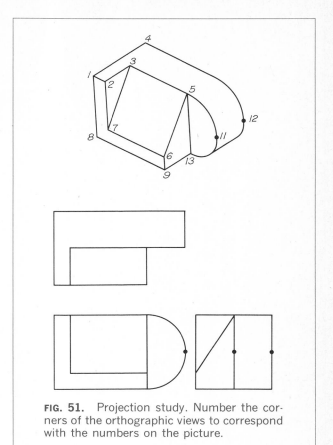

**FIG. 51.** Projection study. Number the corners of the orthographic views to correspond with the numbers on the picture.

**FIG. 50.** Identified corners. Each corner is lettered (or numbered) as an aid in making projections.

**FIG. 52.** Stages in penciling. (*a*) block out the views; (*b*) locate center lines; (*c*) start details, drawing arcs first; (*d*) draw dominant details; (*e*) finish. See text for explanation.

or pick one of the standard drawing-sheet sizes and find a suitable scale.

3. Space the views on the sheet, as described in Sec. 20.

4. Lay off the principal dimensions, and then block in the views with light, sharp, accurate outline and center lines. Draw center lines for the axes of all symmetrical views or parts of views. Every cylindrical part will have a center line—the projection of the axis of the piece. Every circle will have two center lines intersecting at its center.

5. Draw in the details of the part, beginning with the dominant characteristic shape and progressing to the minor details, such as fillets and rounds. Carry the different views along together, projecting a characteristic shape, as shown in one view to the other views, instead of finishing one view before starting another. Use a minimum of construction and draw the lines to finished weight, if possible, as the views are carried along. *Do not make the drawing lightly and then "heavy" the lines later.*

6. Lay out and letter the title.

7. Check the drawing carefully.

## 27. Order of Tracing.

If the drawing is to be traced in ink as an exercise in the use of instruments or for a finished orthographic drawing without dimensions, the order of working is as follows:

1. Place the pencil drawing to be traced on the drawing board, carefully align it with the T square, and put thumbtacks in the two upper corners. Then place the tracing paper or cloth (dull side up) over the drawing. Holding the cloth in position, lift the tacks one at a time and replace them to hold both sheets. Then put tacks in the two lower corners.

2. To remove any oily film, prepare the surface of the cloth or paper by dusting it lightly with prepared pounce or soft white chalk. *Then wipe the surface perfectly clean with a soft cloth.*

3. Carefully set the pen of the compass to the correct line width, and ink all full-line circles and circle arcs, beginning with the smallest. Correct line weights are given in Fig. 57, Chap. 2.

4. Ink dashed circles and arcs in the same order as full-line circles.

5. Carefully set the ruling pen to draw a line

exactly the same width as the line in the full-line circles. The best way to match the straight lines to the circles is to draw with the compass and ruling pen outside the trim line of the sheet or on another sheet of the same kind of paper and adjust the ruling pen until the lines match.

6. Ink irregular curved lines.

7. Ink straight full lines in this order: horizontal (begin at the top of the sheet and work down), vertical (from the left side of the sheet to right), and inclined (uppermost first).

8. Ink straight dashed lines in the same order. Be careful to match these lines with the lines in the dashed circles.

9. Ink center lines.

10. Crosshatch all areas representing cut surfaces.

11. Draw pencil guide lines and letter the title.

12. Ink the border.

13. Check the tracing for errors and omissions.

## 28. Orthographic Freehand Drawing.

Facility in making freehand orthographic drawings is an essential part of the equipment of every engineer, and since ability in sketching presupposes some mastery of other skills (as we have seen in Chaps. 1 to 4), practice should be started early. Although full proficiency in freehand drawing is synonymous with mastery of the graphic language and is gained only after acquiring a background of knowledge and skill in drawing with instruments, sketching is an excellent method for learning the fundamentals of orthographic projection and can be used by the beginner even before he has had much practice with instruments. In training, as in professional work, time can be saved by working freehand instead of with instruments, as with this method more problems can be solved in an allotted amount of time.

Although some experienced teachers advocate the making of freehand sketches before practice in the use of instruments, some knowledge of the use of instruments and especially of applied geometry is a great help because the essentials of line tangents, connections, and intersections as well as the basic geometry of the part should be well defined on a freehand drawing. Drawing freehand is, of course,

an excellent exercise in accuracy of observation. Figure 53 is an example of a good freehand drawing.

## 29. Line Quality for Freehand Work.

Freehand drawings are made on a wide variety of papers, ranging from inexpensive notebook or writing grades to the finer drawing and tracing papers and even pencil cloth. The surface texture—smooth, medium, or rough—combined with the grade of pencil and pressure used will govern the final result. If a bold rough effect is wanted for a scheming or idea sketch, a soft pencil and possibly rough paper would be employed. For a working sketch or for representation of an object with much small or intricate detail, a harder pencil and smoother paper would help to produce the necessary line quality.

Figure 54 is a photograph (reproduced about half size) showing lines made with pencils of various grades with medium pressure on paper of medium texture. Note that the 6B pencil gives a rather wide and rough line. As the hardness increases from 5B to 4B, etc., up to 2H, the line becomes progressively narrower and lighter in color. This does not mean that a wide line cannot be made with a fairly hard pencil, but with ordinary sharpening and with uniform pressure the softer grades wear down much faster than the harder grades. Unless a soft pencil such as 6B, 5B, or 4B is sharpened after each short stroke, fine lines are impossible. Also, the harder grades such as F, H, and 2H, once sharpened, will hold their point for some time, and fairly fine lines can be obtained without much attention to the point. With ordinary pressure and normal use the soft grades give bold rough results, and the harder grades give light smooth lines.

## 30. Range of Pencil Grades.

The 6B grade is the softest pencil made and gives black, rough lines. With normal pressure, the line erases easily but is likely to leave a slight smear. The 6B, 5B, and 4B pencils should be used when a rather rough line is wanted, for example, for scheming or idea sketches, architectural renderings, and illustrations. The range from 3B to HB, inclusive, is usually employed for engineering sketches on medium-textured paper. For example, a sketch

**FIG. 53.** A freehand drawing. Note the "roughness" of lines as compared with instrument work, but that all features are concise and readable.

of a machine part with a normal amount of small detail can be made effectively with a 2B or B pencil. The smoothness and easy response to variations in pressure make these grades stand out as the preferred grades for a wide variety of work. However, for more critical work where there is much detail and also where the smearing of the soft grades is objectionable, the grades from F to 2H are used. For a sketch on fairly smooth paper, F or H are quite satisfactory. Grades harder than 2H are rarely used for freehand work. Incidentally, a fine pencil for ordinary writing is the 2B grade.

### 31. Sharpening the Pencil.
The pencil should be sharpened to a fairly long, conic point, as explained in Sec. 24, Chap. 2. However, for freehand work a point not quite so sharp as for instrument drawing gives the desired line width without too much pressure. If after sharpening in the regular way the point is too fine, it can be rounded off slightly on a piece of scratch paper before use on the drawing.

### 32. Pencil Pressure and Paper Texture.
The pencil grade, pressure, and paper texture all have an effect on the final result. Figure 54 shows the various pencil grades with medium pressure on paper of medium texture. To mark the difference in line quality obtainable by increasing the pressure, Fig. 55 shows the same paper but with firm pressure on the pencil. Note that the lines in Fig. 55 are much blacker than those in Fig. 54. The line quality of Fig. 55 is about right for most engineering sketches. The rather firm opaque lines are preferred to the type in Fig. 54, especially if reproductions, either photographic or by transparency process, are to be made from the sketch.

To show the difference in line quality produced by paper texture, Fig. 56 shows the same firm pressure used in Fig. 55, but this time on smooth paper. In Fig. 57 the same pressure has been used on rough paper. Figure 58 is given to aid in comparing the effect of pressure and texture. On the left is the medium pressure on medium-textured paper; at (*a*), (*b*), and (*c*) firm pressure has been used but at (*a*) on medium-, at (*b*) on smooth-, and at (*c*) on rough-textured paper.

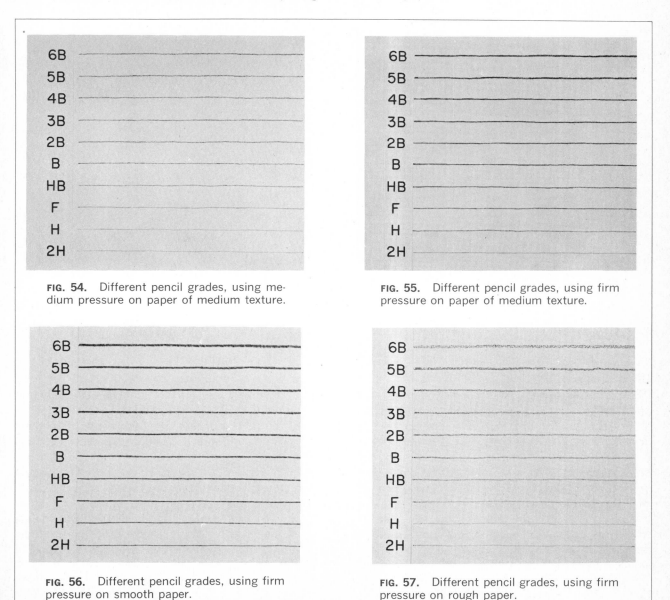

**FIG. 54.** Different pencil grades, using medium pressure on paper of medium texture.

**FIG. 55.** Different pencil grades, using firm pressure on paper of medium texture.

**FIG. 56.** Different pencil grades, using firm pressure on smooth paper.

**FIG. 57.** Different pencil grades, using firm pressure on rough paper.

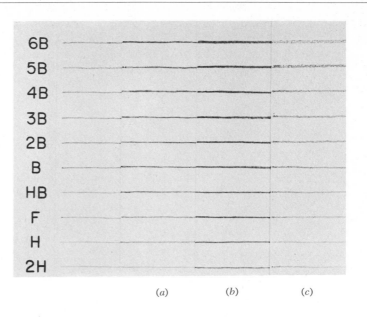

6B
5B
4B
3B
2B
B
HB
F
H
2H

(a)        (b)        (c)

**FIG. 58.** Comparison of different pencil pressures and paper surfaces. Medium pressure on medium paper—left column. Firm pressure on medium paper at (a), on smooth at (b), and on rough at (c).

## 33. Kinds of Paper.

*Plain and Coordinate.* Sketches are made for many purposes and under a variety of circumstances, and as a result on a number of different paper types and surfaces. A field engineer in reporting information to the central office may include a sketch made on notebook paper or a standard letterhead. On the other hand, a sketch made in the home office may be as important as any instrument drawing and for this reason may be made on good-quality drawing or tracing paper and filed and preserved with other drawings in a set. Figure 59 is an example of a sketch made on plain paper, which might be the letterhead paper of the field engineer or a piece of fine drawing or tracing paper. The principal difficulty in using plain paper is that proportions and projec-

tions must be estimated by eye. A good sketch on plain paper requires better-than-average ability and experience. Use of some variety of coordinate paper is a great aid in producing good results. There are many kinds of paper and coordinate divisions available, from smooth to medium texture and coordinate divisions of $\frac{1}{8}$ or $\frac{1}{10}$ to $\frac{1}{2}$ in., printed on tracing paper or various weights of drawing paper. Usually one coordinate size on tracing paper and another (or the same) on drawing paper will supply the needs of an engineering office. Figure 60 is an example of a sketch on paper with coordinate divisions of $\frac{1}{8}$ in. Figure 61 has divisions of $\frac{1}{4}$ in. Figure 62 is a sketch made on $\frac{1}{4}$-in. coordinate paper, actual size.

The paper type, tracing or regular, is another

**FIG. 59.** A freehand drawing on plain paper. This drawing was made by an expert of long experience. More "roughness" or "waviness" of lines is permissible on freehand than on instrument drawings, but care should be taken to keep all details concise and readable.

**FIG. 61.** A freehand drawing on coordinate notebook paper. Coordinates are ¼ in. apart.

**FIG. 60.** A freehand drawing on coordinate tracing paper. The coordinates, on the back of the paper, aid greatly in making projections, in keeping the right proportions, and in drawing straight and accurate lines.

factor to be considered. Reproduction by any of the transparency methods demands the use of tracing paper. If coordinate paper is used, it may be desirable to obtain prints on which the coordinate divisions do not show. Figure 60 is an example of a sketch on tracing paper with the coordinate divisions printed on the back of the paper in faint purplish-blue ink. Since the divisions are on the back, erasures and corrections can be made without erasing the coordinate divisions. Normally the divisions will not reproduce, so prints give the appearance of a sketch made on plain paper. Figure 61 is a sketch on standard, three-hole notebook paper with ¼-in. divisions in pale blue ink.

The use of coordinate paper is a great aid in freehand drawing, and speeds up the work considerably. Projections are much easier to make on it than on plain paper, and to transfer distances from top to side view, the divisions can be counted.

## 34. Technique.

The pencil is held with freedom and not close to the point. Vertical lines are drawn downward with a finger movement in a series of overlapping strokes, the hand somewhat in the position of Fig.

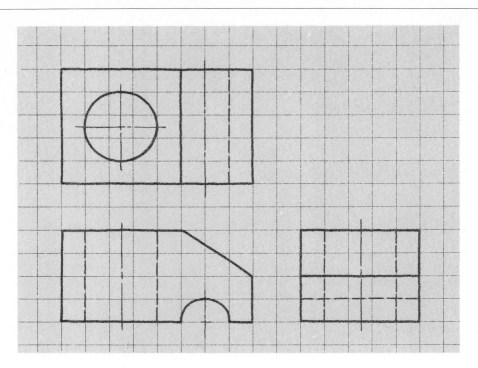

**FIG. 62.** A freehand drawing on coordinate paper (actual size). Note the bold but concise technique.

63. Horizontal lines are drawn with the hand shifted to the position of Fig. 64, using a wrist motion for short lines and a forearm motion for longer ones. In drawing any straight line between two points, *keep your eyes on the point to which the line is to go rather than on the point of the pencil*. Do not try to draw the whole length of a line in a single stroke. It may be helpful to draw a very light line first, as in Fig. 65A, and then to sketch the finished line, correcting the direction of the light line and bringing the line to final width and blackness by using strokes of convenient length, as at (B). The finished line is shown at (C). Do not be disturbed by any nervous waviness. Accuracy of direction is more important than smoothness of line.

## 35. Straight Lines.

Horizontal lines are drawn from left to right as in Fig. 66, vertical lines from top to bottom as in Fig. 67.

Inclined lines running downward from right to left (Fig. 68) are drawn with approximately the same movement as vertical lines, but the paper may be turned and the line drawn as a vertical (Fig. 69).

Inclined lines running downward from left to right (Fig. 70) are the hardest to draw because the hand is in a somewhat awkward position; for this reason, the paper should be turned and the line drawn as a horizontal, as in Fig. 71.

The sketch paper can easily be turned in any direction to facilitate drawing the lines because there is no necessity to fasten the paper to a drawing-table top. The paper may, of course, be taped to a drawing board or attached to a clip board.

It is legitimate in freehand drawing to make long vertical or horizontal lines using the little finger as

FIG. 63. Sketching a vertical line. Draw downward with finger movement, overlapping the strokes for long lines.

FIG. 64. Sketching a horizontal line. Draw from left to right with wrist pivot for short lines and forearm movement for long lines. Overlap strokes if necessary.

FIG. 65. Technique of sketching lines. (*a*) set direction with a *light* construction line; (*b*) first stroke; (*c*) complete line with a series of overlapping strokes.

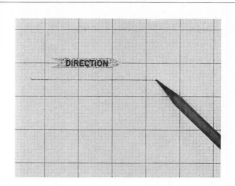

FIG. 66. Sketching a horizontal line. Draw from left to right.

FIG. 67. Sketching a vertical line. Draw from top to bottom.

a guide along the edge of the pad or clip board. The three important things about a straight line are that it (1) be essentially straight, (2) be the right length, and (3) go in the right direction.

## 36. Circles.

Circles can be drawn by marking the radius on each side of the center lines. A more accurate method is to draw two diagonals in addition to the center lines and mark points equidistant from the center of the eight radii; at these points, draw short arcs

**FIG. 68.** Sketching an inclined line sloping downward from right to left. This line may be drawn in either direction, whichever is more convenient by personal preference.

**FIG. 70.** Sketching an inclined line sloping downward from left to right. This is the most awkward position of any line. The paper should be turned as in Fig. 71 to aid in obtaining a smooth, accurate line.

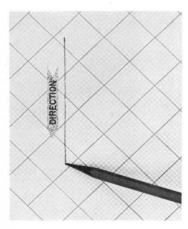

**FIG. 69.** Turning the paper to sketch an inclined line as a vertical line. This is often a great help because of the awkward position of the inclined line. An alternate turn is to the position of Fig. 71.

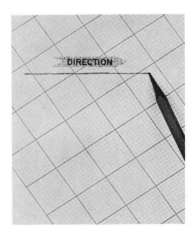

**FIG. 71.** Paper turned to sketch an inclined line as a horizontal line. This should be done especially for the type of line shown in Fig. 70.

perpendicular to the radii, and then complete the circle as shown in Fig. 72. A modification is to use a slip of paper as a trammel. Large circles can be done smoothly, after a little practice, by using the third or fourth finger as a pivot, holding the pencil stationary and rotating the paper under it, or by holding two pencils and using one as a pivot about which to rotate the paper. Another way of drawing a circle is to sketch it in its circumscribing square.

### 37. Projection.

In making an orthographic sketch, remember and apply the principles of projection and applied geometry. Sketches are *not* made to scale but are made to show fair proportions of objects sketched. It is legitimate, however, when coordinate paper is used, to count the spaces or rulings as a means of proportioning the views and as an aid in making projections. Take particular care to have the various details of the views in good projection from view to view. It is an inexcusable mistake to have a detail sketched to a different size on one view from that on another.

When working on plain paper, projections between the top and front views or between the front and side views are easily made by simply "sighting" between the views or using *very* light construction lines, as in Fig. 73. Projections between the top and side views are laid off by judging the distance by

**FIG. 73.** Freehand projections. Judge the projections by eye (aided by coordinate lines if lined paper is used). Judge the measurements (*A* and *B*) by eye; *or* measure the distances with a pencil or by marking on a strip of paper.

eye, by measuring the distance by holding the finger at the correct distance from the end of the pencil and transferring to the view, or by marking the distance on a small piece of paper and transferring to the view. Note in Fig. 73 that distances *A*, *B*, and others could be transferred from the top to the side view by the methods just mentioned.

Even though freehand lines are somewhat "wavy" and not so accurate in position as ruled lines, a good freehand drawing should present the same clean appearance as a good instrument drawing.

### 38. Method.

Practice in orthographic freehand drawing should be started by drawing the three views of a number of simple pieces, developing the technique and the ability to "write" the orthographic language, while exercising the constructive imagination in visualizing the object by looking at the three projections. Observe the following order of working:

1. Study the pictorial sketch and decide what combination of views will best describe the shape of the piece.

2. Block in the views, as in Fig. 74*a*, using a very light stroke of a soft pencil (2B, B, HB, or F) and spacing the views so as to give a well-balanced appearance to the drawing.

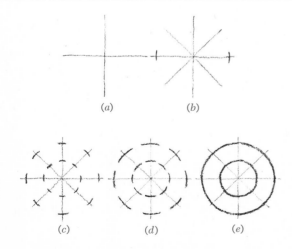

*(a)*      *(b)*

*(c)*      *(d)*      *(e)*

**FIG. 72.** Method of drawing freehand circles. *(a)* draw center lines; *(b)* draw diagonals; *(c)* space points on the circle with *light*, short lines (by eye); *(d)* correct and begin filling in; *(e)* finish.

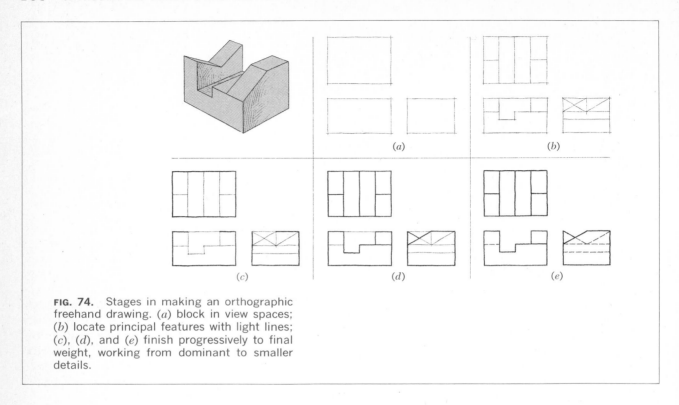

*(a)*

*(b)*

*(c)*

*(d)*

*(e)*

**FIG. 74.** Stages in making an orthographic freehand drawing. (*a*) block in view spaces; (*b*) locate principal features with light lines; (*c*), (*d*), and (*e*) finish progressively to final weight, working from dominant to smaller details.

3. Build up the detail in each view, carrying the three views along together as at (*b*).

4. Brighten the outline of each view with bold strokes as at (*c*).

5. Brighten the detail with bold strokes, thus completing the full lines of the sketch as at (*d*).

6. Sketch in all dashed lines, using a stroke of medium weight and making them lighter than the full lines, as at (*e*), thus completing the shape description of the object.

7. Check the drawing carefully. Then cover the pictorial sketch and visualize the object from the three views.

After drawing a number of simple pieces freehand, try more complicated problems such as Probs. 27 to 55. Faintly ruled coordinate paper (illustrated in Fig. 60) may be used if desired.

### 39. Shop Processes.
Shop processes are properly a part of dimensioning and specification for working drawings and in this text are given in Chap. 9, preceding dimensioning and tolerancing, screw threads and fasteners, and working drawings. However, in order to read the pictorial drawings (problems) to be drawn in orthographic projection, some knowledge of processing fundamentals is necessary. Therefore, Chap. 9, Methods Used in Manufacture, should be studied, especially for a knowledge of hole processing (drilling, reaming, etc.) and for information on fillets, rounds, finished surfaces, and methods of part manufacture. Consult the Glossary for unfamiliar terms.

### 40. Dimensioning Systems.
For the same reasons given in the previous paragraph, some knowledge of dimensioning systems is necessary at this time. Basically, two methods are used: fractional and decimal. These are explained in Sec. 12, Chap. 10. Note especially that a decimally dimensioned part is laid out with a decimal scale.

# READING THE GRAPHIC LANGUAGE

## 41. Orthographic Reading.

The engineer must be able to *read* and *write* the orthographic language. The necessity of learning to read is absolute because everyone connected with technical industry must be able to read a drawing without hesitation or concede technical illiteracy.

Reading the orthographic language is a mental process; a drawing is not read aloud. To describe even a simple object with words is almost impossible. Reading proficiency develops with experience, as similar conditions and shapes occur so often that a person in the field gradually acquires a background of knowledge that enables him to visualize readily the shapes shown. Experienced readers read quickly because they can draw upon their knowledge and recognize familiar shapes and combinations without hesitation. However, reading a drawing should always be done carefully and deliberately, as a whole drawing cannot be read at a glance any more than a whole page of print.

## 42. Prerequisites and Definition.

Before attempting to read a drawing, familiarize yourself with the principles of orthographic projection, as explained in Secs. 1 to 14. Keep constantly in mind the arrangement of views and their projection, the space measurements of height, width, and depth, what each line represents, etc.

*Visualization* is the medium through which the shape information on a drawing is translated to give the reader an understanding of the object represented. The *ability to visualize* is often thought to be a ''gift'' that some people possess and others do not. This, however, is not true. Any person of reasonable intelligence has a visual memory, as can be seen from his ability to recall and describe scenes at home, actions at sporting events, and even details of acting and facial expression in a play or motion picture.

The ability to visualize a shape shown on a drawing is almost completely governed by a person's knowledge of the principles of orthographic projection. The common adage that ''the best way to learn to read a drawing is to learn how to make one'' is quite correct, because in learning to make a drawing you are forced to study and apply the principles of orthographic projection.

*Reading a drawing* can be defined as *the process of recognizing and applying the principles of orthographic projection to interpret the shape of an object from the orthographic views.*

## 43. Method of Reading.

A drawing is read by visualizing units or details one at a time from the orthographic projection and mentally orienting and combining these details to interpret the whole object finally. The form taken in this visualization, however, may not be the same for all readers or for all drawings. Reading is primarily a reversal of the process of making drawings; and inasmuch as drawings are usually first made from a picture of the object, the beginner often attempts to carry the reversal too completely back to the pictorial. The result is that the orthographic views of an object like those shown in Fig. 75 are translated to the accompanying picture, with the thought of the object as positioned in space or placed on a table or similar surface. Another will need only to recognize in the drawing the geometry of the solid, which in the case of Fig. 75 would be a rectangular prism so high, so wide, and so deep with a hole passing vertically through the center of it. This second reader will have read the views just as completely as the first but with much less mental effort.

**FIG. 75.** Orthographic views and picture. Simple objects can be visualized in pictorial form but for complex objects this is difficult.

To most, it is a mental impossibility (and surely unnecessary) to translate more than the simplest set of orthographic views into a complete pictorial form that can be pictured in its entirety. Actually, the reader goes through a routine pattern of procedure (listed in Sec. 44). Much of this is done subconsciously. For example, consider the object in Fig. 76. A visible circle is seen in the top view. Memory of previous projection experience indicates that this must be a hole or the end of a cylinder. The eyes rapidly shift back and forth from the top view to the front view, aligning features of the same size ("in projection"), with the mind assuming the several possibilities and finally accepting the fact that, because of the dashed lines and their extent in the front view, the circle represents a hole that extends through the prism. Following a similar pattern of analysis, the reader will find that Fig. 77 represents a rectangular prism surmounted by a cylinder. This thinking is done so rapidly that the reader is scarcely aware of the steps and processes involved.

The foregoing is the usual method of reading; but how does the beginner develop this ability?

*First,* as stated in Sec. 42, he must have a reasonable knowledge of the principles of orthographic projection.

*Second,* as described in Secs. 45 and 46 he must acquire a complete understanding of the principles behind the meaning of lines, areas, etc., and the mental process involved in interpreting them, as these principles are applied in reading.

There is very little additional learning required. Careful study of all these items plus practice will develop the ability and confidence needed.

## 44. Procedure for Reading.

The actual steps in reading are not always identical because of the wide variety of subject matter (drawings). Nevertheless, the following outline gives the basic procedure and will serve as a guide:

*First,* orient yourself with the views given.

*Second,* obtain a general idea of the over-all shape of the object. Think of each view as the object itself, visualizing yourself in front, above, and at the side, as is done in making the views. Study the dominant features and their relation to one another.

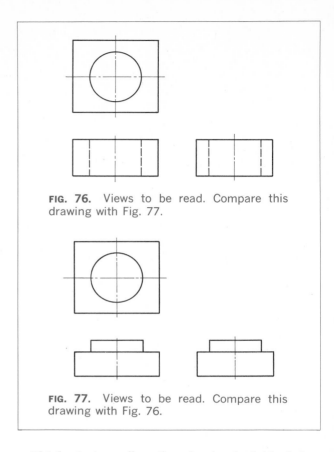

**FIG. 76.** Views to be read. Compare this drawing with Fig. 77.

**FIG. 77.** Views to be read. Compare this drawing with Fig. 76.

*Third,* start reading the simpler individual features, beginning with the most dominant and progressing to the subordinate. Look for familiar shapes or conditions that your memory retains from previous experience. Read all views of these familiar features to note the extent of holes, thickness of ribs and lugs, etc.

*Fourth,* read the unfamiliar or complicated features. Remember that every point, line, surface, and solid appears in every view and that you must find the projection of every detail in the given views to learn the shape.

*Fifth,* as the reading proceeds, note the relationship between the various portions or elements of the object. Such items as the number and spacing of holes, placement of ribs, tangency of surfaces, and the proportions of hubs, etc., should be noted and remembered.

*Sixth,* reread any detail or relationship not clear at the first reading.

## 45. The Meaning of Lines.

As explained in Sec. 11, a line on a drawing indicates (1) the *edge of a surface,* (2) an *intersection of two surfaces,* or (3) a *surface limit.* Because a line on a view may mean any one of these three conditions, the corresponding part of another view must be consulted to determine the meaning. For example, the meaning of line *AB* on the front view of Fig. 78 cannot be determined until the side view is consulted. The line is then found to be the edge view of the horizontal surface of the cutout corner. Similarly, line *CD* on the top view cannot be fully understood without consulting the side view, where it is identified as the edge view of the vertical surface of the cutout corner. Lines *EF* on the top view and *GH* on the front view are identical in appearance. However, the side view shows that line *EF* represents the edge view of the rear surface of the triangular block and that line *GH* is the intersection of the front and rear surfaces of the triangular block.

The top and front views of the objects shown in Figs. 78 and 79 are identical. Nevertheless, lines *AB* and *CD* in Fig. 79 do not represent what they represent in Fig. 78 but are in Fig. 79 the intersection of two surfaces. Also, lines *EF* and *GH* in Fig. 79 are identical in appearance with those in Fig. 78, but in Fig. 79 they represent the surface limits of the circular boss.

From Figs. 78 and 79 it is readily seen that a drawing cannot be read by looking at a single view. Two views are not always enough to describe an object completely, and when three or more views are given, all must be consulted to be sure that the shape has been read correctly. To illustrate with Fig. 80, the front and top views show what appears to be a rectangular projection on the front of the object, but the side view shows this projection to be quarter-round. Similarly, in the front and side views the rear portion of the object appears to be a rectangular prism, but the top view shows that the two vertical rear edges are rounded.

A shape cannot be assumed from one or two views—*all the views must be read carefully.*

The several *lines* representing one feature must be read in all the views. As an exercise in reading the lines on an orthographic drawing, find *all* the lines representing the hole, triangular prism, slot, and cutoff corner in Fig. 81.

## 46. The Meaning of Areas.

The term "area" as used here means the contour

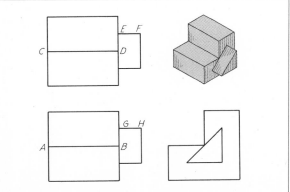

**FIG. 78.** The meaning of lines. *AB, CD,* and *EF* represent the edge views of surfaces. *GH* represents an edge. Study carefully, reading *all* views.

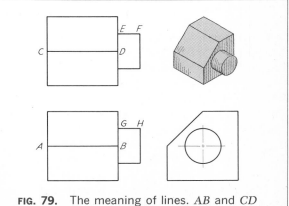

**FIG. 79.** The meaning of lines. *AB* and *CD* represent the intersection of two surfaces. *EF* and *GH* represent a curved-surface limit. Compare this drawing carefully with Fig. 78.

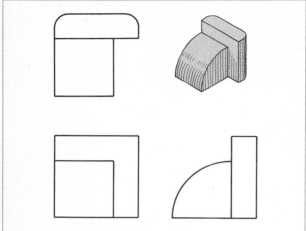

**FIG. 80.** Read all views. Read each feature by looking at all three views; then school yourself to remember all features.

**FIG. 81.** Read all views, features, and lines. This drawing is more complex than Fig. 80. Read all features in all views, by reading all lines, then school yourself to remember all features.

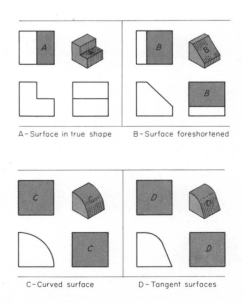

A—Surface in true shape    B—Surface foreshortened

C—Curved surface    D—Tangent surfaces

**FIG. 82.** The meaning of areas. Two views must be read to determine what an area means. Compare *A* with *B* and *C* with *D*.

limits of a surface or combination of tangent surfaces as seen in the different orthographic views. To illustrate, an area of a view as shown in Fig. 82 may represent (1) a surface in true shape as at *A*, (2) a foreshortened surface as at *B*, (3) a curved surface as at *C*, or (4) a combination of tangent surfaces as at *D*.

When a surface is in an oblique position, as surface *E* of Fig. 83, it will appear as an area in all principal views of the surface. A study of the surfaces in Figs. 82, 83, and others will establish with the force of a rule that *a plane surface, whether it is positioned in a horizontal, frontal, profile, or an inclined or skew position, will always appear in a principal orthographic view as a line or an area.* Principal views that show a skew surface as an area in more than one view will always show it in like shape. As an example, surface *A* of Fig. 84 appears as a triangular area in all the principal views; the length of the edges and the angles between the edges may change, but all views show the area with the same number of sides. It should be noted that a plane surface bounded by a certain number of sides can

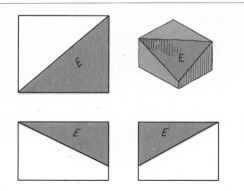

**FIG. 83.** Oblique surface. This will appear as an area in all three principal views.

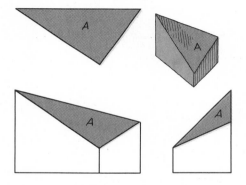

**FIG. 84.** Oblique surface. Note that area *A*, representing this surface, appears in all views as an area of similar shape formed by lines connected in the same order.

**FIG. 85.** Auxiliary surface. Note that the surface appears as an edge in the front view and as an area of similar shape in top and side views.

never appear to have more or fewer sides except when the surface appears as an edge. Moreover, the sides in any view will always connect in the same sequence. For example, in Fig. 85 the front view shows surface *B* as an edge; the top and side views show the surface as an area having a similar shape, the same number of sides, and with the corners in the same sequence.

### 47. Adjacent Areas.

No two adjacent areas can lie in the same plane. It is simple logic that, if two adjacent areas *did* lie in the same plane, there would be no boundary between the areas, and therefore, orthographically, the two adjacent areas would not exist. As an illustration, note that in Fig. 86 areas *A*, *B*, *C*, and *D* are shown in the front and side views to lie in different planes.

Further proof of these principles is given by Fig. 87, in which two top views are shown. By analysis of the projection between top and front views, it is seen that areas *G* and *H* shown in the top views must lie in planes *G* and *H*, respectively, shown in the front view. Also, by projection, it is seen that area *J* of top view *A* must lie in plane *H* and that area *K* must lie in plane *G*. Because areas *H* and *J* lie in plane *H*, and areas *G* and *K* lie in plane *G*, the correct top view, therefore, is top view *B*.

*Hidden areas* may sometimes be confusing to read because the areas may overlap or even coincide with each other. For example, areas *A*, *B*, and *C* in Fig. 88 are not separate areas because they are all formed by the slot on the rear of the object. The apparent separation into separate areas is caused by the dashed lines from the rectangular hole, which is not connected with the slot in any way.

### 48. Reading Lines and Areas.

The foregoing principles regarding the meaning of lines and areas must be used to analyze any given set of views by correlating a surface appearing in one view as a line or an area with its representation in the other views, in which it may appear as a line or an area. Study, for example, Fig. 89, first orienting yourself with the given views. From their arrange-

FIG. 86. Adjacent areas. *A*, *B*, *C*, and *D* all lie in different planes.

FIG. 88. Reading hidden areas. Read all views and note that areas *A*, *B*, and *C* do not represent three different surfaces.

FIG. 87. Adjacent areas. Read all areas and note that *B* is the correct top view.

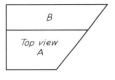

FIG. 89. Reading lines and areas. Read this drawing to determine how areas *A* and *B* of the top view are represented in the front and side views.

ment, the views are evidently top, front, and right-side. An over-all inspection of the views does not reveal a familiar geometric shape, such as a hole or boss, so an analysis of the surfaces is necessary. Beginning with the trapezoidal area *A* in the top view and then moving to the front view, note that a similar-shaped area of the same width is not shown; therefore, the front view of area *A* must appear as an edge, the line *ab*. Next, consider area *B* in the top view. It is shown as a trapezoidal area (four sides) the full width of the view. Again, going to the front view for a mating area or line, the area *abcd* is of similar shape and has the same number of sides with the corners in projection. Area *abcd*, therefore, satisfies the requirements of orthographic

projection and is the front view of area *B*. The side view should be checked along with the other views to see if it agrees. Proceed in this way with additional areas, correlating them one with another and visualizing the shape of the complete object.

Memory and experience aid materially in reading any given drawing. However, every new set of views must be approached with an open mind because sometimes a shape that looks like a previously known condition will crowd the correct interpretation from the mind of the reader. For example, area *E* in Fig. 90 is in a vertical position. The front view in Fig. 91 is identical with the front view in Fig. 90, but in Fig. 91 the surface *F* is inclined to the rear and is not vertical.

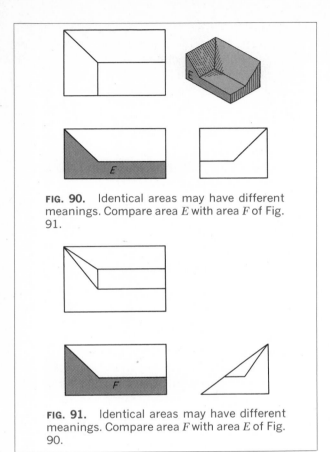

**FIG. 90.** Identical areas may have different meanings. Compare area $E$ with area $F$ of Fig. 91.

**FIG. 91.** Identical areas may have different meanings. Compare area $F$ with area $E$ of Fig. 90.

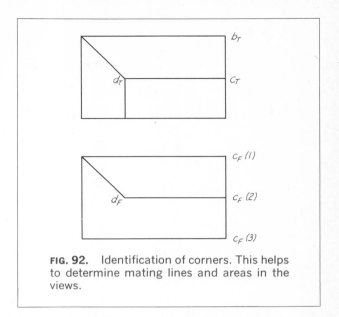

**FIG. 92.** Identification of corners. This helps to determine mating lines and areas in the views.

## 49. Reading Corners and Edges.

The corners and edges of areas may be numbered or lettered to identify them in making additional views or as an aid in reading some complicated shape. If there are no coincident conditions, the corners and edges are easily named by projection; that is, the top view is directly over the front view, and the side view lies on a horizontal projector to the front view. When coincident conditions are present, it may be necessary to coordinate a point with an adjacent point as shown in Fig. 92. Corner $c$ in the front view ($c_F$) may be in projection with $c_T$ at any one of the three positions marked 1, 2, 3. However, the point $c$ is one end of an edge $dc$, and the front view of $c$ must therefore be at position 2. An

experienced reader could probably make the above observations without marking the points, but a beginner can in many cases gain much valuable experience by marking corners and edges, especially if the object he is studying has an unusual combination of surfaces.

## 50. Learning to Read by Sketching.

A drawing is interpreted by mentally understanding the shape of the object represented. You can prove that you have read and understood a drawing by making the object in wood or metal, by modeling it in clay, or by making a pictorial sketch of it. Sketching is the usual method. Before attempting to make a pictorial sketch, make a preliminary study of the method of procedure. Pictorial sketching may be based on a skeleton of three axes, one vertical and two at 30°,[8] representing three mutually perpendicular lines (Fig. 93). On these axes are marked the proportionate width, depth, and height of any rectangular figure. Circles are drawn in their circumscribing squares.

Study the views given in Fig. 94, following the

[8]Isometric position. Oblique or other pictorial methods may also be used. See Chap. 6 for pictorial methods.

**FIG. 93.** Pictorial axes (isometric). This is the "framework" for sketching in isometric. See Fig. 94.

procedure outlined in Sec. 44. Then with a soft pencil (F) and notebook paper make a *very light* pictorial construction sketch of the object, estimating its height, width, and depth and laying the distances off on the axes as at (*a*); sketch the rectangular box that would enclose the piece, or the block from which it could be cut (*b*). On the top face of this box sketch lightly the lines that occur on the top view of the orthographic drawing (*c*). Note that some of the lines in top views may not be in the top plane. Next sketch lightly the lines of the front view on the front face of the box or block, and if a side view is given, outline it similarly (*d*). Now begin

to cut the figure from the block, strengthening the visible edges and adding the lines of intersection where faces of the object meet (*e*). Omit edges that do not appear as visible lines unless they are necessary to describe the piece. Finish the sketch, checking back to the three-view drawing. The construction lines need not be erased unless they confuse the sketch.

### 51. Pictorial Drawing and Sketching.

The foregoing discussion of pictorial sketching should suffice as a guide for making rough sketches as an aid in reading a drawing. However, many experienced teachers like to correlate orthographic drawing, sketching, and reading, with pictorial drawing (which we will take up in Chap. 8). This is valuable training in understanding all methods of graphic expression but also because the making of a pictorial drawing from an orthographic drawing forces the student to read the orthographic drawing.

### 52. Learning to Read by Modeling.

Modeling the object in clay or modeling wax is another interesting, and effective, aid in learning to read a drawing. It is done in much the same way

(*a*)  (*b*)  (*c*)  (*d*)  (*e*)  (*f*)

**FIG. 94.** Stages in making a pictorial sketch. (*a*) draw axes; (*b*) block in the enclosing shape; (*c*) and (*d*) draw outline of detail on top, front, and side; (*e*) and (*f*) finish by completing surfaces represented on the orthographic drawing.

as reading by pictorial sketching. Some shapes are easily modeled by cutting out from the enclosing block; others, by first analyzing and dividing the object into its basic geometric shapes and then combining these shapes.

Starting with a rectangular block of clay, perhaps 1 in. square and 2 in. long, read Fig. 95 by cutting the figure from the solid. With the point of the knife or a scriber, scribe lightly the lines of the three views on the three corresponding faces of the block (Fig. 96$a$). The first cut could be as shown at ($b$) and the second as at ($c$). Successive cuts are indicated at ($d$) and ($e$), and the finished model is shown at ($f$).

Figure 97 illustrates the type of model that can be made by building up the geometric shapes of which the object is composed.

## 53. Calculation of Volume as an Aid in Reading.

To calculate the volume of an object, it must be broken down into its simple geometric elements and the shape of each element must be carefully analyzed before beginning computation. Thus the calculation of volume is primarily an exercise in reading a drawing. Before the computations are completed, the object has usually been visualized, but the mathematical record of the volume of each portion and the correct total volume and weight are proof that the drawing has been read and understood.

The procedure closely follows the usual steps in reading a drawing. Figure 98 illustrates the method.

1. Study the orthographic drawing and pick out the principal masses ($A$, $B$, and $C$ on the breakdown and in the pictorial drawing). Pay no attention in the beginning to holes, rounds, etc., but study each principal over-all shape and its relation to the other masses of the object. Record the dimensions of each of these principal portions and indicate plus volume by placing a check mark in the plus-volume column.

2. Examine each principal mass and find the secondary masses ($D$ and $E$) that must be added to or subtracted from the principal portions. Bosses, lugs, etc., must be added; cutout portions, holes,

**FIG. 95.**    A drawing to be read. This is the object modeled in Fig. 96.

etc., subtracted. Record the dimensions of these secondary masses, being careful to indicate plus or minus volume.

3. Further limit the object to its actual shape by locating smaller details, such as holes, fillets, rounds, etc. (parts $F$, $G$, and $H$). Record the dimensions of these parts.

4. Compute the volume of each portion. This may be done by longhand multiplication or, more conveniently, with a calculator. Record each volume in the proper column, plus or minus. When all unit volumes are completed, find the net volume by subtracting total minus volume from total plus volume.

5. Multiply net volume by the weight per cubic inch of metal to compute the total weight.

The calculations are simplified if all fractional dimensions are converted to decimal form. When a calculator is used, fractions *must* be converted to decimals by calculation or use of conversion tables. A partial conversion table is given in Fig. 98 and a more complete one in the Appendix.

Complete volume and weight calculation not only gives training in recognition of the fundamental geometric portions of an object but also serves to teach neat and concise working methods in the recording of engineering data.

## 54. Exercises in Reading.

Figures 99 and 100 contain a number of three-view drawings of block shapes made for exercises in reading orthographic projection and translating into

**FIG. 96.** Stages in modeling. (*a*) score details on top, front, and side; then make cuts at (*b*), (*c*), (*d*), and (*e*) to finish as at (*f*).

**FIG. 97.** A built-up model. Separate pieces (*a*) are combined to make the finished model (*b*).

**FIG. 98.** Shape breakdown for volume and weight calculations. Each feature of the object is analyzed individually.

WORKING DRAWING

BREAKDOWN

| DECIMAL EQUIVALENTS | | | |
|---|---|---|---|
| 1/2 | 0.5 | 1/16 | 0.0625 |
| 1/4 | 0.25 | 3/16 | 0.1875 |
| 3/4 | 0.75 | 5/16 | 0.3125 |
| | | 7/16 | 0.4375 |
| 1/8 | 0.125 | 9/16 | 0.5625 |
| 3/8 | 0.375 | 11/16 | 0.6875 |
| 5/8 | 0.625 | 13/16 | 0.8125 |
| 7/8 | 0.875 | 15/16 | 0.9375 |

| BEARING REST | | MATERIAL | | WT. — LB. PER CU. IN. | |
|---|---|---|---|---|---|
| PART | DIMENSIONS | | | PLUS VOLUME | MINUS VOLUME |
| A | | | | | |
| B | | | | | |
| C | | | | | |
| D | | | | | |
| E | | | | | |
| F | | | | | |
| G | | | | | |
| H | | | | | |
| | TOTALS | | | | |

| WEIGHT PER CUBIC INCH | | TOTAL NET VOLUME IN CUBIC INCHES | |
|---|---|---|---|
| GRAY CAST IRON | 0.260 LB. | | |
| TOBIN BRONZE | 0.304 LB. | WEIGHT IN POUNDS | |
| CAST STEEL | 0.282 LB. | | |

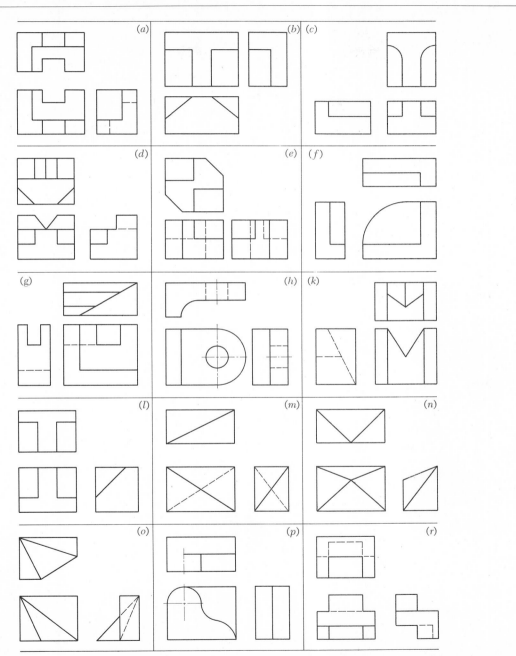

**FIG. 99.** Reading exercises. Read each drawing, (a) to (r). Make pictorial sketches or models if necessary for understanding.

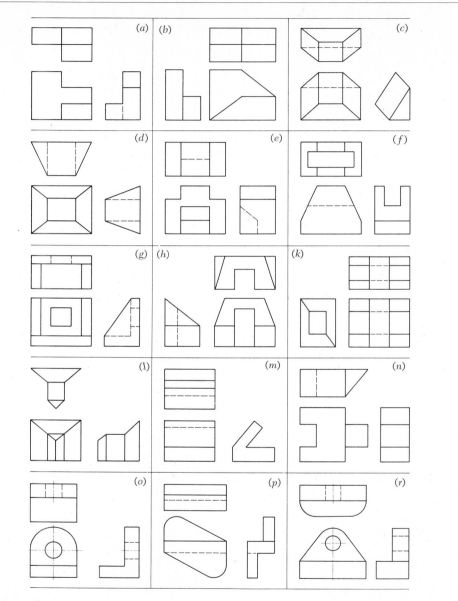

**FIG. 100.** Reading exercises. Read each drawing, (a) to (r). Make pictorial sketches or models if necessary for understanding.

pictorial sketches or models. Proceed as described in the previous paragraphs, making sketches not less than 4 in. over-all. Check each sketch to be sure that all intersections are shown and that the original

**FIG. 101.** Missing-line exercises. Read each drawing, (a) to (r), and sketch the lines missing on the views. Check carefully. Use a model or sketch, if necessary, as an aid in locating all lines.

three-view drawing could be made from the sketch. In each drawing in Fig. 101 some lines have been intentionally omitted. Read the drawings and supply the missing lines.

# PROBLEMS

For practice in orthographic freehand drawing, select problems from the following group.

## Group 1. Freehand Projections from Pictorial Views.

The figures for Probs. 1 to 16 contain a number of pictorial sketches of pieces of various shapes which are to be translated into three-view orthographic freehand drawings. Make the drawings of fairly large size, the front view, say, 2 to 2½ in. in length, and estimate the proportions of the different parts by eye or from the proportionate marks shown but without measuring. The problems are graduated in difficulty for selection depending on ability and experience.

Problem 16 gives a series that can be used for advanced work in freehand drawing or that can be used later on, by adding dimensions, as dimensioning studies or freehand working-drawing problems.

(a)

(b)

(c)

(d)

PROB. 7.

(a)

(b)

(c)

(d)

PROB. 8.

(a)

(b)

(c)

(d)

PROB. 9.

(a)

(b)

(c)

(d)

PROB. 10.

(a)

(b)

(c)

(d)

PROB. 11.

(a)

(b)

(c)

(d)

PROB. 12.

(a)

(b)

(c)

(d)

PROB. 13.

(a)

(b)

(c)

(d)

PROB. 14.

(a)

(b)

(c)

(d)

PROB. 15.

**PROB. 16.** Pieces to be drawn freehand in orthographic projection.

**17.** Make a freehand drawing of the end bracket.
**18.** Make a freehand drawing of the radar wave guide.

Select problems from the following groups for practice in projection drawing. Most of the problems are intended to be drawn with instruments but will give valuable training done freehand, on plain or coordinate paper.

The groups are as follows:

2. Projections from pictorial views

**PROB. 17.** End bracket.

**PROB. 18.** Radar wave guide.

3. Special scales, decimal sizes, projections from photo drawings, references to Appendix material
4. Views to be supplied, freehand

5. Views to be supplied
6. Views to be changed
7. Drawing from memory
8. Volume and weight calculations with slide rule

## Group 2. Projections from Pictorial Views.

**19.** Draw the top, front, and right-side views of the beam support.
**20.** Draw the top, front, and right-side views of the vee rest.
**21.** Draw three views of the saddle bracket.

**22.** Draw three views of the wedge block.
**23.** Draw three views of the slotted wedge.
**24.** Draw three views of the pivot block.
**25.** Draw three views of the inclined support.
**26.** Draw three views of the corner stop.
**27.** Draw three views of the switch base.

**PROB. 19.** Beam support.

**PROB. 20.** Vee rest.

**PROB. 21.** Saddle bracket.

**PROB. 22.** Wedge block.

**PROB. 23.** Slotted wedge.

**PROB. 24.** Pivot block.

**PROB. 25.** Inclined support.

**PROB. 26.** Corner stop.

**PROB. 27.** Switch base.

28.  Draw three views of the adjusting bracket.
29.  Draw three views of the guide base.
30.  Draw three views of the bearing rest.
31.  Draw three views of the swivel yoke.
32.  Draw three views of the truss bearing.
33.  Draw three views of the sliding-pin hanger.
34.  Draw two views of the wire thimble.
35.  Draw three views of the hanger jaw.
36.  Draw three views of the adjustable jaw.
37.  Draw two views of the shifter fork.

PROB. 28.  Adjusting bracket.

PROB. 29.    Guide base.

PROB. 30.    Bearing rest.

PROB. 31.    Swivel yoke.

PROB. 32.    Truss bearing.

PROB. 33.    Sliding-pin hanger.

PROB. 34.    Wire thimble.

PROB. 35.
Hanger jaw.

PROB. 36.    Adjustable jaw.

PROB. 37.
Shifter fork.

PROB. 38.
Mounting bracket.

38. Draw three views of the mounting bracket.
39. Draw three views of the hinged bearing.
40. Draw two views of the clamp lever.
41. Draw three views of the bedplate stop.
42. Draw top, front, and partial side views of the spanner bracket.
43. Draw two views of the sliding stop.
44. Draw three views of the clamp bracket.
45. Draw three views of the tube hanger.
46. Draw three views of the gage holder.
47. Draw three views of the shaft guide.

PROB. 39.  Hinged bearing.

PROB. 40.  Clamp lever.

PROB. 41.  Bedplate stop.

PROB. 42.  Spanner bracket.

PROB. 43.  Sliding stop.

PROB. 44.  Clamp bracket.

PROB. 45.  Tube hanger.

PROB. 46.  Gage holder.

PROB. 47.  Shaft guide.

**48.** Draw three views of the clamp block.
**49.** Draw three views of the offset yoke.
**50.** Draw three views of the angle connector.
**51.** Draw three views of the buckstay clamp.
**52.** Draw three views of the stop base.

**53.** Draw three views of the sliding buttress.
**54.** Draw two views of the end plate.
**55.** Draw three views of the plastic switch base.
**56.** Draw two views of the pawl hook.
**57.** Draw three views of the step-pulley frame.

**PROB. 48.** Clamp block.

**PROB. 49.** Offset yoke.

**PROB. 50.** Angle connector.

**PROB. 51.** Buckstay clamp.

**PROB. 52.** Stop base.

**PROB. 53.** Sliding buttress.

**PROB. 54.** End plate.

**PROB. 55.** Switch base.

**PROB. 56.** Pawl hook.

**PROB. 57.** Step-pulley frame.

## Group 3. Special Scales, Decimal Sizes, Projections from Photo Drawings, References to Appendix Material.

The problems in this group (58 to 73) will give practice in the use of a decimal scale for layout. The projections from the photo drawings of the final chapter are given here not only to serve as problem material, but to acquaint the student with this form of pictorial representation. Details of photo-drawing methods are also given in the final chapter. Wood-screw and bolt sizes are given in the Appendix.

**58.** Draw top, front, and right-side views.
**59.** Draw top, front, and right-side views.
**60.** Draw top, front, and right-side views.
**61.** Draw top, front, and left-side views.
**62.** Draw top, front, and right-side views. Bend radii and setbacks are 0.10 in.
**63.** Draw top, front, and right-side views. Screw and bolt sizes are given in the Appendix. Clearance over bolt and screw diameters are necessary in dimensioning but not in drawing the views.

**PROB. 58.** Motor mount.

**PROB. 59.** Assembly-jig base.

**PROB. 60.** Cargo-hoist tie-down.

**PROB. 61.** Aileron tab-rod servo fitting.

**PROB. 62.** Cover bracket.

**PROB. 63.** Seat-latch support.

**64.** Draw top, front, left-side, and right-side views. Show only *necessary* hidden detail. The 0.44D hole extends through the piece. The limit-dimensioned holes

$$\left(.75 \begin{array}{c} +0.005 \\ -0.002 \end{array} \text{ and } \frac{.86}{.89}\right)$$

are given here as an introduction to precise methods, to be given in detail later. Regardless of dimensional accuracy, features are drawn to basic size with no more scaled accuracy than for other less accurate features.
**65.** Draw top and front views. See Prob. 64 for comment on limit-dimensioned holes.
**66.** Draw top, front, and left-side views. Bolts should have clearance. Undimensioned radii are $\frac{1}{4}R$.
**67.** Draw top and front views. See Prob. 64 for comment on limit-dimensioned holes.
**68.** Draw two views of the conveyor link, 71, final chapter.

**69.** Draw necessary views of lift-strut pivot, 73, final chapter.
**70.** Draw necessary views of hydro-cylinder support (right- and left-hand), Prob. 72, final chapter.
**71.** Draw necessary views of length-adjuster tube, Prob 74, Chap. 23.
**72.** Draw necessary views of third terminal, Prob. 76, final chapter.
**73.** Select a part from the conveyor link assembly, Prob. 77, final chapter, and make an orthographic drawing.
**74.** Make an orthographic drawing of the jet-engine bracket.
**75.** Make an orthographic drawing of the missile gyro support.
**76.** Make an orthographic drawing of the reclining-seat ratchet plate.
**77.** Make an orthographic drawing of the rigging yoke.
**78.** Make an orthographic drawing of the door bracket.

**PROB. 64.** Latch bracket.

**PROB. 65.** Control crank.

**PROB. 66.** Transformer mounting.

**PROB. 67.** Stabilizer link.

**PROB. 74.** Jet-engine bracket.

**PROB. 75.** Missile gyro support.

**79.** Make an orthographic drawing of the missile release pawl.

**80.** Make an orthographic drawing of the jet-engine inner-strut bracket.

**81.** Make an orthographic drawing of the transmission transfer fork.

**82.** Make an orthographic drawing of the reversing fork.

**PROB. 76.** Reclining-seat ratchet plate.

**PROB. 77.** Rigging yoke.

**PROB. 78.** Door bracket.

**PROB. 79.** Missile release pawl.

**PROB. 80.** Jet-engine inner-strut bracket.

**PROB. 81.** Transmission transfer fork.

**PROB. 82.** Reversing fork.

## Group 4. Views to be Supplied Freehand.

These problems (83 to 85) will give valuable training in reading orthographic views, as well as further practice in applying the principles of orthographic projection.

Study the meaning of lines, areas, and adjacent areas. Corners or edges of the object may be numbered or lettered to aid in the reading or to aid later in the projection.

A pictorial sketch may be used, if desired, as an aid in reading the views. This sketch may be made before the views are drawn and completed or at any time during the making of the drawing. For some of the simpler objects, a clay model may be of assistance.

After the views given have been read and drawn, project the third view or complete the views as specified in each case.

Remember that every line representing an edge view of a surface, an intersection of two surfaces, or a surface limit will have a mating projection in the other views. Be careful to represent all hidden features and pay attention to the precedence of lines.

The figures in this group contain a number of objects with two views drawn and the third to be supplied. In addition to helping to develop the ability to draw freehand, this exercise will give valuable practice in reading. These problems may be worked directly in the book or on plain or coordinate paper.

**PROB. 83.** Views to be supplied freehand.

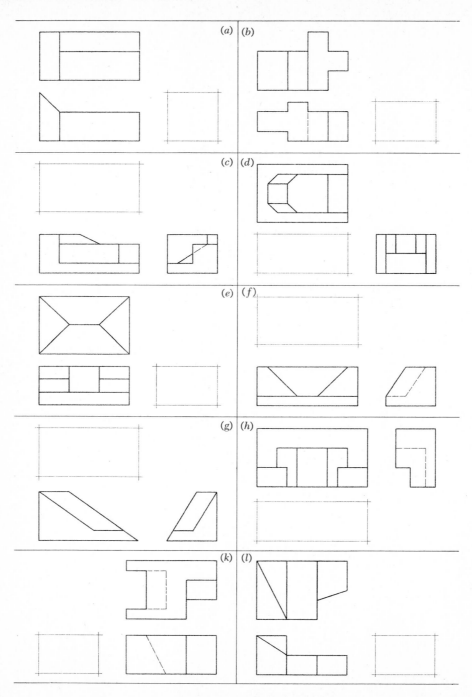

**PROB. 84.** Views to be supplied freehand.

**PROB. 85.** Views to be supplied freehand.

**86.** Given top and front views, add side view. Find at least three solutions. Use tracing paper for the second and third solutions.
**87.** Given front and side views, add top view. Find at least three solutions. Use tracing paper for the second and third solutions.
**88.** Given top and front views, add side view. Find two solutions. Use tracing paper for the second solution.
**89.** Given top and front views, add side view. Find at least three solutions. Use tracing paper for the second and third solutions.

## Group 5. Views to be Supplied.
**90.** Draw the views given, completing the top view from information given on the front and side views. Carry the views along together.
**91.** Given top and front views of the block, add side view. See that dashed lines start and stop correctly.
**92.** Given front and right-side views, add top view.
**93.** Given front and right-side views, add top view.
**94.** Given front and top views, add right-side view.
**95.** Given top and front views, add right-side view.

**96.** Complete the three views given.
**97.** Given front and left-side views, add top view.
**98.** Given front and right-side views, add top view.
**99.** Given front and top views, add right-side view.
**100.** Assume this to be the right-hand part. Draw three views of the left-hand part.
**101.** Given front and top views, add side view.
**102.** Given front and top views, add side view.
**103.** Given top and front views, add left-side view.
**104.** Given front and top views, add side view.
**105.** Given front and top views, add side view.
**106.** Given front and top views, add side view.
**107.** Given top and front views, add side view.
**108.** Given top and front views, add left-side view.
**109.** Given top and front views, add side view.
**110.** Given front and top views, add side view.
**111.** Given front and right-side views of electric-motor support. Add top view.
**112.** Given front and left-side views of master brake cylinder. Add top view.
**113.** Given front and right-side views of end frame for engine starter. Add top view.

PROB. 86.  PROB. 87.  PROB. 88.

PROB. 89.  PROB. 90.  Projection study.  PROB. 91.  Projection study.

**PROB. 92.**   Projection study.

**PROB. 93.**   Bit-point forming die.

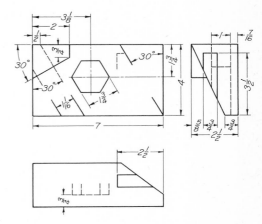

**PROB. 94.**   Rabbeting-plane guide.

**PROB. 95.**   Wedge block.

**PROB. 96.**   Projection study.

**PROB. 97.**   Burner-support key.

**PROB. 98.**   Abutment block.

**PROB. 99.**   Sliding port.

**PROB. 100**
Bumper support and post cap.

**PROB. 101.** Anchor bracket.

**PROB. 102.** Entrance head.

**PROB. 103.** Yoked link.

**PROB. 106.** Tool holder.

**PROB. 104.** Rubber-mounting bracket.

**PROB. 105.** Crosshead.

**PROB. 107.** Lock plate.

**PROB. 108.** Bevel-gear mounting.

**PROB. 109.** Cylinder support.

**PROB. 110.** Rod yoke.

**PROB. 111.** Electric-motor support.

**PROB. 112.** Master brake cylinder.

**PROB. 113.** End frame for engine starter.

## Group 6. Views to be Changed.

These problems (114 to 120) are given to develop the ability to visualize the actual piece in space and from this mental picture to draw the required views as they would appear if the object were looked at in the directions specified.

In addition to providing training in reading orthographic views and in orthographic projection, these problems are valuable exercises in developing graphic technique. Note that all the problems given are castings containing the usual features found on such parts, that is, fillets, rounds, runouts, etc., on unfinished surfaces. Note also that sharp corners are formed by the inter-

section of an unfinished and finished surface or by two finished surfaces. After finishing one of these problems, check the drawing carefully to make sure that all details of construction have been represented correctly.

**114.** Given front and top views, new front, top, and side views are required, turning the block so that the back becomes the front and the top the bottom. The rib contour is straight.

**115.** Given front, left-side, and bottom views, draw front, top, and right-side views.

**116.** Given front, right-side, and bottom views, draw front, top, and left-side views.

**PROB. 114.** Sliding block.

**PROB. 115.** Plunger bracket.

**PROB. 116.** Offset bracket.

**117.** Given front, right-side, and bottom views, draw new front, top, and right-side views, turning the support so that the back becomes the front.

**118.** Given front and left-side views of the left-hand part, draw the right-hand part.

**119.** Given front, left-side, and bottom views, draw front, top, and right-side views.

**120.** Given top and front views of jet-engine hinge plate. Add right- and left-side views.

**PROB. 117.**   Toggle-shaft support.

**PROB. 118.**   Compound link.

**PROB. 119.** Sector bracket.

**PROB. 120.** Jet-engine hinge plate.

## Group 7. Drawing from Memory.

One of the valuable assets of an engineer is a trained memory for form and proportion. A graphic memory can be developed to a surprising degree in accuracy and power by systematic exercises in drawing from memory. It is well to begin this training as soon as you have a knowledge of orthographic projection.

Select an object not previously used; look at it with concentration for a certain time (from 5 sec to $\frac{1}{2}$ min or more), close the book, and make an accurate orthographic sketch. Check with the original and correct any mistakes or omissions. Follow with several different figures. The next day, allow a 2-sec view of one of the objects, and repeat the orthographic views of the previous day.

## Group 8. Volume and Weight Calculations with Hand Calculator

In calculating the weight of a piece from a drawing, the object is broken up into the geometric solids (prisms, cylinders, pyramids, or cones) of which it is composed. The volume of each of these shapes is calculated and the individual volumes are added, or subtracted, to find the total volume. The total volume multiplied by the weight of the material per unit of volume gives the weight of the object.

A table of weights of materials is given in the Appendix.

**121.**  Find the weight of the cast-iron anchor bracket, Prob. 101.

**122.**  Find the weight of the cast-iron bracket, Prob. 104.

**123.**  Find the weight of the wrought-iron tool holder, Prob. 106.

**124.**  Find the weight of the cast-steel cylinder support, Prob. 109.

**125.**  Find the weight of the malleable-iron sliding block, Prob. 114.

# AUXILIARY VIEWS

**6**

Auxiliary Views Are Special Projections Used to Clarify and Complete Orthographic Descriptions of Shape

Items Given in this Chapter:
Basic Concepts • Classification of Surfaces • Directions of Inclined Surfaces • Normal Views of Inclined Surfaces • Purposes of Normal Views • Normal View of Lines • Skew Surfaces • Edge Views of Skew Surfaces • Normal Views of Skew Surfaces • Terms of Reference • Problems

## 1. Auxiliary Views—Basic Concepts.

A plane surface is shown in true shape when the direction of view is perpendicular to the surface; for example, rectangular objects can be placed with their faces parallel to the principal planes of projection and be fully described by the principal views. In Fig. 1 the top, front, and right-side views show the true shape, respectively, of the top, front, and right side of the object. Note especially that the planes of projection are *parallel* to the top, front, and right side of the object and that the directions of observation are *perpendicular* to the object faces and to the planes of projection. Figure 2 is a pictorial of Fig. 1, given here to aid in visualizing the relationship of object faces, planes of projection, and view directions.

Note in Figs. 1 and 2 that each view also shows the *edge* of certain surfaces of the object. For example, the front view shows the edge of the top, bottom, and both sides of the object. For a surface to appear as an edge it must be perpendicular to the plane of projection for the view.

Sometimes an object will have one or more *inclined* surfaces whose true shape it is necessary to show, especially if they are irregular in outline. Figure 3 shows an object with one inclined face, *ABDC*. This face is inclined to the horizontal and profile planes and perpendicular to the frontal plane. The face *ABDC* therefore appears as an edge in the front view, but none of the principal views shows the true size and shape of the surface. To show the true size and shape of *ABDC*, a view is needed that has a direction of observation perpendicular to *ABDC* and is projected on a plane parallel to *ABDC*, as shown in Fig. 4. This view is known as a *normal* view. The top, front, and side views of Fig. 1 are normal views of the top, front, and side surfaces of the object, because they *all* show the true size and shape of a surface by having the direction of observation at right angles to the surface. The dictionary defines a *normal*, in geometry, as "any perpendicular." In graphics, a plane of projection is involved as well as a direction of observation; hence, *a normal view is a projection that has the viewing direction perpendicular to, and made on a plane parallel to, the object face.*

**FIG. 1.** Object faces parallel to the principal projection planes. Top, front, and side views are normal views, respectively, of top, front, and side of the object.

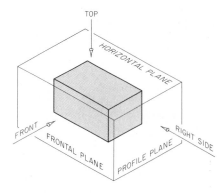

**FIG. 2.** Pictorial of Fig. 1. This is given to aid in visualizing the relationship of principal planes, view directions, and object position.

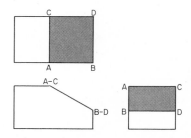

**FIG. 3.** One object face inclined to two principal planes. The inclined face does not appear in its true size and proportions in any view.

Figure 5 is the orthographic counterpart of Fig. 4. To get the normal view of surface *ABDC*, a projec-

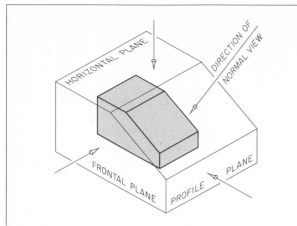

**FIG. 4.** Pictorial of Fig. 3. A view projected perpendicular to the inclined face will show it in true size and proportions.

tion is made perpendicular to *ABDC*. This projection is made *from* the view where the surface shows as an edge, in this case, the front view, and thus perpendicularity from the surface is seen in true relationship. Extra views such as the normal view of Fig. 5 are known as *auxiliary views* to distinguish them from the principal (top, front, side, etc.) views. However, since an auxiliary is made for the purpose of showing the true configuration of a surface, the terms *normal view* or *edge view* state positively what the view is and what it shows.

Detailed instructions for making a normal view of any surface in *any possible position* are given in the sections that follow. Edge views are discussed in Sec. 9.

## 2. Classification of Surfaces.
Surfaces may occur in any of the positions shown in Fig. 6. At (*a*) all the surfaces are aligned with the principal planes of projection, and thus each of the principal views is a normal view. At (*b*) the shaded surface is at an angle to *two* of the principal planes but perpendicular to one plane and is called an *inclined* surface. At (*c*) the shaded surface is at an angle to *all three* principal planes of projection and is known as a *skew*, or *oblique*, surface (that is, one that takes a slanting or oblique course or direction).

**FIG. 5.** Orthographic drawing of the object in Figs. 3 and 4. The normal view gives the true size and proportions of the inclined face.

NORMAL SURFACES
(*a*)

AN INCLINED SURFACE
(*b*)

A SKEW SURFACE
(*c*)

**FIG. 6.** Classification of surfaces.

## 3. Directions of Inclined Surfaces.

Inclined surfaces may occur anywhere on an object, and because other features of the object must also be represented, the inclined surface may be in any one of the twelve positions shown in Fig. 7. The first column, (b) to (e), shows surfaces inclined to the front and side so that the surface may be on the (b) right front, (c) left front, (d) left rear, or (e) right rear. The second column, (f) to (j), shows surfaces inclined to the top and side, so that the surface may be on the (f) upper right, (g) lower right, (h) lower left, or (j) upper left. The third column, (k) to (n), shows surfaces inclined to the top and front, so that the surface may be on the (k) upper front, (l) upper rear, (m) lower rear, or (n) lower front.

Inclined surfaces occur at *angles* of inclination differing from those shown in Fig. 7 but, for general position, no other locations are possible.

## 4. The Normal View of an Inclined Surface.

No matter what the position of an inclined surface may be, the fundamentals of projecting a normal view of the surface are the same, as will be seen from Figs. 8 to 21.

Figure 8 is the orthographic drawing of the object in position (b) of Fig. 7. The inclined surface, identified by the letter A, is on the right front of the object and shows, in Fig. 8, as an *edge* in the top view. Because surface A appears as an edge in the top view, the direction of observation for the normal view is established perpendicular to the edge view, as shown. We are looking, in this case, in a horizontal direction, *directly at* surface A. Projectors parallel to the viewing direction (also, perpendicular to surface A) establish one dimension of the surface needed for the normal view. This is the *horizontal*

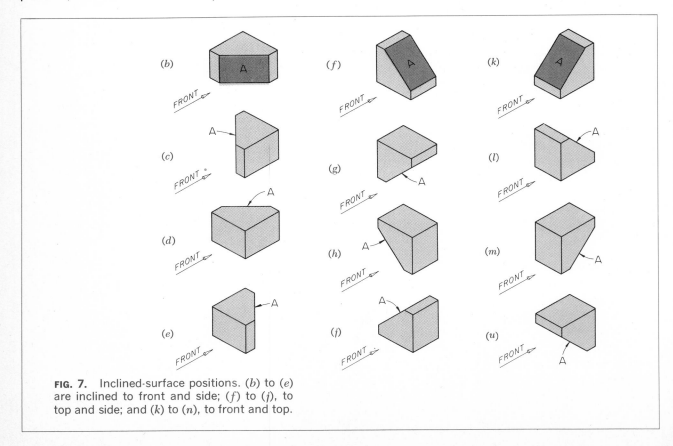

**FIG. 7.** Inclined-surface positions. (b) to (e) are inclined to front and side; (f) to (j), to top and side; and (k) to (n), to front and top.

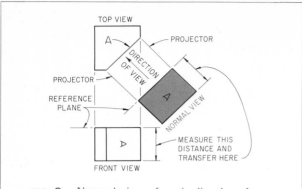

**FIG. 8.** Normal view of an inclined surface. The object is in position (*b*) of Fig. 7.

view, projected directly out from its original coincidence on the object, occurs in the position first shown in Fig. 8. To get the view into coincidence with the plane of the drawing paper, the view is rotated to the position shown in Fig. 10. Carefully study Figs. 8 to 10 and reread the text to fix all relationships.

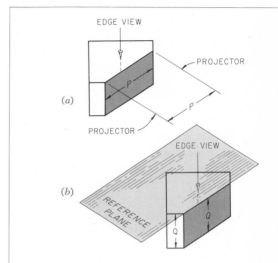

**FIG. 9.** Relationship of projectors, reference plane, and surface dimensions.

distance from the left-front vertical edge to the right-rear vertical edge of surface *A*. Figure 9*a* shows that the distance *between* the projectors (dimension *P*) is equal to dimension *P* on the object.

All orthographic views have two dimensions, and to complete the normal view of surface *A*, we now need its second dimension. That dimension is, in this case, the vertical distance *Q*, shown at (*b*), which is the vertical height of surface *A* and appears in true length *on the front view* (Fig. 8). This distance cannot be projected but must be transferred from the front view to the normal view as shown in Fig. 8. In order to facilitate transferring the distance (this will apply particularly in later problems where the surface may not be rectangular), a *reference plane* (Figs. 9*b* and 10) is established on both the normal view and front view. This reference plane *must* be perpendicular to the distance to be measured and transferred; it is therefore perpendicular to the projectors between the top and front views *and in the front view*, and also perpendicular to the projectors from the top view to the normal view *and in the* normal view, as shown in Fig. 8. The reference plane, here, is at the *top edge* of surface *A* in both views. This can be visualized in the normal view by considering the reference plane as a hinge that will rotate the normal view 90° downward to its original position in coincidence with the top view. This is shown pictorially in Fig. 10 to aid in visualizing the relationship. It must be understood that the normal

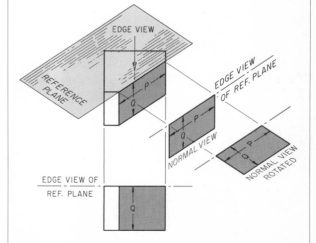

**FIG. 10.** Relationship of projectors, reference plane, surface dimensions, normal view, and rotated position of normal view.

The foregoing constitutes all the geometry and projection needed for drawing *any* normal view of an inclined surface. The projection will in some cases be made from a different view and the measurement transferred will be width or depth instead of height, but the fundamental relationships of the views will *be the same in every case.*

Consider the other positions of the object, (*c*) to (*e*), of Fig. 7. In every case the inclined surface is similarly oriented in space to that of Fig. 8, but turned to a different position. Figure 11 [position (*c*) of Fig. 7] has the inclined surface on the left front of the object. As in the first case, the direction of view must be perpendicular to the inclined face, and is seen in true relationship in the top view, where surface *A* appears as an edge. The projectors for the normal view are parallel to the viewing direction and of course perpendicular to surface *A*. The reference plane is established in the normal view perpendicular to the projectors for the normal view and in the front view perpendicular to the projectors between top and front views. The measurement of height, transferred from the front view to the normal view, completes the normal view.

Note that the reference plane will *always* appear in the two views that are in projection *with* the view that shows the inclined surface as an edge, and that the reference plane is always perpendicular to the projectors between these pairs of views.

Figure 12 shows position (*d*) of Fig. 7. This time the inclined surface is at the left rear of the object. Figure 13 shows position (*e*) of Fig. 7. Read the explanation given for Figs. 8, 9, and 10 while studying these two figures (12 and 13) and note that the directions given for Figs. 8 to 10 apply equally well for Figs. 12 and 13. This shows that the actual position of the inclined surface is simply a variation.

Turning now to the second column of positions (*f*) to (*j*) of Fig. 7, note again that all these positions have the inclined surface at an angle to top and side. Then, looking at Figs. 14 to 17 inclusive (these are all similar because the inclined surface appears in every case as an edge on the front view), observe that Fig. 14 is position (*f*), Fig. 15 position (*g*), Fig. 16 position (*h*), and Fig. 17 position (*j*).

**FIG. 11.**   Normal view of an inclined surface. The object is in position (*c*) of Fig. 7.

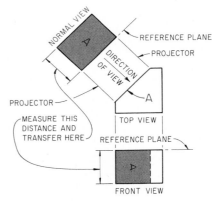

**FIG. 12.**   Normal view of an inclined surface. The object is in position (*d*) of Fig. 7.

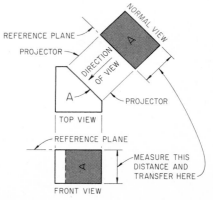

**FIG. 13.**   Normal view of an inclined surface. The object is in position (*e*) of Fig. 7.

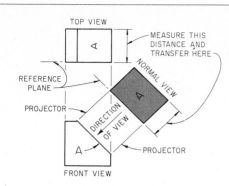

**FIG. 14.** Normal view of an inclined surface. The object is in position (*f*) of Fig. 7.

**FIG. 16.** Normal view of an inclined surface. The object is in position (*h*) of Fig. 7.

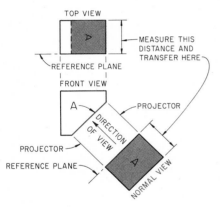

**FIG. 15.** Normal view of an inclined surface. The object is in position (*g*) of Fig. 7.

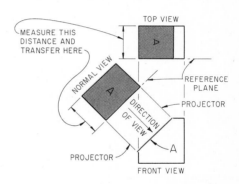

**FIG. 17.** Normal view of an inclined surface. The object is in position (*j*) of Fig. 7.

The projection of the normal view in Figs. 14 to 17 is unchanged geometrically and projectively as compared with Figs. 8 to 13. The *only* difference is that the projectors for the normal view this time emanate from the front view, where the inclined surface appears as an edge. While studying Figs. 14 to 17, consider again the principles of projection involved and the procedure followed: (1) The direction of observation for the normal view is perpendicular to the inclined surface and, on the drawing, perpendicular to the edge view of the surface. (2) The projectors for the normal view are parallel to the viewing direction, perpendicular to the inclined surface and, on the drawing, perpendicular to the edge view of the surface. (3) The projectors (or the spacing between them) determine one dimension of the normal view. (4) The second dimension for the normal view is now needed. (5) The reference plane for making the needed measurement will appear in the normal perpendicular to the projectors between edge view and normal view, and in the principal view (in projection with the edge view) perpendicular to the projectors between edge view and principal view. (6) The measurement perpendicular to the reference plane in the principal view is then transferred to the normal view, perpendicular to the reference plane, thus completing the layout of the normal view.

Figures 18 to 21 show the inclined surface positions of the third column of Fig. 7 [(k) to (n)], in the same order. Note that the projective system is the same as before. The only difference now is that the edge view of the inclined surface appears in the *side* view, and because *either* side view *might* be drawn, the projectors from the edge view to the normal view will have to be drawn from whichever side view, right or left, is used. To illustrate this variation, Figs. 18 and 19 have been drawn with the right-side view and Figs. 20 and 21 with the left-side view. Note also that the reference plane, with measurements taken from it, will in these cases be in the normal view and the front view, since the front view is the view in *direct* projection with the edge view of the inclined surface.

Study Figs 7 to 21 carefully, note the recurrence of certain principles of projection, and study the relationships of views, projectors, directions, and measurements.

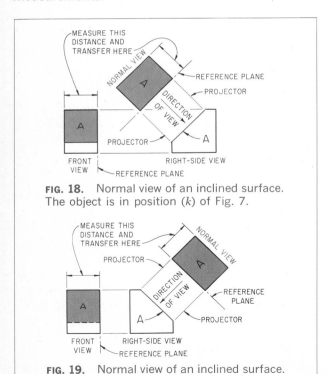

**FIG. 18.** Normal view of an inclined surface. The object is in position (k) of Fig. 7.

**FIG. 19.** Normal view of an inclined surface. The object is in position (l) of Fig. 7.

## 5. Normal Views of Inclined Surfaces on Practical Objects.

So far we have presented only the pertinent principles of projection and the procedure for drawing a normal view, using a rectangular object for illustration. Practical objects are, of course, made up of rectangular, conic, cylindrical, and other shapes, and this is sometimes considered by the uninitiated to be a complication. However, this is not the case; a machine part is often easier to draw than a purely geometric shape because it is more readily visualized.

Figure 22 shows a part with a surface inclined to front and side, about in the same position as the inclined surface in Fig. 8. At (a) the object is shown pictorially, surrounded by the planes of projection, and at (b) the projection planes are opened up into

**FIG. 20.** Normal view of an inclined surface. The object is in position (m) of Fig. 7.

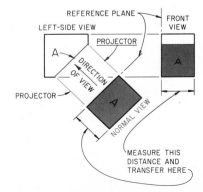

**FIG. 21.** Normal view of an inclined surface. The object is in position (n) of Fig. 7.

**FIG. 22.** Machine part with surface inclined to front and side. (*a*) Planes of projection; (*b*) the planes opened up; (*c*) front, top, and normal views of the inclined surface.

the plane of the paper. Note that the reference plane is placed, in this case, at the *base* of the object because this is a natural reference surface. Thus at (*c*) the reference plane is drawn at the base in the front and normal views, and measurements are made *upward* (dimensions $H$ and $H'$) to needed points. Note that the top view is the normal view

of the reference plane, which appears as an edge in the front and normal views.

Figure 23 illustrates the procedure for drawing the normal view of an inclined surface on a machine part. Steps are as follows:

1. Draw the partial top and front views, as at (*b*), and locate the view direction by drawing projectors

**FIG. 23.** Stages in drawing a machine part with a surface inclined to front and side.

perpendicular to the edge view of the inclined surface, as shown.

2. Locate the reference plane (*RP*) in the front view. The reference plane may be taken above, below, or through the view, and is chosen for convenience in measuring. In this case it is taken through the natural center line of the front view. The reference plane in the normal view will be perpendicular to the projectors already drawn, and is located at a convenient distance from the top view, as shown at (*c*).

3. As shown at (*d*), measure the distance (height) from the reference plane of various points needed, as, for example, $H$ and $H_1$, and transfer these measurements with dividers or scale to the normal view, measuring from the normal-view reference plane.

4. Complete the normal view from specifications of the rounds, etc., as shown at (*e*). Note that any measurement in the front view, made *toward* the top view, is transferred to the *normal view, toward* the top view. Note also that the front view could not be completed without using the normal view.

5. To get the front view of the circular portions, the true shape of which shows only in the normal view as circle arcs, select points in the normal view, project them back to the top view, and then to the front view. On these projectors transfer the heights $H_2$ and $H_3$ from the normal view to find the corresponding points in the front view. $H_4$, $H_5$, and others will complete the curve in the front view. Note that this procedure is exactly reversed from the operation of measuring from the front view to locate a distance in the normal view.

Figure 24 shows an object with *two* inclined surfaces. Note that the normal views are only partial views because the base of the object is fully described in top and front views. The reference plane is taken through the center because the object is symmetrical.

Figure 25 illustrates the steps in drawing the normal view of a face inclined to top and side:

1. Draw the partial top and front views, as at (*b*), and locate the view direction by drawing projectors perpendicular to the inclined surface, as shown.

2. Locate the reference plane in the top view. The reference plane may be taken in front of, through, or to the rear of the view but is here located at the rear flat surface of the object because of convenience in measuring. The reference plane in the normal view will be perpendicular to the projectors already drawn, and is located at a convenient distance from the front view, as shown at (*c*).

**FIG. 24.** A machine part with surfaces inclined to top and side. (*a*) Planes of projection; (*b*) the planes opened up; (*c*) front, top, and normal views of the inclined surfaces.

**FIG. 25.** Stages in drawing a machine part with a surface inclined to top and side.

3. As shown at (d), measure the distance (depths) from the reference plane (of various points needed), and transfer these measurements with dividers or scale to the normal view, measuring from the reference plane in the normal view. Note that the points are in front of the reference plane in the top view and are therefore measured toward the front in the normal view.

4. From specifications, complete the normal view, as shown at (e).

5. Complete the drawing, as shown at (f). In this case the top view could have been completed before the normal view was drawn. However, it is considered better practice to lay out the normal view before completing the view that will show the surface foreshortened.

A part might also be drawn with the inclined surface on the lower or upper front or rear. Orientation of this type is shown in Fig. 26, where the inclined surface is on the upper front. The steps in drawing an object of this type are given in Fig. 27 as follows:

1. Draw the partial front, top, and right-side views, as at (b), and locate the normal view direction by drawing projectors perpendicular to the inclined surface, as shown.

2. Locate the reference plane in the front view. This reference plane is taken at the left side of the object, because both the vertical and inclined portions have a left surface in the same profile plane. The reference plane in the normal view will be perpendicular to the projectors already drawn, and is located at a convenient distance from the right-side view, as shown at (c).

3. As shown at (d), measure the distances (widths) from the reference plane (of various points needed), and transfer these measurements with dividers or scale to the normal view, measuring from the reference plane in the normal view. Note that points to the right of the reference plane in the front view will be measured in a direction *toward* the right-side view in the normal view.

4. From specifications of the surface contour complete the normal view (e).

5. Complete the right-side and front views by projecting and measuring from the normal view. As an example, one intersection of the cut corner is projected to the right-side view and from there to the front view; the other intersection is measured (distance $W$) from the normal view and then laid off in the front view.

## 6. Purposes of Normal Views.

In practical work the chief reason for using a normal view is to show the true shape of an inclined surface.

In a normal view, the inclined surface will be shown in its true shape, but the other faces of the

**FIG. 26.** A machine part with a surface inclined to top and front. (*a*) Planes of projection; (*b*) the planes opened up; (*c*) front, side, and normal views of the inclined surface.

**FIG. 27.** Stages in drawing a machine part with a surface inclined to top and front.

object appearing in the view will be foreshortened. In practical work these foreshortened parts are usually omitted, as in Fig. 28. Views thus drawn are called *partial views*. The exercise of drawing the complete view, however, may aid the student in understanding the subject.

Another important use of a normal view is in the case where a principal view has a part in a fore-shortened position which cannot be drawn without first constructing a normal view in its true shape from which the part can be projected back to the principal view. Figure 29 illustrates this procedure.

**FIG. 28.** Use of partial views.

**FIG. 29.** Use of the normal view for construction of other views.

Note in this figure that the view direction is set up looking along the semihexagonal slot and perpendicular to the face that is at a right angle to the slot. From the normal view showing the true semihexagonal shape, the side view and then the front view can be completed.

In most cases the normal view cannot be projected from the principal views but *must be drawn from dimensional specifications of the surface shape.*

Another practical example is the flanged 45° elbow of Fig. 30, a casting with an irregular inclined face, which not only cannot be shown in true shape in any of the principal views but also is difficult to draw in its foreshortened position. An easier and more practical selection of views for this piece is shown in Fig. 31, where normal views looking in directions perpendicular to the inclined faces show the true shape of the surfaces and allow for simplification. This is because each view can be laid out independently from specifications, and there is not even need for a reference plan, measurements, etc.

### 7. Normal View of a Line.

The normal view of a surface shows the true length of *all* lines in the surface. To prove this, go back to Fig. 8 and visualize *any* line on surface $A$ of the figure. You will find that the line shows in true length in the normal view. Often, however, especially in solving geometric space problems, a line is not a part of a surface but a separate line located in space by a pair of orthographic views. Any line that is parallel to one of the planes of projection will have a principal view that is the normal view. Thus in Fig. 32 where the lines are horizontal ($a$), frontal ($b$), and profile ($c$), the normal views of the lines are at ($a$) the top view, at ($b$) the front view, and at ($c$) the side view.

Often a line is skew to *all* the planes of projection and an extra view is required to give the normal view. To find the normal view of a surface, the direction of observation must be *perpendicular* to the surface; the same direction of observation is needed in finding the normal view of a line—the direction of observation *must* be perpendicular to the line. Thus at ($d$) projectors are drawn in the top view *perpendicular* to the top view of the line. The direction of

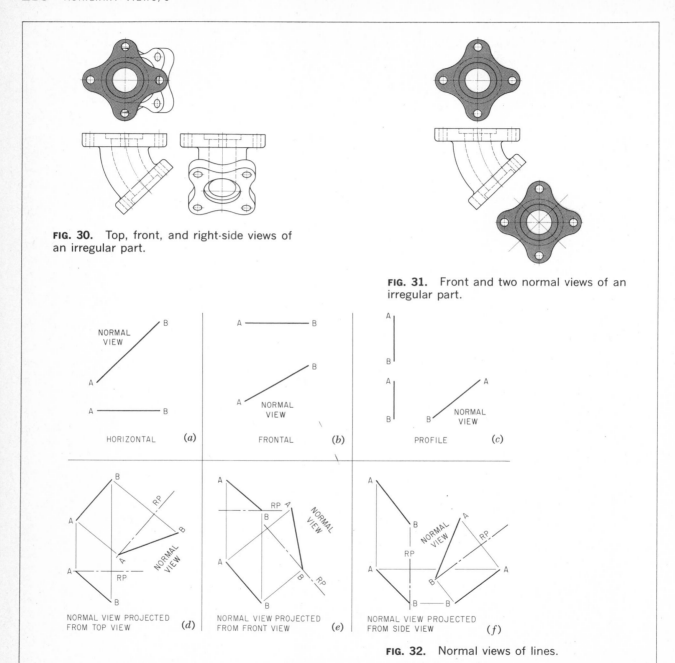

**FIG. 30.**  Top, front, and right-side views of an irregular part.

**FIG. 31.**  Front and two normal views of an irregular part.

**FIG. 32.**  Normal views of lines.

view is now established perpendicular to the line, since the line could be on a surface, the top view being the *edge* view of the assumed surface. Next,

the reference plane will appear perpendicular to the projectors just established for the normal view and will appear again in the front view perpendicular to the projectors between top and front views. Finally, point A is *on* the reference plane and on the projector from the top view and is thus located in the normal view. Point B, then measured from the reference plane in the front view, transferred to the normal view and laid off from the reference plane *on* the projector for point B, completes the normal view. A similar normal view can be projected from the front view as at E, or from the side view as at F. Note in both cases that the projectors for the normal view *must* be perpendicular to the line in the view the normal is projected from.

## 8. Skew Surfaces.

In Sec. 2, in the classification of surfaces, the skew surface was described and illustrated. For convenience, Fig. 33 again shows this type of surface both pictorially and orthographically. Remember that a skew surface is one that occurs at an angle to *all* the principal planes of projection.

The normal view of a skew surface cannot (without resorting to special projective methods) be projected directly from the principal views, but must be projected from an *edge* view of the surface. Also, the

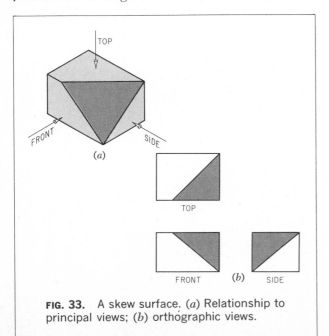

**FIG. 33.** A skew surface. (*a*) Relationship to principal views; (*b*) orthographic views.

edge view is almost always needed to show other features, such as the angle a surface makes with its base. Thus, *two* views are required and we must first learn how to get the edge view.

## 9. Edge Views of Skew Surfaces.

The key to determining the direction of observation needed to obtain the edge view of a surface is to locate a direction that will give the *end view of one line* of the surface. This is easily done if the surface contains a line that is horizontal, frontal, or profile. In Fig. 34a the line is horizontal and a view made looking in a horizontal direction that is *aligned with the line* will give the end view. Note in Fig. 34a that points A and B (the two ends of the line) are both on the same projector and that both lie on the reference plane thus giving coincidence of the two ends of the line in the end view. Also note that the end view must be projected from the normal view of the line, in this case the top view. The end view of a frontal line is shown at (b); it is projected from the front (normal) view. The end view of a profile line is shown at (c); it is projected from the side (normal) view.

As stated before, the edge view of a surface is made by first obtaining the end view of any line of the surface. In Fig. 35 surface ABDC is a skew surface (at an angle to *all* principal planes). However, lines AB and DC are horizontal lines of the surface. Selecting DC as the key line, a projector is drawn aligned with DC and another projector (parallel) is drawn for AB. This establishes the direction of view for the end view of both DC and AB. The reference plane for the edge view will appear as an edge perpendicular to these projectors, and will appear again as an edge in the front view perpendicular to the projectors between top and front views. For convenience, the reference plane is placed through points D and C. Then D and C both project to the edge view on the same projector and both lie on the reference plane, giving DC in the edge view. A and C both project to the edge view on the same projector. The distance from the reference plane in the front view to points A and C is measured and transferred to the edge view, thus locating AC in the edge view. The line connecting

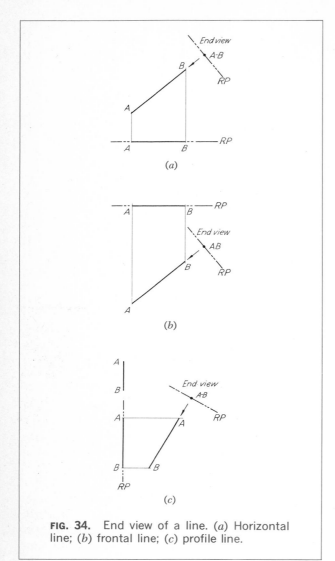

FIG. 34.    End view of a line. (*a*) Horizontal line; (*b*) frontal line; (*c*) profile line.

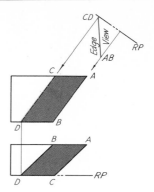

FIG. 35.    Edge view of a skew surface. Projection is in the direction of the horizontal line *DC*.

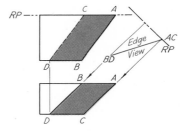

FIG. 36.    Edge view of a skew surface. Projection is in the direction of frontal line *AC*.

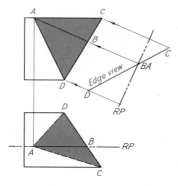

FIG. 37.    Edge view of a skew surface. The direction of the edge view is located by the horizontal line *AB*.

*DC* and *AB* is the edge view of surface *ABDC*.

Figure 36 shows the identical surface *ABDC* of Fig. 35, but this time the edge view has been obtained by looking along the two frontal lines *AC* and *BD*. The edge view is projected in the direction of *AC* and *BD* from the front view because *AC* and *BD* are normal in the front view. The reference plane then located in the edge view and top view, and *AC* and *BD* located with reference to it, complete the edge view.

In Fig. 37 the surface *ACD* does not contain any horizontal, frontal, or profile line, and so to get an

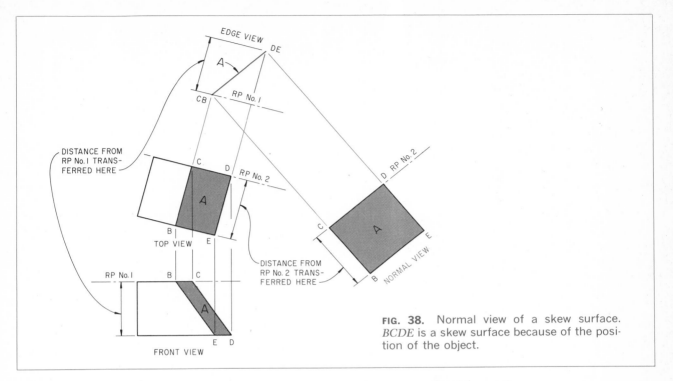

**FIG. 38.** Normal view of a skew surface. *BCDE* is a skew surface because of the position of the object.

end view of one line of the surface, a horizontal, frontal, or profile line must be laid out on the surface. A horizontal line has been selected and is drawn at *AB* in the front view and then projected to the top view. This line (*AB*) is then normal in the top view and projectors parallel to it from *AB*, *C*, and *D* will establish the direction for the edge view of *ACD*. The reference plane is conveniently located through *AB* in front and edge views; then *D* and *C* are measured from the reference plane in the front view and transferred to the edge view. Note that *D* lies *above* the reference plane and *C below* and that they are transferred accordingly to the edge view. To avoid reversing a normal or edge view, remember this simple rule: if a point is on the side of the reference plane *toward* the view projected from in making an edge or normal view (the top view, in this case), the point will be transferred *toward* that same view. As an example of the rule, note in Fig. 37 that *D* is on the side *toward* the top view from the reference plane in both front and edge views while *C* is on the side *away* from the reference plane.

## 10. Normal Views of Skew Surfaces.

A skew surface is produced by an object with an inclined surface that is turned at an angle to the principal planes, as in Fig. 38, or by an object with a face at an angle to three object faces, as in Figs. 39 and 40. Figure 38 is easier to visualize, so we will discuss it first.

Surface *BCDE* of Fig. 38 is a skew surface. *CB* and *DE* are horizontal lines of the surface. The edge view is obtained by looking in the direction of *CB* and *DE*, as shown in Fig. 38 and described in Sec. 9.

The normal view is made with the direction of observation perpendicular to the edge view. To draw the normal view, first draw projectors perpendicular to the edge view (Fig. 38), establishing the direction for the normal view. As in all earlier examples, the reference plane for the normal view (*RP* No. 2 in Fig. 38) is perpendicular to the projectors. The reference plane appears normal in the edge view and appears again as an edge in *any* other view in direct

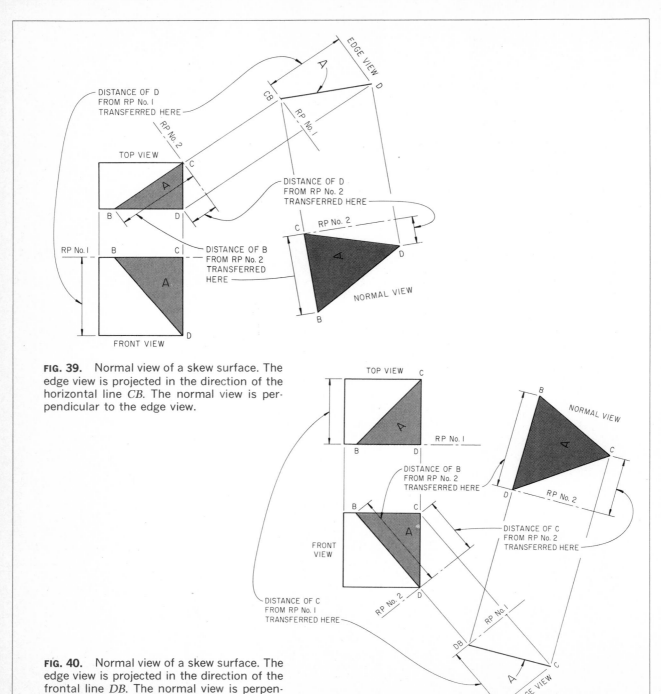

**FIG. 39.** Normal view of a skew surface. The edge view is projected in the direction of the horizontal line *CB*. The normal view is perpendicular to the edge view.

**FIG. 40.** Normal view of a skew surface. The edge view is projected in the direction of the frontal line *DB*. The normal view is perpendicular to the edge view.

projection with the edge view. In this case therefore, the reference plane shows as an edge in the top view, and is perpendicular to the projectors between top and edge views. To prove this relationship to yourself, note that here the top view, edge view, and normal view have *exactly* the same relationship as the front view, top view, and normal view of Fig. 8. Finally, transfer the distance of $B$ and $E$ from the reference plane (No. 2) in the top view to the normal view, as shown.

Figure 39 is similar to Fig. 38, but this time the object faces are parallel to the principal planes, and surface $BCD$ is skew to object faces and principal planes. Again, one line ($BC$) of the skew surface is horizontal and gives the direction for the edge view. In the edge view, $B$ and $C$ are on the reference plane. Point $D$, measured from reference plane No. 1 in the front view, is transferred to the edge view, as shown. Then, projectors perpendicular to the edge view give the direction for the normal view, and reference plane No. 2 is drawn in normal and top views perpendicular to the projectors connecting each with the edge view. Finally the distances of $B$

and $D$ from reference plane No. 2 are transferred to the normal view.

The object of Fig. 40 is similar to that of Fig. 39, but in Fig. 40 the frontal line ($BD$) of surface $BCD$ has been used for the direction of the edge view. The edge view is therefore projected from the front view. All other constructions are the same as before and are evident from the figure.

## 11. Normal Views of Skew Surfaces on Practical Objects.
Figure 41 illustrates the successive steps in drawing a normal view. The pictorial illustration ($a$) shows a typical object with a skew surface. The line of intersection between the skew portion and the horizontal base is line $AB$. In order to get an edge view of the skew surface, a view may be taken looking in the direction of line $AB$, thus giving an end view of $AB$. Because $AB$ is a line of the skew surface, the edge view will result. The reference plane ($RP$ No. 1) for this view will be horizontal. The direction of observation for the normal view will be perpendicular to the skew surface. The reference plane ($RP$

**FIG. 41.** Stages in drawing a machine part with a skew surface.

No. 2) for the normal view will be perpendicular to the edge-view direction and thus perpendicular to edge *AB*, as shown at (*a*).

At (*b*), parital top and front views are shown. The projectors and reference plane for the required edge view are also shown.

At (*c*), the edge view has been drawn. Note that line *AB* appears as a point in the edge view. The angle that the skew surface makes with the base is laid out in this view from specifications.

At (*d*), the normal view is added. The projectors for the view are perpendicular to the edge view. The reference plane is drawn perpendicular to the projectors for the normal view and at a convenient distance from the edge view. The reference plane in the top view is drawn midway between points *A* and *B* in the top view because the skew surface is symmetrical about this reference plane. The normal view is drawn from specifications of the shape. The projection back to the edge view can then be made.

The views thus completed at (*d*) describe the object, but the top and front views may be completed for illustrative purposes or as an exercise in projection. The method is illustrated at (*e*) and (*f*). Any point, say *P*, may be selected and projected back to the edge view. From this view a projector is drawn back to the top view. Then the distance *S* from the normal view is transferred to the reference plane in the top view. A number of points so located will complete the top view of the circular portion, and the straight-line portion can be projected in similar manner. The front view is found by drawing projectors to the front view for the points needed, measuring the heights from the reference plane in the edge view, and transferring these distances to the front view. Note that this procedure for completing the top and front views is the same as for drawing the views originally but in reverse order.

## 12. Terms of Reference for Auxiliary Views.

In this text, emphasis is given to the *purpose* for which auxiliary views are made; namely, to show the *edge view* or *normal view* of a surface. This practice is basic to the representation of solid objects, and, fundamentally, an auxiliary view used to depict a surface needs no further identification than the designation *edge view* or *normal view*. However, auxiliaries are named in various ways in other texts on engineering drawing and graphics, and for this reason the following terms of reference are given:

### Reference Group 1

*Elevation auxiliary* or *auxiliary elevation.* Any view made by looking in a horizontal direction but inclined to frontal and profile planes. These are the views illustrated in Fig. 7*b* to *e*.

*Right auxiliary.* Any view made by looking from the right side in a frontal direction but inclined to horizontal and profile planes. See Fig. 7*f* and *g*.

*Left auxiliary.* Any view made by looking from the left side in a frontal direction but inclined to horizontal and profile planes. See Fig. 7*h* and *j*.

*Front auxiliary.* Any view made by looking from the front in a profile direction but inclined to horizontal and frontal planes. See Fig. 7*k* and *l*.

*Rear auxiliary.* Any view made by looking from the rear in a profile direction but inclined to horizontal and frontal planes. See Fig. 7*m* and *n*.

### Reference Group 2

*Top-adjacent auxiliary.* Any auxiliary view projected from the top view. Such auxiliaries are the same as elevation auxiliaries (group 1). See Fig. 7*b* to *e*.

*Front-adjacent auxiliary.* Any auxiliary view projected from a front view. Such auxiliaries are the same as right-and-left auxiliaries. See Fig. 7*f* to *j*.

*Side-adjacent auxiliary.* Any auxiliary projected from a side view. Such auxiliaries are the same as front-and-rear auxiliaries. See Fig. 7*k* to *n*.

### Reference Group 3

*Oblique view.* This is the normal view of a skew (oblique) surface and is projected from an edge view. See Figs. 38 to 40.

*Reference Group 4*

*Auxiliary-adjacent auxiliary view.* This is the normal view of a skew surface, projected from an edge view. See Figs. 38 to 40.

These terms of reference are used in books on descriptive geometry. However, since the book on descriptive geometry you are using with this text may give other terms, the space below is provided for listing them.

_____

# PROBLEMS

## Group 1. Normal Views of Inclined Surfaces.

**1 to 7.** Draw given views and add normal views of the inclined surface using the reference plane indicated.

**ROB. 1.** Locking wedge.

**PROB. 2.** Statue base.

**PROB. 3.** Rod slide.

**PROB. 4.** Fulcrum.

**PROB. 5.** Adjustable pawl.

**PROB. 6.** Hexagonal-shaft lock.

**PROB. 7.** Molding.

**8.** Draw the front view, partial top view, and normal view of the inclined surface.

**PROB. 8.**   Holder.

**9.** Draw the partial front view, right-side view, partial top view, and normal view of the inclined surface.

**PROB. 9.**   Slotted anchor.

**10.** Draw the front view, partial top view, and normal view of the inclined surfaces.

**PROB. 10.**   Connector strip.

**11.** Draw the top view, partial front view, and normal view of the inclined surfaces.

**PROB. 11.**   Push plate.

**12.** Draw the front view, partial top view, and normal view of the inclined surface.

**PROB. 12.**   Bevel washer.

**13.** Draw the front view, partial right-side view, and normal view of the inclined surface. Draw the normal view before completing the front view.

**PROB. 13.**
Jig angle.

**14.** Draw the front view, parital top view, and normal view of the inclined surface.

**PROB. 14.** Angle clip.

**15.** Draw the front view, partial top view, and normal view of the inclined surface.

**PROB. 15.** Angle swivel.

**16.** Draw the front view, partial top and right-side views, and normal view of the inclined surface.

**PROB. 16.** Corner tie.

**17.** Draw the front view, partial top and left-side views, and normal view of the inclined surface.

**PROB. 17.** Channel support.

**18 and 19.** Determine what views and partial views will best describe the part. Sketch the proposed views before making the drawing with instruments.

**PROB. 18.** Radial swing block.

**PROB. 19.** Angle-shaft base.

**20.** Draw the front view, partial bottom view, and normal view of the inclined surface.

**PROB. 20.** Catenary clip.

**21.** Draw the front view, partial left-side and bottom views, and normal view of the inclined surface.

**PROB. 21.** Bevel-gear housing.

**22 and 23.** Determine what views and partial views will best describe the part.

**PROB. 22.** Slide base.

**PROB. 23.** Idler bracket.

**24.** Draw the given front view and add the views necessary to describe the part.

**PROB. 24.** Corner brace.

**25.** Draw the front view, partial left-side view, and normal view of the inclined surface.

**26 and 27.** This pair of similar objects has the upper lug in two different positions. Layouts are for 11- × 17-in. paper. Draw the views and partial views as indicated on the layouts.

**PROB. 25.** Spindle support.

**PROB. 26.** Spar clip, 90°.

**PROB. 27.** Spar clip, 120°.

**28.** Draw the front view; partial top, right-side, and left-side views; and normal view of the inclined surface. Use decimal scale for layout.

**29.** Draw the views and partial views that will best describe the part.

**30.** Draw top and front views and a normal view of the inclined face. Will the normal view of the inclined face show the true cross section of the square hole?

**31.** Draw front, top, and side views and normal views of the inclined surfaces.

**PROB. 28.** Seat release.

**PROB. 29.** Idler link.

**PROB. 30.** Actuator bracket.

**PROB. 31.** Assembly-fixture base.

**32.** Front and right-side views of the shaft-locator wedge are shown on a layout for 11- × 17-in. paper. Add normal view of the inclined faces at centerline position shown.

## Group 2. Normal Views of Skew Surfaces.

**33.** Draw the partial front and top views, edge view showing the contour of the slot, and normal view of the skew surface.

**34.** Draw the partial front and top views, and edge and normal views of the skew surface. Draw the normal view before completing the edge view.

**35.** Draw the partial front and top views, and edge and normal views of the skew surface.

**PROB. 32.**  Shaft-locator wedge.

**PROB. 33.**  Dovetail clip.

**PROB. 34.**  Anchor base.

**PROB. 35.**  Adjusting clip.

**36.** Draw the views given, omitting the lugs in the top view. Add normal views to describe the lugs.

**37.** Draw the views given, using edge and normal views to obtain the shape of the lugs.

**38.** Draw the partial top, front, and side views. Add edge and normal views to describe the lugs.

**39.** Draw the top and front views and use edge and normal views to describe the slots and skew surfaces. The part is symmetrical about the main axis.

**40.** Draw the spar clip, using the layout shown for 11- × 17-in. paper. Note that an edge and two normal views are required.

**41.** Draw the views given, using edge and normal views to describe the lugs.

**42.** Draw top and front views and auxiliary views that will describe the skew surface.

**PROB. 36.** Bar-strut anchor.

**PROB. 37.** Cable anchor.

**PROB. 38.** Transverse connection.

**PROB. 39.** Chamfer-tool base.

**PROB. 40.** Spar clip.

**43.** Shown on layout for 11- × 17-in. paper are partial front and right-side views and partial auxiliary views that describe the position and shape of the skew surface. Complete front, right-side, and first auxiliary views.

**44.** Shown on layout for 11- × 17-in. paper are partial top and front views and auxiliaries that describe the position and shape of the clamp portion of the part. Complete the top and front views.

**PROB. 41.** Crane-masthead collar and cap.

**PROB. 42.** Bus-bar connector.

**PROB. 43.** Valve control-shaft bracket.

**PROB. 44.** Bipod shaft clamp.

# SECTIONAL VIEWS AND CONVENTIONS

# 7

These Special Views and Practices Are Specific
Aids to Complete and Accurate Orthographic
Representation

Shown in this Chapter:
Definition, Classification • Full, Half, Broken-out,
Rotated, and Removed Sections • Auxiliary
Sections • Assembly Sections • Drawing
Practices • Sections Showing Arms, Ribs,
Lugs • Crosshatching • Aligned Sections •
Conventional Practices • Conventional Breaks
and Symbols • Problems

## 1. Sectional Views Defined.

When the interior of an object is complicated or when the component parts of a machine are drawn assembled, an attempt to show hidden portions by the customary dashed lines in regular orthographic views often results in a confusing network, as shown in Fig. 1 at (*a*), which is difficult to draw and almost impossible to read clearly.

In cases of this kind, to aid in describing the object, one or more views are drawn to show the object as if a portion had been cut away to reveal the interior, as at (*b*). Also, if some detail of the shape of an object is not clear, a cut taken through the portion and then turned up, or turned up and removed, as at (*c*), will describe the shape concisely and often eliminate the need for an extra complete view.

Either of these conventions is called a *section,* which is defined as an imaginary cut made through an object to expose the interior or to reveal the shape of a portion. A view in which all or a substantial portion of the view is sectioned is known as a *sectional view.*

For some simple objects where the orthographic, *unsectioned* views can be easily read, sectional views are often preferable because they show clearly and emphasize the solid portions, the voids, and the shape.

## 2. How Sections are Shown.

The place from which the section is taken must be identifiable on the drawing, and the solid portions and voids must be distinguished on the sectional view. The place from which the section is taken is in many cases obvious, as it is for the sectional view (*b*) in Fig. 1; the section is quite evidently taken through the center (at the center line of the top view). In such cases no further description is needed. If the place from which a section is taken is not obvious, as at (*c*), a *cutting plane,* directional arrows, and identification letters are used to identify it. Whenever there is any doubt of the position where the section is taken, the cutting plane (see alphabet of lines, Sec. 39, Chap. 2) should be shown. A cutting plane is the imaginary medium used to show the

**FIG. 1.** Advantage of sectional views. (*a*) Orthographic view with hidden edges indicated by dashed lines; (*b*) the same view but made as a section to clarify the shape; (*c*) cross-sectional shape of lug shown by removed section.

path of cutting an object to make a section. Cutting planes for each kind of section will be discussed in the following sections.

A sectional view must show which portions of the object are solid material and which are spaces. This is done by section lining, sometimes called "cross-hatching," the solid parts with lines, as shown at (*b*) and (*c*). Section-lining practice is given in Sec. 11, where codes for materials are also discussed.

## 3. Types of Sections.

Although the different sections and sectional views have been named for identification and for specifying the type of view required in a drawing, the names are not shown on the drawing for the same reason that a top, front, or side view would not be so labeled—the views are easily interpreted and a workman does not require the name to read the drawing. The names are assigned by the character of the section or the amount of the *view* in section, *not* by the amount of the object removed.

## 4. Full Section.

A full section is one in which the cutting plane passes entirely across the object, as in Fig. 2, so that the resulting view is completely "in section." The cutting plane may pass straight through, as at (*a*), or be offset, changing direction forward or backward, to pass through features it would otherwise have missed, as at (*b*) and (*c*). Sometimes *two* views are drawn in section on a pair of cutting planes, as at (*a*) in Fig. 3. In such cases each view is considered separately, without reference to what has been removed for another view. Thus (*b*) shows the portion remaining and the cut surface for one sectional view, and (*c*) for the other sectional view. Figure 4 is the orthographic drawing of the object, with indication for sectioning, as shown in Fig. 3. Both the front and side views are full-sectional views.

FIG. 2. Cutting planes for a full section. The plane may cut straight across (*a*) or change direction (*b* and *c*) to pass through features to be shown.

FIG. 3. Two cutting planes. The two planes at (*a*) will produce sections (*b*) and (*c*). Each section is considered separately, without reference to what has been removed for the other view.

FIG. 4. Full sections. This is the orthographic drawing of the object in Fig. 3. The cutting-plane position is obvious for the front-view section and is not identified in the top view. Section *A-A* (side view) has its cutting plane identified because the section might be made elsewhere.

Figure 5 shows a full section made on an offset cutting plane. Note that the *change in plane direction* is *not* shown on the sectional view, for the cut is purely imaginary and *no edge* is present on the object at this position.

## 5. Half Section.

This is a view sometimes used for symmetrical objects in which one half is drawn in section and the other half as a regular exterior view. The cutting plane is imagined to extend halfway across, then forward, as in Fig. 6. A half section has the advantage of showing both the interior and exterior of the object on one view without using dashed lines, as

(a)                (b)

**FIG. 7.** Half sections. Dashed lines are rarely necessary (*a*), but may be used if needed for clarity or to aid in dimensioning (*b*).

(a)

(b)

**FIG. 8.** Cutting planes for broken-out sections. The cutting plane extends only slightly beyond the features to be shown in section.

**FIG. 5.** A full section. The cutting plane is offset to pass through both principal features of the object.

**FIG. 6.** The cutting plane for a half section. The resulting sectional view will be half in section and half an external view.

at (*a*) in Fig. 7. However, a half section thus made is difficult to dimension without ambiguity, and so, if needed for clarity, dashed lines may be added, as at (*b*).

Note particularly that a *center line* separates the exterior and interior portions on the sectional view. This is for the same reason that the change in plane direction for the offset of the cutting plane of Fig. 5 is not shown—no edge exists *on the object* at the center.

**FIG. 9.** A broken-out section. Note that a standard break line terminates the sectional portion.

**FIG. 10.** The cutting plane for a rotated or a removed section. A slice of negligible thickness is taken.

**FIG. 11.** A rotated section. The section is rotated 90° to bring it into the plane of the view.

**FIG. 12.** A rotated section with broken view. The view is broken thus whenever the view outline interferes with the section.

## 6. Broken-out Section.

Often an interior portion must be shown but a full or half section cannot be used because the cutting plane would remove some feature that must be included. For this condition the cutting plane is extended only so far as needed, as illustrated by Fig. 8a and b, and then is thought of as "broken out." Figure 9 is an example. Note the irregular break line that limits the extent of the section.

## 7. Rotated Section.

As indicated in Sec. 1, a section may be a slice of negligible thickness used to show a shape that would otherwise be difficult to see or describe. The cutting plane for such a section is shown in Fig. 10. If the resulting section is then rotated 90° *onto* the view as in Fig. 11, the section is called a rotated section. Whenever the view outline interferes with the section, the view is broken, as in Fig. 12.

## 8. Removed Sections.

These are used for the same purpose as rotated sections, but instead of being drawn *on* the view, they are removed to some adjacent place on the paper, as in Fig. 13.

The cutting plane with reference letters should always be indicated unless the place from which the section has been taken is obvious. Removed sections are used whenever restricted space for the section

**FIG. 13.** Removed sections. The cutting planes and the mating sections must be identified.

**FIG. 14.** Removed sections in projection. Identifications of cutting planes and mating sections are not needed.

or the dimensioning of it prevents the use of an ordinary rotated section. When the shape of a piece changes gradually or is not uniform, several sections may be required (Fig. 14). It is often an advantage to draw the sections to larger scale than that of the main drawing in order to show dimensions more clearly. Sometimes sections are removed to a separate drawing sheet. When this is done, the section must be carefully shown on the main drawing with cutting plane and identifying letters. Often these identifying letters are made as a fraction in a circle, with the numerator a letter identifying the section and the denominator a number identifying the sheet. The sectional view is then marked with the same letters and numbers. The ANSI recommends that, whenever possible, a removed section be drawn in its natural projection position. Note in Fig. 14 that this practice is followed.

## 9. Auxiliary Sections.

These are sectional views conforming to all the principles of edge and normal views given in Chap. 6. The section shows the *normal* view of a cutting plane that is in a position on an inclined feature so as to reveal the interior, as shown in Fig. 15. The edge view of any perpendicular faces will also be seen in the normal view, as in Fig. 15. All types of sections—full, half, broken-out, rotated, and removed—are used on auxiliaries. Figure 16 shows auxiliary partial sections, which are also properly called re-

**FIG. 15.** Auxiliary section. The section is a normal view of the cutting plane.

**FIG. 16.** Auxiliary sections (partial). Identifications of the cutting planes and mating sections are necessary.

**FIG. 17.** The cutting plane for an assembly section. It is often offset to pass through features to be shown in the section.

moved sections in auxiliary position. Note in Fig. 16 that the two sections are normal views of their cutting planes.

## 10. Assembly Sections.

As the name implies, an assembly section is made up of a combination of parts. All the previously mentioned types of sections may be used to increase the clarity and readability of assembly drawings. The cutting plane for an assembly section is often offset, as in Fig. 17, to reveal the separate parts of a machine or structure.

The purpose of an assembly section is to reveal the interior of a machine or structure so that the separate parts can be clearly shown and identified, but the separate parts do not need to be completely described. Thus only such hidden details as are needed for part identification or dimensioning are shown. Also, the small amount of clearance between mating or moving parts is not shown because, if shown, the clearance would have to be greatly exaggerated, thus confusing the drawing. Even the clearance between a bolt and its hole, which may be as much as $\frac{1}{16}$ in., is rarely shown. Figure 18 is an example of an assembly section in tabular form. The component is similar to that in Fig. 17. Note that a half section is used in Fig. 18 on both views.

Crosshatching practice for assembly sections is explained in Sec. 11.

| A | B | C | D | E | F | G | H | I | J | K | L | M |
|---|---|---|---|---|---|---|---|---|---|---|---|---|
| $1\frac{3}{16}$ | $1\frac{1}{4}$ | $1\frac{1}{2}$ | $2\frac{1}{4}$ | $3\frac{3}{4}$ | $2\frac{1}{4}$ | $\frac{1}{2}$ | $\frac{3}{4}$ | $7\frac{3}{4}$ | $1\frac{1}{8}$ | $5\frac{5}{8}$ | $\frac{3}{16}$ | $1\frac{1}{4}$ |
| $1\frac{7}{16}$ | $1\frac{7}{16}$ | $1\frac{3}{4}$ | $2\frac{5}{8}$ | $4\frac{1}{2}$ | $2\frac{5}{8}$ | $\frac{1}{2}$ | $\frac{13}{16}$ | $8$ | $1\frac{1}{4}$ | $5\frac{7}{8}$ | $\frac{3}{16}$ | $1\frac{1}{4}$ |
| $1\frac{11}{16}$ | $1\frac{5}{8}$ | $2\frac{1}{16}$ | $3$ | $5\frac{1}{4}$ | $3$ | $\frac{1}{2}$ | $\frac{7}{8}$ | $8\frac{1}{2}$ | $1\frac{1}{4}$ | $6\frac{3}{8}$ | $\frac{3}{16}$ | $1\frac{1}{4}$ |

**FIG. 18.** A sectional assembly drawing. The sectional views give emphasis to construction and separate parts.

## 11. Drawing Practices for Sectional Views.

In general, the rules of projection are followed in making sectional views. Figure 19 shows the picture of a casting intersected by a cutting plane, giving the appearance that the casting has been cut through by the plane *A-A* and the front part removed, exposing the interior. Figure 20 shows the drawing of the casting with the front view in section. The edge of the cutting plane is shown in the top view by the cutting-plane symbol, with reference letters and arrows to show the direction in which the view is taken. It must be understood that in thus removing the nearer portion of the object to make the sectional view, the portion assumed to be removed is not omitted in making other views. Therefore, the top and right-side views of the object in Fig. 20 are full and complete, and only in the front view has part of the object been represented as removed.

The practices recommended by the ANSI for inclusion of the cutting-plane symbol, of visible and hidden edges, and for crosshatching are as follows:

*The Cutting-plane Symbol.* It may be shown on the orthographic view where the cutting plane appears as an edge and may be more completely identified with reference letters along with arrows to show the direction in which the view is taken. The cutting-plane line symbol is shown in the alphabet of lines, Figs. 56 and 57, Chap. 2. Use of the symbol is illustrated in Figs. 1 and 20. Often when the position of the section is evident, the cutting-plane symbol is omitted (Fig. 21). It is not always desirable to show the symbol through its entire length; so in such cases the beginning and ending of the plane is shown, as in sections *A-A* and *B-B* in Prob. 47 in the final chapter. Removed sections usually need the cutting-plane symbol with arrows to show the direction of sight and with letters to name the resulting sectional view (Fig. 16).

*Unnecessary Hidden Detail.* Hidden edges and surfaces are not shown unless they are needed to describe the object. Much confusion may result if all detail behind the cutting plane is drawn. In Fig. 21, (*a*) shows a sectional view with all the hidden edges and surfaces shown by dashed lines. These

**FIG. 19.** Cutting plane for a sectional view.

**FIG. 20.** A drawing with a sectional view. This is the same object as in Fig. 19.

lines complicate the view and do not add any information. The view at (*b*) is preferred because it is simpler, less time-consuming to draw, and more easily read than the view at (*a*). The holes lie on a circular center line; and where similar details repeat, all may be assumed to be alike.

*Necessary Hidden Detail.* Hidden edges and surfaces are shown if necessary for the description of the object. In Fig. 22 view (*a*) is inadequate since it does not show the thickness of the lugs. The correct treatment is in view (*b*), where the lugs are shown by dashed lines.

*Visible Detail Shown in Sectional Views.* Figure 23 shows an object pictorially with the front half removed, thus exposing edges and surfaces behind the cutting plane. At (*a*) a sectional view of the cut surface only is shown, with the visible elements omitted. This treatment should *never* be used. The view should be drawn as at (*b*) with the visible edges and surfaces behind the cutting plane included in the sectional view.

**FIG. 21.** Hidden edges and surfaces not shown; (b) is preferable to (a) because the dashed lines are not necessary for description of the part.

**FIG. 22.** Hidden edges and surfaces shown. (b) *must* be used instead of (a) because the dashed lines are necessary for description of the part.

**FIG. 23.** Visible edges shown. These are the edges seen behind the plane of the section and must be shown as at (b).

*Visible Detail Not Shown in Sectional Views.* Sometimes confusion results if all visible detail behind the cutting plane is drawn, and it may be omitted if it does not aid in readability. Omission of detail should be carefully considered and may be justified as time saved in drawing. This applies mainly to assembly drawings for showing how the pieces fit together rather than for giving complete information for making the parts (see Fig. 24).

*Section Lining.* Wherever material has been cut by the section plane, the cut surface is indicated by section lining done with fine lines generally at 45° with the principal lines in the view and spaced uniformly to give an even tint. These lines are spaced entirely by eye except when some form of mechanical section liner is used. The pitch, or distance between lines, is governed by the size of the surface. For ordinary working drawings, it will not be much less then $\frac{1}{16}$ in. and rarely more than $\frac{1}{8}$

in. *Very* small pieces may require a spacing closer than $\frac{1}{16}$ in. Take care in setting the pitch by the first two or three lines, and glance back at the first lines often to see that the pitch does not gradually

**FIG. 24.** Omission of detail. Detail behind the plane of the section may be omitted when it does not aid in readability or when it might cause ambiguity.

increase or decrease. Nothing mars the appearance of a drawing more than poor section lining. The alphabet of lines, Figs. 56 and 57, Chap. 2, gives the weight of crosshatch lines.

Two adjacent pieces in an assembly drawing are crosshatched in opposite directions. If three pieces adjoin, one of them may be sectioned at other than 45° (usually 30° or 60°, Fig. 25), or all pieces may be crosshatched at 45° by using a different pitch for each piece. If a part is so shaped that 45° sectioning runs parallel, or nearly so, to its principal outlines, another direction should be chosen (Fig. 26).

Large surfaces are sometimes sectioned around the edge only, as illustrated in Fig. 27.

Very thin sections, as of gaskets, sheet metal, or structural-steel shapes to small scale, may be shown in solid black, with white spaces between the parts where thin pieces are adjacent (Fig. 28).

Section lining for the same piece in different views or for the same piece in different parts of the same view should be identical in spacing and direction[1] (Figs. 18 and 24).

Adjacent pieces are section-lined in opposite directions and are often distinguished more clearly by varying the pitch of the section lines for each piece, using closer spacing for the smaller pieces (Figs. 25 and 29).

**FIG. 25.** Crosshatching of adjacent parts. Adjacent parts are crosshatched in opposite directions and/or different spacing, for emphasis. The same piece in different views or in different parts of the same view should be crosshatched with identical spacing and direction (auxiliaries are a possible exception).

[1]An exception to this rule is made for the crosshatching of an auxiliary view in order to avoid crosshatch lines that are parallel or perpendicular, or nearly so, to the outlines of the view.

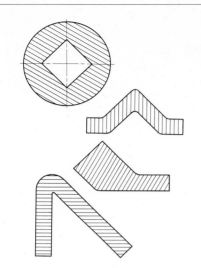

**FIG. 26.** Section-line directions for unusual shapes. Avoid crosshatch directions parallel to the view outlines.

**FIG. 27.** Outline sectioning. This saves drafting time on large views or on large drawings.

**FIG. 28.** Thin material in section. This is drawn solid because there is not enough room for crosshatching.

*Code for Materials in Section.* Symbolic section lining is not commonly used on ordinary working drawings, but in an assembly section it is sometimes useful to show a distinction between materials, and a recognized standard code is an obvious advantage. The ANSI symbols for indicating different materials are given in the Appendix. Code section lining is used only as an aid in reading a drawing and is not to be taken as the official specification of the materials. Exact specifications of the material for each piece are always given on the detail drawing.

## 12. Conventional Practices.

All sections are conventions in that they represent an assumed imaginary cut from which, following the theory of projection rather closely, the sectional views are made. However, the strict rules of projection may be disregarded if by so doing a type of view can be drawn which more accurately depicts the shape of the object. It is impossible to illustrate all the conditions that might occur, but the following discussion shows the principles that are recognized as good practice, since observing them results in added clearness and readability.

## 13. Parts Not Sectioned.

Many machine elements, such as fasteners, pins, and shafts, have no internal construction and, in addition, are more easily recognized by their exterior views. These parts often lie in the path of the section plane, but if they are sectioned (and crosshatched), they are more difficult to read because their typical identifying features (boltheads, rivet heads, chamfers on shafts, etc.) are removed. Thus features of this kind should be left in full view and *not sectioned*. To justify this treatment, the assembly is thought of as being sectioned on a particular plane, with the

**FIG. 29.** Part of a sectional assembly. Shafts, bolts, nuts, rods, rivets, keys, and the like whose axes occur in the plane of the section are left in full (not sectioned).

nonsectioned parts placed in the half holes remaining after the section is made. Figure 17 shows a full section made on an offset cutting plane that passes through a bolt, with the bolt in full view. Figure 29 shows several nonsectioned parts; it is evident that if the shaft, bolts, nuts, rivets, etc., were sectioned, the drawing would be confusing and difficult to read.

## 14. Spoke and Arms in Section.

A basic principle for sectioning circular parts is that any element not continuous (not solid) around the axis of the part should be drawn without crosshatching in order to avoid a misleading effect. For example, consider the two pulleys in Fig. 30. Pulley (a) has a solid web connecting the hub and rim. Pulley (b) has four spokes. Even though the cutting plane passes through two of the spokes, the sectional view of (b) must be made without crosshatching the spokes in order to avoid the appearance of a solid web, as in pulley (a).

Other machine elements treated in this manner are teeth of gears and sprockets, vanes and sup-

**FIG. 30.** Spokes in section. The wheel with spokes (b) is treated as though the cutting plane were in front of the spokes, to avoid misreading the section as a solid web line (a).

**FIG. 31.** Ribs in section. Ribs at (*a*) are treated as though the cutting plane were in front of them, to avoid misreading the section as a solid (*b*).

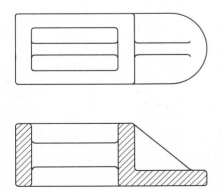

**FIG. 32.** Ribs in section. These are treated as though the cutting plane were in front of the ribs.

porting ribs of cylindrical parts, equally spaced lugs, and similar parts.

## 15. Ribs in Section.
For reasons identical with those given in Sec. 14, when the cutting plane passes longitudinally through the center of a rib or web, as in Fig. 31*a*, the crosshatching is eliminated from the ribs as if the cutting plane were just in front of them or as if they had been temporarily removed and replaced after the section was made. A true sectional view with the ribs crosshatched gives a heavy, misleading effect suggesting a cone shape, as shown at (*b*). The same principle applies to ribs cut longitudinally on rectangular parts (Fig. 32). When the cutting plane cuts a rib transversely, that is, at right angles to its length or axis direction (the direction that shows its thickness), it is always crosshatched (Fig. 24).

## 16. Lugs in Section.
For the same reasons given in Secs. 14 and 15, a lug or projecting ear (Fig. 33*a*), usually of *rectangular* cross section, is not crosshatched; note that crosshatching either of the lugs would suggest a circular flange. However, the somewhat similar condition at (*b*) should have the projecting ears crosshatched as shown because these ears *are* the base of the part.

## 17. Alternate Crosshatching.
In some cases omitting the crosshatching of ribs or similar parts gives an inadequate and sometimes ambiguous treatment. To illustrate, Fig. 34*a* shows a full section of an idler pulley. At (*b*) four ribs have been added. Note that the top surfaces of the ribs are flush with the top of the pulley. Without crosshatching, the section at (*b*) is identical with (*a*) and the ribs of (*b*) are not identified at all on the sectional view. A better treatment in this case is to use alternate crosshatching for the ribs, as at (*c*), where half (alternating) the crosshatch lines are carried through the ribs. Note that the line of demarcation between rib and solid portions is a *dashed* line.

## 18. Aligned Spokes and Arms.
Any part with an odd number (3, 5, 7, etc.) of spokes or ribs will give an unsymmetrical and misleading

**FIG. 33.** Lugs in section. Small lugs (*a*) are treated like spokes (Fig. 30) and ribs. (Figs. 31 and 32). Large lugs (*b*) are considered as the solid base of the part.

**FIG. 34.** Alternate crosshatching. Section (*b*), with ribs flush at the top, looks like section (*a*), with no ribs. Alternate crosshatching (*c*) identifies the ribs.

**FIG. 35.** Aligned spokes. True projection is misleading and difficult to draw. Alignment gives a symmetrical section for a symmetrical part.

section if the principles of true projection are strictly adhered to, as illustrated by the drawing of a handwheel in Fig. 35. The preferred projection is shown in the second sectional view, where one arm is drawn as if aligned, or in other words, the arm is rotated to the path of the vertical cutting plane and then projected to the side view. Note that neither arm should be sectioned, for the reasons given in Sec. 14.

This practice of alignment is well justified logically because a part with an odd number of equally spaced elements is just as symmetrical as a part with an even number and, therefore, should be shown by a symmetrical view. Moreover, the symmetrical view shows the true *relationship* of the elements, while the true projection does not.

### 19. Aligned Ribs, Lugs, and Holes.

Ribs, lugs, and holes often occur in odd numbers and following the principles given in Sec. 18, should be aligned to show the true relationship of the elements. Figure 36 shows several examples of how the cutting plane may pass through a symmetrical

**FIG. 36.** Cutting planes for alignment of holes.

object, permitting the removal of a portion of the object so as to describe the shape better. Note how the cutting planes may change direction so as to pass through the holes. In Fig. 37, true projection

**FIG. 37.** Aligned ribs, lugs, and holes. True projection of ribs (*a*) is misleading. Alignment (*b*) gives a symmetrical section for a symmetrical part. The same is true for lugs (*c*) and holes (*d*).

of the ribs would show the pair on the right fore-shortened, as at (*a*), suggesting in the sectional view that they would not extend to the outer edge of the base. Here, again, the alignment shown at (*b*) gives a symmetrical section of a symmetrical part and shows the ribs in their true relationship to the basic part. To illustrate further, at (*c*) and (*d*) the lugs and holes are aligned, thus showing the holes at their true radial distance from the axis, and, incidently, eliminating some difficult projections.

In all cases of alignment, the element can be thought of as being swung around to a common cutting plane and then projected to the sectional view. Note at (*c*) that because an offset cutting plane is used, each hole is brought separately into position on a common cutting plane before projection to the sectional view. The cutting plane used here is similar to that of (*d*) in Fig. 36.

## 20. Alignment of Elements in Full Views.

In full views, as well as in sectional views, certain violations of the rules of true projection are recognized as good practice because they add to the clearness of the drawing. For example, if a front view shows a hexagonal bolthead "across corners," the theoretical projection of the side view would be "across flats"; but in a working drawing, boltheads are drawn across corners in both views to show better the shape and the space needed. As another example, the slots of screw heads are always drawn at an angle of 45° in all views so that the closely spaced lines will not be confused with horizontal and vertical center lines.

Lugs or parts cast on for holding purposes and to be machined off are shown by "adjacent-part" lines (Figs. 56 and 57, Chap. 2). If such parts are in section, the section lines are dashed. "Alternate-position" lines (Figs. 56 and 57, Chap. 2) are used to indicate the limiting positions of moving parts and to show adjacent parts that aid in locating the position or use of a piece. Note in Figs. 56 and 57, Chap. 2, that the symbol of a long dash and two short dashes, alternating, is used to indicate adjacent parts, alternate positions, and also repeated features.

**FIG. 38.**   Aligned view. This is to avoid drawing in the foreshortened position.

**FIG. 39.**   Developed view. This part is drawn as though it were straightened out in one plane.

## 21. Aligned and Developed Views.

Pieces that have elements at an angle to one another, as the lever of Fig. 38, may be shown straightened out or aligned in one view. Similarly, bent pieces of the type of Fig. 39 should have one view made as a developed view of the *blank* to be punched and formed. Extra metal must be allowed for bends.

## 22. Half Views.

When space is limited, it is allowable to make the top or side view of a symmetrical piece as a half view. If the front is an exterior view, the *front* half of the top or side view would be used, as in Fig. 40; but if the front view is a sectional view, the *rear* half would be used, as in Fig. 41. Figure 42 shows another space-saving combination of a half view with a half section. Examples of half views occur in Probs. 44 to 46, Chap. 24.

## 23. Conventional Practices.

One statement can be made with the force of a rule: *If anything in clearness can be gained by violating a*

**FIG. 40.** Half view. The front half is drawn when the mating view is external.

**FIG. 41.** Half view. The rear half is drawn when the mating view is a full section.

**FIG. 42.** Half view. The rear half is drawn when the mating view is a half section.

*principle of projection, violate it.* This applies to full as well as sectional views. Permissible violations are not readily apparent to the reader, since when they occur the actual conditions are described in a better and usually simpler form than if they were shown in true projection.

However, some care and judgment must be used in applying conventional treatments. Persons trained in their use ordinarily understand their meaning, but workmen often do not. Some unusual convention may be more confusing than helpful, and result in greater ambiguity than true projection.

There are occasions when the true lines of intersection are of no value as aids in reading and should be ignored. Some typical examples are shown in Fig. 43. It must be noted, however, that in certain cases that are similar but where there is a major difference in line position when the true projection is given, as compared with conventional treatment, the true line of intersection should be shown. Compare the treatment of the similar objects in Figs. 43 and 44. It would not be good practice to conventionalize the intersections on the objects in Fig. 44 because the difference between true projection and the convention is too great. To develop your judgment of the use of conventional intersections, carefully study Figs. 43 and 44 and observe the difference between true projection and conventional treatment in each case.

### 24. Fillets and Rounds.

In designing a casting, never leave sharp internal angles because of the liability of fracture at those points. The radius of the fillet depends on the thickness of the metal and other design conditions. When not dimensioned, it is left to the patternmaker. External angles may be rounded for appearance or

**FIG. 43.**   Conventional intersections. These are used when there is a small difference as compared with true projection.

**FIG. 44.**   True intersections. These are used when there is a major difference as compared with conventional treatment.

comfort, with radii ranging from enough merely to remove the sharp edges to an amount nearly equal to the thickness of the piece. An edge made by the intersection of two unfinished surfaces of a casting should always be "broken" by a very small round. A sharp corner on a drawing indicates that one or both of the intersecting surfaces are machined. Small fillets, rounds, and "runouts" are best put in freehand, both in pencil and ink. Runouts, or "die-outs," as they are sometimes called, are conventional indications of filleted intersections where, theoretically, there would be no line because there

is no abrupt change in direction. Figure 45 shows some conventional representations of fillets and rounds with runouts of arms and ribs intersecting other surfaces.

## 25. Conventional Breaks.

In making the detail of a long bar or piece with a uniform cross section, it is rarely necessary to draw its whole length. It may be shown to a larger and therefore better scale by breaking out a piece, moving the ends together, and giving the true length by a dimension, as in Fig. 46. The shape of the cross

**FIG. 45.** Conventional fillets, rounds, and runouts. These are conventionalized intersections.

**FIG. 46.** Broken view with rotated sections. A broken view saves drawing the whole length of the part.

section is indicated by a rotated section or more often by a semipictorial break line, as in Fig. 47.

## 26. Conventional Symbols.

Engineers and draftsmen use conventional representation to indicate many details, such as screw threads, springs, pipe fittings, and electrical apparatus. These have been standardized by the ANSI whose code for materials in section has already been referred to in Sec. 11.

The symbol of two crossed diagonals is used for two distinct purposes: (1) to indicate on a shaft the

**FIG. 47.** Conventional breaks.

**FIG. 48.** Various symbols.

position of finish for a bearing, and (2) to indicate that a certain surface (usually parallel to the picture plane) is flat. These two uses are not apt to be confused (Fig. 48).

Because screw threads recur constantly, the designation for them is one of the most important items under conventional symbols. Up to the time of official standardization by the ANSI, there were a dozen different thread symbols in use. Now a regular symbol and a simplified one have been adopted for American drawings, and both are understood internationally. The symbols for indicating thread on bolts, screws, and tapped holes will be given in a later chapter.

The conventional symbols mentioned are used principally on machine drawings. Architectural drawing, because of the small scales employed, uses many conventional symbols, and topographic drawing is made up almost entirely of symbols. Electrical *diagrams* are completely symbolic.

## PROBLEMS

### Group 1. Single Pieces.

The following problems can be used for practice in shape description only or, by adding dimensions, in making working drawings.

**1.** Draw the top view, and change the front and side views to sectional views as indicated.

**2.** Draw the top view, and make the front and two side views in section on cutting planes as indicated. Scale to suit.

**3 to 5.** Given the side view, draw full front and side views in section. Scale to suit.

**6 and 7.** Change the right-side view to a full section.

**8 and 9.** Change the right-side view to a full section.

**PROB. 1.** Section study.

**PROB. 2.** Section study.

**PROB. 3.** Flanged wheel.

**PROB. 4.** Step pulley.

**PROB. 5.** Double-flanged wheel.

**PROB. 6.** Cap.

**PROB. 7.** Flanged cap.

**PROB. 8.** Pump-rod guide.

**PROB. 9.** Face plate.

**10.** Change the right-side view to a full section.
**11.** Change the right-side view to a sectional view as indicated.

**12 and 13.** Change the front view to a full section.
**14 and 15.** Change the front view to a full section.

**PROB. 10.** Ribbed support.

**PROB. 12.** Filler block.

**PROB. 11.** Housing cover.

**PROB. 13.** Filler block.

**PROB. 14.** Bumper body.

**PROB. 15.** V-belt pulley.

**16 and 17.** Select views that will best describe the piece.

**PROB. 16.** End plate.

**PROB. 17.** Piston cap.

**19.** Draw the top view as illustrated and the front view in half section on *A-A*.

**PROB. 19.** Brake-rod bracket.

**18.** Draw the top view as shown in the illustration and the front view as a full section.

**PROB. 18.** Pump flange.

**20.** Draw the top view as shown, and change the front and side views to sections as indicated.

**PROB. 20.** Bolted anchor block.

**21.** Draw the top view and sectional view (or views) to best describe the object.

**PROB. 21.** Column collar.

**22.** Draw the top view and front view in section.

**PROB. 22.** Stem support.

**23.** Turn the object through 90°, and draw the given front view as the new top view; then make the new front view as section *B-B* and auxiliary section *A-A*. Refer to Sec. 9, for instructions on making auxiliary sections. See also Chap. 7 for the method of projection for the auxiliary section. Note in this case that the new top view, front-view section *B-B*, and auxiliary section *A-A* will completely describe the object. However, if desired, the side view may also be drawn, as shown or as an aligned view (described in Sec. 21).

**23***A*. As an alternate for Prob. 23, draw views as follows: with the object in the position shown, draw the front view as shown, draw the left-side view as section *B-B*, and draw the new top view as an aligned view.

**PROB. 23.** Compound bell-crank.

**24 and 25.** Draw the views and add the sectional views indicated.

**26 and 27.** Draw a view and sectional view to best describe the piece.

**PROB. 24.** Spool base.

**PROB. 26.** Actuator link.

**PROB. 25.** Saddle collar.

**PROB. 27.** Vibrator-drive bearing support.

**PROB. 28.** Cover and valve body.

**28.** Draw the top view and necessary sectional view (or views) to best describe the object.

**29.** Draw three views, making the side view as a section on *B-B*.

**29***A*. Draw three views, making the top view as a half section on *A-A*.

**30.** Select views that will best describe the piece.

**PROB. 30.** Spindle support.

**PROB. 29.** Crosshead.

**31.** Top view and front view in section, of rigging yoke, Prob. 77, Chap. 5.

**32.** Three views with front view in section, of missile gyro support, Prob. 75, Chap. 5.

**33.** Three views with front view in section, of end frame for engine starter, Prob. 113, Chap. 5.

**34.** Three views with front view in section, of master brake cylinder, Prob. 112, Chap. 5.

## Group 2. Assemblies.

**35.** Choose views and sectional views that will best describe the electric-motor support, Prob. 111, Chap. 5.

**36.** Choose views and sectional views that will best describe the jet-engine hinge plate, Prob. 120, Chap. 5.

**37.** Two views with sectional treatment that will best describe the anchor plate, Prob. 64, Chap. 7.

**38.** Draw a half end view and a longitudinal view as full section.

**39.** Draw the top view as shown in the figure and new front view in section. Show the shape (right section) of the link with rotated or removed section. The assembly comprises a cast-steel link, two bronze bushings, steel toggle pin, steel collar, steel taper pin, and part of the cast-steel supporting lug.

**PROB. 38.**    Push rod cap and seal body.

**PROB. 39.**    Link assembly.

**40.** Draw the front view and longitudinal section. The assembly comprises a cast-iron base, a bronze bushing, a bronze disk, and two steel dowel pins.

**PROB. 40.**   Step bearing.

**41.** Draw two half end views and a longitudinal section. The assembly consists of cast-iron body, two bronze bushings, steel shaft, cast-iron pulley, and steel taper pin.

**PROB. 41.**   Pulley-bracket assembly.

**42.** Make an assembly drawing in section. The bracket is cast iron, the wheel is cast steel, the bushing is bronze, and the pin and taper pin are steel. Scale: full size.

**42A.** Make a drawing of the bracket with one view in section. Material is cast iron. Scale: full size.

**42B.** Make a drawing of the wheel with one view in section. Material is cast steel. Scale: full size.

**43.** Make an assembly drawing in section. The assembly comprises two cast-iron brackets, two bronze bushings, steel shaft, cast-steel roller, and cast-iron base. The bushings are pressed into the roller, and the shaft is drilled for lubrication. Scale: full size.

**44A.** Make a drawing of the roller and bushing assembly that gives one view in section. See Prob. 43 for materials. Scale: full size.

**45.** Sectional assembly of sealed shaft unit, Prob. 34 of chapter 24.

**46.** Sectional assembly of crane hook, Prob. 35, chapter 24.

**47.** Sectional assembly of caster, Prob. 36, chapter 24.

**48.** Sectional assembly of antivibration mount, Prob. 53, of chapter 24.

**49.** Sectional assembly of pump valve, Prob. 55, of chapter 24.

**50.** Sectional assembly of boring-bar holder, Prob. 58, chapter 24.

**51.** Sectional assembly of conveyor link unit, Prob. 77, chapter 24.

**52.** Sectional assembly of pressure cell piston, Prob. 62, chapter 24.

**53.** Sectional assembly of pivot nut and adjustable screw, Prob. 63, chapter 24.

**PROB. 42.** Sliding-door guide.

**PROB. 43.** Bell-roller support.

# PICTORIAL DRAWING AND SKETCHING

Pictorial Methods Are Used Either as a Basic
Form of Shape Description or as a Supplement
to Orthographic Depiction

This Chapter Offers:
Pictorial Methods Classified • Isometric,
Dimetric, Trimetric, and Perspective Drawing and
Sketching • Layout of Circles, Arcs, Curves •
Object Orientation • Starting Planes and Picture
Planes • Projection from Orthographic Views •
Graticulation • Pictorial Sketching • Stages in
Sketching • Light and Shade • Rendering •
Problems

## 1. Pictorial Method—Comparison with Orthographic.

In discussing the theory of projection in Chap. 5, we noted that in perspective projection the object is represented as it appears to the eye. However, its lines cannot be measured directly for accurate description of the object; and in orthographic projection the object is shown, in two or more views, as it really is in form and dimensions but interpretation requires experience to visualize the object from the views. To provide a system of drawing that represents the object pictorially and in such a way that its principal lines can be measured directly, several forms of one-plane conventional or projectional picture methods have been devised in which the third dimension is taken care of either by turning the object so that its three dimensions are visible or by employing oblique projection. A knowledge of these picture methods and of perspective projection is extremely desirable as they can all be used to great advantage.

Mechanical or structural details not clear in orthographic views can be drawn pictorially or illustrated by supplementary pictorial views. Pictorial views are used advantageously in technical illustrations, Patent Office drawings, layouts, piping plans, and the like. Pictorial methods are useful also in making freehand sketches, and this is one of the most important reasons for learning them.

## 2. Pictorial Methods.

There are three main divisions of pictorial drawing: (1) axonometric, with its divisions into trimetric, dimetric, and isometric; (2) oblique, with several variations; and (3) perspective. These methods are illustrated in Fig. 1.

The trimetric form gives an effect more pleasing to the eye than the other axonometric and oblique methods and allows almost unlimited freedom in orienting the object, but is difficult to draw. With the dimetric method the result is less pleasing and there is less freedom in orienting the object, but execution is easier than with trimetric. The isometric method gives a result less pleasing than dimetric or trimetric, but it is easier to draw and has the distinct advantage that it is easier to dimension. The oblique method is used principally for objects with circular or curved features only on one face or on parallel faces, and for objects of this type the oblique is easy to draw and dimension. Perspective drawing gives a result most pleasing to the eye, but it is of limited usefulness because many lines are unequally foreshortened; isometric and oblique are the forms most commonly used.

## 3. Axonometric Projection.

Axonometric projection is theoretically orthographic projection in which only one plane is used, the object being turned so that three faces show. Imagine a transparent vertical plane with a cube behind it, one face of the cube being parallel to the plane. The projection on the plane, that is, the front view of the cube, will be a square (Fig. 2). Rotate the cube about a vertical axis through any angle less than

TRIMETRIC        DIMETRIC        ISOMETRIC        OBLIQUE        PERSPECTIVE

**FIG. 1.** Pictorial methods.

90°, and the front view will now show two faces, both foreshortened (Fig. 3). From this position, tilt the cube forward (rotation axis perpendicular to profile) any amount less than 90°. Three faces will now be visible on the front view (Fig. 4). There can be an infinite number of axonometric positions, depending upon the angles through which the cube is rotated. Only a few of these positions are ever used for drawing. The simplest is the isometric (equal-measure) position, in which the three faces are foreshortened equally.

**FIG. 2.** The object face is parallel to the picture plane. One face only is seen in the front view.

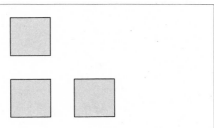

**FIG. 3.** The object is rotated about a vertical axis. Two faces are seen.

**FIG. 4.** The object is rotated about both a vertical and a profile axis. Three faces are seen.

## 4. Isometric Projection.

If the cube in Fig. 5a is rotated about a vertical axis through 45°, as shown at (b), and then tilted forward, as at (c), until the edge $RU$ is foreshortened equally with $RS$ and $RT$, the front view of the cube in this position is said to be an "isometric projection." (The cube has been tilted forward until the body diagonal through $R$ is perpendicular to the front plane. This makes the top face slope approximately 35°16′.[1]) The projections of the three mutu-

[1]The only difference between rotation and auxiliary projection is that in the former the object is moved and in the latter the plane is moved or the observer is considered to have changed his viewing position. Thus an auxiliary view on a plane perpendicular to a body diagonal of the cube in position (b) would be an isometric projection, as illustrated by the dotted view.

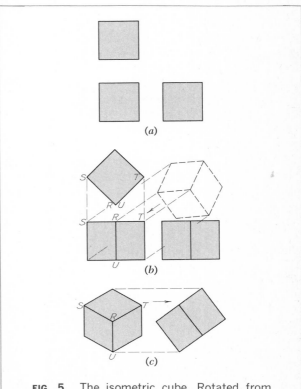

**FIG. 5.** The isometric cube. Rotated from position (a) to (b) then to (c), the three perpendicular edges are now equally foreshortened.

ally perpendicular edges *RS*, *RT*, and *RU* meeting at the front corner *R* make equal angles, 120°, with each other and are called "isometric axes." Since the projections of parallel lines are parallel, the projections of the other edges of the cube will be, respectively, parallel to these axes. Any line parallel to the edge of the cube, whose projection is thus parallel to an isometric axis, is called an "isometric line." The planes of the faces of the cube and all planes parallel to them are called "isometric planes."

The isometric axes *RS*, *RT*, and *RU* are all foreshortened equally because they are at the same angle to the picture plane.

## 5. Isometric Drawing.

In nearly all practical use of the isometric system, this foreshortening of the lines is disregarded, and *their full lengths are laid off on the axes*, as explained in Sec. 6. This gives a figure of exactly the same shape but larger in the proportion of 1.23 to 1, linear, or in optical effect $1.23^3$ to $1.00^3$ (Fig. 6). Except when drawn beside the same piece in orthographic projection, the effect of increased size is usually of no consequence, and since the advantage of measuring the lines directly is of great convenience, isometric drawing is used almost exclusively rather than isometric projection.

In isometric projection the isometric lines have been foreshortened to approximately $81/100$ of their length, and an isometric scale to this proportion can

be made graphically as shown in Fig. 7 if it is necessary to make an isometric projection by the method of isometric drawing.

## 6. To Make an Isometric Drawing.

If the object is rectangular (Fig. 8), start with a point

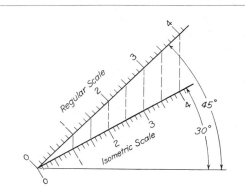

**FIG. 7.** To make an isometric scale.

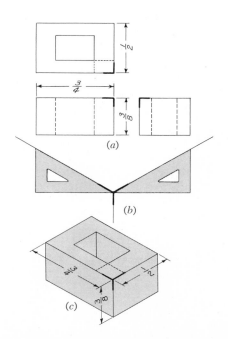

**FIG. 8.** Isometric axes, first position. The starting point is the upper front corner.

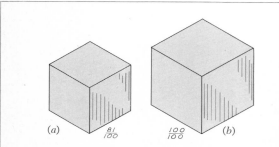

**FIG. 6.** (*a*) isometric projection; (*b*) isometric drawing.

representing a front corner, shown at (*a*) with heavy lines, and draw from it the three isometric axes 120° apart, one vertical (*b*), the other two with the 30° triangle. On these three lines measure the height, width, and depth of the object, as indicated at (*c*); through the points so determined draw lines parallel to the axes, completing the figure. When drawing in isometric, remember the direction of the three principal isometric planes. Hidden lines are omitted except when they are needed to describe the piece.

It is often convenient to build up an isometric drawing from the lower front corner, as illustrated in Fig. 9, starting from axes in what may be called the "second position." The location of the starting corner is again shown by heavy lines at (*a*), (*b*), and (*c*).

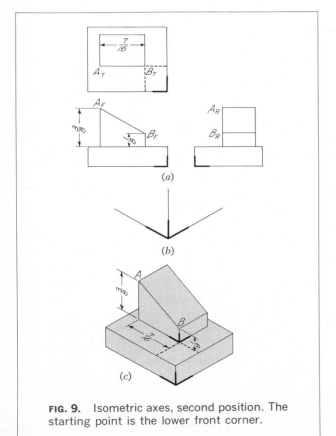

**FIG. 9.** Isometric axes, second position. The starting point is the lower front corner.

## 7. Nonisometric Lines.

Edges whose projections or drawings are not parallel to one of the isometric axes are called "nonisometric lines." The one important rule is that *measurements can be made only on the drawings of isometric lines;* conversely, measurements *cannot* be made on the drawings of *nonisometric* lines. For example, the diagonals of the face of a cube are nonisometric lines; although equal in length, their isometric drawings will not be at all of equal length on the isometric drawing of the cube. Compare the length of the diagonals on the cube in Fig. 6. Since a nonisometric line does not appear in the isometric drawing in its true length, the isometric view of each end of the line must be located and the isometric view of the line found by joining these two points. In Fig. 9*a* and *c*, line *AB* is a nonisometric edge whose true length cannot be measured on the isometric drawing. However, the vertical distances above the base to points *A* and *B* are parallel to the vertical isometric axis. These lines can, therefore, be laid off, as shown at (*c*), to give the isometric view of line *AB*.

## 8. Nonisometric Lines: Boxing Method.

When an object contains many nonisometric lines, it is drawn by the "boxing method" or the "offset method." When the boxing method is used, the object is enclosed in a rectangular box, which is drawn around it in orthographic projection. The box is then drawn in isometric and the object located in it by its points of contact, as in Figs. 10 and 12. It should be noted that the isometric views of lines that are parallel on the object are parallel. This knowledge can often be used to save a large amount of construction, as well as to test for accuracy. Figure 10 might be drawn by putting the top face into isometric and drawing vertical lines equal in length to the edges downward from each corner. It is not always necessary to enclose the whole object in a rectangular "crate." The pyramid (Fig. 11) would have its base enclosed in a rectangle and the apex located by erecting a vertical axis from the center.

The object shown in Fig. 12 is composed almost entirely of nonisometric lines. In such cases the isometric drawing cannot be made without first

**FIG. 10.** Box construction. Points on the orthographic box are transferred to the pictorial box. Identical scale must be used.

**FIG. 11.** Semibox construction. Points on the base are transferred by boxing. The altitude is located by a vertical from the base center. Identical scale must be used.

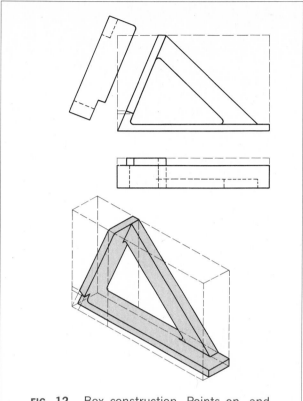

**FIG. 12.** Box construction. Points on, and offsets from, the orthographic box are transferred to the pictorial box. Identical scale must be used.

making the orthographic views necessary for boxing. In general, the boxing method is adapted to objects that have the nonisometric lines in isometric planes.

## 9. Nonisometric Lines: Offset Method.

When an object is made up of planes at different angles, it is better to locate the ends of the edges by the offset method rather than by boxing. When the offset method is used, perpendiculars are ex-tended from each point to an isometric reference plane. These perpendiculars, which are isometric lines, are located on the drawing by isometric coordinates, the dimensions being taken from the orthographic views. In Fig. 13, line *AB* is used as a base line and measurements are made from it as shown, first to locate points on the base; then verticals from these points locate *e*, *f*, and *g*. Figure 14 is another example of offset construction. Here a vertical plane is used as a reference plane. Note that, as in Fig. 13, the *base* of the offset is located first; then the offset distance is measured.

## 10. Angles in Isometric.

The three isometric axes, referred back to the iso-

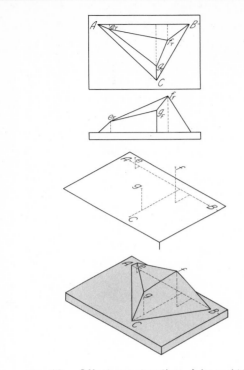

**FIG. 13.** Offset construction. A base (*AB*) is assumed; then all points are located from it.

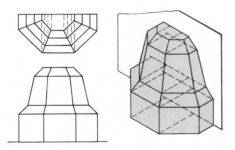

**FIG. 14.** Offset construction. All points are located on a plane or by offsets from the plane. Identical scale must be used.

metric cube, are mutually perpendicular but in an isometric drawing appear at 120° to each other. For this reason, angles specified in degrees do not appear in their true size on an isometric drawing and must be laid off by coordinates that will be parallel

to the isometric axes. Thus if an orthographic drawing has edges specified by angular dimensions, as in Fig. 15A, *a view to the same scale as the isometric drawing* is made as at (*B*); from this view the coordinate dimensions *a*, *b*, and *c* are transferred with dividers or scale to the isometric drawing.

## 11. Curves in Isometric.

For the reasons given in Secs. 7 and 10, a circle or any other curve will not show in its true shape when drawn in isometric. A circle on any isometric plane will be an ellipse, and a curve will be shown as the isometric projection of the true curve.

Any curve can be drawn by plotting points on it from isometric reference lines (coordinates) that are parallel to the isometric axes, as shown in Fig. 16.

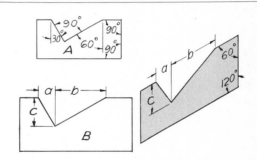

**FIG. 15.** Angles in isometric. These must be laid out by offsets from an orthographic view to the same scale.

**FIG. 16.** Curves in isometric. Points are transferred from the orthographic view to the pictorial by offsets. Identical scale must be used.

A circle plotted in this way is shown in Fig. 17. Note that in both these figures coordinates $a$ and $b$ are parallel to the isometric axes and the coordinate distances must be obtained from an orthographic view drawn to the same scale as the isometric.

## 12. Isometric Circles.

Circles occur so frequently that they are usually drawn by a four-centered approximation, which is sufficiently accurate for ordinary work. Geometrically, the center for any arc tangent to a straight line lies on a *perpendicular from the point of tangency* (Fig. 18a). In isometric, if perpendiculars are drawn from the middle point of each side of the circumscribing square, the intersections of these perpendiculars will be centers for arcs tangent to two sides (b). Two of these intersections will evidently fall at the corners $A$ and $C$ of the isometric square, as the perpendiculars are altitudes of equilateral triangles. Thus the construction at (b) to (d) can be made by simply drawing 60° lines (horizontals also at $C$ and $D$) from the corners $A$ and $C$ and then drawing arcs with radii $R$ and $R_1$, as shown.

Figure 19 shows the method of locating and laying out a hole in isometric from the given orthographic views. First locate and then draw the center lines for the hole by laying out the distances $X$ and $Y$, as shown. On these lines construct an isometric square with sides equal to the diameter of the hole by laying out the radius $R$ in each direction from the intersection of the center lines. Then use the four-center method, as shown in Fig. 18. Should the piece be thin enough, a portion of the backside of the hole will be visible. To determine this, drop the thickness $T$ back on an isometric line and swing the large radius $R_1$ of the isometric circle with this point as center. If the arc thus drawn comes within the boundary of the isometric circle, that portion of the back will be visible. In extra-thin pieces, portions of the small arcs $R_2$ might be visible. This would be determined the same way.

If a true ellipse is plotted by the method of Sec. 11 in the same square, it will be a little longer and narrower and of much more pleasing shape than this four-center approximation, but in most drawings the difference is not sufficient to warrant the extra

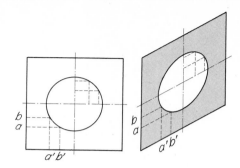

**FIG. 17.** Isometric circle, points plotted. Points are transferred from the orthographic view to the pictorial by offsets. Identical scale must be used.

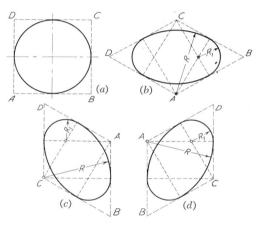

**FIG. 18.** Isometric circles, four-centered method. The ellipse is approximated by circle arcs.

**FIG. 19.** Locating and laying out a hole in isometric. Locate the center, draw the enclosing isometric square, and then draw the circle by the method of Fig. 18.

**FIG. 20.** Isometric quarter circles. The radius center lies on perpendiculars from tangent points that are the radius distance from the corner.

**FIG. 21.** Isometric with reversed axes. The bottom and two sides are shown. Construction methods are the same as for regular position.

expenditure of time required in execution.

The isometric drawing of a *sphere* is a circle with its diameter equal to the long axis of the ellipse that is inscribed in the isometric square of a great circle of the sphere. It would thus be 1.23/1.00 of the actual diameter (the isometric *projection* of a sphere would be a circle of the actual diameter of the sphere).

### 13. Isometric Circle Arcs.

To draw any circle arc, draw the isometric square of its diameter in the plane of its face, with as much of the four-center construction as is necessary to find centers for the part of the circle needed, as illustrated in Fig. 20. The arc occurring most frequently is the quarter circle. Note that in illustrations (*d*) and (*e*) only two construction lines are needed to find the center of a quarter circle in an isometric

plane. Measure the true radius $R$ of the circle from the corner on the two isometric lines as shown, and draw *actual* perpendiculars from these points. Their intersection will be the required center for radius $R_1$ or $R_2$ of the isometric quadrant. (*f*) illustrates the construction for the two vertical isometric planes.

### 14. Reversed Isometric.

It is often desirable to show the lower face of an object by tilting it *back* instead of *forward*, thus reversing the usual position so as to show the underside. The construction is the same as when the top is shown, but the directions of the principal isometric planes must be kept clearly in mind. Figure 21 shows the reference cube and the position of the axes, as well as the application of reversed-isometric construction to circle arcs. A practical use of this construction is in the representation of such architectural features as are naturally viewed from below. Figure 22 is an example of this use.

Sometimes a piece is shown to better advantage with the main axis horizontal, as in Fig. 23.

## 15. Isometric Sections.

Isometric drawings are, from their pictorial nature, usually outside views, but sometimes a sectional view (see Chap. 8) is used to good advantage to show a detail of shape or interior construction. The cutting planes are taken as isometric planes, and the section lining is done in the direction that gives the best effect; this is, in almost all cases, the direction of the long diagonal of a square drawn on the surface. As a general rule, a half section is made by outlining the figure in full and then cutting out the front quarter, as in Fig. 24; for a full section, the cut face is drawn first and then the part of the object behind it is added (Fig. 25).

## 16. Dimetric Projection.

The reference cube can be rotated into any number of positions in which two edges are equally foreshortened, and the direction of axes and ratio of foreshortening for any one of these positions might be taken as the basis for a system of dimetric draw-

FIG. 23. Isometric with the main axis horizontal. It is used when the object looks more natural in this position.

FIG. 24. Isometric half section. One-fourth of the object is removed to reveal interior construction.

FIG. 25. Isometric full section. Half of object is removed to reveal object shape.

FIG. 22. An architectural detail on reversed axes.

ing. A simple dimetric position is one with the ratios 1 to 1 to ½. In this position the tangents of the angles are ⅛ and ⅞, making the angles approximately 7° and 41°. Figure 26 shows a drawing in this system. Dimetric is seldom used because of the difficulty of drawing circles in this projection.

## 17. Trimetric Projection.

Any position in which all three axes are unequally foreshortened is called "trimetric." Compared with isometric and dimetric, distortion is reduced in trimetric projection, and even this effect can be lessened with some positions. However, because it is slower to execute than isometric or dimetric, it is seldom used except when done by projection. Axonometric projection from orthographic views is given in Sec. 26.

## 18. Oblique Projection.

When the projectors make an angle other than 90° with the picture plane, the resulting projection is called "oblique." The name "cavalier projection" is given to the special and most used type of oblique projection in which the projectors make an angle of 45° with the plane of projection. Cavalier projection is often called by the general name "oblique projection," or "oblique drawing." The principle is as follows: Imagine a vertical plane with a rectangular block behind it, having its long edges parallel

**FIG. 26.** Dimetric drawing. It is used principally for rectangular objects.

to the plane. Assume a system of parallel projecting lines in any direction making an angle of 45° with the picture plane (they could be parallel to any one of the elements of a 45° cone with its base in the picture plane). Then that face of the block which is parallel to the plane is projected in its true size, and the edges perpendicular to the plane are projected in their true length. Figure 27 illustrates the principle. The first panel shows the regular orthographic projection of a rectangular block with its front face in the frontal plane. An oblique projector

**FIG. 27.** Oblique projection. Projectors are at an oblique angle to the picture plane.

from the back corner $B$ is the hypotenuse of a 45°
right triangle of which $AB$ is one side and the projection of $AB$ on the plane is the other side. When this
triangle is horizontal, the projection on the plane
will be $AC$. If the triangle is rotated about $AB$ through
any angle $\beta$, $C$ will revolve to $C'$ and $A_F C_F'$ will be
the oblique projection of $AB$.

### 19. To Make an Oblique Drawing.

Oblique drawing is similar to isometric drawing in
that it has three axes that represent three mutually
perpendicular edges and upon which measurements
can be made. Two of the axes are always at right
angles to each other, as they are in a plane parallel
to the picture plane. The third, or depth, axis may
be at any angle to the horizontal, 30° or 45° being
generally used (Fig. 28). Oblique drawing is thus
more flexible than isometric drawing. To draw a
rectangular object (Fig. 29) start with a point representing a front corner ($A$) and draw from it the three
oblique axes, one vertical, one horizontal, and one
at an angle. On these three axes measure the height,
width, and depth of the object. In this case the width
is made up of the $2\frac{1}{2}$-in. distance and the $1\frac{5}{16}$-in.
radius. Locate the center of the arc, and draw it as
shown. The center for the arc of the hole in the
figure will be at the same point as the center for
the outside arc on the front face. The center for the
rear arc of the hole will be $1\frac{1}{8}$-in. rearward on a
depth-axis line through the front center.

### 20. Object Orientation for Oblique.

Any face parallel to the picture plane will evidently
be projected without distortion. In this, oblique projection has an advantage over isometric that is of
particular value in representing objects with circular
or irregular outline.

The *first rule* for oblique projection is to *place the
object with the irregular outline or contour parallel to
the picture plane.* Note in Fig. 30 the greater distortion at (*b*) and (*c*) than at (*a*).

One of the greatest disadvantages in the use of
isometric or oblique drawing is the effect of distortion produced by the lack of convergence in the
receding lines—a violation of perspective. In some

**FIG. 28.**   Various oblique positions: (*a*) up, to
the right at 30°; (*b*) up, to the right at 45°;
(*c*) up, to the left at 45°; (*d*) down, to the
right at 30°; (*e*) down, to the left at 30°.

cases, particularly with large objects, this becomes
so painful as practically to preclude the use of these
methods. This is perhaps even more noticeable in
oblique than in isometric and of course increases
with the length of the depth dimension.

Hence the *second rule: preferably, the longest dimension should be parallel to the picture plane.* In Fig.
31, (*a*) is preferable to (*b*).

In case of conflict between these two rules, *the
first always takes precedence,* as the advantage of
having the irregular face without distortion is greater
than that gained by the second rule, as illustrated
in Fig. 32. The first rule should be given precedence

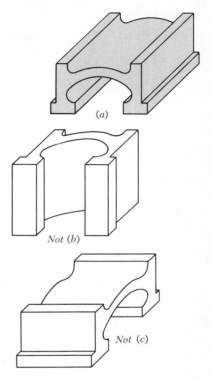

(a)

Not (b)

Not (c)

**FIG. 29.** Oblique drawing. The front face, parallel to the picture plane, is identical with an orthographic view.

**FIG. 30.** Illustration of the first rule. Note the distortion at (b) and (c).

(a)

Not (b)

**FIG. 31.** Illustration of the second rule. Note the exaggerated depth at (b).

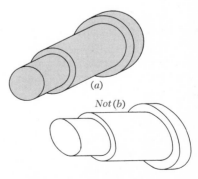

(a)

Not (b)

**FIG. 32.** Precedence of the first rule. (a), following the first rule, is easier to draw and also shows less distortion than (b).

even with shapes that are not irregular if, in the draftsman's judgment, the distortion can be lessened, as in Fig. 33, where (*b*) is perhaps preferable to (*a*).

## 21. Starting Plane.

Note that as long as the front of the object is in one plane parallel to the plane of projection, the front face of the oblique projection is *exactly the same as in the orthographic front view*. When the front is made up of more than one plane, take care to preserve the relationship between the planes by selecting one as the starting plane and working from it. In a piece such as the link in Fig. 34, the front bosses can be imagined as cut off on the plane *A-A*, and the front view, that is, the section on *A-A*, drawn as the front of the oblique projection. Then lay off depth axes through the centers *C* and *D*, the distances, for example, *CE* behind and *CF* in front of the plane *A-A*.

When an object has no face perpendicular to its base, it can be drawn in a similar way by cutting a right section and measuring offsets from it, as in Fig. 35. This offset method, previously illustrated in the isometric drawings in Figs. 13, 14, and 16, is a rapid and convenient way of drawing almost any figure, and it should be studied carefully.

## 22. Circles in Oblique.

When it is necessary to draw circles that lie on oblique faces, they can be drawn as circle arcs, with the compasses, on the same principle as the four-center isometric approximation shown in Fig. 18. In isometric it happens that *two of the four intersections of the perpendiculars from the middle points* of the containing square fall at the corner of the square, and advantage is taken of the fact. In oblique, the position of the corresponding points depends on the angle of the depth axis. Figure 36 shows three squares in oblique positions at different angles and the construction of their inscribed circles. The important point to remember is that the circle arcs *must* be tangent at the mid-points of the sides of the oblique square.

FIG. 33.  Choice of position. (*b*) is preferable to (*a*).

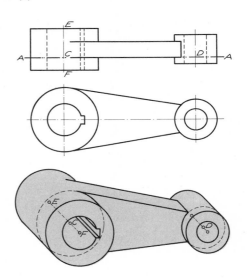

FIG. 34.  Offsets from reference plane. Distances forward and rearward are measured from the frontal plane.

## 23. Arcs in Oblique.

Circle arcs representing rounded corners, etc., are drawn in oblique by the same method given for isometric arcs in Sec. 13. The only difference is that the angle of the sides tangent to the arc will vary according to the angle of the depth axis chosen.

## 24. Cabinet Drawing.

This is a type of oblique projection in which the parallel projectors make an angle with the picture plane of such a value that distances measured parallel to the depth axis are reduced one-half that of

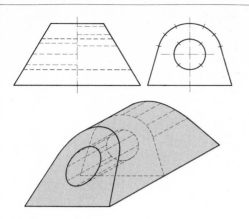

**FIG. 35.** Offsets from right section. Measurements forward and rearward are made from the frontal plane.

**FIG. 36.** Oblique circle construction. Note that the tangent points of arcs must be at the midpoints of the enclosing oblique square.

**FIG. 37.** Oblique (*a*) and cabinet drawing (*b*). Note less exaggeration of depth at (*b*).

cavalier projection. The appearance of excessive thickness that is so disagreeable in cavalier projection is entirely overcome in cabinet projection. The depth axis may be at any angle with the horizontal but is usually taken at 30° or 45°. The appearance of cavalier and cabinet drawing is shown in Fig. 37.

## 25. Other Forms.
Cabinet drawing is popular because of the easy ratio, but the effect is often too thin. Other oblique drawing ratios, such as 2 to 3 or 3 to 4, may be used with pleasing effect.

## 26. Axonometric Projection from Orthographic Views.
In making pictorial drawings of complicated parts, especially whenever curves are plotted, axonometric projection from orthographic views may give an advantage in speed and ease of drawing over axonometric projection made directly from the object. Any position—isometric, dimetric, or trimetric—may be used.

The three axes of an axonometric drawing are *three mutually perpendicular edges* in space. If the angle of rotation and the angle of tilt of the object are known or decided upon, the three axes for the pictorial drawing and the location of the orthographic views for projection to the pictorial can easily be found. Figure 38 illustrates the procedure. The three orthographic views of a cube are shown at (*g*). The three mutually perpendicular edges *OA*, *OB*, and *OC* will be foreshortened differently when the cube is rotated in space for some axonometric position, but the ends of the axes *A*, *B*, and *C* will always lie on the surface of a sphere whose radius is *OA* = *OB* = *OC*, as illustrated by (*G'*). At any particular angle of tilt of the cube, the axis ends *A* and *B* will describe an ellipse, as shown, if the cube is rotated about the axis *OC*. The axis *OC* will appear foreshortened at *oc'*. Thus for any particular position of the cube in space, representing some desired axonometric position, the axes can be located and their relative amounts of foreshortening found.

Moreover, if a face of the cube is rotated about a *frontal axis perpendicular to the axis that is at right angles to the face*, an orthographic view of the face, in projection with the axonometric view, will result. Thus, the top and right-side views may be located as at (*j*) and projected as at (*k*) to give the axonometric drawing.

The drawings at (*h*), (*j*), and (*k*) illustrate the practical use of the theory of rotation just described. The

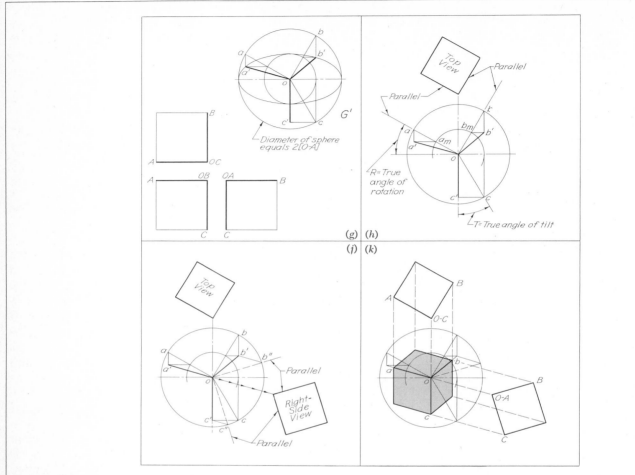

**FIG. 38.** Axonometric projection from orthographic views. (g) Construction to locate axes; (h) location of top view; (j) location of side view; (k) projection.

actual size of the sphere is unimportant, as it is used only to establish the direction of the axes. First, the desired angle of rotation $R$ and the angle of tilt $T$ are decided upon and laid out as at (h). The minor diameter for the ellipse upon which $A$ and $B$ will lie is found by projecting vertically from $c$ and drawing the circle as shown. $A$ and $B$ on the major-diameter circle of the ellipse will be at $a$ and $b$; on the minor-diameter circle, they will be at $a_m$ and $b_m$; and they are found in the axonometric position by projecting,

as in the concentric-circle ellipse method, to $a'$ and $b'$. The foreshortened position of $C$ is found by projecting horizontally across from $c$ to $c'$.

The top orthographic view of the cube (or object) will be parallel to $oa$ and $ob$, and projection from the orthographic view to the axonometric will be vertical (parallel to $aa'$ and $bb'$).

Projection from an orthographic right-side view would be as shown at (j). The right side of the cube, containing axes $OC$ and $OB$, is found by projecting

ORDER OF DRAWING:
1. DRAW CIRCLE OF ANY CONVENIENT RADIUS (IN THIS CASE WITH CENTER ON FRONT OBJECT CORNER EXTENDED).

2. LOCATE P-I (R=ANGLE OF ROTATION).

3. LOCATE P-2 (T=ANGLE OF TILT).

4. COMPLETE CONSTRUCTION, AS SHOWN, TO LOCATE LINES P-3 AND P 4.

5. LOCATE TOP VIEW WITH A RECEDING EDGE PARALLEL TO P-1.

6. LOCATE RIGHT-SIDE VIEW WITH HORIZONTAL EDGE PARALLEL TO P-3.

7. LOCATE FRONT VIEW WITH HORIZONTAL EDGE PARALLEL TO P-4.

8. DRAW AXONOMETRIC BY PROJECTION FROM ORTHOGRAPHIC VIEWS.

Top View

Right-Side View

Front View

**FIG. 39.** An axonometric drawing by projection from orthographic views. Note the order of drawing and refer back to Fig. 38 for details of construction.

from $b'$ and $c'$, parallel to $oa'$, to locate $b''$ and $c''$ on the circle representing the sphere. The sides of the cube (or object) are parallel to $ob''$ and $oc''$, as shown at ($j$). Projection from the right-side view to the axonometric view is in the direction of $oa'$, as indicated.

The axonometric drawing is shown projected at ($k$). The dashed lines indicate the actual projectors, and the light solid lines and circles show the necessary construction just described.

One advantage of this method is that the angle of rotation and tilt can be decided upon so that the object will be shown in the best position. Figure 39 is an example of an axonometric drawing made by projection from orthographic views. The curved faces are plotted by projecting points as shown.

## 27. Isometric Projection From Orthographic Views.

Isometric is, of course, a special type of axonometric projection in which all three axes are foreshortened equally. The work of finding the axes for isometric projection from orthographic views is reduced if the views are located by angle, as illustrated in Fig. 40.

## 28. Oblique Projection From Orthographic Views.

In oblique projection the projectors make some oblique angle with the picture plane. It should be noted that the actual angle of the projectors (with horizontal and frontal planes) is not critical, so that a variety of angles may therefore be used. The

**FIG. 40.** Isometric by projection. Top view turned 45° and side view turned 15° locates the views for projection to the isometric.

making of an oblique drawing by projection from orthographic views is simple, as illustrated by Fig. 41. The picture plane is located, and one face of the object is made coincident with the picture plane. The front view is located at a convenient place on the paper. The angle of the projectors in the top view may be assumed (in this case 45°) and projections made to the picture plane as shown. The angle of the projectors in the front view can than be assumed (in this case 30°). Projection from the front view at the assumed angle and vertically from the picture plane, as shown, will locate the necessary lines and points for the oblique view.

Reversed axes can be obtained by projecting downward from the front view. An axis to the left can be located by changing the direction of the projectors in the top view. Any desired oblique axes can be located by altering the angles (top and front) for the projectors.

### 29. Perspective Drawing.
Perspective drawing represents an object as it appears to an observer stationed at a particular posi-

tion relative to it. The object is seen as the figure resulting when visual rays from the eye to the object are cut by a picture plane. There is a difference between an artist's use of perspective and geometric perspective. The artist often disregards true perspective since he draws the object as he sees it through his creative imagination, while geometric perspective is projected instrumentally on a plane from views or measurements of the object represented. Projected geometric perspective is, theoretically, very similar to the optical system in photography.

In a technical way, perspective is used more in architecture and in illustration than in other fields, but every engineer will find it useful to know the principles of the subject.

### 30. Fundamental Concepts.
Imagine an observer standing on the sidewalk of a city street, as in Fig. 42, with the picture plane erected between him and the street scene ahead. Visual rays from his eye to the ends of lamppost $A$ intercept a distance $aa'$ on the picture plane. Similarly, rays from post $B$ intercept $bb'$, a smaller distance than $aa'$. This apparent diminution in the size of like objects as the distance from the objects to the eye increases agrees with our everyday experience and is the keynote of perspective drawing. It is evident from the figure that succeeding lampposts will intercept shorter distances on the picture plane than the preceding ones, and that a post at infinity would show only as a point $o$ at the level of the observer's eye.

In Fig. 43 the plane of the paper is the picture plane, and the intercepts $aa'$, $bb'$, etc., show as the heights of the respective lampposts as they diminish in their projected size and finally disappear at a point on the horizon. In a similar way the curbings and balustrade appear to converge at the same point $O$. Thus a system of parallel horizontal lines will vanish at a single point on the horizon, and all horizontal planes will vanish on the horizon. Verticals such as the lampposts and the edges of the buildings, being parallel to the picture plane, pierce the picture plane at an infinite distance and therefore show as vertical lines in the picture.

**FIG. 41.** Oblique by projection. Projection at an angle from the top view to the picture plane and directly from the side view gives the oblique view.

**FIG. 42.** Theory of perspective illustrated. Rays from the objects to the observer's eye intersect the picture plane.

**FIG. 43.** The perspective drawing. This is the image formed on the picture plane of Fig. 42.

## 31. Definitions and Nomenclature.

Figure 44 illustrates perspective theory and names the points, lines, and planes used. An observer in viewing an object selects his *station point* and thereby determines the *horizon plane*, as the horizontal plane is at eye level. This horizon plane is normally above the horizontal *ground plane* upon which the object is assumed to rest. The *picture plane* is usually located between the station point and the object being viewed and is ordinarily a vertical plane perpendicular to the horizontal projection of the line of sight to the object's center of interest. The *horizon line* is the intersection of the horizon plane and picture plane, and the *ground line* is the intersection of the ground plane and picture plane. The *axis of vision* is the line through the station point which is perpendicular to the picture plane. The piercing point of the axis of vision with the picture plane is the *center of vision*.

## 32. Selection of the Station Point.

In beginning a perspective drawing, take care in selecting the station point, as an indiscriminate choice may result in a distorted drawing. If the station point is placed to one side of the drawing, the same effect is obtained as when a theater screen is viewed from a position close to the front and well off to one side: heights are seen properly but not horizontal distances. Therefore, *the center of vision should be somewhere near the picture's center of interest.*

Wide angles of view result in a violent convergence of horizontal lines and so should be avoided. The angle of view is the included angle $\theta$ between the widest visual rays (Fig. 45). Figure 46 shows the difference in perspective foreshortening for different lateral angles of view. In general, an angle of about 20° gives the most natural picture.

The station point should be located at the point from which the object is seen to best advantage. For this reason, for large objects such as buildings, the station point is usually taken at a normal standing height of about 5 ft above the ground plane; for small objects, the best representation demands that the top, as well as the lateral surfaces, be seen,

**FIG. 44.** Perspective nomenclature.

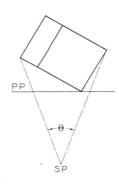

**FIG. 45.** Lateral angle of view.

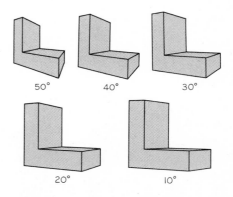

**FIG. 46.** Comparative lateral angles of view. Angles greater than 30° give an unpleasing perspective.

and the station point must be elevated accordingly. Figure 47 shows the angle of elevation Ω between the horizon plane and the extreme visual ray. By illustrating several different angles of elevation (Ω), Fig. 48 shows the effect of elevation of the station point. In general, the best picturization is obtained at an angle of about 20° to 30°.

Accordingly, *the visual rays to the object should be kept within a right-circular cone whose elements make an angle of not more than 15° with the cone axis* (total included angle of 30°).

In choosing the station point, see that its position is always offset to one side and also that it is offset vertically from the exact middle of the object, or a rather stiff and awkward perspective will result. Similarly, in locating the object with reference to the picture plane, avoid having the faces make identical angles with the picture plane, or the same stiffness will appear.

## 33. To Draw a Perspective.

Perspective projection is based on the theory that visual rays from the object to the eye pierce the picture plane and form an image of the object on the plane. Thus in Fig. 49, the image of line $YZ$ is formed by the piercing points $y$ and $z$ of the rays. Several projective methods may be used. The simplest method, basically, but the most laborious to draw is illustrated by the purely orthographic method of Fig. 50, in which the top and side views are drawn in orthographic. The picture plane (edge view) and the station point are located in each view. Assuming that the line $YZ$ in Fig. 49 is one edge of the L-shaped block in Fig. 50, visual rays from $Y$ and $Z$ will intersect the picture plane in the top view, thus locating the perspective of the points laterally. Similarly, the intersections of the rays in the side view give the perspective heights of $Y$ and $Z$. Projection from the top and side views of the picture plane gives the perspective of $YZ$, and a repetition of the process for the other lines will complete the drawing. Note that *any* point such as $Y$ or $Z$ can be located on the perspective, and thus the perspective is actually plotted, by projection, point after point.

**FIG. 47.** Elevation angle of view.

**FIG. 48.** Comparative elevation angles of view. Angles greater than 30° give an unpleasing perspective.

**FIG. 49.** Perspective of a line.

**FIG. 50.** Perspective drawing (orthographic method). Points are plotted from the intersection of rays with the picture plane.

## 34. The Use of Vanishing Points and Measuring Lines.

These facilitate the projections. Let it be required to make a perspective of the sliding block in Fig. 51. The edge view of the picture plane (plan view) is drawn (Fig. 52), and behind it the top view of the object is located and drawn. In this case, one side of the object is oriented at 30° to the picture plane in order to emphasize the L shape more than the end of the block. The station point is located a little to the left of center and far enough in front of the picture plane to give a good angle of view. The ground line is then drawn, and on it is placed the front view of the block from Fig. 51. The height of the station point is then decided—in this case, well

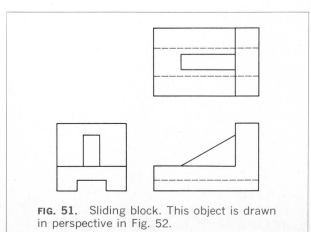

**FIG. 51.** Sliding block. This object is drawn in perspective in Fig. 52.

**FIG. 52.** Use of vanishing points and measuring lines. This saves time and extra construction, as compared with the orthographic method used in Fig. 50.

above the block so that the top surfaces will be seen—and the horizon line is drawn at the station-point height.

To avoid the labor of redrawing the top and front views in the positions just described, the views can be cut from the orthographic drawing, oriented in position, and fastened with tacks or tape.

The *vanishing point* for any horizontal line can be found by drawing a visual ray from the station point *parallel* to the horizontal line and finding the piercing point of this visual ray with the picture plane. Thus, in Fig. 52, the line *SP* to *R* is parallel to the edge *AB* of the object, and *R* is the piercing point. Point

*R* is then projected to the horizon line, locating *VR*, the vanishing point for *AB* and all edges parallel to *AB*. The vanishing point *VL* for *AC* and edges parallel to *AC* is found similarly, as shown.

In visualizing the location of a vanishing point, imagine that the edge, for example, *AB*, is moved to the right along the ground, still making the same angle with the picture plane; the intercept of *AB* will become less and less until, when *A* is in coincidence with *R*, the intercept will be zero. *R* then must be the top view of the vanishing point for all lines parallel to *AB*.

Point *A* lies in both the picture plane and the

ground plane and will therefore be shown in the perspective at *a*, on the ground line, and in direct projection with the top view. The perspective of *AB* is determined by drawing a line from *a* to *VR* (the perspective *direction* of *AB*) and then projecting the intercept *Z* (of the visual rays *SP* to *B*) to the line, thus locating *b*.

All lines behind the picture plane are foreshortened in the picture, and only those lying in the picture plane will appear in their true length. For this reason, *all measurements must be made in the picture plane*. Since *AD* is in the picture plane, it will show in its actual height as *ad*.

A *measuring line* will be needed for any verticals such as *BF* that do not lie in the picture plane. If a vertical is brought forward to the picture plane along some established line, the true height can be measured in the picture plane. If, in Fig. 52, *BF* is imagined as moved forward along *ab* until *b* is in coincidence with *a*, the true height can be measured vertically from *a*. This vertical line at *a* is then the measuring line for all heights in the vertical plane containing *a* and *b*. The height of *f* is measured from *a*, and from this height point, a vanishing line is drawn to *VR*; then from *Z* (the piercing point in the picture plane of the visual ray to *F*), *f* can be projected to the perspective.

The measuring line can also be thought of as the intersection of the picture plane with a vertical plane that contains the distance to be found. Thus *ad*, extended, is the measuring line for all heights in surface *ABFEGD*. The triangular rib in Fig. 52 is located by continuing surface *HJK* until it intersects the picture plane at *XY*, thereby establishing *xy* as the measuring line for all heights in *HJK*. In the figure, the height of *J* is measured on the measuring line *xy*, and *j* is found as described for *f*.

Note that heights can be measured with a scale on the measuring line or they can be projected from the front view, as indicated in Fig. 52.

*To Make a Perspective Drawing:*

1. Draw the top view (edge of the picture plane).
2. Orient the object relative to the picture plane so that the object will appear to advantage, and draw the top view of the object.

3. Select a station point that will best show the shape of the object.

4. Draw the horizon and ground line.

5. Find the top view of the vanishing points for the principal horizontal edges by drawing lines parallel to the edges, through the station point, and to the picture plane.

6. Project from the top views of the vanishing points to the horizon line, thus locating the vanishing points for the perspective.

7. Draw the visual rays from the station point to the corners of the object in the top view, locating the piercing point of each ray with the picture plane.

8. Start the picture, building from the ground up and from the nearest corner to the more distant ones.

## 35. Planes Parallel to the Picture Plane.

Objects with circles or other curves in a vertical plane can be oriented with their curved faces parallel to the picture plane. The curves will then appear in true shape. This method, often called "parallel perspective," is also suitable for interiors and for street vistas and similar scenes where considerable depth is to be represented.

The object in Fig. 53 has been placed so that the planes containing the circular contours are parallel to the picture plane. The horizontal edges parallel to the picture plane will appear horizontal in the picture and will have no vanishing point. Horizontals perpendicular to the picture plane are parallel to the axis of vision and will vanish at the center of vision *CV*. Except for architectural interiors, the station point is usually located above the object and either to the right or left, yet not so far in any direction as to cause unpleasant distortion. For convenience, one face of the object is usually placed in the picture plane and is therefore not reduced in size in the perspective.

In Fig. 53, the end of the hub is in the picture plane; thus the center *o* is projected from *O* in the top view, and the circular edges are drawn in their true size. The center line *ox* is vanished from *o* to *CV*. To find the perspective of center line *MN*, a

**FIG. 53.** Planes parallel to the picture plane. Compare the position of this object with that of Fig. 52.

vertical plane is passed through $MN$ intersecting the picture plane in measuring line $gh$. A horizontal line from $o$ intersecting $gh$ locates $m$, and $m$ vanished to $CV$ is the required line.

By using the two center lines from $o$ and $m$ as a framework, the remaining construction is simplified. A ray from the station point to $B$ pierces the picture plane at $J$, which, projected to $mn$, locates $b$. The horizontal line $bz$ is the center line of the front face of the nearer arm, and the intercept $IJ$ gives the perspective radius $ab$. The circular hole having a radius $CB$ has an intercept $PJ$, giving $cb$

as the perspective radius. The arc $qy$ has its center on $ox$ at $z$. On drawing the tangents $lq$ and $ky$, the face "$F$" is completed.

The remaining construction for the arms is exactly the same as that for "$F$." The centers are moved back on the center lines, and the radii are found from their corresponding intercepts on the picture plane.

### 36. Circles in Perspective.

The perspective of a circle is a circle only when its plane is parallel to the picture plane; the circle appears as a straight line when its plane is receding from the station point. In all other positions the circle projects as an ellipse whose major and minor diameters are not readily determinable. The major diameter of the ellipse will be at some odd angle except when a vertical circle has its center on the horizon plane; then the major diameter will be vertical. Also, when a horizontal circle has its center directly above, below, or on the center of vision, the major diameter will be horizontal. It should be noted that in all cases the center of the circle is not coincident with the center of the ellipse representing the circle and that concentric circles are not represented by concentric ellipses. The major and minor diameters of the ellipses for concentric circles are not even parallel except in special cases.

The perspective of a circle can be plotted point by point, but the most rapid solution is had by enclosing the circle in a square, as shown in Fig. 54, and plotting points at the tangent points and at the intersections of the diagonals. The eight points thus determined are usually sufficient to give an accurate curve. The square, with its diagonals, is first drawn in the perspective. From the intersection of the diagonals, the vertical and horizontal center lines of the circle are established; where these center lines cross the sides of the square are four points on the curve. In the orthographic view, the measurement $X$ is made, then laid out *in the picture plane* and vanished, crossing the diagonals at four additional points.

Note that the curve is tangent to the lines enclosing it and that the *direction* of the curve is established by these tangent lines; if the lines completing the circumscribing octagon are projected and drawn, the direction of the curve is established at eight points.

## 37. Graticulation.

The perspectives of irregular curves can be drawn by projecting a sufficient number of points to establish the curve, but if the curve is complicated, the method of graticulation may be used to advantage. A square grid is overlaid on the orthographic view as shown in Fig. 55; then the grid is drawn in per-

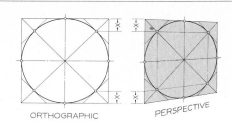

**FIG. 54.** Perspective of a circle. Points are plotted.

ORTHOGRAPHIC

PERSPECTIVE

**FIG. 55.** Graticulation. Points are plotted.

spective and the outlines of the curve are transferred by inspection from the orthographic view.

**FIG. 56.** Use of measuring points. An edge of the object is rotated into the picture plane.

## 38. Measuring Points.

It has been shown that all lines lying in the picture plane will be their own perspectives and can be scaled directly on the perspective drawing. The adaptation of this principle has an advantage in laying off a series of measurements, such as a row of pilasters, because it avoids a confusion of intercepts on the picture plane and the inaccuracies due to long projection lines.

In the measuring-points method, a surface, such as the wall between $A$ and $B$ in Fig. 56, is rotated into the picture plane for the purpose of making measurements, as shown at $AB'$. While in the picture plane, the entire surface can be laid out directly to the same scale as the top view; therefore, $ab'$ and other horizontal dimensions of the surface are established along the ground line as shown. The counterrotation of the wall to its actual position on the building and the necessary projections in the perspective are based on the principle that the rotation has been made about a vertical axis and that any point has traveled in a horizontal plane. By drawing,

as usual, a line parallel to $BB'$, from the station point to the picture plane, and then projecting to the horizon, the vanishing point $MR$ is found. This vanishing point is termed a *measuring point* and may be defined as the vanishing point for lines joining corresponding points of the actual and rotated positions of the face considered. The divisions on $ab'$ are therefore vanished to $MR$; where this construction intersects $ab$ (the perspective of $AB$), the lateral position of the pilasters, in the perspective, is determined. Heights are scaled on the vertical edge through $a$, as this edge lies in the picture plane. The perspective of the wall between $A$ and $B$ is completed by the regular methods previously described. For work on the end of the building, the end wall is rotated as indicated, measuring point $ML$ is found, and the projections are continued as described for the front wall.

Measuring points can be more readily located if the draftsman recognizes that the triangles $ABB'$ and $R\,O\,SP$ are similar. Therefore, a measuring point is as far from its corresponding vanishing point as the station point is from the picture plane, measuring the latter parallel to the face concerned. *MR* can then be found by measuring the distance from the station point to *R* and laying off *RO* equal to the measurement, or by swinging an arc, with *R* as center, from the station point to *O*, as shown. The measuring point *MR* is then projected from *O*.

### 39. Inclined Lines.

Any line that is not parallel or perpendicular to either the picture plane or the horizon plane is termed an inclined line. Any line may have a vertical plane passed through it, and if the vanishing line of the plane is found, a line in the plane will vanish at some point on the vanishing line of the plane. Vertical planes will vanish on vertical lines, just as horizontal planes vanish on a horizontal line, the horizon. In Fig. 57, the points *a* to *e* have all been found by regular methods previously described. The vanishing point of the horizontal *ab* is *VR*. The vertical line through *VR* is the vanishing line of the plane of *abc* and all planes *parallel to abc*. This vanishing line is intersected by the extension of *de* at *UR*, thereby determining the vanishing point for *de* and all edges *parallel to de*.

The vanishing point for inclined lines can also be located on the theory that the vanishing point for any line can be determined by moving the line until it appears as a point, while still retaining its original angle with the picture plane. The vanishing point of *de* can therefore be located by drawing a line through the station point parallel to *DE* and finding its piercing point with the picture plane. This is done by laying out *SP T* at the angle $\beta$ to *SP R* and erecting *RT* perpendicular to *SP R*. Then *RT* is the height of the vanishing point *UR* above *VR*.

If measuring points are used for the initial work on the perspective, it will be an advantage to recog-

**FIG. 57.** Vanishing point of inclined lines. This simplifies construction when there are many parallel inclined lines.

nize which one of the measuring points was used for determining horizontal measurements in the parallel vertical planes containing the inclined lines; at that measuring point, the angle $\beta$ is laid out, above or below the horizon depending upon whether the lines slope up or down as they go into the distance. Where this construction intersects the vanishing line for the vertical planes containing the inclined lines, the vanishing point is located.

## 40. Inclined Planes.

An inclined plane is any plane not parallel or perpendicular to either the picture plane or the horizon plane. The vanishing line for an inclined plane can be found by locating the vanishing points for any two systems of parallel lines in the inclined plane. To determine the vanishing line of plane *ABCD* in Fig. 58, the vanishing point *VL* of the horizontal edges *AD* and *BC* is one point, and the vanishing point *UR* for the inclined edges *AB* and *DC* gives a second point on the vanishing line *VL UR* for plane *ABCD*.

It is often necessary to draw the line of intersection of two inclined planes. The intersection will vanish at the point of intersection of the vanishing lines of both planes. The intersection *J* of the two vanishing lines of the roof planes in Fig. 58 is the vanishing point of the line of intersection of the two planes.

## 41. Pictorial Sketching.

The need for the engineer to be trained in freehand sketching was emphasized in Chap. 5, where the discussion referred particularly to sketching in orthographic projection. Before he can be said to have a command of the graphic language, his training in freehand drawing must include also acquiring the ability to sketch *pictorially* with skill and facility.

In designing and inventing, the first ideas come into the mind in pictorial form, and sketches made in this form preserve the ideas as visualized. From this record the preliminary orthographic design sketches are made. A pictorial sketch of an object or of some detail of construction can often be used

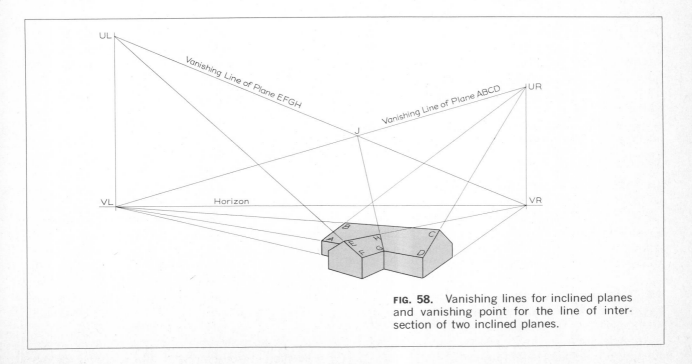

**FIG. 58.** Vanishing lines for inclined planes and vanishing point for the line of intersection of two inclined planes.

to explain it to a client or workman who cannot read the orthographic projection intelligently. One of the best ways of reading a working drawing that is difficult to understand is to start a pictorial sketch of it. Usually before the sketch is finished, the orthographic drawing is perfectly clear. Often a pictorial sketch can be made more quickly and serve as a better record than orthographic views of the same piece. A young engineer should not be deterred by any fancied lack of "artistic ability." An engineer's sketch is a record of information, not a work of art. The one requirement is *good proportion*.

## 42. Methods.
Although this is not a complete classification, there may be said to be three pictorial methods of sketching: axonometric, oblique, and perspective. The mechanical construction has been explained in detail

## 43. Prerequisites.
It should be clearly understood at the outset that pictorial sketching means the making of a pictorial drawing *freehand*. The same construction that is used for locating points and lines and for drawing circles and arcs with instruments will be used in pictorial sketching. From this standpoint, a knowledge of the constructions already given is necessary before attempting pictorial sketching. Note in Figs. 69 to 75 that the ellipses representing holes and rounded contours have, before being drawn, been boxed in with construction lines representing the enclosing square in exactly the same manner as for an instrument drawing.

## 44. Materials and Technique.
The same materials, pencil grades, etc., used for orthographic freehand drawing, described in Chap. 5, are employed for pictorial sketching. The directions given there for drawing straight lines, circles, and arcs will apply here also.

## 45. Pictorial Sketching: Choice of Type and Direction of View.
After a clear visualization of the object, the first step is to select the type of pictorial—axonometric, oblique, or perspective—to be used.

Isometric is the simplest axonometric position, and it will serve admirably for representing most objects. Although dimetric or trimetric may be definitely advantageous for an object with some feature that is obscured or misleading in isometric, it is best to try isometric first, especially if there is doubt that another form will be superior. This is principally because proportions are easier to judge in isometric.

Oblique forms (cavalier or cabinet) may be used to advantage for cylindrical objects or for objects with a number of circular features in parallel planes. Nevertheless, a true circle, representing a circular feature parallel to the picture plane in oblique, is much harder to sketch than an ellipse, representing the same feature in isometric, as the slightest deviation from a circle is evident, while the same deviation in an ellipse is unnoticed. Also, inherently, there is more distortion in the oblique than in axonometric forms. Therefore, especially for the sake of professional appearance, an axonometric form has the advantage.

Perspective is the best form for pictorial sketching because it is free from any distortion. A perspective is not much more difficult to sketch than an axonometric or an oblique, but attention must be paid to the convergence of the lines and to keeping good proportion. However, do not discard axonometric and oblique from consideration. As will be seen in Secs. 46 and 47, these forms can be handled as successfully as perspective by some flattening of the axes and by converging the lines properly.

Choose carefully the *direction* in which the object is to be viewed. There are many possibilities. The object may be turned so that any lateral face will be represented on the right or the left side of the pictorial. Orient the object so that the two *principal* faces will show to advantage. Use reversed axes if necessary. Do this by mentally visualizing and turning the object in every possible position in order to arrive at the best representation for all features. Be alert to see that some feature will not be hidden by a portion in front of it. The proper choice of direction is an important factor in pictorial sketching.

Fig. 59 illustrates these points. At (*a*) an object is sketched in isometric; all features are clearly rep-

**FIG. 59.**  Choice of axes and object position.

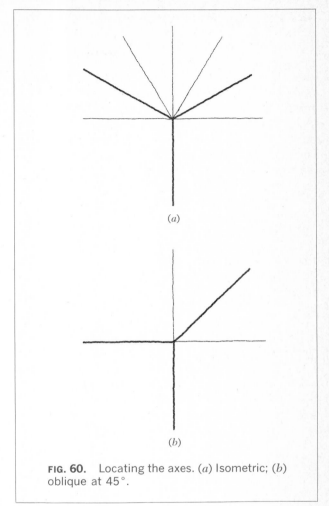

**FIG. 60.**  Locating the axes. (*a*) Isometric; (*b*) oblique at 45°.

resented and the object appears natural in this form. At (*b*) trimetric has been used so that the slot in the lower portion is not obscured by the horizontal middle portion. At (*c*) another trimetric position has been chosen to present the semicircular ear as the definite front face of the object. For further illustration of the possibilities, study the pictorial drawings in Chap. 5.

## 46. Sketching the Axes.

After the type of pictorial and the position of the object have been decided upon, the first step in making the sketch is to draw the axes.

*In isometric,* the three axes should be located as nearly as possible 120° from one another (one vertical and two at 30° with the horizontal). Because no triangles are used in sketching, the angles must be located by judgment. Figure 60*a* shows a satis-

factory method of judging the position of the lines. First draw a *light* horizontal and vertical and then divide both upper quadrants into thirds. The lines at the top of the lower thirds are then the two axes at 30° to the horizontal, and the third axis is the vertical. This method is simple and accurate because it is easy to estimate equal thirds of a quadrant.

*In dimetric,* the standard angles are 7° and 41°. For the 7° axis, again referring to Fig. 60*a*. draw the bisector of the 30° axis for isometric to get 15°, and then bisect again to get 7½°, which will be quite satisfactory and can be done fairly accurately. For the 41° axis, take the mid-line of the second 30°

section, which is 45°, and shade it a little to approximate 41°.

*In trimetric*, almost any combination representing three mutually perpendicular lines is possible. However, a pronounced distortion will occur if the two transverse axes are more than 30° from the horizontal. Read the cautions in Sec. 47.

*In oblique*, the common angles are 30° and 45° although theoretically any angle is possible. To prevent violent distortion, never make the depth axis greater than 45°. To locate a 45° axis, sketch a vertical and horizontal as in Fig. 60b, and then sketch the bisector of the quadrant (see oblique positions) where the axis is wanted. For an axis at 30°, proceed as in Fig. 60a for isometric.

*For perspective*, as explained in Sec. 32, take care to have a reasonable included angle of view; otherwise a violent convergence will occur. In order to prevent difficulty, first locate two vanishing points, as in Fig. 61, as widely separated as the paper will allow (attach extra paper with scotch tape if necessary). Then sketch the *bottom* edges of the object (the heavier lines in Fig. 61) and on these sketch a rectangular shape, as shown. It will be immediately evident that the arrangement is satisfactory or that the vanishing points must be moved *or* that the base lines must be altered. Remember that the two vanishing points *must* be at the same level, the horizon.

### 47. Sketching the Principal Lines.
Almost without exception, the first lines sketched should be those that box in the whole object, or at least its major portion. These first lines are all-important to the success of the sketch, for mistakes at the outset are difficult to correct later. Observe carefully the following three points:

1. *Verticals must be parallel to the vertical axis.* In Fig. 62 a cube is sketched correctly at (a). At (b) and (c) the same cube is sketched but at (b) the verticals converge downward and at (c), upward. It is evident that (b) and (c) do not look cubical at all, but like frustums of pyramids. This is proof that verticals *must be kept accurately vertical.* Be critical of the verticals throughout the construction. Accurate verticals add a stability and crispness not attained in any other way.

2. *Transverse lines must be parallel or converging.* In Fig. 63 a cube is sketched at (a) with the transverse lines (receding right and left edges) made accurately parallel. At (b) these lines are made to converge as they recede. Note that (b) looks more natural than (a) because of the effect of perspective foreshortening. The monstrosity at (c) is produced by the separating of the receding lines as they recede. In his attempt to get the lines parallel, the beginner often makes a mistake like that at (c). Converge the lines deliberately, as in (b), to avoid the results of (c)!

3. *Axes must be kept flat to avoid distortion.* In Fig. 64 an accurate isometric sketch is shown at (a). At (b) the axes have been flattened to less than 30° with the horizontal. (b) possibly looks more natural than (a). At (c), however, the axes are somewhat more than 30° to the horizontal. Note the definite

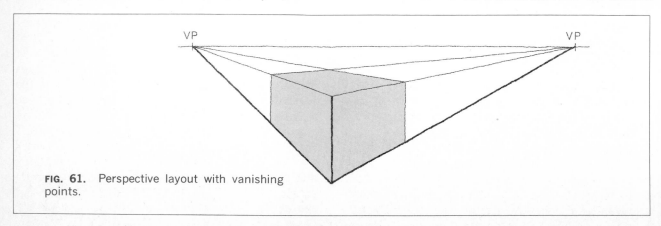

**FIG. 61.**  Perspective layout with vanishing points.

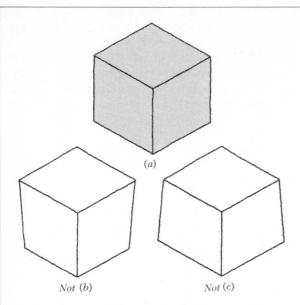

**FIG. 62.** Sketching vertical lines. These must be accurately vertical to define object shape.

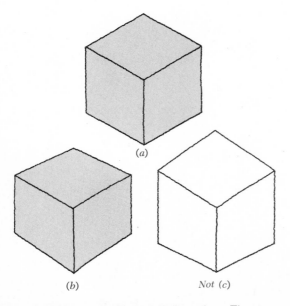

**FIG. 63.** Sketching the receding edges. These must be parallel (*a*); or converging (*b*); *never* separating (*c*).

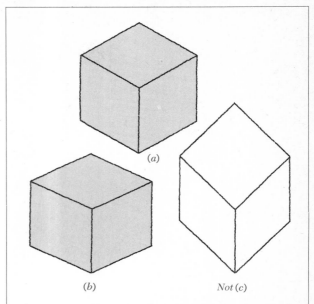

**FIG. 64.** Angle of axes. Isometric position (*a*) or flattened (*b*) gives a natural appearance. Distortion is inherent in steep axes as at (*c*).

distortion and awkward appearance of (*c*). Therefore, especially in isometric, but also in other forms, keep the axes at their correct angle or flatter than the normal angle.

The foregoing three points must be kept constantly in mind. Hold your sketch at arm's length often during the work, so that you will see errors that are not so evident in the normal working position. Become critical of your own work and you will soon develop confidence and a good sense of line direction.

## 48. Divisions for Symmetry.

Continually in sketching, centers must be located and divisions of a face must be made into thirds, fourths, fifths, etc. Division into halves to locate a center line is the simplest and most common, and is easily accomplished by judging the mid-point along one of the sides, as indicated on the top face of the rectangular shape in Fig. 65. Also shown on this top face are additional divisions into quarters and eights. Practice this a few times to develop your

**FIG. 65.** Judging equal spaces. This can be done satisfactorily by eye.

**FIG. 66.** Location of centers. Use diagonals or judge the position of center lines.

**FIG. 67.** Use of diagonals for increasing or decreasing rectangular shapes.

judgment of equal spacings. It is really quite easy to do.

Division into thirds is a little harder but after short practice is readily done. The left face of Fig. 65 shows third divisions and one space divided in half to give a sixth point. The right face of Fig. 65 shows division into fifths.

Practice dividing lines or spaces into various equal units. The experience will be valuable in later work.

## 49. Uses of the Diagonal.

The two diagonals of a square or rectangle will locate its geometric center, as indicated on the left face of Fig. 66. The center is also readily located by drawing two center lines, estimating the middle of the space as shown on the top and right faces.

The diagonals of a rectangular face can also be used to increase or decrease the rectangle symmetrically about the same center and in proportion as shown on the left face of Fig. 67, or with two sides coincident as shown on the right face. To increase or decrease by equal units, the distance (or space) between lines must be judged as explained in Sec. 48.

## 50. Sketching Circular Features.

A circle in pictorial is an ellipse whose major diameter is always perpendicular to the *rotation axis*. Thus its minor diameter coincides on the drawing with the rotation axis (Fig. 68). These facts can be used to advantage when drawing an object principally made up of cylinders on the same axis. Note particularly from the above that *all* circles on horizontal planes are drawn as ellipses *with the major diameter horizontal*, as shown in Fig. 69.

Most objects, however, are made up of combina-

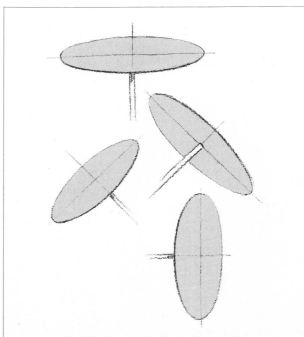

**FIG. 68.** Circles in pictorial. The major diameter of the ellipse is perpendicular to the axis of rotation.

tions of rectangular and circular features, and for this reason it is best to draw the enclosing pictorial square for all circular features. Figure 70 shows circles on all three axonometric planes. Note particularly that the ellipses must be tangent to the sides of the pictorial square at the mid-points of the sides; accordingly, it is best always to draw center lines, also shown. Always sketch the enclosing pictorial square for *all* circular features because by this method the size of the ellipse and the thickness of the cylindrical portion are easily judged. Figure 71 illustrates the boxing of circular features. This object would be difficult to sketch without first boxing in the circular portions.

## 51. Proportioning the Distances.

The ability to make divisions into equal units, discussed in Sec. 48, is needed in proportioning distances on a sketch. The average object does not have distances that are easily divisible into even *inches, half inches,* etc., but as sketches are *not* made to scale, only to good proportions, great accuracy is not necessary. Also, most objects, dimensioned as they invariably are in various odd distances, are difficult to judge from the standpoint of one distance being $\frac{1}{2}$, $\frac{1}{3}$, $\frac{1}{4}$, etc., of another. Nevertheless, the proportioning can be done easily and quickly by the method we are about to discuss.

Figure 72 shows a simple rectangular object, but the distances are not multiples of any simple unit. The best way to proportion this object (and others)

**FIG. 69.** Circular features on horizontal planes. The major diameter of the ellipse is horizontal.

**FIG. 70.** Circles in isometric. Ellipses are tangent to the enclosing isometric squares at the mid-points of the sides.

**FIG. 71.** Boxing construction for circular features. This is necessary to assure correct shape and proportion.

is to lay off on *one* axis a distance that is to represent one side of the object. This *sets the size* of the sketch. Then divide this first side into some unit that can be used easily to proportion other distances. At (*a*) the *left* side has been laid off and divided into three parts. Each of these parts now represents *approxi-*

ORTHOGRAPHIC

(*a*)

(*b*)

**FIG. 72.** Proportioning the distances. By dividing one side into units, all other distances are proportioned.

*mately* a half inch. To get the distance for the vertical axis, the last third (at the rear) has been divided in half, and again in half, so that the dimension shown is approximately $1\frac{1}{8}$ in. Transfer this distance to the vertical axis (1) by judging by eye, (2) by measuring with the finger on the pencil, or (3) by marking the distance on a piece of scratch paper. The method to be used will suggest itself according to the relative accuracy needed. The right-axis distance is obtained similarly, by first transferring the whole left distance (representing approximately $1\frac{1}{2}$ in.) and then adding two-thirds of the left distance (approx. 1 in.), which gives a total of $2\frac{1}{2}$ in., close enough to the actual $2\frac{5}{8}$ in. dimension of the object. Any side may be chosen to start the sketch. At (*b*) the right side has been laid off and divided, this time, into five parts so that again each unit is approximately $\frac{1}{2}$ in. The procedure is then similar to that described for (*a*).

Remember that great accuracy is unnecessary. Do

**FIG. 73.** Proportioning the distances. After one side is divided into units, distances are projected or transferred.

not make the proportioning a burden. It is simply a method of getting distances reasonably close to the actual distance and eliminating the need for wild guesses or extra construction on the sketch.

Figure 73 shows the method applied to a more complex object. The left axis is divided into four parts. In this case the upper details have been located by projection upward from the left axis divisions, a method that is often used. Note the boxing of circular features, as described in Sec. 50. Study this figure carefully, with particular attention to the proportioning and construction.

## 52. Steps in Making a Pictorial Sketch.

Because a variety of objects are sketched, the order of procedure will not always be the same, but the following will serve as a guide:

A. Visualize the shape and proportions of the object from the orthographic views, a model, or other source.
B. Mentally picture the object in space and decide the pictorial position that will best describe its shape.
C. Decide on the type of pictorial to use—axonometric, oblique, or perspective.
D. Pick a suitable paper size.
E. Then proceed as shown in Fig. 74.

The numbers that follow refer to the figure; 1 through 6 shows light construction, 7, 8, and 9 completion to final weight.

**FIG. 74.** Steps in making a pictorial sketch. Progressive layout, (1) to (6); finishing, (7) to (9).

1. Sketch the axes.
2. Lay off the proportions of an enclosing rectangular box for the whole object or a principal portion of it.
3. Sketch the enclosing box.
4. Divide one axis for proportioning distances and sketch the most dominant detail of the object.
5. Proportion smaller details by reference to the divided axis and sketch the enclosing boxes, center lines, or outlines.
6. Complete the boxes for circular features. Check to be sure that all features are in good proportion and that there are no errors in representation (see Sec. 47). In working entirely with light lines, corrections are easily made.
7. Start sketching to the final line width. Begin with the most dominant feature.
8. Sketch the smaller details.
9. Remove construction.

## 53. Axonometric Sketching.

The methods presented thus far have been directed toward the making of axonometric sketches because they are the type most used. However, the practices given apply to oblique and perspective sketching, for which additional help is given in the two sections that follow.

## 54. Oblique Sketching.

The advantage of oblique projection in preserving one face without distortion is of particular value in sketching, as illustrated by Fig. 75. The painful effect of distortion in oblique drawing that is done instrumentally can be greatly lessened in sketching by foreshortening the depth axis to a pleasing proportion. The effect of perspective is obtained by converging the lines parallel to the depth axis. This converging in axonometric or oblique is sometimes called "fake perspective."

## 55. Perspective Sketching.

A sketch made in perspective gives a better effect than in axonometric or oblique. For constructing a perspective drawing of a proposed structure from its plans and elevations, a knowledge of the principles of perspective drawing is required, but for making a perspective sketch from the object, you can get along by observing the ordinary phenomena of perspective which affect everything we see: the fact that objects appear proportionately smaller as their distance from the eye increases, that parallel lines appear to converge as they recede, and that horizontal lines and planes appear to "vanish" on the horizon.

In perspective sketching from the model, make the drawing simply by observation, estimating the directions and proportionate lengths of lines by sighting and measuring on the pencil held at arm's length and use your knowledge of perspective phenomena as a check. With the drawing board or sketch pad held in a comfortable drawing position perpendicular to the line of sight from the eye to the object, test the direction of a line by holding the pencil at arm's length parallel to the board, rotating the arm until the pencil appears to coincide with the line on the model, and then moving it parallel to this position back to the board. Estimate the apparent lengths of lines in the same way; holding the pencil in a plane perpendicular to the line of sight, mark with the thumb the length of pencil which covers the line of the model, rotate the arm with the thumb held in position until the pencil coincides with another line, and then estimate the proportion of this measurement to the second line (Fig. 76).

Make the sketch lightly, with free sketchy lines, and do not erase any lines until the whole sketch has been blocked in. *Do not make the mistake of getting the sketch too small.*

In starting a sketch from the object, set it in a position to give the most advantageous view, and sketch the directions of the principal lines, running them past the limits of the figure toward their vanishing points. Block in the enclosing squares for all circles and circle arcs and proceed with the figure, drawing the main outlines first and adding details later; then brighten the sketch with heavier lines.

The drawing in Fig. 77 shows the general appearance of a "one-point" perspective sketch before the construction lines have been erased. Figure 78 is an example showing the object turned at an angle to the picture plane.

FIG. 75. An oblique sketch. Note boxing of circular features.

FIG. 76. Estimating distances and proportion.

FIG. 77. A perspective sketch. The front face of the object is parallel to the picture plane.

## 56. Pictorial Illustration.

Pictorial illustration combines any one of the regular pictorial methods with some method of shading or ''rendering.'' In considering a specific problem, decide upon the pictorial form—axonometric, oblique, or perspective—and then choose a method of shading that is suited to the method of reproduction and the general effect desired.

## 57. Light and Shade.

The conventional position of the light in light-and-shade drawing is the same as that used for orthographic line shading, that is, a position to the left, in front of, and above the object. Any surface

FIG. 78. A perspective sketch. The object is in an angular position.

or portion of a surface perpendicular to the light direction and directly illuminated by the light would receive the greatest amount of light and be lightest in tone on the drawing; any face not illuminated by the light would be "in shade" and darkest on the drawing. Other surfaces, receiving less light than the "high" light but more than a shade portion, would be intermediate in tone.

An understanding of the simple one-light method of illumination is needed at the outset, as well as some artistic appreciation for the illumination on various surfaces of the object. Figure 79 shows a sphere, cylinder, cone, and cube illuminated as described and shaded accordingly. Study the tone values in this illustration.

## 58. Shade Lines.

Shade lines, by their contrast with other lines, add some effect of light and shade to the drawing. These lines used alone, without other shading, give the simplest possible shading method. Usually the best effect is obtained by using heavy lines only for the left vertical and upper horizontal edges of the dark faces (Fig. 80). Holes and other circular features are drawn with heavy lines on the shade side. Shade

lines should be used sparingly as the inclusion of too many heavy lines simply adds weight to the drawing and does not give the best effect.

## 59. Pencil Rendering.

There are two general methods of pencil shading—continuous tone and line tone. Continuous-tone shading is done with a fairly soft pencil with its point flattened. A medium-rough paper is best for the purpose. Start with a light, over-all tone and then build the middle tones and shade portions gradually. Figure 81 is an example. Clean high lights with an eraser.

Line-tone shading requires a little more skill, as the tones are produced by line spacing and weight. Light lines at wide spacing produce the lightest tone, and heavy lines at close spacing make the darkest shade. Leave high lights perfectly white. Pure black may be used sparingly for deep shade or shadow. Figure 82 is an example, drawn with only a light outline.

**FIG. 79.** Light and shade. The light source is from the upper left front.

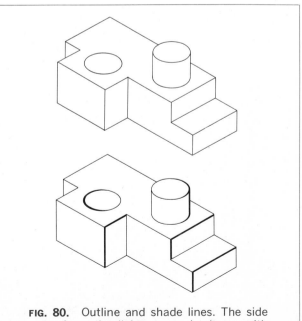

**FIG. 80.** Outline and shade lines. The side away from the light source is drawn with heavier lines.

FIG. 81.   Continuous-tone shading.

FIG. 82.   Line-tone shading.

FIG. 83.   Line-shading technique in pencil.

FIG. 84.   Line-shading technique in ink.

Complete over-all shading is somewhat heavy, and a lighter, more open treatment is usually desired. To achieve this, leave light portions of the object with little or no shading, and line middle tones and shade sparingly. The few lines used strongly suggest light, shade, and surface finish (Fig. 83). There are many variations which can be made in this type of rendering.

## 60. Pen-and-ink Rendering.

Pen-and-ink methods follow the same general pattern as work in pencil, with the exception that no continuous tone is possible. However, there are some variations not ordinarily used in pencil work. Figure 84 shows line techniques. As in pencil work, the common and usually the most pleasing method is the partially shaded, suggestive system.

# PROBLEMS

The following problems are intended to furnish practice (1) in the various methods of pictorial representation and (2) in reading and translating orthographic projections.

In reading a drawing, remember that a line on any view always means an edge or a change in direction of the surface of the object and always look at another view to interpret the meaning of the line.

## Group 1. Isometric Drawings.

**PROB. 1.**  Jig block.

**PROB. 2.**  Frustum of pyramid.

**PROB. 3.**  Hopper.

**PROB. 4.**  Stop block.

**PROB. 5.**  Skid mount.

**PROB. 6.**  Guide block.

**PROB. 7.**  Dovetail stop.

**PROB. 8.**  Bracket.

**PROB. 9.**  Hinged catch.

**PROB. 10.**  Bearing.

**PROB. 11.**  Cross link

**PROB. 12.** Wedge block.

**PROB. 13.** Head attachment.

**PROB. 14.** Slide stop.

**PROB. 15.** Dovetail bracket.

**PROB. 16.** Offset bracket.

**PROB. 17.** Cradle bracket.

**PROB. 18.** Dovetail hinge.

**PROB. 19.** Cable clip.

**PROB. 20.** Strut anchor.

**PROB. 21.** Strut swivel.

**PROB. 22.** Tie plate.

**PROB. 23.** Forming punch.

## Group 2. Isometric Sections.

**PROB. 24.** Springing stone.

**PROB. 25.** Column base. **PROB. 26.** Base plate.

**PROB. 27.** Gland. **PROB. 28.** Squared collar. **PROB. 29.** Blank for gland.

**PROB. 30.**  Sliding cover.

**PROB. 31.**  Rod support.

**PROB. 32.**  Side-beam bracket.

**PROB. 33.**  Head yoke.

**PROB. 34.**  Trunnion plate.

**PROB. 35.**  Section study.

**PROB. 36.**  Section study.

Flange A

Tube

2 15/16 Rad.

4 15/16 Rad.

90°±1°

1

1/2

3.476 ±.001 I.D.

3.9/6 ±.001 O.D.

6" Dia.

.150 ±.001

#4-40 UNC-2B
1/8 Deep-2 Holes

30°

Flange B

1/8

1 3/4

1/8

1/16

1/2

1/16

2 1/4

1/8

Note: Flanges A & B Are Identical

60°

9/16

1/2

3/4

3/16

9/16

1/2 Drill, 6 Holes
On 5 1/8 Dia. B.C.

**PROB. 37.** Spectrometer elbow.

## Group 3. Oblique Drawings.

5 5/8

9/16

3 3/4

3/4

3/4

1/2

3/4

Draw 45° to left

**PROB. 38.** Letter die.

Draw 45° to right
3/4 Drill
4 Holes

3 3/4

9/16

1/8

5 5/8

9/16

Draw 45° to right

**PROB. 39.** Guide plate.

2 7/8

2 5/8

7/8

4 1/4

5

1/4 Sq.

1

1/2
2 7/8

Draw 45° to right

**PROB. 40.** Brace base.

1/8

2 1/4 sq.

Involute
of square

3 3/8

11/16

7/16

1/8

1/2

Draw half size and 30° to right

**PROB. 41.** Heart cam.

Draw 45° to left

**PROB. 42.** Ratchet wheel.

Draw 30° to left

**PROB. 43.** Slotted link.

Draw 45° to right

**PROB. 44.** Swivel plate.

Draw 45° to right

**PROB. 45.** Slide bracket.

Draw 45° to right

**PROB. 46.** Jaw bracket.

**PROB. 47.** Bell crank.

Draw 30° to right

**PROB. 48.** Stop plate.

Draw 30° to right

**PROB. 49.** Hook brace.

Draw 30° to right

**PROB. 50.** Adjusting-rod support.

**PROB. 51.**   Link.

**PROB. 52.**   Pawl.

Offsets from right section, 30° to right

**PROB. 53.**   Culvert model.

Offsets from right section, 30° to right

**PROB. 54.**   Slotted guide.

Draw 30° to right

**PROB. 55.**   Support bracket.

Draw 45° to right

**PROB. 56.**   Port cover.

## Group 4.  Oblique Sections.

**57.**   Oblique full section of sliding cone.
**57A.**   Oblique half section of sliding cone.
**58.**   Oblique full section of conveyor-trough end.

**58A.**   Oblique half section of conveyor-trough end.
**59.**   Oblique full section of ceiling flange.
**59A.**   Oblique half section of ceiling flange.

Draw 30° to right

**PROB. 57.**   Sliding cone.

Draw 30° to right

**PROB. 58.**   Conveyor-trough end.

Draw 45° to right

**PROB. 59.**   Ceiling flange.

**60.**   Oblique full section of hanger flange.
**60A.**   Oblique half section of hanger flange.
**61.**   Oblique half section of the base plate of Prob. 26.
**62.**   Oblique half section of the gland of Prob. 27.
**63.**   Oblique half section of regulator level.
**64.**   Oblique full section of anchor plate.

**PROB. 60.**   Hanger flange.

**PROB. 63.**   Regulator level.

**PROB. 64.**   Anchor plate.

**65.** Oblique drawing of shaft guide. Use sections, as needed, to describe the part.

**66.** Oblique half section of rod support. Partial sections and phantom lines can also be used to describe the part.

## Group 5. Dimetric and Cabinet Drawing.

**67.** Make a dimetric drawing of the jig block, Prob. 1.
**68.** Make a dimetric drawing of the guide block, Prob. 6.
**69.** Make a cabinet drawing of the gland, Prob. 27.
**70.** Make a cabinet drawing of the ceiling flange as shown in Prob. 59.

## Group 6. Perspective Drawings.

The following are a variety of different objects to be drawn in perspective. A further selection can be made from the orthographic drawings in other chapters.

**71.** Double wedge block.
**72.** Notched holder.
**73.** Crank.
**74.** Corner lug.
**75.** House.
**76.** Church.

PROB. 65. Shaft guide.

PROB. 66. Rod support.

**PROB. 71.** Double wedge block.

**PROB. 72.** Notched holder.

**PROB. 73.** Crank.

**PROB. 74.** Corner lug.

**PROB. 75.** House.

**PROB. 76.** Church.

## Group 7. Pictorial Sketching.

The following problems are planned to develop skill not only in pictorial sketching but also in reading orthographic drawings. Make the sketches to suitable size on 8½- by 11-in. paper, choosing the most appropriate form of representation—axonometric, oblique, or perspective—with partial, full, or half sections as needed. Small fillets and rounds may be ignored in these problems and shown as sharp corners.

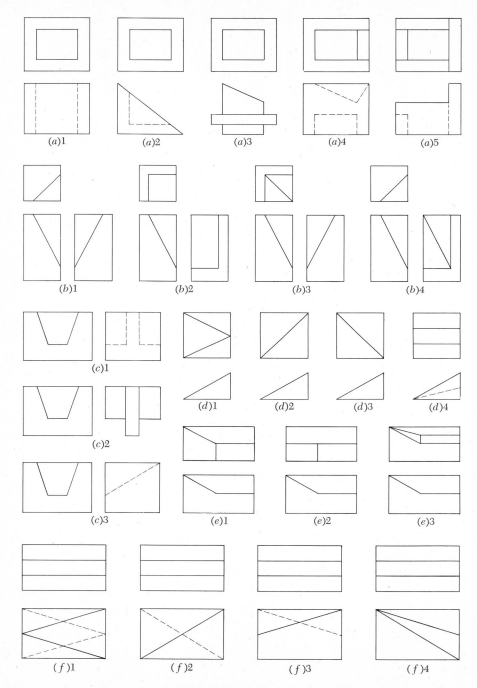

**PROB. 77.** Objects to be sketched.

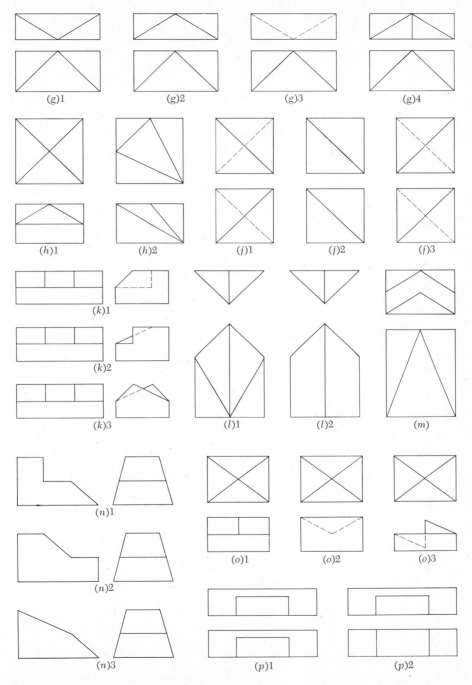

**PROB. 78.**  Objects to be sketched.

**PROB. 79.** Objects to be sketched.

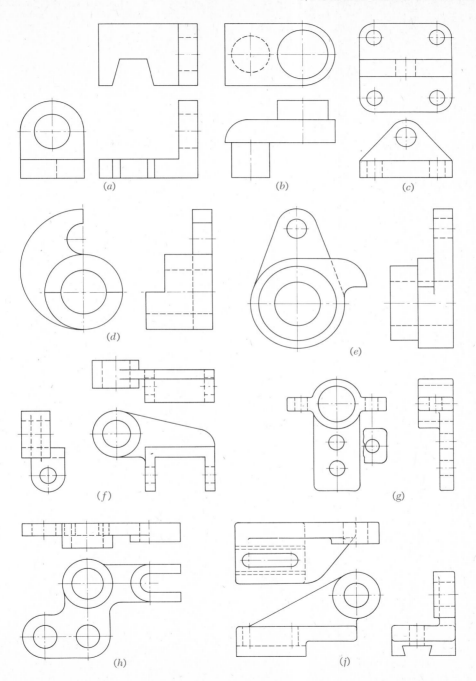

(a)   (b)   (c)

(d)   (e)

(f)   (g)

(h)   (j)

**PROB. 80.** Objects to be sketched.

(k)

(l)

(m)

(n)

(o)

(p)

(q)

(r)

**PROB. 81.** Objects to be sketched.

## Group 8. Oblique Sketching.

Select an object from Group 3 not previously drawn with instruments, and make an oblique sketch.

## Group 9. Perspective Sketching.

Select an object not previously drawn with instruments, and make a perspective sketch.

## Group 10. Axonometric and Oblique Projection from Orthographic Views.

Any of the problems given in this chapter may be used for making axonometric or oblique projections from the orthographic views by first drawing the orthographic views to suitable scale and then employing these views as described in Secs. 26 to 28 to obtain the pictorial projection.

## Group 11. Pictorial Drawings from Machine Parts.

Machine parts, either rough castings and forgings or finished parts, offer valuable practice in making pictorial drawings. Choose pieces to give practice in isometric and oblique drawing. Use the most appropriate form of representation, and employ section and half-section treatments where necessary to give clearer description.

## Group 12. Pictorial Working Drawings.

Any of the problems in this chapter offer practice in making complete pictorial working drawings. Follow the principles of dimensioning in Chap. 11. The form and placement of the dimension figures are given in Sec. 37, Chap. 9.

**82.** Pictorial working drawing of dovetail stop, Prob. 7.

**83.** Pictorial working drawing of hinged catch, Prob. 9.

**84.** Pictorial working drawing of tie plate, Prob. 22.

**85.** Pictorial working drawing of head yoke, Prob. 33.

**86.** Pictorial working drawing of jaw bracket, Prob. 46.

**87.** Pictorial working drawing of adjusting-rod support, Prob. 50.

**88.** Pictorial working drawing of port cover, Prob. 56.

**89.** Pictorial working drawing of ceiling flange, Prob. 59.

**90.** Make a pictorial working drawing of the extension block. Use any reasonable technique such as exploding the part (with note to explain), partial sections, extra views, or phantom lines to completely describe the part.

**PROB. 90.** Extension block.

**91.** Make a pictorial working drawing of the double sleeve clamp. Show the two halves as an exploded assembly.

PROB. **91.** Double sleeve clamp.

# METHODS USED IN MANUFACTURE

Concise and Complete Specification of Size
Requires a Knowledge of Manufacturing
Procedures

In this Chapter:
The Effect of the Basic Manufacturing Method
on the Drawing • Drawings for Sand Castings,
Pattern Shop, Foundry • Forgings and Drawings
of Forgings • Fundamentals of Machining • The
Lathe, Drill Press, Shaper, Planer • Parts
Machined from Stock • The Turret Lathe, Milling
Machine, Grinder • Lapping, Honing, and
Superfinishing • Broaching, Small Tools, Welded
Parts, Parts Made from Sheet Stock, Plastics •
Heat Treatment, Tools for Mass Production, Jigs
and Fixtures, Inspection, Assembly

## 1. Drawings—Their Relationship to Manufacturing.

The test of any working drawing for legibility, completeness, and accuracy is the production of the object or assembly by the shop without further information than that given on the drawing. A knowledge of shop methods will, to a great extent, govern the effectiveness and completeness of the drawing. Study the glossary of shop terms to become familiar with the terms and the form of designation in notes. This chapter is given as an introduction to those on dimensioning (Chap. 12) and working drawings (final chapter).

The relation of drawings and the prints made from them to the operations of production is illustrated in the accompanying chart (Fig. 1). This chart shows in diagrammatic form the different steps in the development of drawings and their distribution and use in connection with shop operations from the time the order is received in the plant until the finished machine is delivered to the shipping room.

## 2. Effect of the Basic Manufacturing Method on the Drawing.

In drawing any machine part, consider first the manufacturing process to be used, as on this depends the representation of the detailed features of the part and, to some extent, the choice of dimensions. Special or unusual methods may occasionally be used, but most machine parts are produced by (1) casting, (2) forging, (3) machining from standard stock, (4) welding, or (5) forming from sheet stock.

Each of the different methods produces a characteristic detailed shape and appearance of the parts, and these features must be shown on the drawing. Figure 2 shows and lists typical features of each method and indicates the differences in drawing practice.

## 3. The Drawings.

For the production of any part, a detail working drawing is necessary, complete with shape and size description and giving, where needed, the opera-

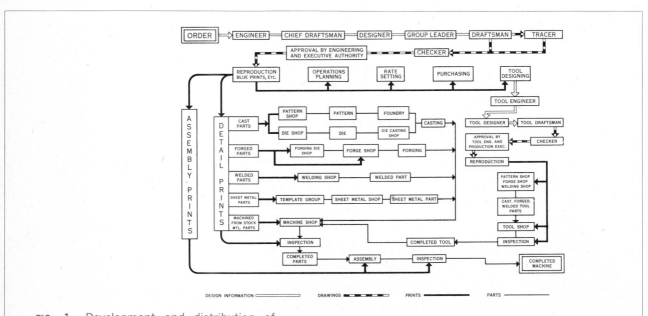

**FIG. 1.** Development and distribution of drawings. This shows the routing of drawings, prints, and parts through all design and manufacturing units.

**FIG. 2.** Drawing requirements for different manufacturing methods.

tions that are to be performed by the shop. Machined surfaces must be clearly indicated, with dimensions chosen and placed so as to be useful to the various shops without the necessity of adding or subtracting dimensions or scaling the drawing.

Two general practices are followed: (1) the "single-drawing" system, in which only one drawing, showing the finished part, is made to be used by all the shops involved in producing the part; and (2) the "multiple-drawing" system, in which different drawings are prepared, one for each shop, giving only the information required by the shop for which the drawing is made.

The second practice is recommended, as the drawings are easier to dimension without ambiguity, somewhat simpler and more direct, and therefore easier for the shop to use. Figure 3 is a single drawing, to be used by the pattern shop and the machine shop. Figures 107 and 108, Chap. 12, are multiple drawings, Fig. 107 for the patternmaker and Fig. 108 for the machine shop.

## 4. Sand Castings.

Figure 3 shows (in the title strip) that the material to be used is cast iron (*C.I.*), indicating that the part will be formed by pouring molten iron into a mold (in this case a "sand mold"), resulting in a sand casting.

After casting, subsequent operations produce the finished part. Figure 4 illustrates the shop interpretation of the casting drawing in Fig. 3 and indicates the order of operations to be performed.

| DWN | C.H.L. | 7-9-46 | LIMITS, UNLESS OTHERWISE NOTED:<br>FRACTIONAL ±1/64, DECIMAL ±.010, ANGULAR ±1/2° | | | | REVISION | DATE | CHANGE OR ADDITION | | |
|---|---|---|---|---|---|---|---|---|---|---|---|
| TCD | JONES | 7-15-46 | | | | | MATERIAL | STOCK SIZE | HEAT TR. | | FINISH |
| | | | PART NO. | REQD | N A M E | | C.I. | | | | As Shown |
| CKD | H.J.J. | 7-21-46 | 107 | 2 | HOUSING SUPPORT | | SCALE | | COMPANY NAME | | DRAWING NO. |
| APPD | J. Smith. | 7-29-46 | | | | | Full | | CITY | | NO. OF |
| | NAME | DATE | UNIT OR<br>ASSEMBLY | | No. 7 SCREENER | | ASSY. DR. NO.<br>74200 | | | | 74238 |

**FIG. 3.** A working drawing of a cast part. Complete information is given for all shops and all operations.

**FIG. 4.** Interpretation of the casting drawing in Fig. 3. These are the operations that must be performed to produce the part.

## 5. The Pattern Shop.

The drawing is first used by the patternmaker, who makes a pattern, or "model," of the part in wood. From this, if a large quantity of castings is required, a metal pattern, often of aluminum, is made. The patternmaker provides for the shrinkage of the casting by making the pattern oversize, using a "shrink rule" for his measurements. He also provides additional metal (machining allowance) for the machined surfaces, indicated on the drawing by (1) finish marks, (2) dimensions indicating a degree of precision attainable only by machining, or (3) notes giving machining operations. The patternmaker also provides the "draft," or slight taper, not shown on the drawing, so that the pattern can be withdrawn easily from the sand. A "core box," for making sand cores for the hollow parts of the casting, is also made in the pattern shop. A knowledge of patternmaking

MYER'S PATTERN WORKS
COLUMBUS

| CUSTOMER | CUSTOMER'S DRAWING NO. |
|----------|------------------------|
| Mine Equipment Co. | 74238 |
| TO BE CAST IN | PATTERN MATERIAL |
| Cast Iron | White Pine |

| DRAWN BY | DATE | SCALE | DRAWING NO |
|----------|------|-------|------------|
| R.E.M. | 8-23-46 | Full | A-4702 |

1 - Rounds $\frac{1}{16}$ R.

2 - Use Shrink Rule for Cast Iron.

3 - Provide $\frac{1}{16}$" per ft. Draft where required.

**FIG. 5.** A pattern drawing. This is a drawing of the pattern for the casting in Fig. 3.

is a great aid in dimensioning, as almost all the dimensions are used by the patternmaker, while only the dimensions for finished features are used by the machine shop.

## 6. Drawings of Castings.

A casting drawing is usually made as a single drawing of the machined casting, with dimensions for both the patternmaker and the machinist (Fig. 3). If the multiple-drawing system is followed, a drawing of the unmachined casting, with allowances for machining accounted for and with no finish marks or finish dimensions, is made for the patternmaker; then a second drawing for the machinist shows the finished shape and gives machining dimensions.

For complicated or difficult castings, a special "pattern drawing" may be made (Fig. 5), showing every detail of the pattern, including the amount of draft, the parting line, "core prints" for supporting the cores in the mold, and the pattern material. Similar detail drawings may also be made for the core boxes.

## 7. The Foundry.

The pattern and core box or boxes are sent to the foundry, and sand molds are made so that molten metal can be poured into the molds and allowed to cool, forming the completed rough casting. Figure 6 is a cross section of a two-part mold, showing the space left by the pattern and the core in place. Only

**FIG. 6.** Cross section of a two-part mold. This shows the cavity produced by the pattern in Fig. 5.

in occasional instances does the foundryman call for assistance from the drawing, as his job is simply to reproduce the pattern in metal.

*Permanent molds,* made of cast iron coated on the molding surfaces with a refractory material, are sometimes an advantage in that the mold can be used over and over again, thus saving the time to make an individual sand mold for each casting. This method is usually limited to small castings.

*Die castings* are made by forcing molten metal under pressure into a steel die mounted in a special die-casting machine. Alloys with a low melting point are used in order to avoid damaging the die. Because of the accuracy possible in making a die, a fine finish and accurate dimensions of the part can be obtained; thus machining may be unnecessary.

## 8. Forgings.

Forgings are made by heating metal to make it plastic and then forming it to shape on a power hammer with or without the aid of special steel dies. Large parts are often hammered with dies of generalized all-purpose shape. Smaller parts in quantity may warrant the expense of making special dies. Some small forgings are made with the metal cold.

*Drop forgings* are the most common and are made in dies of the kind shown in Fig. 7. The lower die is held on the bed of the drop hammer, and the upper die is raised by the hammer mechanism. The hot metal is placed between the dies, and the upper die is dropped several times, causing the metal to flow into the cavity of the dies. The slight excess of material will form a thin fin, or "flash," surrounding the forging at the parting plane of the dies (Fig. 7). This flash is then removed in a "trimming" die made for the purpose. Considerable draft must be provided for release of the forging from the dies.

## 9. Drawings of Forgings.

Forging drawings are prepared according to the multiple-drawing system, one drawing for the diemaker and one for the machinist (Fig. 110, Chap. 10); or the single-drawing system, one drawing for both (Fig. 8). In either case the parting line and draft should be shown and the amount of draft specified.

**FIG. 7.** Drop-forging dies and the forged part. These are the dies for the part shown in Fig. 8.

On the single drawing (Fig. 8), the shape of the finished forging is shown in full outline, and the machining allowance is indicated by "alternate-position" lines, thus completing the shape of the rough forging. This single drawing combines two drawings in one, with complete dimensions for both diemaker and machinist.

Figure 9 illustrates the shop interpretation of the forging drawing in Fig. 8 and indicates the order of operations to be performed.

## 10. The Machine Shop.

The machine shop produces parts machined from stock material and finishes castings, forgings, etc., requiring machined surfaces. Cylindrical and conic surfaces are machined on a lathe. Flat or plane surfaces are machined on a planer, shaper, milling machine, broaching machine, or in some cases (facing) a lathe. Holes are drilled, reamed, counterbored, and countersunk on a drill press or lathe; holes are bored on a boring mill or lathe. For exact work, grinding machines with wheels of abrasive

**FIG. 17.** Working drawing of a part machined from stock. Note the specification of material and stock size in the title block.

## 16. The Turret Lathe.

The *quill* of Fig. 17, produced in quantity, may be made on a turret lathe, except for the rack teeth and the outside-diameter grinding. The stock is held in the collet chuck of the lathe. First the end surface

**FIG. 18.** Interpretation of the drawing in Fig. 17. This lists machining operations.

is faced, and then the cylindrical surface (OD) is turned. The work piece is then ready for drilling and reaming. The turret holds the various tools and swings them around into position as needed. A center drill starts a small hole to align the larger drill, and then the drill and reamer are brought successively into position. The drill provides a hold slightly undersize, and then the reamer, cutting with its fluted sides, cleans out the hole and gives a smooth surface finished to a size within the dimensional limits on the drawing. Figure 19 shows the turret indexed so that the drill is out of the way and the reamer in position. At the right is seen the cutoff tool ready to cut the piece to the length shown on the drawing.

## 17. The Milling Machine.

The dimensions of the rack teeth (Fig. 17) give the depth and spacing of the cuts and the specifications for the cutter to be used. This type of work may be done on a milling machine. The work piece is

**FIG. 19.** Turret-lathe operations (drilling and reaming). The chuck (*A*) holds, indexes (for length), and rotates the stock (*B*) in the direction shown. The turret (*C*) swings the drill (*D*) and reamer (*E*) successively into position.

**FIG. 20.** Milling. The vise (*A*) holds the part (*B*) to the table (*C*), which moves laterally to index cuts and longitudinally, in the direction shown, for individual cuts. The milling cutter (*D*) revolves in the direction shown.

**FIG. 21.** Milling cutters.

held in a vise and moved horizontally into the rotating milling cutter, which, in profile, is the shape of the space between the teeth (Fig. 20). The cuts are spaced by moving the table of the machine to correspond with the distance shown on the drawing. Note that this operation is a forming process, as the shape depends upon the contour of the cutter. With several cutters mounted together (gang milling), a number of teeth can be cut at the same time.

There are many types of milling cutters made to cut on their peripheries, their sides, or their ends, for forming flat, curved, or special surfaces. Three milling cutters are shown in Fig. 21.

## 18. The Grinder.

The general purpose of grinding is to make a smoother and more accurate surface than can be obtained by turning, planing, milling, etc. In many cases, pieces hardened by heat-treatment will warp slightly; and as ordinary machining methods are impractical with hardened materials, such parts are finish-ground after hardening.

The limit dimensions for the outside diameter of the quill (Fig. 17) indicate a grinding operation on a cylindrical grinder (Fig. 22). The abrasive wheel rotates at high speed, while the work piece, mounted on a mandrel between conic centers, rotates slowly in the opposite direction. The wheel usually moves laterally to cover the surface of the work piece. The work piece is gaged carefully during the operation to bring the size within the dimensional limits shown on the drawing and to check for a cylindrical surface without taper. The machine for flat surfaces, called a "surface grinder," holds the work piece on a flat table moving back and forth under the abrasive

**FIG. 22.** Grinding. The cone centers (*A*) hold the mandrel, which mounts and rotates the part (*B*) in the direction shown. The wheel (*C*) moves laterally to traverse the entire surface of (*B*).

wheel. The table "indexes" laterally after each pass under the work.

### 19. Lapping, Honing, and Superfinishing.

These are methods of producing smooth, accurate, mirrorlike surfaces after grinding. All three methods use fine abrasives (1) powdered and carried in oil on a piece of formed soft metal (lapping) or (2) in the form of fine-grained compact stones (honing and superfinishing) to rub against the surface to be finished and reduce scratches and waviness.

### 20. The Broaching Machine.

A broach is a long, tapered bar with a series of cutting edges (teeth), each successively removing a small amount of material until the last edge forms the shape desired. For flat or irregular external surfaces, the broach and work piece are held by the broaching machine, and the broach is passed across the surface of the work piece. For internal surfaces, the broach is pulled or pushed through a hole to give the finished size and shape.

Some machined shapes can be more economically produced by broaching than by any other method.

Figure 23 shows several forms of broaches and the shapes they produce.

### 21. Small Tools.

The shop uses a variety of small tools, both in powered machines and as hand tools. Figure 24 shows a *twist drill*, available in a variety of sizes (numbered, lettered, fractional, and metric) for producing holes in almost any material; a *reamer*, used to enlarge and smooth a previously existing hole and to give greater accuracy than is possible by drilling alone; a *counterbore* and a *countersink*, both used to enlarge and alter the end of a hole (usually for screwheads). A *spot-facing tool* is similar to a counterbore. *Taper, plug*, and *bottoming taps* for cutting the thread of a tapped hole and a die for threading a rod or shaft are also shown.

### 22. Welded Parts.

Simple shapes cut from standard rod, bar, or plate stock can be combined by welding to form a finished part. Some machining after welding is frequently necessary.

### 23. Parts from Standard Sheet.

A relatively thin sheet or strip of standard thickness may first be cut to size "in the flat" and then bent, formed, punched, etc., to form the final required part. The drawing should be made so as to give information for the "template maker" and the information required for bending and forming the sheet. Sometimes separate developments (Chap. 10) are made. The thickness of sheet stock is specified by giving (1) the gage (see table, Appendix) and the equivalent thickness in decimals of an inch or (2) only the decimal thickness (the practice followed in specifying aluminum sheet). Figure 111, Chap. 10, shows a working drawing of a sheet-metal part.

### 24. Plastics.

Plastics are available either in standard bar, rod, tubing, sheet, etc., from which parts can be made by machining, or in granular form to be used in "molding," a process similar to die-casting, in which the material is heated to a plastic state and com-

Broaching a Keyway

Hexagon Broach

Surface Broach

**FIG. 23.** Broaches. Each tooth takes its small cut as the broach is forced across or through the work piece.

Drill   Reamer   Counterbore   Countersink   Taper Tap   Plug Tap   Bottom Tap   Die

**FIG. 24.** Small tools.

pressed by a die (compression molding) or injected under pressure into a die (injection molding). Metal inserts for threads, wear bushings, etc., are sometimes cast into the part. Consideration should be given the diemaker when dimensioning the drawing.

## 25. Heat-treatment.

This is a general term applied to the processing of metals by heat and chemicals to change the physical properties of the material.

The glossary of shop terms gives definitions of such heat-treatment processes as annealing, carburizing, casehardening, hardening, normalizing, and tempering.

The specification of heat-treatment may be given on the drawing in several ways: (1) by a general note listing the steps, temperatures, and baths to be used, (2) by a standard heat-treatment number (SAE or company standard) in the space provided in the title block, (3) by giving the Brinell or Rockwell hardness number to be attained, or (4) by giving the tensile strength, in pounds per square inch, to be attained through heat-treatment.

Figure 8, and Figs. 67, 106, and 110 in Chap. 12, illustrate these methods.

## 26. Tools for Mass Production.

Many special machine tools, both semi-automatic and fully automatic, are used in modern factories. These machines are basically the same as ordinary lathes, grinders, etc., but contain mechanisms to control the movements of cutting tools and produce identical parts with little attention from the operator once the machine has been "tooled up." Automatic screw machines and centerless grinders are examples.

## 27. Jigs and Fixtures.

Jigs for holding the work and guiding the tool and fixtures for holding the work greatly extend the production rate for general-purpose machine tools.

## 28. Inspection.

Careful inspection is an important feature of modern production. Good practice requires inspection after each operation. For production in quantity, special gages are usually employed, but in small-quantity production, the usual measuring instruments, calipers and scale, micrometers, dial gages, etc., are used. For greater precision in gaging, electrical, air, or optical gages, etc., are often employed.

## 29. Assembly.

The finished separate pieces come to the assembly department to be put together according to the assembly drawings. Sometimes it is desirable or necessary to perform some small machining operation during assembly, often drilling, reaming, or hand-finishing. In such cases the assembly drawing should carry a note explaining the required operation and give dimensions for the alignment or location of the pieces. If some parts are to be combined before final assembly, a subassembly drawing or the detail drawings of each piece will give the required information; "$\frac{1}{8}$ drill in assembly with piece No. 107" is a typical note form for an assembly machining operation.

Cover $\frac{13}{16}$-18N-2
Cast Brass 1 Req'd.

$\frac{9}{16}$-18N

$1\frac{1}{4}$

$\frac{1}{4}$  $\frac{7}{16}$

Annular
Groove

$3\frac{9}{16}$

$1\frac{1}{2}$

$1\frac{5}{16}$

2

$1\frac{9}{16}$

$\frac{5}{32}$  $\frac{3}{16}$

$\frac{1}{4}$

$2\frac{1}{16}$

$\frac{7}{16}$

$2\frac{1}{4}$

$\frac{7}{32}$

$2\frac{5}{16}$

$1\frac{5}{16}$

$\frac{1}{4}$-20NC-2

$1\frac{7}{8}$-16N16-2

Body
Cast Brass 1 Req'd.

$\frac{1}{4}$-20NC-2

$\frac{3}{8}$  $\frac{1}{8}$  $\frac{11}{32}$

$\frac{11}{16}$

40°

Lock Screw
Hex, Rolled Brass
1 Req'd.

$\frac{9}{32}$ Drill

$\frac{7}{8}$-20N-2

90°

$1\frac{3}{8}$

$1\frac{3}{16}$

1

$\frac{9}{16}$

120°

$\frac{1}{4}$

# DIMENSIONS, NOTES, LIMITS, AND PRECISION

# 10

These Are the Elements Used To Describe Size,
which along with a Description of Shape
Comprise a Complete Graphic Representation

Size Description Requires a Knowledge of:
Lines and Symbols • Selection of Distances •
Placement of Dimensions • The Dimensioning of
Standard Features • Precision and Tolerance •
Production Methods • Metric Dimensioning •
Problems

## 1. Size Description— First Considerations.

After the shape of an object has been described by orthographic (or pictorial) views, the value of the drawing for the construction of the object depends upon dimensions and notes that describe the *size*. In general, the description of shape and size together gives complete information for producing the object represented.

The dimensions put on the drawing are *not necessarily* those used in making the drawing but are those required for the proper functioning of the part after assembly, selected so as to be readily usable by the workers who are to make the piece. Before dimensioning the drawing, study the machine and understand its functional requirements; then put yourself in the place of the patternmaker, diemaker, machinist, etc., and mentally construct the object to discover which dimensions would best give the information.

## 2. Method.

The basic factors in dimensioning practice are:

1. *Lines and Symbols*. The first requisite is a thorough knowledge of the elements used for dimensions and notes and of the weight and spacing of the lines on the drawing. These lines, symbols, and techniques are the "tools" for clear, concise representation of size.

2. *Selection of Distances*. The most important consideration for the ultimate operation of a machine and the proper working of the individual parts is the selection of distances to be given. This selection is based upon the functional requirements, the "breakdown" of the part into its geometric elements, and the requirements of the shop for production.

3. *Placement of Dimensions*. After the distances to be given have been selected, the next step is the actual placement of the dimensions showing these distances on the drawing. The dimensions should be placed in an orderly arrangement that is easy to read and in positions where they can be readily found.

4. *Dimensioning Standard Features*. These include angles, chamfers, standard notes, specifications of holes, spherical shapes, round-end shapes, tapers, and others for which, through long usage and study, dimensioning practice has been standardized.

5. *Precision and Tolerance*. The ultimate operation of any device depends upon the proper interrelationship of the various parts so that they operate as planned. In quantity production, each part must meet standards of size and position to assure assembly and proper functioning. Through the dimensioning of the individual parts, the limits of size are controlled.

6. *Production Methods*. The method of manufacturing (casting, forging, etc.) affects the detailed information given on the drawing proper, and in notes and specifications. The operations of various shops must be known, in order to give concise information.

# LINES AND SYMBOLS

## 3. Dimension Forms.

Two basic methods are used to give a distance on a drawing: a *dimension* (Fig. 1) or a *note* (Fig. 2).

A dimension is used to give the distance between two points, lines, or planes, or between some combination of points, lines, and planes. The numerical value gives the actual distance, the dimension line indicates the direction in which the value applies, and the arrowheads indicate the points between which the value applies. Extension lines refer the dimension to the view when the dimension is placed outside the view.

A note provides a means of giving explanatory information along with a size. The leader and arrowhead refer the word statement of the note to the proper place on the drawing. Notes applying to the object as a whole are given without a leader in some convenient place on the drawing.

The lines and symbols used in dimensioning are dimension lines, arrowheads, extension lines, leaders, numerical values, notes, finish marks, etc.

## 4. Line Weights.

Dimension lines, extension lines, and leaders are made with fine full lines the same width as center

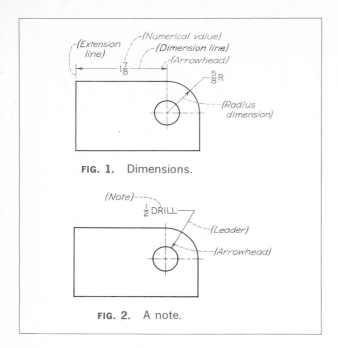

FIG. 1.  Dimensions.

FIG. 2.  A note.

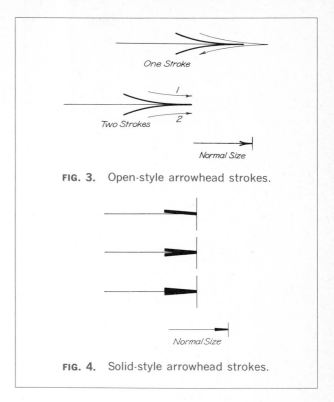

FIG. 3.  Open-style arrowhead strokes.

FIG. 4.  Solid-style arrowhead strokes.

lines so as to contrast with the heavier outlines of the views. Note the line widths given in the alphabet of lines, Figs. 56 and 57, Chap. 2.

## 5. Arrowheads.

These are carefully drawn freehand. The sides of the arrowhead are made either in one stroke, toward the point and then away from it, or in two strokes toward the point, as shown in enlarged form in Fig. 3. The general preference is for the solid head, as in Fig. 4. The solid head is usually made narrower and slightly longer than the open head and has practically no curvature to the sides. It is made in one stroke and then filled, if necessary, without lifting the pen or pencil; a rather blunt pencil or pen is required for this style of head. The bases of arrowheads should not be made wider than one-third the length. All arrowheads on the same drawing should be the same type, open or solid, and the same size, except in restricted spaces. Arrowhead lengths vary somewhat depending upon the size of the drawing. One-eighth inch is a good general length for small drawings and $\frac{3}{16}$ in. for larger drawings.

Poor arrowheads ruin the appearance of an otherwise carefully made drawing. Avoid the incorrect shapes and placements shown in Fig. 5.

## 6. Extension Lines.

These extend from the view to a dimension placed outside the view. They should not touch the outline of the view but should start about $\frac{1}{16}$ in. from it and extend about $\frac{1}{8}$ in. beyond the last dimension line (Fig. 6a). This example is printed approximately one-half size.

Dimensions may also terminate at *center lines* or *visible outlines of the view*. Where a measurement between centers is to be shown, as at (*b*), the center lines are continued to serve as extension lines, extending about $\frac{1}{8}$ in. beyond the last dimension line. Usually the outline of the view becomes the terminal for arrowheads, as at (*c*), when a dimension must be placed inside the view. This might occur because of limited space, when extension lines crossing parts

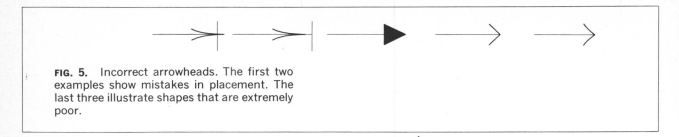

**FIG. 5.** Incorrect arrowheads. The first two examples show mistakes in placement. The last three illustrate shapes that are extremely poor.

of the view would cause confusion, or when long extension lines would make the dimension difficult to read.

*(a)*

*(b)*      *(c)*

**FIG. 6.** Dimension terminals.

*(a)*

*(b)*

*(c)*

**FIG. 7.** Technique of drawing extension lines.

Extension lines for an angular dimension are shown in Fig. 7*a*, which also shows one of the extension lines used for a linear dimension, a common occurrence.

Extension lines should not be broken where they cross each other or an outline of a view, as shown in Figs. 6*a* and 7*a*. However, when space is restricted and the extension lines come close to arrowheads, the extension lines may be broken for clarity, as in Fig. 7*b*.

Where a point is located by extension lines alone, the extension lines should pass through the point, as at (*c*).

**7. Leaders.**
These are *straight* (not curved) lines leading from a dimension value or an explanatory note to the feature on the drawing to which the note applies (Fig. 8). An arrowhead is used at the pointing end of the leader but never at the note end. The note end of the leader should terminate with a short horizontal bar at the mid-height of the lettering and should run to the beginning or the end of the note, never to the middle.

Leaders should be drawn at an angle to contrast with the principal lines of the drawing, which are mainly horizontal and vertical. Thus leaders are usually drawn at 30°, 45°, or 60° to the horizontal; 60° looks best. When several leaders are used, the appearance of the drawing is improved if the leaders can be kept parallel.

The ANSI allows the use of a "dot" for termination of a leader, as in Fig. 9*a*, when the dot is considered to be a clearer representation than an arrowhead. The dot should fall within the outline of the object.

**FIG. 8.** Leaders for notes. Observe that the leader emanates from the *beginning* or the *end* of the lettered information.

**FIG. 9.** Leaders. (*a*) Special terminal; (*b*) and (*c*) proper angle.

When dimensioning a circular feature with a note, the leader should be *radial* as shown at (*b*). A leader directed to a flat surface should meet the surface at an angle of *at least* 30°; an angle of 45° or 60° as at (*c*) is best.

If possible *avoid* crossing leaders; *avoid* long leaders; *avoid* leaders in a horizontal or vertical direction; *avoid* leaders parallel to adjacent dimension lines, extension lines, or crosshatching; and *avoid* small angles between leaders and the lines on which they terminate.

### 8. Figures.

For dimension values, figures must be carefully lettered in vertical or inclined style. In an effort to achieve neatness, the beginner often gets them too small. One-eighth inch for small drawings and $\frac{5}{32}$ in. for larger drawings are good general heights.

The general practice is to leave a space in the dimension line for the dimension value (Fig. 10). It is universal in structural practice and common in

architectural practice to place the value above a continuous dimension line (Fig. 10).

### 9. Common Fractions.

These should be made with the fraction bar parallel to the guide lines for making the figure and with the numerator and denominator each somewhat smaller than the height of the whole number so that the *total* fraction height is twice that of the integer (Fig. 11). Avoid the incorrect forms shown. The figures should not touch the fraction bar.

### 10. Feet and Inches.

Indicate these thus: 9′-6″. When there are no inches, it should be so indicated, as 9′-0″, 9′-0½″. When dimensions are all in inches, the inch mark is preferably omitted from all the dimensions and notes unless there is some possibility of misunderstanding; if that is the case, "1 bore," for example, should be given for clarity as "1″ bore."

In some machine industries, all dimensions are given in inches. In others where feet and inches are used, the ANSI recommends that dimensions up to and including 72 in. be given in inches and greater lengths in feet and inches.

In structural drawing, length dimensions should be given in feet and inches. Plate widths, beam sizes, etc., are given in inches. Inch marks are omitted, even though the dimension is in feet and inches (Fig. 10).

**FIG. 10.** Placement of dimension values.

**FIG. 11.** Technique for lettering values with common fractions.

In the United States, if no foot or inch marks appear on a drawing, the dimension values indicate inches unless a different unit of measurement is indicated by a general note. Drawings made in foreign countries employing the metric system are commonly dimensioned in millimeters.

### 11. Reading Direction of Figures.

This is arranged according to the aligned system or the unidirectional system.

The *aligned system* is the older of the two methods. The figures are oriented to be read from a position *perpendicular* to the dimension line; thus the guide lines for the figures will be parallel to the dimension line, and the fraction bar in line with the dimension line (Fig. 12). The figures should be arranged so as to be read from the *bottom* or *right side* of the drawing. Avoid running dimensions in the directions included in the shaded area of Fig. 12; if this is unavoidable, they should read downward with the line.

The *unidirectional system* originated in the automotive and aircraft industries, and is sometimes called the "horizontal system." All figures are oriented to read from the bottom of the drawing. Thus the guide lines and fraction bars are horizontal regardless of the direction of the dimension (Fig. 13). The "avoid" zone of Fig. 12 has no significance with this system.

*Notes must be lettered horizontally and read from the bottom of the drawing in either system.*

### 12. Systems of Writing Dimension Values.

Dimension values may be given as common fractions, ¼, ⅜, etc., or as decimal fractions, 0.25, 0.375, etc.; and from these, three systems are evolved.

The *common-fraction system*, used in general drawing practice, including architectural and structural work, has all dimension values written as units and common fractions, as $3\frac{1}{2}$, $1\frac{1}{4}$, $\frac{3}{8}$, $\frac{1}{16}$, $\frac{3}{32}$, $\frac{1}{64}$. Values thus written can be laid out with a steel tape or scale graduated in sixty-fourths of an inch.

The *common-fraction, decimal-fraction system* is used principally in machine drawing whenever the degree of precision required calls for fractions of an

**FIG. 12.** Reading direction of values (aligned system). The drawing is read from the bottom and right side.

**FIG. 13.** Reading direction of values (unidirectional system). The drawing is read from the bottom.

inch smaller than those on the ordinary steel scale. To continue the use of common fractions below $\frac{1}{64}$, such as $\frac{1}{128}$ or $\frac{1}{256}$, is considered impractical. The method followed is to give values: (1) in units and common fractions for distances not requiring an accuracy closer than $\frac{1}{64}$ in.; and (2) in units and decimal fractions, as 2.375, 1.250, 0.1875, etc., for distances requiring greater precision. The decimal fractions are given to as many decimal places as needed for the degree of precision required.

The *complete decimal system* uses decimal fractions exclusively for all dimension values. This system has the advantages of the metric system but uses the inch as its basis, thus making it possible to use present measuring equipment.

The ANSI complete decimal system[1] uses a two-place decimal for all values where common fractions would ordinarily be used. The digits after the decimal point are preferably written to even fiftieths, .02, .10, .36, etc., so that when halved, as for radii, etc., two-place decimals will result. Writing the values in even fiftieths allows the use of scales divided in

[1] Y14.5—1964.

**FIG. 14.**  Decimal scale. Graduations are in fiftieths.

fiftieths (Fig. 14), which are much easier to read than scales divided in hundredths.

Dimension values for distances requiring greater precision than that expressed by the two-place decimal are written to three, four, or more decimal places as needed for precision.

Figure 15 is a detail drawing dimensioned according to the ANSI decimal system. The advantage of this system in calculating, adding, and checking and

**FIG. 15.**  A drawing dimensioned in the decimal system.

in doing away with all conversion tables, as well as in lessening chances for error, is apparent.

Designers and draftsmen working in the complete decimal system find it necessary to think in terms of tenths and hundredths of inches instead of common fractions. New designs must be made in decimal sizes without reference to common fractional sizes. However, until standard-stock materials, tools, and commercial parts are available in decimal sizes, some dimensions will have to be given as the decimal equivalent of a common fraction. Thus, for example, a standard $\frac{3}{8}$-16UNC-2A thread would be given as 0.375-16UNC-2A.

*Decimal equivalents* of some common fractions come out to a greater number of decimal places (significant digits) than is necessary or desirable for use as a dimension value, and in such cases the decimal should be adjusted, or "rounded off," to a smaller number of decimal places. The following procedure from the SAE Aerospace–Automotive Drafting Standard[2] is recommended:

When the figure beyond the last figure to be retained is less than 5, the last figure retained should not be changed. Example: 3.46325, if cut off to three places, should be 3.463.

When the figure beyond the last figure to be retained is more than 5, the last figure retained should be increased by 1. Example: 8.37652, if cut off to three places, should be 8.377.

When the figure beyond the last place to be retained is exactly 5 with only zeros following, the preceding number, if even, should be unchanged; if odd, should be increased by 1. Example: 4.365 becomes 4.36 when cut off to two places. Also, 4.355 becomes 4.36 when cut off to two places.

### 13. Finish Marks.
These are used to indicate that certain surfaces of metal parts are to be machined and that allowance must therefore be provided for finish. Finish marks need not be used for parts made by machining from rolled stock, as the surfaces are necessarily machined. Neither are they necessary on drilled, reamed, or counterbored holes or on similar machined features when the machining operation is specified by note. Also, when limit dimensions are given, the accuracy required is thereby stated, and a finish mark on the surface is unnecessary.

The standard finish mark recommended by the ANSI is a 60° **V** with its point touching the line representing the edge view of the surface to be machined. The **V** is placed on the "air side" of the surface. Figure 16 shows the normal size of the **V** and its position for lines in various directions as given on a drawing.

Finish marks should be placed on all views in which the surface to be machined appears as a line, including dashed lines. If the part is to be machined on all surfaces, the note "Finish all over," or "FAO," is used, and the marks on the views are omitted.

In addition to using the finish mark to indicate a machined surface, it may be necessary in some cases to indicate the degree of smoothness of the surface. The ANSI gives a set of symbols to indicate the degree of smoothness of a surface, the various conditions of *surface quality*. These symbols are explained and illustrated in Art. 79.

### 14. Scale of the Drawing.
Even though a workman is never expected to scale a distance on a drawing to obtain a dimension value, the scale to which the drawing is made should be stated in the title block. Standard scales are listed in Fig. 34, Chap. 2.

### 15. Revision of Dimensions.
As a project is being developed, changes in design, in engineering methods, etc., may make it necessary to change some drawings either before or after they have been released to the shop. If the change is a major one, the drawing may have to be remade. But in many cases it is merely a question of altering the dimension values and leaving the shape description unchanged. For changes in dimensions, the out-of-scale dimensions should be indicated by one of the methods in Fig. 17. Drawing changes should be listed in tabular form, in connection with the title block, or in the upper right corner, with

[2]Section A.6 (1963).

**FIG. 16.** The ANSI finish mark. It indicates removal of material to produce the desired surface.

**FIG. 17.** Out-of-scale dimensions.

reference letters and the date, as explained in Sec. 19.

## 16. Abbreviations.

Many abbreviations are used on drawings to save time and space. The common ones such as DIA, THD, MAX, MIN, ID, and OD are universally understood. Uncommon abbreviations such as PH BRZ (Phosphor bronze) should be used cautiously because of the possibility of misinterpretation. Abbreviations should conform to the SAE Aerospace–Automotive Drafting Standard.

## 17. Decimal Points.

These should be made distinctly, in a full letter space, and should be aligned with the bottom edges of digits and letters.

## 18. Dashes.

These are used in many standard expressions such as ½—20 and should be made clearly, one letter space in length, at the mid-height of letters or digits and parallel to the direction of the expression.

# SELECTION OF DISTANCES

## 19. Theory of Dimensioning.

Any object can be broken down into a combination of basic geometric shapes, principally prisms and cylinders. Occasionally, however, there will be parts of pyramids and cones, now and then a double-curved surface, and rarely, except for surfaces of screw threads, a warped surface. Any of the basic shapes may be positive or negative, in the sense that a hole is a negative cylinder.

If the *size* of each of these elementary shapes is dimensioned and the relative position of each is given, measuring from center to center, from base lines, or from surfaces, the dimensioning of any piece can be done systematically. Dimensions can thus be classified as dimensions of *size* and dimensions of *position*.

## 20. Dimensions of Size.

Since every solid has three dimensions, each of the geometric shapes making up the object must have its height, width, and depth indicated in the dimensioning.

The *prism*, often in plinth or flat form, is the most common shape and requires three dimensions for square, rectangular, or triangular (Fig. 18*a*). For regular hexagonal or octagonal types, usually only two dimensions are given, either the distance "across corners" and the length or the distance "across flats" and the length.

The *cylinder*, found on nearly all mechanical pieces as a shaft, boss, or hole, is the second most common shape. A cylinder obviously requires only two dimensions, diameter and length (*b*). Partial cylinders, such as fillets and rounds, are dimensioned by radius instead of diameter. A good general rule is to dimension complete circles with the diameter and circle arcs (partial circles) with the radius.

*Right cones* can be dimensioned by giving the altitude and the diameter of the base. They usually occur as frustums, however, and require the diameters of the ends and the length (*c*). Sometimes it is desirable to dimension cone frustums as *tapers* or with an angular dimension, as described in Art. 41.

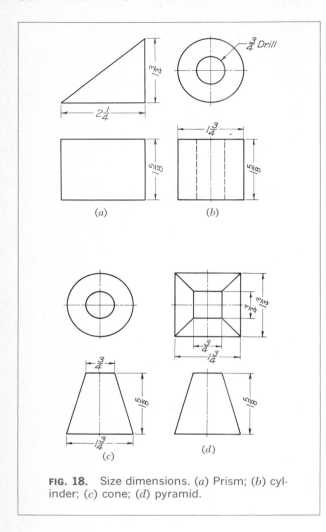

**FIG. 18.** Size dimensions. (*a*) Prism; (*b*) cylinder; (*c*) cone; (*d*) pyramid.

*Right pyramids* are dimensioned by giving the dimensions of the base and the altitude. Right pyramids are often frustums, requiring dimensions of both bases (*d*).

*Oblique cones* and *pyramids* are dimensioned in the same way as right cones and pyramids but with an additional dimension parallel to the base to give the offset of the vertex.

*Spheres* are dimensioned by giving the diameter and other surfaces of revolution by dimensioning the generating curve.

*Warped surfaces* are dimensioned according to their method of generation; and as their representation requires numerous sections, each of these must be fully dimensioned by ordinate and abscissa dimensions.

## 21. Dimensions of Position.

After the basic geometric shapes have been dimensioned for size, the position of each relative to the others must be given. *Position must be established in height, width, and depth directions.* Rectangular shapes are positioned with reference to their faces, cylindrical and conic shapes with reference to their center lines and their ends.

One basic shape will often coincide or align with another on one or more of its faces. In such cases the alignment serves partially to locate the parts and eliminates the need of a dimension of position in a direction perpendicular to the line of coincidence. Thus in Fig. 19, prism *A* requires only one dimension for complete positioning with respect to prism *B*, as two surfaces are in alignment and two in contact.

Coincident center lines often eliminate the need for position dimensions. In the cylinder in Fig. 18*b* the center lines of the hole and of the cylinder coincide, and no dimensions of position are needed. The two holes of Fig. 19 are on the same center line, and the dimension perpendicular to the common center line positions both holes in that direction.

## 22. Selection of Dimensions.

The dimensions arrived at by reducing the part to its basic geometric shapes will, in general, fulfill the requirements of practical dimensioning. However, sometimes other dimensions are required to ensure satisfactory functioning of the part and to give the information in the best way from the standpoint of production.

*The draftsman must therefore correlate the dimensions on drawings of mating parts to ensure satisfactory functioning and, at the same time, select dimensions convenient for the workmen to use.*

Here our study of graphics as a language must be supplemented by a knowledge of shop methods.

**FIG. 19.** Dimensions of size and position. *S* indicates size, *P* position.

process is involved, such as drilling, reaming, counterboring, and punching.

*Selecting dimensions of position* ordinarily requires more consideration than selecting dimensions of size because there are usually several ways in which a position might be given. In general, positional dimensions are given between finished surfaces, center lines, or a combination thereof (Fig. 22). Remember that rough castings or forgings vary in size; so do not position machined surfaces from unfinished surfaces—except in one situation: A machined surface may be positioned from an unmachined surface only when the *initial*, or *starting*, dimension gives the position for the first surface to be machined—a position from which the other machined surfaces will in turn be positioned. *Coinciding center lines of unfinished and finished surfaces often take the place of a starting dimension.*

The position of a point or center by offset dimensions from two center lines or surfaces (Fig. 20) is preferable to angular dimensions (Fig. 21) unless

To be successful, the machine draftsman must have an intimate knowledge of patternmaking, foundry practice, forging, and machine-shop practice, as well as, in some cases, sheet-metal working, metal and plastic die casting, welding, and structural-steel fabrication.

The beginning student who is without this knowledge should not depend upon his instructor alone but should learn by observing work going through the shops and reading books and periodicals on methods used in modern production work.

The *selection of dimensions of size* arrived at by shape breakdown will usually meet the requirements of the shop since the basic shapes result from the fundamental shop operations. However, a shop often prefers to receive dimensions of size in note form rather than as regular dimensions when a shop

**FIG. 20.** Position by offsets.

**FIG. 21.** Position by angle.

the angular dimension is more practical from the standpoint of construction.

### 23. Correlation of Dimensions.

Mating parts must have their dimensions correlated so that the two parts will fit and function as intended. Figure 22 illustrates this principle. Note that the tongue of the bracket is to fit the groove in the body, and that the drilled holes in both pieces must align. Study the dimensioning of both pieces and observe that the dimensions of position ($P$) are correlated so that the intended alignment and fitting of the parts will be accomplished.

Not only must dimensions be correlated with the dimensions of the mating part, but the accuracy to which these distances are produced must meet certain requirements, or the parts still may not fit and function properly. Distances between the surfaces or center lines of finished features of an object must usually be more accurately made than distances between unfinished features. In Fig. 23, note that the position dimensions between center lines or surfaces of finished features are given as three-place decimals, as dimension $A$. Dimensions of position for unfinished features are given as common frac-

tions, as dimension $B$. The decimal dimensions call for greater precision in manufacture than do the common fractions. Dimension $B$ is in this case used by the patternmaker to position cylinder $C$ from the right end of the piece. The machinist will first locate the finished hole in this cylinder, making it concentric with the cylinder; then all other machined surfaces will be positioned from this hole, as, for example, the spline position by dimension $A$. The four spot-faced holes are positioned with reference to each other with fractional dimensions since the holes are oversize for the fastenings used, allowing enough shifting of the fastenings in the holes so that great accuracy in location is not necessary. The mating part, with its holes to receive the screws, would be similarly dimensioned.

Study Fig. 23 and note the classification, size, and position of each dimension.

### 24. Superfluous Dimensions.

*Duplicate* or *unnecessary* dimensions are to be avoided because they may cause confusion and delay. When a drawing is changed or revised, a duplicate dimension may not be noticed and changed along with its counterpart; hence a distance

BRACKET                              BODY

**FIG. 22.** Correlation of dimensions. Dimensions must be given so that parts will assemble properly.

**FIG. 23.** An example of dimensioning. See text for detailed explanation.

will have two different values, one incorrect. An unnecessary dimension is any dimension, other than a duplicate, that is not essential in making the piece. Because of the allowable variation permitted the manufacturer on each dimension (see Art. 54), difficulties will be encountered if unnecessary dimensions occur when parts are to be interchangeable. Actually, if the proper dimensions have been selected, it will be possible to establish a point on the object in any given direction with only one dimension. Unnecessary dimensions always occur when all the individual dimensions are given, in addition to the over-all dimension (Fig. 24). One dimension of the series must be omitted if the over-all dimension is used, thus allowing only one possible positioning from each dimension (Fig. 25).

In architectural and structural work, where the interchangeability of parts is usually irrelevant, unnecessary dimensions cause no difficulty and all dimensions are given.

**FIG. 24.** One unnecessary dimension.

**FIG. 25.** Unnecessary dimension omitted.

Although it is important not to "over-dimension" a part, it is equally important to give all the dimensions that are needed to locate every point, line, or surface of the object. Every dimension that the workman will require in making the part must be given.

Dimensions for similar features, such as the thickness of several ribs obviously of the same size, need not be repeated (Fig. 26). Also, such details as the size of fillets and rounds can be provided for with a general note. Any superfluous or omitted dimensions can be easily discovered by mentally going through the manufacture, or even the drawing, of the part, checking each dimension as it is needed.

### 25. Reference Dimensions.

Occasionally it is useful, for reference and checking purposes, to give all dimensions in a series as well as the over-all dimension. In such cases one dimension is marked with the abbreviation "REF" as shown in Fig. 26. According to the American Standard definition, a reference dimension is "a dimension without tolerance, used for informational purposes only, and does not govern machining or inspection operations."

### 26. Dimensions from Datum.

Datum points, lines, and edges of surfaces of a part are features that are assumed to be exact for purposes of computation or reference, and *from* which the position of other features is established. In Fig. 27a the left side and bottom surfaces of the part are the datum surfaces, and at (*b*) the center lines of the central hole in the part are datum lines. Where positions are specified by dimensions from a datum, different features are always positioned from this datum and *not* with respect to one another.

### 27. Selection of a Datum.

A feature selected to serve as a datum must be clearly identified and readily recognizable. Note in Fig. 27 that the datum lines and edges are obvious. On an actual part, the datum features must be accessible during manufacture so that there will be no difficulty in making measurements. In addition, *corresponding* features on mating parts must be

ALL FILLLETS $\frac{3}{16}$ R.

**FIG. 26.**   One reference dimension.

(*a*)

(*b*)

**FIG. 27.**   Dimensions from datum. (*a*) Datum surfaces; (*b*) datum center.

employed as datum features to assure assembly and proper functioning.

A datum surface (on a physical part) must be more accurate than the allowable variation on any dimension of position which is referred to the datum. Thus, it may be necessary to specify the perfection of datum surfaces for flatness, straightness, roundness, etc. See Arts. 77 and 78.

### 28. Basic Dimensions.

Any dimension on a drawing specified as BASIC is a *theoretical* value used to describe the exact size, shape, or position of a feature. It is used as a reference from which permissible variations are established.

### 29. Maximum and Minimum Sizes.

In some cases a maximum or a minimum size represents a limit beyond which any variation in size cannot be permitted, but in the *opposite* direction there is no difficulty. To illustrate, Fig. 28*a* shows a shaft and hub assembly. If the change in diameter of the shaft (rounded) is too large or the hub edge (chamfer) too small, the hub will not seat against the shaft shoulder; but if the shaft radius is somewhat smaller or the hub edge somewhat larger, no difficulty exists. Therefore, in the dimensioning of both, at (*b*) and (*c*), MAX and MIN are applied to the dimensions. This method is often used in connection with depths of holes, lengths of threads, chamfers, and radii.

## PLACEMENT OF DIMENSIONS

### 30. Placement of Dimensions.

After the distances have been selected, it is possible to decide (1) the *view* on which the distance will be indicated, (2) the particular *place* on that view, and (3) the *form* of the dimension itself. Numerous principles, some with the force of a rule, can be given, but in all cases the important consideration is *clarity*.

### 31. Views.

*The Contour Principle.* One of the views of an object will usually describe the shape of some detailed feature better than will the other view or views, and the feature is then said to be "characteristic" in that particular view. In reading a drawing, it is natural to look for the dimensions of a given feature wherever that feature appears most characteristic, and an advantage in clarity and in ease of reading will certainly result if the dimension is placed there. In Fig. 29 the rounded corner, the drilled hole, and the lower notched corner are all characteristic in, and dimensioned on, the front view. The projecting shape on the front of the object is more characteristic in the top view and is dimensioned there.

**FIG. 28.** Maximum and minimum sizes. Mating edges at (*a*) are dimensioned as at (*b*) and (*c*) so that there will be no assembly interference.

**FIG. 29.** The contour principle. Dimensions are placed where the feature dimensioned is most easily recognized.

*Dimensions for prisms* should be placed so that two of the three dimensions are on the view showing the contour shape and the third on one of the other views (Figs. 18 and 29).

*Dimensions for cylinders,* the diameter and length, are usually best placed on the noncircular view (Fig.

**FIG. 30.** Dimensions for cylinders. (*a*) shows usual practice for cylindrical parts; (*b*), (*c*), and (*d*), for holes.

30*a*). This practice keeps the dimensions on one view, a convenience for the workman. Occasionally a cylindrical hole is dimensioned with the diameter at an angle on the circular view, as indicated at (*b*). This practice should never be used unless there is a clear space for the dimension value. In some cases, however, the value can be carried outside the view, as at (*c*). When a round hole is specified by a note, as at (*d*), the leader should point to the circular view if possible. The note has an advantage in that the diameter, operation, and depth can all be given together. Giving the diameter on the circular view as at (*b*), (*c*), and (*d*) may make for ease of reading, as the dimensions of position will probably be given there also, as indicated at (*d*). When it is not obvious from the drawing, a dimension may be indicated as a diameter by following the value with the letter *D*, as shown in Fig. 31.

*Principles for the Placement of Dimensions:*

1. Dimensions outside the view are preferred, unless added clearness, simplicity, and ease of reading will result from placing some of them inside. For good appearance, dimensions should be kept off the cut surfaces of sections. When it is not possible to do this, the section lining is omitted around the numbers, as shown in Fig. 31 (see Art. 35).

2. Dimensions between the views are preferred unless there is some reason for placing them elsewhere, as in Fig. 29 where the dimensions for the lower notched corner and the position of the hole must come at the bottom of the front view.

3. Dimensions should be applied to one view only; that is, with dimensions between views, the extension lines should be drawn from one view, not from both views (Fig. 32).

4. Dimensions should be placed on the view that shows the distance in its true length (Fig. 33).

5. Dimension lines should be spaced, in general, $\frac{1}{2}$ in. away from the outlines of the view. This applies to a single dimension or to the first dimension of several in a series.

6. Parallel dimension lines should be spaced uniformly with at least $\frac{3}{8}$ in. between lines.

7. Values should be midway between the arrowheads, except when a center line interferes (Fig. 34) or when the values of several parallel dimensions are staggered (Fig. 35).

8. Continuous or staggered dimension lines may be used, depending upon convenience and readability. Continuous dimension lines are preferred where possible (Figs. 36 and 37).

9. Always place a longer dimension line outside a shorter one to avoid crossing dimension lines with the extension lines of other dimensions. Thus an over-all dimension (maximum size of piece in a given direction) will be outside all other dimensions.

10. Dimensions should never be crowded. If the space is small, follow one of the methods given in Art. 32.

11. Center lines are used to indicate the sym-

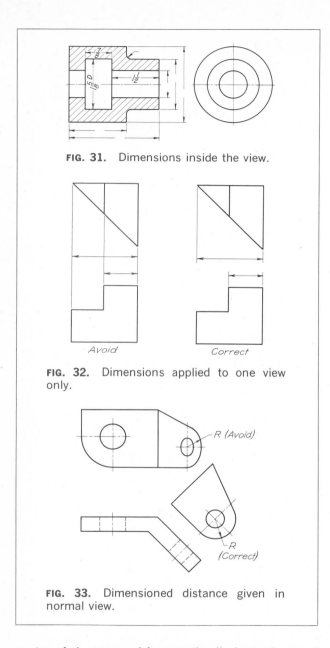

FIG. 31.  Dimensions inside the view.

FIG. 32.  Dimensions applied to one view only.

*Avoid*  *Correct*

*R (Avoid)*

*R (Correct)*

FIG. 33.  Dimensioned distance given in normal view.

FIG. 34.  Values midway between arrow-heads.

FIG. 35.  Values staggered for clarity.

FIG. 36.  Dimensions arranged in continuous form.

FIG. 37.  Dimensions staggered.

metry of shapes, and frequently eliminate the need for a positioning dimension. They should be considered as part of the dimensioning and drawn in finished form at the time of dimensioning. They should extend about $\frac{1}{8}$ in. beyond the shape for which they indicate symmetry unless they are car-

ried further to serve as extension lines. Center lines should not be continued between views.

12. All notes must read horizontally (from the bottom of the drawing).

*Cautions:*

1. Never use a center line, a line of a view, or an

extension line as a dimension line.

2. Never place a dimension line on a center line or place a dimension line where a center line should properly be.

3. Never allow a line of any kind to pass through a dimension figure.

4. Never allow the crossing of two dimension lines or an extension line and a dimension line.

5. Avoid dimensioning to dashed lines if possible.

## 32. Dimensioning in Limited Space.

Dimensions should never be crowded into a space too small to contain them. One of the methods in Fig. 38 can be used where space is limited. Sometimes a note is appropriate. If the space is small and crowded, an enlarged removed section or part view can be used (Fig. 39).

FIG. 38.   Dimensions in limited space.

FIG. 39.   Use of enlarged view to clarify dimensions.

## 33. Order of Dimensioning.

A systematic order of working is a great help in placing dimensions. Figure 40 illustrates the procedure. First complete the shape description (*a*). Then place the extension lines and extend the center lines where necessary (*b*), thus planning for the location of both size and position dimensions; study the placement of each dimension and make alterations if desirable or necessary. Add the dimension lines (*c*). Draw arrowheads and leaders for notes (*d*). Then add values and letter notes (*e* and *f*).

It is desirable to add the notes *after* the dimensions have been placed. If the notes are placed first, they may occupy a space needed for a dimension. Because of the freedom allowed in the use of leaders, notes may be given in almost any available space.

## 34. Dimensioning of Auxiliary Views.

In placing dimensions on an auxiliary view, the same principles of dimensioning apply as for any other drawing, but special attention is paid to the contour principle given in Art. 31. An auxiliary view is made for the purpose of showing the normal view (size) of some inclined or skew face, and for this reason the dimensioning of the face should be placed where it is easiest to read, *which will be on the normal view.* Note in Fig. 41 that the spacing and size of holes as well as the size of the inclined face are dimensioned on the auxiliary view. Note further that the angle and dimension of position tying the inclined face to the rest of the object could not be placed on the auxiliary view.

## 35. Dimensioning of Sectional Views.

Dimensions that must be placed on sectional views are usually placed outside the view so as not to be crowded within crosshatched areas. However, sometimes a dimension *must* be placed across a crosshatched area. When this is the case, the crosshatching is left out around the dimension figures, as illustrated in Fig. 31. Examples showing dimensioning practice on sectional views are given in Probs. 36, 37, and 49 of the final chapter.

**FIG. 40.** Order of dimensioning. See text for details.

**FIG. 41.** Dimensioning on an auxiliary (normal) view.

## 36. Dimensioning a Half Section.

In general, the half section is difficult to dimension clearly without some possibility of crowding and giving misleading or ambiguous information. Generous use of notes and careful placement of dimension lines, leaders, and figures will in most cases make the dimensioning clear; but if a half section cannot be clearly dimensioned, an extra view or part view should be added on which to describe the size.

Inside diameters should be followed by the letter *D* and the dimension line carried over the center line, as in Fig. 42, to prevent the possibility of reading the dimension as a radius. Sometimes the view and the dimensioning can both be clarified by showing the dashed lines on the unsectioned side. Dimensions of internal parts, if placed inside the view, will prevent confusion between extension lines and the outline of external portions.

## 37. Dimensioning of Pictorial Drawings.

Pictorial drawings are often more difficult to dimension than orthographic drawings because there is one view instead of several, and the dimensioning may become crowded unless the placement is care-

fully planned. In general, the principles of dimensioning for orthographic drawings should be followed whenever possible. The following rules should be observed:

1. Dimension and extension lines should be placed so as to lie *in* or *perpendicular* to the face on which the dimension applies. See Probs. 4 and 5, Chap. 24.

2. Dimension numerals should be placed so as to lie in the plane in which the dimension and extension lines lie. See Probs. 8 and 14, final chapter.

3. Leaders for notes and the lettered note should be placed so as to lie in a plane parallel or perpendicular to the face on which the note applies. See Probs. 18 and 59 in Chap. 24.

4. Finish may be indicated by the standard finish symbol (**V**). The symbol is applied *perpendicular to* the face with its point touching a short line lying in the surface. The symbol and line should be parallel to one of the principal axes. If the symbol cannot be applied in the above manner, the finish symbol may be attached to a leader pointing to the face. See Probs. 5 and 18 in Chap. 24.

5. Lettering of dimension values and of notes

**FIG. 42.** Dimensioning a half section.

should be made so that the lettering appears to lie in or parallel to one of the principal faces of the pictorial drawing. To do this, the lettering must be the *pictorial representation of vertical figures*. Note the placement and lettering of the dimension values and the notes in Probs. 54 and 56, Chap. 24.

The American Standard permits lettering of pictorial drawings according to the unidirectional system, with either vertical or slant lettering. This is principally to make possible the use of mechanical lettering devices. If the unidirectional system is employed,

1. Dimension values are lettered to read from the bottom of the sheet.

2. Notes are lettered so that they lie *in the picture plane,* to read from the bottom of the sheet. Notes should be kept off the view, if possible.

## DIMENSIONING STANDARD FEATURES

### 38. Notes.

These are word statements giving information that cannot be given by the views and dimensions. They almost always specify some standard shape, operation, or material, and are classified as *general* or *specific*. A general note applies to the entire part, and a specific note applies to an individual feature. Occasionally a note will save making an additional view or even an entire drawing, for example, by indicating right- and left-hand parts.

Do not be afraid to put notes on drawings. Supplement the graphic language with the English language whenever added information can be conveyed by so doing, but be careful to word the note so clearly that the meaning cannot possibly be misunderstood.

*General notes* do not require the use of a leader and should be grouped together above the title block. Examples are "Finish all over," "Fillets ¼R, rounds ⅛R, unless otherwise specified," "All draft angles 7°," "Remove burs."

Much of the information provided in the title strip of a machine drawing is a grouping of general notes. Stock size, material, heat-treatment, etc., are general notes in the title of the drawing of Fig. 106.

*Specific notes* almost always require a leader and should therefore be placed fairly close to the feature to which they apply. Most common are notes giving an operation with a size, as "½ Drill, 4 holes."

Recommended wordings for notes occurring more or less frequently are given in Fig. 43.

When lower-case lettering is used, capitalization of words in notes depends largely on company policy. One common practice is to capitalize all important words. However, for long notes, as on civil engineering or architectural drawings, the grammatical rules for capitalization usually prevail.

### 39. Angles.

The dimension line for an angle is a circle arc with its center at the intersection of the sides of the angle (Fig. 44). The value is placed to read horizontally, with the exception that in the aligned system large arcs have the value aligned with the dimension arc. Angular values should be written in the form 35°7′ with no dash between the degrees and the minutes.

### 40. Chamfers.

Chamfers may be dimensioned by note, as in Fig. 45a, if the angle is 45°. The linear size is understood to be a short side of the chamfer triangle; the dimensioning without a note shown at (b) is in conformity with this. If the chamfer angle is other than 45°, it is dimensioned as at (c).

### 41. Tapers.

The term "taper" as used in machine work usually means the surface of a cone frustum. The dimensioning will depend on the method of manufacture and the accuracy required. If a standardized taper (see Appendix) is used, the specification should be accompanied by one diameter and the length as shown at (a) in Fig. 46. The general method of giving the diameters of both ends and the taper per foot is illustrated at (b). An alternate method is to give one diameter, the length, and the taper per foot. Taper per foot is defined as the difference in diameter in inches for 1 ft of length. The method of dimensioning for precision work, where a close fit between the parts as well as a control of entry dis-

-GENERAL NOTES-
10. All draft angles 7° unless otherwise noted.
11. Fillets $\frac{1}{2}$ R, Rounds $\frac{1}{8}$ R. unless otherwise noted.
12. Unspecified radii .12
13. Finish all over.
14. Finish all over except____
15. Heat-treat to Rockwell C 52-56.
16. Use pattern No. 72141.
17. Pickle before machining.
18. Dimensions to be met after plating.
19. Polish after plating.
20. Sandblast before painting.
21. Locate holes from template No. ____
22. Tumble after removing sprue.

| NOTE (appears on) | PC. No. | NOTE (appears on) | PC. No. | NOTE (appears on) | PC. No. | NOTE (appears on) | PC. No. |
|---|---|---|---|---|---|---|---|
| Bore | 4 | Fillet | 11,12 | Peen | 9 | Section | 2 |
| Broach | 9 | Finish | 8(∨),9(f),13,14 | Pickle | 17 | Spot | 7 |
| Chamfer | 8 | Grind | 6 | Plate | 2,18,19 | Spotface | 3 |
| Core | 3,8 | Heat-treat | 6,15 | Polish | 1,19 | Tap | 3,9 |
| Counterbore | 9 | Keyseat | 1,7 | Ream | 1,3 | Taper | 4,6 |
| Countersink | 2 | Keyway | 4 | Relief (thread) | 7 | Template | 21 |
| Crown | 4 | Knurl | 6 | Rounds | 11,12 | Thread | 1,5,9 |
| Drill | 1,2,3,9 | Neck | 6 | Sandblast | 20 | Tumble | 22 |
| Draft | 10 | Pattern | 16 | Sawcut | 9 | Undercut | 8 |

FIG. 43. Recommended wording of notes.

FIG. 44.   Dimensions for angles.

FIG. 45.   Dimensioning of chamfers.

FIG. 46.   Dimensioning of tapers.

FIG. 47.   Dimensioning of batters and slopes.

## 42. Batters, Slopes, and Grade.

*Batter* is a deviation from the vertical, such as is found on the sides of retaining walls, piers, etc., and *slope* is a deviation from the horizontal. Both are expressed as a ratio with one factor equal to unity, as illustrated in Fig. 47. *Grade* is identical with slope but is expressed in percentage, the inclination in feet per hundred feet. In structural work angular measurements are shown by giving the ratio of run to rise with the larger side 12 in.

## 43. Arcs.

Arcs should be dimensioned by giving the radius on the view that shows the true shape of the curve. The dimension line for a radius should always be drawn as a radial line at an angle (Fig. 48), never horizontal or vertical; and only one arrowhead is used. There is no arrowhead at the arc center. The numerical

tance is required, is shown at (c). Because of the inaccuracy resulting from measuring at one of the ends, a gage line is established where the diameter is to be measured. The entry distance is controlled through the allowable variation in the location of the gage line, and the fit of the taper is controlled by the accuracy called for in the specification of the angle.

FIG. 48.　Dimensioning of radii.

FIG. 49.　Dimensioning of radii having inaccessible centers. Note that the positioning dimension is marked out of scale.

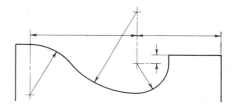

FIG. 50.　Dimensioning a curve made up of radii. The position of centers must be given.

FIG. 51.　A curve dimensioned by offsets.

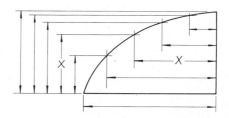

FIG. 52.　A curve dimensioned from datum edges.

value should be followed by the letter $R$. Depending upon the size of the radius and the available space for the value, the dimension line and value are both inside the arc, or the line is inside and the value outside, or, for small arcs, both are outside, as shown in the illustration.

When the center of an arc lies outside the limits of the drawing, the center is moved closer along a center line of the arc and the dimension line is jogged to meet the new center (Fig. 49). The portion of the dimension line adjacent to the arc is a radial line of the true center.

A curved line made up of circle arcs is dimensioned by radii with the centers located, as in Fig. 50.

## 44. Curves.

Curves for which great accuracy is not required are dimensioned by offsets, as in Fig. 51. For greater accuracy, dimensions from datum features, as in Fig. 52, are recommended. Note in Fig. 52 that *any* pair of dimensions (indicated at $x$) could be given to greater accuracy than the others, in order to position a point for which greater accuracy is required.

## 45. Shapes with Rounded Ends.

These should be dimensioned according to their method of manufacture. Figure 53 shows several similar contours and the typical dimensioning for each. The link (*a*), to be cut from thin material, has the radius of the ends and the center distance given as it would be laid out. At (*b*) is shown a cast pad dimensioned as at (*a*), with the dimensions most usable for the patternmaker. The drawing at (*c*) shows a slot machined from solid stock with an end-milling cutter. The dimensions give the diameter of the cutter and the travel of the milling-machine table. The slot at (*d*) is similar to that at (*c*) but it is dimensioned for quantity production, where, instead of the table travel, the over-all length is wanted for gaging purposes. Pratt and Whitney keys and key seats are dimensioned by the method shown at (*d*).

## 46. Dimensions and Specifications for Holes.

Drilled, reamed, bored, punched, or cored holes are usually specified by note giving the diameter, operation, and depth if required. If there is more than one hole of the same kind, the leader needs to point to but one hole, and the number of holes is stated in the note (Fig. 54). Several operations involving one hole may be grouped in a common note. Figures 54 to 57 show typical dimensioning practice for drilled and reamed, counterbored, countersunk, and spot-faced holes.

The ANSI specifies that standard drill sizes be given as decimal fractions, such as 0.250, 0.375, 0.750, and 1.500. If the size is given as a common fraction, the decimal equivalent should be added.

The leader to a hole should point to the *circular* view if possible. The pointing direction is toward the center (Fig. 9*b*). With concentric circles, the arrowhead should touch the inner circle (usually the first operation) unless an outer circle would pass through the arrowhead. In such a case, the arrow should be drawn to touch the outer circle.

Holes made up of several diameters and involving several stages of manufacture may be dimensioned

**FIG. 53.**   Dimensioning of round-end shapes.

$\frac{31}{64}$ (.484) Drill
.500 Ream
2 Holes

**FIG. 54.** Dimensioning of drilled and reamed holes.

$\frac{3}{8}$ Drill, $\frac{3}{4}$ C'bore
2 Holes

.375 Drill, .75 C'bore
.40 Deep, 2 Holes

$\frac{3}{8}$

**FIG. 55.** Dimensioning of counterbored holes.

$\frac{1}{2}$ Drill
82° Csk
to $\frac{5}{8}$ D.

.531 Drill

$\frac{17}{32}$ Drill, Csk for
$\frac{1}{2}$ Flat Head
Cap Screw

78°
.88

**FIG. 56.** Dimensioning of countersunk holes.

$\frac{1}{2}$ Drill
$\frac{7}{16}$ D. Spotface
$\frac{1}{16}$ Deep

$\frac{1}{2}$ Drill
$\frac{7}{16}$ D. Spotface
to clean up

**FIG. 57.** Dimensioning of spot-faced holes.

as shown in Fig. 58. This method of dimensioning combines notes with the regular dimensions.

## 47. Positioning of Holes.

Mating parts held together by bolts, screws, rivets, etc., must have holes for fastenings positioned from common datum surfaces or lines in order to assure matching of the holes. When two or more holes are on an established center line, the holes will require a dimension of position in one direction only (Fig. 59). If the holes are not on a common center line, they will require positioning in two directions, as in Fig. 60. The method at (b) is preferred when it is

$2\frac{1}{2}$ Core
$2\frac{3}{4}$ Bore

$1\frac{3}{4}$ Core
$\frac{1.970}{1.990}$ Rough Bore
$\frac{2.000}{2.003}$ Finish Bore

**FIG. 58.** Method of specifying several operations on a hole.

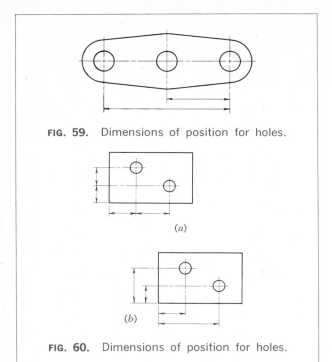

**FIG. 59.** Dimensions of position for holes.

*(a)*

*(b)*

**FIG. 60.** Dimensions of position for holes.

circle is often drawn and its diameter given for reference purposes, as in the figure. The diameter of a hole circle is invariably given on the circular view. The datum lines in this case are the center lines of the part.

*Hole circles* are circular center lines, often called "bolt circles," on which the centers of a number of holes are located.

One practice is to give the diameter of the hole circle and a note specifying the size of the holes, the number required, and the spacing, as in Fig. 61*b*. If one or more holes are not in the regular equally spaced position, their location may be given by an offset dimension, as shown at (*c*).

Figure 62 shows holes located by polar coordinates. This method should be used only when modern, accurate shop equipment is available for locating holes by angle.

### 48. Cylindrical Surfaces.
Cylindrical surfaces on a drawing having an end view may be dimensioned as in Fig. 63*a*; if there is no end view, "DIA" should be placed after each value as in Fig. 63*b*.

### 49. Spherical Surfaces.
Spherical surfaces should be dimensioned as in Fig. 64, that is, by placing the abbreviation "SPHER" after the dimension value.

important to have the positions of both holes established from datum features (left and lower sides of the part).

The coordinate method for the positioning of holes (Fig. 61*a*) is preferred in precision work. The hole

**FIG. 61.** Positioning of holes on "hole circles." (*a*) From datum lines; (*b*) equally spaced; (*c*) one hole offset.

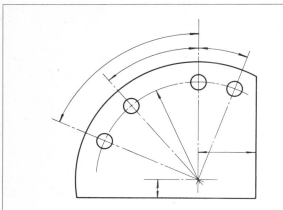

**FIG. 62.** Holes positioned by angle from datum line.

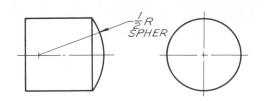

**FIG. 64.** Dimensioning a spherical surface.

**FIG. 65.** Dimensioning a position on a curved surface.

### 50. Curved Surfaces.

A position on a curved part may be misconstrued unless there is a specification showing the *surface* of the part to which the dimension applies, as in Fig. 65.

### 51. Threads, Fasteners, Keyways, and Keyseats.

The dimensioning of these will be given in a later section.

### 52. The Metric System.

A knowledge of the metric system is advantageous, as it is used on all drawings from countries where this system is standard and with increasing frequency on drawings made in the United States. The first instance of international standardization of a mechanical device is that of ball bearings, which have been standardized in the metric system.

Scale drawings in the metric system are not made to English or American scales, but are based on divisions of 10 as full size and then 1 to 2, 1 to 2½, 1 to 5, 1 to 10, 1 to 20, 1 to 50, and 1 to 100. The unit of measurement is the millimeter (mm), and the figures are all understood to be millimeters, without any indicating marks. Figure 66 is an example of metric dimensioning. A table of metric equivalents is given in the Appendix.

**FIG. 63.** Dimensioning of cylinders. (*a*) End view shown; (*b*) no end view.

**FIG. 66.** A metric drawing. Dimensions are in millimeters.

## 53. Standard Sizes, Parts, and Tools.

In dimensioning any machine part, it is often necessary to specify some standard thickness or diameter or the size produced by some standard tool. The American Standard, prevailing company standard, or manufacturer's standard should be consulted in order to assure giving correct information.

*Wire and sheet-metal gages* are given by number and are followed by the equivalent thickness or diameter in decimal form.

*Bolts and screws* are supplied in fractional and numbered sizes.

*Keys* are available in manufacturer's numbered sizes or, for square and flat keys, in fractional sizes.

*Rivets*, depending upon the variety, are supplied in fractional or numbered sizes.

*Drills* are available in numbered, lettered, fractional, and metric sizes.

*Reamers, milling cutters*, and other standard tools are available in a variety of standard sizes.

The Appendix gives tables of standard wire and metal gages, bolt and screw sizes, key sizes, etc. ANSI or manufacturer's standards will give further information required.

## PRECISION AND TOLERANCE

### 54. Precision and Tolerance.

In the manufacture of any machine or structure, quality is a primary consideration. The manufacturing care put into the product determines its quality in relation to competitive products on the market and, in part, its accompanying relative cost and selling price.

*Precision* is the degree of accuracy necessary to ensure the functioning of a part as intended. As an example, a cast part usually has two types of surfaces: mating surfaces and nonmating surfaces. The mating surfaces are machined to the proper smoothness and to be at the correct distance from each other. The nonmating surfaces, exposed to the air and with no important relationship to other parts or surfaces, are left in their original rough-cast form. Thus mating surfaces ordinarily require much greater manufacturing precision than nonmating surfaces. The dimensions on a drawing must indicate which surfaces are to be finished and the degree of precision required in finishing. However, because it is impossible to produce any distance to an absolute size, some variation must be allowed in manufacture.

*Tolerance* is the allowable variation for any given size and provides a practical means of achieving the precision required. The tolerance on any given dimension varies according to the degree of precision necessary for the particular surface. For nonmating surfaces, the tolerance may vary from 0.01 in. for small parts to as much as 1 in. on very large parts. For mating surfaces, tolerances as small as a few millionths of an inch are sometimes necessary (for extremely close-fitting surfaces), but usually surfaces are finished to an accuracy of 0.001 to 0.010, depending upon the function of the part. Figure 67 shows variously toleranced dimensions on a machine drawing. Methods of expressing tolerances are given in Art. 61.

In some cases, particularly in structural and architectural work, tolerances are not stated on the drawing but are given in a set of specifications or are understood to be of an order standard for the industry.

**FIG. 67.** Toleranced dimensions. The title strip gives tolerances on fractional, decimal, and angular dimensions. All other tolerances are specified directly as limits.

## 55. Fits of Mating Parts.

The working parts of any machine have some definite relationship to their mating parts in order to achieve a particular function, such as free rotation, free longitudinal movement, clamping action, permanent fixed position, etc. To ensure the proper relationship, the old practice was to mark the drawings of both parts with the same fractional dimension and add a note such as "running fit" or "drive fit," leaving the difference in size required (the allowance) to the experience and judgment of the machinist.

The tongue of Fig. 68 is to slide longitudinally in the slot. Thus, if the slot is machined first and measures 1.499 in. and the machinist, from his experience, assumes an allowance of 0.004 in., he will carefully machine the tongue to 1.495 in.; the parts will fit and function as desired. In making up a second machine, if the slot measured, say, 1.504

in. after machining, the tongue would be made 1.500 in. and an identical fit obtained; but the tongue of the first machine would be much too loose in the slot of the second machine, and the tongue of the second would not enter the slot of the first. The parts would, therefore, not be interchangeable.

Since it is not possible to work to absolute sizes, it is necessary where interchangeable assembly is required to give the dimensions of mating parts with "limits," that is, the maximum and minimum sizes within which the actual measurements must fall in order for the part to be accepted. The dimensions for each piece are given by three- or four-place decimals, the engineering department taking all the responsibility for the correctness of fit required.

Figure 69 shows the same tongue and slot as in Fig. 68 but dimensioned for interchangeability of parts. In this case, for satisfactory functioning, it has been decided that the tongue must be at least

**FIG. 68.** Dimensioning a fit. According to this practice the machinist fits one part to the other.

**FIG. 69.** Dimensioning a fit with limits. Tolerance on tongue, 0.002 in.; tolerance on groove, 0.002 in.; allowance, 0.002 in.

**FIG. 70.** An example of limit dimensioning. Tolerance on tongue, 0.002 in.; tolerance on groove, 0.001 in.; allowance, 0.004 in.

0.002 in. smaller than the slot but not more than 0.006 in. smaller. This would provide an average fit similar to that used in the previous example. The maximum and minimum sizes acceptable for each part are then figured.

The value 1.500 in. has been assigned as the size of the minimum acceptable slot. This value minus the minimum clearance, 0.002 in., gives a size for the maximum tongue of 1.498. The maximum allowable clearance, 0.006, minus the minimum allowable clearance, 0.002, gives the amount, 0.004, available as the total manufacturing tolerance for both parts. This has been evenly divided and applied as 0.002 to the slot and 0.002 to the tongue. Thus the size of the maximum slot will be the size of the minimum slot *plus* the slot tolerance, or 1.500 + 0.002 = 1.502. The size of the minimum tongue will be the size of the maximum tongue *minus* the tongue tolerance, or 1.498 − 0.002 = 1.496.

A study of Fig. 69 will show that, made in any quantity, the two parts will allow interchangeable assembly and that any pair will fit approximately as any other pair, as planned. This system is essential in modern quantity-production manufacture. Quantity-production and unit-production methods are discussed in Sec. 80.

**56. Nomenclature.**
The terms used in limit dimensioning are so interconnected that their meaning should be clearly understood before a detailed study of the method is attempted.

The following are adapted from ANSI definitions:
*Nominal Size.* The nominal size is the designation that is used for the purpose of general identification.

*Dimension.* A dimension is a geometric characteristic such as diameter, length, angle, or center distance.

*Size.* Size is a designation of magnitude. When a value is assigned to a dimension, it is referred to as the size of that dimension.

*Allowance.* An allowance is an intentional difference between the maximum material limits of mating parts. (See definition of "Fit.") It is a minimum clearance (positive allowance) or maximum interference (negative allowance) between mating parts.

*Tolerance.* A tolerance is the total permissible variation of a size. The tolerance is the difference between the limits of size.

*Basic Size.*   The basic size is that size from which the limits of size are derived by the application of allowances and tolerances.

*Design Size.*   The design size is that size from which the limits of size are derived by the application of tolerances. When there is no allowance, the design size is the same as the basic size.

*Actual Size.*   An actual size is a measured size.

*Limits of Size.*   The limits of size are the applicable maximum and minimum sizes.

*Maximum Material Limit.*   A maximum material limit is the maximum limit of size of an external dimension or the minimum limit of size of an internal dimension.

*Minimum Material Limit.*   A minimum material limit is the minimum limit of size of an external dimension or the maximum limit of size of an internal dimension.

*Tolerance Limit.*   A tolerance limit is the variation, positive or negative, by which a size is permitted to depart from the design size.

*Unilateral Tolerance.*   A unilateral tolerance is a tolerance in which variation is permitted only in one direction from the design size.

*Bilateral Tolerance.*   A bilateral tolerance is a tolerance in which variation is permitted in both directions from the design size.

*Unilateral Tolerance System.*   A design plan that uses only unilateral tolerance is known as a Unilateral Tolerance System.

*Bilateral Tolerance System.*   A design plan that uses only bilateral tolerances is known as a Bilateral Tolerance System.

*Fit.*   Fit is the general term used to signify the range of tightness which may result from the application of a specific combination of allowances and tolerances in the design of mating parts.

*Actual Fit.*   The actual fit between two mating parts is the relation existing between them with respect to the amount of clearance or interference that is present when they are assembled.

*Clearance Fit.*   A clearance fit is one having limits of size so prescribed that a clearance always results when mating parts are assembled.

*Interference Fit.*   An interference fit is one having limits of size so prescribed that an interference always results when mating parts are assembled.

*Transition Fit.*   A transition fit is one having limits of size so prescribed that either a clearance or an interference may result when mating parts are assembled.

*Basic-hole System.*   A basic-hole system is a system of fits in which the design size of the hole is the basic size and the allowance is applied to the shaft.

*Basic-shaft System.*   A basic-shaft system is a system of fits in which the design size of the shaft is the basic size and the allowance is applied to the hole.

In illustration of some of these terms, a pair of mating parts is dimensioned in Fig. 70. In this example the *nominal size* is $1\frac{1}{2}$ in. The *basic size* is 1.500. The *allowance* is 0.004. The *tolerance* on the tongue is 0.002, and on the slot it is 0.001. The *limits* are, for the tongue, 1.496 (maximum) and 1.494 (minimum) and, for the slot, 1.501 (maximum) and 1.500 (minimum).

## 57. General Fit Classes.

The fits established on machine parts are classified as follows:

A *clearance fit* is the condition in which the internal part is smaller than the external part, as illustrated by the dimensioning in Fig. 71. In this case the largest shaft is 1.495 in. and the smallest hole 1.500 in., leaving a clearance of 0.005 for the tightest possible fit.

An *interference fit* is the opposite of a clearance fit, having a definite interference of metal for all possible conditions. The parts must be assembled by pressure or by heat expansion of the external member. Figure 72 is an illustration. The shaft is 0.001 in. larger than the hole for the loosest possible fit. The allowance in this case is 0.003 in. interference.

A *transition fit* is the condition in which either a clearance fit or an interference fit may be had. Figure 73 illustrates a transition fit; the smallest shaft in the largest hole results in 0.0003 in. clearance and the largest shaft in the smallest hole results in 0.0007 in. interference.

**FIG. 71.** A clearance fit. The tightest fit is 0.005 in. clearance; the loosest, 0.009 in. clearance.

**FIG. 72.** An interference fit. The loosest fit is 0.001 in. interference; the tightest, 0.003 in. interference.

**FIG. 73.** A transition fit. The loosest fit is 0.0003 in. clearance; the tightest, 0.0007 in. interference.

### 58. Selective Assembly.

Sometimes the fit desired may be so close and the tolerances so small that the cost of producing interchangeable parts is prohibitive. In such cases tolerances as small as practical are established; then the parts are gaged and graded as, say, *small, medium,* and *large.* A small shaft in a small hole, medium in medium, or large in large will produce approximately the same fit allowance. Transition and inter-

ference fits often require a selection of parts in order to get the amount of clearance or interference desired. Antifriction bearings are usually assembled selectively.

### 59. Basic-hole and Basic-shaft Systems.

Production economy depends to some extent upon which mating part is taken as a standard size. In the *basic-hole system*, the minimum size of the hole is taken as a base from which all variations are made; the hole can often be made with a standard tool.

Where a number of different fits of the same nominal size are required on one shaft, as, for example, when bearings are fitted to line shafting, the *basic-shaft system* is employed, in which the maximum shaft size is taken as the basic size.

### 60. Unilateral and Bilateral Tolerances.

A unilateral tolerance is one in which the total allowable variation is in *one* direction, plus or minus (not both) from the basic value. A bilateral tolerance is one in which the tolerance is divided, with part plus and the remainder minus from the basic value.

### 61. Methods of Expressing Tolerances.

Tolerances may be *specific*, given with the dimension value; or *general*, given as a note in the title block. The general tolerances apply to all dimensions not carrying a specific tolerance. The general tolerance should be allowed to apply whenever possible, using specific tolerances only when necessary. If no tolerances are specified, the value usually assumed for fractional dimensions is $\pm\frac{1}{64}$ in.; for angular dimensions $\pm\frac{1}{2}°$; and for decimal dimensions plus or minus the nearest significant figure, as, for example, $\pm 0.01$ in. for a two-place decimal and $\pm 0.001$ in. for a three-place decimal.

There are several methods of expressing tolerances. The method preferred in quantity-production work, where gages are employed extensively, is to write the two limits representing the maximum and minimum acceptable sizes as in Fig. 74. An internal dimension has the *minimum* size above the line, and an external dimension has the *maximum* size above

$$\begin{array}{c} 1.250 \\ \hline 1.252 \end{array}$$

**FIG. 74.** A tolerance expressed as limits.

$$1.250 \ ^{+.002}_{-.000} \qquad\qquad 1.250 \ ^{}_{+.002}$$

(a)                              (b)

**FIG. 75.** Tolerances plus and minus.

the line. This arrangement is for convenience in machining.

Another method is to give the basic size followed by the tolerance, plus and minus (with the plus above the minus), as in Fig. 75. If only one tolerance value is given, as at (b), the other value is assumed to be zero.

*Unilateral* tolerances are expressed by giving the two limits, as in Fig. 74, or by giving one limiting size and the tolerance, as $2.750 + 0.005$ or $2.750$

$^{+0.005}_{-0.000}$; for fractional dimensions, $\frac{1}{2} - \frac{1}{32}$ or $\frac{1}{2}$ $^{+0}_{-\frac{1}{32}}$; for angular dimensions, $64°15'30'' +$

$0°45'0''$ or $64°15'30'' \ ^{+0°45'0''}_{-0°0'0''}$

*Bilateral* tolerances are expressed by giving the basic value followed by the divided tolerance, both plus and minus (commonly equal in amount), as

$1.500 \ ^{+0.002}_{-0.002}$ or $1.500 \pm 0.002$; for fractional dimensions, $1\frac{1}{2} \ ^{+\frac{1}{64}}_{-\frac{1}{64}}$ or $1\frac{1}{2} \pm \frac{1}{64}$; for angular dimensions, $30°0' \ ^{+0°10'}_{-0°10'}$ or $30°0' \pm 0°10'$.

## 62. Decimal Places.

A dimension value should be carried to the same number of decimal places as the tolerance. For example, with a tolerance of 0.0005 on a nominal dimension of $1\frac{1}{2}$ in., the basic value should be written 1.5000. Tolerances for common fractional values should be given as common fractions, as $\frac{7}{8} \pm \frac{1}{64}$. Tolerances for decimal values should be given as decimal fractions, as $0.750 \pm 0.010$.

## 63. Fundamentals for Tolerance Selection.

Before the engineer can decide on the precision necessary for a particular part and specify the proper fits and tolerances, he must have experience in the manufacturing process used and understand the particular mechanism involved. The following quotation from the ANSI Standard is pertinent: "Many factors, such as length of engagement, bearing load, speed, lubrication, temperature, humidity, and materials, must be taken into consideration in the selection of fits for a particular application."

A table of fits, such as the ANSI table of cylindrical fits (Appendix) explained in Sec. 67, may be taken as a guide for ordinary work.

In many cases practical experience is necessary in determining the fit conditions guaranteeing proper performance. Often it is difficult to determine the definite size at which performance fails, and critical tolerances are sometimes determined through exhaustive testing of experimental models.

It is essential to know the precision attainable with various machine tools and machining methods. As an example, holes to be produced by drilling must not be specified to a smaller tolerance than can be attained by drilling. Attainable manufacturing precision is discussed in Sec. 64. A knowledge of kinds and types of equipment is needed to assure that the tolerances specified can be attained.

## 64. Manufacturing Precision.

The different manufacturing processes all have inherent minimum possible accuracies, depending upon the size of the work, the condition of the equipment, and, to some extent, the skill of the workmen. The following *minimum* tolerances are given as a guide and are based on the assumption that the work is to be done on a quantity-production basis with equipment in good condition. Greater precision can be attained by highly skilled workmen on a unit-production basis.

In general, the following are recommended as tolerances for dimensions having *no effect on the function of the part;* for sizes of 0 to 6 in., $\pm\frac{1}{64}$; 6 to 18 in., $\pm\frac{1}{32}$; 18 in. and larger, $\pm\frac{1}{16}$ (or more).

*Sand Castings.* For unmachined surfaces, a tolerance of $\pm\frac{1}{32}$ is recommended for small castings and a tolerance of $\pm\frac{1}{16}$ for medium-sized castings. On larger castings the tolerance should be increased to suit the size. Small- and medium-sized castings are rarely below the nominal size since the pattern is "rapped" for easy removal from the sand, and this tends to increase the size.

*Die Castings and Plastic Moulding.* A tolerance of $\pm\frac{1}{64}$ or less can easily be held with small and medium-sized parts; for large parts, the tolerance should be increased slightly. Hole-center distances can be maintained within 0.005 to 0.010, depending on the distance of separation. Certain alloys can be die-cast to tolerances of 0.001 or less.

*Forgings.* The rough surfaces of drop forgings weighing 1 lb or less can be held to $\pm\frac{1}{32}$; for weights up to 10 lb, $\pm\frac{1}{16}$; for weights up to 60 lb, $\pm\frac{1}{8}$. Because of die wear, drop forgings tend to increase in size as production from the die increases.

*Drilling.* For drills from no. 60 to no. 30, allow a tolerance of $+0.002 - 0.000$; from no. 29 to no. 1, $+0.0004 - 0.000$; from $\frac{1}{4}$ to $\frac{1}{2}$ in., $+0.005 - 0.000$; from $\frac{1}{2}$ to $\frac{3}{4}$ in., $+0.008 - 0.000$; from $\frac{3}{4}$ to 1 in., $+0.010 - 0.000$; from 1 to 2 in., a tolerance of $+0.015 - 0.000$.

*Reaming.* In general, a tolerance of $+0.0005 - 0.0000$ can be held with diameters up to $\frac{1}{2}$ in. For diameters from $\frac{1}{2}$ to 1 in., $+0.001 - 0.000$; from 1 in. and larger, $+0.0015 - 0.0000$.

*Lathe Turning: Rough Work.* For diameters of $\frac{1}{4}$ to $\frac{1}{2}$ in., allow a total tolerance of 0.005; for diameters of $\frac{1}{2}$ to 1 in., 0.007; for diameters of 1 to 2 in., 0.010; for diameters of 2 in. and larger, 0.015.

*Finish Turning.* For diameters of $\frac{1}{4}$ to $\frac{1}{2}$ in., allow a total tolerance of 0.002; for diameters of $\frac{1}{2}$ to 1 in., 0.003; for diameters of 1 to 2 in., 0.005; for diameters of 2 in. and larger, 0.007.

*Milling.* When single surfaces are to be milled, tolerances of 0.002 to 0.003 can be maintained. When two or more surfaces are to be milled, the most important can be toleranced to 0.002 and the remainder to 0.005. In general, 0.005 is a good value to use with most milling work.

*Planing and Shaping.* These operations are not commonly used with small parts in quantity-production work. For larger parts, tolerances of 0.005 to 0.010 can be maintained.

*Broaching.* Diameters up to 1 in. can be held within 0.001; diameters of 1 to 2 in., 0.002; diameters of 2 to 4 in., 0.003. Surfaces up to 1 in. apart can be held within 0.002; 1 to 4 in. apart, 0.003; 4 in. apart and over, 0.004.

*Threads.* Tolerances for ANSI threads are provided on the pitch diameter through the thread class given with the specification. For a given class, the tolerances increase as the size of the thread increases.

*Grinding.* For both cylindrical and surface grinding, a tolerance of 0.0005 can be maintained.

## 65. Selection of Tolerances.

A common method of determining and applying tolerances is to determine at the outset how much clearance or interference there can be between the mating parts *without impeding their proper functioning.* The difference between the tightest and loosest conditions will be the *sum* of the tolerances of both parts. To obtain tolerances for the individual parts, take half of this value; or if it seems desirable because of easier machining on one part, use slightly less tolerance for that part, with a proportionately larger tolerance for the part more difficult to machine. The following example will illustrate the procedure:

Assume that a running fit is to be arranged between a 2-in. shaft and bearing. It has been determined that in order to provide clearance for a film of oil, the parts cannot fit closer than 0.002 in., and in order to prevent excessive looseness and radial movement of the shaft, the parts cannot be looser than 0.007 in. Then we have the following calculations:

| | | |
|---|---|---|
| Loosest fit | $= 0.007$ | (maximum clearance) |
| Tightest fit | $= 0.002$ | (minimum clearance—allowance) |
| Difference | $= 0.005$ | (sum of tolerances) |

$\frac{1}{2}$ of difference = 0.0025   (possible value for each tolerance)

Assuming that the shaft will be ground and the bearing reamed, 0.002 can be used for the shaft tolerance and 0.003 for the bearing tolerance since these values conform better to the precision attainable by these methods of production.

Figure 76 illustrates the completed dimensions. Note that the minimum hole is taken as the basic size of 2.000 and the tolerance of 0.003 applied. Then the largest shaft size will be the basic size minus the value for the tightest fit (the allowance); the shaft tolerance then subtracted from the maximum shaft size gives the minimum shaft size. From the figure, the 2.003 bearing minus the 1.996 shaft gives 0.007, the loosest fit; the 2.000 bearing minus the 1.998 shaft gives 0.002, the tightest fit.

## 66. ANSI Preferred Limits and Fits for Cylindrical Parts.

This standard conforms with the recommendations of American-British-Canadian conferences. Agreement has been reached for diameters up to 20 in., and larger diameters are under study.

## 67. Designation of ANSI Fits.

The standard ANSI fits are designated by symbols that facilitate reference for educational purposes. These symbols are not to be shown on manufac-

**FIG. 76.** Limits calculated from maximum and minimum clearances. See text for details.

turing drawings; instead, sizes should be specified. The letter symbols used are as follows:

RC, running or sliding fit
LC, locational clearance fit
LT, transition fit
LN, locational interference fit
FN, force or shrink fit

These letter symbols are used in conjunction with numbers for the class of fit; thus FN4 represents a class 4 force fit. Each symbol (two letters and a number) represents a complete fit, for which the minimum and maximum clearance or interference and the limits of size for the mating parts are given in the Appendix.

## 68. Description of Fits.

The following is a description of each class of fit, with a reference in each case to the tables in the Appendix.

*Running and Sliding Fits (Table 1).* These fits provide a similar running performance, with suitable lubrication allowance, throughout the range of sizes. The clearances for the first two classes, used chiefly as slide fits, increase more slowly than for the other classes, so that accurate location is maintained, even when this is at the expense of free relative motion.

RC 1, *close sliding fits* accurately locate parts that must assemble without perceptible play.

RC 2, *sliding fits* are for accurate location, but with greater maximum clearance than RC 1. Parts move and turn easily but do not run freely, and in the larger sizes may seize with small temperature changes.

RC 3, *precision running fits* are about the closest fits expected to run freely, and are for precision work at slow speeds and light journal pressures. They are not suitable under appreciable temperature differences.

RC 4, *close running fits* are chiefly for running fits on accurate machinery with moderate surface speeds and journal pressures, where accurate location and minimum play are desired.

RC 5 and RC 6, *medium running fits* are for higher running speeds, heavy journal pressures, or both.

RC 7, *free running fits* are for use where accuracy is not essential, where large temperature variations are likely, or under both these conditions.

RC 8 and RC 9, *loose running fits* are for materials such as cold-rolled shafting and tubing, made to commercial tolerances.

*Locational Fits* (Tables 2, 3, and 4). These fits determine only the location of mating parts and may provide rigid or accurate location—as in interference fits—or some freedom of location—as in clearance fits. They fall into three groups:

LC, *locational clearance fits* are for normally stationary parts that can be freely assembled or disassembled. They run from snug fits for parts requiring accuracy of location, through the medium clearance fits for parts such as spigots, to the looser fastener fits where freedom of assembly is of prime importance.

LT, *transition locational fits* fall between clearance and interference fits for application where accuracy of location is important, but a small amount of clearance or interference is permissible.

LN, *locational interference fits* are used where accuracy of location is of prime importance, and for parts needing rigidity and alignment with no special requirements for bore pressure. Such fits are not for parts that transmit frictional loads from one part to another by virtue of the tightness of fit; these conditions are met by force fits.

*Force Fits* (Table 5). A force fit is a special type of interference fit, normally characterized by maintenance of constant bore pressures throughout the range of sizes. Thus the interference varies almost directly with diameter, and to maintain the resulting pressures within reasonable limits, the difference between its minimum and maximum value is small.

FN 1, *light drive fits* require light assembly pressures and produce more or less permanent assemblies. They are suitable for thin sections, long fits, or cast-iron external members.

FN 2, *medium drive fits* are for ordinary steel parts or for shrink fits on light sections. They are about the tightest fits that can be used with high-grade cast-iron external members.

FN 3, *heavy drive fits* are suitable for heavier steel parts or for shrink fits in medium sections.

FN 4 and FN 5, *force fits* are for parts that can be highly stressed, or for shrink fits where heavy pressing forces are impractical.

### 69. Examples of Dimensioning Standard ANSI Fits.

Assume that a 1-in. shaft is to run with moderate speed but with a fairly heavy journal pressure. The fit class chosen is RC 6. From Table 1 (Appendix) the limits given are:

Hole: $\begin{array}{l} +1.2 \\ -0 \end{array}$

Shaft: $\begin{array}{l} -1.6 \\ -2.8 \end{array}$

These values are thousandths of an inch. The basic size is 1.0000. Therefore, for the hole, 1.0000 + 0.0012 gives 1.0012 as the maximum limit, and 1.0000 + 0.0000 gives 1.0000 for the minimum limit. For the shaft, 1.0000 − 0.0016 gives 0.9984 for the maximum limit, and 1.0000 − 0.0028 gives 0.9972 for the minimum limit. The dimensioning is shown in Fig. 77. To analyze the fit, the tightest condition (largest shaft in smallest hole) is 1.0000 − 0.9984, or 0.0016, clearance. This is the *allowance*. The loosest condition (smallest shaft in largest hole) is 1.0012 − 0.9972, or 0.0040, clearance. The val-

**FIG. 77.** An ANSI clearance fit. This one is class RC 6. Tolerance on shaft, 0.0012 in.; tolerance on hole, 0.0012 in; allowance, 0.0016 in. clearance.

ues of these two limits are given in the table under "limits of clearance."

To illustrate further, suppose a 2-in. shaft and hub are to be fastened permanently with a drive fit. The hub is high-grade steel. Fit FN 4 has been chosen. Table 5 gives:

Hole: $\begin{array}{l} +1.2 \\ -0 \end{array}$

Shaft: $\begin{array}{l} +4.2 \\ +3.5 \end{array}$

These values are in thousandths of an inch. The basic size is 2.0000. Thus, for the hole 2.0000 + 0.0012 gives 2.0012 for the maximum limit, and 2.0000 − 0.0000 gives 2.0000 as the minimum limit. For the shaft, 2.0000 + 0.0042 gives 2.0042 for the maximum limit, and 2.0000 + 0.0035 gives 2.0035 for the minimum limit. The limits of fit are 0.0023 minimum interference and 0.0042 maximum interference. The allowance is 0.0042. The dimensioning of this fit is shown in Fig. 78.

Note in both the above cases that, in dimensioning, the maximum limit of the hole and the minimum limit of the shaft are placed below the dimension line. This is for convenience in reading the drawing and to help prevent mistakes by the machinist.

The ANSI fits are based on standard hole practice. Note in the tables (1 to 5) that the minimum hole size is always the basic size.

### 70. Cumulative Tolerances.

Tolerances are said to be cumulative when a position in a given direction is controlled by more than one tolerance. In Fig. 79 the holes are positioned one from another. Thus, the distance between two holes separated by two, three, or four dimensions will vary in position by the sum of the tolerances on all the dimensions. This difficulty can be eliminated by dimensioning from *one* position, which is used as a datum for all dimensions, as shown in Fig. 80. This system is commonly called base-line dimensioning.

Figure 81 is a further example of the effect of cumulative tolerances. The position of surface $Y$ with respect to surface $W$ is controlled by the additive

**FIG. 78.** An ANSI interference fit. This one is class FN 4. Tolerance on shaft, 0.0007 in.; tolerance on hole, 0.0012 in.; allowance, 0.0042 in. interference.

**FIG. 79.** Successive dimensioning. Tolerances accumulate.

**FIG. 80.** Dimensions from datum. Position from datum is subject to only one tolerance, but the center distance between two holes positioned from datum is subject to variation of two tolerances.

tolerances on dimensions $A$ and $B$. If it is important, functionally, to hold surface $Y$ with respect to surface $X$, the dimensioning used is good. If, however, it is more important to hold surface $Y$ with respect to surface $W$, the harmful effect of cumulative tolerances can be avoided by dimensioning as in Fig. 82. Cumulative tolerance, however, is always present; in Fig. 82 the position of surface $Y$ with respect

**FIG. 81.** Control of surface position through different toleranced dimensions. Here one dimension is successive.

**FIG. 82.** Control of surface position through different toleranced dimensions. All dimensions are from datum.

**FIG. 83.** Control of surface position through different toleranced dimensions. This drawing is overdimensioned.

to surface $X$ is now subject to the cumulative tolerances of dimensions $A$ and $C$.

In machine drawing, overdimensioning a drawing may cause confusion in the shop due to cumulative tolerances. This is illustrated in Fig. 83, where one of the surfaces will be positioned by two dimensions, both of which are subject to a tolerance. Thus surface $Z$ may be positioned with respect to surface $W$ by means of dimensions $A$, $B$, and $D$ and be within $\pm0.003$ in. of the basic position; this variation is inconsistent with the tolerance on dimension $E$. The situation can be clarified by assigning smaller

tolerances to dimensions $A$, $B$, and $D$ so that, cumulatively, they will be equal to $\pm0.001$ or less. This is poor practice, however, since it will probably increase the production cost. Another solution is to increase the tolerance on dimension $E$ to $\pm0.003$ if the function of the part will permit. The best solution, however, is to eliminate one of the four dimensions, since one dimension is superfluous. If all four dimensions are given, one should be marked "REF," and its tolerance thus eliminated.

**71. Tolerance between Centers.**

In all cases where centers are arranged for interchangeable assembly, the tolerance on shafts, pins, etc., and the tolerance on bearings or holes in the mating pieces will affect the possible tolerance between centers. In Fig. 84, observe that smaller tolerances on the holes would necessitate a smaller tolerance on the center distances. A smaller allowance for the fit of the pins would make a tighter fit and reduce the possible tolerances for the center-to-center dimensions. Study carefully the dimensions of both pieces.

**72. Tolerance of Concentricity.**

Tolerance of concentricity is a special case of tolerance in which there is a coincidence of centers. In most cases, concentric cylinders, cones, etc., gener-

**FIG. 84.** Tolerance on centers. The parts will not assemble unless all features are held within the limits shown.

ated about common axes in manufacture, will be concentric to a degree of precision more than adequate for functional requirements, and no statement is required on the drawing concerning the allowable variation. However, mating pairs of two (or more) precise, close-fitting, machined cylindrical surfaces must have the axes of adjoined cylinders closely coinciding in order to permit assembly of the parts; thus it is sometimes necessary to give the permissible deviation from concentricity. Since the center lines of adjoining cylinders coincide on a drawing, the tolerance cannot be given as a dimension. One method of indicating it is to mark the diameters with reference letters and give the tolerance in note form as in Fig. 85a. The reference letters can be dispensed with if the note is applied directly to the surfaces, as at (b).

### 73. Tolerance for Angular Dimensions.
When it is necessary to give the limits of an angular

dimension, the tolerance is generally bilateral, as $32 \pm \frac{1}{2}°$. When the tolerance is given in minutes, it is written $\pm 0°10'$; and when it is given in seconds, it is written $\pm 0'30''$. Where the location of a hole or other feature depends upon an angular dimension, the length along the leg of the angle governs the angular tolerance permitted. A tolerance of $\pm 1°$ gives a variation of 0.035 in. for a length of 1 in. and can be used as a basis for computing the tolerance in any given problem.

As an example, assume an allowable variation of 0.007 in.; then $(0.007/0.035) \times 1° = \frac{1}{5}°$ is the angular tolerance at 1 in. If the length is assumed as 2 in., the tolerance would be one-half the tolerance computed for 1 in., or $\frac{1}{5}° \times \frac{1}{2} = \frac{1}{10}°$, or $0°6'$.

### 74. Coinciding Center Lines and Dimensions.
In many cases the center lines for two different features of a part will coincide. Often one center line

**FIG. 85.** Tolerance of concentricity. (a) Shown by general note; (b) shown by note with leaders to surfaces.

FIG. 86. Coinciding center lines and dimensions. Clarification is obtained by giving dimensions for separate operations.

the patternmaker. If holes are to be machined in this link, the drawing would be as at (b); the patternmaker would not use the dimension between centers, as shown, but would assume the nominal dimension of $1\frac{1}{2}$ in. with the usual pattern tolerance of $\pm\frac{1}{32}$ in. The clearest dimensioning in this case would be as at (c).

In any instance where there is a coincidence of centers, it may be difficult to indicate the limits within which the coincidence must be maintained. At (c) there are actually two horizontal center lines, one for the cast link and another for the hole centers. One method of controlling the deviation from coincidence is to give the wall thickness as a minimum, which is understood to apply in all radial directions.

In cases where the coincident center lines are both for finished features with differing tolerances, there may be a serious ambiguity on the drawing unless the dimensioning is specially arranged. Figure 87a shows a milled slot with nominal dimensions and, on the coincident center lines, two accurate holes with a closely toleranced center distance. Unless the dimensioning is cleared by two separate dimensions, as shown, the machinist would not know the difference in tolerance. A somewhat more difficult case is shown at (b) where pairing holes are diagonally opposite. Unless all the holes are to be toleranced the same on their center distance, the dimensioning must be made clear with notes, as shown.

is for an unfinished feature and the other (and coincident) center line for a finished feature. Figure 86a shows the drawing of a link dimensioned for

FIG. 87. Coinciding center lines and dimensions. Clarification is obtained by giving dimensions for separate features.

## 75. Positional Tolerances.

Figures 61 and 62 show holes positioned by rectangular or polar coordinates. This method of positioning by tolerances on two dimensions produces a square tolerance zone for the common case where the positioning dimensions are at right angles to each other. The engineering intent can often be specified more accurately by giving information that states the *true position*, with tolerances to indicate how far actual position on the part can vary from true position. True position denotes the basic or theoretically exact position of a feature. This method results in a circular tolerance zone when the tolerance applies in all directions from true position.

Features such as holes may be allowed to vary in any direction from the specified true position. Features such as slots may be allowed to vary from the specified true position on either side of the true-position plane. Thus there are two methods of applying true-position tolerances. Both methods are shown in Fig. 88.

When a feature is allowed to vary in any direction, a note in one of the following forms should be used:

(*a*) 6 holes located at true position within 0.010 dia.

(*b*) 6 holes located within 0.005R of true position.

When features are allowed to vary from a true-position plane, a note in one of the following forms should be used:

(*a*) 6 slots located at true position within 0.010 wide zone.

(*b*) 6 slots located within 0.005 either side of true position.

When a feature is allowed to vary in any direction from true position, the position of the feature is given by untoleranced dimensions, as shown in Figs. 88 and 89. The fact that there is no allowable variation on the positioning dimensions should be indicated either on the drawing, in the title block, or in a separate specification, by a note—"Dimensions locating true position are basic." If this is not done, the word "basic" should be shown on each dimension subject to true-position tolerance.

## 76. Maximum Material Condition.

Since all features of a part have allowable variations in size, for *mating* parts the least favorable assembly condition exists when the mating parts are both at their maximum material condition. This means that a hole is at its minimum size and a shaft at its maximum. In terms of the cylindrical surface of a hole, it means that no point on the surface will be inside a cylinder having a diameter equal to the *actual* diameter of the hole, minus the true-position tolerance (diameter or twice the radius of the tolerance circle), the axis of the cylinder being at true position.

Where the maximum material condition applies, it is stated by the addition of "maximum material condition" to the true-position note, or by using the abbreviation "MMC," as shown in Fig. 89. Also, MMC may be stated in a general note or a specification.

In some cases it may be necessary to state a positional tolerance without reference to MMC. This is done by the reference "regardless of feature size" (abbreviated RFS), as shown on Fig. 89, or by a general note or specification.

**FIG. 88.**  True-position dimensioning (ANSI).

**FIG. 89.**  True-position dimensioning (ANSI).

In most cases the datum for true position is obvious from the dimensioning itself, but where there may be any doubt, the positional tolerance note should read "XX holes located within .xxxR of true position in relation to datum *A*." Of course, datum *A* must be clearly marked on the drawing.

For a complete discussion of true-position tolerancing and related details, see ANSI Y14.5 (tentative) or any subsequent revisions thereof.

## 77. Zone Tolerances.

A zone tolerance may be specified where a uniform variation can be permitted along a contour, as in Fig. 90. For complete coverage, see ANSI Y14.5 or any subsequent revisions thereof.

## 78. Form Tolerances.

Tolerances of form state how far actual surfaces may vary from the perfect geometry implied by the drawing. The methods of indicating straightness, flatness, and parallelism are shown in Fig. 91; squareness in Fig. 92; angularity, symmetry, concentricity, and roundness in Fig. 93; and parallelism with a surface and another hole in Fig. 94. For complete details on form tolerances, see ANSI Y14.5 (tentative).

## 79. Surface Quality.

The proper functioning and wear life of a part frequently depend upon the smoothness quality of its surfaces. ANSI B46.1 defines the factors of surface quality and describes the meaning and use of symbols on drawings. Any surface, despite its apparent smoothness, has minute peaks and valleys, the height of which is termed "surface roughness" and which may or may not be superimposed on a more general *"waviness."* The most prominent direction of tool marks and minute scratches is called "lay."

*Roughness,* produced principally by cutting edges and tool feed, is expressed as the arithmetical average from the mean in microinches (Fig. 95).

*Roughness width* is rated in inches as the maximum permissible spacing between repetitive units of the surface pattern (Fig. 95).

FIG. 90.    Zone tolerances (ANSI).

FIG. 91.    Form tolerances. Straightness, flatness, and parallelism (ANSI).

*Roughness-width cutoff* is the maximum width in inches of surface irregularities to be included in the measurement of roughness height (Fig. 95).

*Waviness* designates irregularities of greater spacing than the roughness, resulting from factors such as deflection and vibration (Fig. 95). The height is rated in inches as peak-to-valley height. The width is rated in inches as the spacing of adjacent waves.

*Lay* is the direction of the predominant surface

**FIG. 92.** Form tolerances. Squareness (ANSI).

**FIG. 93.** Form tolerances. Angularity, symmetry, concentricity, and roundness (ANSI).

pattern, produced by tool marks or grains of the surface ordinarily determined by the production method used (Fig. 95).

Although other instruments are used, the common method of measuring surface roughness employs electrical amplification of the motion of a stylus over the surface.

*The symbol* for indicating surface irregularities is a "check mark" with the horizontal extension as shown in Figs. 96 to 103. Symbols for lay and the application of the symbols on a drawing are shown in Fig. 104. The caption explains each case.

*Values* for roughness height, waviness height, and roughness-width cutoff are shown in Table 1.

Figure 105 is a chart adapted from several sources, giving the range of surface roughness for various methods.

## Table 1

| Recommended roughness-height values (microinches) | | | | |
|---|---|---|---|---|
| 0.25 | 5 | 20 | 80 | 320 |
| 0.5 | 6 | 25 | 100 | 400 |
| 1.0 | 8 | 32 | 125 | 500 |
| 2.0 | 10 | 40 | 160 | 600 |
| 3.0 | 13 | 50 | 200 | 800 |
| 4.0 | 16 | 63 | 250 | 1000 |

| Recommended waviness-height values (inches) | | | | | |
|---|---|---|---|---|---|
| 0.00002 | 0.00008 | 0.0003 | 0.001 | 0.005 | 0.015 |
| 0.00003 | 0.0001 | 0.0005 | 0.002 | 0.008 | 0.020 |
| 0.00005 | 0.0002 | 0.0008 | 0.003 | 0.010 | |

| Recommended roughness-width-cutoff values (inches) | | | | | |
|---|---|---|---|---|---|
| 0.003 | 0.010 | 0.030 | 0.100 | 0.300 | 1.0000 |

# PRODUCTION METHODS

### 80. Production Methods and Dimensioning Practice.

Production methods can be classified as (1) *unit production*, when one or only a few devices or structures are to be built, and (2) *quantity* or *mass production*, when a large number of practically identical machines or devices are to be made with the parts interchangeable from one machine to another.

*Unit-production* methods almost always apply to large machines and structures, especially if they are custom-made. The large size to some extent eliminates the need for great accuracy. Each individual part is produced to fit or is fitted to the adjacent parts, frequently on the job, by experienced workmen in accordance with common fractional dimensions and directions given on the drawings. Since interchangeability of parts is no object, tolerances are not ordinarily used.

Similar methods are employed for unit production of smaller machines and mechanical devices. The drawings may have common fractional dimensions exclusively, on the assumption that the parts will be individually fitted in the shop. If this practice is followed, the manufacturing group accepts the re-

**FIG. 94.** Form tolerances. Parallel to surface and parallel to hole (ANSI).

**FIG. 95.** Surface roughness (ANSI).

**FIG. 96.** Symbol. The roughness height rating is placed at the left of the long leg.

**FIG. 97.** Symbol. The specification of maximum and minimum roughness height indicates the allowable range.

**FIG. 98.** Symbol. The maximum waviness height rating is placed above the horizontal extension.

**FIG. 99.** Symbol. The maximum waviness width rating is placed to the right of the waviness height rating.

**FIG. 100.** Symbol. To specify the contact area, when required, the percentage value is placed above the extension line.

**FIG. 101.** Symbol. Lay designation is given by the lay symbol placed to the right of the long leg.

**FIG. 102.** Symbol. The roughness width cutoff rating is placed below the horizontal extension.

**FIG. 103.** Symbol. When it is required, the maximum roughness width rating is placed at the right of the lay symbol.

sponsibility for the proper functioning of the machine, and in some cases even for designing some of the parts. Skilled workmen are employed for this work. Usually one machine is completed before another is started, and the parts are not interchangeable.

*Quantity-production* methods are used whenever a great many identical products are made. After a part has been detailed, the operations-planning group of the engineering department plans the shop operations step by step. Then special tools are designed by the tool-design group so that, in production, semiskilled workmen can perform operations that would otherwise require skilled workmen. These

tools, built by a highly skilled toolmaker, simplify and greatly increased the rate of production.

One workman performs a single operation on a part; then it is passed to a second workman, who performs another operation; and so on until the part has been completed. Specially designed tools and equipment make it possible to produce economical parts that have dimensional exactness consistent with the requirements for interchangeability. The assembly may also be made by semiskilled workers using special assembly fixtures and tools.

With this system nothing can be left to the judgment of the workman. In preparing drawings intended for quantity production, the engineering department must assume full responsibility for the success of the machine by making the drawings so exact and complete that, if followed to the letter, the resulting parts cannot fail to be satisfactory. The engineering department alone is in a position to correlate corresponding dimensions of mating parts, establish dimensional tolerances, and give complete directions for the entire manufacturing job.

It is sometimes expedient for concerns doing unit or small-production work to follow the methods of the quantity-production system. The advantage they gain is interchangeability of parts, which can be produced without reference or fitting to mating parts.

## 81. Principles for the Selection of Dimensions.
Systematic selection of dimensions demands consideration of the *use* or *function* of the part and the *manufacturing process* to be used in producing it.

SYMBOLS FOR LAY

APPLICATION

**FIG. 104.** Lay symbols and application of surface-quality symbols on the drawing.

The *functional principle* recognizes that it is essential to dimension between points or surfaces associated through their functional relationship with points or surfaces of mating parts. This is accomplished by correlating the dimensions on a drawing of one part with the mating dimensions on the drawing of a mating part and arranging the tolerances of these dimensions to ensure interchangeability and proper functioning.

The *process principle,* or "workman's rule," as it is sometimes called, recognizes that the work of manufacture can be made easier by giving directly the dimensions the shop will find most convenient to "work to" in producing the part. Here a knowledge of manufacturing processes and procedure is necessary, as explained in Chap. 9.

In some cases there may be a conflict between these two principles. Whenever there is, the functional principle must take precedence, as any attempt to satisfy both principles would result in overdimensioning, as described in Sec. 24, causing confusion for the workmen and possible malfunctioning of the part. With few exceptions, however, dimensions can be chosen to satisfy both principles.

## 82. Procedure for the Selection of Dimensions.
A systematic procedure for selecting dimensions is of course desirable. The following steps can serve as a guide:

1. Carefully study the part along with its mating part or parts. Pay particular attention to the mating and controlling surfaces. Plan dimensions meeting functional requirements before placing any dimensions on the drawing so that you can correlate dimensions of mating parts.

2. Study the part to determine whether or not the manufacturing processes can be simplified by an alteration of any of the functional dimensions. Do not make changes if the functioning of the part would be impaired in any way.

3. Select the nonfunctional dimensions, being guided by the process principle, so that the dimensions will be readily usable by the workmen. Avoid overdimensioning and duplication.

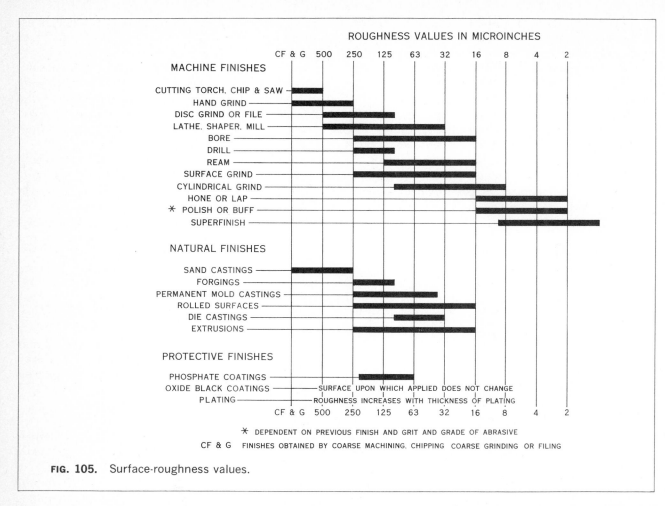

**FIG. 105.** Surface-roughness values.

In general, dimensions for mating surfaces are governed by the functional and process principles, and the dimensions for nonmating surfaces are governed by the process principle only.

Occasionally the manufacturing process will not be known at the time of dimensioning. This may happen when there are optional methods of manufacture, all equally good, or when the details of the manufacturing equipment of a contracting firm are not known. When this is the case, the dimensions should be selected and toleranced in a logical manner so that the part, regardless of the method by which it is produced, cannot fail to be satisfactory.

The size and location dimensions arrived at by the shape-breakdown system described in Art. 22 apply here to a great extent since these dimensions fulfill most production requirements. Contracting firms often redraw incoming part drawings, dimensioning them for the most economical production with their own shop equipment.

### 83. Methods of Part Production.

In following the process principle, the basic methods of part production—casting, forging, etc.—as described in Chap. 9, must be known. The manufacturing procedures followed with the particular type of part involved are then considered in selecting the dimensions. The only workman to be considered in dimensioning a part cut from solid stock is the machinist. For parts produced by casting, the workmen to be considered are the patternmaker (for sand castings) or the diemaker (for die castings) and, for finishing, the machinist. Forged parts subject to quantity production are dimensioned for the diemaker and machinist. For parts produced from sheet stock, the template maker, the diemaker, and the machinist must be considered; information for making the template and for forming the blank is obtained from a detail drawing showing the part as is should be when completed. In every case one drawing, appropriately dimensioned, must show the finished part.

The sections that follow give examples of the dimensioning of machine parts for quantity production.

## 84. Dimensioning a Part Machined from Stock.

Figure 106 is a detail drawing of the stud from the rail-transport hanger in Prob. 51, final chapter. Study the drawing to determine the function of the stud. The stud is produced by machining on a lathe. Cold-rolled steel stock, $1\frac{1}{4}$ in. in diameter, is used. The stock diameter is the same as the large end of the stud, thus eliminating one machining operation.

Shape breakdown of the part results in a series of cylinders each requiring two dimensions—diameter and length. The important functional dimensions have been marked (on Fig. 106) with the

FIG. 106. Dimensioning a part machined from stock.

letters $A$, $B$, $C$, and $D$. Diameter $A$ is given to correlate with the bore of the bearings; a four-place decimal limit provides for the desired fit. Dimension $B$ is a three-place decimal limit to correlate with a similar dimension for the hole in the hanger. Dimension $C$ is made 0.03 in. larger than the combined width of the two bearings in order to allow the inner races of the bearings to "creep." Dimension $C$, a two-place decimal, can vary ±0.030 in., but clearance for the bearings is assured under all conditions. Dimension $D$ is made approximately 0.05 in. less than the length of the hanger hub to ensure that the nut will bear against the hanger rather than on the shoulder of the stud.

Functional dimensions need not always be extremely accurate. Note that dimensions $C$ and $D$, with the relatively broad tolerance of two-place decimals, will allow the part to function as intended.

The thread specification can be considered as a functional dimension with the tolerance provided through the thread class.

The remainder of the dimensions selected are those that best suit shop requirements. Note that the thread length and over-all dimension cannot both be given, or the part would be overdimensioned.

## 85. Dimensioning a Casting.

The dimensions required for sand castings can be classified as those used by the patternmaker and those used by the machinist. Since a cast part has two distinct phases in its manufacture, we will discuss a particular case first with the drawings made according to the multiple system explained in Chap. 9, one for the patternmaker (Fig. 107) and one for the machinist (Fig. 108).

**FIG. 107.** Dimensioning an unmachined casting.

For "Casting" see Dwg. 87342.

5.000 ±.005

3.500 +.003

2.000 −.005

$\frac{3}{8}$ (A)

$\frac{1}{2}$ Drill

3.000 ±.005

3.000 +.003 Bore

| DWN | R.S. | 6-7-47 | LIMITS, UNLESS OTHERWISE NOTED: FRACTIONAL ±$\frac{1}{64}$. DECIMAL ±.010, ANGULAR ±$\frac{1}{2}$° | | | REVISION | DATE | | CHANGE OR ADDITION | | |
|---|---|---|---|---|---|---|---|---|---|---|---|
| TCD | R.S. | 6-7-47 | PART NO. | REQD | NAME | MATERIAL | | STOCK SIZE | | HEAT TR. | FINISH |
| CKD | L.K.D. | 6-8-47 | 142 | 4 | ROCKER ARM | C. I. | 87342 Casting | | | | As Shown |
| APPD | James | 6-10-47 | | | | SCALE $\frac{1}{2}$ Size | COMPANY NAME CITY | | | | DRAWING NO. |
| | NAME | DATE | UNIT OR ASSEMBLY | | MODEL 47 DIESEL | ASSY. DR. NO. 87200 | | | | NO.   OF | 87343 |

FIG. 108.    Dimensions for machining a casting.

The *casting drawing* gives the shape of the un-machined casting and carries dimensions for the patternmaker only. Shape breakdown will show that each geometric shape has been dimensioned for size and then positioned, thus providing dimensions easily usable by the workman. Some of the dimensions might be altered, depending upon how the pattern is made; the most logical and easily usable combination should be given. Note that the main central shape is dimensioned as it would be laid out on a board. Note also that several of the dimensions have been selected to agree with required functional dimensions of the machined part although the dimensions employed have been selected so as to be directly usable by the patternmaker; they also achieve the *main objective*, which is to state *the sizes that the unmachined casting must fulfill when produced.*

The *machining drawing* shows only the dimensions required by the machinist. These are almost all functional dimensions and have been selected to correlate with mating parts. It is important to note that a starting point must be established in each of the three principal directions for machining the casting. In this case a starting point is provided by (1) the coincidence of the center lines of the large hole and cylinder (positioning in two directions) and (2) dimension $A$ to position the machined surface on the back, from which is positioned the drilled hole. Dimension $A$ is a common fraction carrying the broad tolerance of ±$\frac{1}{64}$ in., as there is no functional reason for working to greater precision.

Figure 109 is a drawing of the same part used in Figs. 107 and 108 but with the casting drawing dispensed with and the patternmaker's dimensions incorporated in the drawing of the finished part. In combining the two drawings, some dimensions have been eliminated, as the inclusion of all the dimensions of both drawings would result in over-dimensioning; thus the patternmaker must make use of certain machining dimensions in his work. In working from the drawing in Fig. 109, the patternmaker provides for machining allowance, being guided by the finish marks. In the drawing in Fig. 107 the engineering department provides for the machining allowance by showing and dimensioning the rough casting oversize where necessary for machining, and no finish marks are used.

## 86. Dimensioning a Drop Forging.

Figure 110 is a drawing of a drop forging showing, at the left, the unmachined forging and, at the right, the machined forging. The drawing of the unmachined forging carries the dimensions it must fulfill when produced; these dimensions have been selected so as to be most useful to the diemaker for producing the forging dies. As the draft on drop forgings is considerable, it is shown on the drawing and dimensioned (usually) by a note. If the draft varies for different portions of the part, the angles may be given on the views. The dimensions parallel to the horizontal surfaces of the die are usually given so as to specify the size at the *bottom of the die cavity*. Thus, in dimensioning, visualize the draft as stripped off; then its apparent complication will no longer be a difficulty.

The machining drawing shows the dimensions for finishing. These are all functional, selected from the standpoint of the required function of the part. Study the illustration carefully.

## 87. Dimensioning a Sheet-metal Part.

Parts to be made of thin materials are usually drawn showing the part in its finished form, as in Fig. 111.

**FIG. 109.** All dimensions for a casting.

10/DIMENSIONS, NOTES, LIMITS, AND PRECISION 393

**FIG. 110.** Dimensioning a drop forging.

Draft angle 7°
Unspecified radii .10

**FIG. 111.** Dimensioning a sheet-metal part.

L.H. Bracket shown.
Make R.H. opposite.
Bend radii ⅛

.130 +.005 Punch
8 Holes

The template maker first uses the drawing to lay out a flat pattern of the part. If only a few parts are to be made, this template will serve as a pattern for cutting the blanks. Then the part will be formed and completed by hand. If a large number of parts are to be made, the diemaker will use the template and drawing in making up the necessary dies for blanking, punching, and forming. The work of both template maker and diemaker is simplified by giving the dimensions to the same side of the material inside or outside, whichever is more important from the functional standpoint, as shown in Fig. 111. Dimensions to rounded edges (bends) are given to the theoretical sharp edges, which are called *mold lines.* The thickness of the material is given in the "stock" block of the title strip. Note in the figure that the holes are positioned in groups (because of functional requirements) and that important functional dimensions are three-place decimals.

### 88. Metric Dimensioning.

There are no significant differences (except for details) between the practices of (1) metric and (2) inch dimensioning. The only important item to specify is the *units* in which the dimensions are given. Thus, a prominent note on the drawing must state: All dimensions in [give unit] unless otherwise specified. The units given may be *millimeters, centimeters* or *meters.*

The following, adapted from the ANSI standards on dimensioning, (ANSI Y14.5-1973), gives instructions for metric dimensioning and, *specifically, dual* dimensioning (both *inch* and *metric* units).

## DUAL DIMENSIONING[1]

### General

Dual dimensioning is a procedure where both U.S. customary (inch) units and SI (metric) units of measurement are shown on the same engineering drawing. This procedure permits the use of certain ISO (International Organization for Standardization) recommendations related

[1]Caution: This form of dimensioning is applicable where interchangeable parts are to be manufactured in both US customary (inch) and SI (metric) units of measurement. It is not recommended as a practice for implementing the transition from inch to metric units.

to drawing practices in conjunction with SI units. The International System of Units (SI) is described in the NBS (National Bureau of Standards) Special Publication 330 and in ISO/R1000.

### Interchangeability

Both U.S. customary and SI dimensions on a drawing must achieve the required degree of interchangeability of parts from a drawing. Interchangeability is determined at the time of dimensional conversion by the number of decimal places retained when rounding off a converted dimension as well as to what degree the tolerance limits of the conversion are allowed to violate the limits of the original dimension. Suggested sources for conversion principles and conversion tables are:

1. Rules for Conversion and Rounding—American National Standard Z210.1-1973, Standard Metric Practice Guide.

2. Conversion of Toleranced Linear Dimensions—Society of Automotive Engineers J390, Dual Dimensioning Standard.

3. National Bureau of Standards Miscellaneous Publication 286, Units of Weights and Measure, International (Metric) and U.S. Customary.

Examples herein demonstrate placement and presentation practices only and do not exemplify conversion practice.

### Identification of Units

Each drawing must identify which units are U.S. customary and which are SI. Two methods are illustrated to distinguish the U.S. customary unit from the SI unit, either the position or the bracket (symbol) method.

**Position method.**   Units may be displayed in either of two ways:

(a) Show the millimetre dimension above the inch dimension, separated by a horizontal line, or to the left of the inch dimension, separated by a virgule (slash line).

*Examples:*

(b) Same as (a) except inch dimension above or to the left of the millimetre dimension.

**Bracket method.** Units may be displayed in either of two ways:

(a) Show the millimetre dimension enclosed in square brackets (symbol [ ]). Location of the millimetre dimension is optional but shall be uniform on any drawing; that is, above or below, or to the left or the right of the inch dimension.

*Examples:*

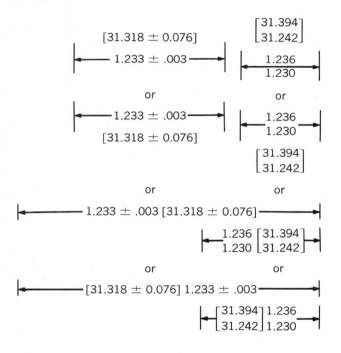

(b) Same as (a), except enclose the inch dimension in square brackets instead of the millimetre dimension.

Computer prepared drawings may require the use of parentheses rather than brackets due to computer character limitations. In such cases, parentheses shall not be used to identify reference dimensions.

Only one method, position (a) or (b) or bracket (a) or (b), shall be used to identify the inch and millimetre dimensions on a single drawing. Each drawing shall illustrate or make note of how the inch and millimetre dimensions can be identified. An illustration can apply to the position or bracket method as:

$$\frac{\text{MILLIMETRE}}{\text{INCH}}; \text{MILLIMETRE/INCH}$$

or

$$\frac{[\text{MILLIMETRE}]}{\text{INCH}}; [\text{MILLIMETRE}]\ \text{INCH}$$

Note form may be used for the bracket method, as:

DIMENSIONS IN [ ] ARE MILLIMETRES

or

DIMENSIONS IN [ ] ARE INCHES

**Other units.** Units other than linear dimensions and all units in notes or text on a drawing are shown in the same manner as that for dimensions, except the appropriate measurement unit symbol is specified.

*Example:*

DENSITY [32 kg/m³] 2 LB/FT³

**Common units.** The callout of some unit quantities can satisfy both systems of measurement, that is, 0.006 inch per inch of taper and 0.006 millimetre per millimetre have the same rate and can be expressed as a ratio, .006 : 1. A note could read, TAPER .006 : 1.

**Angles.** Angles stated in degrees, minutes, and seconds, or in degrees and decimals of a degree are common to the inch and metric systems of measurement.

**Nominal designations.** Designations such as nominal thread sizes, pipe sizes, and wood cross-sectional sizes, are not converted. In such cases, the origin of the drawing controls which measurement system is the basis of such designations.

**Practices Applying to SI Values.**

In general, SI numerical values appear on engineering drawings in the manner prescribed by international drawing practices. The following paragraphs point out the manner in which zeros are used. Two other details are shown that apply equally to both the inch and the SI values.

**Use of zeros.** A zero precedes a decimal point in an SI value of less than unity.

*Example:* 0.13

Zeros do not follow a significant digit of an SI value to the right of the decimal point. Where equal plus and minus tolerancing is used, the millimetre dimension and its tolerance may not have the same number of decimal places. In addition, if a millimetre dimension is a whole number, neither the decimal point nor a zero is shown.

Where unilateral tolerancing is used and either the plus or minus tolerance is nil, a single zero is shown without a plus or minus sign.

*Example:*

$$32 \quad {}^{0}_{-0.02} \quad \text{or} \quad 32 \quad {}^{+\ 0.02}_{0}$$

Where bilateral tolerancing is used, both the plus and minus tolerance have the same number of decimal places, using zeros where necessary.

*Example:*

$$32 \quad {}^{+0.25}_{-0.10} \quad \text{not} \quad 32 \quad {}^{+0.25}_{-0.1}$$

Where limit dimensioning is used and either the maximum or minimum value has digits following a decimal point, the other value has zeros added for uniformity.

*Example:*

$$\begin{matrix} 25.45 \\ 25.00 \end{matrix} \quad \text{not} \quad \begin{matrix} 25.45 \\ 25 \end{matrix}$$

**Commas.** Commas shall not be used to denote thousands with either U.S. customary or SI values because it is the practice in many countries to use the comma as a decimal point sign. Another practice is to leave a space where a comma would otherwise be used to denote thousands. On engineering drawings, a space may be interpreted as a place where a decimal point was missing, particularly when a reproduction of the drawing was made. Therefore, spaces and commas are not permitted in numerical expressions. See following example:

*Example:*

$$32541 \text{ not } 32{,}541 \quad \text{nor} \quad 32\ 541$$

**Application.**
Some procedures related to the metric drawing practice and individual company drawing practices could cause difficulty in producing a dual dimensioned drawing. The following paragraphs indicate a uniform procedure that will eliminate a drafting problem or a language translation problem and will clarify some of the questions that seem to be frequently asked.

**Decimal point.** A period is used for the decimal point sign for both U.S. customary and SI units of measurement. (The comma, which is used as the decimal point sign in many metric or SI oriented countries, could have been used with SI units without prejudice to the principles established herein.)

**Angle of projection.** Many metric or SI oriented countries use first angle projection while third angle projection is normal U.S. practice. To avoid confusion, dual dimensioned drawings shall specify the angle of projection. See American National Standard for Projections, Y14.3.

**New drawings.** Dual dimensioning of new drawings is facilitated if all dimensions are shown in decimals except where fractions normally designate nominal sizes.

**Existing drawings with fractional dimensions.** Existing drawings with dimensions in fractions may be dual dimensioned by adding the millimetre conversion adjacent to the fractional inch dimension. Fractions are considered the same magnitude as two place decimal inch dimensions when determining the number of places to retain in the millimetre conversion.

*Examples:*

$$[61.91 \pm 0.4] \qquad [61.91 \pm 0.4]$$

$$\longleftarrow 2^{7}/_{16} \pm {}^{1}/_{64} \longrightarrow \quad | \quad \longleftarrow 2^{7}/_{16} \pm {}^{1}/_{64} \longrightarrow$$

**General tolerances.** General tolerances expressed as part of the drawing format or in a general note shall be dual identified.

# PROBLEMS

The following problems are given as studies in dimensioning in which the principles presented in this chapter are to be applied. Attention should be given to the methods of manufacture. Assume a function for the part in order to fix the position of finished surfaces and limit possibilities in the selection of dimensions.

### Group 1. Dimensioned Drawings from Models.

Excellent practice in dimensioning is afforded by making a detail drawing from a pattern, casting, or forging or from a model made for the purpose. Old or obsolete patterns can often be obtained from companies manufacturing a variety of small parts, and "throw-out" castings or forgings are occasionally available. In taking measurements from a pattern, a shrink rule should be used; allowance must be made for finished surfaces.

### Group 2. Pieces to be Drawn and Dimensioned.

The problem illustrations are printed to scale, as indicated in each problem. Transfer distances with dividers or by scaling, and draw the objects to a convenient scale on paper of the size to suit. For proper placement of dimensions, more space should be provided between the views than is shown in the illustrations.

Use the aligned or horizontal dimensioning system, as desired. It is suggested that some problems be dimensioned in the complete decimal system.

**13.** Stud shaft, shown half size. Machined from steel-bar stock.

**PROB. 12.**   Desurger case. (*Courtesy of Westinghouse Air Brake Co.*)

**PROB. 13.**   Stud shaft.

**14.**   Shaft bracket, shown half size. Malleable iron. Hole in base is drilled and counterbored for a socket-head cap screw. Base slot and front surface of hub are finished. Hole in hub is bored and reamed. The function of this part is to support a shaft at a fixed distance from a machine bed, as indicated by the small pictorial view.

**15.**   Idler bracket, shown half size. Cast iron. Hole is bored and reamed. Slot is milled.

**15A.**   See Prob. 15. Draw and dimension the right-hand part.

**PROB. 14.**   Shaft bracket.

**PROB. 15.**   Idler bracket, left hand.

**16.**   Filter flange, shown half size. Cast aluminum. The small holes are drilled. Add spot faces.

**17.**   Boom-pin rest. Steel drop forging. Shown half size; draw half size or full size. Add top view if desired. Show machining allowance with alternate position lines. All draft angles 7°. Holes are drilled, corner notches milled.

**17A.**   Same as Prob. 17, but make two drawings; (*a*) the unmachined forging dimensioned for the diemaker and (*b*) the machined forging dimensioned for the machinist. Reference: Fig. 110.

**PROB. 16.**   Filter flange.

**PROB. 17.**   Boom-pin rest.

**18.** Clutch lever. Aluminum drop forging. Shown half size, draw full size or twice size. Add top view if desired. Holes are drilled and reamed; ends of hub are finished; left-end lug is straddle milled; slot in lower lug is milled. Show machining allowance with alternate position lines, and dimension as in Fig. 8, Chap. 9. Draft angles 7°.

**18A.** Same as Prob. 18, but make two drawings: (*a*) the unmachined forging dimensioned for the diemaker and (*b*) the machined forging dimensioned for the machinist. Reference: Fig. 110.

**19.** Radiator mounting clip, LH, no. 16 (0.0625) steel sheet. Shown half size. Holes and slot are punched. Reference: Fig. 111.

**PROB. 18.** Clutch lever.

**PROB. 19.** Radiator mounting clip, left hand.

**20.** Pulley bracket. Shown half size. Aluminum sheet, 24-ST, 0.032 in. thick.

**21.** Using the scale shown, draw and dimension the check-valve body.

**22.** Using the scale shown, draw and dimension the hydrostatic pressure housing. Draw to convenient scale on paper of a size adequate for clear representation. Use either the fractional or decimal system.

**23.** Using the scale shown, draw and dimension the universal-joint housing. Use a scale and paper size adequate for clear representation. Either the fractional, metric, or decimal system may be used.

**24.** Using dividers on the scale shown, draw and dimension the universal joint. Note that the flexing-rod portions are helical. Use fractional or decimal system.

**PROB. 20.**    Pulley bracket.

SECTION A-A

1 N.P.T.

A

A

0    1    2    3    4

SCALE - INCHES

**PROB. 21.**    Check-valve body.

DETAIL A

SECTION A-A

DETAIL A
View Enlarged Four Times

Case To Be Hydrostatically
Tested At 6000 P.S.I.

1½ -12 UNF-2B THD.

SECTION B-B

⅛ Drill × ¼ Dp
4 Holes, for
Ident. Plate

⅜-24 UNF-2B THD.

0 1 2 3 4 5 6
SCALE – INCHES

**PROB. 22.** Hydrostatic pressure housing.
(*Courtesy of Westinghouse Air Brake Co.*)

0 1 2 3 4 5 6
SCALE – INCHES

**PROB. 23.** Universal-joint housing.

**PROB. 24.**   Universal joint.

## Group 3. Dimensioned Drawings from an Assembly or Design Drawing.

The assembly drawings given in the problem section of the final chapter are well suited for exercises in dimensioning detail working drawings. The assembly shows the position of each part, and the function can be understood by a study of the motion, relationship, etc., of the different parts. Note particularly the mating and controlling surfaces and the logical reference surfaces for dimensions of position. Inasmuch as the dimensioning of an assembly drawing is crowded, and it will probably not even have the same views as the detail drawing of one of the parts, disregard the position and selection of the assembly dimensions, and use them only to obtain sizes.

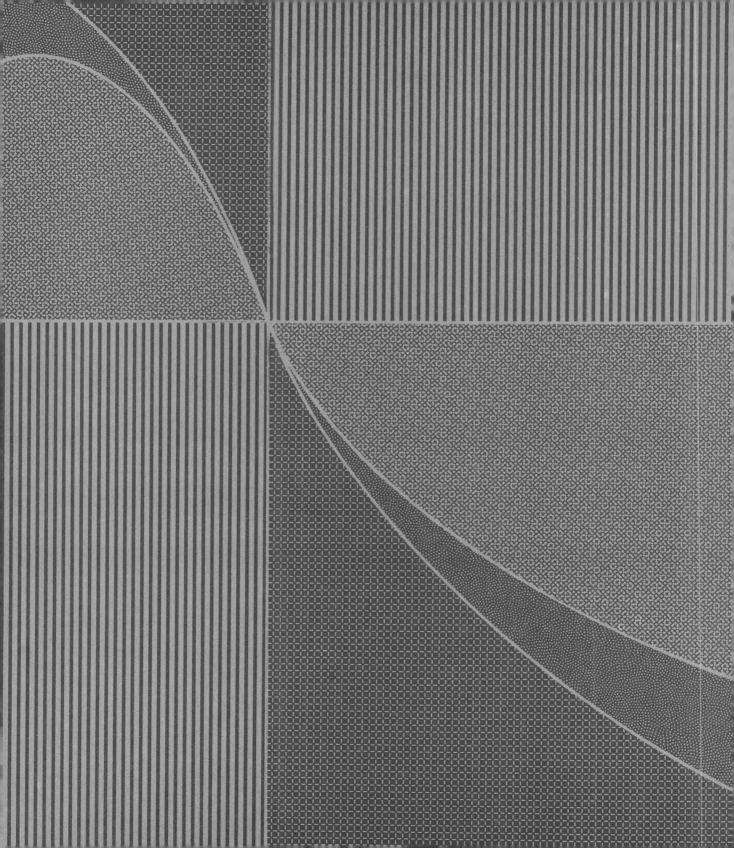

# CHARTS, GRAPHS AND DIAGRAMS

# 11

These Are Graphic Representations of Data and
Are Fundamental to All of Science and
Engineering

Graphic Presentation of Facts:
Representation of Data • Rectilinear Charts •
The Drawing of Curves • Titles and Notation •
To Draw a Chart • Standards • Logarithmic,
Semilogarithmic, Polar, and Trilinear Charts •
Empirical Data and Charts • Popular, Bar, Pie,
Strata, Volume, and Pictorial Charts • Use of
Color for Charts • Problems

## 1. Graphic Presentations—
## Their Importance.

This chapter is given as an introduction to the use of graphic methods in tabulating data for analysis, solving problems, and presenting facts. We will discuss the value of this application of graphics in engineering and suggest ways of studying the subject further.

The graphic chart is an excellent method for presenting a series of quantitative facts quickly. When properly constructed, charts, graphs, and diagrams constitute a powerful tool for computation, analysis of engineering data, and the presentation of statistics for comparison or prediction.

## 2. Classification.

Charts, graphs, and diagrams fall roughly into two classes: (1) those used for purely technical purposes and (2) those used in advertising or in presenting information in a way that will have popular appeal. The engineer is concerned mainly with those of the first class, but he should be acquainted also with the preparation of those of the second and understand their potential influence. The aim here is to give a short study of the types of charts, graphs, and diagrams with which engineers and those in allied professions should be familiar.

It is assumed that the student is familiar with the use of rectangular coordinates and that such terms as "axes," "ordinates," "abscissas," "coordinates," and "variables," are understood.

## 3. Representation of Data:
## Rectilinear Charts.

Because the preliminary chart work in experimental engineering is done on rectilinear graph paper, the student should become familiar with this form of chart early in his course. The rectilinear chart is made on a sheet ruled with equispaced horizontal lines crossing equispaced vertical lines. The spacing is optional. One commercial graph paper is divided into squares of $\frac{1}{20}$ in., with every fifth line heavier, to aid in plotting and reading. Sheets are available with various other rulings, such as 4, 6, 8, 10, 12, and 16 divisions per inch.

It is universal practice to use the upper right-hand quadrant for plotting experimental-data curves, making the lower left-hand corner the origin. In case both positive and negative values of a function are to be plotted, as occurs with many mathematical curves, the origin must be placed so as to include all desired values.

Figure 1 shows a usual form of rectilinear chart, such as might be made for inclusion in a report.

**FIG. 1.** A rectilinear chart. This type of chart shows how one variable changes with respect to another. Corresponding values can be determined.

## 4. Drawing the Curve.

In drawing graphs from experimental data, it is often a question whether the curve should pass through all the points plotted or strike a mean between them. In general, observed data not backed up by definite theory or mathematical law are shown by connecting the points plotted with straight lines, as in Fig. 2a. An empirical relationship between curve and plotted points may be used, as at (b), when, in the opinion of the engineer, the curve should exactly follow some points and go to one side of others. Consistency of observation is indicated at (c), in which case the curve should closely follow a true theoretical curve.

## 5. Titles and Notation.

The title is an important part of a chart, and its wording should be clear and concise. In every case, it should contain sufficient description to tell what the chart is, the source or authority, the name of the observer, and the date. Approved practice places the title at the top of the sheet, arranged in phrases symmetrically about a center line. If it is placed within the ruled space, a border line or box should set it apart from the sheet. Each sheet of curves should have a title; and when more than one curve is shown on a sheet, the different curves should be drawn so as to be easily distinguishable. This can be done by varying the character of the lines, using full, dashed, and dot-and-dash lines, with a tabular key for identification, or by lettering the names of the curves directly along them. When the charts are not intended for reproduction, inks of different colors can be used.

## 6. To Draw a Chart.

In drawing a coordinate chart, the general order is: (1) compute and assemble all data; (2) determine the size and kind of chart best adapted and whether to use printed or plain paper; (3) determine, from the limits of the data, the scales for abscissas and ordinates to give the best effect to the resulting curve; (4) lay off the independent variable (often *time*) on the horizontal, or $X$, axis and the dependent variable on the vertical, or $Y$, axis; (5) plot points from the data and pencil the curves; (6) ink the curve; (7) compose and letter the title and coordinates.

(a)

(b)

(c)

**FIG. 2.** Methods of drawing curves. (a) "Curve" for data not supported by theory and therefore not necessarily producing a smooth curve; (b) and (c) curves for data known to produce smooth curves.

The construction of a graphic chart requires good draftsmanship, especially for the lettering, but in engineering and scientific work the primary considerations are judgment in the proper selection of coordinates, accuracy in plotting points and drawing the graph, and an understanding of the functions and limitations of the resulting chart.

When the chart is drawn on a printed form, to be blueprinted, the curve can be drawn on the reverse side of the paper, enabling erasures to be made without injuring the ruled surface.

Green is becoming the standard color for printed forms. Blue will not print or photograph; red is trying on the eyes.

If the curve is for purposes of computation, it should be drawn with a fine, accurate line; if for demonstration, it should be fairly heavy, for contrast and effect.

The following rules are adapted from ANSI Y15 (formerly Z15):

*Standards for Graphic Presentations*

1. A graph should be free of all lines and lettering that are not essential to the reader's clear understanding of its message.

2. All lettering and numbers on a graph should be placed so as to be easily read from the bottom and from the right-hand side of the graph, not the left-hand side.

3. Standard abbreviations should be used where space is limited as, for example, in denoting the unit of measurement in scale captions.

4. The range of scales should be chosen so as to ensure effective and efficient use of the coordinate area in attaining the objective of the chart.

5. The zero line should be included if visual comparison of plotted magnitudes is desired.

6. When it is desired to show whether the rate of change of the dependent variable is increasing, constant, or decreasing, a logarithmic vertical scale should be used in conjunction with an arithmetical horizontal scale.

7. The horizontal (independent variable) scale values should usually increase from left to right and the vertical (dependent variable) scale values from bottom to top.

8. Scale values and scale captions should be placed outside the grid area, normally at the bottom for the horizontal scale and at the left side for the vertical scale. On wide graphs it may be desirable to repeat the vertical scale at the right side.

9. For arithmetical scales, the scale numbers shown on the graph and the space between coordinate rulings should preferably correspond to 1, 2, or 5 units of measurement, multiplied or divided by 1, 10, 100, etc.

10. The use of many digits in scale numbers should be avoided.

11. The scale caption should indicate both the variable measured and the unit of measurement. For example: EXPOSURE TIME IN DAYS.

12. Coordinate rulings should be limited in number to those necessary to guide the eye in making a reading to the desired degree of approximation. Closely spaced coordinate rulings are appropriate for computation charts but not for graphs intended primarily to show relationship.

13. Curves should preferably be represented by solid lines.

14. When more than one curve is presented on a graph, relative emphasis or differentiation of the curves may be secured by using different types of line, that is, solid, dashed, dotted, etc., or by different widths of line. A solid line is recommended for the most important curve.

15. The observed points should preferably be designated by circles.

16. Circles, squares, and triangles should be used rather than crosses or filled-in symbols to differentiate observed points of several curves on a graph.

17. Curves should, if practicable, be designated by brief labels placed close to the curves (horizontally or along the curves) rather than by letters, numbers, or other devices requiring a key.

18. If a key is used, it should preferably be placed within the grid in an isolated position, and enclosed by a light line border—grid lines, if convenient.

19. The title should be as clear and concise as possible. Explanatory material, if necessary to ensure clearness, should be added as a subtitle.

20. Scale captions, designations, curves, and blank spaces should, so far as practicable, be ar-

ranged to give a sense of balance around vertical and horizontal axes.

21. The appearance and the effectiveness of a graph depend in large measure on the relative widths of line used for its component parts. The widest line should be used for the principal curve. If several curves are presented on the same graph, the line width used for the curves should be less than that used when a single curve is presented.

22. A simple style of lettering, such as Gothic with its uniform line width and no serifs, should in general be used.

### 7. Logarithmic Scales.

An important type of chart is that in which the divisions are made proportional to the logarithms of the numbers instead of to the numbers themselves, and accordingly are not equally spaced. When ruled logarithmically in one direction with equal spacing at right angles, the spacing is called "semilogarithmic."

Logarithmic spacing may be done directly from the graduations on one of the scales of a slide rule. Logarithmic paper is sold in various combinations of ruling. It is available in one, two, three, or more cycles; in multiples of 10; and also in part-cycle and split-cycle form. In using logarithmic paper, interpolations should be made logarithmically; arithmetical interpolation with coarse divisions might lead to considerable error.

### 8. Semilogarithmic Charts.

These charts have equal spacing on one axis, usually the $X$ axis, and logarithmic spacing on the other axis. Owing to a property by virtue of which the slope of the curve at any point is an exact measure of the rate of increase or decrease in the data plotted, the semilogarithmic chart is frequently called a "ratio chart." Often called the "rate-of-change" chart as distinguished from the rectilinear, or "amount-of-change," chart, it is extremely useful in statistical work as it shows at a glance the rate at which a variable changes. By the use of this chart it is possible to predict a trend, such as the future increase of a business, growth of population, etc.

In choosing between rectilinear ruling and semi-logarithmic ruling, the important point to consider is whether the chart is to represent *numerical* increases and decreases or *percentage* increases and decreases. In many cases it is desirable to emphasize the percentage, or rate, change, not the numerical change; for these a semilogarithmic chart should be used.

An example of a semilogarithmic chart is given in Fig. 3. This curve was drawn from data compiled for the *World Almanac*. The dashed line shows the actual production by years, and the full line is the trend curve, the extension of which predicts future production.

Exponential equations of the form $y = ae^{bx}$ plot on semilogarithmic coordinates as a straight line. This represents the type of relationship in which a quantity increases or decreases at a rate proportional to the amount present at any time. For example, the passage of light through a translucent substance varies (in intensity) exponentially with the thickness of material.

### 9. Logarithmic Charts.

These are charts in which both abscissas and ordinates are spaced logarithmically. Any equation of the form $y = ax^b$, in which one quantity varies directly as some power of another, will plot as a straight line on logarithmic coordinates. Thus, multiplication, division, powers, and roots are examples. Figure 4 shows a logarithmic plot of sound intensity (and the power required to produce it) which varies as a power of the distance from the source. Other examples are the distance-time relationship of a falling body and the period of a simple pendulum.

### 10. Polar Charts.

The use of polar coordinate paper for representing intensity of illumination, intensity of heat, polar forms of curves, etc., is common. Figure 5 shows a candle-power distribution curve for an ordinary Mazda B lamp, and Fig. 6 the curve for a certain type of reflector. The candle power in any given direction is determined by reading off the distance from the origin to the curve. Use of polar forms of curves enables the determination of the foot-candle intensity at any point.

**FIG. 3.** A semilogarithmic chart. The ordinate axis (in this case) is logarithmic, the abscissa uniform.

## 11. Trilinear Charts.

The trilinear chart, or "triaxial diagram," as it is sometimes called, affords a valuable means of studying the properties of chemical compounds consisting of three elements, alloys of three metals or compounds, and mixtures containing three variables. The chart has the form of an equilateral triangle, the altitude of which represents 100 per cent of each of the three constituents. Figure 7, showing the ultimate tensile strength of copper-tin-zinc alloys, is a typical example of its application. The usefulness of such diagrams depends upon the geometric principle that the sum of the perpendiculars to the sides from any point within an equilateral triangle is a constant and is equal to the altitude.

## 12. Choice of Type and Presentation.

The function of a chart is to reveal facts. It may be entirely misleading if wrong paper or coordinates are chosen. The growth of an operation plotted on a rectilinear chart might, for example, entirely mislead an owner analyzing the trend of his business, while if plotted on a semilogarithmic chart, it would give a true picture of conditions. Intentionally misleading charts have been used many times in advertising; the commonest form is the chart with a greatly exaggerated vertical scale. In engineering

**FIG. 4.** A logarithmic chart. Both ordinate and abscissa are divided logarithmically.

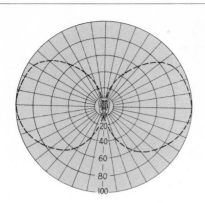

**FIG. 5.** A polar chart. This type is used when direction from a point (pole) must be shown.

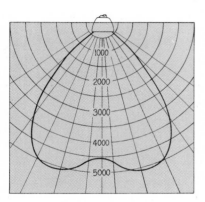

**FIG. 6.** A polar chart. This one shows the area of illumination from a reflector.

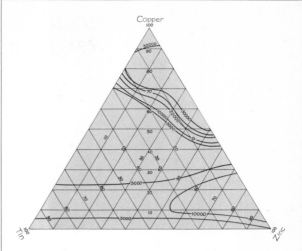

**FIG. 7.** A trilinear chart. This type is used to represent properties of three combined elements.

work it is essential to present the facts honestly and with scientific accuracy.

## 13. Analysis of Experimental Data.

So far we have discussed only the representation of data. In many cases representation alone on suitable coordinates gives an accurate description—for, from the plot, the *value* of one variable relative to a second variable can be ascertained. Nevertheless, the equation of a relationship can often be obtained from the data. To do this, the data must be plotted on coordinates that will "justify" the curve to a straight line. The equation of a straight

line on any system of coordinates is easily determined. Equations thus resolved are known as *empirical* equations.

The standard procedure is first to plot the data on rectangular coordinates. If a straight line results, it is known that the equation is of the first degree (linear). If a curve results from the plot, the equation is of a higher form, and semilogarithmic coordinates or logarithmic coordinates should be tried; if the data justifies to a straight line on either system, the equation can be determined.

## 14. Rectangular Plots: Linear Equations.

If a plot of data results in a straight line on rectilinear coordinates, the equation is of the first degree. Figure 8 shows a plot that justifies to a straight line on rectilinear coordinates. To determine the equation of the line, select two points, such as $y_1 x_1$ and $y_2 x_2$ shown in Fig. 8. Then by simple proportion,

$$\frac{y - y_1}{x - x_1} = \frac{y_2 - y_1}{x_2 - x_1}$$

and from Fig. 8,

**FIG. 8.** A straight-line plot on rectilinear co-ordinates. The equation of the line is linear, $y = a + bx$.

**FIG. 9.** A straight-line plot on logarithmic coordinates. The equation of the line is a power form, $y = ax^b$.

**FIG. 10.** A straight-line plot on semilogarithmic coordinates. The equation of the line is an exponential form, $y = ae^{bx}$.

$y_1 = 13.8$, $y_2 = 43.5$; $x_1 = 8$, $x_2 = 35.0$

substituting,

$$\frac{y - 13.8}{x - 8.0} = \frac{43.5 - 13.8}{35.0 - 8.0}$$

whence,

$$\frac{y - 13.8}{x - 8.0} = 1.1$$

and $y = 5.0 + 1.1x$

Thus, this equation is of a straight line, $y = a + bx$, where $a = 5.0$ and $b = 1.1$.

The coefficient 1.1 is the slope of the line and the constant 5.0 is the ordinate-axis intercept.

### 15. Logarithmic Plots: Power Equations.
If the data justifies to a straight line on logarithmic coordinates, the equation is a power relationship of the form $y = ax^b$. Figure 9 shows a plot that justifies to a straight line on logarithmic coordinates. Two selected points give

$x_1 = 2.5$    $y_1 = 3.8$
$x_2 = 55.0$   $y_2 = 26.5$

Because the plot is logarithmic,

$$\frac{\log y - \log 3.8}{\log x - \log 2.5} = \frac{\log 26.5 - \log 3.8}{\log 55.0 - \log 2.5}$$

whence,

$$\log y = 0.330 + 0.628 \log x$$

and

$$y = 2.14x^{0.628}$$

## 16. Semilogarithmic Plots: Exponential Equations.

If plotted data justifies to a straight line on semi-logarithmic coordinates, the equation is an exponential form, $y = ae^{bx}$ or $y = a10^{bx}$ (for base 10 logarithms). Figure 10 shows a typical plot. Two selected points give

$$x_1 = 0.8 \qquad y_1 = 3.8$$
$$x_2 = 4.5 \qquad y_2 = 50.0$$

Writing the equation of the straight line,

$$\frac{\ln y - \ln 3.8}{x - 0.8} = \frac{\ln 50.0 - \ln 3.8}{4.5 - 0.8}$$

whence,

$$\ln y = 0.777 + 0.697x$$

and

$$y = 2.18e^{0.697x}$$

## 17. Nomography.

After the equation of a relationship has been determined, specific values of the variables can be found by substitution in the equation. To facilitate such calculations, a chart called a nomogram can be made in which the variables are represented by plotted scales arranged so that a straightedge placed across the scales will give corresponding values of the variables. A simple example of a nomogram (conversion chart) is shown in Fig. 11.

The construction of a nomogram requires a knowledge of functional scales, chart forms, and the necessary mathematics. Thus, nomography is a special graphic field and is beyond the scope of this book. For a discussion of it, see "Graphic Science and Design" by Thomas E. French and Charles J. Vierck.[1]

## 18. Graphic Calculus.

Considered in the same light as nomography, this is a specialized field of chart construction in which the rates of change of variables are determined graphically. This field is also beyond the scope of this book. For a discussion of graphic calculus, see "Graphic Science and Design" by Thomas E. French and Charles J. Vierck.[1]

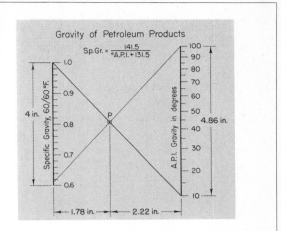

**FIG. 11.** A nomogram. This one is a conversion chart for specific gravity to degrees API.

## 19. Classification Charts, Route Charts, and Flow Sheets.

The uses to which these three classes of charts can be put are widely different; but their underlying principles are similar, and so they have been grouped together for convenience.

A *classification chart*, illustrated in Fig. 12, is intended to show the subdivisions of a whole and the interrelation of its parts with one another. Such a chart often takes the place of a written outline as it gives a better visualization of the facts than can be achieved with words alone. A common application is an organization chart of a corporation or business. It is customary to enclose the names of the divisions in rectangles, although circles or other shapes may be used. The rectangle has the advantage of being convenient for lettering, while the circle can be drawn more quickly and possesses greater popular appeal. Often a combination of both is used.

The *route chart* is used mainly for the purpose of showing the various steps in a process, either of manufacturing or other business. The *flow sheet*, illustrated in Fig. 13, is an example of a route chart applied to a chemical process. Charts of this type show in a dynamic way facts that might require

[1] McGraw-Hill, New York, 3d ed., 1970.

**FIG. 12.** A classification chart. This type of chart shows the relationship of parts of a whole.

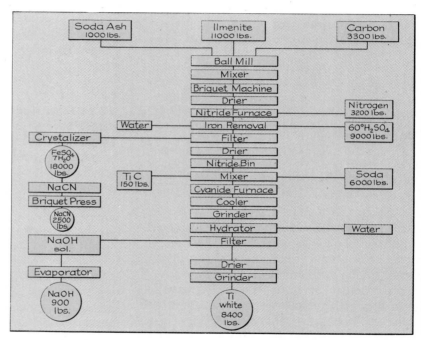

**FIG. 13.** A flow sheet. This is an effective way to show the fundamentals of a process or method.

considerable study to be understood from a written description. Figure 1, Chap. 9, showing the course of a drawing through the shops, illustrates a different form of route chart.

## 20. Popular Charts.

Engineers and draftsmen are frequently called upon to prepare charts and diagrams that will be under-

stood by nontechnical readers. For such charts it is often advisable to present the facts, not by means of curves on coordinate paper but in a way that is more easily read, even though the representation may be somewhat less accurate. Charts for popular use must be made so that the impression they produce is both quick and accurate. They are seldom studied critically but are taken in at a glance; hence the method of presentation requires the exercise of careful judgment and the application of a certain amount of psychology.

## 21. Bar Charts.

The bar chart is a type of chart easily understood by nontechnical readers. One of its simplest forms is the *100 per cent bar* for showing the relations of the constituents to a given total. Figure 14 is an example. The different segments should be cross-hatched, shaded, or distinguished in some other effective manner, the percentage represented should be placed on the related portion of the diagram or directly opposite, and the meaning of each segment should be clearly stated. Bars may be placed vertically or horizontally, the vertical position giving an advantage for lettering and the horizontal an advantage in readability, as the eye judges horizontal distances readily.

Figure 15 is an example of a *multiple-bar chart,* in which the length of each bar is proportional to the magnitude of the quantity represented. Means should be provided for reading numerical values represented by the bars. If it is necessary to give the exact value represented by the individual bars, the values should not be lettered at the ends of the bars as this would increase their apparent length. This type of chart is made horizontally, with the description at the base, and vertically. The vertical form is sometimes called the "pipe-organ chart." When vertical bars are drawn close together, touching along the sides, the diagram is called a "staircase chart." This chart is more often made as the "staircase curve," a line plotted on coordinate paper representing the profile of the tops of the bars.

A *compound-bar chart* is made when it is desirable to show two or more components in each bar. It is really a set of 100 per cent bars of different lengths set together in pipe-organ or horizontal form.

## 22. Pie Charts.

The "pie diagram," or 100 per cent circle (Fig. 16) is much inferior to the bar chart but is used constantly because of its popular appeal. It is a simple form of chart and, except for the lettering, is easily constructed. It can be regarded as a 100 per cent

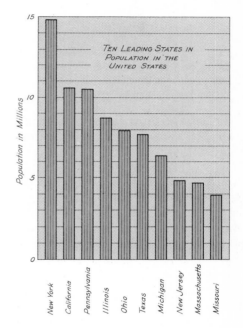

**FIG. 14.** A 100 per cent bar chart. This is a simple and impressive method of showing percentages of a whole.

**FIG. 15.** A multiple-bar chart. Each bar gives (graphically) the magnitude of the item represented.

bar bent into circular form. The circumference of the circle is divided into 100 parts, and sectors are used to represent percentages of the total.

To be effective, this diagram must be carefully lettered and the percentages marked on the sectors or at the circumference opposite the sectors. For contrast, it is best to crosshatch or shade the individual sectors. If the original drawing is to be displayed, the sectors may be colored and the diagram supplied with a key showing the meaning of each

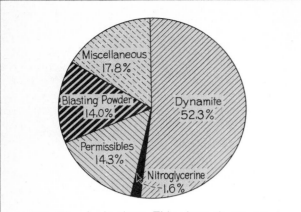

**FIG. 16.** A pie chart. This chart shows percentages of a whole. Compare it with a 100 per cent bar chart.

color. The percentage notation should always be placed where it can be read without removing the eyes from the diagram.

### 23. Strata and Volume Diagrams.

The use of strata and volume diagrams has been common although they are usually the most deceptive of the graphic methods of representation. Figure 17 is a strata chart showing a change in amounts over a period of time. The area between the curves is shaded to emphasize each variable represented. However, in reading it should be remembered that the areas *between* curves have no significance.

### 24. Pictorial Charts.

Pictorial charts were formerly much used for comparisons, for example of costs, populations, standing armies, livestock, and various products. It was customary to represent the data by human or other figures whose heights were proportional to numerical values or by silhouettes of the animals or products concerned whose heights or sometimes areas were proportional. Since volumes vary as the cubes of the linear dimensions, such charts are grossly misleading. Bar charts or charts such as the one shown in Fig. 18, where the diameter of the column is constant, should be used for the type of comparisons that are shown in pictorial charts.

### 25. Charts for Reproduction.

Charts for reproduction by the zinc-etching process should be carefully penciled to about twice the size of the required cut. Observe the following order in inking: First, ink the circles around plotted points; second, ink the curves with strong lines. A border pen is useful for heavy lines, and a Leroy pen (in socket holder) may be used to advantage, particularly for dashed lines. Third, ink the title box and

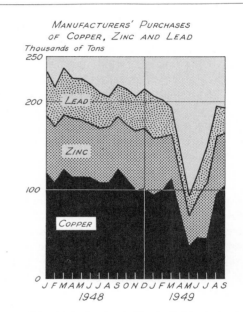

**FIG. 17.** A strata chart. The individual charts are superimposed.

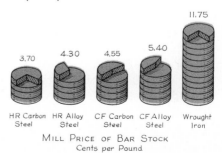

**FIG. 18.** A pictorial chart. Pictorialism adds to popular interest but detracts from accuracy of representation.

all lettering; fourth, ink the coordinates with fine black lines, putting in only as many as are necessary for easy reading and breaking them wherever they interfere with title or lettering or where they cross plotted points.

### 26. Charts for Display.

Large charts for demonstration purposes are sometimes required. These can be drawn on sheets 22 by 28 in. or 28 by 44 in. known as "printer's blanks." The quickest way to make them is with show-card colors and single-stroke sign-writer's brushes. Large bar charts can be made with strips of black adhesive tape. Lettering can be done with the brush or with gummed letters.

### 27. Standards.

Many standards are available for graphic presentations of data. Of particular interest are the ANSI standards for graphic symbols (Y32), letter symbols (Y10), and time-series charts (Y15.2—1960), and ANSI Y15.1 "Guide for Preparing Technical Illustrations for Publications and Projections."

### 28. The Advantage of Color.

Even though "black-and-white" charts, graphs, and diagrams are satisfactory for statistical and scientific purposes, the use of color will add greatly to readability, separation, and emphasis. In black and white, line codes (dashes, etc.) must be used and areas can be emphasized only by crosshatching. In color, lines stand out prominently and areas are easily distinguished. Especially for popular charts, color is almost mandatory. Today, many of the charts in good periodicals, corporate reports, promotional charts, and many others are being done in color. The use of color is increasing rapidly. Note the use of color techniques in all of the illustrations in this chapter, especially Figs. 12, 13, 15, and Figs. 19 to 23.[2] Also, compare Fig. 19 with Fig. 20.

### 29. Principles for the Use of Color.

The use of color always requires some artistic judg-

ment and consequently it is difficult to give definitive rules. Nevertheless, the following principles will serve as a guide.

*Backgrounds.* A pure white background for a chart is not nearly so effective as a background made of a light tint of color. This is because a white background, even if it is outlined, is the same value as the page itself. A background made as a light tint adds cohesiveness and vigor to the charts. Notice the use of tinted backgrounds in all the charts in this chapter.

*White lines and areas.* White lines and areas can be produced only if a tinted background is employed. This adds one more color to any series used and is quite effective. Notice the white area in Fig. 23 and the white line in Fig. 22.

*Values of tints.* Tints of a color are made either by diluting the color with white or, in printing, by the use of screens which reduce the color intensity. In printing, the screens are rated in percentage, which means that a 75% screen will print (in actual minute areas) 75% of the color and 25% white (blank area). Thus a 30% screen would print 30% of the color and 70% as blank area. For the dilution of a color with white when applying color by hand, it is advantageous to think of the color *value* also in percentage because it is easy to judge color intensity, for example, as approximately $\frac{3}{4}$ the intensity of the pure color. The actual dilution *will not* be the same as the value of the intensity; the dilution must be made by carefully adding white and then judging the intensity. Note the various intensities of colors used in the illustrations of this chapter.

*Contrast.* Contrast may be defined as the degree of dissimilarity. Thus, high contrast is had when a color of strong intensity is placed adjacent to a color of very weak intensity. Low contrast is had when two colors of about the same intensity are placed together. High contrast will accentuate the areas and low contrast will diminish the impact of the areas. Notice the gradual (low-contrast) changes in both color and value of the areas in Fig. 19 and also the high contrast and accentuation of the darkest and white areas of Fig. 20. If emphasis is needed, use high contrast. If a subtle change is desirable, use low contrast.

[2] Figures 19 to 23 are supplied through the courtesy of Mr. W. Anthony Dowell of the Palm Beach office to Merrill, Lynch, Pierce, Fenner, and Smith and are adapted from the U.S. Industrial Outlook, 1965, by the U.S. Department of Commerce.

**FIG. 19.** A multiple-bar chart in color. Two basic colors and tints of each are used. Compare with Fig. 20.

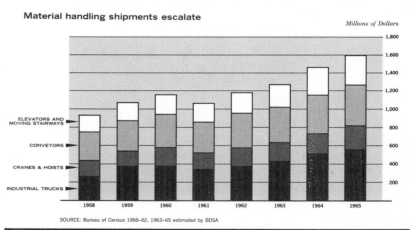

**FIG. 20.** The chart of Fig. 19 in "black and white" and tints. Compare with Fig. 19.

*Adjacent areas.* Adjacent areas are well defined and easily read if they are made of different colors. Notice in Fig. 21 that adjacent areas are different colors and also that different values of the colors have been used. Maximum separation of areas occurs when a strong color and value is placed next to a weak color and value. In some cases maximum separation is needed, in others a more gradual change is desirable. For example, in Fig. 20 the stronger colors are used at the bottom of the bars and the weaker colors at the top. This is done not only for the sake of stability of the bars, but because, if adjacent areas of the bars are made as high-contrast areas, the bars become "spotty" in ap-

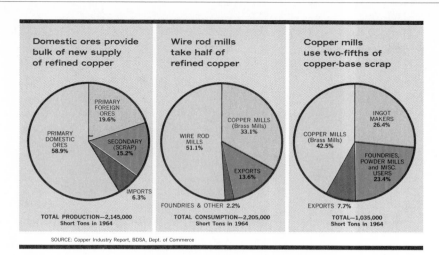

Domestic ores provide bulk of new supply of refined copper

PRIMARY FOREIGN ORES 19.6%

PRIMARY DOMESTIC ORES 58.9%

SECONDARY (SCRAP) 15.2%

IMPORTS 6.3%

TOTAL PRODUCTION—2,145,000 Short Tons in 1964

Wire rod mills take half of refined copper

COPPER MILLS (Brass Mills) 33.1%

WIRE ROD MILLS 51.1%

EXPORTS 13.6%

FOUNDRIES & OTHER 2.2%

TOTAL CONSUMPTION—2,205,000 Short Tons in 1964

Copper mills use two-fifths of copper-base scrap

INGOT MAKERS 26.4%

COPPER MILLS (Brass Mills) 42.5%

FOUNDRIES, POWDER MILLS and MISC. USERS 23.4%

EXPORTS 7.7%

TOTAL—1,035,000 Short Tons in 1964

SOURCE: Copper Industry Report, BDSA, Dept. of Commerce

**FIG. 21.** A series of pie charts in color. Readability and appeal is superior to single-color rendition.

pearance and the value of a continuous bar is lost.

*Adjacent lines.* The best separation of lines occurs when adjacent lines are contrasty, either in color or value, or both. Note the use of alternate brown and green lines in Fig. 22 and the contrasting white line at the top.

*Choice of Colors.* The actual choice of colors is almost entirely one of artistic judgment, but this is often tempered by other factors. Colors at the high (wave-length) end of the spectrum are known as "cold" colors. Colors at the low end of the spectrum are known as "warm" colors. The colors of nature are the blue of the sky; the greens of trees, grass, and shrubbery; and the browns of tree trunks, soil, and rock. Note that these colors comprise a cold color, a medium color, and a warm color. However, even though we view the colors of nature daily and are not disturbed by them, such colors are not commonly put together in a chart. Also, strange as it may seem to the layman, a study of the colors of the spectrum and other scientific aspects of color transmission and reflectance will be of little help in the choice of colors to be used for a chart, graph, or diagram. This is because the *only* important considerations are those of pleasing and visually effec-

tive combinations, contrasts, and color values. Colors that are not pleasingly compatible should not be used together unless a startling or garish effect is wanted. For example, violet, purple, and yellow are almost never used for charts. The colors used most are blues, greens, reds, browns, white, and black. The symbolic colors of a company or organization may also play a part in the choice, as for example, the blue and white of the Pure Oil Company or the red, white, and blue of The Standard Oil Company. Of interest is the fact that blue has predominated recently in the annual reports of many large companies. In addition, many of the charts and graphs in such magazines as *Scientific American* have been printed in black, shades of black (gray), red, blue, and green with occasionally some yellow, but black, gray, and red are predominant.

The best practice is to lay out the chart and then try several colors by holding colored sheets or strips over the chart to judge finally the most pleasing and effective combination.

## 30. Color for Primary Charts.

By the term "primary charts" is meant the chart that is made by hand for use directly. This excludes

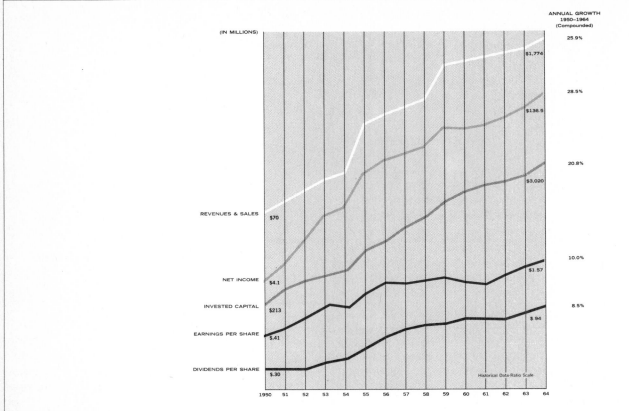

**FIG. 22.** A graph in color. Note that one more colored line is produced by employing a tinted background with the white line on it.

reproduction. Color may be applied in the form of colored inks, water color, or tempera. All of these may be used in a ruling pen for making lines or applied with a brush for areas. Chart-Pak (Chart-Pak Inc., River Road, Leeds, Mass.) colored acetate fiber strip, available in many colors and widths, is a convenient method of coloring for line graphs and bar charts. Colored paper can be used for backgrounds, drawing the chart *on* the colored paper.

## 31. Color for Reproduced Charts.
When the chart is made exclusively for reproduction, the basic chart is outlined and drawn in india ink as would be done for a black-and-white chart, with the exception that areas are not blacked-in. Then, an overlay sheet of tracing paper or drafting film such as Mylar is placed over the chart and fastened with drafting (or transparent) tape. On this overlay sheet directions are lettered or written, designating the colors to be used for lines and areas. The engraver, following these directions, will then make plates for the printing of each color. In giving the directions for color and color value, either an approximate value must be given, such as *light blue*, or an exact value must be designated, such as 30% blue.

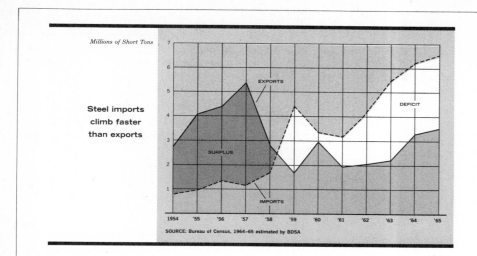

**FIG. 23.** A strata chart in color. Note that the overlap of colored strata produces readability for the *surplus* area and the separation of strata emphasizes the *deficit* area.

# PROBLEMS

These problems are given for practice in preparing various types of charts, graphs, and diagrams for technical or popular presentation.

**1.** The data given below were obtained in a tension test of a machine-steel bar. Plot the data on rectangular coordinates, using the elongation as the independent variable, the applied load as the dependent variable.

| Applied load, lb per sq in. | Elongation per in. of length |
|---|---|
| 0 | 0 |
| 3,000 | 0.00011 |
| 5,000 | 0.00018 |
| 10,000 | 0.00033 |
| 15,000 | 0.00051 |
| 20,000 | 0.00067 |
| 25,000 | 0.00083 |
| 30,000 | 0.00099 |
| 35,000 | 0.00115 |
| 40,000 | 0.00134 |
| 42,000 | 0.00142 |

**2.** A test of the corrosive effect of 5 per cent sulfuric acid, both air-free and air-saturated, on 70 per cent nickel, 30 per cent copper alloy (Monel) over a temperature range from 20 to 120°C resulted in the data tabulated below. Plot this data on rectangular coordinates with corrosion rate as ordinate versus temperature as abscissa.

| Temp., °C | Corrosion rate, MDD (mg per sq dm per day) | |
|---|---|---|
| | Acid sat. with air | Acid air-free, sat. with $N_2$ |
| 20 | 195 | 35 |
| 30 | 240 | 45 |
| 40 | 315 | 50 |
| 50 | 425 | 63 |
| 60 | 565 | 70 |
| 70 | 670 | 74 |
| 80 | 725 | 72 |
| 83 | 715 | 70 |
| 85 | 700 | 68 |
| 92 | 580 | 57 |
| 95 | 470 | 42 |
| 101 | 60 | 12 |

**3.** In testing a small 1-kw transformer for efficiency at various loads, the following data were obtained: Watts delivered: 948, 728, 458, 252, 000. Losses: 73, 62, 53, 49, 47.

Plot curves on rectangular coordinate paper showing the relation between percentage of load and efficiency, using watts delivered as the independent variable and remembering that efficiency = output ÷ (output + losses).

**4.** The following data were obtained from a test of an automobile engine:

| Rpm | Length of run, min | Fuel per run, lb | Bhp |
|---|---|---|---|
| 1,006 | 11.08 | 1.0 | 5.5 |
| 1,001 | 4.25 | 0.5 | 8.5 |
| 997 | 7.53 | 1.0 | 13.0 |
| 1,000 | 5.77 | 1.0 | 16.3 |
| 1,002 | 2.38 | 0.5 | 21.1 |

Plot curves on rectangular coordinate paper showing the relation between percentage of load and efficiency, using watts delivered as the independent variable between thermal efficiency and brake horsepower developed, assuming the heat value of the gasoline to be 19,000 Btu per lb.

**5.** During the year 1950 the consumption of wood pulp by various grades in the United States was as shown in the table. Show these facts by means of a 100 per cent bar, a pie diagram, and a multiple-bar chart.

| Grade of pulp | Consumption in paper and board manufacture, tons |
|---|---|
| Sulfate | 8,380,864 |
| Sulfite | 3,331,668 |
| Groundwood | 2,483,980 |
| Defibrated, exploded, etc. | 971,912 |
| Soda | 564,355 |
| All other | 754,331 |
| Total (all grades) | 16,487,110 |

**6.** Put the data given in Fig. 16 into 100 per cent bar form.

**7.** Put the data of Fig. 14 into pie-chart form.

**8.** A test of the resistance of alloy steels to high-temperature steam resulted in the data tabulated below. Represent the results of the test graphically by means of a multiple-bar chart.

## Corrosion of Steel Bars in Contact with Steam at 1100°F for 2,000 Hr

| Steel | Average penetration, in. |
|---|---|
| SAE 1010 | 0.001700 |
| 1.25 Cr-Mo | 0.001169 |
| 4.6 Cr-Mo | 0.000956 |
| 9 Cr 1.22 Mo | 0.000694 |
| 12 Cr | 0.000045 |
| 18-8-Cb | 0.000012 |

**9.** From the data below, plot curves showing the "thinking distance" and "braking distance." From these curves, plot the sum curve "total distance." Title "Automobile Minimum Travel Distances when Stopping—Average Driver."

| Mph | Ft per sec | Thinking distance, ft | Braking distance, ft |
|---|---|---|---|
| 20 | 29 | 22 | 18 |
| 30 | 44 | 33 | 40 |
| 40 | 59 | 44 | 71 |
| 50 | 74 | 55 | 111 |
| 60 | 88 | 66 | 160 |
| 70 | 103 | 77 | 218 |

**10.** Make a semilogarithmic chart showing the comparative rate of growth of the five largest American cities from 1880 to 1950. Data for this chart are in the table in the next column.

## Population

| City | 1880 | 1890 | 1900 | 1910 |
|---|---|---|---|---|
| New York, N.Y. | 1,911,698 | 2,507,414 | 3,437,202 | 4,766,883 |
| Chicago, Ill. | 503,185 | 1,099,850 | 1,698,757 | 2,185,283 |
| Philadelphia, Pa. | 847,170 | 1,046,964 | 1,293,697 | 1,549,008 |
| Los Angeles, Calif. | 11,181 | 50,395 | 102,479 | 319,198 |
| Detroit, Mich. | 116,340 | 205,876 | 285,704 | 465,766 |

| City | 1920 | 1930 | 1940 | 1950 |
|---|---|---|---|---|
| New York, N.Y. | 5,620,048 | 6,930,446 | 7,454,995 | 7,835,099 |
| Chicago, Ill. | 2,701,705 | 3,376,348 | 3,396,808 | 3,606,436 |
| Philadelphia, Pa. | 1,823,779 | 1,950,961 | 1,931,334 | 2,064,794 |
| Los Angeles, Calif. | 576,673 | 1,238,048 | 1,504,277 | 1,957,692 |
| Detroit, Mich. | 993,678 | 1,568,662 | 1,623,452 | 1,838,517 |

| Angle,° | (1) Mazda lamp and porcelain enameled reflector, candle power | (2) Type S-1 sun lamp, candle power |
|---|---|---|
| 0 | 600 | 1,000 |
| 5 | 655 | 925 |
| 10 | 695 | 790 |
| 15 | 725 | 640 |
| 20 | 760 | 460 |
| 25 | 800 | 350 |
| 30 | 815 | 260 |
| 35 | 825 | 205 |
| 40 | 830 | 170 |
| 45 | 815 | 150 |
| 50 | 785 | 130 |
| 55 | 740 | 120 |
| 60 | 640 | 0 |
| 65 | 485 | |
| 70 | 330 | |
| 75 | 200 | |
| 80 | 90 | |
| 85 | 25 | |
| 90 | 0 | |

**11.** On polar coordinate paper, plot a curve for the first set of data given below as a solid line (Fig. 6) and a curve for the second set as a broken line (Fig. 5). The lower end of the vertical center line is to be taken as the zero-degree line. The curves will be symmetrical about this line, two points being plotted for each angle, one to the left and one to the right. Mark the candle power along this center line also.

**12.** On trilinear coordinate paper plot the data given below. Complete the chart, identifying the curves and lettering the title below the coordinate lines (Fig. 7).

### Freezing Points of Solutions of Glycerol and Methanol in Water

| Weight, per cent | | | Freezing points, °F |
|---|---|---|---|
| **Water** | **Methanol** | **Glycerol** | |
| 86 | 14 | 0 | +14 |
| 82 | 8 | 10 | +14 |
| 78 | 6 | 16 | +14 |
| 76 | 24 | . . . | −4 |
| 74 | 2 | 24 | +14 |
| 71 | 17 | 12 | −4 |
| 70 | 0 | 30 | +14 |
| 68 | 32 | 0 | −22 |
| 65 | 10 | 25 | −4 |
| 62 | 23 | 15 | −22 |
| 60 | 40 | 0 | −40 |
| 58 | 5 | 37 | −4 |
| 57 | 17 | 26 | −22 |
| 55 | 0 | 45 | −4 |
| 54 | 28 | 18 | −40 |
| 53 | 47 | 0 | −58 |
| 52 | 10 | 38 | −22 |
| 50 | 20 | 30 | −40 |
| 47 | 31 | 22 | −58 |
| 45 | 0 | 55 | −22 |
| 42 | 7 | 51 | −40 |
| 39 | 12 | 49 | −58 |
| 37 | 0 | 63 | −40 |
| 36 | 8 | 56 | −58 |
| 33 | 0 | 67 | −58 |

**13.** Thermal conductivities of a bonded-asbestos-fiber insulating material at various mean temperatures are tabulated below:

| Mean temperature (hot to cold surface) $T$, °F | Thermal conductivity $K$, Btu per hr-sq ft-°F temp diff-in. thickness |
|---|---|
| 100 | 0.365 |
| 200 | 0.415 |
| 320 | 0.470 |
| 400 | 0.520 |
| 460 | 0.563 |
| 600 | 0.628 |
| 730 | 0.688 |
| 900 | 0.780 |

Plotting the data on uniform rectangular coordinates with thermal conductivity as ordinate and mean temperature as abscissa, evaluate the constants $a$ and $b$ in the linear equation $K = a + bT$ relating the variables.

**14.** Corrosion tests on specimens of pure magnesium resulted in values for weight increase in pure oxygen at 525°C tabulated below:

| Elapsed time $T$, hr | Weight increase $W$, mg per sq cm |
|---|---|
| 0 | 0 |
| 2.0 | 0.17 |
| 4.0 | 0.36 |
| 8.2 | 0.65 |
| 11.5 | 1.00 |
| 20.0 | 1.68 |
| 24.1 | 2.09 |
| 30.0 | 2.57 |

Plotting the data on uniform rectangular coordinates with weight increase as ordinate and elapsed time as abscissa, evaluate the constant $a$ in the linear equation $W = aT$ relating the variables.

**15.** Approximate rates of discharge of water under various heads of fall through a 1,000-ft length of 4-in. pipe having an average number of bends and fittings are tabulated on the next page:

| Head of fall H, ft | Discharge Q, gal per min |
|---|---|
| 1 | 35.8 |
| 2 | 50.6 |
| 4 | 71.6 |
| 6 | 87.7 |
| 9 | 107.5 |
| 12 | 123.7 |
| 16 | 142.9 |
| 20 | 159.7 |
| 25 | 178.9 |
| 30 | 195.8 |
| 40 | 225.8 |
| 50 | 252.2 |
| 75 | 309.8 |
| 100 | 357.9 |

Plotting the data on logarithmic coordinates with discharge as ordinate and head as abscissa, evaluate the constants $a$ and $b$ in the power equation $Q = aH^b$ relating the variables.

**16.** Creep-strength tests of a high-chromium (23 to 27 per cent) ferritic steel used in high-temperature service resulted in the stress values, to produce a 1 per cent deformation in 10,000 hr at various temperatures, tabulated below:

| Temperature T, °F | Stress S, lb per sq in. |
|---|---|
| 1,000 | 6,450 |
| 1,100 | 2,700 |
| 1,200 | 1,250 |
| 1,300 | 570 |
| 1,400 | 270 |

Plotting the data on semilogarithmic coordinates with stress as ordinate on a logarithmic scale and temperature as abscissa on a uniform scale, evaluate the constants $a$ and $b$ in the exponential equation $S = a10^{bT}$ relating the variables.

**17.  COLORED CHARTS.** Select any of Probs. 1 to 16, and draw in color, following the suggestions given in Sec. 30.

# SURFACE INTERSECTIONS

# 12

Geometric Components of an Object or Assembly Meet in Intersections Which Must Be Shown in order to Complete the Graphic Description

Included Here:
Definitions and Classifications • Intersections of Lines and Planes • Intersections of Prisms • Intersections of Pyramids • Intersections of Curved Surfaces • Intersections of Cylinders and Cones with Lines and Planes • Intersections of Cylinders and Cones • Problems

# 1. Definition and Classifications.

In making orthographic drawings, there is the repeated necessity to represent the *lines of intersection* between the various surfaces of a wide variety of objects. Nearly every line on a drawing is a line of intersection, generally the intersection of two planes, giving a straight line, or of a cylinder and a plane, giving a circle or an ellipse. The term "intersection of surfaces" refers, however, to the more complicated lines that occur when geometric surfaces such as planes, cylinders, and cones intersect one another. These lines of intersection are shown by one of two basic methods: (1) *conventional intersections*, ordinarily used to represent a fillet, round, or runout, as explained in Sec. 23, Chap. 7, and shown in Fig. 43 of that chapter, or (2) *plotted intersections*, used when an intersection must be located accurately for purposes of dimensioning or for development of the surfaces. In sheet-metal combinations the intersection *must* be found before the piece can be developed. In this chapter we are concerned solely with the methods of projecting plotted intersections.

*Classification of Surfaces.*   A surface may be considered to be generated by the motion of a line: the generatrix. Surfaces are thus divided into two general classes: (1) those that can be generated by a moving *straight* line, and (2) those that can be generated by a moving *curved* line. The first are called *ruled surfaces;* the second, *double-curved surfaces.* Any position of the generatrix is called an *element* of the surface.

*Ruled surfaces* are divided into (*a*) *the plane,* (*b*) *single-curved surfaces,* and (*c*) *warped surfaces.*

The plane is generated by a straight line moving so as to touch two other intersecting or parallel straight lines or a plane curve.

*Single-curved surfaces* have their elements parallel or intersecting. In this class are the cylinder and the cone and also a third surface, which we shall not consider, known as the "convolute," in which only consecutive elements intersect.

*Warped surfaces* have no two consecutive elements that are parallel or intersecting. There are a great variety of warped surfaces. The surface of a screw

thread and that of an airplane wing are two examples.

*Double-curved surfaces* are generated by a curved line moving according to some law. The commonest forms are *surfaces of revolution,* made by revolving a curve about an axis in the same plane, such as the sphere, torus or ring, ellipsoid, paraboloid, hyperboloid. Illustrations of various surfaces may be found in Fig. 101, Chap. 3.

*Definitions and Details of Common Surfaces.*   A *prism* is a polyhedron whose bases or ends are equal parallel polygons and whose lateral faces are parallelograms. A right prism is one whose lateral faces are rectangles; all others are called oblique prisms. The axis of a prism is a straight line connecting the centers of the bases. A truncated prism is that portion of a prism lying between one of its bases and a plane which cuts all its lateral edges.

A *pyramid* is a polyhedron whose base is a polygonal plane and whose other surfaces are triangular planes meeting at a point called the "vertex." The axis is a line passing through the vertex and the mid-point of the base. The altitude is a perpendicular from the vertex to the base. A pyramid is *right* if the altitude coincides with the axis; it is *oblique* if they do not coincide. A truncated pyramid is that portion of a pyramid lying between the base and a cutting plane which cuts all the lateral edges. The frustum of a pyramid is that portion of a pyramid lying between the base and a cutting plane parallel to the base which cuts all the lateral edges.

A *cylinder* is a single-curved surface generated by the motion of a straight-line generatrix remaining parallel to itself and constantly intersecting a curved directrix. The various positions of the generatrix are elements of the surface. It is a *right* cylinder when the elements are perpendicular to the bases, an *oblique* cylinder when they are not. A truncated cylinder is that portion of a cylinder which lies between one of its bases and a cutting plane which cuts all the elements. The axis is the line joining the centers of the bases.

A *cone* is a single-curved surface generated by the movement, along a curved directrix, of a straight-line generatrix, one point of which is fixed. The directrix is the base, and the fixed point is the vertex of the cone. Each position of the generatrix is an element of the surface. The axis is a line connecting the vertex and the center of the base. The altitude is a perpendicular dropped from the vertex to the base. A cone is *right* if the axis and altitude coincide; it is *oblique* is they do not coincide. A truncated cone is that portion of a cone lying between the base and a cutting plane which cuts all the elements. The frustum of a cone is that portion of a cone lying between the base and a cutting plane parallel to the base which cuts all the elements.

## 2. Intersections of Plane Surfaces.

The intersection of a line and a plane is a *point* common to both. The intersection of two planes is a *line* common to both. After mastering the graphic methods of locating intersections (Secs. 3 to 10), apply the principles you have learned to finding the line of intersection between objects made up of plane surfaces (prisms and pyramids, Secs. 11 to 13). The method of solution for the intersection of other polyhedrons should follow logically from the examples given.

## 3. Intersection of a Line and a Plane, Both in Principal Positions.

Principal positions of planes are horizontal, frontal, or profile. Principal positions of lines occur when the lines are *not* inclined to any principal plane; thus a line is (1) horizontal-frontal, (2) horizontal-profile, or (3) frontal-profile. Therefore, if the line is parallel to, or lies in, the principal plane, no single point of intersection is possible, but in other positions when the line is perpendicular to the plane, a single point exists. To illustrate, Fig. 1 shows a rectangular object made up of horizontal, frontal, and profile planes, with an accompanying horizontal-profile line $AB$. Because the line is parallel to horizontal and profile planes, the line and these planes do not intersect except at infinity. However, the line is perpendicular to frontal planes and there is a single point of intersection, observed in the top view at $P_T$ and in the side view at $P_R$, where the front of the object appears as an edge. The front view of the intersection is

coincident with $A_F B_F$ at $P_F$ because this is the end view of line $AB$.

Thus it is established that *the intersection of a line and plane is a point on the line coincident with the edge view of the plane.* The point of intersection of a line and plane is often called a "piercing point."

## 4. Intersection of an Inclined Line with Planes in Principal Positions.

Two points of intersection are possible, one with each plane to which the line is not parallel. For example, Fig. 2 shows a frontal line $AB$ and an object made up of horizontal, frontal, and profile planes. The line will not intersect any frontal plane (except

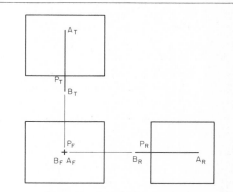

**FIG. 1.** Intersection of a line and plane. The line is perpendicular to the plane.

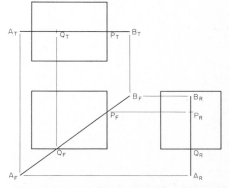

**FIG. 2.** Intersection of an inclined line with planes. In this case the line intersects the horizontal and profile sides of an object.

at infinity). The intersection with the (profile) right side is seen at $P_T$ and $P_F$, where the plane appears as an edge, and is then easily projected to $P_R$. The intersection with the horizontal bottom is observed at $Q_F$ and $Q_R$, where the plane appears as an edge, and is projected from $Q_F$ to $Q_T$.

Observe that line $AB$ does not intersect the top or left side of the *object*, but if the top and side planes were extended, a point of intersection would exist.

## 5. Intersection of a Skew Line with Planes in Principal Positions.

Theoretically, a skew line (a line inclined to *all* principal planes) will intersect all three principal planes. However, on a *rectangular object*, depending upon the length and position of the line, only one or two intersections exist. Figure 3 illustrates all possibilities. At (*a*), the skew line from point $A$ inside the object emanates upward, backward, and to the right, and obviously intersects the top (horizontal) surface of the object at $P$, observed first at $P_F$ and $P_R$, where the top surface appears as an edge, and then projected to $P_T$. But line $AB$ does not intersect any other surface of the object.

At (*b*), line $AB$ is seen to intersect two surfaces of the object, the right side at $Q$ and the front at $P$. Note that in each case the points are found first where the surface appears as an edge, $Q_T$ and $Q_F$ for point $Q$, and $P_T$ and $P_R$ for point $P$.

At (*c*), the top surface of the object is extended so that three intersections are possible. Physically this occurs on an object when two rectangular shapes are offset or when a base or lug extends. The line $AB$ intersects the right side at $P$ and the front at $Q$. The top surface and the line give an intersection at $R$, shown on the figure with dashed lines for the extensions.

Note in every case, (*a*), (*b*), and (*c*), of Fig. 3, that the intersection is found by observing where the line intersects the surface *in the edge view* of the surface. For horizontal, frontal, and profile surfaces, two views always show the surface as an edge. Thus it is easily seen that, for example, at (*b*) the line crosses the *edge* view of the *right side* of the object in the top view at $Q_T$ and in the front view at $Q_F$,

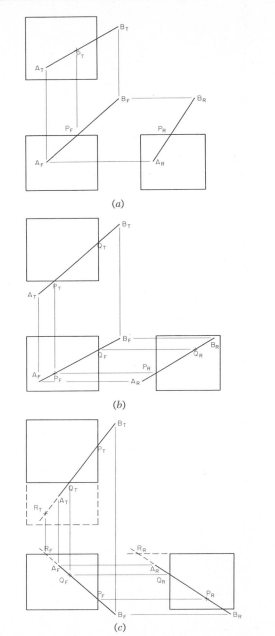

**FIG. 3.** Intersection of a skew line with planes. (*a*) Intersection with horizontal; (*b*) intersection with frontal and profile; (*c*) intersection with horizontal, frontal, and profile.

and because point $Q$ falls *within the confines* of the surface, a real point of intersection has been found. Also observe at (*a*) that line $AB$ *apparently* crosses the edge view of the right side of the object *in the top view*. Nevertheless, reading the front view, line $AB$ is seen to miss the right side and there is therefore no intersection of $AB$ with the right side. To illustrate further, line $AB$ of Fig. 4*a* apparently inter-

sects the edge view of the top surface, as observed in the front view. But projecting $Q_F$ to the top view at $Q_T$ on $AB$, we see that $Q_T$ falls within the confines of the top surface of the object and is therefore an intersection. However, at (*b*) when $Q_F$ is projected to the top view, $Q_T$ is outside the confines of the top surface and there is no intersection. From this it is clear that to determine a real point of intersection, (1) *two views must be consulted*, and (2) *the apparent point of intersection in one view must project within the confines of the surface in another view.*

### 6. Intersection of a Skew Line with Inclined Planes.

An inclined plane appears as an edge in one view and, because of this, its intersection with any line is easily found. In Fig. 5*a* the inclined surface appears as an edge in the front view. The apparent intersection $P_F$ is projected to $P_T$ on $A_TB_T$, where it is seen that $P_T$ falls within the confines of the inclined surface and therefore is a real point of intersection. Figure 5 shows the principle applied at (*b*) to a surface inclined to front and side and at (*c*) to a surface inclined to top and front. Note that at (*b*) the intersection will be located first in the top view, and at (*c*) it will be found first in the side view.

### 7. Intersection of a Skew Line with a Skew Surface.

A skew surface does not appear as an edge in any principal view, so if the methods of Secs. 3 to 6 are used, an edge view will have to be made. This is illustrated in Fig. 6 where the edge view has been made by looking in the direction of $SR$, a horizontal line of the skew surface. Line $AB$ is now projected into the edge view, and the intersection $P$ is located and then projected back to top, front, and side views.

It is not necessary, however, to employ an extra (edge) view. Figure 7 shows pictorially a line and skew plane. If any edge-view plane is passed through the line, the intersection of this plane with the skew plane will contain the point of intersection between the line and the skew plane. The intersection of the edge-view plane and the skew plane is easily located,

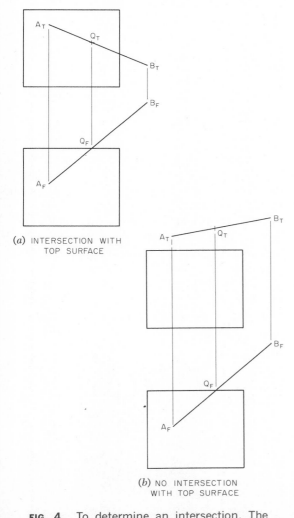

(*a*) INTERSECTION WITH TOP SURFACE

(*b*) NO INTERSECTION WITH TOP SURFACE

**FIG. 4.** To determine an intersection. The intersection (piercing point) must fall within the surface area in two (or more) views.

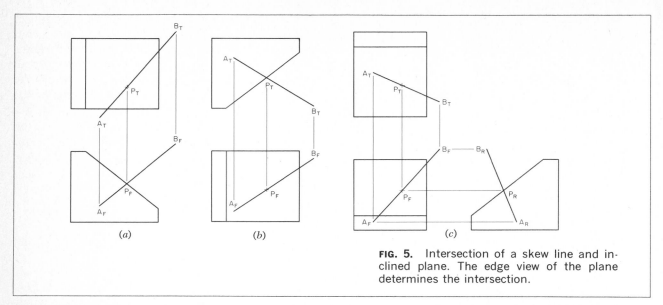

**FIG. 5.** Intersection of a skew line and inclined plane. The edge view of the plane determines the intersection.

as shown in Fig. 8, where $AB$ is a line and $RST$ a skew plane. A plane *appearing as an edge in the top*

*view*, passed through $AB$, will intersect $RS$ at $Z_T$ and $RT$ at $X_T$, and these points projected to the front view at $Z_F$ and $X_F$ establish $ZX$ as the line of intersection between the edge-view plane and $RST$, the skew plane. The front view shows the intersection of $AB$ with $ZX$ at $P_F$, the point common to line, skew plane, and edge-view plane. To complete, $P$ is projected to the top view at $P_T$.

It often happens that one edge of a skew surface is profile, as is $ST$ in Figs. 8 and 9; and when the line, in this case $AB$, is in a position like that of Fig.

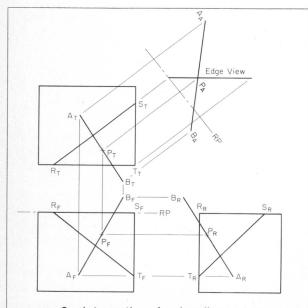

**FIG. 6.** Intersection of a skew line and skew plane. An auxiliary (edge view) determines the intersection.

**FIG. 7.** A receding plane passed through a line.

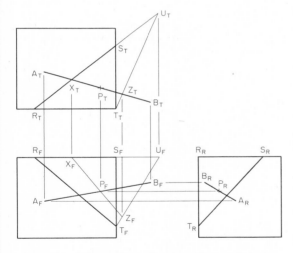

**FIG. 8.** Intersection of a skew line and skew plane. The intersection is determined by a receding plane (edge view plane in top view) passed through line *AB*.

**FIG. 9.** Intersection of a skew line and skew plane. This illustrates enlargement of the skew plane to eliminate difficult projection.

9, the intersection (with the skew plane) of an edge-view plane through the line is not easily projected. An accepted method in such cases is to *extend* the skew plane to eliminate the profile edge. In Fig. 9, *RS* has been extended to *U*, making the skew sur-

face *RUT*, still the *same* surface but larger. An edge-view plane (in the top view) passed as before gives *XZ* as its line of intersection with *RUT*. Point *P*, the point common to edge-view plane, skew plane, and line, is then located first in the front view at $P_F$ and then projected to top and side views.

## 8. Intersection of a Plane with a Plane in a Principal Position.

Any line of a given plane will intersect a second plane *on the line of intersection* between two planes. Therefore, to establish the line of intersection between two planes, the intersection of either of two lines of one plane with the other plane (or one line of each plane with the opposite plane) will establish two points on the line of intersection of the planes. The simplest case is when one of the planes is in a principal position as is the top surface of the object in Fig. 10. Lines *A-A'* and *B-B'* are edges of a skew plane, which, extended, are seen to intersect the top surface of the rectangular object at *P* and *Q*, as explained in Sec. 5.

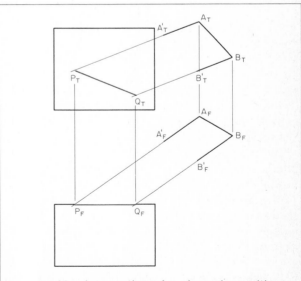

**FIG. 10.** Intersection of a skew plane with a plane in principal position. The edge view of the principal plane determines the intersection.

## 9. Intersection of a Plane with an Inclined Plane.

Figure 11 is similar to Fig. 10 but the intersection this time is with the inclined surface of the object. The intersection is again found by extending $A$-$A'$ and $B$-$B'$ to the edge view of the inclined surface. Refer to Sec. 6.

## 10. Intersection of Two Skew Planes.

As before, the points of intersection of two lines determine the intersection of the planes. In Fig. 12 two edge-view (in the front view) planes have been passed through $A$-$A'$ and $B$-$B'$ extended. The edge-view planes then intersect the skew plane $RSTU$ in lines $WX$ and $YZ$. Then $P_T$ and $Q_T$, the intersections of $A$-$A'$ and $B$-$B'$ extended, in the top view determine the top view of the line of intersection, which, projected to the front view at $P_F$ and $Q_F$, completes the solution. Compare Fig. 12 (edge-view plane in front view is used) with Fig. 8 (edge-view plane in top view is used). An edge-view plane can also be passed in a side view whenever it is convenient to do so.

## 11. To Find the Intersection of Two Prisms (Fig. 13).

In general, find the line of intersection of a surface on one prism with all surfaces on the other. Then take a surface adjacent to the first surface, and find its intersection with the other prism. Continue in this manner until the complete line of intersection of the prisms is determined.

The method of locating end points on the line of intersection of two surfaces depends upon the position of the surfaces, as follows:

*Both Surfaces Receding (Edge View).* Their intersection appears as a point in the view in which they recede. Project the intersection to an adjacent view, locating the two ends of the intersection on the edges of one or both intersecting surfaces so that they will lie within the boundaries of the other surface. The intersection 4-5 of surfaces $QRST$ and $EF$-3 was obtained in this manner.

*One Surface Receding, the Other Skew.* An edge of the oblique surface may appear to pierce the

receding surface in a view in which these conditions exist. If, in an adjacent view, the piercing point lies on the edge of the oblique surface and within the

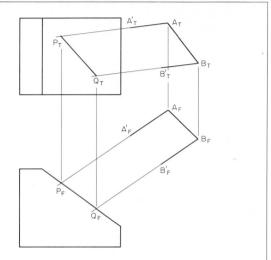

**FIG. 11.** Intersection of a skew plane with an inclined plane. The edge view of the inclined plane determines the intersection.

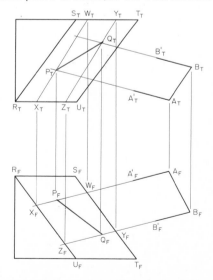

**FIG. 12.** Intersection of two skew planes. Receding planes, passed through two lines of one plane, determine the intersection.

boundaries of the other surface, it is an end point on the intersection of the surfaces. Point 5, lying on edge F-5 of the oblique surface FG-1-5 and the surface QRST, is located in the top view in this manner. Point 1 was similarly established. Point 6, lying on edge ST, is found by passing a vertical plane AA through edge ST. Plane AA cuts line 1-c from plane GF-5-1, giving point 6 where line 1-c crosses ST.

*Both Surfaces Skew.* Find the piercing point of an edge of one surface with the other surface, as follows: Pass a receding plane through an edge of one surface. Find the line of intersection of the receding plane and the other surface, as explained above. The piercing point of the edge and surface is located where the line of intersection, just found, and the edge intersect. Repeat this operation to establish the other end of the line of intersection of the surfaces. Point 3, on the line of intersection 2-3 of the oblique surfaces NORS and EG-1-3, was found in this manner by passing the receding plane BB through edge E-3, finding the intersection b-4 of the surfaces, and then locating point 3 at the intersection of b-4 and E-3.

### 12. To Find the Intersection of Two Pyramids.

In general, find the point where one edge on one pyramid pierces a surface of the other pyramid. Then find where a second edge pierces, and so on. To complete the line of intersection, the piercing points of the edges of the second pyramid with surfaces of the first will probably also have to be found. Figure 14 illustrates the method. Find where edge AD pierces plane EHG, by assuming a vertical cutting plane through edge AD. This plane cuts line 1-2 from plane EHG, and the piercing point is point P, located first on the front view and then projected to the top view. Next, find where AD pierces plane EFG, by using a vertical cutting plane through AD. This plane cuts line 3-4 from plane EFG, and the piercing point is point Q.

After finding point P on plane EHG and point Q on plane EFG, locate the piercing point of edge EG with plane ABD in order to draw lines of intersection.

A vertical plane through EG cuts line 5-6 from plane ABD, and the intersection is point R on edge EG.

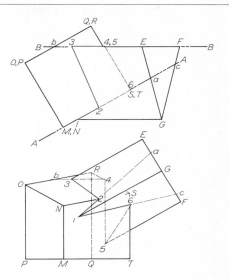

FIG. 13. Intersection of two prisms. The edges of one, intersecting the faces of the other, determine the complete "line" of intersection.

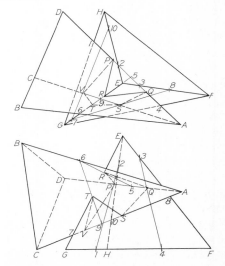

FIG. 14. Intersection of two pyramids. The edges of one, intersecting the faces of the other, determine the intersection.

Thus edges of the "first" pyramid pierce surfaces of the "second," and edges of the second pierce surfaces of the first. Continue in this manner until the complete line of intersection *PRQSTV* has been found.

The use of a vertical cutting plane to obtain the piercing points is perhaps the simplest method and the easiest to visualize. Nevertheless, it should be noted that a plane receding from the frontal or profile plane could also be used. As an example of the use of a plane receding from the frontal, consider that such a plane has been passed through line *CA* in the front view. This plane cuts line 7-8 from plane *EFG*, and the point of intersection is *S* on line *CA*. The use of a plane receding from the profile plane would be basically the same but would require a side view.

### 13. To Find the Line of Intersection between a Prism and a Pyramid (Fig. 15).

The method is basically the same as for two pyramids. Thus a vertical plane through edge *G* cuts line 1-2 from surface *AED*, and a vertical plane through edge *K* cuts line 3-4 from surface *AED*, giving the two piercing points *P* and *Q* on surface *AED*. A vertical plane through edge *AE* cuts elements 7 and 8 from the prism and gives piercing points *R* and *T*. Continue in this manner until the complete line of intersection *PQSTVR* is found.

### 14. Intersections of Curved Surfaces.

The intersections of single- and double-curved surfaces with lines and planes and with each other, especially when one or both of the surfaces are in a skew position, are properly a part of the study of engineering geometry and therefore not included here. However, the intersections of cylinders and cones in simple positions occur more or less frequently on machine parts, and so are discussed in the paragraphs that follow.

### 15. Intersection of a Cylinder and Line.

The common case occurs when the cylinder is in a simple position, as in Fig. 16, with one view the

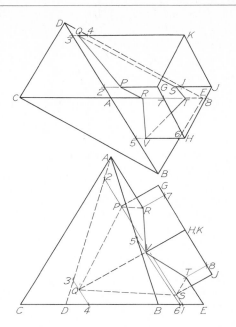

**FIG. 15.** Intersection of a prism and pyramid. The edges of one, intersecting the faces of the other, determine the intersection.

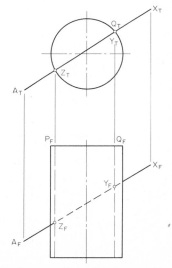

**FIG. 16.** Intersection of a line and cylinder. The edge view of the cylinder determines the two piercing points.

end view—the *edge* view of the surface and the other view showing the normal view of all the elements. Any line such as $AX$ is therefore seen to intersect the cylinder where the surface appears as an edge, in this case the top view at $Y_T$ and $Z_T$. These two points lie on elements of the cylinder $P$ and $Q$, which are projected to the front view thus locating $Y_F$ and $Z_F$.

## 16. Intersection of a Cylinder and Plane.

Planes theoretically contain an infinite number of lines, and so, by the method of Sec. 15, the intersection of a number of *lines* of a plane will determine the intersection of a plane with a cylinder. In Fig. 17 the intersection of $ABC$ with the cylinder is to be located. Any line $AX$ is drawn on the plane and the intersections $Z$ and $Y$ found, giving two points on the curved line of intersection. The process repeated with a number of lines similar to $AX$ will give points to complete the curve.

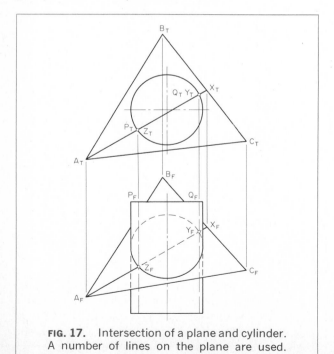

**FIG. 17.**  Intersection of a plane and cylinder. A number of lines on the plane are used.

## 17. Intersection of a Cone and a Line.

Figure 18*a* illustrates pictorially the problem of finding the points where a line pierces a cone. Select a cutting plane, $VXY$, containing the given line $AB$ and passing through the apex, $V$, of the cone. A plane passing through the apex of a cone will intersect the cone in straight lines. The lines of intersection of the cutting plane and the cone are lines $V$-1 and $V$-2. These lines are determined by finding the line of intersection, $RS$, between the cutting plane and the base plane of the cone and then finding the points 1 and 2 where this line crosses

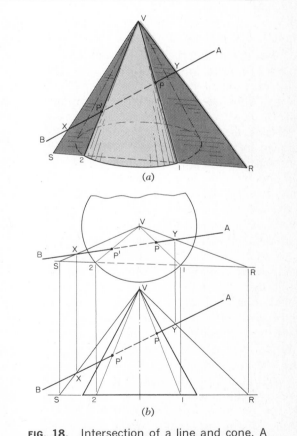

**FIG. 18.**  Intersection of a line and cone. A plane passed through the line and simultaneously through the apex of the cone is the basis for determination.

the base curve. Line $AB$ intersects both lines $V$-1 and $V$-2, establishing the points $P$ and $P'$ where line $AB$ pierces the cone.

The orthographic solution is illustrated at ($b$). Convenient points $X$ and $Y$ on line $AB$ are selected so that plane $VXY$ (a plane containing the apex of the cone and line $AB$) extended will intersect the base plane of the cone in the line $RS$. In the top view $RS$ cuts the base curve at 1 and 2, thus locating elements $V$-1 and $V$-2 *in* the cutting plane and *on* the surface of the cone. $V$-1 and $V$-2 then intersect $BA$ at $P$ and $P'$, the piercing points of line $AB$ and the cone.

### 18. Intersection of a Cone and a Plane.
Figure 19$a$ illustrates pictorially the determination of the line of intersection between a cone and a plane. Lines (elements) of the cone, such as $V$-1, $V$-2, etc., will intersect the plane at $a$, $b$, etc., respectively. If the edge view of the plane appears in one of the views, as at ($b$), the solution is quite simple. Selected elements of the cone, $V$-1, $V$-2, etc., are drawn in both views. These elements are seen to intersect the plane in the front view at $a$, $b$, etc. Projection of these points to the top view then gives points through which a smooth curve is drawn to complete the solution. The edge view of the plane (within the confines of the cone) is, of course, the intersection in the front view.

If the plane intersecting the cone is skew, as in Fig. 20, the cutting-plane method is probably preferable because of the simplicity of the solution. Any plane passed through the apex of the cone will cut straight lines from the cone. Therefore, as shown at ($a$), if vertical cutting planes are used, all passing through $V$, the apex, these planes will cut lines such as $V$-2 and $V$-8 from the cone. At the same time, a vertical cutting plane will cut $XZ$ from the plane. $XZ$ then intersects $V$-2 at $b$ and $V$-8 at $h$, giving two points on the line of intersection. A series of planes thus passed and points found will complete the solution.

Even though the above plane method is simple and requires only the given views for solution, it *might* be advantageous to draw an extra view (auxil-

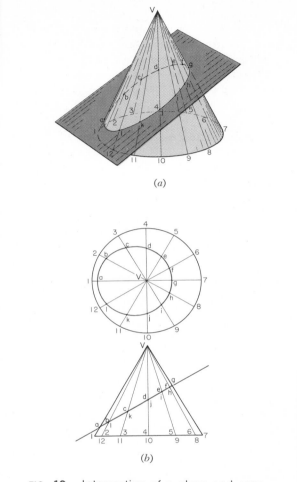

($a$)

($b$)

**FIG. 19.** Intersection of a plane and cone. The edge view of the plane determines the intersection.

iary), as at ($b$), where, by projecting parallel to a horizontal line of the plane, *the plane appears as an edge*. In this event, selected lines of the cone are seen to intersect the plane in the auxiliary view (where the plane appears as an edge), and the solution becomes basically the same as in Fig. 19$b$. In addition, points located first in the auxiliary view and then in the top view will have to be found in the front view by projection from the top view and

FIG. 20.  Intersection of a plane and cone. (a) A series of planes, passed through the cone apex, locate points common to plane and cone; (b) an auxiliary (edge view of plane) determines points common to cone and plane.

measurement from the auxiliary, indicated by distance $X$ on the figure.

## 19. Intersection of Two Cylinders (Fig. 21).

Cutting planes parallel to the axis of a cylinder will cut straight-line elements from the cylinder. The frontal cutting planes $A$, $B$, $C$, and $D$, *parallel to the axis of each cylinder*, cut elements from each cylinder, the intersections of which are points on the curve. The pictorial sketch shows a slice cut by a plane from the object, which has been treated as a solid in order to illustrate the method more easily.

When the axes of the cylinders do not intersect, as in Fig. 22, the same method is used. Certain "critical planes" give the limits and turning points of the curve. Such planes should always be taken through the contour elements. For the position shown, planes $A$ and $D$ give the depth of the curve, the plane $B$ the extreme height, and the plane $C$ the tangent or turning points on the contour element of the vertical cylinder. After the critical points

have been determined, a sufficient number of other cutting planes are used to give an accurate curve.

## 20. Intersection of a Cylinder and a Cone (Fig. 23).

Cutting planes may be taken, as at ($a$), so as to pass through the vertex of the cone and parallel to the axis of the cylinder, thus cutting the straight-line elements from both cylinder and cone; or, as at ($b$), with a right-circular cone, when the cylinder's axis is parallel or perpendicular to the cone's axis, cutting planes may be taken parallel to the base so as to cut circles from the cone. Both systems of planes are illustrated in the figure. The pictorial sketches show slices taken by each plane through the objects, which have been treated as solids in order to illustrate the method more easily. Some judgment is necessary in the selection of both the direction and the number of cutting planes. More points need to be found at the places of sudden curvature or change of direction of the projections of the line of intersections.

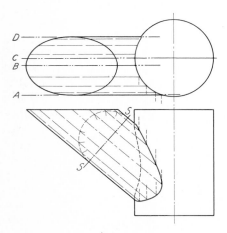

**FIG. 21.**   Intersection of two cylinders. Planes cut straight-line elements from both.

**FIG. 22.**   Intersection of two cylinders, axes not intersecting. Planes, parallel to each of the axes, cut straight-line elements from both surfaces.

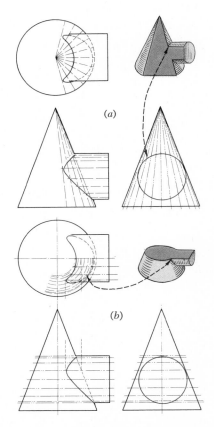

**FIG. 23.**   Intersection of a cylinder and cone. (*a*) Elements of the cone are observed to intersect the cylinder in the side view (edge view of cylinder); or (*b*) horizontal planes cut circles from both surfaces and locate points common to both.

In Fig. 23a the cutting planes appear as edges in the right-side view, where the cylinder surface also appears as an edge. The observed intersections in the side view are then projected to the cone elements formed by the cutting planes, in the top and front views, to complete the solution.

At (b), horizontal cutting planes cut circles from the cone and straight-line elements from the cylinder. The top view then reveals intersections of cone circles and cylinder elements. Projection to the front view will complete the solution.

# PROBLEMS

Selections may be made from the following problems. Construct the figures accurately in pencil without inking. Any practical problem can be resolved into some combi-

nation of the "type solids," and the exercises given illustrate the principles involved in the various combinations.

The following problems may be drawn on $8\frac{1}{2}$- $\times$ 11-in. or 11- $\times$ 17-in. sheets. Assume the objects to be made of thin metal with open ends unless otherwise specified.

## Group 1. Intersections of Prismatic Ducts.
**1. to 3.** Find the line of intersection, considering the prisms as pipes opening into each other. Use care in indicating visible and invisible parts of the line of intersection.

**4 to 6.** Find the line of intersection, indicating visible and invisible parts and considering prisms as pipes opening into each other.

**7.** Find the intersection between the two prismatic ducts.

PROB. 1.

PROB. 2.

PROB. 3.

PROB. 4.

PROB. 5.

PROB. 6.

PROB. 7.  Prismatic ducts.

**PROB. 9.**

**PROB. 10.**

**PROB. 8.**   Tower and conveyor-belt galleys.

**PROB. 11.**         **PROB. 12.**         **PROB. 13.**

**8.**   The layout as illustrated shows a tower and proposed conveyor-belt galleys. Find the intersection between the tower and the main supply galley.

**8A.**   Same layout as Prob. 8. After solving Prob. 8., find the intersection between the two galleys.

## Group 2. Intersections of Pyramidal Objects.

**9 to 13.**   Find the lines of intersection.

**14.**   Find the line of intersection between the prismatic duct and hopper.

**PROB. 14.**   Prismatic duct and hopper.

**PROB. 15.**    **PROB. 16.**

## Group 3. Intersections of Cylindrical Ducts.

**15 to 17.**    Find the line of intersection, indicating visible and invisible portions and considering cylinders as pipes opening into each other.

**18.**    The layout shows a pump casing composed of a cylinder and elbow. Find the intersection between the cylinder and elbow.

**19.**    The layout shows a portion of a liquid oxygen pumping system consisting of pipes $A$, $B$, $C$, and $D$. Find the intersection of the pipes.

**PROB. 17.**

**PROB. 18.**    Pump casing.

**PROB. 19.**    Lox pump outlet.

**20.** The layout shows an oblique pipe meeting a cylindrical inlet in a power-station piping system. Find the intersection between the inlet and pipe.

## Group 4. Intersections of Conic Objects and Ducts.

**21 to 31.**   Find the line of intersection.

**32.**   The layout shows the plan view of an intake manifold consisting of a conical connector intersected by two cylindrical pipes. Find the intersection of the pipe that

VIEW A-A

ELEVATION

Mk-13   1- REQ'D     **PROB. 20.**   Power-station piping.

**PROB. 21.**

**PROB. 22.**

**PROB. 23.**

**PROB. 24.**

**PROB. 25.**

**PROB. 26.**

**PROB. 27.**

**PROB. 28.**

**PROB. 29.**

enters the side of the conical connector. The axes of the two shapes intersect. Solve this problem using only the plan view.

**33.** The layout shows the cylindrical outlet pipe and conical inlet pipe of a "Kaplan-system" outlet. Find the intersection of the two shapes, using the cutting-sphere method.

**34.** The layout shows a right circular cylinder and an elliptical-base cone. Find the intersection of the two surfaces.

**PROB. 30.**          **PROB. 31.**

**PROB. 32.**    Discharge manifold.

**PROB. 33.**    "Kaplan-system" outlet.

**PROB. 34.**    Cylinder and cone.

# DEVELOPED VIEWS

# 13

These Show a Representation of the Shape and Size of Thin Material Used to Make an Object by Folding, Rolling, or Forming

This is a Discussion and Illustration of: Developments Defined and Classified • Basic Considerations • Developments of Prisms, Pyramids • Triangulation • Developments of Connectors and Transition Pieces • Development of a Sphere • Joints, Connectors, Hems • Problems

## 1. Developments, Defined.

In many different kinds of construction full-size patterns of some or all of the faces of an object are required; for example, in stonecutting, a template or pattern giving the shape of an irregular face, or in sheet-metal work, a pattern to which a sheet may be cut so that when rolled, folded, or formed it will make the object.

The complete surface laid out in a plane is called the "development" of the surface.

Surfaces about which a thin sheet of flexible material (such as paper or tin) can be wrapped smoothly are said to be developable; these include objects made up of planes and single-curved surfaces only. Warped and double-curved surfaces are nondevelopable; and when patterns are required for their construction, they can be made only by methods that are approximate, but, assisted by the ductility or pliability of the material, they give the required form. Thus while a ball cannot be wrapped smoothly, a two-piece pattern developed approximately and cut from leather can be stretched and sewed on in a smooth cover, or a flat disk of metal can be die-stamped, formed, or spun to a hemispherical or other shape.

## 2. Basic Considerations.

We have learned the method of finding the true size of a plane surface by projecting its normal view. If the true size of all the plane faces of an object are found and joined in order at their common edges so that all faces lie in a common plane, the result will be the developed surface. Usually this may be done to the best advantage by finding the true length of the edges.

The development of a right cylinder is evidently a rectangle whose width is the altitude and length the rectified circumference (Fig. 1); and the development of a right-circular cone is a circular sector with a radius equal to the slant height of the cone and an arc equal in length to the circumference of its base (Fig. 1).

As illustrated in Fig. 1, developments are drawn with the inside face up. This is primarily the result of working to inside rather than outside dimensions of ducts. This procedure also facilitates the use of fold lines, identified by punch marks at each end, along which the metal is folded in forming the object.

In laying out real sheet-metal designs, an allowance must be made for seams and lap and, in heavy sheets, for the thickness and crowding of the metal; there is also the consideration of the commercial sizes of material as well as the question of economy in cutting. In all of this some practical shop knowledge is necessary. Figure 17 and Sec. 14 indicate the usage of some of the more common joints, although the developments given in this chapter will be confined to the principles.

## 3. To Develop a Truncated Hexagonal Prism (Fig. 2).

First draw two projections of the prism: (1) a normal view of a right section (a section or cut obtained by a plane perpendicular to the axis) and (2) a normal view of the lateral edges. The base *ABCDEF* is a right section shown in true size in the bottom view. Lay off the perimeter of the base on line *AA* of the development. This line is called by sheet-metal workers the "stretchout" or "girth" line. At points *A*, *B*, *C*, etc., erect perpendiculars called "measuring lines" or "bend lines," representing the lateral edges along which the pattern is folded to form the prism. Lay off on each of these its length *A*-1, *B*-2, *C*-3, etc., as given on the front view. Connect the points 1, 2, 3, etc., in succession, to complete the development of the lateral surfaces. Note on the pattern that the inside of the lateral faces is toward the observer. For the development of the entire surface in one piece, attach the true sizes of the upper end and the base as shown, finding the true size of the upper end by an auxiliary (normal) view as described in Sec. 4, Chap. 6. For economy of solder or rivets and time, it is customary to make the seam on the shortest edge or surface. In seaming along the intersection of surfaces whose dihedral angle is other than 90°, as is the case here, the lap seam lends itself to convenient assembling. The flat lock could be used if the seam were made on one of the lateral faces.

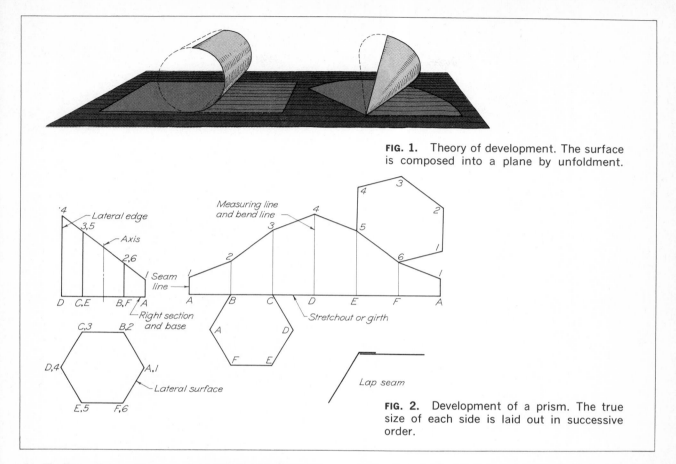

**FIG. 1.** Theory of development. The surface is composed into a plane by unfoldment.

**FIG. 2.** Development of a prism. The true size of each side is laid out in successive order.

## 4. To Develop a Truncated Right Pyramid (Fig. 3).

Draw the projections of the pyramid which show (1) a normal view of the base or right section and (2) a normal view of the axis. Lay out the pattern for the pyramid, and then superimpose the pattern of the truncation.

Since this is a portion of a right regular pyramid, the lateral edges are all of equal length. The lateral edges $OA$ and $OD$ are parallel to the frontal plane and consequently show in their true length on the front view. With center $O_1$, taken at any convenient place, and a radius $O_F A_F$, draw an arc that is the stretchout of the pattern. On it step off the six equal sides of the hexagonal base, obtained from the top view, and connect these points successively with

each other and with the vertex $O_1$, thus forming the pattern for the pyramid.

The intersection of the cutting plane and lateral surfaces is developed by laying off the true length of the intercept of each lateral edge on the corresponding line of the development. The true length of each of these intercepts, such as $OH$ and $OJ$, is found by rotating it about the axis of the pyramid until they coincide with $O_F A_F$. The path of any point, as $H$, will be projected on the front view as a horizontal line. To obtain the development of the entire surface of the truncated pyramid, attach the base; also find the true size of the cut face and attach it on a common line.

The lap seam is suggested for use here also for convenient assembling.

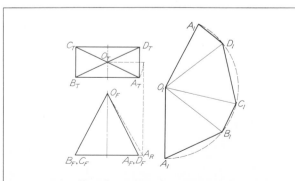

**FIG. 3.** Development of a pyramid. This is a right pyramid. The true lengths of all edges (from the vertex to the base) are equal.

*The right-rectangular pyramid,* Fig. 4, is developed in a similar way, but as the edge $OA$ is not parallel to the plane of projection, it must be rotated to $O_F A_R$ to obtain its true length.

## 5. To Develop an Oblique Pyramid (Fig. 5).

Since the lateral edges are unequal in length, the true length of each must be found separately by rotating it parallel to the frontal plane. With $O_1$ taken at any convenient place, lay off the seam line $O_1 A_1$ equal to $O_F A_R$. With $A_1$ as center and radius $A_1 B_1$ equal to the true length of $AB$, describe an arc. With $O_1$ as center and radius $O_1 B_1$ equal to $O_F B_R$, describe a second arc intersecting the first in vertex $B_1$. Connect the vertices $O_1$, $A_1$, and $B_1$, thus forming the pattern for the lateral surface $OAB$. Similarly, lay out the patterns for the remaining three lateral surfaces, joining them on their common edges. The stretchout is equal to the summation of the base edges. If the complete development is required, attach the base on a common line. The lap seam is suggested as the most suitable for the given conditions.

## 6. To Develop a Truncated Right Cylinder (Fig. 6).

The development of a cylinder is similar to the development of a prism. Draw two projections of the cylinder: (1) a normal view of a right section and (2) a normal view of the elements. In rolling the cylinder out on a tangent plane, the base or right section, being perpendicular to the axis, will develop

**FIG. 4.** Development of a pyramid. The true length from vertex to base is found by rotating one edge until frontal.

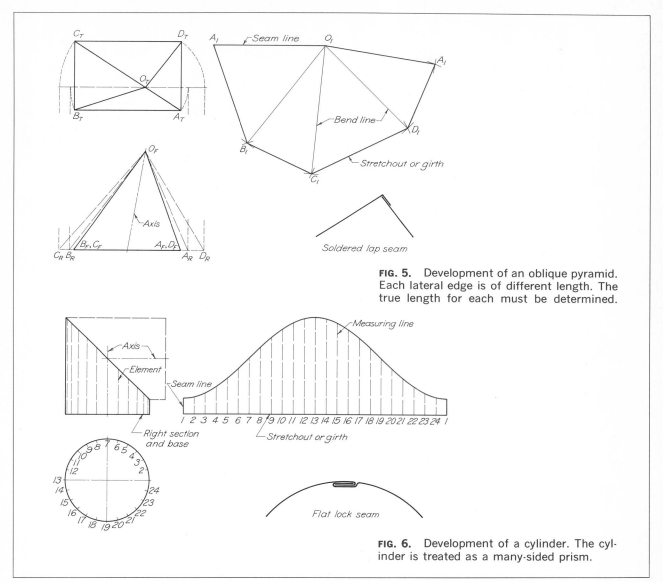

**FIG. 5.** Development of an oblique pyramid. Each lateral edge is of different length. The true length for each must be determined.

**FIG. 6.** Development of a cylinder. The cylinder is treated as a many-sided prism.

into a straight line. For convenience in drawing, divide the normal view of the base, here shown in the bottom view, into a number of equal parts by points that represent elements. These divisions should be spaced so that the chordal distances closely enough approximate the arc to make the stretchout practically equal to the periphery of the base or right section. Project these elements to the front view. Draw the stretchout and measuring lines as in Fig. 2, the cylinder now being treated as a many-sided prism. Transfer the lengths of the elements in order, by projection or with dividers, and join the points thus found by a smooth curve, sketching it in freehand very lightly before fitting the French curve to it. This development might be the pattern of one-half of a two-piece elbow. Three-piece, four-piece, or five-piece elbows can be drawn similarly, as illustrated in Fig. 7. As the base is symmetrical, only one-half of it need be drawn. In these cases, the intermediate pieces, as $B$, $C$, and $D$, are developed on a stretchout line formed by laying off the perimeter of a right section. If the right section is taken through the middle of the piece, the stretchout line becomes the center line of the development.

Evidently any elbow could be cut from a single

sheet without waste if the seams were made alternately on the long and short sides. The flat lock seam is recommended for Figs. 6 and 7, although other types could be used.

*The octagonal dome* (Fig. 8) illustrates an application of the development of cylinders. Each piece is a portion of a cylinder. The elements are parallel to the base of the dome and show in their true lengths in the top view. The true length of the stretchout line for sections $A$ and $A'$ shows in the

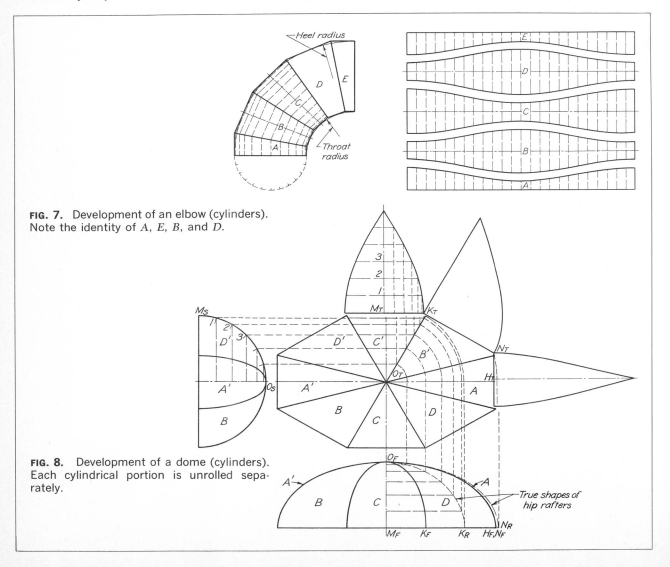

**FIG. 7.** Development of an elbow (cylinders). Note the identity of $A$, $E$, $B$, and $D$.

**FIG. 8.** Development of a dome (cylinders). Each cylindrical portion is unrolled separately.

front view at $O_F H_F$. By considering $O_T H_T$ as the edge of a plane cutting a right section, the problem is identical with the preceding problem.

Similarly, the stretchout line for sections $B$, $B'$, $D$, and $D'$ shows in true length at $O_F K_R$ in the front view, and for section $C$ and $C'$ at $O_S M_S$ in the side view.

The true shape of hip rafter $ON$ is found by rotating it until it is parallel to the frontal plane, as at $O_F N_R$, in the same manner as in finding the true length of any line. A sufficient number of points should be taken to give a smooth curve.

## 7. To Develop a Truncated Right-Circular Cone (Fig. 9).
Draw the projections of the cone which will show (1) a normal view of the base or right section and (2) a normal view of the axis. First develop the surface of the complete cone and then superimpose the pattern for the truncation.

Divide the top view of the base into a sufficient number of equal parts so that the sum of the resulting chordal distances will closely approximate the periphery of the base. Project these points to the

front view, and draw front views of the elements through them. With center $A_1$ and a radius equal to the slant height $A_F \cdot 1_F$, which is the true length of all the elements, draw an arc, which is the stretchout, and lay off on it the chordal divisions of the base, obtained from the top view. Connect these points $1_1$, 2, 3, etc., with $A \cdot 1$, thus forming the pattern for the cone. Find the true length of each element from vertex to cutting plane by rotating it to coincide with the contour element $A \cdot 1$, and lay off this distance on the corresponding line of the development. Draw a smooth curve through these points. The flat lock seam along element $S \cdot 1$ is recommended, although other types could be employed. The pattern for the inclined surface is obtained from the auxiliary (normal) view.

## 8. Triangulation.
Nondevelopable surfaces are developed approximately by assuming them to be made of narrow sections of developable surfaces. The commonest and best method for approximate development is that of triangulation, that is, the surface is assumed to be made up of a large number of triangular strips,

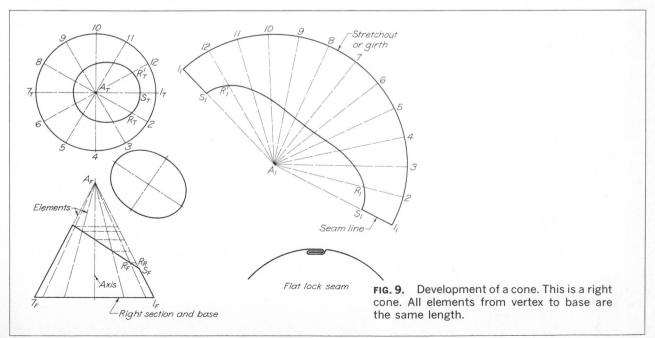

**FIG. 9.** Development of a cone. This is a right cone. All elements from vertex to base are the same length.

or plane triangles with very short bases. This method is used for all warped surfaces and also for oblique cones. Oblique cones are single-curved surfaces and thus are theoretically capable of true development, but they can be developed much more easily and accurately by triangulation, a simple method which consists merely of dividing the surface into triangles, finding the true lengths of the sides of each, and constructing them one at a time, joining these triangles on their common sides.

### 9. To Develop an Oblique Cone (Fig. 10).

An oblique cone differs from a cone of revolution in that the elements have different lengths. The development of the right-circular cone is, practically, made up of a number of equal triangles which meet at the vertex and whose sides are elements and whose bases are the chords of short arcs of the base of the cone. In the oblique cone each triangle must be found separately.

Draw two views of the cone showing (1) a normal view of the base and (2) a normal view of the altitude. Divide the true size of the base, here shown in the top view, into a sufficient number of equal parts so that the sum of the chordal distances will closely approximate the length of the base curve. Project these points to the front view of the base.

Through these points and the vertex, draw the elements in each view. Since this cone is symmetrical about a frontal plane through the vertex, the elements are shown only on the front half of it. Also, only one-half of the development is drawn. With the seam on the shortest element, the element $OC$ will be the center line of the development and can be drawn directly at $O_1C_1$, as its true length is given at $O_FC_F$. Find the true length of the elements by rotating them until parallel to the frontal plane or by constructing a "true-length diagram." The true length of any element would be the hypotenuse of a triangle, one leg being the length of the projected element as seen in the top view and the other leg being equal to the altitude of the cone. Thus to make the diagram, draw the leg $OD$ coinciding with or parallel to $O_FD_F$. At $D$ and perpendicular to $OD$, draw the other leg, on which lay off the lengths $D$-1, $D$-2, etc., equal to $D_T1_T$, $D_T2_T$, etc., respectively. Distances from $O$ to points on the base of the diagram are the true lengths of the elements.

Construct the pattern for the front half of the cone as follows: With $O_1$ as center and radius $O$-1, draw an arc. With $C_1$ as center and radius $C_T1_T$, draw a second arc intersecting the first at $1_1$; then $O_1$-$1_1$ will be the developed position of the element $O$-1. With $1_1$ as center and radius $1_T$-$2_T$, draw an arc

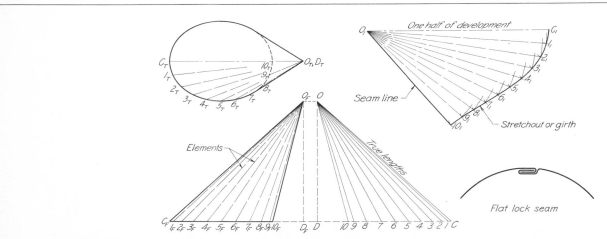

**FIG. 10.** Development of an oblique cone. A true-length diagram is used to obtain the lengths of the elements.

intersecting a second arc with $O_1$ as center and radius $O$-2, thus locating $2_1$. Continue this procedure until all the elements have been transferred to the development. Connect the points $C_1$, $1_1$, $2_1$, etc., with a smooth curve, the stretchout line, to complete the development. The flat lock seam is recommended for joining the ends to form the cone.

## 10. A Conic Connection between Two Parallel Cylindrical Pipes of Different Diameters.

This is shown in Fig. 11. The method used in drawing the pattern is an application of the development of an oblique cone. One-half of the elliptical base is shown in true size in an auxiliary view, here attached to the front view. Find the true size of the base from its major and minor diameters, divide it into a number of equal parts so that the sum of these chordal distances closely approximates the periphery of the curve, and project these points to the front and top views. Draw the elements in each view through these points, and find the vertex $O$ by extending the contour elements until they intersect. The true length of each element is found by using the vertical distance between its ends as the vertical leg of the diagram and its horizontal projection as the other leg. As each true length from vertex to base is found, project the upper end of the intercept horizontally across from the front view to the true

length of the corresponding element to find the true length of the intercept. The development is drawn by laying out each triangle in turn, from vertex to base, as in Sec. 9, starting on the center line $O_1C_1$ and then measuring on each element its intercept length. Draw smooth curves through these points to complete the pattern. Join the ends with a flat lock seam.

## 11. Transition Pieces.

These are used to connect pipes or openings of different shapes of cross section. Figure 12, showing a transition piece for connecting a round pipe and a rectangular pipe, is typical. Transition pieces are always developed by triangulation. The piece shown in Fig. 12. is made up of four triangular planes, whose bases are the sides of the rectangle, and four parts of oblique cones, whose common bases are arcs of the circle and whose vertices are at the corners of the rectangle. To develop it, make a true-length diagram as in Fig. 10. When the true length of $O$-1 is found, all the sides of triangle $A$ will be known. Attach the development of cones $B$ and $B'$, then those of triangles $C$ and $C'$, and so on.

Figure 13 is another transition piece joining a rectangular to a circular pipe whose axes are nonparallel. By using a partial right-side view of the round opening, the divisions of the bases of the oblique cones can be found (as the object is sym-

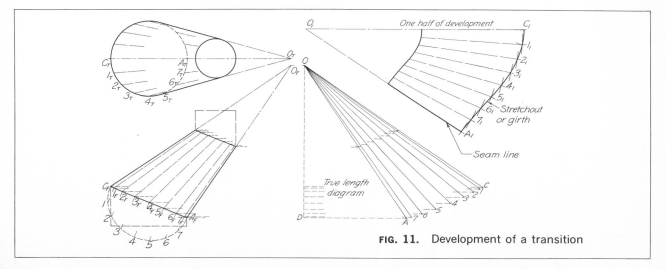

**FIG. 11.**   Development of a transition

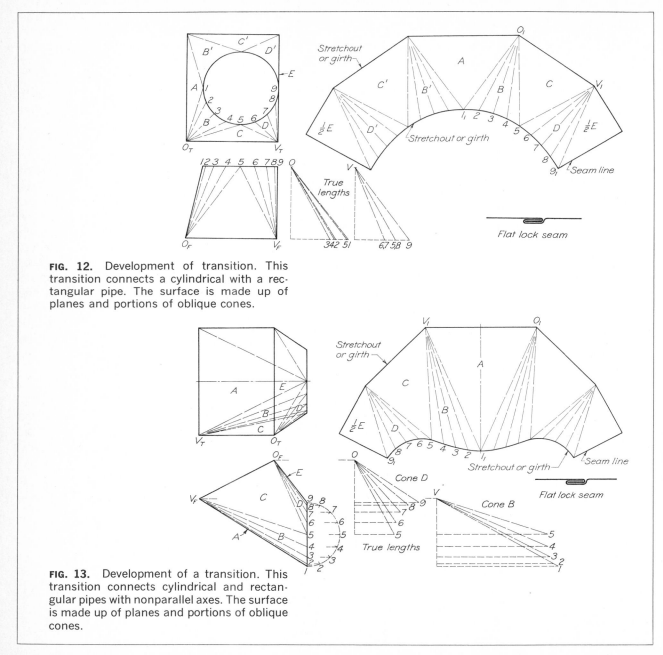

**FIG. 12.** Development of transition. This transition connects a cylindrical with a rectangular pipe. The surface is made up of planes and portions of oblique cones.

**FIG. 13.** Development of a transition. This transition connects cylindrical and rectangular pipes with nonparallel axes. The surface is made up of planes and portions of oblique cones.

metrical, one-half only of the opening need be divided). The true lengths of the elements are obtained as in Fig. 11.

With the seam line the center line of the plane $E$ in Figs. 12 and 13, the flat lock is recommended for joining the ends of the development.

## 12. Triangulation of Warped Surfaces.

The approximate development of a warped surface is made by dividing it into a number of narrow quadrilaterals and then splitting each of these quadrilaterals into two triangles by a diagonal, which is assumed to be a straight line, although really a curve. Figure 14 shows a warped transition piece to connect an ovular (upper) pipe with a right-circular cylindrical pipe (lower). Find the true size of one-half the elliptical base by rotating it until horizontal about an axis through 1, when its true shape appears on the top view. The major diameter is $1\text{-}7_R$, and the minor diameter through $4_R$ will equal the diameter of the lower pipe. Divide the semiellipse into a sufficient number of equal parts, and project these to the top and front views. Divide the top semicircle into the same number of equal parts, and connect similar points on each end, thus dividing the surface into approximate quadrilaterals. Cut each into two triangles by a diagonal. On true-length diagrams find the lengths of the elements and the diagonals, and draw the development by constructing the true sizes of the triangles in regualr order. The flat-lock seam is recommended for joining the ends of the development.

## 13. To Develop a Sphere.

The sphere may be taken as typical of double-curved surfaces, which can be developed only approximately. It may be cut into a number of equal meridian sections, or lunes, as in Fig. 15, and these may be considered to be sections of cylinders. One of these sections, developed as the cylinder in Fig. 15, will give a pattern for the others.

Another method is to cut the sphere into horizontal sections, or zones, each of which may be

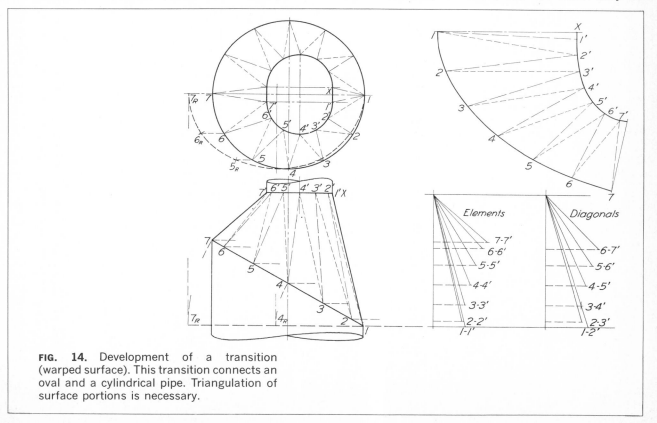

**FIG. 14.** Development of a transition (warped surface). This transition connects an oval and a cylindrical pipe. Triangulation of surface portions is necessary.

taken as the frustum of a cone whose vertex is at the intersection of the extended chords (Fig. 16).

## 14. Joints, Connectors, and Hems.

There are numerous joints used in seaming sheet-metal ducts and in connecting one duct to another. Figure 17 illustrates some of the more common types, which may be formed by hand on a break or by special seaming machines. No attempt to dimension the various seams and connections has been made here because of the variation in sizes for different gages of metal and in the forming machines of manufacturers.

Hemming is used in finishing the raw edges of the end of the duct. In wire hemming, an extra allowance of about $2\frac{1}{2}$ times the diameter of the wire is made for wrapping around the wire. In flat hemming, the end of the duct is bent over once or twice to relieve the sharp edge of the metal.

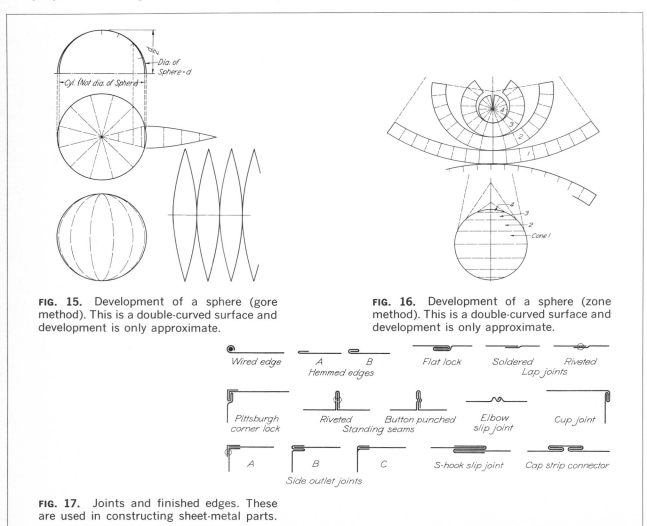

**FIG. 15.** Development of a sphere (gore method). This is a double-curved surface and development is only approximate.

**FIG. 16.** Development of a sphere (zone method). This is a double-curved surface and development is only approximate.

**FIG. 17.** Joints and finished edges. These are used in constructing sheet-metal parts.

# PROBLEMS

Selections may be made from the following problems. Construct the figures accurately in pencil without inking. Any practical problem can be resolved into some combination of the "type solids," and the exercises given illustrate the principles involved in the various combinations.

An added interest in developments may be found by working problems on suitable paper, allowing for fastenings and lap, and cutting them out. It is recommended that at least one or two models be constructed in this manner.

In sheet-metal shops, development problems, unless very complicated, are usually laid out directly on the metal.

The following problems may be drawn on $8\frac{1}{2}$- $\times$ 11-in. or 11- $\times$ 17-in. sheets. Assume the objects to be made of thin metal with open ends unless otherwise specified.

## Group 1. Developments of Prisms.
**1 to 6.** Develop lateral surfaces of the prisms.

## Group 2. Developments of Pyramids.
**7 to 9.** Develop lateral surfaces of the hoppers.
**10 and 11.** Develop lateral surfaces of the pyramids.

## Group 3. Developments of Cylinders.
**12 to 18.** Develop lateral surfaces of the cylinders.
**19.** Develop the oblique pipe section of the power-station piping of Prob. 20, Chap. 12. Also develop the cylindrical inlet, showing the opening to be made for the oblique pipe.

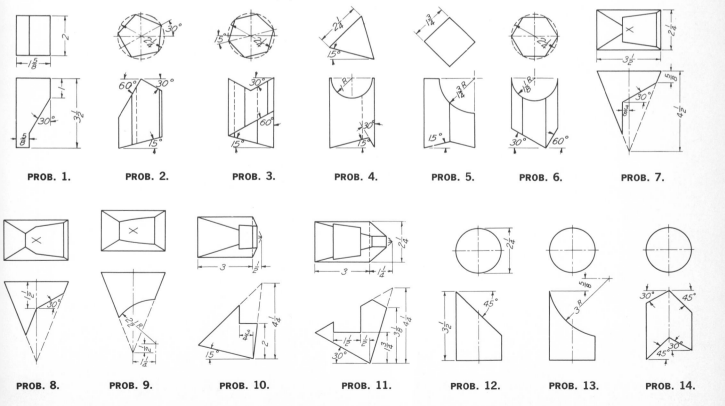

PROB. 1.  PROB. 2.  PROB. 3.  PROB. 4.  PROB. 5.  PROB. 6.  PROB. 7.

PROB. 8.  PROB. 9.  PROB. 10.  PROB. 11.  PROB. 12.  PROB. 13.  PROB. 14.

PROB. 15.          PROB. 16.          PROB. 17.          PROB. 18.

PROB. 20.

## Group 4. Developments of Combinations of Prisms and Cylinders.

**20 to 22.**   Develop lateral surfaces.

## Group 5. Developments of Cones.

**23 to 27.**   Develop lateral surfaces.

**28.**   Develop the concentric reducer for the power-station piping of Prob. 20, Chap. 12.

PROB. 21.          PROB. 22.          PROB. 23.          PROB. 24.

PROB. 25.          PROB. 26.          PROB. 27.          PROB. 29.

## Group 6. Developments of Combinations of Surfaces.

**29 to 32.** Develop lateral surfaces of the objects.

**33.** Develop the cylindrical pipe for the pump casing of Prob. 18, Chap. 12.

**34.** Develop the elbow sections, *A*, *B*, and *C*, of the pump casing of Prob. 18, Chap. 12.

## Group 7. Developments of Cones and Transition Pieces.

**35 to 42.** Develop lateral surfaces of the objects (one-half in Probs. 35, 36, 38, 39, and 41).

PROB. 30.

PROB. 31.

PROB. 32.

PROB. 35.

PROB. 36.

PROB. 37.

PROB. 38.

PROB. 39.

PROB. 40.

PROB. 41.

PROB. 42.

## Group 8. Developments of Furnace-pipe Fittings.

**43 to 50.**   Develop surfaces and make paper models.

**PROB. 43.**

**PROB. 45.**

**PROB. 44.**

## Group 9. Specialities.

There are innumerable combinations of geometric shapes and, sometimes, nongeometric forms used in modern engineering work that, fundamentally, are problems of layout and intersection and then either development or representation. The following three problems are typical examples.

**51.**   The diverter-duct system for a VTOL jet consists essentially of a formed elbow and a cylinder. A typical layout is shown in Prob. 51. Find the intersection between the cylinder and elbow. Develop the cylinder and make an accurate representational drawing of the elbow.

**52.**   The evaporator shown in the layout consists of a combustion sphere and cone. Connected to this combination is a steam outlet *A* and an exhaust pipe *B*. Find all intersections and then develop the parts that can be developed.

**53.**   The microwave horn reflector antenna consists essentially of three surfaces: a cone, a cylinder, and a paraboloid of revolution, which, if chosen correctly, should intersect one another on the same curve. To verify this construction, lay out the complete horn according to the dimensions shown in the illustration. Then, find the intersection between the cone and cylinder and between the cylinder and paraboloid. Develop all parts that can be developed.

**PROB. 46.**

**PROB. 47.**

**PROB. 49.**

**PROB. 48.**

**PROB. 50.**

**PROB. 51.** Diverter duct system.

**PROB. 52.** Evaporator.

**PROB. 53.** Reflector horn.

# COMPUTER GRAPHICS AND NUMERICAL CONTROL

# 14

A discussion of the computer and its importance in graphics

This Chapter Presents
Computer Graphics and Numerically Controlled
Machines • Flatbed Drum Line and Analog
Plotters • Matrix Printers • Animation •
Cathode-Ray Tube Displays • Storage Cathode-
Ray Tubes • Refresh Cathode-Ray Tubes •
Plasma Panel Displays • Hard-Copy Devices •
Computer Programming for Displays •
Mathematical Models • Computer Graphics Lists,
Graphs, Maps, and Pictures • Wireframe
Pictures • Interactive Graphics and Terminals •
Light Pens • Computer-assisted Instruction •
Numerically Controlled Machine Tools

## 1. Introduction.

Digital and analog computers were developed in the 1940s. Today computers may perform several million calculations per second and may be programmed to execute sophisticated tasks such as guiding rockets to the moon. Development of the computer dramatically affected all engineering and science. As we shall see, the computer has enhanced the technical capabilities of the graphic language.

## 2. Computer Graphics.

Computer graphics combines the methods of graphic arts and the technical capabilities of the computer to convert scientific data into descriptive pictures. The most common and least expensive form of computer graphic representation is the familiar line drawing. All the engineering drawings in this text are basically line drawings. The actual drawing may be displayed on paper or on a cathode-ray tube. The second and more complicated form of computer graphic representation is a lifelike picture drawn according to the mathematical model of an object, as shown in Figs. 1 and 2. In this process a camera produces a pattern of light and dark on a television screen according to mathematical calculations. This form of computer graphics is the most dramatic because it is possible to produce a picture of a nonexistent object with nothing more than a mathematical model.

Computer graphics involves six basic steps: (1) a design is conceptualized or planned; (2) the design is translated into data that is acceptable to the computer; (3) the data describing the design concept is fed into the computer along with the programming that will direct the graphics equipment (this is the *input* stage, in which the data and program make up the *software* of the graphics process); (4) the computer operates the graphics equipment and produces an image according to the data describing the design concept (this is the *output* stage, in which the computer and graphics equipment make up the *hardware* of the computer process); (5) the graphics designer makes changes and refines the design; and (6) reproduction of the final

**FIG. 1**  A computer-generated picture of an object which does not exist except as a set of numbers in a computer program. This was drawn on a matrix type display.

**FIG. 2**  Another object also computer-drawn which also does not exist except as a set of numbers in a computer program.

design is implemented by any of the standard methods such as photoprocessing.

## 3. Numerically Controlled Machine Tools.

Numerically controlled machine tools are run by a numerical control program rather than by a human operator. (See Fig. 14.) A computer produces the numerical control program.

Computer graphics and numerically controlled

machining have many points in common, so much so that paths to be taken by the numerically controlled machine tools are often drawn using computer graphics to verify whether or not the tools will cut the correct contours.

## 4. Computer Graphics and Numerically Controlled Machines.

Computer graphics and numerically controlled machine tools basically perform the same task of producing a line with a given function at a specified place on a drawing or machine part, according to a computer-generated program. Both are forms of automation in which the enormous calculating power of the computer is used to direct a stylus which will draw pictures or produce machined parts more rapidly than could be done by conventional hand methods. A strong advantage of both processes is that they may quickly and accurately produce a curve or contour matching a mathematical function. This latter process, known as plotting a curve, requires a large number of calculations of $x$ and $y$ coordinates.

## 5. Line Plotters or Drafting Machines.

The graphic devices which draw lines on paper are called plotters, line plotters, or automated drafting machines.

## 6. The Flatbed Plotter.

As shown in Fig. 3, the flatbed plotter supports the drawing with a flat surface. Bed dimensions of four or five feet are fairly common. The plotter pen is mounted on a bridge that moves as an entire unit in both $x$ and $y$ directions corresponding to the $x$ and $y$ coordinates of a drawing. The pen module can lift or drop the pen so that lines can be drawn or not drawn as the pen moves across the paper.

Pen movements may be controlled by two different types of motors: stepping motors or variable-speed motors.

Plotters controlled by stepping motors are called incremental or digital plotters. The majority of plotters are of this type. The advantage of these plotters is that voltage fluctuations within the system do not

**FIG. 3** A small flat bed plotter

affect the pen movement. Plotters of this type have eight basic movements: positive and negative $x$, positive and negative $y$, and the four combinations of simultaneous $x$ and $y$ movements.

Alternatives to stepping motors are variable-speed plotters, which are controlled by two motors, one for the $x$ function and one for the $y$ function. Thus, a line with a slope of one unit in the $y$ direction and three units in the $x$ direction would be drawn with the $x$ motor running at three times the speed of the $y$ motor.

The pen bridge flatbed plotter is constructed so that the pen module remains stationary while a linear motor, which is similar to a variable-speed motor, moves the flatbed in $x$ and $y$ directions.

## 7. Drum Line Plotters.

As seen in Fig. 4, the drum line plotter also has a stationary pen bridge. Paper is moved underneath it on a rotating drum. The drum acts as a sprocket wheel. The edge of the paper is perforated with sprocket holes. Sprocket teeth on the drum pull the paper in the desired direction. A typical drum plotting system can handle ten feet of drawing. Drum plotters are usually driven by stepping motors and are therefore of the incremental type. Because the steps around the drum are produced by an angular motion of the drum, maintaining accuracy on a long drawing is more of a problem with a drum plotter

**FIG. 4**   A drum-type plotter

than with a flatbed plotter. However, most drum plotters have an adjustment to permit drawing a line the correct length by a trial-and-error manipulation of the drum drive. Drum plotters do not require different programming than that used for flatbed plotters. One advantage for the user is that the paper can be advanced to provide a new starting point without the need to remove the current drawing and replace the paper.

### 8. Analog Plotters.

The plotters that have been described so far are normally controlled by digital computers. Input voltages to the analog plotter are usually controlled by an analog computer. The analog plotter is a vector or continuous-line plotter that resembles the flatbed plotter. The input to such a plotter is time-dependent. The $x$ and $y$ voltages must change with respect to each other in such a fashion that the pen traces the desired line, whether it be a straight line or a curve.

### 9. The Matrix Printer or Matrix Plotter.

The next type of plotter may be referred to as a matrix plotter or a matrix printer. For convenience we shall refer to it as a matrix printer. The matrix printer places dots at positions on a matrix, closely approximating the lines of a drawing. The printing heads used in many computer card punches are

matrix printers. This type of printer has a pen bridge that remains stationary while the paper moves. The printer operates either by causing a tiny wire to press a printing ribbon against a paper at the correct spot or by heating a wire to cause a sensitized paper to turn dark at the heated spot. The hardware of such a printer must translate a program into the correct dot pattern to be placed on each line of printing. Since typical matrix alphabets have letters and numbers that are six to ten dots high, a single line of letters or numbers is really six to ten lines of printing. Matrix printers are quiet, inexpensive, and fast in comparison to the drum line plotter. Two major disadvantages are noticed by the user of such a device. The first is that the entire drawing must be developed inside the computer before any of it is printed. The second is that lines are not continuous but are strings of dots. Consequently, such a plotter cannot be used to produce precise patterns, such as those used to construct a printed circuit board.

### 10. Animation.

The plotters that have been described up to this point are all line plotters. There are two distinct disadvantages to use of the line plotter. Line plotters cannot produce shading, and motion can be produced only by drawing each frame of the picture, photographing it, and repeating the process until a film is produced. This is the process that was used to produce the animated cartoons that Disney made famous. Although realistic motion can be given to a computer graphics picture by this technique, it is a slow and expensive process. Animation of mechanical motions is very valuable to designers, and there are alternative methods of producing animated computer graphics.

### 11. Cathode-ray Tube Displays.

The cathode-ray tube, such as that used in black-and-white television, may be used as a computer graphics display device. (See Fig. 5.) The cathode-ray tube may display drawings of the continuous-line, matrix, or incremental type. Cathode-ray tubes offer an advantage over pen plotters in that the letters and numbers are formed by electronic cir-

**FIG. 5** A cathode ray tube display device of the storage tube type.

cuitry and not by a program in the computer. This allows for speed and flexibility.

There are two major types of cathode-ray tubes: refresh and storage.

### 12. Storage Cathode-ray Tube.
The storage-type cathode-ray tube has on the tube face a phosphor which is continually bombarded with a "fog" of electrons. To produce a picture the phosphor is lit by additional electrons according to the specifications of a program. The picture is, then, stored on the face of the tube by the fog of electrons. Unfortunately, once a line is drawn, it cannot be erased individually. The only way to remove part of a picture is to erase the entire thing and redraw it correctly.

### 13. Refresh Cathode-ray Tube
Like the television tube, the picture on the refresh-type cathode-ray tube must be continually redrawn, or "refreshed," by the electron beam that produces the image. This means that to erase a line on a refresh tube one simply does not redraw it. Either the display equipment must have an internal memory where the graphics program can be stored so that it may be used to refresh the picture, or the

computer memory must be used for this purpose. The impermanence of the refresh cathode-ray tube allows for the creation of animation. Changing the position of a given line on the display every time the picture is refreshed gives the illusion of motion in a cartoon style. When a computer graphics system allows for motion to be simulated on a screen, it is said to be operating *in real time*. This cartoon technique of producing motion on a refresh cathode-ray tube is the most complex, difficult, and expensive of the computer graphics techniques discussed so far. However, the result is dramatic and provides information about moving objects which cannot be obtained in any other fashion. For instance, watching computer simulation of the wing of a jet aircraft undulating under the stresses of flight can provide invaluable information to the aircraft designer.

### 14. The Raster Cathode-ray Tube.
The raster refresh tube is actually a computer-driven television set. The picture on the screen is produced as the result of the computer providing information which is interpreted by the television tube as a picture. Television pictures are formations of raster lines. Raster lines are successions of horizontal lines. The quality of the picture depends on the number of lines per inch. Each line in a raster also has a certain brightness level that must be preprogrammed. A black-and-white television picture could easily require a million bits of picture information. To produce a realistic series of pictures, the picture information must be changed often enough to give the illusion of motion. Obviously, the raster cathode-ray tube requires a fantastically fast computer. The appropriate computer is correspondingly expensive and requires equally sophisticated programming.

### 15. Plasma Panel Displays.
The screen of the plasma panel display consists of glowing neon gas dots spaced in a matrix fashion. The dots are spaced about 60 to the inch. (See Fig. 6.) Clever electronic circuitry permits any dot to be lit or unlit. Plasma panels are storage-type display panels. The display is stable as long as power is

**FIG. 6**  A matrix or dot-type ionized gas display known as a plasma panel.

applied to the panel. Since individual dots can be erased, the display can be selectively updated without the requirement of erasing the entire picture to redraw those portions which are not desired. Thus, the plasma panel has some of the advantages of the refresh tube. However, an unwanted line on a plasma panel must be erased abruptly. This causes a noticeable flickering of the picture when attempting to produce motion. Although this effect may be annoying, the plasma panel still produces the illusion of motion and at a cost that is considerably under that of typical refresh tubes. The dot pattern of the screen resembles the dots of a halftone illustration and, as a result, various shading patterns can be used to produce realistic pictures. Shading patterns can be formed by various combinations of lighted and unlighted dots, providing some of the shading capabilities of the raster refresh cathode-ray tube.

The plasma panel is the thickness of two sheets of glass and is essentially transparent. A rear-projection screen can be placed behind the display, permitting slides or motion pictures to be projected onto the screen while the plasma panel displays computer-drawn pictures on the front of the screen. Various combinations of actual pictures and computer-drawn displays are possible. The plasma panel

operates in the matrix mode and is programmed quite differently than vector or incremental types.

### 16. Hard-copy Devices.

Hard-copy devices seem to intrigue inventors, and the variety of devices which have been developed is staggering. Any attempt to do more than simply describe a few popular types is an exercise in futility.

The obvious advantages of cathode-ray tubes are that they can produce pictures much faster than permanent copy devices, the illusion of motion can be produced, and a great number of trial-and-error solutions can be tested at little cost to the user. However, when the user would like to retain a picture, some method must be used to produce a permanent copy of the cathode-ray tube display. One of the earliest methods that is still used is to convert the display information into data for use by a hard-copy plotter. Unfortunately, hard-copy plotters are slow, adding to the time required to convert the picture information. More versatile methods are now available.

The xerographic process is frequently used to copy pictures from the face of the cathode-ray screen. This is the process pioneered and made famous by the Xerox Corporation. Two newer processes have become popular. One process places an electrostatic image on paper using a charged stylus. The paper is then treated with a toner, which reacts with the electrical charge on the paper to produce a permanent picture. Various companies make copiers of this type, and each company uses some variation of the process to make theirs unique. Another new process is called thermal printing, which uses specially treated paper which turns dark when heated. The paper is drawn over a printing head which consists of a line of wires whose ends face the paper. The wires, often spaced 100 to the inch, can be heated very rapidly and cooled just as rapidly. The picture on the cathode-ray tube is drawn matrix-style as the paper is reeled past the thermal printing head. Information in incremental or vector mode must be converted to matrix information. A picture may be drawn by a thermal printer much more rapidly than by the usual line plotter. In addition, thermal printers are silent.

### 17. Computer Programming for Displays.

The user of the computer graphics system is generally spared much of the logical difficulties associated with computer programming. Fortunately, the software and hardware necessary to produce a drawing according to number input is usually furnished by computer manufacturers, because only they are familiar enough with the system to provide efficient algorithms for these actions. The most basic operations of computer graphics programming may sometimes be wired into electronic circuitry. Examples of basic operations are drawing a line and forming an alphabet.

### 18. Standard Graphics Programs.

The computer graphics system manufacturers often supply programs called graphics subroutine packages to perform standardized drawing operations. For instance, normally a circle must be programmed by describing a series of straight lines shaped into a polygon of at least 180 sides. However, the graphics subroutine user may program a circle by calling for a radius $R$ and a center at $X$ and $Y$. These programs can be added to, to provide for additional operations. However, because every engineering operation is unique, many engineers write their own subroutines to fit their individual needs.

### 19. User Graphics Programs.

Let us assume that we have received a computer graphics system of the line drawing type for either a line plotter or a cathode-ray tube. The manufacturer has provided three basic drawing operations: DRAW, MOVE, and ERASE. For the moment we shall ignore the need for characters on the drawing. We might decide that it would be nice to be able to draw rectangles without having to move the drawing operations to a specified point to plot four straight lines from that point. So we might elect to produce an operation called RECT $(X,Y,L,W)$ which would produce a rectangle of length $L$ and width $W$ with

the lower left corner at point $X,Y$. The program to do so might look something like this:

| MOVE $(X,Y)$ | move drawing operation to point $X, Y$ |
| DRAW $(X\&L,Y)$ | draw bottom edge of rectangle of length $L$ |
| DRAW $(X\&L, Y\&W)$ | draw right hand vertical edge of length $W$ |
| DRAW $(X,Y\&W)$ | draw top horizontal edge |
| DRAW $(X,Y)$ | return to starting point, drawing left vertical edge |

To draw a rectangle of length 7.5 and width 3.75 with the starting corner at $X = 2.75$ and $Y = 1.35$ would require only this command in the RECT program:

RECT (2.75, 1.35, 7.5, 3.75)

This program would actually execute the previous instructions, but the user would be spared the task of actually writing out five commands. It can be seen that the user has ample opportunity to make mistakes in placing the correct numbers into statement commands. Therefore, it is desirable to reduce the number of places where the user needs to enter numbers or perform mathematical operations.

In an attempt to reduce the amount of busywork needed to produce drawings on computer graphics systems, special computer languages are being developed especially for graphics work. They will be oriented to easy usage and will try to provide computer power without the need for computer knowledge. The ultimate goal is to produce computer-generated pictures by describing the picture in strictly layman's terms without the limitations imposed by the computer. The fact that this is not yet present technology is shown by the qualifications necessary to be a computer graphics professional.

### 20. Mathematical Models.

Practical objects to be drawn by a computer graphics system are seldom as simple as the foregoing rectangle illustrations. Even an elementary object like a block letter E would require drawing four rectangles of varying sizes and orientations and then eliminating the three lines which are at the junction of the rectangles. Drawing practical objects becomes a major problem because of the sheer volume of data which must be handled. A formidable stumbling block to development of inexpensive, easy-to-use computer graphics programs is the need to describe designs by mathematical equations. This mathematical model of the object is so difficult to create that producing a new computer graphics picture can be likened to the process of producing a new automobile model. It is not a casual, overnight process.

## 21. Graphics Program Complexities.

Early attempts to draw pictures of solid objects usually consisted of producing a drawing made as though the object were constructed of wires representing each of the edges. (See Fig. 7.) Even today, drawing of solid objects is largely of the wire-frame type. Drawing objects of the wire-frame model in orthographic views, perspective views, or axonomet-

**FIG. 8**  A slightly different view of the object of Fig. 7 with the "hidden" edges really hidden.

**FIG. 9**  The object of Figs. 7 and 8 reoriented and drawn with the "hidden" edges represented by dashed lines.

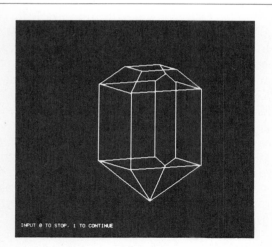

**FIG. 7**  A "wire frame" computer generated picture of a polyhedron. Note that the far side or hidden edges are drawn as though visible.

ric views is relatively simple. Because a perspective drawing of a wire-frame model of an object can be processed very rapidly by the computer, it is possible to give a cartoon-type motion to wire-frame drawings on cathode-ray tubes. However, wire-frame models of objects provide poor pictures because the

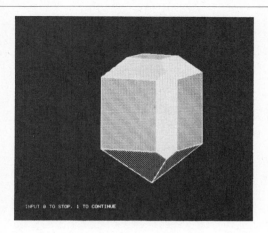

**FIG. 10** The object of Fig. 9 with the faces "lit" by a light from the upper right.

**FIG. 11** Orthographic views produced by the same program that drew the objects in Fig. 9-12.

parts of the object which should be hidden from our view are just as visible as features on the front of the object. Unfortunately, a wire-frame object that can be drawn with 200 or 300 statements in a computer program might suddenly require 1,500 or 2,000 statements if drawn using a hidden-line elimination addition to the program. (See Figs. 7 to 11.) In addition to the complexity added by the hidden-line part of the program, the description of the object itself becomes much more complex. Any attempt to make a wire-frame drawing of an object in which the faces or surfaces of the object are opaque adds several magnitudes of complexity to the drawing. In addition, hiding the edge of an object by a plane surface is one thing, but hiding it by a cylindrical surface is another vastly more difficult. It is so difficult that a general purpose picture-drawing program is not now available for computer graphics.

## 22. Uses of Computer Graphics.

It should not be construed from the foregoing that useful pictures of solid objects cannot be drawn by a computer graphics system, because that is not true. Very useful and dramatic pictures can be drawn, but each one becomes something of a special case. This leads one to ask, "Just what is computer graphics good for?" The answer is, "Any design for which you are willing to take the time to develop a special program." A simplification is that computer graphic systems are used to draw curves of equations. Computer graphics techniques are best suited for those calculations that are the most complex and time-consuming. A good rule of thumb is that if the development of a design requires a great number of repetitive calculations, it is advantageous to use a computer graphics system; if its development requires a great deal of analysis of existing drawings, it is not advantageous.

## 23. Computer Graphics-Generated Lists.

One major use of display devices for graphics is to exhibit lists requiring large quantities of characters or alphanumeric material, such as metric-to-English conversion tables. The display devices best suited to this usage are the cathode-ray tubes because of their drawing speed and because many of them have a variety of alphabets built into their operating systems. Of course, unless the cathode-ray tube has a hard-copy capability, the table must be read from the face of the display while it is kept alive. Electrostatic plotters and thermal printers are also desirable because they produce permanent copy. The mechanical plotter is not desirable for list production. The mechanical plotter is unable to obtain high drawing speeds because of the need to stop, start, or change direction constantly in drawing letters, numbers, and other characters. An electron beam is easier to move jerkily than a mechanical device having a weighable mass. However, all the display devices are superior to hand lettering, so the real choice is whether or not it is economical to develop a computer program to do a specific tabular chore.

## 24. Computer Graphics Graphs.

Graphing a function is suitable for any of the display devices. (See Fig. 12.) Most commercial graphics systems include a graph-drawing program developed to utilize the best features of that particular system. Usually the graphing program includes drawing the curve and the grid or coordinate system, and putting in the calibrations, legends, and other labels of the graph. Specialized types of charts such as bar charts (See Fig. 13.) and pie charts are also ideally suited to computer graphics display devices. Except for speed or hard-copy considerations, there is little reason to choose one display device over another.

## 25. Computer Graphics Maps.

Map makers might disagree, but the development of maps of all varieties presents special cases of the process of plotting points and drawing curves in a grid or coordinate system. To the computer, the choice of a coordinate system makes no real difference, since all coordinate systems must be described in mathematical terms. Complex coordinate systems require complex equations of description which, in turn, require more calculations to locate items in those coordinate systems. But to a com-

**FIG. 12**   A line graph drawn from data points without any equations being known about the data.

**FIG. 13**   A bar chart with textured bars drawn on a matrix-type display.

puter capable of doing hundreds of thousands or even millions of calculations per second, the additional complexity does not seem to have a serious practical effect. Maps requiring contour lines are easily produced by computer graphics systems although the programs which must calculate the contour lines are complex. Flatbed plotters are usually the only ones capable of producing a picture of sufficiently large physical size to be suitable for most maps. Developing a map-drawing program is not an easy task, so the choice of a suitable computer graphics system often depends upon whether or not the manufacturer of the system has programs available to produce the kind of drawing needed. Civil engineers use computer graphics systems extensively in producing maps of highways.

### 26. Computer Graphic Pictures.

Solid objects are the most difficult to produce for computer graphic representation. In this discussion we shall redefine orthographic projection drawings as *special orientations of a pictorial projection of an object;* the purpose is to show that the problems of drawing pictorials of a solid object are also the problems of orthographic projection, and that there are no problems of orthographic projection that do not have counterparts in general pictorial projection. In other words, an orthographic drawing of an object and a pictorial view of an object become much the same project. The computer graphics designer may attempt to shade or otherwise give three-dimensional qualities to a picture, but the basic picture is drawn as either an orthographic or pictorial representation.

### 27. Wire-Frame Pictures.

A major problem in developing a picture of an object is describing the object in computer language. One common way is to imagine the object as composed of wires which represent each edge of the object. This is known as the wire-frame model of the object. (See Figs. 7 to 11.) The description of a wire-frame model must consist of the $x,y,z$ coordinates of each junction point and an explanation of the manner in which corners, nodes, or vertices are connected. Any type of object composed of plane surfaces is easily described by this method, and pictorial drawings of wire-frame objects are quite easy to produce. Figure 7 illustrates a simple geometric solid drawn in wire-frame style. Understanding this object is not an easy task. Nevertheless, wire-frame models are often drawn because of the simplicity of the computer program which draws them.

### 28. Hidden Line Removal.

To convert the wire-frame model of an object into a picture of a solid object we must remove those lines or portions of lines which are on the back side of the object and which should be hidden from view. This is a most difficult task. Figure 8 represents the object of Fig. 7 with the hidden lines deleted. The picture is far easier to read and understand. However, the object seems flat and the illusion of depth is not yet pronounced.

### 29. Shaded or Lighted Objects.

The next step is to shade or color the surfaces of the object to give the illusion of depth. One technique is to assume the object is lighted from a certain angle. With a raster cathode-ray tube it is possible to duplicate the lighting by making some portions of the object brighter than others. All mechanical plotters must duplicate the illusion by cross-hatching, which is a less than desirable illusion. Thermal printers, plasma panels, storage cathode-ray tubes, and nonraster cathode-ray tubes must use a density of dots to produce shading. Electrostatic graphics devices can also produce shaded areas with some realism. Figure 10 shows the object of Figs. 7 to 9 with shading which produces the effect of light shining on the object from a certain direction. The illusion which is created gives the object a more solid look and obviously adds to the ease of reading and understanding the object.

The more realistic the pictorial representation of the object, the greater the complexity of the computer program. In addition, curved surfaces add a very real degree of complexity which is not easy to solve.

### 30. Orthographic Views.

To illustrate that orthographic projection drawings are no more than special cases of general pictorial drawing programs, the object of Figs. 7 to 10 is shown in Fig. 11 from the top, front, and right-side orthographic views produced with the identical program that drew Fig. 7, changing only the angles used to view the object. Once the general pictorial case has been solved, a program is capable of drawing all perspectives of a single object.

### 31. Summary Orthographic Views.

The efficient use of computer graphics requires a thorough knowledge of the principles of graphics and graphic projection. The computer graphics routines do not relieve us of the responsibility of knowledge. They provide us with a tireless robot which will do our drawing for us, once we know what the drawing is to be. We must command the robot, the robot cannot command us. An old computer principle reminds us that input mistakes create output mistakes. The picture will be only as good as our ideas.

### 32. Interactive Graphics.

A designer might wish to draw a picture, examine it, and have some parts of the picture changed. This can be done in various ways. One is to change the computer program in use and resubmit it for processing. This type of operation is commonly known as *batch mode*. Another method is to write the computer graphics program so that it can accept changes as it is being executed. This is commonly known as *interactive graphics* and is the best way of doing what might be called trial-and-error drawing.

### 33. Interactive Terminals.

The most common form of interactive graphics utilizes a cathode-ray tube display connected to a typewriter-like keyboard for typing information into the computer program. So popular and common is this form of interaction, that the very word "terminal" has come to mean that type of device which includes a cathode-ray tube display for printing out answers from the computer. Terminals of this kind can be operated just like a regular typewriter; the typing appears on the display instead of on paper. Since a terminal is a two-way device, the computer program can type out questions and responses to user actions on the display. If the display equipment is capable of producing pictorial graphics, the user can type in parameters to produce a drawing, have it produced on the display, examine it, and type in changes in the parameters to produce another drawing. An advantage of using cathode-ray tubes for this operation is that hard copy is not obtained

until the drawing is correct. When using mechanical plotters, for instance, an incorrect drawing takes as long to draw as a correct one. In addition, cathode-ray tube displays are usually much faster than mechanical plotters, and mistakes take less time to find. Any graphics display device could be used for interactive graphics.

### 34. Light Pens.

A unique device known as a light pen is used to identify information on a display panel. The light pen looks like an oversized pen with a cable attached. It is used to point to a particular spot on the display picture. When the electron beam which is drawing a picture on the cathode-ray tube passes the end of the light pen, a sensor in the pen triggers special circuits in the display, which then identifies the screen coordinates of the position of the electron beam at that instant. Thus, the position of the pen is known. Special graphics programs have built-in tests which are capable of interpreting this light pen as a signal to produce specific actions. For example, if the cathode-ray tube has a list of printed items on the display, the light pen can be used to point to one of the lines of print. If the display has a drawing on it, the light pen can be used to drag some portion of the design from one point to another on the screen. If the picture on the display is composed of separate representations of electronic components such as transistors, capacitors, or resistors, the light pen can be used to indicate how wiring connections are to be made with these components. The user gets the impression that one can draw a line on the cathode-ray tube display with a pen. Obviously, considerable hardware is involved in this type of operation, so that light pens with cathode-ray tubes are expensive.

### 35. Computer-assisted Instruction.

This computer-assisted instruction program utilizes the interactive capabilities of a cathode-ray tube or plasma panel display to ask questions of students and respond to their answers. Because pictures are helpful in learning, computer-assisted instruction displays usually utilize picture-graphics capabilities extensively. Here, the picture is not the end result but an aid to understanding a topic.

### 36. Numerical Control of Machine Tools.

Numerically controlled machine tools are usually used to produce complex objects. The programs which run the tools are commonly produced with the help of a computer. In most cases, a computer does not run the machine tool directly. Note that we are referring to numerically controlled machining, not computer-controlled machining. Numerically controlled machine tools are run by programs similar to computer programs. Through a program code, the programs describe each move the machine must make. The numerical control program is usually punched on tape. The tape is then read by the machine as it produces the desired object. Many of the numerically controlled tools merely substitute electric or hydraulic motors for the machine operator.

### 37. Numerically Controlled Machine Types.

Almost any machine tool is capable of being designed for numerical control operation. Some common machines which have been built as numerical control models are lathes, drill presses, and boring mills. Specialized machine tools have also been developed strictly for numerical control use. Two-axis machines are capable of moving forward, backward, to the left, and to the right. A three-axis machine also has an up-and-down movement capability. The two-and-a-half–axis machine has an up-and-down motion that must be set manually (See Fig. 14.) Note that the rotation of the drill bit or the milling cutter is not considered an axis, so that the normal lathe might be described as a two-axis machine. Very special capability tools require more axes of control. For example, if the basic three-axis drill press were designed so that the drilling head could be tipped forward or backward, or to the left or the right, we would have a five-axis machine. Unlike the usual drill press which moves the drill bit in and out of the drill head, this five-axis machine would move the drilling head along the drilling axis to produce that axis of motion. Thus, any hole could be drilled at any angle from vertical to horizontal and at any point on the upper surface of an object. By replacing the drill bit with a milling cutter, very complex shapes could be produced using the five axes of motion. It is axiomatic that cost follows function and that five-axis machines are very expensive.

### 38. Two-axis Numerically Controlled Machines.

The most familiar two-axis numerically controlled machine is the engine lathe. The lathe, which has been called the king of machine tools, basically produces surfaces of revolution. However, some items such as screw threads are produced on lathes and do not conform to the common definition of a surface of revolution. Lathes for commercial production are usually of the turret variety where a series of cutting tools is held in a rotating assembly on the carriages. The proper cutting tools for various machining operations such as cylinder turning, facing, and threading can be easily rotated into cutting position by turning the turret.

The functions of a lathe which are under the control of a human operator include the rotational speed of the workpiece, tool position of the turret, direction of coolant to the workpiece, longitudinal feed, cross feed, and gearing together of workpiece and longitudinal feed to produce threads. A numerically controlled machining program can control all these items automatically. Because of both the amount of information and the precise timing needed to produce tool actions correctly, a computer would be helpful in producing the master tapes which run the tool.

### 39. Two-and-a-half and Three-axis Machines.

Another common machine is the drilling machine, also known as the drill press. It consists of a table upon which the workpiece is mounted, a drilling head which can rotate the drilling tool or bit, a mechanism for moving the drill bit up and down, and a sufficiently strong frame to connect all these items. The drilling table can be designed to move to the left or right and from front to rear, giving movement in the $x$- and $y$- coordinate directions. This accounts for two axes of motion. Moving the drill bit down to drill a hole gives a third axis of motion. However, if the up-and-down motion is operated by mechanical stops instead of by the numerically controlled program, the machine is called a two-and-a-half–axis machine. The machine shown in Figs. 14 and 15 has the third axis stops built into a sort of turret so that a variety of stops can be chosen by rotating the turret. This choice of stops is under control of the numerically controlled program. By moving the table with the workpiece attached, holes can be drilled at selected places on the workpiece. The functions under numerical control in a machine of this type include $x$-position of the workpiece; $y$-position of the workpiece; rotational speed of the drilling tool; choice of up-down stop pairs, of coolant use (on or off), and of drill up, down, or feed; and halting the machine. The use of a computer to prepare the program for the numerical control of a two-and-a-half–axis machine is helpful but not mandatory.

Three-axis machines have full control over the $X$, $Y$, and $Z$ axes of motion. The machine of Fig. 14 could be converted to a three-axis machine by replacing the mechanical stops with numerical control of the exact up-and-down positions of the drilling tool. Other types of machines which might be classed as three-axis machines include milling machines, shapers, and planers. Three-axis machines require complex numerical control that must be prepared by a computer.

## 40. Five-axis Machines.

The five-axis machines are the most versatile of the numerically controlled machines. The five-axis machine gives the kind of flexibility that the wrist gives our hand and arm. The five-axis machine has all the capabilities of a three-axis machine. In addition, the

FIG. 14  A two-and-half axis numerically controlled drilling machine.

FIG. 15  Boring and drilling tools for holes of different diameters and depths to be used in the machine of Fig. 14.

drilling head of this machine is modified so that it can operate with any rotary-cutting tool. The table on which the workpiece is mounted can be moved in $x$- and $y$- coordinate directions as well as in $z$-coordinate directions. The head may also be moved up and down while the table remains stationary in the $Z$ axis. An additional axis of motion is obtained by hinging the drilling or milling head for rotation about the $X$ axis so that it can tip forward or backward. The other additional axis of motion is rotation about the $Y$ axis, causing the drilling head to lean from side to side. With the proper machining head and tool, practically any form can be machined. For example, wing sections for supersonic aircraft are often machined out of slab stock using a five-axis machine. The use of a computer to prepare a five-axis numerical control program is mandatory if any practical work is to be done by machine. The five-axis machine could be called a sculpting machine, as it is capable of producing surfaces which are neither planar nor surfaces of revolution. Numerically controlled machine tools can produce objects which heretofore were the province of casting methods. The main reasons that five-axis machines do not replace all sculpting methods of manufacture is that they cost hundreds of thousands of dollars and need a computer to keep them running.

## 41. Monitoring Numerically Controlled Machines.

Numerically controlled machines must have the machining axis movements under precise control so that the object being machined will be exactly the desired design. Electric motors driving either mechanical linkages or hydraulic movements are often used to produce the required motions. Two major methods are used to monitor the magnitude of the motion. One is to use a basic increment of motion which can be counted to produce the desired total motion. Thus a machining motion increment of 0.1 mm would require 283 increments to produce a total machining motion of 28.3 mm. Problems exist with this method, because small increments of motion require a large number of increments to pro-

duce a large motion, and large increments of motion produce objects of less precision than might be desirable.

A second method of monitoring motion is by means of feedback. In essence, the position of the numerically controlled cutting is detected by a mechanism, usually an electronic sensing circuit, so that the numerically controlled tool can have its position verified and modified with split-second timing. Neither of these methods is a trivial design project. Even if precise motions can be monitored by the numerically controlled tool, it is necessary that it produce these motions from precisely located reference points. This means that the numerically controlled machine must be able to locate the tool very accurately so that the programmer can designate a desired location of the tool easily and accurately. A combination of electrical, mechanical, hydraulic, and electronic sensing devices is used to achieve this.

Despite accurate sensing and locating methods, cutting bits are not held in tool holders as accurately as desired, and cutting bits become dull and need sharpening. For these reasons it is difficult to locate the cutting edge of the cutting bit with respect to the coordinates of the tool holder. To solve this problem, numerically controlled machines are normally equipped with devices which permit the operator to dial in a compensating movement if it is discovered that the cutting bit is not starting the machining cut at the proper place. Additional compensation can be added if it is discovered that tool wear has not produced parts to specification.

## 43. Programming Numerically Controlled Machines.

The basic problem of monitoring numerically controlled machines is that of producing the proper motions along the axes of motion to produce an object of desired configuration. Add to these requirements the need to control cutting speeds, coolants, and object geometry, and it will be seen that a computer is a helpful device in performing all the calculations needed. One aspect of object geometry will illustrate the need for high-level assistance. Suppose a milling cutter has finished a cut at point *A* and is to be moved to point *B* on the object. Unlike the mechanical pen plotter, the milling cutter does not perform this movement simply. First, it must be determined if there is any part of the object which lies in the path from *A* to *B*. If the cutter collides with a part of the object during the move, it may damage the numerically controlled machine or the object being machined. A useful program for numerically controlled machines includes the ability to detect such collisions and to prevent them by moving the cutting tool with an alternative route. So complex are the motions of a five-axis machine that only a computer program of enormous capability can handle the calculations needed. A number of special numerically controlled machine languages have been developed, such as APT, SPLIT, UNIAPT, and COMPACT. In each of these, the programmer must describe the machine being used and the geometry of the object being machined. The programmer and the computer program interact to produce a final tape of the desired motions the tool is to take.

## 44. Uses of Numerically Controlled Machines.

Numerically controlled machines fill a gap in machine tools between hand-controlled production and mass production methods. Operating with the versatility of the hand-operated machine and the precise operation qualities of the special machine, the numerically controlled machine can produce a moderate number of essentially identical objects. The five-axis numerically controlled machine is able to produce objects that were previously impossible to manufacture. Numerically controlled machines form an important part of the industrial world.

## 45. Impact on Engineering Graphics.

The major impact of both computer graphics and numerically controlled machining upon engineering drawing is the methods by which the geometry of an object is described. Dimensioning practice will be different in specific industries using numerically controlled machine tools because each industry will wish to provide their users with information in the most convenient form for their special circumstances. However, at the present time there are no dimensioning standards for numerically controlled machining. It is hoped that ANSI will soon develop guidelines and recommended procedures.

# DRAWINGS FOR ENGINEERING DESIGN AND CONSTRUCTION

# 15

This Chapter Discusses Professional Practices Employed in the Making of Design, Detail, Assembly, Production, Construction, and Other Drawings

In this Chapter:
Definitions • Engineering Design Procedure • Assembly and Detail Drawings • Tabular and Standard Drawings • Bills of Material and Parts Lists • Drawings • Order of Drawing, Penciling, Inking • Titles, Checking, Zoning • Sketches • Measuring and Dimensioning • Reproduction • Filing and Storage • Simplified Practices • Templates, Overlays • Uses of Reproduction • Problems

## 1. Drawings—Fundamental Concepts.

A "production," "construction," or "working drawing" is any drawing used to give information for the manufacture or construction of a machine or the erection of a structure. Complete knowledge for the production of a machine or structure is given by a *set* of drawings conveying all the facts fully and explicitly so that further instructions are not required.

The description given by the set of drawings will include:

1. The full graphic representation of the shape of each part (shape description).

2. Dimensions of each part (size description).

3. Explanatory notes, general and specific, on the individual drawings, giving the specifications of material, heat-treatment, finish, etc. Often, particularly in architectural and structural work, the notes of explanation and information concerning details of materials and workmanship are too extensive to be included on the drawings and so are made up separately in typed or printed form and called the "specifications"—thus the term "drawings and specifications."

4. A descriptive title on each drawing.

5. A description of the relationship of each part to the others (assembly).

6. A parts list or bill of material.

A set of drawings will include, in general, two classes of drawings: *detail drawings,* giving the information included in items 1 to 4; and an *assembly drawing,* giving the location and relationship of the parts, item 5.

## 2. Engineering Procedure.

When a new machine or structure is designed, the first drawings are usually in the form of freehand sketches on which the original ideas, scheming, and inventing are worked out. These drawings are accompanied or followed by calculations to prove the suitability of the design. Working from the sketches and calculations, the design department produces a *design assembly* (also called a "design layout" or "design drawing")(Fig. 1). This is a preliminary pencil drawing on which more details of the design are worked out. It is accurately made with instruments, full size if possible, and shows the shape and position of the various parts. Little attempt is made to show all the intricate detail; only the essential dimensions, such as basic calculated sizes, are given. On the drawing, or separately as a set of written notes, will be the designer's general specifications for materials, heat-treatments, finishes, clearances or interferences, etc., and any other information needed by the draftsman in making up the individual drawings of the separate parts.

Working from the design drawing and notes, draftsmen (detailers) then make up the individual detail drawings. Figure 2 shows a detail drawing taken from the design drawing of Fig. 1. On a detail drawing, all the views necessary for complete shape description of a part are provided, and all the necessary dimensions and manufacturing directions are given. Dimension values for nonmating surfaces are obtained by scaling the design drawing, and the more critical values are determined from the design notes and from drafting-room standards. The detailer correlates the dimensions of mating parts and gives all necessary manufacturing information.

The set of drawings is completed with the addition of an assembly drawing and a parts list or bill of material.

If a machine is to be quantity produced, "operation" or "job" sheets will be prepared giving the separate manufacturing steps required and indicating the use and kind of special tools, jigs, fixtures, etc. The tool-design group, working from the detail drawings and the operation sheets, designs and makes the drawings for the special tools needed.

## 3. Assembly Drawings.

An *assembly drawing* is, as its name implies, a drawing of the machine or structure put together, showing the relative positions of the different parts.

The term "assembly drawings" includes preliminary design drawings and layouts, piping plans, unit assembly drawings, installation diagrams, and final complete drawings used for assembling or erecting the machine or structure.

The design drawing is the preliminary layout on which the scheming, inventing, and designing are accurately worked out. The assembly drawing is in

**FIG. 1.** A portion of a design drawing. Notes and specifications accompany the drawing.

| DWN | D.P.V. | 1-8-47 | LIMITS. UNLESS OTHERWISE NOTED:<br>FRACTIONAL ±1/64.   DECIMAL ±.010,   ANGULAR ±1/2° | | REVISION | DATE | CHANGE OR ADDITION | | |
|---|---|---|---|---|---|---|---|---|---|
| TCD | B.D.F. | 2-1-47 | | | MATERIAL | | STOCK SIZE | HEAT TR. | FINISH |
| CKD | Mc AL | 2-10-47 | PART NO. | REQD | C.I. | | | | Noted |
| APPD | Thomas | 2-11-47 | 12 | 1 | NAME | | | | DRAWING NO. |
| | NAME | DATE | | | DRIVING FLANGE | | COMPANY NAME | | NO.     OF |
| | | | UNIT OR<br>ASSEMBLY | | SCALE<br>Full size | | CITY | | |
| | | | | | ASSY. DR. NO. | | | | 1005 |
| | | | BENCH DRILL | | 1026 | | | | |

**FIG. 2.** A detail drawing. This gives complete information for the production of a single part.

some cases made by tracing from the design drawing. More often it is drawn from the dimensions of the detail drawings; this provides a valuable check on the correctness of the detail drawings.

The assembly drawing sometimes gives the overall dimensions and the distances between centers or from part to part of the different pieces, thus fixing the relation of the parts to each other and aiding in erecting the machine. However, many assembly drawings need no dimensions. An assembly drawing should not be overloaded with detail, particularly hidden detail. Unnecessary dashed lines should not be used on any drawing, least of all on an assembly drawing.

Assembly drawings usually have reference letters or numbers designating the different parts. These "piece numbers," sometimes enclosed in circles (called "balloons" by draftsmen) with a leader pointing to the piece (Fig. 3), are used in connection with the details and bill of material.

A *unit assembly drawing* or subassembly (Fig. 3) is a drawing of a related group of parts used to show the assembly of complicated machinery where it would be practically impossible to show all the features on one drawing. Thus, in the drawing of a lathe, there would be included unit assemblies of groups of parts such as the headstock, tailstock, gearbox, etc.

An *outline assembly drawing* is used to give a general idea of the exterior shape of a machine or structure and contains only the principal dimensions (Fig. 4). When it is made for catalogues or

**FIG. 3.** A unit assembly drawing. For complex machines, a number of units may be used in place of one complete assembly drawing.

| CODE WORD | PIPE SIZES DISCH. | SUCT'N | A | B | C | D | E | F | G | H | J | K | L | M | N | O | P | Q | R | S | T | U | W | X |
|---|---|---|---|---|---|---|---|---|---|---|---|---|---|---|---|---|---|---|---|---|---|---|---|---|
| JAY | 1½" | 2" | TAPPED OPENINGS | | | | | | 5½ | 6⅝ | 6½ | 7¾ | 20⅛ | 3⅛ | 8⅜ | ⅜ | | | | | | | | |
| JABBER | 2" | 2½" | 6 | 4½ | 4-⅜ | 7 | 5½ | 4-⅜ | 5⅝ | 7¼ | 6½ | 7⅞ | 20¹⁄₁₆ | 3 | 8¼ | ⅜ | | | | | | | | |
| JACKET | 3" | 3" | 7½ | 6 | 4-⅝ | 7½ | 6 | 4-⅝ | 6¼ | 7⅞ | 7 | 7¾ | 20 | 3¾ | 8½ | | ABOVE DIMENSIONS DETERMINED BY SIZE | | | | | | | |
| JACK | 4" | 4" | 9 | 7½ | 8-⅜ | 9 | 7½ | 8-⅜ | 7⅞ | 8⅜ | 7¾ | 8½ | 22½ | 4½ | 10 | | OF MOTOR AND BASE USED | | | | | | | |
| JACOBUS | 5" | 5" | 10 | 8½ | 8-¾ | 10 | 8½ | 8-¾ | 8⅝ | 9⅞ | 8¼ | 8½ | 22½ | 5 | 10¼ | ½ | | | | | | | | |

**FIG. 4.** An outline assembly drawing (tabular). Tabulation gives the dimensions for different-sized units.

other illustrative purposes, dimensions are often omitted. Outline assembly drawings are frequently used to give the information required for the installation or erection of equipment and are then called *installation drawings.*

An *assembly working drawing* gives complete information for producing a machine or structure on one drawing. This is done by providing adequate orthographic views together with dimensions, notes, and a descriptive title. The figure for Prob. 45 is an example.

A *diagram drawing* is an assembly showing, symbolically, the erection or installation of equipment. Erection and piping and wiring diagrams are examples. Diagram drawings are often made in pictorial form.

## 4. Detail Drawings.

A *detail* drawing is the drawing of a single piece, giving a complete and exact description of its form, dimensions, and construction. A successful detail drawing will tell the workman *simply* and *directly* the shape, size, material, and finish of a part; what shop

operations are necessary; what limits of accuracy must be observed; the number of parts wanted; etc. It should be so exact in description that, if followed, a satisfactory part will result. Figure 5 illustrates a commercial detail drawing.

Detailing practice varies somewhat according to the industry and the requirements of the shop system. For example, in structural work, details are often grouped together on one sheet, while in modern mechanical practice a separate sheet is used for each part.

If the parts are grouped on one sheet, the detailed pieces should be set, if possible, in the same relative position as on the assembly and, to facilitate reading, placed as nearly as possible in natural relationship. Parts of the same material or character are usually grouped together, as, for example, forgings on one sheet, castings on another, and parts machined from stock on another. A subtitle must be provided for each part, giving the part number, material, number required for each assembly, etc.

The accepted and best system in mechanical work is to have each piece, no matter how small, on a

**FIG. 5.**  A detail drawing. Note the completeness of manufacturing information given on the drawing.

separate sheet. As described in Sec. 3, Chap. 11, if the single-drawing system is followed, one drawing will be used by all shops. If the multiple system is used, a separate drawing must be made for each shop; thus there may be a *pattern drawing*, a *casting drawing*, and a *machining drawing*, all for a single cast part. A detail drawing should be a complete unit for the purpose intended and should not be dependent in any way upon any other detail drawing.

## 5. Tabular Drawings.

A tabular drawing, either assembly or detail, is one on which the dimension values are replaced by reference letters, an accompanying table on the drawing listing the corresponding dimensions for a series of sizes of the machine or part, thus making one drawing serve for the range covered. Some companies manufacturing parts in a variety of sizes use this tabular system of size description, but a serious danger with it is the possibility of misreading the

table. Figure 4 illustrates a tabular assembly drawing.

## 6. Standard Drawings.

To avoid the difficulties experienced with tabular drawings, some companies make a "standard drawing," complete except for the actual figured dimensions. This drawing is reproduced by offset printing or black-and-white reproduction on vellum paper, and the reproductions are dimensioned separately for the various sizes of parts. This method gives a separate complete drawing for each size of part; and when a new size is needed, the drawing is easily and quickly made. Figure 6 shows a standard drawing and Fig. 7, the drawing filled in, making a completed working drawing.

## 7. Standard Parts.

Purchased or company standard parts are specified by name and size or by number and, consequently,

COLLAR _____
Material _____
Heat-treat. _____
No. Req'd. _____

**FIG. 6.** A standard drawing. This is reproduced on tracing paper to provide the shape description needed for the drawing.

COLLAR _X 6___
Material _S.A.E. 1020_
Heat-treat. _S.A.E. 2_
No. Req'd. __50__

**FIG. 7.** A filled-in standard drawing. With the addition of dimensions, a standard drawing becomes a completed drawing to be used for manufacturing purposes.

do not need to be detailed. All standard parts, such as bolts, screws, antifriction bearings, etc., are shown on the assembly drawing and are given a part number. The complete specifications for their purchase are given in the parts list.

Sometimes, however, a part is made by *altering* a standard or previously produced part. In this case a detail drawing is made, showing and specifying the original part with changes and dimensions for the alteration.

### 8. The Bill of Material, or Parts List.

This is a tabulated statement, usually placed on a separate sheet in the case of quantity production (as in Prob. 37) or on the assembly drawing in other cases (as in Prob. 50). This table gives the piece number, name, quantity, material, sometimes the stock size of raw material, detail drawing numbers, weight of each piece, etc. A final column is usually left for remarks. The term "bill of material" is ordinarily used in structural and architectural drawing. The term "parts list" applies more accurately in machine-drawing practice. In general, the parts are listed in the order of their importance, with the larger parts first and the standard parts such as screws, pins, etc., at the end.

The blank ruling for a bill of material should not be crowded. Lines should never be spaced closer than $\frac{1}{4}$ in.; $\frac{5}{16}$ or $\frac{3}{8}$ in. spacing is better, with the height of the lettering not more than half the space and centered between the lines. Instead of being lettered, bills of material are frequently typed on forms printed on thin paper. Intensifying the impression by typing with carbon paper on the back increases the opacity of the letters, and a clearer blueprint will result.

### 9. Set of Drawings.

A *complete set* of drawings consists of detail sheets and assembly sheets, the former giving all necessary information for the manufacture of each of the individual parts which cannot be purchased and the latter showing the parts assembled as a finished unit or machine. The set includes the bill of material or parts list and may also contain special drawings for the purchaser, such as foundation plans or oiling diagrams.

### 10. Making a Working Drawing: Basic Concepts.

Although pictorial drawings are used to some extent in special cases, the basis of all working drawings is orthographic projection. Thus, to represent an object completely, at least two views are ordinarily needed, often more. The only general rule is: *make as many views as are necessary to describe the object clearly—and no more.* Instances may occur in which the third dimension is so evident as to make one view sufficient, as, for example, in the drawing of a shaft or bolt. In other cases, perhaps a half-dozen views will be required to describe a piece completely. Sometimes half, partial, or broken views can be used to advantage.

Select for the front view the face showing the largest dimension, preferably the obvious front of the object when in its functioning position, and then decide what other views are necessary. A vertical cylindrical piece, for example, would require only a front and a top view; a horizontal cylindrical piece, only a front and a side view. Determine which side view to use or whether both are needed. The one with the fewest dashed lines should be preferred. In some cases the best representation will be obtained by using *both* side views with all unnecessary dashed lines omitted. See whether an auxiliary view or a note will eliminate one or more other views and whether a section is better than an exterior view. One statement can be made with the force of a rule:

*If anything in clearness can be gained by violating a principle of projection, violate it.*

Arts. 12 to 26, Chap. 7, give a number of examples of conventions that are in violation of theoretical representation but are in the interest of clearness. The draftsman must remember that his responsibility is to the reader of the drawing and that he is not justified in saving himself any time or trouble at the expense of the drawing, by making it less plain or easy to read. The time so saved by the draftsman may be lost to the company a hundredfold in the shop, where the drawing is used not once but repeatedly.

There is a *style* in drawing, just as there is in literature, which indicates itself in one way by ease of reading. Some drawings stand out, while others, which may contain all the information, are difficult to decipher. Although dealing with mechanical thought, there is a place for some artistic sense in

mechanical drawing. The number, selection, and disposition of views; the omission of anything unnecessary, ambiguous, or misleading; the size and placement of dimensions and lettering; and the contrast of lines are all elements in the style.

In commercial drafting, *accuracy* and *speed* are the two requirements. The drafting room is an expensive department, and time is an important element. The draftsman must therefore have a ready knowledge not only of the principles of drawing but of the conventional methods and abbreviations and of any device or system that will save time without sacrificing clearness.

The usual criticism of the beginner by the employer is the result of the former's lack of appreciation of the necessity for speed.

## 11. Materials Used for Working Drawings.

Working drawings go to the shop in the form of blueprints, black-line prints, or other similar forms of reproduction, and the drawings must therefore be made on translucent material, either directly or as tracings. Pencil drawings may be made on tracing paper or on pencil cloth; inked drawings, on tracing paper or on tracing cloth.

*Tracing paper* is a thin translucent material, commonly called "vellum." Considerable time and expense may be saved by making the original pencil drawing on this material. Excellent prints can be obtained if the lines are of sufficient blackness and intensity.

*Pencil cloth* is a transparentized fabric with one or both sides of its surface prepared to take pencil so that the original drawing can be made on it and prints made either from the pencil drawing directly or after it has been inked. Some of the newer cloths are moisture resistant, others are really waterproof. Pencil cloth is made for pencil drawings, and perfect blueprints can be made from drawings made on it with sharp, hard pencils. Ink lines, however, do not adhere well and have a tendency to chip or rub off in cleaning.

*Tracing cloth* is a fine-thread fabric sized and transparentized with a starch preparation. The smooth side is considered by the makers as the working side, but most draftsmen prefer to work on the dull side, which will take pencil marks. The cloth should be fastened down smoothly over the pencil drawing and its selvage torn off. To remove the traces of grease that sometimes prevent the flow of ink, dust the tracing cloth with chalk or prepared pounce (a blackboard eraser may be used) and then rub it off with a cloth. Carbon tetrachloride is an effective cleaning agent. Rub a moistened cloth over the surface—any excess will evaporate in a moment.

A *plastic material,* known as "Mylar,"[1] is now available for both pencil and ink drawings. It is evenly translucent, has a fine matte surface, good lasting qualities, and requires no special storage precautions.

## 12. Drawing Sizes.

Drawing paper and cloth are available in rolls of various widths and in standard trimmed sizes. Most drafting rooms use standard sheets, printed with border and title block. The recommended sizes shown in Table 1, based on multiples of 8½ by 11 in. and 9 by 12 in., permit the filing of prints in a standard letter file.

Figure 8 shows the most common trimmed sizes. Larger drawings can be made on rolled stock of

[1]DuPont Manufacturing Co.

**Table 1.   Finished Flat-sheet Sizes.**

| Designation | Width | Length | Designation | Width | Length |
|---|---|---|---|---|---|
| A | 8½ | 11 | A | 9 | 12 |
| B | 11 | 17 | B | 12 | 18 |
| C | 17 | 22 | C | 18 | 24 |
| D | 22 | 34 | D | 24 | 36 |
| E | 34 | 44 | E | 36 | 48 |
| F* | 28 | 40 | F* | 28 | 40 |

*Not a multiple of the basic size. To be used when width of E size is not adaptable.

**FIG. 8.** ANSI trimmed sizes of paper and cloth. ANSI sizes are based on multiples of $8^{1}/_{12}$ by 11 in. and 9 by 12 in. (see Table 1).

standard width, with the length as a multiple of 11 or 12 in., not to exceed 144 in.

### 13. Order of Penciling.

After the scheming, inventing, and calculating have been done and the design drawing has been completed, the order of procedure for making the detail drawings is:

1. Select a suitable standard sheet or lay off a sheet to standard size, with the excess paper to the right, as a convenient space for making sketches and calculations, and block out the space for the title.

2. Decide what scale to use, choosing one large enough to show all dimensions without crowding, and plan the arrangement of the sheet by making a little preliminary freehand sketch, estimating the space each view will occupy and placing the views to the best advantage for preserving, if possible, a balance in the appearance of the sheet. Be sure to leave sufficient space *between* views for the dimensions.

3. Draw the center lines for each view, and on these "block in" the views by laying off the principal dimensions and outlines, using *light, sharp, accurate* pencil lines. Center lines are drawn for the axes of symmetry of all symmetrical views or parts of views. Thus every cylindrical part should have a center line—the projection of the axis of the piece. Every circle should have two center lines intersecting at its center.

4. Draw the views, beginning with the most dominant features and progressing to the subordinate. Carry the different views on together, projecting a characteristic shape as shown on one view to the other views and *not* finishing one view before starting another. Draw the lines to the final finished weight (wherever possible), using a minimum of construction. *Never* make a drawing lightly and "heavy" it later.

5. Finish the projections, putting in last the minor details. Check the projections and make sure that all views are complete and correct.

6. Draw all necessary dimension lines; then put in the dimension values.

7. Draw guide lines for the notes, and then letter them.

8. Lay out the title.

9. Check the drawing carefully.

The overrunning lines of the constructive stage should not be erased before tracing or inking. These extensions are often convenient in showing the stopping points. Avoid unnecessary erasing as it abrades the surface of the paper so that it catches dirt more readily.

As an aid in stopping tangent arcs in inking, it is desirable to mark the tangent point on the pencil drawing with a short piece of the normal to the curve at the point of tangency.

Figure 9 illustrates the stages of penciling.

### 14. Order of Inking.

To ensure good printing, the ink should be perfectly black and the ruling pens in good condition. Red ink should not be used unless it is desired to have some lines inconspicuous on the print. Blue ink will not print well. Sometimes, on maps, diagrams, etc., it is desirable to use colored inks on the tracing to avoid confusion of lines; in such cases, add a little Chinese white to the colored inks and it will render them sufficiently opaque to print.

Ink lines can be removed from tracing cloth by rubbing with a hard pencil eraser, slipping a triangle under the tracing to give a harder surface. The rubbed surface should afterward be burnished with a burnisher or fingernail. In tracing a part that has been section-lined, a piece of white paper can be slipped under the cloth and the section lining done without reference to the section lines underneath.

Tracing cloth is sensitive to atmospheric variations, often changing overnight so as to require restretching. If a large tracing cannot be finished in one day, some views should be finished and no figure left only partly traced.

In making a large tracing, it is well to cut the required piece from the roll and lay it exposed, flat, for a short time before fastening it down.

Water will ruin a tracing on starch-coated cloth, and moist hands or arms should not come in contact with it. Form the habit of keeping the hands off

**FIG. 9.** Order of penciling. (*a*) Layout; (*b*) details progress to (*c*); completion at (*d*). See Sec. 13.

drawings. In both drawing and tracing on large sheets, it is a good plan to cut a mask of drawing paper to cover all but the view being worked on. Unfinished drawings should be covered overnight.

Tracings can be cleaned of pencil marks and dirt by wiping with a cloth moistened with benzine or carbon tetrachloride. To prevent smearing when using this method of cleaning, borders and titles should be printed in an ink not affected by benzine.

*Order of Inking*

1. Ink all full-line circles, beginning with the smallest, and then circle arcs.

2. Ink dashed circles and arcs in the same order as full-line circles.

3. Ink any irregular curved lines.

4. Ink straight full lines in this order: horizontal, vertical, and inclined.

5. Ink straight dashed lines in the same order.

6. Ink center lines.

7. Ink extension and dimension lines.

8. Ink arrowheads and dimensions.

9. Section-line all areas representing cut surfaces.

10. Letter notes and titles. (On tracings, draw pencil guide lines first.)

11. Ink the border.

12. Check the inked drawing.

Figure 10 shows the stages of inking.

## 15. Title Blocks.

The title of a working drawing is usually placed in the lower right-hand corner of the sheet, the size of the space varying with the amount of information to be given. The spacing and arrangement are designed to provide the information most helpful in a particular line of work.

In general, the title of a machine drawing should contain the following information:

1. Name of company and its location

**FIG. 10.** Order of inking. (*a*) Layout; (*b*) details progress to (*c*); completion at (*d*). See Sec. 14.

2. Name of machine or unit
3. Name of part (if a detail drawing)
4. Drawing number
5. Part number (if a detail drawing)
6. Number of parts required (for each assembly)
7. Scale
8. Assembly-drawing number (given on a detail drawing to identify the part in assembly)
9. Drafting-room record: names or initials of draftsman, tracer, checker, approving authority; each with date
10. Material

To these, depending upon the need for the information, may be added:

11. Stock size
12. Heat-treatment

13. Finish
14. Name of purchaser, if special machine
15. The notation that the drawing "supersedes" and is "superseded by"

*Form of Title.* Every drafting room has its own standard form for titles. In large offices the blank form is often printed in type on paper, plastic, film or cloth (see Figs. 11 and 12 for characteristic examples.)

A form of title that is used to some extent is the *record strip*, a strip marked off across the lower part or right end of the sheet, containing the information required in the title and space for the recording of orders, revisions, changes, etc., that should be noted, with the date, as they occur. Figure 13 illustrates one form.

FIG. 11. A printed title form. This one contains a "change-record" block.

FIG. 12. A printed title form. This one is for an electrical department.

FIG. 13. A strip title. This type extends across one side of the drawing.

It is sometimes desired to keep the records of orders and other private information on the tracing but not to have them appear on the print. In such cases a record strip is put outside the border and trimmed off the print before sending it out.

*To Letter a Title.* The title should be lettered freehand in single-stroke capitals, vertical or inclined—not both styles in the same title. Write out the contents on a separate piece of paper; then refer to Art. 15, Chap. 4, where full instructions have been given.

## 16. Zoning.
As an aid in locating some item on a large drawing, the lower and right borders may be ruled and marked as shown on the *D*-size drawing in Fig. 8. Item *Q* would be located in zone *b*2. A separate column in the change-record block is often used to indicate the position of each drawing change.

## 17. Checking.
When a working drawing is finished, it must be checked by an experienced person, who, in signing his name to it, becomes responsible for any errors. This is the final "proofreading" and cannot be done by the one who has made the drawing nearly so well as by another person. In small offices all the work is checked by the chief draftsman, or sometimes draftsmen check one another's work; in large drafting rooms one or more checkers are employed who devote all their time to this kind of work. All notes, computations, and checking layouts should be preserved for future reference.

Students can gain experience in this work by checking one another's drawings.

To be effective, checking must be done in an absolutely systematic way and with thorough concentration.

## 18. Order of Checking.
Each of the following items should be gone through separately. As each dimension or feature is verified, a check mark in colored pencil should be placed on or above it and corrections indicated with a different-colored pencil.

1. Put yourself in the position of those who are to read the drawing, and find out whether it is easy to read and tells a straight story. Always do this before checking any individual features, in other words, before you have had time to become accustomed to the contents.

2. See that each piece is correctly designed and illustrated and that all necessary views are shown but none that is not necessary.

3. Check all the dimensions by scaling and, where advisable, also by calculation. Preserve the calculations.

4. See that dimensions for the shop are given as required by the shop and that the shop is not left to do any adding or subtracting in order to get a needed dimension.

5. Check for tolerances. See that they are neither too "fine" nor too "coarse" for the particular conditions of the machine, so that they will not, on the one hand, increase unnecessarily the cost of production or, on the other, impair accuracy of operation or duplication.

6. Go over each piece and see that finishes are properly specified.

7. See that every specification of material is correct and that all necessary ones are given.

8. Look out for "interferences." This means that each detail must be checked with the parts that will be adjacent to it in the assembled machine to see that proper clearances have been allowed.

9. When checking for clearances in connection with a mechanical movement, lay out the movement to scale, figure the principal angles of motion, and see that proper clearances are maintained in all positions, drawing small mechanisms to double size or larger.

10. Check to see that all the small details such as screws, bolts, pins, and rivets are standard and that, where it is possible, stock sizes have been employed.

11. Check every feature of the title or record strip and bill of material.

12. Review the drawing in its entirety, adding such explanatory notes as will increase its efficiency.

## 19. Alterations.

Once a drawing has been printed and the prints have been released to the shop, any alterations or changes should be recorded on the drawing and new prints issued. If the changes are extensive, the drawing may be *obsoleted* and a new drawing made which *supersedes* the old drawing. Many drawing rooms have "change-record" blocks printed in conjunction with the title, where minor changes are recorded (Fig. 11). The change is identified in the record and on the face of the drawing by a letter.

New designs may be changed so often that the alterations cannot be made fast enough to reach the shop when needed. In this case sketches showing the changes are rapidly made, reproduced, and sent to the shop, where they are fastened to each print of the drawing. These sketches, commonly known as "engineering orders," are later incorporated on the drawing.

Portions of a drawing can be canceled by drawing closely spaced parallel lines, usually at 45°, over the area to be voided.

## 20. Working Sketches.

Facility in making freehand orthographic drawings is an essential part of the equipment of every engineer. Routine men such as tracers and detailers may get along with skill and speed in mechanical execution, but the designer must be able to record his ideas freehand. In all inventive mechanical thinking, in all preliminary designing, in all explanations and instructions to draftsmen, freehand sketching is the mode of expression. Its mastery means mastery of the language, and it is gained only after full proficiency in drawing with instruments has been acquired. It is mastery which the engineer, inventor, designer, chief draftsman, and contractor, with all of whom time is too valuable to spend in mechanical execution, must have.

Working sketches may be made in orthographic or pictorial form. Chapter 5 gives the fundamentals for orthographic freehand drawings, and Chap. 6 discusses pictorial sketching.

## 21. Kinds of Working Sketches.

Working sketches can be divided into two general classes: (1) those made before the structure is built and (2) those made after the structure is built.

In the first class are included the sketches made in connection with the designing of the structure. These can be classified as (*a*) *scheme* or *idea* sketches, used in studying and developing the arrangement and proportion of parts; (*b*) *computation sketches*, made in connection with the figured calculations for motion and strength; (*c*) *executive sketches*, made by the chief engineer, inventor, or consulting engineer to give instructions for special arrangements or ideas that must be embodied in the design; (*d*) *design sketches*, used in working up the schemes and ideas into suitable form so that the design drawing can be started; and (*e*) *detail sketches*, made as substitutes for detail drawings.

The second class includes (*a*) *detail sketches*, drawn from existing models or parts, with complete notes and dimensions, from which duplicate parts can be constructed directly or from which working drawings can be made (Fig. 14); (*b*) *assembly sketches*, made from an assembled machine to show the relative positions of the various parts, with center and location dimensions, or sometimes for a simple machine, with complete dimensions and specifications; and (*c*) *outline* or *diagrammatic sketches*, generally made for the purpose of location: sometimes, for example, to give the size and location of pulleys and shafting, piping, or wiring, that is, information for use in connection with the setting up of machinery; sometimes to locate a single machine, giving the over-all dimensions, sizes, and center distances for foundation bolts and other necessary information.

## 22. Making a Working Sketch.

In making a working sketch, the principles of projection and the rules of practice for working drawings are to be remembered and applied. A systematic order should be followed for both idea sketches and sketches from objects, as listed below:

1. Visualize the object.

**FIG. 14.** A detail sketch. This is a sketch of two parts, the guide and block parts for a leveling device.

2. Decide on the treatment, orthographic or pictorial.

3. Determine the view or views.

4. Determine the size of the sketch.

5. Locate the center lines.

6. Block in the main outlines.

7. Complete the detail.

8. Add dimension lines and arrowheads.

9. Put on the dimension values.

10. Letter notes and title, with date.

11. Check the drawing.

Before a good graphic description of an object or idea can be developed, it is essential that the mental image of it be definite and clear. The clearness of the sketch is a direct function of this mental picture. Hence the first step is to concentrate on visualization, which leads directly to the second step: determination of the best method of representation.

The method of representation will probably not be just the same as would be used in a scale drawing. For example, a note in regard to thickness or shape of section will often be used in a working drawing to save a view (Fig. 15); thus one view of a piece that is circular in cross section would be

sufficient. In other cases additional part views and extra sections may be sketched rather than complicate the regular views with added lines that might confuse the sketch, although the same lines might be perfectly clear in a measured drawing.

The third step is to determine the view (pictorial) or views (orthographic). Draw the object in its functioning position, if possible, but if another position will show the features to better advantage, use it. A machine should, of course, be represented right-side up, in its natural working position. If symmet-

**FIG. 15.** A one-view sketch. Notes and the dimensioning details make another view unnecessary.

rical about an axis, often only one-half need be sketched. If a whole view cannot be made on one sheet, it may be put on two, each part being drawn up to a break line used as a datum line.

The fourth step is to proportion the size of the sketch to the sheet. Have it large enough to show all detail clearly, but allow plenty of room for dimensions, notes, and memorandums. Small parts can be sketched larger than full size. Do not try to crowd all the views on one sheet of paper. Use as many sheets as may be required, but name each view, indicating the direction in which it is taken. Sometimes one view alone will require a whole sheet.

*In drawing on plain paper*, the location of the principal points, centers, etc., should be so judged that the sketches will fit on the sheet, and the whole sketch, with as many views, sections, and auxiliary views as are necessary to describe the piece, will be drawn in as nearly correct proportion as the eye can determine, *without making any measurements*.

*Cross-section Paper*. Sketches are often made on coordinate paper ruled faintly in $\frac{1}{16}$, $\frac{1}{8}$, or $\frac{1}{4}$ in., used simply as an aid in drawing straight lines and judging proportions, or for drawing to approximate scale by assigning suitable values to the unit spaces. The latter use is more applicable to design sketches than to sketches from the object (Fig. 16).

In order to gain skill through practice, sketches should be made entirely freehand. However, in commercial work the engineer often saves time by making a hybrid sketch, drawing circles with the compass or even with a coin from his pocket, ruling some lines with a pocket scale or a triangle and making some freehand but always keeping a workmanlike quality and good proportion.

### 23. Dimensioning and Finishing.

After completing the views of a piece, go over it and add *dimension lines* for all the dimensions needed for its construction, drawing extension lines and arrowheads carefully and checking to see that none is omitted.

The dimension values are now added. If the sketch is made from reference drawings and specifications, these sources give the dimension values.

**FIG. 16.** A sketch on coordinate paper.

If the sketch is of an existing part or model, measurements must be made to determine the dimension values, as explained in Sec. 24.

Add all remarks and notes that seem to be of possible value. The title should be written or lettered on the sketch.

*Always Date a Sketch.* Valuable inventions have been lost through inability to prove priority because the first sketches were not dated. In commercial work the draftsman's notebook with sketches and calculations is preserved as a permanent record, and its sketches should be made so as to stand the test of time and be legible after the details of their making have been forgotten.

### 24. Measuring and Dimensioning.

Before adding values, *if the sketch is of an existing model or part*, the object has not been handled, and so the drawing has been kept clean. The measurements for the dimensions indicated on the drawing are now needed. A flexible rule or steel scale will serve for getting most of the dimensions. Never use a draftsman's scale for measuring castings, as it would become soiled and its edges marred. The diameter of cylindrical shapes or the distance between outside surfaces can be measured by using outside calipers and scale (Fig. 17); and the sizes of holes or internal surfaces, by using inside calipers. Figure 18 illustrates the inside transfer caliper, used when a projecting portion prevents removing

FIG. 17. Outside caliper. It is used for measuring diameters or thicknesses.

FIG. 18. Inside transfer caliper. It is used for measuring internal distances when a shoulder prevents direct removal.

FIG. 19. Gages. (*a*) Depth gage; (*b*) screw-pitch gage; (*c*) fillet-and-round gage.

FIG. 20. Making a measurement.

the ordinary caliper. The outside transfer caliper is used for a similar condition occurring with an outside measurement. The depth of a hole is easily measured with the depth gage (Fig. 19*a*). Screw threads are measured by calipering the body diameter and either counting the number of threads per inch or gaging with a screw-pitch gage (Fig. 19*b*). A fillet-and-round gage measures radii, the gage fitting to the circular contour (Fig. 19*c*). It is often necessary to lay a straightedge across a surface, as in Fig. 20. This type of measurement could be made conveniently with a combination square or with a surface gage. The combination square has two different heads, the regular 90-45° head (Fig. 21*a*) for a variety of measurements, and the protractor

head (Fig. 21*b*) for measuring or laying out angles. For accurate measurements, outside or inside micrometer calipers are necessary. The outside type is illustrated in Fig. 22. Readings to 0.001 in. are easily obtained. Accurate measurements of holes can be made with a telescopic gage in conjunction with an outside micrometer.

A variety of gages made for special purposes, such

**FIG. 21.** Combination square. (*a*) Standard head; (*b*) adjustable head.

**FIG. 22.** Outside micrometer caliper.

as a wire gage, gage for sheet metal, can be used as occasion demands. With some ingenuity, measurements can often be made with the simpler instruments when special ones are not available.

Always measure from finished surfaces, if possible. Judgment must be exercised in measuring rough castings so as not to record inequalities.

In finding the distance between centers of two holes of the same size, measure from the edge of one to the corresponding edge of the other. Curves are measured by coordinates or offsets, as shown in Figs. 51 and 52, Chap. 10. A curved outline can be recorded by laying a sheet of paper on it and making a rubbing.

## 25. Checking the Sketch.

The final step is to check the sketch. It is a curious fact that when a beginner omits a dimension, it is usually a basic, vital one, such as that of the center height of a machine or an over-all length.

Sketches are not always made on paper with a printed title, but essentially the same information as for an instrument drawing should be recorded in some convenient place on the sheet. All notes and special directions should be checked for accuracy along with the drawing proper. In general, follow the order of checking given in Sec. 17.

## 26. Reproduction of Drawings.

Working drawings go to the shop in the form of prints made from the original drawings. Several different printing processes are in use, all of which give the best results from tracings inked on tracing cloth or paper. However, quite satisfactory prints can be obtained from pencil drawings on translucent paper when the penciling is done skillfully, with uniform, opaque lines. In fact, most of the drawings of industry are not inked; only those of a permanent nature, such as maps, charts, etc., and tracings that must be printed a great many times, are inked.

*Blueprints.* The simplest and most generally used copying process is the blueprinting process, in which the prints are made by exposing a piece of sensitized paper and a tracing in close surface contact with each other to sunlight or electric light in a printing frame or machine made for the purpose. On exposure to the light, a chemical action takes place, which, when fixed by washing in water, gives a strong blue color. The parts protected from the light by the black lines of the tracing wash out, leaving the white paper.

*Vandyke Paper.* This is a thin sensitized paper that turns dark brown when exposed to light and properly "fixed." A reversed negative of a tracing can be made on it by exposing it to light with the inked side of the drawing next to the sensitized side of the paper; then this negative can be printed on blueprint paper, giving a positive print with blue lines on white.

*BW Prints and Directo Prints.* These have black lines on a white ground and are made directly from the original tracing, in a blueprinting machine (and developed by hand) or in a special machine made for the purpose. They are used extensively when positive prints are desired.

*Ozalid Prints.* This process is based on the chemical action of light-sensitive diazo compounds. It is a contact method of reproduction in which the exposure is made in a regular blueprinting machine or an ozalid "whiteprint" machine, and the exposed print is developed dry with ammonia vapors in a developing machine. Standard papers giving black, blue, and maroon lines on a white ground are available. Dry developing has the distinct advantage of giving prints without distortion, and it also makes possible the use of transparent papers, cloth, and foils, which effects savings in drawing time, as these transparent replicas can be changed by additions or erasures and prints made from them without altering the original tracing.

*Photostat Prints.* These are extensively used by large corporations. By this method, a print with white lines on a dark background is made directly from any drawing or tracing, to any desired reduction or enlargement, through the use of a large, specially designed camera. This print can be again photostated, giving a brown-black line on a white ground. This method is extremely useful to engineers for drawings to be included in reports and for matching drawings of different scales which may have to be combined into one.

*Duplicating Tracings.* Tracings with all the qualities of ordinary inked ones are made photographically from pencil drawings by using a sensitized tracing cloth.

*Lithoprinting.* When a number of copies of a drawing (50 or more) are needed, they may be reproduced by lithoprinting, a simplified form of photolithography, at comparatively small cost.

*Copying Methods.* Copying methods such as the mimeograph, ditto machine, and other forms of the hectograph or gelatin pad are often used for small drawings.

*Photo-copying Methods.* Negative copies on high-contrast film are reproduced on paper or cloth either to the same size as the original or to a reduced or increased size. Micromaster copies (Keuffel & Esser Co.) are well known. The reproductions are

very good, and when the process is used for reproduction of an old drawing, wrinkles and other blemishes can be eliminated by using opaque on the film negative.

## 27. Filing and Storing Drawings.

Drawings are filed in steel or wooden cabinets made for the purpose. Many engineering offices store their drawings in fireproof vaults and remove them only for making alterations or for printing. Photographic copies are sometimes made as a separate permanent record. Drawings are always filed flat or rolled. Prints, however, are folded for filing or mailing. The usual method is the "accordion" fold illustrated in Fig. 8. To aid in the filing of accordion-folded prints, a supplementary number block may be added, as shown.

## 28. Simplified Practices.

The bulk of commercial design and representation is done in the conventional manner, that is, by drawing orthographic or pictorial views and dimensioning them as described in the preceding chapters. However, there are many modifications of standard practice which result in economies both in making and in using drawings. These economies can be classified as (1) simplified practices, (2) elimination of drafting effort by the use of templates and overlays, (3) special uses of reproduction, (4) photo drawing, (5) model making,[2] (6) illustrated drawings, and (7) use of special methods and equipment. The purpose in this chapter is to describe and illustrate modifications of standard practice and to suggest ways in which they can be applied.

## 29. "Simplified" Drawing: Economical Drafting Practices.

The drafting of any engineering creation is generally considered by management to be an expensive part of the complete operation of producing a finished, salable product. Furthermore, the percentage of total cost relegated to research, development, draft-

[2]The material on photo drawing and model-making in this chapter is adapted from original subject matter and a selection of illustrations by G. P. Hammond of Procter & Gamble.

ing, tooling, factory, labor, and other items of expense can be judged only on the economy of each individual department. Thus the drafting department comes under the same critical scrutiny as any other division of operation. Any saving in time and expense that can be made without injuring the effectiveness and efficiency of the department is true economy.

This question of thrift has been studied by many executives for years, but W. L. Healy and A. H. Rau,[3] of the General Electric Company, are credited with the first publication recommending and describing methods by which economies can be effected in the drafting room. There are a number of economical practices in making drawings and others that are important only to management. The principles on which economical drafting is based are discussed in the paragraphs that follow.

### 30. Unnecessary Views.

Any view that is not necessary for clear description is a waste of drafting time, paper, reproduction (area and cost), and filing space. Figure 23 is a conventional drawing of a pulley in which the top view is used to describe the cylindrical shapes. This pulley *can* be represented economically, as in Fig. 24, by eliminating the top view and using the letter $D$ on the dimensions of the front view to show that the shapes represented are cylinders. Some drafting economists argue that even the letter $D$ is not necessary because the title block will tell that the object is a pulley and therefore a combination of cylinders. Each drafting executive must judge the extent to which extra views, letters, etc., can be eliminated without making the interpretation of the drawing obscure or misleading. Nevertheless, the following principles should be observed: (1) Never draw more views than are necessary for clear description; (2) examine the object to see if a view can be eliminated by use of a descriptive title, note, symbols, letters, or numbers; (3) examine the object to determine whether a word statement can eliminate *all* views (this is usually possible only for very simple parts).

3 "Simplified Drafting Practice" (New York: John Wiley & Sons, Inc., 1953).

**FIG. 23.** Conventional drawing of a pulley. The two views represent the shape; the dimensions give size description (magnitudes).

**FIG. 24.** Simplified drawing of a pulley. The single view describes the cross-sectional shape and the letter $D$ on the dimensions specifies that the shape is cylindrical. The figured dimensions give size description (magnitudes).

### 31. Unnecessary Scale.

The reader of the drawing does not establish distances by scaling the drawing, and most commercial drawings carry a note stating that the drawing is not to be scaled. Drafting time should not be used in drawing to scale *when it is not necessary*. A good example is given in Fig. 25, a structural detail *not to any scale* at all but just as effective and readable as if drawn carefully and meticulously to scale. The

**FIG. 25.** A structural drawing. This drawing is "simplified" in that it is not drawn to scale; thus some drafting effort is eliminated.

cardinal points are: (1) do not draw to scale unless it is necessary for proper layout and readability; (2) do not cautiously draw to scale when approximations will suffice.

In sketching, *proportions* need not be meticulously maintained. Note in Fig. 26 that the two holes ($\frac{7}{8}$ and $\frac{3}{8}$ in. diameter) are *not* in good proportion and that the two $\frac{1}{2}$ in. radii for the bends are not shown centered according to the dimensioning. Nevertheless, the drawing is readable, and if the dimensions are followed—and of course they *must* be—the correct object will be manufactured. Figure 26 is not an extreme case—proportions could be *much* poorer than in Fig. 26 and the drawing still be perfectly usable.

## 32. Unnecessary Lines and Elaboration.

Every line made on a drawing costs in drafting time and the company's money. Therefore, any line that does not accomplish a useful purpose should be eliminated. Such lines will include extra lines (on any view) that are not necessary for describing the object. Meticulous crosshatching and embellishment, especially, can be omitted without seriously affecting readability. Compare the two drawings in Figs. 27 and 28. Painstaking and complete representation of sections such as is given in Fig. 27 is a waste of time and could be advantageously simplified, as in Fig. 28. The drawing in Fig. 28 actually gives more structural information than the detailed representation in Fig. 27.

## 33. Arrowless Dimensioning.

Time can be saved by eliminating unnecessary arrows on dimension lines and leaders, as has been done in Figs. 24, 25, 26, 28, and 30. To the uninitiated, these drawings may appear to be unfinished, but careful examination will show that arrows are not really needed.

## 34. Simplified Base-line Dimensioning.

If the base-line system of dimensioning is used,

**FIG. 26.** A simplified sketch. In this example, proportions have not been carefully maintained, thus saving time. The object is built by following the general shape shown and adhering to the dimensions; therefore, careful proportioning is not necessary.

**FIG. 27.** A conventional drawing. Because of the drafting effort required to produce the detail and embellishment, this drawing is expensive.

**FIG. 28.** A simplified drawing. This drawing represents the same structure as in Fig. 27. Note the absence of extra detail and embellishment, which reduces drawing cost.

even the dimension *lines* can be eliminated. As shown in Fig. 29, two base lines, one horizontal and the other vertical, are established. The location of base lines may be at the edge of the part, on the center of symmetry, or through some important feature. Distances are then given, as shown in the figure, *from* the base lines (corresponding to any coordinate system) and include all the information

**FIG. 29.** Simplified base-line dimensioning. This system accomplishes the same purpose as standard dimensioning from datum, but eliminates all unnecessary lines and arrowheads, thus reducing drafting time.

**FIG. 30.** A simplified sketch. This drawing, made freehand, is economical, but economy is carried further by elimination of extra views, embellishment, and arrowheads on the dimensions.

necessary for construction.

### 35. Freehand Drawing.

The greatest saving of drafting time can often be realized by using the methods just described and by making the drawing freehand. Compare the simplified sketch in Fig. 30 with the conventional sketch in Fig. 15. See also Art. 31 and Fig. 26 where proportioning is discussed and illustrated.

### 36. Cautions in the Use of Simplified Practices.

Simplified drafting practices are comparatively new and therefore may not be understood by some engineers and manufacturers. For example, for one

unfamiliar with simplified practices, the drawing in Fig. 29 *may* be more difficult to read than a conventional drawing. Further, if mistakes occur in manufacture because of misreading a simplified drawing, the saving in drafting time is offset many times over in costly rejected parts. The common, and safest, use of simplified practices occurs *within an organization* where everyone concerned is thoroughly trained in all company practices. It is potentially dangerous to send a simplified drawing to a subcontractor or supplier *unless it is known* that *all* special drafting practices will be understood.

### 37. Abbreviations.

Another economy results from using abbreviations where this does not detract from the readability of the drawing. American Standard abbreviations are given in the Appendix. Note the use of abbreviations in Figs. 24 to 30.

## TEMPLATES AND OVERLAYS

### 38. Templates.

Use of modern aids such as the drafting machine, pencil sharpener and pointer, and special instruments such as an ellipsograph (Fig. 62) conserves drafting time. For features such as circles, ellipses, boltheads, nuts, and symbols which must be drawn repeatedly, a template greatly shortens the time required for layout. Figure 31 shows several typical templates: (*a*) and (*b*), Rapidesign templates for ellipses and circles; (*c*), a Tilt-hex template for boltheads; (*d*), a Du-all template for a variety of operations; (*e*), a Holometer for boltheads and nuts; (*f*), a K & E thread template; and (*g*), a Rapidesign template for piping symbols.

A number of symbol templates are available for use with Leroy lettering equipment; Fig. 32 shows some of them. Figure 33 is an example of a symbolic drawing.

### 39. Overlays.

An overlay is typed or printed detail reproduced on tracing paper, Mylar, or some other translucent or transparent material. Overlays are fastened to a

**FIG. 31.** Templates. A wide variety are available for many purposes. Their use conserves drafting time.

drawing by means of transparent tape or the adhesive backing on the overlay.

Most companies have standard items that must be repeated on drawings. The use of overlays eliminates the need for redrawing these items each time they appear. One prominent processing company, for example, is continually building new plants and expanding existing plants, and it must make plant layout drawings showing the location of equipment. The plan views of tanks, pumps, filters, and other items are drawn and identified for size and scale, and then reproduced on a suitable overlay material.

**FIG. 32.** Leroy symbol templates (Keuffel & Esser Co.). A wide variety of these templates, for use with standard Leroy equipment, are available. Special templates of a user's design also can be obtained.

Needed items are cut from this sheet and applied to the drawing with tape. The outline of the overlays and the tape will show faintly on the reproduced drawing but will not impair its readability. Extra information may be lettered on the overlays.

Many standard items are available from companies that manufacture a general line of overlays. For details that are time-consuming to draw and that occur again and again, use an overlay.

Parts lists, tables, references, and specifications must often appear on drawings, and it is time-consuming to letter these. Laborious lettering can be eliminated by employing a standard overlay. If a standard overlay cannot be used, information can be typed on blank stanpat or some other overlay material and applied to the drawing.

The use of overlays is not restricted to prepared details and lettering. Black tape is used to block out unwanted areas or in place of drawn material. It is especially useful for wide detail. Tape can be cut with a sharp blade to any width needed. Tapes in widths from $1/32$ in. to 1 in. for full, dashed, and special lines and for areas, and a "tape pen" for application are manufactured by Chart-Pac, Inc., Leeds, Massachusetts.

**FIG. 33.** A symbolic drawing. Components are shown by symbols. Magnitudes and other details are given by lettered information. (*Courtesy of Radio Corporation of America.*)

# PROBLEMS

## Group 1. Detail Drawings.

This group of problems gives practice in making drawings with sectional views, auxiliaries, and conventional representation. Several methods of part production are included: casting, forging, forming of sheet metal, and plastic molding.

**1.** Make complete working drawing with necessary sectional views. Cast iron.

**2.** Working drawing of friction-shaft bearing. Cast iron.

**3.** Working drawing of mixing-valve body. Cast brass.

**PROB. 1.** Gear-shifter bracket.

**PROB. 2.** Friction-shaft bearing.

**PROB. 3.** Mixing-valve body.

**PROB. 4.** Conveyor hanger.

**4.** Working drawing of conveyor hanger. Determine what views and part views will best describe the piece. Cast steel.

**5.** Working drawing of strut base. Determine what views and part views will best describe the piece. Cast aluminum.

**6.** Draw given front view. Add part views and auxiliaries to describe the piece best. Cast aluminum.

**7.** Working drawing of valve cage. Cast bronze.

**8.** Determine what views and part views will most adequately describe the piece. Malleable iron.

**9.** Working drawing of water-pump cover. Aluminum alloy die casting.

**10.** Working drawing of slotted spider. Malleable iron.

**11.** Make working drawing of relief-valve body. Cast brass.

**12.** Working drawing of breaker. Steel.

**13.** Working drawing of meter case. Molded Bakelite.

**14.** (*a*) Make detail working drawings on same sheet, one for *rough* forging and one for *machining;* or (*b*) make one detail drawing for forging and machining. Alloy steel.

**PROB. 5.** Strut base.

**PROB. 7.** Valve cage.

**PROB. 6.** Hinge bracket.

**PROB. 8.** Brace plate.

**PROB. 9.** Water-pump cover.

**PROB. 10.** Slotted spider.

**PROB. 11.** Relief-valve body.

**PROB. 12.** Breaker.

**PROB. 13.** Meter case.

**PROB. 14.** Steering knuckle.

**15.** Working drawing of automotive connecting rod. Drop forging, alloy steel. Drawings same as Prob. 14.
**16.** Working drawing of puller body. Press forging, steel. Drawings same as Prob. 14.
**17.** Make working drawing of buffer stand. Steel drop forging. Drawings same as Prob. 14.
**18.** Working drawing of torque-tube support. Drop forging, aluminum alloy. Drawings same as Prob. 14.
**19.** Working drawing of supply head. Cast iron.

**PROB. 16.**  Puller body.

**PROB. 15.**  Automotive connecting rod.

**PROB. 17.**  Buffer stand.

**PROB. 18.**  Torque-tube support.

**PROB. 19.**  Supply head.

**20.** Working drawing of bearing block, cast iron.

**21.** Working drawing of elevator control bracket. Aluminum alloy, welded.

**22.** Working drawing of flap link. High-strength aluminum alloy.

**23.** Working drawing of wing fitting. High-strength aluminum alloy.

**24.** Working drawing of micro-switch bracket. Aluminum sheet, 0.15 thick.

**PROB. 20.** Bearing block.

**PROB. 21.** Elevator control bracket.

**PROB. 23.** Wing fitting.

**PROB. 22.** Flap link.

**PROB. 24.** Micro-switch bracket.

**25.** Working drawing of tab link. High-strength aluminum alloy.

**26.** Working drawing of drag strut. High-strength aluminum alloy.

**27.** Make a complete working drawing of the 1½-in., no. 2000 desurger case head of Prob. 12, Chap. 12. Material is ductile iron.

**28.** Make a complete working drawing of the desurger loading-valve gage connector. This part is made of ¾-in. hexagon steel (AISI C1212).

**29.** Make a complete working drawing of the needle-valve needle. Material is stainless steel.

**30.** Make a complete working drawing of the wheel-cylinder boot.

**PROB. 25.** Tab link.

**PROB. 26.** Drag strut.

**PROB. 28.** Desurger loading-valve gage connector. (*Courtesy of Westinghouse Air Brake Co.*)

**PROB. 29.** Needle-valve needle. (*Courtesy of Westinghouse Air Brake Co.*)

**PROB. 30.** Wheel-cylinder boot and piston. (*Courtesy of Bendix Corp.*)

**31.** Make a complete drawing of the wheel-cylinder piston, Prob. 30.

**32.** Make a complete working drawing of the desurger loading-valve body. Material is brass.

**33.** Make a complete working drawing of the wheel-cylinder cup.

## Group 2. An Assembly Drawing from the Details.

**34.** Assembly drawing of sealed shaft unit. Bracket and gland are cast iron. Shaft, collar, and studs are steel. Bushing is bronze. Drill bushing in assembly with bracket.

**PROB. 32.** Desurger loading-valve body. (*Courtesy of Westinghouse Air Brake Co.*)

**PROB. 33.** Wheel-cylinder cup. (*Courtesy of Bendix Corp.*)

**PROB. 34.** Sealed shaft unit.

**35.** Make an assembly drawing of the crane hook from details given. Standard parts 7 to 10 are not detailed; see Appendix or handbook for sizes.

**36.** Make an assembly drawing, front view in section, of caster.

**36A.** Redesign caster for ball-bearing installation.

**37.** Make an assembly drawing of the Brown and Sharpe

rotary geared pump, with top view, longitudinal section, and side view. Show direction of rotation of shafts and flow of liquid with arrows. Give dimensions for base holes to be used in setting; also give distance from base to center of driving shaft and size of shaft and key. For the parts that are not detailed, see the parts list.

**38.** Make assembly drawing of parallel-index drive.

| PC.NO. | NAME | QUAN. | NOTES |
|---|---|---|---|
| 1 | SIDE PLATE | 2 | S.A.E.-1030-PLATE |
| 2 | PULLEY | 2 | " 1020-CASTING |
| 3 | PULLEY PIN | 2 | " 1045-FORGING |
| 4 | NUT LOCK PIN | 1 | " 1030-CR |
| 5 | CRANE HOOK | 1 | " 3140-FORGING |
| 6 | HOOK TRUNION | 1 | " 1045 " |
| 7 | SEMIFIN HEX NUT | 6 | 2½-4UNC-2B |
| 8 | HEX SLOTTED NUT | 1 | 3-6NS-2B |
| 9 | WASHER | 2 | 3" HEAVY |
| 10 | COTTER PIN | 2 | NO. 11 × 1"LONG |

**PROB. 35.** Crane hook.

**PROB. 36.** Caster.

| PC. NO. | DRAW SIZE | NAME | QUAN. | MAT. | STOCK | | USED ON | | REMARKS |
|---|---|---|---|---|---|---|---|---|---|
| | | | | | DIA. | LGTH. | NAME | PC. NO. | |
| | | | | PARTS LIST | | | | | |
| | | | | BROWN AND SHARPE No. 1 ROTARY GEARED PUMP | | | | | |
| 101 | | Base | 1 | C.I. | | | | | |
| 102 | | Body | 1 | C.I. | | | | | |
| 103 | | Cover | 1 | C.I. | | | | | |
| 104 | | Pulley | 1 | C.I. | | | | | |
| 105 | | Gland | 1 | C.I. | | | | | |
| 106 | | Gland Bushing | 1 | Bro. | | | | | |
| 107 | | Gear Bushing | 4 | Bro. | | | | | |
| 108 | | Driving Gear | 1 | S.A.E. #1045 | 1 9/16 | 5 7/8 | | | |
| 109 | | Driven Gear | 1 | S.A.E. #1045 | 1 9/16 | 2 9/16 | | | |
| 110 | | Gasket | 2 | Sheet Copper | | | Body | 102 | #26 B&S Gage (0.0159) |
| 111 | | #10-32 x 1 5/8 Slotted Hex. Hd. Mach. Scr. & Nut | 4 | | | | Cover | 103 | |
| 112 | | #10-32 x 1 5/8 Slotted Hex. Hd. Cap Scr. | 2 | | | | Cover | 103 | |
| 113 | | #10-32 x 7/8 Slotted Hex. Hd. Cap Scr. | 2 | | | | Gland | 105 | |
| 114 | | Woodruff Key #405 | 1 | | | | Driving Gear | 108 | |
| 115 | | 3/8 x 3/8 Headless Set Scr., 3/8-16NC-2 | 1 | | | | Pulley | 104 | |
| 116 | | 3/16 x 1 7/16 Dowel Pin | 2 | C.R.S. | | | Cover | 103 | |
| | | Packing | To Suit | | | | | | Garlock Rotopac #239 |

**PROB. 37.** Rotary geared pump and parts list.

**PROB. 38.** Parallel-index drive. (*Courtesy of Commercial Cam and Machine Co.*)

## Group 3. Detail Drawings from the Assembly.

**39.** Make detail drawings of the jig table. Parts, cast iron.

**40.** Make detail drawings of the door catch.

**41.** Make detail drawings of belt drive. The pulley and bracket are cast iron; the gear and shaft, steel. The bushing is bronze.

**PROB. 40.** Door catch.

**PROB. 39.** Jig table.

**PROB. 41.** Belt drive.

**42.** Make detail drawings of swing table.

**43.** Make detail drawings of sealed ball joint.

**44.** Make detail drawings of belt tightener. The bracket, pulley, and collar are cast iron. The bushing is bronze; the shaft, steel.

| PC. NO. | NAME | MAT. | QUAN. | NOTES |
|---|---|---|---|---|
| 101 | Base | C.I. | 1 | |
| 102 | Table | C.I. | 1 | |
| 103 | Trunnion Stud | Steel | 1 | |

**PROB. 42.**  Swing table.

| PC. NO. | NAME | MAT. | QUAN | NOTES |
|---|---|---|---|---|
| 101 | Base | Mal. I. | 1 | |
| 102 | Cover | Mal. I. | 1 | |
| 103 | Sleeve Ball | Steel | 1 | |
| 104 | Stud Ball | Steel | 1 | |
| 105 | Diaphragm | Fabric | 1 | Purchased |

**PROB. 43.**  Sealed ball joint.

**PROB. 44.**  Belt tightener.

**45.** Make detail drawings of rotary pressure joint.
**46.** Make detail drawings of ball-bearing idler pulley.

**PROB. 45.** Rotary pressure joint.

| PC.NO. | NAME | MAT. | QUAN. | | | | |
|---|---|---|---|---|---|---|---|
| 1 | Body | C.I. | 1 | 5 | Thrust Collar | C.I. | 1 |
| 2 | Head | C.I. | 1 | 6 | Seal Ring | No.61 Graphitar | 2 |
| 3 | Nipple Tube | Steel | 1 | 7 | Spring | Stainless Steel | 1 |
| 4 | Nipple Body | C.I. | 1 | 8 | Gasket | Durabla | 1 |
| | | | | 9 | Key | Steel | 1 |

| PC.NO. | NAME | MAT. | QUAN. | NOTES |
|---|---|---|---|---|
| 101 | Bracket | C.I. | 1 | |
| 102 | Pulley | C.I. | 1 | |
| 103 | Flange | C.I. | 2 | |
| 104 | Shaft | Steel | 1 | |
| 105 | Sleeve | Steel | 1 | |
| 106 | Nut | Steel | 1 | |

**PROB. 46.** Ball-bearing idler pulley.

**47.**   Make detail drawings of butterfly valve.

**48.**   Make detail drawings for V-belt drive.

| PC. NO. | NAME | | MAT. | QUAN. | NOTES |
|---------|------|--|------|-------|-------|
| 101 | Butterfly Housing | | Mal. I. | 1 | |
| 102 | Rack Housing | | Mal. I. | 1 | |
| 103 | Butterfly | (.0625″ thick) | Steel | 1 | #16 U.S.S. Gage |
| 104 | Butterfly Shaft | | Steel | 1 | |
| 105 | Pinion | | Steel | 1 | |
| 106 | Rack | | Steel | 1 | 16 Pitch |
| 107 | Key Screw | | Steel | 1 | |
| 108 | Cover | (.0625″ thick) | Steel | 1 | #16 U.S.S. Gage |

**PROB. 47.**   Butterfly valve.

| PC. NO. | NAME | MAT. | QUAN. | NOTES |
|---------|------|------|-------|-------|
| 101 | Bracket | C. I. | 1 | |
| 102 | Gear | C. I. | 1 | |
| 103 | Shaft | Steel | 1 | |
| 104 | Gland | C. I. | 1 | |
| 105 | Pulley | Al. | 1 | Die Cast |
| 106 | Bushing | Bro. | 1 | |

**PROB. 48.**   V-belt drive.

**49.**   Make detail drawings of hydraulic punch. In action, the punch assembly proper advances until the cap, piece 109, comes against the work. The assembly (piece 106 and attached parts) is then stationary, and the tension of the punch spring (piece 114) holds the work as the punch advances through the work and returns.

| PC. NO. | NAME | MAT. | QUAN. | NOTES |
|---------|------|------|-------|-------|
| 101 | Bracket | C.I. | 1 | |
| 102 | Cylinder | C.I. | 1 | |
| 103 | Piston | C.I. | 1 | |
| 104 | Cylinder Head | C.I. | 1 | |
| 105 | Sleeve Stop | C.I. | 1 | |
| 106 | Sleeve | Steel | 1 | |
| 107 | Sleeve Nut | Steel | 1 | |
| 108 | Punch | Steel | 1 | |
| 109 | Punch Cap | Steel | 1 | |
| 110 | Packing Plate | Steel | 1 | |
| 111 | Packing | Leather | 1 | Purchase |
| 112 | Piston Rod | Steel | 1 | |
| 113 | Sleeve Spring | Steel | 1 | Purchase |
| 114 | Punch Spring | Steel | 1 | Purchase |

**PROB. 49.**   Hydraulic punch.

**50.** Make detail drawings of double-acting air cylinder. Length of stroke to be assigned. Fix length of cylinder to allow for clearance of 1 in. at ends of stroke. Note that pieces 101 and 102 are identical except for the extra machining of the central hole in piece 101 for the shaft, packing, and gland. Make separate drawings for this piece, one for the pattern shop and two for the machine shop.

| PC.NO. | NAME | MAT. | QUAN. | NOTES |
|---|---|---|---|---|
| 101 | Cylinder Head, Front | C.I. | 1 | |
| 102 | Cylinder Head, Rear | C.I. | 1 | Use Cast. 101 |
| 103 | Packing Gland | C.I. | 1 | |
| 104 | Piston | C.I. | 1 | |
| 105 | Piston Rod | Steel | 1 | |
| 106 | Piston Plate | Steel | 2 | |
| 107 | Cylinder | Steel | 1 | Shelby Tubing |
| 108 | Tie Rod | Steel | 4 | |
| 109 | Stop Collar | Steel | 1 | |
| 110 | Piston Packing | Leath. | 2 | Purchase |
| 111 | Gasket | Fiber | 2 | Purchase |

**PROB. 50.** Double-acting air cylinder.

**51.** Make detail drawings of rail-transport hanger. Rail is 10-lb ASCE.

| PC. NO. | NAME | MAT. | QUAN. | NOTES |
|---|---|---|---|---|
| 101 | Hanger | C.S. | 1 | |
| 102 | Wheel | C.I. | 1 | |
| 103 | Stud | Steel | 1 | |
| 104 | Plate | Steel | 2 | #13 U.S. Gage (.0938) |
| 105 | Eyebolt | D.F. | 1 | Purchase |

**PROB. 51.**   Rail-transport hanger.

**52.** Make detail drawings of the laboratory pump. Materials are shown on the assembly drawing.

**PROB. 52.** Laboratory pump.

## Group 4.  A Set of Drawings from an Exploded Pictorial.

The problems that are given in this group have been arranged as complete exercises in making a set of working drawings. Remember that the dimensions given on the pictorial views are to be used only to obtain distances or information needed. In some cases the data needed for a particular part may have to be obtained from the mating part.

The detail drawings should be made with each part on a separate sheet. Drawings of cast or forged parts may be made in the single-drawing or the multiple-drawing system described in Chap. 10.

The assembly drawing should include any necessary dimensions, such as number, size, and spacing of mounting holes, that might be required by a purchaser or needed for checking with the machine on which a subassembly is used.

For the style and items to be included in the parts list, see Prob. 37.

**53.**  Make a complete set of working drawings for the antivibration mount.

**54.**  Make a complete set of drawings for pivot hanger, including detail drawings, assembly drawing, and parts list. All parts are steel. This assembly consists of a yoke, base, collar, and standard parts.

**PROB. 54.**  Pivot hanger.

**PROB. 53.**  Antivibration mount.

**55.** Make a complete set of drawings for pump valve. The valve seat, stem, and spring are brass; the disk is hard-rubber composition. In operation, pressure of a fluid upward against the disk raises it and allows flow. Pressure downward forces the disk tighter against the seat and prevents flow.

**56.** Make a complete set of working drawings for the cartridge-case trimmer. Note that some of the dimensions will have to be obtained from the mating part. A tabular drawing may be made of the case holder, covering holders for various cartridge cases. The dimensions to be tabulated are (*a*) holder length, (*b*) diameter, and (*c*) taper of hole.

**57.** Make a complete set of working drawings of the hydraulic check valve. Spring is stainless steel; gasket is soft aluminum; all other parts are steel.

**58.** Make a complete set of working drawings of the boring-bar holder. All parts are steel. Note that the *body* is made in one piece, then split with a ⅛-in.-wide cut (exaggerated in the picture). The holder can be seen in use in Fig. 13, Chap. 9.

**58***A.* Design three boring bars (see Fig. 13, Chap. 9) to fit the boring-bar holder, and make a tabular drawing.

| NAME | MAT. |
|---|---|
| Base | 1020 |
| Base Rod | 1020 |
| Stop Screw | B-1113 |
| Stop Nut | B-1113 |
| Case Holder | B-1113 |
| Cutter Bearing | B-1113 |
| Cutter (Harden) | 1.05-1.10 Carbon Tool Steel |
| Handle | 1020 |
| Ball | Bro |
| Fibre Disc | |
| Slotted Flat Pt Setscrew | |
| Fil Hd Mach Screw | |
| Semifin Hex Nut | |
| Lock Nut (alter std. nut) | |

**PROB. 55.**  Pump valve.

**PROB. 56.**  Cartridge-case trimmer.

**PROB. 57.**  Hydraulic check valve.

**PROB. 58.**  Boring-bar holder.

**59.** Make a complete set of working drawings of the ratchet wrench.

**60.** Make detail and assembly drawings of landing-gear

door hinge. High-strength aluminum alloy.

**61.** Make a complete set of working drawings for access-door support.

**PROB. 59.** Ratchet wrench.

| NAME | MAT. | REQ | NOTES |
|------|------|-----|-------|
| Body | 3240 | 1 | Drop forging, Heat-treat |
| Cover | 1025 | 1 | |
| Ratchet Wheel | 2315 | 1 | Caseharden |
| Ratchet | 2315 | 1 | Caseharden |
| Lever | 1025 | 1 | |
| Spring, 5/32 OD | Music Wire | 2 | No.8 Ga, 5 Coils, 5/16 F.L. |
| Ball | Steel | 2 | 5/32 D, Hardened |
| Flat Hd Mach Scr | | 2 | Recess drive |

**PROB. 60.** Landing-gear door hinge.

**PROB. 61.** Access-door support.

**62.** Make detail and assembly drawings of the pressure-cell piston. Body parts are high-strength aluminum alloy. O rings are Teflon.

**PROB. 62.** Pressure-cell piston.

**63.** Make a complete set of working drawings of the pivot nut and adjusting screw.

## Group 5. A Set of Drawings from a Pictorial Assembly.

**64.** Make a complete set of working drawings of the high-tension coil mount.

**65.** Make a complete set of working drawings of the pipe clamp. The flange is cast steel.

**PROB. 63.** Pivot nut and adjusting screw. *(Courtesy of Bendix Corp., Products Division.)*

**PROB. 64.** High-tension coil mount.

**PROB. 65.** Pipe clamp.

**66.** Make a complete set of working drawings of the stay-rod pivot. Parts are malleable iron.

**67.** Make a complete set of working drawings of the tool post. All parts are steel.

**68.** Make a unit-assembly working drawing of the wing-nose rib.

**PROB. 66.** Stay-rod pivot.

**PROB. 67.** Tool post.

**PROB. 68.** Wing-nose rib.

**69.** Make a complete set of working drawings of the valve-poppet assembly. Flexible washer made of Koroseal 67-77 Durometer (B. F. Goodrich grade 116). Draw ten times actual size.

**70.** Make a complete set of working drawings of the diaphragm assembly. Draw four times actual size.

## Group 6. Working Drawings from Photo-drawings.

Photo-drawings are coming into use commercially, especially in the processing industries. These problems give practice in making detail and assembly drawings, and illustrate the possibilities of drawings produced from photographs of models or test parts.

**71.** Detail drawing of conveyor link. SAE 1040 steel.
**72.** Detail drawing of hydro-cylinder support. Aluminum sheet.

**PROB. 69.** Valve-poppet assembly. (*Courtesy of Bellows-Valvair, Division of International Basic Economy Corporation.*)

**PROB. 70.** Diaphragm assembly. (*Courtesy of Bellows-Valvair, Division of International Basic Economy Corporation.*)

**PROB. 71.** Conveyor link.

**PROB. 72.** Hydro-cylinder support.

**73.**  Detail drawing of lift-strut pivot. Forged aluminum.
**74.**  Detail drawing of length adjuster-tube. SAE 1220 steel.
**75.**  Detail and unit assembly of pivot shaft. Shaft is bronze; nuts are steel.
**76.**  Detail drawing of third terminal (temperature control). Material is red brass (85% CU).
**77.**  Set of working drawings of conveyor link unit. All parts are steel.

**PROB. 73.**  Lift-strut pivot.

**PROB. 74.**  Length adjuster-tube.

**PROB. 75.**  Pivot shaft.

**PROB. 76.**  Third terminal.

**PROB. 77.**  Conveyor link unit.

# PROBLEM DRAWINGS IN METRIC DIMENSIONS

The problems following are assembled as a collection from which the student, teacher, or graphics director may choose in order to gain some experience in the metric system of measurement. All dimensions on all drawings are given in the metric system; the exception are screws and threads.

One of the outstanding learning values in the solution of graphics problems is the transfer from a pictorial representation to an orthographic presentation and vice versa. The following group of problems provides the student with an opportunity to make this transfer within the metric system.

For further experience in the metric system, use a hand calculator to determine any metric values that must be inserted in the drawing during the transfer from one presentation to another. Also use the calculator to convert inches to metric equivalents in the drawings.

The general assignments for the students in doing these problems are:

Produce a complete graphic representation (orthographic or pictorial) of the part, group of parts, or machine.

Describe the part, group of parts, or machine completely in the metric system.

Be so complete and thorough in your graphic description that the part, group of parts, or machine could certainly be manufactured from your drawings and specifications.

Students may check with the teacher on the choices of method of manufacture but must produce the drawings completely by themselves.

PROB. 1. Boat winch pawl.

PROB. 2. Handwheel.

PROB. 3. Gasket.

PROB. 4. Inclined support.

PROB. 5. Wire thimble.

PROB. 6. Splice plate.

PROB. 7. Bearing rest.

538

PROB. 8. Rubber-mounting
bracket.

PROB. 9. Crosshead.

PROB. 10. Slotted wedge.

PROB. 11. Parking-brake cam.

116 Drill
Counterbore 22D×6Dp

PROB. 12. Switch base.

PROB. 13. Latch pocket.

PROB. 14. Guide base.

PROB. 15. Corner block.

PROB. 16. Inclined support.

PROB. 17. Leveling wedge.

PROB. 18. Step-pulley frame.

PROB. 19. Sliding socket.

540

PROB. 20. Flanged pocket.

PROB. 21. Bumper.

PROB. 22. Slotted wedge.

PROB. 23. Clamp frame.

PROB. 24. Truss bearing.

PROB. 25. Coupling.

PROB. 26. Toggle-shaft support.

10 Drill for oil

20.8 Drill, Spotface 35 D x 1.5 deep, 3 holes

25 Drill Thru
Counterbore 35D x 5 Dp.
Both Ends

PROB. 27. Shifter fork.

12.5 Drill - 4 Holes

PROB. 28. Mounting bracket.

PROB. 29. Skew slide.

542

PROB. 30. Rubber support anchor.

PROB. 31. Convector strip.

PROB. 32. Push plate.

PROB. 33. Circuit-breaker stop.

PROB. 34. Transverse connection.

PROB. 35. Chamber tool base.

PROB. 36. Dovetail clip.

PROB. 37. Anchor base.

PROB. 38. Connector strip.

PROB. 39. Push plate.

PROB. 40. Channel support.

PROB. 41. Corner tie.

PROB. 42. Bevel-gear housing.

544

PROB. 43. Bevel washer.

PROB. 44. Angle bracket.

PROB. 45. Catenary clip.

PROB. 46. Cable anchor.

10 Drill, Spotface
20 D x 2 deep,
3 holes.
10 Drill, 2 hls.

Lug makes angle of 60° with base

Calculator push button (nylon)

PROB. 47. Calculator push button (nylon).

PROB. 48. Carrier base.

PROB. 49. Selector ring.

PROB. 50. Impeller.

Draw 45° to right

PROB. 51. Slide bracket.

Draw 45° to right

PROB. 52. Jaw bracket.

PROB. 53. Bracket.

PROB. 54. Hinged catch.

PROB. 55. Ceiling flange.

PROB. 56. Hanger flange.

PROB. 57. Bell crank.

PROB. 58. Stop plate.

Offsets from right section 30° to right

PROB. 59. Culvert model.

Offsets from right section, 30° to right

PROB. 60. Slotted guide.

PROB. 61. Slotted link.

Draw 30° to left

PROB. 62. Swivel plate.

Draw 45° to right

PROB. 63. Column base.

PROB. 64. Base plate.

PROB. 65. Gland.

PROB. 66. Head yoke.

PROB. 67. Trunnion plate.

548

PROB. 68. Squared collar.

PROB. 69. Blank for gland.

PROB. 70. Sliding cover.

PROB. 71. Side-beam bracket.

PROB. 72. Rod support.

PROB. 73. Strut swivel.

PROB. 74. Tie plate.

PROB. 75. Heat sink for transistor radio (1 mm aluminum).

PROB. 76. Dovetail bracket.

PROB. 77. Offset bracket.

PROB. 79. Notched holder.

PROB. 78. Double wedge block.

PROB. 80. Bracket.

PROB. 81. Face plate.

PROB. 82. Spindle carrier.

PROB. 83. Bed bracket.

PROB. 85. Compound gear arm.

PROB. 84. Drawbar pivot.

PROB. 86. Fan bracket.

PROB. 87. Valve flange.

PROB. 88. Switch-box cover.

PROB. 89. Adjustable base.

| PC.NO. | NAME | MAT. | QUAN. | NOTES |
|---|---|---|---|---|
| 101 | Ball Fitting | Mal. I. | 1 | |
| 102 | Threaded Fitting | Mal. I. | 1 | |
| 103 | Gland | Mal.I. | 1 | |
| 104 | Packing | — | 2 | Purchased |

PROB. 90. Screw-end ball joint.

552

| PC.No. | NAME | MAT. | QUAN | NOTES |
|--------|------|------|------|-------|
| 101 | Base | C.I. | 1 | |
| 102 | Support | C.R.S. | 2 | |
| 103 | Housing | C.I. | 1 | Make with #104 |
| 104 | " Cap | C.I. | 1 | " " #103 |
| 105 | Bearing Sleeve | C.I. | 1 | |
| 106 | " | W.M. | 1 | Die Casting |
| 107 | Stud | C.R.S. | 2 | |

Annular Groove, 8 wide
8.3 Drill, ⅛ Am. Std. pipe tap
155
16
R→13
90°
16R
0.2×45
1/2
5
75
5D×25 Pin
6R
160
18R
45
30
2
107
92
130
20
102
32
3
101
16
12
32
45

Oil Groove
6 wide x 2 deep
M12-1.75 Stud, Hex Nut
13 Drill, Spotface
25D × 2Dp, 2 Req'd.
Annular Oil Groove
8 wide x 2 deep

104
22
103
105
3
65D
85
55
42D
106
3 Sawcut
45
70
75
3
20
12
12

PROB. 91. Adjustable midbearing.

19.8D., 2 Holes
50R
38R
65 Pitch
6.5 Thick

PROB. 92. Conveyor link.

8R
(2)
64
4.0 Thk
4.8 Drill,
2 Holes
38
5
30
1.0R
12
Inside (near) surface
50
30
30
6R
3.7 Drill,
3 Holes
37
30
6R
6R
6
38
6
55

PROB. 93. Hydrocylinder support.

PROB. 94. Independent face-plate chuck.

554

St'd Eyebolt 12.8 Drill, C'bore 22 × 2 Deep,
C'sink 3 Deep × 90° Rivet over
at end to loose fit.

10D

32D

6R

35
6   25D

12

54

67 Core

50

10

160

50

25

103

83

40R

102      101

$\frac{5}{8}$-18UNF-2B Semi-
fin. Hex Jam Nut

M12-1.75 × 25
Soc, Cone Pt.
Set Scr

105

40

45R

85R      11

25R

12R

12R

15°     10     29

M6.3-1 × 12
Round Hd.
Cap Scr, 8 Req.

50

25
50      6

3        30     2.5

3R

64

90      79

38

20

33

35

16   16

32

5

45

52.0

116

12

103      104

L#8505 Sealed Ball
Bearing  2 Req'd.

3R

| PC. NO. | NAME | MAT. | QUAN. | NOTES |
|---|---|---|---|---|
| 101 | Hanger | C.S. | 1 | |
| 102 | Wheel | C.I. | 1 | |
| 103 | Stud | Steel | 1 | |
| 104 | Plate | Steel | 2 | |
| 105 | Eyebolt | D.F. | 1 | Purchase |

PROB. 95. Rail-transport hanger.

PROB. 96. Crane hook.

| PC.NO. | NAME | QUAN | NOTES |
|---|---|---|---|
| 1 | SIDE PLATE | 2 | SAE-1030-PLATE |
| 2 | PULLEY | 2 | " 1020-CASTING |
| 3 | PULLEY PIN | 2 | " 1045-FORGING |
| 4 | NUT LOCK PIN | 1 | " 1030-CR |
| 5 | CRANE HOOK | 1 | " 3140-FORGING |
| 6 | HOOK TRUNION | 1 | " 1045 " |
| 7 | SEMIFIN HEX NUT | 6 | M64-6 |
| 8 | HEX SLOTTED NUT | 1 | M80-6 |
| 9 | WASHER | 2 | M80-HEAVY |
| 10 | COTTER PIN | 2 | 3 × 25 |

| PC.NO. | NAME | MAT. | QUAN. | NOTES |
|---|---|---|---|---|
| 101 | Cylinder Head, Front | C.I. | 1 | |
| 102 | Cylinder Head, Rear | C.I. | 1 | Use Cast.101 |
| 103 | Packing Gland | C.I. | 1 | |
| 104 | Piston | C.I. | 1 | |
| 105 | Piston Rod | Steel | 1 | |
| 106 | Piston Plate | Steel | 2 | |
| 107 | Cylinder | Steel | 1 | Shelby Tubing |
| 108 | Tie Rod | Steel | 4 | |
| 109 | Stop Collar | Steel | 1 | |
| 110 | Piston Packing | Leath. | 2 | Purchase |
| 111 | Gasket | Fiber | 2 | Purchase |

PROB. 97. Double-acting air cylinder.

| PC. NO. | NAME | MAT. | QUAN. | NOTES |
|---|---|---|---|---|
| 101 | Bracket | C.I. | 1 | |
| 102 | Cylinder | C.I. | 1 | |
| 103 | Piston | C.I. | 1 | |
| 104 | Cylinder Head | C.I. | 1 | |
| 105 | Sleeve Stop | C.I. | 1 | |
| 106 | Sleeve | Steel | 1 | |
| 107 | Sleeve Nut | Steel | 1 | |
| 108 | Punch | Steel | 1 | |
| 109 | Punch Cap | Steel | 1 | |
| 110 | Packing Plate | Steel | 1 | |
| 111 | Packing | Leather | 1 | Purchase |
| 112 | Piston Rod | Steel | 1 | |
| 113 | Sleeve Spring | Steel | 1 | Purchase |
| 114 | Punch Spring | Steel | 1 | Purchase |

PROB. 98. Hydraulic punch.

PROB. 99. Belt tightener.

| PC. NO. | NAME | MAT. | QUAN. | NOTES |
|---|---|---|---|---|
| 101 | Base | C.I. | 1 | |
| 102 | Table | C.I. | 1 | |
| 103 | Trunnion Stud | Steel | 1 | |

PROB. 100. Swing table.

PROB. 101. Slide base.

PROB. 102. Idler bracket.

558

Stake at 2 places
each end

10 ø Drill × 11 Dp.
8 O.D. Spring × 25 lg.
8 Coils × .5 wire

5 × 12 × 10 I.D.
Split spring band

PROB. 103. Screw-driving attachment for electric drill.

3D

6 Hex

16  6  16

9

200

190

180

12.50

32D

6D

12.50

1.5

19  3  6  25  6  3  9  16

All chamfer 1 × 45°

Screw-driving attachment
for electric drill

Detail of dog clutch
Cut across end of each
clutch, 3 times equally spaced
using cutter of profile

Directions of
cutter travel

90°
¢

1.5

8

M24-3 Hex
Jam Nut,
Both ends

Oil Groove,
6 Wide x
1.5 Deep

83

6

Oil Pocket, 18 ø

Woodruff Key
#100B, Both
ends

38°

1.5R

20

.75R

25

175

49

6R

100

3

46

175

162

111

50

70

35

28

28

70.D

175

28

216.D

45

6 ø

54

28

[10058]-24T-6P

140

120.5

21.5

8

45

28

50

95.6

9.5

48

95.6

9.5

Oil Groove, 10 Wide x 1.5 Deep

M12-175 x 45 Hex Hd
Cap Scr, 4 Req.

17.5  17.5  8

23  5

56

PROB. 104. Belt drive.

45

27

64

102

15°

51

15°

PROB. 105. Inclined support.

PROB. 106. Propeller hub.

## Part A. Shop Terms

**anneal** (*v.*)  To soften a metal piece and remove internal stresses by heating to its critical temperature and allowing to cool very slowly.

**arc-weld** (*v.*)  To weld by electric-arc process.

**bore** (*v.*)  To enlarge a hole with a boring tool, as in a lathe or boring mill. Distinguished from *drill*.

**boss** (*n.*)  A projection of circular cross section, as on a casting or forging.

**braze** (*v.*)  To join by the use of hard solder.

**broach** (*v.*)  To finish the inside of a hole to a shape usually other than round.  (*n.*) A tool with serrated edges pushed or pulled through a hole to enlarge it to a required shape.

**buff** (*v.*)  To polish with abrasive on a cloth wheel or other soft carrier.

**burnish** (*v.*)  To smooth or polish by a rolling or sliding tool under pressure.

**bushing** (*n.*)  A removable sleeve or liner for a bearing; also a guide for a tool in a jig or fixture.

**carburize** (*v.*)  To prepare a low-carbon steel for heat-treatment by packing in a box with carbonizing material, such as wood charcoal, and heating to about 2000°F for several hours, then allowing to cool slowly.

**caseharden** (*v.*)  To harden the surface of carburized steel by heating to critical temperature and quenching, as in an oil or lead bath.

**castellate** (*v.*)  To form into a shape resembling a castle battlement, as castellated nut. Often applied to a shaft with multiple integral keys milled on it.

**chamfer** (*v.*)  To bevel a sharp external edge. (*n.*) A beveled edge.

**chase** (*v.*)  To cut threads in a lathe, as distinguished from cutting threads with a die. (*n.*) A slot or groove.

**chill** (*v.*)  To harden the surface of cast iron by sudden cooling against a metal mold.

**chip** (*v.*)  To cut or clean with a chisel.

**coin** (*v.*)  To stamp and form a metal piece in one operation, usually with a surface design.

Boss

Bushing

Chamfer

**cold-work** (*v.*)   To deform metal stock by hammering, forming, drawing, etc., while the metal is at ordinary room temperature.

**color-harden** (*v.*)   To caseharden to a very shallow depth, chiefly for appearance.

**core** (*v.*)   To form the hollow part of a casting, using a solid form made of sand, shaped in a core box, baked, and placed in the mold. After cooling, the core is easily broken up, leaving the casting hollow.

**counterbore** (*v.*)   To enlarge a hole to a given depth. (*n.*) 1. The cylindrical enlargement of the end of a drilled or bored hole. 2. A cutting tool for counterboring, having a piloted end the size of the drilled hole.

**countersink** (*v.*)   To form a depression to fit the conic head of a screw or the thickness of a plate so that the face will be level with the surface. (*n.*) A conic tool for countersinking.

**crown** (*n.*)   Angular or rounded contour, as on the face of a pulley.

**die** (*n.*)   1. One of a pair of hardened metal blocks for forming, impressing, or cutting out a desired shape. 2 (thread). A tool for cutting external threads. Opposite of *tap*.

**die casting** (*n.*)   A very accurate and smooth casting made by pouring a molten alloy (or composition, as Bakelite) usually under pressure into a metal mold or die. Distinguished from a casting made in sand.

**die stamping** (*n.*)   A piece, usually of sheet metal, formed or cut out by a die.

**draw** (*v.*)   1. To form by a distorting or stretching process. 2. To temper steel by gradual or intermittent quenching.

**drill** (*v.*)   To sink a hole with a drill, usually a twist drill. (*n.*) A pointed cutting tool rotated under pressure.

**drop forging** (*n.*)   A wrought piece formed hot between dies under a drop hammer, or by pressure.

**face** (*v.*)   To machine a flat surface perpendicular to the axis of rotation on a lathe. Distinguished from *turn*.

**feather** (*n.*)   A flat sliding key, usually fastened to the hub.

**fettle** (*v.*)   To remove fins and smooth the corners on unfired ceramic products.

**file** (*v.*)   To finish or trim with a file.

**fillet** (*n.*)   A rounded filling of the internal angle between two surfaces.

**fin** (*n.*)   A thin projecting rib. Also, excess ridge of material.

**fit** (*n.*)   The kind of contact between two machined surfaces. 1. *Drive, force,* or *press:* When the shaft is slightly larger than the hole and must be forced in with sledge or power press. 2. *Shrink:* When the shaft is slightly larger than the hole, the piece containing the hole is heated, thereby expanding the hole sufficiently to slip over the shaft. On cooling, the shaft will be seized firmly if the fit allowances have been correctly proportioned. 3. *Running* or *sliding:* When sufficient allowance has been made between sizes of shaft and hole to allow free running without seizing or heating. 4. *Wringing:* When the allowance is smaller than a running fit and the shaft will enter the hole by twisting the shaft by hand.

**flange** (*n.*)   A projecting rim or edge for fastening or stiffening.

**forge** (*v.*)   To shape metal while hot and plastic by a hammering or forcing process either by hand or by machine.

**galvanize** (*v.*)   To divide a scale or dial into regular spaces.

**grind** (*v.*)   To finish or polish a surface by means of an abrasive wheel.

Counterbore

Countersink

Fillet

Flange

Kerf

**harden** (*v.*)  To heat hardenable steel above critical temperature and quench in bath.

**hot-work** (*v.*)  To deform metal stock by hammering, forming, drawing, etc., while the metal is heated to a plastic state.

**kerf** (*n.*)  The channel or groove cut by a saw or other tool.

**key** (*n.*)  A small block or wedge inserted between shaft and hub to prevent circumferential movement.

**keyway, key seat** (*n.*)  A groove or slot cut to fit a key. A key fits into a key seat and slides in a keyway.

**knurl** (*v.*)  To roughen or indent a turned surface, as a knob or handle.

**lap** (*n.*)  A piece of soft metal, wood, or leather charged with abrasive material, used for obtaining an accurate finish. (*v.*) To finish by lapping.

**lug** (*n.*)  A projecting "ear," usually rectangular in cross section. Distinguished from *boss*.

**malleable casting** (*n.*)  An ordinary casting toughened by annealing. Applicable to small castings with uniform metal thicknesses.

**mill** (*v.*)  To machine with rotating toothed cutters on a milling machine.

**neck** (*v.*)  To cut a groove around a shaft, usually near the end or at a change in diameter. (*n.*) A portion reduced in diameter between the ends of a shaft.

**normalize** (*v.*)  To remove internal stresses by heating a metal piece to its critical temperature and allowing it to cool very slowly.

**pack-harden** (*v.*)  To carburize and caseharden.

**pad** (*n.*)  A shallow projection. Distinguished from *boss* by shape or size.

**peen** (*v.*)  To stretch, rivet, or clinch over by strokes with the peen of a hammer. (*n.*) The end of a hammer head opposite the face, as *ball peen.*

**pickle** (*v.*)  To clean castings or forgings in a hot weak sulfuric acid bath.

**plane** (*v.*)  To machine work on a planer having a fixed tool and reciprocating bed.

**planish** (*v.*)  To finish sheet metal by hammering with polished-faced hammers.

**plate** (*v.*)  The electrochemical coating of a metal piece with a different metal.

**polish** (*v.*)  To make smooth or lustrous by friction with a very fine abrasive.

**profile** (*v.*)  To machine an outline with a rotary cutter usually controlled by a master cam or die.

**punch** (*v.*)  To perforate by pressing a nonrotating tool through the work.

**ream** (*v.*)  To finish a drilled or punched hole very accurately with a rotating fluted tool of the required diameter.

**relief** (*n.*)  The amount one plane surface of a piece is set below or above another plane, usually for clearance or for economy in machining.

**rivet** (*v.*)  1. To fasten with rivets. 2. To batter the headless end of a pin used as a permanent fastening.

**round** (*n.*)  A rounded exterior corner between two surfaces. Compare with *fillet*.

**sandblast** (*v.*)  To clean castings or forgings by means of sand driven through a nozzle by compressed air.

Keyway

Key and seat

Lug

Neck

Pad

Round

Spline

Spot-face

Trepan          Undercut

**shape** (*v.*) To machine with a shaper, a machine tool differing from a planer in that the work is stationary and the tool reciprocating.

**shear** (*v.*) To cut off sheet or bar metal between two blades.

**sherardize** (*v.*) To galvanize with zinc by a dry heating process.

**shim** (*n.*) A thin spacer of sheet metal used for adjusting.

**shoulder** (*n.*) A plane surface on a shaft, normal to the axis and formed by a difference in diameter.

**spin** (*v.*) To shape sheet metal by forcing it against a form as it revolves.

**spline** (*v.*) A long keyway. Sometimes also a flat key.

**spot-face** (*v.*) To finish a round spot on a rough surface, usually around a drilled hole, to give a good seat to a screw or bolthead, cut, usually $\frac{1}{16}$ in. deep, by a rotating milling cutter.

**spot-weld** (*v.*) To weld in spots by means of the heat of resistance to an electric current. Not applicable to sheet copper or brass.

**steel casting** (*n.*) Material used in machine construction. It is ordinary cast iron into which varying amounts of scrap steel have been added in the melting.

**swage** (*v.*) To shape metal by hammering or pressure with the aid of a form or anvil called a "swage block."

**sweat** (*v.*) To join metal pieces by clamping together with solder between and applying heat.

**tack-weld** (*v.*) To join at the edge by welding in short intermittent sections.

**tap** (*v.*) To cut threads in a hole with a rotating tool called a "tap," having threads on it and fluted to give cutting edges.

**temper** (*v.*) To change the physical characteristics of hardened steel by reheating to a temperature below the critical point and allowing to cool.

**template, templet** (*n.*) A flat pattern for laying out shapes, location of holes, etc.

**trepan** (*v.*) To cut an outside annular groove around a hole.

**tumble** (*v.*) To clean, smooth, or polish castings or forgings in a rotating barrel or drum by friction with each other, assisted by added mediums, as scraps, "jacks," balls, sawdust, etc.

**turn** (*v.*) To machine on a lathe. Distinguished from *face*.

**undercut** (*v.*) To cut, leaving an overhanging edge. (*n.*) A cut having inwardly sloping sides.

**upset** (*v.*) To forge a larger diameter or shoulder on a bar.

**weld** (*v.*) To join two pieces by heating them to the fusing point and pressing or hammering together.

## Part B. Structural Terms

**bar** Square or round rod; also flat steel up to 6 in. in width.

**batten plate** A small plate used to hold two parts in their proper position when made up as one member.

**batter** A deviation from the vertical in upright members.

**bay** The distance between two trusses or transverse bents.

**beam** A horizontal member forming part of the frame of a building or structure.

**bearing plate** A steel plate, usually at the base of a column, used to distribute a load over a larger area.

**bent** A vertical framework usually consisting of a truss or beam supported at the ends on columns.

**brace** A diagonal member used to stiffen a framework.

**buckle plate** A flat plate with dished depression pressed into it to give transverse strength.

**builtup member** A member built from standard shapes to give one single stronger member.

**camber**  Slight upward curve given to trusses and girders to avoid effect of sag.

**cantilever**  A beam, girder, or truss overhanging one or both supports.

**chord**  The principal member of a truss on either the top or bottom.

**clearance**  Rivet driving clearance is distance from center of rivet to obstruction. Erection clearance is amount of space left between members for ease in assembly.

**clevis**  U-shaped shackle for connecting a rod to a pin.

**clip angle**  A small angle used for fastening various members together.

**column**  A vertical compression member.

**cope**  To cut out top or bottom of flanges and web so that one member will frame into another.

**coping**  A projecting top course of concrete or stone.

**counters**  Diagonal members in a truss to provide for reversal of shear due to live load.

**cover plate**  A plate used in building up flanges, in a builtup member, to give greater strength and area or for protection.

**crimp**  To offset the end of a stiffener to fit over the leg of an angle.

**diagonals**  Diagonal members used for stiffening and wind bracing.

**dowel**  An iron or wooden pin extending into, but not through, two timbers to connect them.

**driftpin**  A tapered steel pin used to bring rivet holes fair in assembling steel work.

**edge distance**  The distance from center of rivet to edge of plate or flange.

**fabricate**  To cut, punch, and subassemble members in the shop.

**fillers**  Plate or ring fills used to take up space in riveting two members where a gusset is not used.

**flange**  The projecting portion of a beam, channel, or column.

**gage line**  The center line for rivet holes

**gin pole**  A guyed mast with block at the top for hoisting.

**girder**  A horizontal member, either single or built up, acting as a principal beam.

**girt**  A beam usually bolted to columns to support the side covering or serve as window lintels.

**gusset plate**  A plate used to connect various members, such as in a truss.

**hip**  The intersection between two sloping surfaces forming an exterior angle.

**knee brace**  A corner brace used to prevent angular movement.

**lacing or lattice bars**  Bars used diagonally to space and stiffen two parallel members, such as in a builtup column.

**laterals**  Members used to prevent lateral deflection.

**lintel**  A horizontal member used to carry a wall over an opening.

**louvers**  Metal slats either movable of fixed, as in a monitor ventilator.

**monitor ventilator**  A framework that carries fixed or movable louvers at the top of the roof.

**panel**  The space between adjacent floor supports or purlins in a roof.

**pitch**  Center distance between rivets parallel to axis of member. Also, for roofs, the ratio of rise to span.

**plate**  Flat steel over 6 in. in width and $\frac{1}{4}$ in. or more in thickness.

**purlins**  Horizontal members extending between trusses, used as beams for supporting the roof.

**rafters**  Beams or truss members supporting the purlins.

**sag ties**  Tie rods between purlins in the plane of the roof to carry the component of the roof load parallel to the roof.

**separator**  Either a cast-iron spacer or wrought-iron pipe on a bolt for the purpose of holding members a fixed distance apart.

**sheet**  Flat steel over 6 in. in width and less than $\frac{1}{4}$ in. in thickness.

**shim**  A thin piece of wood or steel placed under a member to bring it to a desired elevation.

**sleeve nut**  A long nut with right and left threads for connecting two rods to make an adjustable member.

**span**  Distance between centers of supports of a truss, beam, or girder.

**splice**   A longitudinal connection between the parts of a continuous member.

**stiffener**   Angle, plate, or channel riveted to a member to prevent buckling.

**stringer**   A longitudinal member used to support loads directly.

**strut**   A compression member in a framework.

**truss**   A rigid framework for carrying loads, formed in a series of triangles.

**turnbuckle**   A coupling, threaded right and left or swiveled on one end, for adjustably connecting two rods.

**valley**   The intersection between two sloping surfaces, forming a reentrant angle.

**web**   The part of a channel, I beam, or girder between the flanges.

## Part C. Architectural Terms

**apron**   The finished board placed against the wall surface, immediately below a window stool.

**ashlar**   Thin, squared, and dressed stone facing of a masonry wall.

**backing**   The inner portion of a wall—that which is used behind the facing.

**batten**   A strip of wood used for nailing across two other pieces of wood to hold them together and cover a crack.

**batter boards**   Boards set up at the corners of a proposed building from which the lines marking off the walls are stretched.

**bearing wall**   A wall that supports loads other than its own weight.

**bond**   The joining together of building materials to ensure solidity.

**bridging**   The braces or system of bracing used between joists or other structural members to stiffen them and to distribute the load.

**centering**   A substructure of temporary nature, usually of timber or planks, on which a masonry arch or vault is built.

**coffer**   An ornamental panel deeply recessed, usually in a dome or portico ceiling.

**corbel**   A bracket formed on a wall by building out successive courses of masonry.

**curtain wall**   A wall that carries no building load other than its own weight.

**fenestration**   The arrangement and proportioning of window and door openings.

**flashing**   The sheet metal built into the joints of a wall, or covering the valleys, ridges, and hips of a roof for the purpose of preventing leakage.

**footing**   A course or series of courses projecting at the base of a wall for the purpose of distributing the load from above over a greater area, thereby preventing excessive settlement.

**furring**   The application of thin wood, metal, or other building material to a wall, beam, ceiling, or the like to level a surface for lathing, boarding, etc., or to make an air space within a wall.

**glazing**   The act of furnishing or fitting with glass.

**ground**   Strips of wood, flush with the plastering, to which moldings, etc., are attached. Grounds are usually installed first and the plastering floated flush with them.

**grout**   A thin mortar used for filling up spaces where heavier mortar will not penetrate.

**head**   The horizontal piece forming the top of a wall opening, as a door or window.

**hip**   The intersection of two roof surfaces, which form an external angle on the plan.

**jamb**   The vertical piece forming the side of a wall opening.

**lintel**   The horizontal structural member that supports the wall over an opening.

**millwork**   The finish wordwork, machined and in some cases partly assembled at the mill.

**miter**   To match together, as two pieces of molding, on a line bisecting the angle of junction.

**mullion**   A vertical division of a window opening.

**muntin**   The thin members that separate the individual lights of glass in a window frame.

**party wall**   A division wall common to two adjacent pieces of property.

**plate**   A horizontal member that carries other structural members; usually the top timber of a wall that carries the roof trusses or rafters directly.

**rail**   A horizontal piece in a frame or paneling.

**return**   The continuation in a different direction, most often at right angles, of the face of a building or any member, as a colonnade or molding; applied to the shorter in contradistinction to the longer.

**reveal** The side of a wall opening; the whole thickness of the wall; the jamb.

**riser** The upright piece of a step, from tread to tread.

**saddle** A small double-sloping roof to carry water away from behind chimneys, etc.

**scratch coat** The first coat in platering, roughened by scratching or scoring so that the next coat will firmly adhere to it.

**screeds** A strip of plaster of the thickness proposed for the work, applied to the wall at intervals of 4 or 5 ft, to serve as guides.

**shoring** A prop, as a timber, placed against the side of a structure; a prop placed beneath anything, as a beam, to prevent sinking or sagging.

**sill** The horizontal piece, as a timber, which forms the lowest member of a frame.

**sleepers** The timbers laid on a firm foundation to carry and secure the superstructure.

**soffit** The underside of subordinate parts and members of buildings, such as staircases, beams, arches, etc.

**stile** A vertical piece in a frame or paneling.

**stool** The narrow shelf fitted on the inside of a window against the actual sill.

**threshold** The stone, wood, or metal piece that lies directly under a door.

**trap** A water seal in a sewage system to prevent sewer gas from entering the building.

**tread** The upper horizontal piece of a step, on which the foot is placed.

**valley** The intersection of two roof surfaces which form, on the plan, a reentrant angle.

## Part D. Welding Terms

**air-acetylene welding** A gas-welding process in which coalescence is produced by heating with a gas flame or flames obtained from the combustion of acetylene with air, without the application of pressure and with or without the use of filler metal.

**arc cutting** A group of cutting processes in which the severing of metals is effected by melting with the heat of an arc between an electrode and the base metal.

**arc welding** A group of welding processes in which coalescence is produced by heating with an electric arc or arcs, with or without the application of pressure and with or without the use of filler metal.

**As-welded** The condition of weld metal, welded joints, and weldments after welding prior to any subsequent thermal or mechanical treatment.

**atomic-hydrogen welding** An arc-welding process in which coalescence is produced by heating with an electric arc maintained between two metal electrodes in an atmosphere of hydrogen. Shielding is obtained from the hydrogen. Pressure may or may not be used, and filler metal may or may not be used.

**automatic welding** Welding with equipment which performs the entire welding operation without constant observation and adjustment of the controls by an operator. The equipment may or may not perform the loading and unloading of the work.

**axis of a weld** A line through the length of a weld, perpendicular to the cross section at its center of gravity.

**backing** Material (metal, weld metal, asbestos, carbon, granular flux, etc.) backing up the joint during welding to facilitate obtaining a sound weld at the root.

**bevel** A type of edge preparation.

**braze** A weld in which coalescence is produced by heating to suitable temperatures above 800°F and by using a nonferrous filler metal having a melting point below that of the base metals. The filler metal is distributed between the closely fitted surfaces of the joint by capillary attraction.

**butt joint** A joint between two members lying approximately in the same plane.

**coalesce** To unite or merge into a single body or mass. Fusion.

**die welding** A forge-welding process in which coalescence is produced by heating in a furnace and by applying pressure by means of dies.

**edge joint**   A joint between the edges of two or more parallel or nearly parallel members.

**filler metal**   Metal to be added in making a weld.

**fillet weld**   A weld of approximately triangular cross section joining two surfaces approximately at right angles to each other in a lap joint, tee joint, or corner joint.

**forge welding**   A group of welding processes in which coalescence is produced by heating in a forge or other furnace and by applying pressure or blows.

**fusion**   The melting together of filler metal and base metal, or of base metal only, which results in coalescence.

**gas welding**   A group of welding processes in which coalescence is produced by heating with a gas flame or flames, with or without the application of pressure, and with or without the use of filler metal.

**groove weld**   A weld made in the groove between two members to be joined. The standard types of groove welds are: square-groove weld; single-V-groove weld; single-bevel-groove weld; single-U-groove weld; single-J-groove weld; double-V-groove weld; double-bevel-groove weld; double-U-groove weld; double-J-groove weld.

**hammer welding**   A forge-welding process in which coalescence is produced by heating in a forge or other furnace and by applying pressure by means of hammer blows.

**intermittent welding**   Welding in which the continuity is broken by recurring unwelded spaces.

**joint penetration**   The minimum depth a groove weld extends from its face into a joint, exclusive of reinforcement.

**kerf**   The space from which metal has been removed by a cutting process.

**lap joint**   A joint between two overlapping members.

**oxy-acetylene welding**   A gas-welding process in which coalescence is produced by heating with a gas flame or flames obtained from a combustion of acetylene with oxygen, with or without the application of pressure and with or without the use of filler metal.

**peening**   The mechanical working of metals by means of hammer blows.

**plug weld**   A circular weld made by either arc or gas welding through one member of a lap or tee joint joining that member to the other. The weld may or may not be made through a hole in the first member.

**pressure welding**   Any welding process or method in which pressure is used to complete the weld.

**projection welding**   A resistance-welding process in which coalescence is produced by the heat obtained from resistance to the flow of electric current through the work parts held together under pressure by electrodes.

**root opening**   The separation between the members to be joined, at the root of the joint.

**seam weld**   A weld consisting of a series of overlapping spot welds, made by seam welding or spot welding.

**slot weld**   A weld made in an elongated hole in one member of a lap or tee joint joining that member to the portion of the surface of the other member which is exposed through the hole.

**spot welding**   A resistance-welding process in which coalescence is produced by the heat obtained from resistance to the flow of electric current through the work parts, which are held together under pressure by electrodes.

**tack weld**   A weld made to hold parts of a weldment in proper alignment until the final welds are made.

**tee joint**   A joint between two members located approximately at right angles to each other in the form of a T.

**upset welding**   A resistance-welding process in which coalescence is produced, simultaneously over the entire area of abutting surfaces or progressively along a joint, by the heat obtained from resistance to the flow of electric current through the area of contact of those surfaces. Pressure is applied before heating is started and is maintained throughout the heating period.

# APPENDIX

## The Hand Calculator

### 1. Introduction.

Early engineers were forced to make calculations by long-hand methods with some assistance from the slide rule. The situation was not aided by the fact that most slide rules could not calculate values of more than three or four digits. Such feats as landing a spacecraft on the moon would not have been possible if slide rules were still our primary calculating tool.

### 2. Mechanical Calculators.

For many years, mechanical calculators have been available to perform addition, subtraction, multiplication, and division of numbers with twelve or more digits. Such calculators were relatively slow, quite expensive, and about the size of an office typewriter. Complex mathematical entities could be handled only by entering the proper equation values from a handbook. Names such as Friden, Marchant, and Monroe are associated with such calculators.

### 3. Digital Computers.

Any type of engineering calculation can be easily solved using a digital computer. For example, the calculations required to define the orbit of a space-

craft can now be handled easily because the computer can make millions of calculations per second.

### 4. Scientific or Engineering Calculators.

The development of integrated circuits has made it possible to produce small computers known as minicomputers, microcomputers, microprocessors, or hand calculators. Descriptions of how calculators actually perform their operations is beyond the scope of this appendix. However, calculators are basically small preprogrammed or programmable computers.

The calculators most suitable for engineering calculations are usually described as scientific calculators. (See Figs. 1 to 3.) Scientific calculators usually include trigonometric functions, logarithms, roots of numbers, powers of numbers, built-in constants like pi, and other operations suitable for engineering calculations. However, even the simple four-function calculator is of such value that it can be considered a suitable device for beginning students.

### 5. Calculator Accuracy.

Each manufacturer of a calculator designs its operating circuits to solve the desired functions with a certain accuracy in a reasonable time. The more expensive calculators usually have more digits of accuracy and solve functions in less time. Cheaper calculators often seem faster than the more expensive ones because they have fewer digits in the answer. Repeated calculations will accumulate errors more rapidly on a calculator with fewer numbers of accurate digits than a model with a higher number of accurate digits. A calculator of less than five-digit accuracy would be inappropriate for scientific or engineering calculations.

## 6. Number Representation.

The internal form of numbers used by a calculator is far different from the form actually displayed. This only affects the user if he or she tries to use a calculator for a purpose that the designers did not intend it for. Most scientific calculators use floating decimal point notation. This means that there is not a fixed number of fractional digits in an answer unless the answer is represented by exponential form. The number 131.1 will not be displayed as

**FIG. 1** Hewlett-Packard's HP-45 battery-operated scientific calculator.

**FIG. 2** Texas Instruments SR-51 scientific calculator.

**FIG. 3** A typical scientific function keyboard. Note the dual function keys for trigonometric functions. The secondary function printed above each key is invoked by first pushing the ARC key and then the desired function key. Note the DEG-RAD (degrees-radians) selector switch.

131.10 or 131.100 despite the fact that the machine might be listed as a five- or six-digit machine. To the user doing scientific calculations this is eminently satisfactory. However, for a user doing banking problems, the inability to show dollars and cents is annoying.

Calculators can display as many numbers as they have in their capacity. For example, a typical eight-digit calculator will display all of the whole number digits in a number, as long as that number does not exceed eight digits. For all whole numbers of less than eight digits, any fractional digits that exist will be displayed to a maximum display size of eight digits. When a number to be displayed or entered exceeds eight whole number digits, the calculator uses the exponential mode. For example, an eight-digit display would now show five digits of the number plus the two-digit power-of-ten value required to give the number the proper magnitude. The number 1,575,671,283,916 has too many digits to be represented by an eight-digit display. It would be displayed as:

$$1.5757 \quad 12$$

This is the calculator's way of showing scientific notation which says the number is:

$$1.5757 \times 10^{12}$$

This notation usually permits calculations using numbers of up to $10^{99}$. Figure 4 shows an eight-digit calculator representing an eight-digit and a nine-digit number. The jump to exponential form is noticeable.

Many calculators carry more digits in their internal calculations than can be entered or displayed, so the displayed value is usually quite accurate. An eight-digit machine might do its calculations in ten- or eleven-digit form. This is especially true of the more expensive calculators which can be said to display answers of almost 100 per cent accuracy. Doubtful digits at the end of a number are never displayed.

## 7. Choosing a Calculator.

There are several items that a buyer of a scientific calculator should examine:

**FIG. 4**   A calculator display changing from regular notation to exponential notation as more digits are entered than can be shown. This is an eight-digit display that changes to exponential when a ninth digit is entered.

1. *Display.*

   *a.* Does the display have enough digits to satisfy your needs? Less than five is not recommended.

   *b.* Displays are either red or blue-green. In general, the construction of the red display produces smaller digits than that of the green display. The red display places the digits in a kind of cave, which makes the red digits readable only from a straightforward position.

2. *Calculator functions.* Will the calculator handle the functions you need? See Sec. 9 on calculator functions to determine some of the available functions. Be especially careful on the $Y^x$. In many calculators, the $x$th root of $Y$ can be found only by first calculating $P = 1/x$ and then calculating $Y^P$. This represents extra steps and the chance that you will inadvertently calculate the $x$th power of $Y$ instead of the $x$th root of $Y$. Hyperbolic sines and cosines are often omitted from the less expensive scientific calculator. These functions (sinh and cosh) can be calculated using the $e^x$ function if needed only occa-

sionally. Currently, the inclusion of sinh and cosh in a calculator raises its price to as much as double that of other full-function calculators.

3. *Calculator logic.* To the user, calculator logic defines the sequence in which numbers and operations must be entered in order to obtain the correct answer. The three major forms of logic are algebraic, algebraic with hierarchy, and reverse Polish. Section 8 describes the various types of logic and gives some examples of how logic differences affect the user.

Generally speaking, calculators with algebraic logic and parenthesis control solve problems in the order the numbers and operations appear in a problem. Operators of calculators using reverse Polish notation logic must rearrange the numbers and operations before entering them into the calculator. Since few calculators can handle enough parentheses to solve all problems, algebraic logic with parenthesis control is not enough reason to select that type of calculator. However, reverse Polish logic calculators seldom have enough memory in the reverse Polish stack to solve many problems without some juggling of numbers and operations; therefore, that type of logic also has its flaws. Never choose a calculator without trying it on practical problems. Avoid the problems generally used in advertising the calculators, since these problems are specifically chosen to prove the superiority of the machine's type of logic.

4. *Keyboard.* Various calculator manufacturers have different ideas about the entry keys on the keyboard. Two major points about entry keys should be noted, and the user should determine if either one would be annoying to use.

Many calculators use double- or even triple-function keys. The primary function is obtained by depressing the function key directly. The secondary function is obtained by first depressing a function shift key and then the proper function key. Usually the primary function is marked on the key and the secondary function is marked on the face plate directly above the key. The only problem with this type of operation is that a person who is prone to forget details may forget to depress the function

shift key and get the primary operations instead of the desired secondary operation. Triple-function keys are rare but represent a noticeable increase in complexity of operation.

Some calculators have operation keys with an "oil can" spring which creates a distinct snap which can be felt in the fingertip as the key is depressed. This type of key cannot be depressed a second time unless it is first completely released; a second snap is felt upon a second depression. However, many calculators do not have this feature, and the only way an operator can be sure that a digit has been entered is to watch the display. This can be time-consuming. In addition, because an insufficient depression of the key will not activate the entry, it is possible to fail to enter a number and not notice it. A second quirk of this type of key is that it is easy to enter two or more items upon one depression of the key. A nervous person or one who has been distracted can depress the key in a kind of stuttering action. Each "stutter" enters the digit once.

## 8. Calculator Logic.

The specific logic of a calculator affects the user only by requiring that the numbers and operations be entered in a specific order to produce the correct answers. Each method of logic has its adherents among both manufacturers and users. There is no *right* method since each logic method produces the correct answers when correctly used. The majority of calculators use a straight algebraic logic. In this logic numbers and operations are ordered the way they would be written down in long-hand calculations. Numbers and operations need not and cannot be stacked or held back once they have been entered into the calculator.

To illustrate this point use the example:

$$X = 1.45 + 7.28 \div 2.961$$

The entry of this problem as it appears from left to right would give the following results for each operation:

| Operation | Display | Action |
|---|---|---|
| 1. Enter 1.45. | 1.45 | |
| 2. Depress +. | 1.45 | |

3. Enter 7.28.   7.28
4. Depress ÷.   8.73          1.45 and 7.28
                              have now been added.
5. Enter 2.961.  2.961
6. Depress =.   2.9483282  8.73 has been
                              divided by
                              2.961.

While this answer is correct, it is correct for a problem which would normally have been written as:

$$X = (1.45 + 7.28) \div 2.961$$

But the problem which we solved could just as easily be interpreted as:

$$X = 1.45 + (7.28 \div 2.961)$$

Depending upon which of these interpretations one uses on the problem, the straight entry from left to right of the items in the problems does or does not give the proper answer with straight algebraic logic. If we are to interpret the problem by the second interpretation, the order of entry must be something like this:

| Operation | Display | Action |
|---|---|---|
| 1. Enter 7.28. | 7.28 | |
| 2. Depress ÷. | 7.28 | |
| 3. Enter 2.961. | 2.961 | |
| 4. Depress +. | 2.4586288 | 7.28 divided by 2.961. |
| 5. Enter 1.45. | 1.45 | |
| 6. Depress =. | 3.9086288 | 2.4586288 added to 1.45. |

In algebraic logic the operation entered will be performed as soon as the *next* function key is depressed. Calculators which operate like this are also said to have *chain calculations* because the = key is *not* needed to cause operations except as the last operation of the problem. Some early calculators needed a depression of the execute (=) key for each operation entered. Obviously, considerable rearrangement of problems may be needed if the calculator uses straight algebraic logic. The problem has been resolved in more recent and more expensive calculators by adding parenthesis control to straight algebraic logic. This is discussed in later paragraphs.

Some calculators, notably most models from Texas Instruments, use algebraic logic with hierarchy. In this mode, the operation or function just entered into the calculator may be held in abeyance or stacked if the next entry is of higher priority. The Texas Instruments models assign ÷ and × equally as having a higher priority than + and −. Now let's do that example problem again.

$$X = 1.45 + 7.28 \div 2.961$$

| Operation | Display | Action |
|---|---|---|
| 1. Enter 1.45 | 1.45 | |
| 2. Depress +. | 1.45 | |
| 3. Enter 7.28. | 7.28 | |
| 4. Depress ÷. | 7.28 | Stack 1.45 +. |
| 5. Enter 2.961. | 2.961 | |
| 6. Depress =. | 3.9086288 | Execute 7.28 ÷ 2.961, then unstack and add 1.45 to that answer. |

This calculator tried to calculate the answer as though we wanted the second interpretation of the problem.

If the problem had been written as:

$$7.28 \div 2.961 + 1.45$$

both types of calculators would have produced the same answer, 3.9086288.

Writing the problem as $(1.45 + 7.28) \div 2.961$ and executing the parentheses would cause the answer to be 2.9483282. Thus the use of parenthesis control permits the user to control the order in which operations are performed without changing the order in which the arithmetic functions are entered. Calculators differ in the number of open left parentheses which can be placed in a problem before a right parenthesis or = must be used. Thus a calculator which has only two levels of parentheses could handle the problem:

$$1.25 + (6.34 \times (9.83 + .87))$$

but not the problem:

$$1.25 + (6.34 \times (9.38 + (5.41 \div 2.87)))$$

The latter problem would cause the first (+), second (×) and third (+) operations to be stacked until the ÷ operation is performed. Since only two operations can be stacked with a calculator having two levels of parenthesis control, depressing the third left parenthesis would cause either an error indication on the display or, in some calculators, a truncating of the first operation and left parenthesis.

Calculators vary in the number of parenthesis levels which can be handled. Two to ten levels seem to be common numbers for the more popular moderately priced calculators.

Many Texas Instruments calculators include algebraic logic with both hierarchy and parenthesis control so that operations can be stacked either by hierarchy or by parenthesis control. However, the total number of stacked operations would be the same whether hierarchy or parentheses were used to force the stacking. This ability to stack operations is a major characteristic of the scientific calculators, as opposed to the less versatile slide rule calculators.

The third major method uses a type of logic developed to assist mathematicians in determining the order of operations in a long equation. The logic is also used by computers in determining the order for performing operations in a computer program equation. The name of the system is reverse Polish notation. Hewlett-Packard calculators are a major brand that uses this type of logic. Here, numbers to be used in the problem are entered first, then the operations in the order they are to be used. In the example we have been using, $1.45 + 7.28 ÷ 2.961$, a calculator using reverse Polish notation would have the three numbers entered into a stack as follows:

Enter 1.45.
Enter 7.28.
Enter 2.961.

If we were to look at the numbers stored in the stack, we would find that the first number entered is now at the bottom of the stack and that the last is at the top of the stack:

2.961
7.28
1.45

Since we desire to have the ÷ operation performed before the + operation, we would enter the ÷. The calculator would take the top two numbers in the stack, 2.961 and 7.28, and divide the lower one, 7.28, by the upper one, 2.961. The answer to this operation, 2.4586288, would now be placed back into the stack, replacing both of the numbers used in this operation.

2.4586288
1.45

If we now enter the + operation, the top two numbers in the stack are added together, and the answer, 3.9086288, replaces both of these numbers in the stack.

It can be seen that the ability to use this logic without setting up some intermediate answers is directly related to the amount of information that can be stacked. Most Hewlett-Packard calculators can stack four levels of operations.

This form of logic is totally satisfactory to those persons who can transform the normal algebraic equation into the reverse Polish notation form, but many individuals have some trouble with this. No one should buy a reverse Polish notation machine without using it on a trial basis to see if the logic bothers them. The problem

$$1.45 + 2.76$$

would be solved as follows in algebraic logic and reverse Polish notation logic:

| *Algebraic* | *RPN* |
|---|---|
| 1. Enter 1.45. | 1. Enter 1.45. |
| 2. Depress +. | 2. Depress ENTER. |
| 3. Enter 2.76. | 3. Enter 2.76. |
| 4. Depress =. | 4. Depress +. |

The answer 4.21 would appear in the display of both machines.

## 9. Operations and Functions.
The operations and functions of calculators seem to be added to the machines in somewhat the order to

be listed here. The cheaper, less versatile machines will have operations only from the first part of the list, while the more expensive, more versatile machines will have operations from a much greater part of the list. Full scientific calculators will usually have all items in the list plus some special extras peculiar to the individual manufacturer.

*a*  All calculators in normal use have an eleven-key number-entering keyboard consisting of the digits 0 through 9 and a decimal point key. Digits are entered in left to right order from the problem.

*b*  The standard four arithmetic functions are included in all calculators. In most cases, five keys operate these functions: add ($+$), subtract ($-$), multiply ($\times$), divide ($\div$), and execute ($=$).

*c*  The change sign key reverses the algebraic sign of the entered number without causing any other operations to occur. It is often marked ($+/-$). Normally, the subtract key will not put a negative sign in front of an entered number but will attempt to subtract it from something.

*d*  A percent key (%) appears on many calculators. It usually has one of two actions. In some cases depressing the key changes the position of the decimal point in the displayed number to a position two places to the left. Thus the entry 10.82, which represents 10.82 per cent, would be changed to .1082 if the % key were depressed under this system. This permits entering numbers as percent values while changing them to actual fractional values for the calculations.

In some cases, the % key is a special multiplier key which permits entering the multiplier as per cent value instead of decimal value. Multiplying 34.5 by 15.86 per cent would be done as follows under the two systems:

| *First case* | *Second case* |
|---|---|
| 1. Enter 34.5. | 1. Enter 34.5. |
| 2. Depress $\times$. | 2. Depress %. |
| 3. Enter 15.86. | 3. Enter 15.86. |
| 4. Depress $=$. | 4. Depress $=$. |
| 5. Depress %. | |

The same answer, 5.4717, would appear on both displays.

*e*  A reciprocal key ($1/X$) is used to obtain the re-ciprocal of a number in the display of the machine. The normal sequence to use is:

1. Enter number.
2. Depress $1/X$.

*f*  A squaring key ($X^2$) is used to shorten the process of squaring a number. In some calculators a number can be squared by the sequence:

1. Enter number.
2. Depress $\times$.
3. Depress $=$.

In a calculator with an $X^2$ key, the sequence for squaring a number would be:

1. Enter number.
2. Depress $X^2$.

*g*  A square root key ($\sqrt{\phantom{x}}$) is used to extract the square root of an entered number. The sequence to use is:

1. Enter number.
2. Depress $\sqrt{\phantom{x}}$.

*h*  An exponential entry key (EE) appears on most scientific machines. This is a necessity if numbers larger than the maximum display size are to be handled. Basically the operation is as follows: if the display is an eight-digit display, numbers larger than eight digits cannot be entered into the calculator except by the EE notation. The eight-digit display now becomes a five-digit display (sometimes called the mantissa) *plus* a two-digit display for a power of 10 which scales the total number. The number 3,145,678,912 could not be entered in an eight-digit display. The number would be entered as follows:

1. Enter 31456 or 31457 depending on whether you wish to round or not.
2. Depress EE.
3. Enter 5.

The number entered reads on the display as 31457 05, denoting that the correct number in the calculator is really $31457 \times 10^5$.

Very small numbers can be entered by using negative powers of 10. The number .0000005673 could also be expressed as $5.673 \times 10^{-7}$. This would be entered into the calculator as follows:

1. Enter 5.673.
2. Depress EE.
3. Depress 7.
4. Depress $+/-$.

The number in the display now reads 5.673 $-07$. Most EE keys also perform other minor operations, but they differ with each calculator and no general rules can be stated.

*i* Logarithms to the base 10 and the base *e* are available using keys marked LOG and LN. Because the calculation of the logarithms requires considerable calculation, the answer usually takes a noticeable amount of time before it appears. Since logarithms are not defined for negative numbers, an attempt to obtain the logarithm of a negative number will cause some sort of error notation to appear in the display.

*j* Antilogarithms are obtained by evaluating $10^x$ or $e^x$ where $x$ is the logarithm to the base 10 or the base *e*. Many calculators have separate keys for these functions. However, some calculators—Texas Instruments models, for example—use a key which reverses the action of the LOG or LN keys. This is the INV key. Obtaining the antilogarithm in base 10 to the number .30103 would proceed as follows under the two types of operations:

*Regular action*
1. Enter .30103.
2. Depress $10^x$.

*Reverse action*
1. Enter .30103.
2. Depress INV.
3. Depress LOG.

In both operations, the answer in the display is 2.

*k* Numbers can be raised to any power by using logarithms, but most calculators which have logarithms also have a $Y^x$ key for raising $Y$ to the $x$th power. The operating sequence is usually:

1. Enter number $Y$.
2. Depress $Y^x$.
3. Enter power $x$ (may be a fractional value).
4. Depress $=$.

Some calculators can perform the reverse operation of extracting the $x$th root of $Y$ by using the INV key. Other calculators without an INV key but with a parenthesis control can perform the operation in a much different way. Calculators without reverse action or parenthesis control but with a memory can use a third method. The three methods can be contrasted as follows for the problem of finding the 2.7th root of the number 45.67:

| *Reverse action* | *Parenthesis control* | *Memory only* |
|---|---|---|
| 1. Enter 45.67. | 1. Enter 45.67. | 1. Enter 2.7. |
| 2. Depress INV. | 2. Depress $Y^x$. | 2. Depress $1/X$. |
| 3. Depress $Y^x$. | 3. Depress (. | 3. Store in memory. |
| 4. Depress 2.7. | 4. Enter 2.7. | 4. Enter 45.67. |
| 5. Depress $=$. | 5. Depress $1/X$. | 5. Depress $Y^x$. |
| | 6. Depress ) or $=$. | 6. Recall memory. |
| | | 7. Depress $=$. |

It is fairly obvious which method is most efficient, but it is just as obvious that any calculator with one of the three arrangements of functions will satisfactorily compute the $x$th root of $Y$ without writing down any intermediate answers.

*l* Trigonometric functions are usually the sine, cosine, and tangent of an angle. The internal calculating hardware of the machine requires that the angle be measured in radians (2 radians = 360 degrees), but most calculators permit entries in either radians or degrees. Most calculators have switches indicating the mode of entry. If the switch is set to degrees, the entry must be made in degrees; if set to radians, it must be made in radians. Occasionally a calculator has a special function key which permits reversing the entry mode without changing the entry mode switch. To find the sine, cosine, or tangent of an angle:

1. Select the mode of entry as radians or degrees.
2. Set the entry mode switch to correspond.
3. Enter angle value.
4. Depress function key (SIN, COS, or TAN).

The arc trigonometric functions return the angle whose sine, cosine, or tangent is entered. The answers are displayed in the units set by the entry mode switch discussed in the trigonometric functions paragraph. Almost all calculators use either a INV key or second function key to perform the arc functions. Sometimes marked as the ARC key, the INV key merely reverses the action of the SIN, COS, or TAN keys. The second function key is used to choose a second function for which a key can be utilized. For example, if SIN is marked on the key itself, the sine function is the first or primary function. Directly above or below the key, the faceplate of the calculator might be marked $SIN^{-1}$. The second function of this key is, then, the arc sine. The key which chooses the second function is often marked 2ND, F, or FUNC. The process of obtaining the angle whose sine, cosine, or tangent is known would be:

1. Choose form of angular measurement desired for answer (radians or degrees).
2. Set entry mode switch to correspond.
3. Enter function value (sine, cosine, or tangent).
4. Depress INV or 2ND key.
5. Depress trigonometric function key.

$m$  Special constants or functions are often included on special keys. A very common constant is pi. A typical calculator having a $\pi$ key would calculate the area of a circle of radius 1.55 as follows:

1. Enter 1.55.
2. Depress $X^2$.
3. Depress $\times$.
4. Depress $\pi$.
5. Depress $=$.

The answer 7.5476763 would then be in the display.

A less common function found on some calculators is the N!, the factorial function.

Hyperbolic sines and cosines are also found on a few calculators.

$n$  Parenthesis control is available on some calculators. The keys which operate these functions are generally marked ( and ). The left and right parenthesis functions permit stacking or setting aside some operation just entered until a later operation has been entered and executed. Parenthesis control permits entering numbers and operations in the

exact order they appear in the written problem. The problem

$$1.45 \times (7.12 + (3.12 \div 1.71))$$

would be executed on many calculators as follows:

1. Enter 1.45.
2. Depress $\times$.
3. Depress (.
4. Enter 7.12.
5. Depress $+$.
6. Depress (.
7. Enter 3.12.
8. Depress $\div$.
9. Enter 1.71.

Two types of execution are often found on calculators. In one type, typified by the Texas Instruments calculators, depressing the $=$ key will cause all operations to be executed. The other type of common operation requires a right parenthesis to be executed for every left parenthesis that has been entered. Fortunately, this type of operation also uses the right parenthesis, ), as an execution key so that simply depressing the ) key until no further change occurs in the display will usually guarantee that all pending operations have been executed. The answer, 12.969614, will then appear in the display.

A major problem with parenthesis control is that the number of operations and numbers that can be stacked or set aside is limited. The usual number of pending operations is from two to ten on popular calculators. The important feature in most problems is the number of pending operations that can be handled, not the number of parentheses. The problem

$$((1.45) \times ((7.12) + (((3.12) \div (1.71))))$$

has as many as five levels of parentheses, most of them redundant and confusing, open at times, but only three operations pending. Only two levels of parentheses are needed in this problem as seen in the earlier example, which is the clean version of this problem, but three levels of pending operations cannot be reduced without rearranging the problem. Calculators without parenthesis control would perform the operations in the following order to obtain the same answer:

1. Enter 3.12.
2. Depress ÷.
3. Enter 1.71.
4. Depress + to execute
   $3.12 \div 1.71 = 1.8245614$.
5. Enter 7.12.
6. Depress × to execute
   $1.8245614 + 7.12 = 8.9445614$.
7. Enter 1.45.
8. Depress = to execute
   $8.9445614 \times 1.45 = 12.969614$.

o Many calculators have one or more storage memories. Simple storage-only memories may be found in very elementary machines. However, versatile memories are usually provided on only the more expensive machines. The memories may be used for storing various temporary values which will be used several times in a given problem. One particular use of a single memory is in accumulating the sum of a series. Some calculators offer a special key which permits adding the value in the display to the value in the memory. A typical problem involving a memory is that of evaluating the equation $1.23x^3 + .79x^2 - 2.51x + .56$. If the calculator has parenthesis control, the value of $x$ might be stored in the memory and the sum of the individual terms accumulated in the display register. If the calculator does not have parenthesis control, the sum of the individual terms might be accumulated in the memory. The procedures for the two types of operations might go as follows:

*Parenthesis control*
1. Enter 1.3
   (the value of $x$).
2. Depress STO
   (store in memory).
3. Depress $Y^x$.

4. Enter 3.
5. Depress ×.
6. Enter 1.23.

7. Depress + to execute $1.23x^3$.
8. Depress (.

9. Depress RCL to recall $x$ from memory.
10. Depress $X^2$.
11. Depress ×.
12. Enter .79.
13. Depress ) to execute ×.
14. Depress = to execute +.
15. Depress −.

16. Depress (.
17. Depress RCL to recall $x$ from memory.
18. Depress ×.
19. Enter 2.51.

20. Depress ) to execute ×.

21. Depress = to execute −.
22. Depress +.
23. Enter .56.

24. Depress = to execute +.

*No parenthesis control*
1. Enter 0 or clear display.
2. Depress STO (zero memory).
3. Enter 1.3 (the value of $x$).
4. Depress $Y^x$.
5. Enter 3.
6. Depress × to execute $x^3$.

7. Enter 1.23.
8. Depress = to execute $1.23x^3$.
9. Depress M+ to add display to memory.
10. Enter 1.3.
11. Depress $X^2$.
12. Depress ×.
13. Enter .79.
14. Depress = to execute $.79x^2$.
15. Depress M+ to add display to memory.
16. Enter 1.3.
17. Depress ×.

18. Enter 2.51.
19. Depress = to execute $2.51x$.
20. Depress +/− to change display to −.
21. Depress M+ to add display to memory.
22. Enter .56.
23. Depress M+ to add display to memory.
24. Depress RCL to recall answer from memory.

The answer is 1.33441.

The add-to-memory key used in the second version of the problem is sometimes marked M+, SUM, ACC, or Σ (sigma, the Greek letter for summation). Either of these problems would have required some additional steps if the memory had not been available. The first version of the problem is especially useful if $x$ is an eight- or ten-digit number, for it avoids entering that large value repeatedly with the chance of a mistake. If the second version of the problem were done on a calculator with two or more memories, the value for $x$ would not have to be entered each time it is used, but could be recalled from the second memory, avoiding the possibility of an error in entry.

Some calculators also include functions so that one can subtract the display from the memory, multiply the memory by the display, and divide the memory by the display. This permits some parallel calculations in both the memory and the display.

Some calculators also include an exchange key for exchanging or trading the values in the memory and the display. This is especially useful if the value in the memory is to be divided by the value in the display and there is not a memory-divide function. The value in the display can only be divided by the value in the memory by the following steps:

1. Depress ÷.
2. Depress RCL to recall value from memory.
3. Depress = to execute divide display by memory.

By exchanging the memory and display values, the divisor ends up in the memory, where it is entered second in the operation.

The answer can be obtained by the following sequence without exchanging the memory and display:

1. Depress $1/X$ to get reciprocal of divisor in display.
2. Depress ×.
3. Depress RCL to recall number to be divided from memory.
4. Depress = to execute $N \times 1/X$ which is, obviously, division.

$p$   All calculators have two operating registers for performing the various arithmetic operations requiring two numbers or operands. The first register is the one we have always called the display. The second register is hidden from the user. This register is used in any operation involving two numbers. The sequence for such a situation is something like this:

| | |
|---|---|
| 1. Enter first number. | Value appears in display register. |
| 2. Enter operation. | Display value is transferred into hidden register. |
| 3. Enter second number. | Second value now replaces first value in display register. |

| | |
|---|---|
| 4. Depress =. | First value (in hidden register) is operated on by second value (in display register). |

This sequence of operations explains why some calculators can perform certain operations when only one number has been entered. The operations of +, −, ×, and ÷ can be used as follows in these calculators:

| | |
|---|---|
| 1. Enter value. | Value in display register. |
| 2. Depress operation key. | Value goes into hidden register. |
| 3. Depress =. | Operation performed on value in hidden register using value in display register. |

This means that the following operations provide the following answers:

| Operation | Answer |
|---|---|
| Add. | Value doubled in display. |
| Subtract. | Zero in display. |
| Multiply. | Value squared in display. |
| Divide. | One in display. |

Some calculators do not permit this type of operation; a second value must be entered into the display, or the operation will not be performed. Care must be taken to read the instructions carefully to determine the special register operations provided by the calculator.

$q$   Keys for the clearing of registers are provided on all calculators. Their operations are very much the same but the names of the keys are not. The two major clearing operations are:

1. Clear the entry in the display register.
2. Clear both the hidden register and the display register.

The keys for clearing the display are often marked CD, CK, CE, or some similar abbreviation for "clear display," or "keyboard," or "entry." The key for clearing both registers is generally marked C or CLR. Some calculators use a double-purpose key for clearing in which the first depression of the key clears the display register and the second depression of the key clears the hidden register.

The operation which has been entered is always voided by the clearing of both registers. In some cases, using the display-clearing key will *also* clear the operation; in other cases, first clearing the display will permit a different operation to be reentered if desired. If it is not reentered, the previously entered operation still will be executed. Careful reading of the calculator instruction book is needed to determine the exact mode of operation of these keys.

## 10. Sample Problems.

Solving a problem on a hand calculator is dependent upon the type of calculator logic and operations which are available. We will consider three major types of calculators, each having a single memory:

1. Standard algebraic logic with no parenthesis control.
2. Standard algebraic logic with parenthesis control to four levels.
3. Reverse Polish notation with a four-level stack.

Solving the following problem on each of these calculators could proceed as shown in the sample calculations below. Here is the problem:

$$\frac{3.5 \times (2.7 - (\sin 57° \div 1.65))}{5.61 \times \log (9.6 \div (4.7 + \cos 35°))}$$

When using a standard algebraic logic calculator with memory but no parenthesis control, the general approach is to calculate the divisor first and store it in the memory.

The final stage of the problem is to divide the number in the display register by the number in the memory. Problems involving parentheses are best handled by starting with the innermost parenthesis pair for the calculation.

1. Enter 4.7.
2. Depress +.
3. Enter 35.
4. Depress COS.
5. Depress = to execute 4.7 + sin 35°.
6. Depress STO.

7. Enter 9.6.
8. Depress ÷.
9. Depress RCL to recall value from memory.
10. Depress = to execute 9.6 ÷ (4.7 + cos 35°).
11. Depress LOG to take logarithm to base 10 of above value.
12. Depress ×.
13. Enter 5.61.
14. Depress = to execute 5.61 × log(9.6 ÷ (4.7 + cos 35°)).
15. Depress STO to store divisor in memory.
16. Enter 57.
17. Depress SIN.
18. Depress ÷.
19. Enter 1.65.
20. Depress = to execute sin 57° ÷ 1.65.
21. Depress +/−.
22. Depress +.
23. Enter 2.7.
24. Depress × to execute 2.7 + (− (sin 57° ÷ 1.65)).
25. Enter 3.5.
26. Depress = to execute 3.5 × (2.7 − (sin 57° ÷ 1.65)).
27. Depress ÷.
28. Depress RCL to recall divisor from memory.
29. Depress = to execute final division.

The answer is 5.6879634.

Note steps 21, 22, and 23, where the numbers are in the wrong order for subtraction so that the number doing the subtraction must be changed to a negative value and the other value added to this.

When using a standard algebraic logic calculator with four levels of parenthesis control plus memory, the problem must be rearranged if parenthesis control is to be used. However, full parenthesis control does not need to be used. This would shorten the problem execution somewhat. Rewrite the problem as:

$$3.5 \times (2.7 - (\sin 57° \div 1.65))$$
$$\div [5.61 \times \log (9.6 \div (4.7 + \cos 35°))]$$

Note the extra brackets in this problem. The execution of such a problem is done on a left-to-right basis, entering the numbers and operations in exactly the order they appear in the rewritten problem.

1. Enter 3.5.
2. Depress ×.
3. Depress (.
4. Enter 2.7.
5. Depress −.
6. Depress (.
7. Enter 57.
8. Depress SIN.
9. Depress ÷.
10. Enter 1.65.
11. Depress ) to execute sin 57° ÷ 1.65.
12. Depress ) to execute 2.7 − value from step 11.
13. Depress = to execute 3.5 × value from step 12.
14. Depress ÷.
15. Depress (.
16. Enter 5.61.
17. Depress ×.
18. Depress (.
19. Enter 9.6.
20. Depress ÷.
21. Depress (.
22. Enter 4.7.
23. Depress +.
24. Enter 35.
25. Depress COS.
26. Depress ) to execute 4.7 + cos 35°.
27. Depress ) to execute 9.6 ÷ value from step 26.
28. Depress LOG to obtain logarithm to base 10 of value from step 27.
29. Depress ) to execute 5.61 × value from step 28.
30. Depress = to execute value of step 13 ÷ value of step 29.

This problem could be shortened by partly using nonparenthesis logic and partly using parenthesis logic.

When using a reverse Polish notation machine with a four-level stack, the major rule is to start the calculations with the innermost set of parentheses.

Because the items in the stack can be reversed in most machines, it is usual to start with either the dividend or the divisor.

1. Enter 57.
2. Depress SIN.
3. Depress ENTER to push value down one level in stack.
4. Enter 1.65.
5. Depress ÷ to execute sin 57° ÷ 1.65.
6. Enter 2.7.
7. Depress EXC to exchange values in top two levels of stack.
8. Depress − to execute 2.7 − value of step 5.
9. Enter 3.5.
10. Depress × to execute 3.5 × value of step 8.
11. Enter 4.7.
12. Depress ENTER to put value of step 10 in third level of stack, 4.7 in second level of stack.
13. Enter 35.
14. Depress COS to put cos 35° in first level of stack.
15. Depress + to execute 4.7 + cos 35°.
16. Enter 9.6.
17. Depress EXC to exchange top two levels of stack, putting 9.6 on second level and value from step 15 on first level.
18. Depress ÷ to execute 9.6 ÷ (4.7 + cos 35°).
19. Depress LOG to obtain logarithm to base 10 of value from step 18.
20. Enter 5.61.
21. Depress × to execute 5.61 × value from step 19.
22. Depress ÷ to execute dividend ÷ divisor now in top two levels of the stack.

Note that the reverse Polish notation machine needs fewer keystrokes than the other two types of calculators. It should be noted that manufacturers of each type of machine always chose a problem to illustrate that their machine works best with their logic and poorly with the competitors' logic.

# APPENDIX B

## Mathematical Tables

Selected Mathematical Values, A22
Length of Chord for Circle Arcs of 1-in. Radius, A23
Decimal Equivalents of Inch Fractions, A24
Metric Equivalents, A24
Conversion Table: Fractions, Decimals, and Millimeters, A25

Conversion Factors as Exact Numerical Multiples of SI Units, A26
ISO Metric Screw Threads, A38
Dual (Inch-metric) Dimensioning, A46

## Selected Mathematical Values

| | | | | | |
|---|---|---|---|---|---|
| $\pi =$ | 3.141593 | $\sqrt{\pi} =$ | 1.772454 | $\sqrt{2} =$ | 1.414214 |
| $2\pi =$ | 6.283185 | $\sqrt[3]{\pi} =$ | 1.464592 | $\sqrt{3} =$ | 1.732055 |
| $3\pi =$ | 9.424778 | $\log_e \pi* =$ | 1.144730 | | |
| $4\pi =$ | 12.566371 | $\dfrac{1}{\pi} =$ | 0.318310 | $\dfrac{\pi}{\sqrt{g}} =$ | 0.55399 |
| $\dfrac{\pi}{2} =$ | 1.570796 | | | $\dfrac{1}{\sqrt{g}} =$ | 0.17634 |
| $\dfrac{\pi}{3} =$ | 1.047198 | $\dfrac{1}{\pi^2} =$ | 0.101321 | $\sqrt{2g} =$ | 8.01998 |
| $\dfrac{\pi}{4} =$ | 0.785398 | $\dfrac{1}{\sqrt{\pi}} =$ | 0.564190 | $\dfrac{1}{2g} =$ | 0.01555 |
| | | $\log_{10} \pi =$ | 0.497150 | $g^2 =$ | 1034.266 |
| $\pi^2 =$ | 9.869604 | $\dfrac{1}{e} =$ | 0.367879 | $2g =$ | 64.32 |
| $\pi^3 =$ | 31.006277 | | | $g =$ | 32.16 |

$*e = 2.718282$

## Length of Chord for Circle Arcs of 1-in. Radius

| ° | 0′ | 10′ | 20′ | 30′ | 40′ | 50′ |
|---|-----|-----|-----|-----|-----|-----|
| 0 | 0.0000 | 0.0029 | 0.0058 | 0.0087 | 0.0116 | 0.0145 |
| 1 | 0.0175 | 0.0204 | 0.0233 | 0.0262 | 0.0291 | 0.0320 |
| 2 | 0.0349 | 0.0378 | 0.0407 | 0.0436 | 0.0465 | 0.0494 |
| 3 | 0.0524 | 0.0553 | 0.0582 | 0.0611 | 0.0640 | 0.0669 |
| 4 | 0.0698 | 0.0727 | 0.0756 | 0.0785 | 0.0814 | 0.0843 |
| 5 | 0.0872 | 0.0901 | 0.0931 | 0.0960 | 0.0989 | 0.1018 |
| 6 | 0.1047 | 0.1076 | 0.1105 | 0.1134 | 0.1163 | 0.1192 |
| 7 | 0.1221 | 0.1250 | 0.1279 | 0.1308 | 0.1337 | 0.1366 |
| 8 | 0.1395 | 0.1424 | 0.1453 | 0.1482 | 0.1511 | 0.1540 |
| 9 | 0.1569 | 0.1598 | 0.1627 | 0.1656 | 0.1685 | 0.1714 |
| 10 | 0.1743 | 0.1772 | 0.1801 | 0.1830 | 0.1859 | 0.1888 |
| 11 | 0.1917 | 0.1946 | 0.1975 | 0.2004 | 0.2033 | 0.2062 |
| 12 | 0.2091 | 0.2119 | 0.2148 | 0.2177 | 0.2206 | 0.2235 |
| 13 | 0.2264 | 0.2293 | 0.2322 | 0.2351 | 0.2380 | 0.2409 |
| 14 | 0.2437 | 0.2466 | 0.2495 | 0.2524 | 0.2553 | 0.2582 |
| 15 | 0.2611 | 0.2639 | 0.2668 | 0.2697 | 0.2726 | 0.2755 |
| 16 | 0.2783 | 0.2812 | 0.2841 | 0.2870 | 0.2899 | 0.2927 |
| 17 | 0.2956 | 0.2985 | 0.3014 | 0.3042 | 0.3071 | 0.3100 |
| 18 | 0.3129 | 0.3157 | 0.3186 | 0.3215 | 0.3244 | 0.3272 |
| 19 | 0.3301 | 0.3330 | 0.3358 | 0.3387 | 0.3416 | 0.3444 |
| 20 | 0.3473 | 0.3502 | 0.3530 | 0.3559 | 0.3587 | 0.3616 |
| 21 | 0.3645 | 0.3673 | 0.3702 | 0.3730 | 0.3759 | 0.3788 |
| 22 | 0.3816 | 0.3845 | 0.3873 | 0.3902 | 0.3930 | 0.3959 |
| 23 | 0.3987 | 0.4016 | 0.4044 | 0.4073 | 0.4101 | 0.4130 |
| 24 | 0.4158 | 0.4187 | 0.4215 | 0.4244 | 0.4272 | 0.4300 |
| 25 | 0.4329 | 0.4357 | 0.4386 | 0.4414 | 0.4442 | 0.4471 |
| 26 | 0.4499 | 0.4527 | 0.4556 | 0.4584 | 0.4612 | 0.4641 |
| 27 | 0.4669 | 0.4697 | 0.4725 | 0.4754 | 0.4782 | 0.4810 |
| 28 | 0.4838 | 0.4867 | 0.4895 | 0.4923 | 0.4951 | 0.4979 |
| 29 | 0.5008 | 0.5036 | 0.5064 | 0.5092 | 0.5120 | 0.5148 |
| 30 | 0.5176 | 0.5204 | 0.5233 | 0.5251 | 0.5289 | 0.5317 |
| 31 | 0.5345 | 0.5373 | 0.5401 | 0.5429 | 0.5457 | 0.5485 |
| 32 | 0.5513 | 0.5541 | 0.5569 | 0.5597 | 0.5625 | 0.5652 |
| 33 | 0.5680 | 0.5708 | 0.5736 | 0.5764 | 0.5792 | 0.5820 |
| 34 | 0.5847 | 0.5875 | 0.5903 | 0.5931 | 0.5959 | 0.5986 |
| 35 | 0.6014 | 0.6042 | 0.6070 | 0.6097 | 0.6125 | 0.6153 |
| 36 | 0.6180 | 0.6208 | 0.6236 | 0.6263 | 0.6291 | 0.6319 |
| 37 | 0.6346 | 0.6374 | 0.6401 | 0.6429 | 0.6456 | 0.6484 |
| 38 | 0.6511 | 0.6539 | 0.6566 | 0.6594 | 0.6621 | 0.6649 |
| 39 | 0.6676 | 0.6704 | 0.6731 | 0.6758 | 0.6786 | 0.6813 |
| 40 | 0.6840 | 0.6868 | 0.6895 | 0.6922 | 0.6950 | 0.6977 |
| 41 | 0.7004 | 0.7031 | 0.7059 | 0.7086 | 0.7113 | 0.7140 |
| 42 | 0.7167 | 0.7195 | 0.7222 | 0.7249 | 0.7276 | 0.7303 |
| 43 | 0.7330 | 0.7357 | 0.7384 | 0.7411 | 0.7438 | 0.7465 |
| 44 | 0.7492 | 0.7519 | 0.7546 | 0.7573 | 0.7600 | 0.7627 |
| 45 * | 0.7654 | 0.7681 | 0.7707 | 0.7734 | 0.7761 | 0.7788 |

* For angles between 45° and 90°, draw 90° angle and lay off complement from 90° line.

## Decimal Equivalents of Inch Fractions

| Fraction | Equiv. | Fraction | Equiv. | Fraction | Equiv. | Fraction | Equiv. |
|---|---|---|---|---|---|---|---|
| 1/64 | 0.015625 | 17/64 | 0.265625 | 33/64 | 0.515625 | 49/64 | 0.765625 |
| 1/32 | 0.03125 | 9/32 | 0.28125 | 17/32 | 0.53125 | 25/32 | 0.78125 |
| 3/64 | 0.046875 | 19/64 | 0.296875 | 35/64 | 0.546875 | 51/64 | 0.796875 |
| 1/16 | 0.0625 | 5/16 | 0.3125 | 9/16 | 0.5625 | 13/16 | 0.8125 |
| 5/64 | 0.078125 | 21/64 | 0.328125 | 37/64 | 0.578125 | 53/64 | 0.828125 |
| 3/32 | 0.09375 | 11/32 | 0.34375 | 19/32 | 0.59375 | 27/32 | 0.84375 |
| 7/64 | 0.109375 | 23/64 | 0.359375 | 39/64 | 0.609375 | 55/64 | 0.859375 |
| 1/8 | 0.1250 | 3/8 | 0.3750 | 5/8 | 0.6250 | 7/8 | 0.8750 |
| 9/64 | 0.140625 | 25/64 | 0.390625 | 41/64 | 0.640625 | 57/64 | 0.890625 |
| 5/32 | 0.15625 | 13/32 | 0.40625 | 21/32 | 0.65625 | 29/32 | 0.90625 |
| 11/64 | 0.171875 | 27/64 | 0.421875 | 43/64 | 0.671875 | 59/64 | 0.921875 |
| 3/16 | 0.1875 | 7/16 | 0.4375 | 11/16 | 0.6875 | 15/16 | 0.9375 |
| 13/64 | 0.203125 | 29/64 | 0.453125 | 45/64 | 0.703125 | 61/64 | 0.953125 |
| 7/32 | 0.21875 | 15/32 | 0.46875 | 23/32 | 0.71875 | 31/32 | 0.96875 |
| 15/64 | 0.234375 | 31/64 | 0.484375 | 47/64 | 0.734375 | 63/64 | 0.984375 |
| 1/4 | 0.2500 | 1/2 | 0.5000 | 3/4 | 0.7500 | 1 | 1.0000 |

## Metric Equivalents

| Mm | In.* | Mm | In. | In. | Mm † | In. | Mm |
|---|---|---|---|---|---|---|---|
| 1 = 0.0394 | | 17 = 0.6693 | | 1/32 = 0.794 | | 17/32 = 13.493 | |
| 2 = 0.0787 | | 18 = 0.7087 | | 1/16 = 1.587 | | 9/16 = 14.287 | |
| 3 = 0.1181 | | 19 = 0.7480 | | 3/32 = 2.381 | | 19/32 = 15.081 | |
| 4 = 0.1575 | | 20 = 0.7874 | | 1/8 = 3.175 | | 5/8 = 15.875 | |
| 5 = 0.1969 | | 21 = 0.8268 | | 5/32 = 3.968 | | 21/32 = 16.668 | |
| 6 = 0.2362 | | 22 = 0.8662 | | 3/16 = 4.762 | | 11/16 = 17.462 | |
| 7 = 0.2756 | | 23 = 0.9055 | | 7/32 = 5.556 | | 23/32 = 18.256 | |
| 8 = 0.3150 | | 24 = 0.9449 | | 1/4 = 6.349 | | 3/4 = 19.050 | |
| 9 = 0.3543 | | 25 = 0.9843 | | 9/32 = 7.144 | | 25/32 = 19.843 | |
| 10 = 0.3937 | | 26 = 1.0236 | | 5/16 = 7.937 | | 13/16 = 20.637 | |
| 11 = 0.4331 | | 27 = 1.0630 | | 11/32 = 8.731 | | 27/32 = 21.431 | |
| 12 = 0.4724 | | 28 = 1.1024 | | 3/8 = 9.525 | | 7/8 = 22.225 | |
| 13 = 0.5118 | | 29 = 1.1418 | | 13/32 = 10.319 | | 29/32 = 23.018 | |
| 14 = 0.5512 | | 30 = 1.1811 | | 7/16 = 11.112 | | 15/16 = 23.812 | |
| 15 = 0.5906 | | 31 = 1.2205 | | 15/32 = 11.906 | | 31/32 = 24.606 | |
| 16 = 0.6299 | | 32 = 1.2599 | | 1/2 = 12.699 | | 1 = 25.400 | |

\* Calculated to *nearest* fourth decimal place.
† Calculated to *nearest* third decimal place.

# Conversion Table: Fractions, Decimals, and Millimeters*

| inches fractions | inches decimals | m m | inches fractions | inches decimals | m m | inches fractions | inches decimals | m m | inches fractions | inches decimals | m m |
|---|---|---|---|---|---|---|---|---|---|---|---|
| — | .0004 | .01 | 25/32 | .781 | 19.844 | — | 2.165 | 55. | 3-11/16 | 3.6875 | 93.663 |
| — | .004 | .10 | — | .7874 | 20. | 2-3/16 | 2.1875 | 55.563 | — | 3.7008 | 94. |
| 1/64 | .01 | .25 | 51/64 | .797 | 20.241 | — | 2.2047 | 56. | 3-23/32 | 3.719 | 94.456 |
| 1/64 | .0156 | .397 | 13/16 | .8125 | 20.638 | 2-7/32 | 2.219 | 56.356 | — | 3.7401 | 95. |
| — | .0197 | .50 | — | .8268 | 21. | — | 2.244 | 57. | 3-3/4 | 3.750 | 95.250 |
| — | .0295 | .75 | 53/64 | .828 | 21.034 | 2-1/4 | 2.250 | 57.150 | — | 3.7795 | 96. |
| 1/32 | .03125 | .794 | 27/32 | .844 | 21.431 | 2-9/32 | 2.281 | 57.944 | 3-25/32 | 3.781 | 96.044 |
| — | .0394 | 1. | 55/64 | .859 | 21.828 | — | 2.2835 | 58. | 3-13/16 | 3.8125 | 96.838 |
| 3/64 | .0469 | 1.191 | — | .8661 | 22. | 2-5/16 | 2.312 | 58.738 | — | 3.8189 | 97. |
| — | .059 | 1.5 | 7/8 | .875 | 22.225 | — | 2.3228 | 59. | 3-27/32 | 3.844 | 97.631 |
| 1/16 | .062 | 1.588 | 57/64 | .8906 | 22.622 | 2-11/32 | 2.344 | 59.531 | — | 3.8583 | 98. |
| 5/64 | .0781 | 1.984 | — | .9055 | 23. | — | 2.3622 | 60. | 3-7/8 | 3.875 | 98.425 |
| — | .0787 | 2. | 29/32 | .9062 | 23.019 | 2-3/8 | 2.375 | 60.325 | — | 3.8976 | 99. |
| 3/32 | .094 | 2.381 | 59/64 | .922 | 23.416 | — | 2.4016 | 61. | 3-29/32 | 3.9062 | 99.219 |
| — | .0984 | 2.5 | 15/16 | .9375 | 23.813 | 2-13/32 | 2.406 | 61.119 | — | 3.9370 | 100. |
| 7/64 | .109 | 2.778 | — | .9449 | 24. | 2-7/16 | 2.438 | 61.913 | 3-15/16 | 3.9375 | 100.013 |
| — | .1181 | 3. | 61/64 | .953 | 24.209 | — | 2.4409 | 62. | 3-31/32 | 3.969 | 100.806 |
| 1/8 | .125 | 3.175 | 31/32 | .969 | 24.606 | 2-15/32 | 2.469 | 62.706 | — | 3.9764 | 101. |
| — | .1378 | 3.5 | — | .9843 | 25. | — | 2.4803 | 63. | 4 | 4.000 | 101.600 |
| 9/64 | .141 | 3.572 | 63/64 | .9844 | 25.003 | 2-1/2 | 2.500 | 63.500 | 4-1/16 | 4.062 | 103.188 |
| 5/32 | .156 | 3.969 | 1 | 1.000 | 25.400 | — | 2.5197 | 64. | 4-1/8 | 4.125 | 104.775 |
| — | .1575 | 4. | — | 1.0236 | 26. | 2-17/32 | 2.531 | 64.294 | — | 4.1338 | 105. |
| 11/64 | .172 | 4.366 | 1-1/32 | 1.0312 | 26.194 | — | 2.559 | 65. | 4-3/16 | 4.1875 | 106.363 |
| — | .177 | 4.5 | 1-1/16 | 1.062 | 26.988 | 2-9/16 | 2.562 | 65.088 | 4-1/4 | 4.250 | 107.950 |
| 3/16 | .1875 | 4.763 | — | 1.063 | 27. | 2-19/32 | 2.594 | 65.881 | 4-5/16 | 4.312 | 109.538 |
| — | .1969 | 5. | 1-3/32 | 1.094 | 27.781 | — | 2.5984 | 66. | — | 4.3307 | 110. |
| 13/64 | .203 | 5.159 | — | 1.1024 | 28. | 2-5/8 | 2.625 | 66.675 | 4-3/8 | 4.375 | 111.125 |
| — | .2165 | 5.5 | 1-1/8 | 1.125 | 28.575 | — | 2.638 | 67. | 4-7/16 | 4.438 | 112.713 |
| 7/32 | .219 | 5.556 | — | 1.1417 | 29. | 2-21/32 | 2.656 | 67.469 | 4-1/2 | 4.500 | 114.300 |
| 15/64 | .234 | 5.953 | 1-5/32 | 1.156 | 29.369 | — | 2.6772 | 68. | — | 4.5275 | 115. |
| — | .2362 | 6. | — | 1.1811 | 30. | 2-11/16 | 2.6875 | 68.263 | 4-9/16 | 4.562 | 115.888 |
| 1/4 | .250 | 6.350 | 1-3/16 | 1.1875 | 30.163 | — | 2.7165 | 69. | 4-5/8 | 4.625 | 117.475 |
| — | .2559 | 6.5 | 1-7/32 | 1.219 | 30.956 | 2-23/32 | 2.719 | 69.056 | 4-11/16 | 4.6875 | 119.063 |
| 17/64 | .2656 | 6.747 | — | 1.2205 | 31. | 2-3/4 | 2.750 | 69.850 | — | 4.7244 | 120. |
| — | .2756 | 7. | 1-1/4 | 1.250 | 31.750 | — | 2.7559 | 70. | 4-3/4 | 4.750 | 120.650 |
| 9/32 | .281 | 7.144 | — | 1.2598 | 32. | 2-25/32 | 2.781 | 70.6439 | 4-13/16 | 4.8125 | 122.238 |
| — | .2953 | 7.5 | 1-9/32 | 1.281 | 32.544 | — | 2.7953 | 71. | 4-7/8 | 4.875 | 123.825 |
| 19/64 | .297 | 7.541 | — | 1.2992 | 33. | 2-13/16 | 2.8125 | 71.4376 | — | 4.9212 | 125. |
| 5/16 | .312 | 7.938 | 1-5/16 | 1.312 | 33.338 | — | 2.8346 | 72. | 4-15/16 | 4.9375 | 125.413 |
| — | .315 | 8. | — | 1.3386 | 34. | 2-27/32 | 2.844 | 72.2314 | 5 | 5.000 | 127.000 |
| 21/64 | .328 | 8.334 | 1-11/32 | 1.344 | 34.131 | — | 2.8740 | 73. | — | 5.1181 | 130. |
| — | .335 | 8.5 | 1-3/8 | 1.375 | 34.925 | 2-7/8 | 2.875 | 73.025 | 5-1/4 | 5.250 | 133.350 |
| 11/32 | .344 | 8.731 | — | 1.3779 | 35. | 2-29/32 | 2.9062 | 73.819 | 5-1/2 | 5.500 | 139.700 |
| — | .3543 | 9. | 1-13/32 | 1.406 | 35.719 | — | 2.9134 | 74. | — | 5.5118 | 140. |
| 23/64 | .359 | 9.128 | — | 1.4173 | 36. | 2-15/16 | 2.9375 | 74.613 | 5-3/4 | 5.750 | 146.050 |
| — | .374 | 9.5 | 1-7/16 | 1.438 | 36.513 | — | 2.9527 | 75. | — | 5.9055 | 150. |
| 3/8 | .375 | 9.525 | — | 1.4567 | 37. | 2-31/32 | 2.969 | 75.406 | 6 | 6.000 | 152.400 |
| 25/64 | .391 | 9.922 | 1-15/32 | 1.469 | 37.306 | — | 2.9921 | 76. | 6-1/4 | 6.250 | 158.750 |
| — | .3937 | 10. | — | 1.4961 | 38. | 3 | 3.000 | 76.200 | — | 6.2992 | 160. |
| 13/32 | .406 | 10.319 | 1-1/2 | 1.500 | 38.100 | 3-1/32 | 3.0312 | 76.994 | 6-1/2 | 6.500 | 165.100 |
| — | .413 | 10.5 | 1-17/32 | 1.531 | 38.894 | — | 3.0315 | 77. | — | 6.6929 | 170. |
| 27/64 | .422 | 10.716 | — | 1.5354 | 39. | 3-1/16 | 3.062 | 77.788 | 6-3/4 | 6.750 | 171.450 |
| — | .4331 | 11. | 1-9/16 | 1.562 | 39.688 | — | 3.0709 | 78. | 7 | 7.000 | 177.800 |
| 7/16 | .438 | 11.113 | — | 1.5748 | 40. | 3-3/32 | 3.094 | 78.581 | — | 7.0866 | 180. |
| 29/64 | .453 | 11.509 | 1-19/32 | 1.594 | 40.481 | — | 3.1102 | 79. | — | 7.4803 | 190. |
| 15/32 | .469 | 11.906 | — | 1.6142 | 41. | 3-1/8 | 3.125 | 79.375 | 7-1/2 | 7.500 | 190.500 |
| — | .4724 | 12. | 1-5/8 | 1.625 | 41.275 | — | 3.1496 | 80. | — | 7.8740 | 200. |
| 31/64 | .484 | 12.303 | — | 1.6535 | 42. | 3-5/32 | 3.156 | 80.169 | 8 | 8.000 | 203.200 |
| — | .492 | 12.5 | 1-21/32 | 1.6562 | 42.069 | 3-3/16 | 3.1875 | 80.963 | — | 8.2677 | 210. |
| 1/2 | .500 | 12.700 | 1-11/16 | 1.6875 | 42.863 | — | 3.1890 | 81. | 8-1/2 | 8.500 | 215.900 |
| — | .5118 | 13. | — | 1.6929 | 43. | 3-7/32 | 3.219 | 81.756 | — | 8.6614 | 220. |
| 33/64 | .5156 | 13.097 | 1-23/32 | 1.719 | 43.656 | — | 3.2283 | 82. | 9 | 9.000 | 228.600 |
| 17/32 | .531 | 13.494 | — | 1.7323 | 44. | 3-1/4 | 3.250 | 82.550 | — | 9.0551 | 230. |
| 35/64 | .547 | 13.891 | 1-3/4 | 1.750 | 44.450 | — | 3.2677 | 83. | — | 9.4488 | 240. |
| — | .5512 | 14. | — | 1.7717 | 45. | 3-9/32 | 3.281 | 83.344 | 9-1/2 | 9.500 | 241.300 |
| 9/16 | .563 | 14.288 | 1-25/32 | 1.781 | 45.244 | — | 3.3071 | 84. | — | 9.8425 | 250. |
| — | .571 | 14.5 | — | 1.8110 | 46. | 3-5/16 | 3.312 | 84.1377 | 10 | 10.000 | 254.001 |
| 37/64 | .578 | 14.684 | 1-13/16 | 1.8125 | 46.038 | 3-11/32 | 3.344 | 84.9314 | — | 10.2362 | 260. |
| — | .5906 | 15. | 1-27/32 | 1.844 | 46.831 | — | 3.3464 | 85. | — | 10.6299 | 270. |
| 19/32 | .594 | 15.081 | — | 1.8504 | 47. | 3-3/8 | 3.375 | 85.725 | 11 | 11.000 | 279.401 |
| 39/64 | .609 | 15.478 | 1-7/8 | 1.875 | 47.625 | — | 3.3858 | 86. | — | 11.0236 | 280. |
| 5/8 | .625 | 15.875 | — | 1.8898 | 48. | 3-13/32 | 3.406 | 86.519 | — | 11.4173 | 290. |
| — | .6299 | 16. | 1-29/32 | 1.9062 | 48.419 | — | 3.4252 | 87. | — | 11.8110 | 300. |
| 41/64 | .6406 | 16.272 | — | 1.9291 | 49. | 3-7/16 | 3.438 | 87.313 | 12 | 12.000 | 304.801 |
| — | .6496 | 16.5 | 1-15/16 | 1.9375 | 49.213 | — | 3.4646 | 88. | 13 | 13.000 | 330.201 |
| 21/32 | .656 | 16.669 | — | 1.9685 | 50. | 3-15/32 | 3.469 | 88.106 | — | 13.7795 | 350. |
| — | .6693 | 17. | 1-31/32 | 1.969 | 50.006 | 3-1/2 | 3.500 | 88.900 | 14 | 14.000 | 355.601 |
| 43/64 | .672 | 17.066 | 2 | 2.000 | 50.800 | — | 3.5039 | 89. | 15 | 15.000 | 381.001 |
| 11/16 | .6875 | 17.463 | — | 2.0079 | 51. | 3-17/32 | 3.531 | 89.694 | — | 15.7480 | 400. |
| 45/64 | .703 | 17.859 | 2-1/32 | 2.03125 | 51.594 | — | 3.5433 | 90. | 16 | 16.000 | 406.401 |
| — | .7087 | 18. | — | 2.0472 | 52. | 3-9/16 | 3.562 | 90.4877 | 17 | 17.000 | 431.801 |
| 23/32 | .719 | 18.256 | 2-1/16 | 2.062 | 52.388 | — | 3.5827 | 91. | — | 17.7165 | 450. |
| — | .7283 | 18.5 | — | 2.0866 | 53. | 3-19/32 | 3.594 | 91.281 | 18 | 18.000 | 457.201 |
| 47/64 | .734 | 18.653 | 2-3/32 | 2.094 | 53.181 | — | 3.622 | 92. | 19 | 19.000 | 482.601 |
| — | .7480 | 19. | 2-1/8 | 2.125 | 53.975 | 3-5/8 | 3.625 | 92.075 | — | 19.6850 | 500. |
| 3/4 | .750 | 19.050 | — | 2.126 | 54. | 3-21/32 | 3.656 | 92.869 | 20 | 20.000 | 508.001 |
| 49/64 | .7656 | 19.447 | 2-5/32 | 2.156 | 54.769 | — | 3.6614 | 93. | | | |

* Adapted from "TRUARC Technical Manual of Retaining Rings and Assembly Tools."

# CONVERSION FACTORS AS EXACT NUMERICAL MULTIPLES OF SI UNITS

The following tables express the definitions of various units of measure as exact numerical multiples of coherent SI units, and provide multiplying factors for converting numbers and miscellaneous units to corresponding new numbers and SI units.

The first two digits of each numerical entry represent a power of 10. An asterisk follows each number which expresses an exact definition. For example, the entry "−2 2.54*" expresses the fact that 1 inch = 2.54 × 10$^{-2}$ meter, exactly, by definition. Most of the definitions are extracted from National Bureau of Standards documents. Numbers not followed by an asterisk are only approximate representations of definitions or are the results of physical measurements. The conversion factors are listed alphabetically and by physical quantity.

The Listing by Physical Quantity includes only relationships which are frequently encountered and deliberately omits the great multiplicity of combinations of units which are used for more specialized purposes. Conversion factors for combinations of units are easily generated from numbers given in the Alphabetical Listing by the technique of direct substitution or by other well-known rules for manipulating units. These rules are adequately discussed in many science and engineering textbooks and are not repeated here.

## Alphabetical listing

| To convert from | To | Multiply by |
|---|---|---|
| Abampere | ampere | +01 1.00* |
| Abcoulomb | coulomb | +01 1.00* |
| Abfarad | farad | +09 1.00* |
| Abhenry | henry | −09 1.00* |
| Abmho | mho | +09 1.00* |
| Abohm | ohm | −09 1.00* |
| Abvolt | volt | −08 1.00* |
| Acre | meter$^2$ | +03 4.046 856 422 4* |
| Ampere | ampere | −01 9.998 35 |
| Angstrom | meter | −10 1.00* |
| Are | meter$^2$ | +02 1.00* |
| Astronomical unit | meter | +11 1.495 978 9 |
| Atmosphere | newton/meter$^2$ | +05 1.013 25* |
| Bar | newton/meter$^2$ | +05 1.00* |
| Barn | meter$^2$ | −28 1.00* |
| Barrel (petroleum, 42 gal) | meter$^3$ | −01 1.589 873 |
| Barye | newton/meter$^2$ | −01 1.00* |
| British thermal unit (ISO/TC 12) | joule | +03 1.055 06 |
| British thermal unit (International Steam Table) | joule | +03 1.055 04 |
| British thermal unit (mean) | joule | +03 1.055 87 |
| British thermal unit (thermochemical) | joule | +03 1.054 350 264 488 |
| British thermal unit (39°F) | joule | +03 1.059 67 |
| British thermal unit (60°F) | joule | +03 1.054 68 |
| Bushel (U.S.) | meter$^3$ | −02 3.523 907 016 688* |

| To convert from | To | Multiply by |
|---|---|---|
| Cable | meter | $+02$ 2.194 56* |
| Caliber | meter | $-04$ 2.54* |
| Calorie (International Steam Table) | joule | $+00$ 4.1868 |
| Calorie (mean) | joule | $+00$ 4.190 02 |
| Calorie (thermochemical) | joule | $+00$ 4.184* |
| Calorie (15°C) | joule | $+00$ 4.185 80 |
| Calorie (20°C) | joule | $+00$ 4.181 90 |
| Calorie (kilogram, International Steam Table) | joule | $+03$ 4.1868 |
| Calorie (kilogram, mean) | joule | $+03$ 4.190 02 |
| Calorie (kilogram, thermochemical) | joule | $+03$ 4.184* |
| Carat (metric) | kilogram | $-04$ 2.00* |
| Celsius (temperature) | kelvin | $t_K = t_C + 273.15$ |
| Centimeter of mercury (0°C) | newton/meter$^2$ | $+03$ 1.333 22 |
| Centimeter of water (4°C) | newton/meter$^2$ | $+01$ 9.806 38 |
| Chain (engineer or ramden) | meter | $+01$ 3.048* |
| Chain (surveyor or gunter) | meter | $+01$ 2.011 68* |
| Circular mil | meter$^2$ | $-10$ 5.067 074 8 |
| Cord | meter$^3$ | $+00$ 3.624 556 3 |
| Coulomb | coulomb | $-01$ 9.998 35 |
| Cubit | meter | $-01$ 4.572* |
| Cup | meter$^3$ | $-04$ 2.365 882 365* |
| Curie | disintegration/second | $+10$ 3.70* |
| Day (mean solar) | second (mean solar) | $+04$ 8.64* |
| Day (sidereal) | second (mean solar) | $+04$ 8.616 409 0 |
| Degree (angle) | radian | $-02$ 1.745 329 251 994 3 |
| Denier (international) | kilogram/meter | $-07$ 1.00* |
| Dram (avoirdupois) | kilogram | $-03$ 1.771 845 195 312 5* |
| Dram (troy or apothecary) | kilogram | $-03$ 3.887 934 6* |
| Dram (U.S. fluid) | meter$^3$ | $-06$ 3.696 691 195 312 5* |
| Dyne | newton | $-05$ 1.00* |
| Electron volt | joule | $-19$ 1.602 10 |
| Erg | joule | $-07$ 1.00* |
| Fahrenheit (temperature) | kelvin | $t_K = (5/9)(t_F + 459.67)$ |
| Fahrenheit (temperature) | Celsius | $t_C = (5/9)(t_F - 32)$ |
| Farad | farad | $-01$ 9.995 05 |
| Faraday (based on carbon 12) | coulomb | $+04$ 9.648 70 |
| Faraday (chemical) | coulomb | $+04$ 9.649 57 |
| Faraday (physical) | coulomb | $+04$ 9.652 19 |
| Fathom | meter | $+00$ 1.828 8* |
| Fermi (femtometer) | meter | $-15$ 1.00* |
| Fluid ounce (U.S.) | meter$^3$ | $-05$ 2.957 352 956 25* |
| Foot | meter | $-01$ 3.048* |
| Foot (U.S. survey) | meter | $+00$ 1200/3937* |

| To convert from | To | Multiply by |
|---|---|---|
| Foot (U.S. survey) | meter | −01 3.048 006 096 |
| Foot of water (39.2°F) | newton/meter$^2$ | +03 2.988 98 |
| Foot-candle | lumen/meter$^2$ | +01 1.076 391 0 |
| Foot-lambert | candela/meter$^2$ | +00 3.426 259 |
| Furlong | meter | +02 2.011 68* |
| Gal (galileo) | meter/second$^2$ | −02 1.00* |
| Gallon (U.K. liquid) | meter$^3$ | −03 4.546 087 |
| Gallon (U.S. dry) | meter$^3$ | −03 4.404 883 770 86* |
| Gallon (U.S. liquid) | meter$^3$ | −03 3.785 411 784* |
| Gamma | tesla | −09 1.00* |
| Gauss | tesla | −04 1.00* |
| Gilbert | ampere turn | −01 7.957 747 2 |
| Gill (U.K.) | meter$^3$ | −04 1.420 652 |
| Gill (U.S.) | meter$^3$ | −04 1.182 941 2 |
| Grad | degree (angular) | −01 9.00* |
| Grad | radian | −02 1.570 796 3 |
| Grain | kilogram | −05 6.479 891* |
| Gram | kilogram | −03 1.00* |
| Hand | meter | −01 1.016* |
| Hectare | meter$^2$ | +04 1.00* |
| Henry | henry | +00 1.000 495 |
| Hogshead (U.S.) | meter$^3$ | −01 2.384 809 423 92* |
| Horsepower (550 ft lbf/s) | watt | +02 7.456 998 7 |
| Horsepower (boiler) | watt | +03 9.809 50 |
| Horsepower (electric) | watt | +02 7.46* |
| Horsepower (metric) | watt | +02 7.354 99 |
| Horsepower (U.K.) | watt | +02 7.457 |
| Horsepower (water) | watt | +02 7.460 43 |
| Hour (mean solar) | second (mean solar) | +03 3.60* |
| Hour (sidereal) | second (mean solar) | +03 3.590 170 4 |
| Hundredweight (long) | kilogram | +01 5.080 234 544* |
| Hundredweight (short) | kilogram | +01 4.535 923 7* |
| Inch | meter | −02 2.54* |
| Inch of mercury (32°F) | newton/meter$^2$ | +03 3.386 389 |
| Inch of mercury (60°F) | newton/meter$^2$ | +03 3.376 85 |
| Inch of water (39.2°F) | newton/meter$^2$ | +02 2.490 82 |
| Inch of water (60°F) | newton/meter$^2$ | +02 2.4884 |
| Joule | joule | +00 1.000 165 |
| Kayser | 1/meter | +02 1.00* |
| Kilocalorie (International Steam Table) | joule | +03 4.186 74 |
| Kilocalorie (mean) | joule | +03 4.190 02 |
| Kilocalorie (thermochemical) | joule | +03 4.184* |
| Kilogram mass | kilogram | +00 1.00* |
| Kilogram force (kgf) | newton | +00 9.806 65* |

| To convert from | To | Multiply by |
|---|---|---|
| Kilopond force | newton | +00 9.806 65* |
| Kip | newton | +03 4.448 221 615 260 5* |
| Knot (international) | meter/second | −01 5.144 444 444 |
| Lambert | candela/meter$^2$ | +04 1/$\pi$* |
| Lambert | candela/meter$^2$ | +03 3.183 098 8 |
| Langley | joule/meter$^2$ | +04 4.184$^2$ |
| lbf (pound force, avoirdupois) | newton | +00 4.448 221 615 260 5* |
| lbm (pound mass, avoirdupois) | kilogram | −01 4.535 923 7* |
| League (British nautical) | meter | +03 5.559 552* |
| League (international nautical) | meter | +03 5.556* |
| League (statute) | meter | +03 4.828 032* |
| Light year | meter | +15 9.460 55 |
| Link (engineer or ramden) | meter | −01 3.048* |
| Link (surveyor or gunter) | meter | −01 2.011 68* |
| Liter | meter$^3$ | −03 1.00* |
| Lux | lumen/meter$^2$ | +00 1.00* |
| Maxwell | weber | −08 1.00* |
| Meter | wavelengths Kr 86 | +06 1.650 763 73* |
| Micron | meter | −06 1.00* |
| Mil | meter | −05 2.54* |
| Mile (U.S. statute) | meter | +03 1.609 344* |
| Mile (U.K. nautical) | meter | +03 1.853 184* |
| Mile (international nautical) | meter | +03 1.852* |
| Mile (U.S. nautical) | meter | +03 1.852* |
| Millibar | newton/meter$^2$ | +02 1.00* |
| Millimeter of mercury (0°C) | newton/meter$^2$ | +02 1.333 224 |
| Minute (angle) | radian | −04 2.908 882 086 66 |
| Minute (mean solar) | second (mean solar) | +01 6.00* |
| Minute (sidereal) | second (mean solar) | +01 5.983 617 4 |
| Month (mean calendar) | second (mean solar) | +06 2.628* |
| Nautical mile (international) | meter | +03 1.852* |
| Nautical mile (U.S.) | meter | +03 1.852* |
| Nautical mile (U.K.) | meter | +03 1.853 184* |
| Oersted | ampere/meter | +01 7.957 747 2 |
| Ohm | ohm | +00 1.000 495 |
| Ounce force (avoirdupois) | newton | −01 2.780 138 5 |
| Ounce mass (avoirdupois) | kilogram | −02 2.834 952 312 5* |
| Ounce mass (troy or apothecary) | kilogram | −02 3.110 347 68* |
| Ounce (U.S. fluid) | meter$^2$ | −05 2.957 352 956 25* |
| Pace | meter | −01 7.62* |
| Parsec | meter | +16 3.083 74 |
| Pascal | newton/meter$^2$ | +00 1.00* |
| Peck (U.S.) | meter$^3$ | −03 8.809 767 541 72* |
| Pennyweight | kilogram | −03 1.555 173 84* |
| Perch | meter | +00 5.0292* |

| To convert from | To | Multiply by |
|---|---|---|
| Phot | lumen/meter$^2$ | +04 1.00 |
| Pica (printers) | meter | −03 4.217 517 6* |
| Pint (U.S. dry) | meter$^3$ | −04 5.506 104 713 575* |
| Pint (U.S. liquid) | meter$^3$ | −04 4.731 764 73* |
| Point (printers) | meter | −04 3.514 598* |
| Poise | newton second/meter$^2$ | −01 1.00* |
| Pole | meter | +00 5.0292* |
| Pound force (lbf avoirdupois) | newton | +00 4.448 221 615 260 5* |
| Pound mass (lbm avoirdupois) | kilogram | −01 4.535 923 7* |
| Pound mass (troy or apothecary) | kilogram | −01 3.732 417 216* |
| Poundal | newton | −01 1.382 549 543 76* |
| Quart (U.S. dry) | meter$^3$ | −03 1.101 220 942 715* |
| Quart (U.S. liquid) | meter$^3$ | −04 9.463 529 5 |
| Rad (radiation dose absorbed) | joule/kilogram | −02 1.00* |
| Rankine (temperature) | kelvin | $t_K = (5/9)t_R$ |
| Rayleigh (rate of photon emission) | 1/second meter$^2$ | +10 1.00* |
| Rhe | meter$^2$/newton second | +01 1.00* |
| Rod | meter | +00 5.0292* |
| Roentgen | coulomb/kilogram | −04 2.579 76* |
| Rutherford | disintegration/second | +06 1.00* |
| Second (angle) | radian | −06 4.848 136 811 |
| Second (ephemeris) | second | +00 1.000 000 000 |
| Second (mean solar) | second (ephemeris) | Consult American Ephemeris and Nautical Almanac |
| Second (sidereal) | second (mean solar) | −01 9.972 695 7 |
| Section | meter$^2$ | +06 2.589 988 110 336* |
| Scruple (apothecary) | kilogram | −03 1.295 978 2* |
| Shake | second | −08 1.00 |
| Skein | meter | +02 1.097 28* |
| Slug | kilogram | +01 1.459 390 29 |
| Span | meter | −01 2.286* |
| Statampere | ampere | −10 3.335 640 |
| Statcoulomb | coulomb | −10 3.335 640 |
| Statfarad | farad | −12 1.112 650 |
| Stathenry | henry | +11 8.987 554 |
| Statmho | mho | −12 1.112 650 |
| Statohm | ohm | +11 8.987 554 |
| Statute mile (U.S.) | meter | +03 1.609 344* |
| Statvolt | volt | +02 2.997 925 |
| Stere | meter$^3$ | +00 1.00* |
| Stilb | candela/meter$^2$ | +04 1.00 |
| Stoke | meter$^2$/second | −04 1.00* |
| Tablespoon | meter$^3$ | −05 1.478 676 478 125* |

| To convert from | To | Multiply by |
|---|---|---|
| Teaspoon | meter$^3$ | $-06$ 4.928 921 593 75* |
| Ton (assay) | kilogram | $-02$ 2.916 666 6 |
| Ton (long) | kilogram | $+03$ 1.016 046 908 8* |
| Ton (metric) | kilogram | $+03$ 1.00* |
| Ton (nuclear equivalent of TNT) | joule | $+09$ 4.20 |
| Ton (register) | meter$^3$ | $+00$ 2.831 684 659 2* |
| Ton (short, 2000 lb) | kilogram | $+02$ 9.071 847 4* |
| Tonne | kilogram | $+03$ 1.00* |
| Torr (0°C) | newton/meter$^2$ | $+02$ 1.333 22 |
| Township | meter$^2$ | $+07$ 9.323 957 2 |
| Unit pole | weber | $-07$ 1.256 637 |
| Volt | volt | $+00$ 1.000 330 |
| Watt | watt | $+00$ 1.000 165 |
| Yard | meter | $-01$ 9.144* |
| Year (calendar) | second (mean solar) | $+07$ 3.1536* |
| Year (sidereal) | second (mean solar) | $+07$ 3.155 815 0 |
| Year (tropical) | second (mean solar) | $+07$ 3.155 692 6 |
| Year 1900, tropical, Jan., day 0, hour 12 | second (ephemeris) | $+07$ 3.155 692 597 47* |
| Year 1900, tropical, Jan., day 0, hour 12 | second | $+07$ 3.155 692 597 47 |

## Listing by physical quantity

### Acceleration

| | | |
|---|---|---|
| Foot/second$^2$ | meter/second$^2$ | $-01$ 3.048* |
| Free fall, standard | meter/second$^2$ | $+00$ 9.806 65* |
| Gal (galileo) | meter/second$^2$ | $-02$ 1.00* |
| Inch/second$^2$ | meter/second$^2$ | $-02$ 2.54* |

### Area

| | | |
|---|---|---|
| Acre | meter$^2$ | $+03$ 4.046 856 422 4* |
| Are | meter$^2$ | $+02$ 1.00* |
| Barn | meter$^2$ | $-28$ 1.00* |
| Circular mil | meter$^2$ | $-10$ 5.067 074 8 |
| Foot$^2$ | meter$^2$ | $-02$ 9.290 304* |
| Hectare | meter$^2$ | $+04$ 1.00* |
| Inch$^2$ | meter$^2$ | $-04$ 6.4516* |
| Mile$^2$ (U.S. statute) | meter$^2$ | $+06$ 2.589 988 110 336* |
| Section | meter$^2$ | $+06$ 2.589 988 110 336* |
| Township | meter$^2$ | $+07$ 9.323 957 2 |
| Yard$^2$ | meter$^2$ | $-01$ 8.361 273 6* |

| To convert from | To | Multiply by |
|---|---|---|
| **Density** | | |
| Gram/centimeter$^3$ | kilogram/meter$^3$ | +03 1.00* |
| lbm/inch$^3$ | kilogram/meter$^3$ | +04 2.767 990 5 |
| lbm/foot$^3$ | kilogram/meter$^3$ | +01 1.601 846 3 |
| Slug/foot$^3$ | kilogram/meter$^3$ | +02 5.153 79 |
| **Energy** | | |
| British thermal unit (ISO/TC 12) | joule | +03 1.055 06 |
| British thermal unit (International Steam Table) | joule | +03 1.055 04 |
| British thermal unit (mean) | joule | +03 1.055 87 |
| British thermal unit (thermochemical) | joule | +03 1.054 350 264 488 |
| British thermal unit (39°F) | joule | +03 1.059 67 |
| British thermal unit (60°F) | joule | +03 1.054 68 |
| Calorie (International Steam Table) | joule | +00 4.1868 |
| Calorie (mean) | joule | +00 4.190 02 |
| Calorie (thermochemical) | joule | +00 4.184* |
| Calorie (15°C) | joule | +00 4.185 80 |
| Calorie (20°C) | joule | +00 4.181 90 |
| Calorie (kilogram, International Steam Table) | joule | +03 4.1868 |
| Calorie (kilogram, mean) | joule | +03 4.190 02 |
| Calorie (kilogram, thermochemical) | joule | +03 4.184* |
| Electron volt | joule | −19 1.602 10 |
| Erg | joule | −07 1.00* |
| Foot lbf | joule | +00 1.355 817 9 |
| Foot poundal | joule | −02 4.214 011 0 |
| Joule (international of 1948) | joule | +00 1.000 165 |
| Kilocalorie (International Steam Table) | joule | +03 4.1868 |
| Kilocalorie (mean) | joule | +03 4.190 02 |
| Kilocalorie (thermochemical) | joule | +03 4.184* |
| Kilowatt hour | joule | +06 3.60* |
| Kilowatt hour | joule | +06 3.600 59 |
| Ton (nuclear equivalent of TNT) | joule | +09 4.20 |
| Watt hour | joule | +03 3.60* |
| **Energy/area time** | | |
| Btu (thermochemical)/foot$^2$ second | watt/meter$^2$ | +04 1.134 893 1 |
| Btu (thermochemical)/foot$^2$ minute | watt/meter$^2$ | +02 1.891 488 5 |
| Btu (thermochemical)/foot$^2$ hour | watt/meter$^2$ | +00 3.152 480 8 |
| Btu (thermochemical)/inch$^2$ second | watt/meter$^2$ | +06 1.634 246 2 |

| To convert from | To | Multiply by |
|---|---|---|
| Calorie (thermochemical)/cm$^2$ minute | watt/meter$^2$ | +02 6.973 333 3 |
| Erg/centimeter$^2$ second | watt/meter$^2$ | −03 1.00* |
| Watt/centimeter$^2$ | watt/meter$^2$ | +04 1.00* |

## Force

| | | |
|---|---|---|
| Dyne | newton | −05 1.00* |
| Kilogram force (kgf) | newton | +00 9.806 65* |
| Kilopond force | newton | +00 9.806 65* |
| Kip | newton | +03 4.448 221 615 260 5* |
| lbf (pound force, avoirdupois) | newton | +00 4.448 221 615 260 5* |
| Ounce force (avoirdupois) | newton | −01 2.780 138 5 |
| Pound force, lbf (avoirdupois) | newton | +00 4.448 221 615 260 5* |
| Poundal | newton | −01 1.382 549 543 76* |

## Length

| | | |
|---|---|---|
| Angstrom | meter | −10 1.00* |
| Astronomical unit | meter | +11 1.495 978 9 |
| Cable | meter | +02 2.194 56* |
| Caliber | meter | −04 2.54* |
| Chain (surveyor or gunter) | meter | +01 2.011 68* |
| Chain (engineer or ramden) | meter | +01 3.048* |
| Cubit | meter | −01 4.572* |
| Fathom | meter | +00 1.8288* |
| Fermi (femtometer) | meter | −15 1.00* |
| Foot | meter | −01 3.048* |
| Foot (U.S. survey) | meter | +00 1200/3937* |
| Foot (U.S. survey) | meter | −01 3.048 006 096 |
| Furlong | meter | +02 2.011 68* |
| Hand | meter | −01 1.016* |
| Inch | meter | −02 2.54* |
| League (U.K. nautical) | meter | +03 5.559 552* |
| League (international nautical) | meter | +03 5.556* |
| League (statute) | meter | +03 4.828 032* |
| Light year | meter | +15 9.460 55 |
| Link (engineer or ramden) | meter | −01 3.048* |
| Link (surveyor or gunter) | meter | −01 2.011 68* |
| Meter | wavelengths Kr 86 | +06 1.650 763 73* |
| Micron | meter | −06 1.00* |
| Mil | meter | −05 2.54* |
| Mile (U.S. statute) | meter | +03 1.609 344* |
| Mile (U.K. nautical) | meter | +03 1.853 184* |
| Mile (international nautical) | meter | +03 1.852* |
| Mile (U.S. nautical) | meter | +03 1.852* |
| Nautical mile (U.K.) | meter | +03 1.853 184* |

| To convert from | To | Multiply by |
|---|---|---|
| Nautical mile (international) | meter | +03 1.852* |
| Nautical mile (U.S.) | meter | +03 1.852* |
| Pace | meter | −01 7.62* |
| Parsec | meter | +16 3.083 74 |
| Perch | meter | +00 5.0292* |
| Pica (printers) | meter | −03 4.217 517 6* |
| Point (printers) | meter | −04 3.514 598* |
| Pole | meter | +00 5.0292* |
| Rod | meter | +00 5.0292* |
| Skein | meter | +02 1.097 28* |
| Span | meter | −01 2.286* |
| Statute mile (U.S.) | meter | +03 1.609 344* |
| Yard | meter | −01 9.144* |

## Mass

| | | |
|---|---|---|
| Carat (metric) | kilogram | −04 2.00* |
| Dram (avoirdupois) | kilogram | −03 1.771 845 195 312 5* |
| Dram (troy or apothecary) | kilogram | −03 3.887 934 6* |
| Grain | kilogram | −05 6.479 891* |
| Gram | kilogram | −03 1.00* |
| Hundredweight (long) | kilogram | +01 5.080 234 544* |
| Hundredweight (short) | kilogram | +01 4.535 923 7* |
| kgf second$^2$ meter (mass) | kilogram | +00 9.806 65* |
| Kilogram mass | kilogram | +00 1.00* |
| lbm (pound mass, avoirdupois) | kilogram | −01 4.535.923 7* |
| Ounce mass (avoirdupois) | kilogram | −02 2.834 952 312 5* |
| Ounce mass (troy or apothecary) | kilogram | −02 3.110 347 68* |
| Pennyweight | kilogram | −03 1.555 173 84* |
| Pound mass, lbm (avoirdupois) | kilogram | −01 4.535 923 7* |
| Pound mass (troy or apothecary) | kilogram | −01 3.732 417 216* |
| Scruple (apothecary) | kilogram | −03 1.295 978 2* |
| Slug | kilogram | +01 1.459 390 29 |
| Ton (assay) | kilogram | −02 2.916 666 6 |
| Ton (long) | kilogram | +03 1.016 046 908 8* |
| Ton (metric) | kilogram | +03 1.00* |
| Ton (short, 2000 lb) | kilogram | +02 9.071 847 4* |
| Tonne | kilogram | +03 1.00* |

## Power

| | | |
|---|---|---|
| Btu (thermochemical/second | watt | +03 1.054 350 264 488 |
| Btu (thermochemical/minute | watt | +01 1.757 250 4 |
| Calorie (thermochemical)/second | watt | +00 4.184* |
| Calorie (thermochemical)/minute | watt | −02 6.973 333 3 |
| Foot lbf/hour | watt | −04 3.766 161 0 |
| Foot lbf/minute | watt | −02 2.259 696 6 |

| To convert from | To | Multiply by |
|---|---|---|
| Foot lbf/second | watt | +00 1.355 817 9 |
| Horsepower (550 ft lbf/s) | watt | +02 7.456 998 7 |
| Horsepower (boiler) | watt | +03 9.809 50 |
| Horsepower (electric) | watt | +02 7.46* |
| Horsepower (metric) | watt | +02 7.354 99 |
| Horsepower (U.K.) | watt | +02 7.457 |
| Horsepower (water) | watt | +02 7.460 43 |
| Kilocalorie (thermochemical)/ minute | watt | +01 6.973 333 3 |
| Kilocalorie (thermochemical)/ second | watt | +03 4.184* |
| Watt | watt | +00 1.000 165 |

**Pressure**

| To convert from | To | Multiply by |
|---|---|---|
| Atmosphere | newton/meter$^2$ | +05 1.013 25* |
| Bar | newton/meter$^2$ | +05 1.00* |
| Barye | newton/meter$^2$ | −01 1.00* |
| Centimeter of mercury (0°C) | newton/meter$^2$ | +03 1.333 22 |
| Centimeter of water (4°C) | newton/meter$^2$ | +01 9.806 38 |
| Dyne/centimeter$^2$ | newton/meter$^2$ | −01 1.00* |
| Foot of water (39.2°F) | newton/meter$^2$ | +03 2.988 98 |
| Inch of mercury (32°F) | newton/meter$^2$ | +03 3.386 389 |
| Inch of mercury (60°F) | newton/meter$^2$ | +03 3.376 85 |
| Inch of water (39.2°F) | newton/meter$^2$ | +02 2.490 82 |
| Inch of water (60°F) | newton/meter$^2$ | +02 2.4884 |
| kgf/centimeter$^2$ | newton/meter$^2$ | +04 9.806 65* |
| kgf/meter$^2$ | newton/meter$^2$ | +00 9.806 65* |
| lbf/foot$^2$ | newton/meter$^2$ | +01 4.788 025 8 |
| lbf/inch$^2$ (psi) | newton/meter$^2$ | +03 6.894 757 2 |
| Millibar | newton/meter$^2$ | +02 1.00* |
| Millimeter of mercury (0°C) | newton/meter$^2$ | +02 1.333 224 |
| Pascal | newton/meter$^2$ | +00 1.00* |
| psi (lbf/inch$^2$) | newton/meter$^2$ | +03 6.894 757 2 |
| Torr (0°C) | newton/meter$^2$ | +02 1.333 22 |

**Speed**

| To convert from | To | Multiply by |
|---|---|---|
| Foot/hour | meter/second | −05 8.466 666 6 |
| Foot/minute | meter/second | −03 5.08* |
| Foot/second | meter/second | −01 3.048* |
| Inch/second | meter/second | −02 2.54* |
| Kilometer/hour | meter/second | −01 2.777 777 8 |
| Knot (international) | meter/second | −01 5.144 444 444 |
| Mile/hour (U.S. statute) | meter/second | −01 4.4704* |
| Mile/minute (U.S. statute) | meter/second | +01 2.682 24* |
| Mile/second (U.S. statute) | meter/second | +03 1.609 344* |

| To convert from | To | Multiply by |
|---|---|---|

## Temperature

| | | |
|---|---|---|
| Celsius | kelvin | $t_K = t_C + 273.15$ |
| Fahrenheit | kelvin | $t_K = (5/9)(t_F + 459.67)$ |
| Fahrenheit | Celsius | $t_C = (5/9)(t_F - 32)$ |
| Rankine | kelvin | $t_K = (5/9)t_R$ |

## Time

| | | |
|---|---|---|
| Day (mean solar) | second (mean solar) | +04 8.64* |
| Day (sidereal) | second (mean solar) | +04 8.616 409 0 |
| Hour (mean solar) | second (mean solar) | +03 3.60* |
| Hour (sidereal) | second (mean solar) | +03 3.590 170 4 |
| Minute (mean solar) | second (mean solar) | +01 6.00* |
| Minute (sidereal) | second (mean solar) | +01 5.983 617 4 |
| Month (mean calendar) | second (mean solar) | +06 2.628* |
| Second (ephemeris) | second | +00 1.000 000 000 |
| Second (mean solar) | second (ephemeris) | Consult American Ephemeris and Nautical Almanac |
| Second (sidereal) | second (mean solar) | −01 9.972 695 7 |
| Year (calendar) | second (mean solar) | +07 3.1536* |
| Year (sidereal) | second (mean solar) | +07 3.155 815 0 |
| Year (tropical) | second (mean solar) | +07 3.155 692 6 |
| Year 1900, tropical, Jan., day 0, hour 12 | second (ephemeris) | +07 3.155 692 597 47* |
| Year 1900, tropical, Jan., day 0, hour 12 | second | +07 3.155 692 597 47 |

## Viscosity

| | | |
|---|---|---|
| Centistoke | meter$^2$/second | −06 1.00* |
| Stoke | meter$^2$/second | −04 1.00* |
| Foot$^2$/second | meter$^2$/second | −02 9.290 304* |
| Centipoise | newton second/meter$^2$ | −03 1.00* |
| lbm/foot second | newton second/meter$^2$ | +00 1.488 163 9 |
| lbf second/foot$^2$ | newton second/meter$^2$ | +01 4.788 025 8 |
| Poise | newton second/meter$^2$ | −01 1.00* |
| Poundal second/foot$^2$ | newton second/meter$^2$ | +00 1.488 163 9 |
| Slug/foot second | newton second/meter$^2$ | +01 4.788 025 8 |
| Rhe | meter$^2$/newton second | +01 1.00* |

## Volume

| | | |
|---|---|---|
| Acre foot | meter$^3$ | +03 1.233 481 9 |
| Barrel (petroleum, 42 gal) | meter$^3$ | −01 1.589 873 |
| Board foot | meter$^3$ | −03 2.359 737 216* |
| Bushel (U.S.) | meter$^3$ | −02 3.523 907 016 688* |

| To convert from | To | Multiply by |
|---|---|---|
| Cord | meter$^3$ | $+00$ 3.624 556 3 |
| Cup | meter$^3$ | $-04$ 2.365 882 365 * |
| Dram (U.S. fluid) | meter$^3$ | $-06$ 3.696 691 195 312 5 * |
| Fluid ounce (U.S.) | meter$^3$ | $-05$ 2.957 352 956 25 * |
| Foot$^3$ | meter$^3$ | $-02$ 2.831 684 659 2 * |
| Gallon (U.K. liquid) | meter$^3$ | $-03$ 4.546 087 |
| Gallon (U.S. dry) | meter$^3$ | $-03$ 4.404 883 770 86 * |
| Gallon (U.S. liquid) | meter$^3$ | $-03$ 3.785 411 784 * |
| Gill (U.K.) | meter$^3$ | $-04$ 1.420 652 |
| Gill (U.S.) | meter$^3$ | $-04$ 1.182 941 2 |
| Hogshead (U.S.) | meter$^3$ | $-01$ 2.384 809 423 92 * |
| Inch$^3$ | meter$^3$ | $-05$ 1.638 706 4 * |
| Liter | meter$^3$ | $-03$ 1.00 * |
| Ounce (U.S. fluid) | meter$^3$ | $-05$ 2.957 352 956 25 * |
| Peck (U.S.) | meter$^3$ | $-03$ 8.809 767 541 72 * |
| Pint (U.S. dry) | meter$^3$ | $-04$ 5.506 104 713 575 * |
| Pint (U.S. liquid) | meter$^3$ | $-04$ 4.731 764 73 * |
| Quart (U.S. dry) | meter$^3$ | $-03$ 1.101 220 942 715 * |
| Quart (U.S. liquid) | meter$^3$ | $-04$ 9.463 529 5 |
| Stere | meter$^3$ | $+00$ 1.00 * |
| Tablespoon | meter$^3$ | $-05$ 1.478 676 478 125 * |
| Teaspoon | meter$^3$ | $-06$ 4.928 921 593 75 * |
| Ton (register) | meter$^3$ | $+00$ 2.831 684 659 2 * |
| Yard$^3$ | meter$^3$ | $-01$ 7.645 548 579 84 * |

# ISO METRIC SCREW THREADS

| Nominal size | Pitch $P$ | Basic major diameter $d$ and $D$, mm | Basic pitch diameter $d_2$ and $D_2$, mm | Basic minor diameter $d_1$ and $D_1$, mm | Minor diameter of external thread $d_{3max}$, mm | Basic major diameter $d$ and $D$, in | Basic pitch diameter $d_2$ and $D_2$, in | Basic minor diameter $d_1$ and $D_1$, in | Minor diameter of external thread $d_{3max}$, in |
|---|---|---|---|---|---|---|---|---|---|
| 0.25 | 0.075 | 0.250000 | 0.201286 | 0.168810 | 0.157985 | 0.009842 | 0.007925 | 0.006646 | 0.006220 |
| 0.3 | 0.08 | 0.300000 | 0.248038 | 0.213397 | 0.201850 | 0.011811 | 0.009765 | 0.008401 | 0.007947 |
| 0.35 | 0.09 | 0.350000 | 0.291543 | 0.252572 | 0.239582 | 0.013779 | 0.011478 | 0.009944 | 0.009432 |
| 0.4 | 0.1 | 0.400000 | 0.335048 | 0.291747 | 0.277313 | 0.015748 | 0.013191 | 0.011486 | 0.010918 |
| 0.45 | 0.1 | 0.450000 | 0.385048 | 0.341747 | 0.327313 | 0.017716 | 0.015159 | 0.013454 | 0.012886 |
| 0.5 | 0.125 | 0.500000 | 0.418810 | 0.364684 | 0.346641 | 0.019685 | 0.016488 | 0.014358 | 0.013647 |
| 0.55 | 0.125 | 0.550000 | 0.468810 | 0.414684 | 0.396641 | 0.021653 | 0.018457 | 0.016326 | 0.015616 |
| 0.6 | 0.15 | 0.600000 | 0.502572 | 0.437620 | 0.415970 | 0.023622 | 0.019786 | 0.017229 | 0.016377 |
| 0.7 | 0.175 | 0.700000 | 0.586334 | 0.510557 | 0.485298 | 0.027559 | 0.023084 | 0.020101 | 0.019106 |
| 0.8 | 0.2 | 0.800000 | 0.670096 | 0.583494 | 0.554626 | 0.031496 | 0.026382 | 0.022972 | 0.021836 |
| 0.9 | 0.225 | 0.900000 | 0.753858 | 0.656430 | 0.623954 | 0.035433 | 0.029679 | 0.025844 | 0.024565 |
| 1 | 0.25 | 1.000000 | 0.837620 | 0.729367 | 0.693283 | 0.039370 | 0.032977 | 0.028715 | 0.027294 |
| 1 | 0.2 | 1.000000 | 0.870096 | 0.783494 | 0.754626 | 0.039370 | 0.034256 | 0.030846 | 0.029710 |
| 1.1 | 0.25 | 1.100000 | 0.937620 | 0.829367 | 0.793283 | 0.043307 | 0.036914 | 0.032652 | 0.031231 |
| 1.1 | 0.2 | 1.100000 | 0.970096 | 0.883494 | 0.854626 | 0.043307 | 0.038193 | 0.034783 | 0.033647 |
| 1.2 | 0.25 | 1.200000 | 1.037620 | 0.929367 | 0.893283 | 0.047244 | 0.040851 | 0.036589 | 0.035168 |
| 1.2 | 0.2 | 1.200000 | 1.070096 | 0.983494 | 0.954626 | 0.047244 | 0.042130 | 0.038720 | 0.037584 |
| 1.4 | 0.3 | 1.400000 | 1.205144 | 1.075240 | 1.031939 | 0.055118 | 0.047446 | 0.042332 | 0.040627 |
| 1.4 | 0.2 | 1.400000 | 1.270096 | 1.183494 | 1.154626 | 0.055118 | 0.050004 | 0.046594 | 0.045458 |
| 1.6 | 0.35 | 1.600000 | 1.372668 | 1.221114 | 1.170596 | 0.062992 | 0.054042 | 0.048075 | 0.046086 |
| 1.6 | 0.2 | 1.600000 | 1.470096 | 1.383494 | 1.354626 | 0.062992 | 0.057878 | 0.054468 | 0.053332 |
| 1.8 | 0.35 | 1.800000 | 1.572668 | 1.421114 | 1.370596 | 0.070866 | 0.061916 | 0.055949 | 0.053960 |
| 1.8 | 0.2 | 1.800000 | 1.670096 | 1.583494 | 1.554626 | 0.070866 | 0.065752 | 0.062342 | 0.061206 |
| 2 | 0.4 | 2.000000 | 1.740192 | 1.566987 | 1.509252 | 0.078740 | 0.068511 | 0.061692 | 0.059419 |
| 2 | 0.25 | 2.000000 | 1.837620 | 1.729367 | 1.693283 | 0.078740 | 0.072347 | 0.068085 | 0.066664 |
| 2.2 | 0.45 | 2.200000 | 1.907716 | 1.712861 | 1.647909 | 0.086614 | 0.075107 | 0.067435 | 0.064878 |
| 2.2 | 0.25 | 2.200000 | 2.037620 | 1.929367 | 1.893283 | 0.086614 | 0.080221 | 0.075955 | 0.074538 |
| 2.5 | 0.45 | 2.500000 | 2.207716 | 2.012861 | 1.947909 | 0.098425 | 0.086918 | 0.079246 | 0.076689 |
| 2.5 | 0.35 | 2.500000 | 2.272668 | 1.121114 | 2.070596 | 0.098425 | 0.089475 | 0.083508 | 0.081519 |
| 3 | 0.5 | 3.000000 | 2.675240 | 2.458734 | 2.386565 | 0.118110 | 0.105324 | 0.096800 | 0.093959 |
| 3 | 0.35 | 3.000000 | 2.772668 | 2.621114 | 2.570596 | 0.118110 | 0.109160 | 0.103193 | 0.101204 |
| 3.5 | 0.6 | 3.500000 | 3.110289 | 2.850481 | 2.763878 | 0.137795 | 0.122452 | 0.112223 | 0.108814 |
| 3.5 | 0.35 | 3.500000 | 3.272668 | 3.121114 | 3.070596 | 0.137795 | 0.128845 | 0.122878 | 0.120889 |
| 4 | 0.7 | 4.000000 | 3.545337 | 3.242228 | 3.141191 | 0.157480 | 0.139580 | 0.127646 | 0.123669 |
| 4 | 0.5 | 4.000000 | 3.675240 | 3.458734 | 3.386565 | 0.157480 | 0.144694 | 0.136170 | 0.133329 |
| 4.5 | 0.75 | 4.500000 | 4.012861 | 3.688101 | 3.579848 | 0.177165 | 0.157986 | 0.145200 | 0.140939 |
| 4.5 | 0.5 | 4.500000 | 4.175240 | 3.958734 | 3.886565 | 0.177165 | 0.164379 | 0.155855 | 0.153014 |
| 5 | 0.8 | 5.000000 | 4.480385 | 4.133975 | 4.018505 | 0.196850 | 0.176393 | 0.162755 | 0.158208 |
| 5 | 0.5 | 5.000000 | 4.675240 | 4.458734 | 4.386565 | 0.196850 | 0.184064 | 0.175540 | 0.172699 |
| 5.5 | 0.5 | 5.500000 | 5.175240 | 4.958734 | 4.886565 | 0.216535 | 0.203749 | 0.195225 | 0.192384 |
| 6 | 1 | 6.000000 | 5.350481 | 4.917468 | 4.773131 | 0.236220 | 0.210648 | 0.193601 | 0.187918 |
| 6 | 0.75 | 6.000000 | 5.512861 | 5.188101 | 5.079848 | 0.236220 | 0.217041 | 0.204255 | 0.199994 |

| Nominal size | Pitch $P$ | Basic major diameter $d$ and $D$, mm | Basic pitch diameter $d_2$ and $D_2$, mm | Basic minor diameter $d_1$ and $D_1$, mm | Minor diameter of external thread $d_{3max}$, mm | Basic major diameter $d$ and $D$, in | Basic pitch diameter $d_2$ and $D_2$, in | Basic minor diameter $d_1$ and $D_1$, in | Minor diameter of external thread $d_{3max}$, in |
|---|---|---|---|---|---|---|---|---|---|
| 7 | 1 | 7.000000 | 6.350481 | 5.917468 | 5.773131 | 0.275590 | 0.250018 | 0.232971 | 0.227288 |
|  | 0.75 | 7.000000 | 6.512861 | 6.188101 | 6.079848 | 0.275590 | 0.256411 | 0.243625 | 0.239364 |
|  | 1.25 | 8.000000 | 7.188101 | 6.646835 | 6.466413 | 0.314960 | 0.282995 | 0.261686 | 0.254583 |
| 8 | 1 | 8.000000 | 7.350481 | 6.917468 | 6.773131 | 0.314960 | 0.289388 | 0.272341 | 0.266658 |
|  | 0.75 | 8.000000 | 7.512861 | 7.188101 | 7.079848 | 0.314960 | 0.295781 | 0.282995 | 0.278734 |
|  | 1.25 | 9.000000 | 8.188101 | 7.646835 | 7.466413 | 0.354330 | 0.322365 | 0.301056 | 0.293953 |
| 9 | 1 | 9.000000 | 8.350481 | 7.917468 | 7.773131 | 0.354330 | 0.328758 | 0.311711 | 0.306028 |
|  | 0.75 | 9.000000 | 8.512861 | 8.188101 | 8.079848 | 0.354330 | 0.335151 | 0.322365 | 0.318104 |
|  | 1.5 | 10.000000 | 9.025721 | 8.376202 | 8.159696 | 0.393700 | 0.355343 | 0.329771 | 0.321247 |
| 10 | 1.25 | 10.000000 | 9.188101 | 8.646835 | 8.466413 | 0.393700 | 0.361735 | 0.340426 | 0.333322 |
|  | 1 | 10.000000 | 9.350481 | 8.917468 | 8.773131 | 0.393700 | 0.368128 | 0.351081 | 0.345398 |
|  | 0.75 | 10.000000 | 9.512861 | 9.188101 | 9.079848 | 0.393700 | 0.374521 | 0.361735 | 0.357474 |
|  | 1.5 | 11.000000 | 10.025721 | 9.376202 | 9.159696 | 0.423070 | 0.394713 | 0.369141 | 0.360617 |
| 11 | 1 | 11.000000 | 10.350481 | 9.917468 | 9.773131 | 0.433070 | 0.407498 | 0.390451 | 9.384768 |
|  | 0.75 | 11.000000 | 10.512861 | 10.188101 | 10.079848 | 0.433070 | 0.413891 | 0.401105 | 0.396844 |
|  | 1.75 | 12.000000 | 10.863342 | 10.105569 | 9.852979 | 0.472440 | 0.427690 | 0.397856 | 0.387912 |
| 12 | 1.5 | 12.000000 | 11.025721 | 10.376202 | 10.159696 | 0.472440 | 0.434083 | 0.408511 | 0.399987 |
|  | 1.25 | 12.000000 | 11.188101 | 10.646835 | 10.466413 | 0.472440 | 0.440475 | 0.419166 | 0.412063 |
|  | 1 | 12.000000 | 11.350481 | 10.917468 | 10.773131 | 0.472440 | 0.446868 | 0.429821 | 0.424138 |
|  | 2 | 14.000000 | 12.700962 | 11.834936 | 11.546261 | 0.551180 | 0.500037 | 0.465941 | 0.454576 |
| 14 | 1.5 | 14.000000 | 13.025721 | 12.376202 | 12.159696 | 0.551180 | 0.512823 | 0.487251 | 0.478727 |
|  | 1.25 | 14.000000 | 13.188101 | 12.646835 | 12.466413 | 0.551180 | 0.519215 | 0.497906 | 0.490803 |
|  | 1 | 14.000000 | 13.350481 | 12.917468 | 12.773131 | 0.551180 | 0.525608 | 0.508561 | 0.502878 |
| 15 | 1.5 | 15.000000 | 14.025721 | 13.376202 | 13.159696 | 0.590550 | 0.552193 | 0.526621 | 0.518097 |
|  | 1 | 15.000000 | 14.350481 | 13.917468 | 13.773131 | 0.590550 | 0.564978 | 0.547931 | 0.542248 |
|  | 2 | 16.000000 | 14.700962 | 13.834936 | 13.546261 | 0.629920 | 0.578777 | 0.544681 | 0.533316 |
| 16 | 1.5 | 16.000000 | 15.025721 | 14.376202 | 14.159696 | 0.629920 | 0.591563 | 0.565991 | 0.557467 |
|  | 1 | 16.000000 | 15.350481 | 14.917468 | 14.773131 | 0.629920 | 0.604348 | 0.587301 | 0.581618 |
| 17 | 1.5 | 17.000000 | 16.025721 | 15.376202 | 15.159696 | 0.669290 | 0.630933 | 0.605361 | 0.596837 |
|  | 1 | 17.000000 | 16.350481 | 15.917468 | 15.773131 | 0.669290 | 0.643718 | 0.626671 | 0.620988 |
|  | 2.5 | 18.000000 | 16.376202 | 15.293671 | 14.932827 | 0.708660 | 0.644731 | 0.602112 | 0.587905 |
| 18 | 2 | 18.000000 | 16.700962 | 15.834936 | 15.546261 | 0.708660 | 0.657517 | 0.623421 | 0.612056 |
|  | 1.5 | 18.000000 | 17.025721 | 16.376202 | 16.159696 | 0.708660 | 0.670303 | 0.644731 | 0.596837 |
|  | 1 | 18.000000 | 17.350481 | 16.917468 | 16.773131 | 0.708660 | 0.683088 | 0.666041 | 0.660358 |
|  | 2.5 | 20.000000 | 18.376202 | 17.293671 | 16.932827 | 0.787400 | 0.723471 | 0.680852 | 0.666645 |
| 20 | 2 | 20.000000 | 18.700962 | 17.834936 | 17.546261 | 0.787400 | 0.736257 | 0.702163 | 0.690796 |
|  | 1.5 | 20.000000 | 19.925721 | 18.376202 | 18.159696 | 0.787400 | 0.749043 | 0.723471 | 0.714947 |
|  | 1 | 20.000000 | 19.350481 | 18.917468 | 18.773131 | 0.787400 | 0.761828 | 0.744781 | 0.739098 |
|  | 2.5 | 22.000000 | 20.376202 | 19.293671 | 18.932827 | 0.866140 | 0.802211 | 0.759592 | 0.745385 |
| 22 | 2 | 22.000000 | 20.700962 | 19.834936 | 19.546261 | 0.866140 | 0.814997 | 0.780901 | 0.769536 |
|  | 1.5 | 22.000000 | 21.025721 | 20.376202 | 20.159696 | 0.866140 | 0.827783 | 0.802211 | 0.793687 |
|  | 1 | 22.000000 | 21.350481 | 20.917468 | 20.773131 | 0.866140 | 0.840568 | 0.823521 | 0.817838 |
|  | 3 | 24.000000 | 22.051443 | 20.752405 | 20.319392 | 0.944880 | 0.868165 | 0.817022 | 0.799974 |
| 24 | 2 | 24.000000 | 22.700962 | 21.834936 | 21.546261 | 0.944880 | 0.893737 | 0.859641 | 0.848276 |
|  | 1.5 | 24.000000 | 23.025721 | 22.376202 | 22.159696 | 0.944880 | 0.906523 | 0.880951 | 0.872427 |
|  | 1 | 24.000000 | 23.350481 | 22.917468 | 22.773131 | 0.944880 | 0.919308 | 0.902261 | 0.896578 |

| Nominal size | Pitch $P$ | Basic major diameter $d$ and $D$, mm | Basic pitch diameter $d_2$ and $D_2$, mm | Basic minor diameter $d_1$ and $D_1$, mm | Minor diameter of external thread $d_{3max}$, mm | Basic major diameter $d$ and $D$, in | Basic pitch diameter $d_2$ and $D_2$, in | Basic minor diameter $d_1$ and $D_1$, in | Minor diameter of external thread $d_{3max}$, in |
|---|---|---|---|---|---|---|---|---|---|
|  | 2 | 25.000000 | 23.700962 | 22.834936 | 22.546261 | 0.984250 | 0.933107 | 0.899011 | 0.887646 |
| 25 | 1.5 | 25.000000 | 24.025721 | 23.376202 | 23.159696 | 0.984250 | 0.945893 | 0.920321 | 0.911797 |
|  | 1 | 25.000000 | 24.350481 | 23.917468 | 23.773131 | 0.984250 | 0.958678 | 0.941631 | 0.935948 |
| 26 | 1.5 | 26.000000 | 25.025721 | 24.376202 | 24.159696 | 1.023620 | 0.985263 | 0.959691 | 0.951167 |
|  | 3 | 27.000000 | 25.051443 | 23.752405 | 23.319392 | 1.062990 | 0.986275 | 0.935132 | 0.918084 |
| 27 | 2 | 27.000000 | 25.700962 | 24.834936 | 24.546261 | 1.062990 | 1.011847 | 0.977751 | 0.966386 |
|  | 1.5 | 27.000000 | 26.025721 | 25.376202 | 25.159696 | 1.062990 | 1.024633 | 0.999061 | 0.990537 |
|  | 1 | 27.000000 | 26.350481 | 25.917468 | 25.773131 | 1.062990 | 1.037418 | 1.020371 | 1.014688 |
|  | 2 | 28.000000 | 26.700962 | 25.834936 | 25.546261 | 1.102360 | 1.051217 | 1.017121 | 1.005756 |
| 28 | 1.5 | 28.000000 | 27.025721 | 26.376202 | 26.159696 | 1.102360 | 1.064003 | 1.038431 | 1.029907 |
|  | 1 | 28.000000 | 27.350481 | 26.917468 | 26.773131 | 1.102360 | 1.076788 | 1.059741 | 1.054058 |
|  | 3.5 | 30.000000 | 27.726683 | 26.211139 | 25.705957 | 1.181100 | 1.091599 | 1.031932 | 1.012043 |
|  | 3 | 30.000000 | 28.051443 | 26.752405 | 26.319392 | 1.181100 | 1.104385 | 1.053242 | 1.036194 |
| 30 | 2 | 30.000000 | 28.700962 | 27.834936 | 27.546261 | 1.181100 | 1.129957 | 1.095861 | 1.084496 |
|  | 1.5 | 30.000000 | 29.025721 | 28.376202 | 28.159696 | 1.181100 | 1.142743 | 1.117171 | 1.108647 |
|  | 1 | 30.000000 | 29.350481 | 28.917468 | 28.773131 | 1.181100 | 1.155528 | 1.138481 | 1.132798 |
| 32 | 2 | 32.000000 | 30.700962 | 29.834936 | 29.546261 | 1.259840 | 1.208697 | 1.174601 | 1.163236 |
|  | 1.5 | 32.000000 | 31.025721 | 30.376202 | 30.259696 | 1.259840 | 1.221483 | 1.195911 | 1.191324 |
|  | 3.5 | 33.000000 | 30.726683 | 29.211139 | 28.705957 | 1.299210 | 1.209709 | 1.150047 | 1.130153 |
| 33 | 3 | 33.000000 | 31.051443 | 29.752405 | 29.319392 | 1.299210 | 1.222495 | 1.171357 | 1.154304 |
|  | 2 | 33.000000 | 31.700962 | 30.834936 | 30.546261 | 1.299210 | 1.248067 | 1.213971 | 1.202606 |
|  | 1.5 | 33.000000 | 32.025721 | 31.376202 | 31.159696 | 1.299210 | 1.260852 | 1.235281 | 1.226757 |
| 35 | 1.5 | 35.000000 | 34.025721 | 33.376202 | 33.159696 | 1.377950 | 1.339593 | 1.314021 | 1.305497 |
|  | 4 | 36.000000 | 33.401924 | 31.669873 | 31.092523 | 1.417320 | 1.315034 | 1.246843 | 1.224113 |
| 36 | 3 | 36.000000 | 34.051443 | 32.752405 | 32.319392 | 1.417320 | 1.340605 | 1.289462 | 1.272414 |
|  | 2 | 36.000000 | 34.700962 | 33.834936 | 33.546261 | 1.417320 | 1.366177 | 1.332081 | 1.320716 |
|  | 1.5 | 36.000000 | 35.025721 | 34.376202 | 34.159696 | 1.417320 | 1.378963 | 1.353391 | 1.344867 |
| 38 | 1.5 | 38.000000 | 37.025721 | 36.376202 | 36.159696 | 1.496060 | 1.457703 | 1.432131 | 1.423607 |
|  | 4 | 39.000000 | 36.401924 | 34.669873 | 34.092523 | 1.535430 | 1.433144 | 1.364953 | 1.342223 |
| 39 | 3 | 39.000000 | 37.051443 | 35.752405 | 35.319392 | 1.535430 | 1.458715 | 1.407572 | 1.390524 |
|  | 2 | 39.000000 | 37.700962 | 36.834936 | 36.546261 | 1.535430 | 1.484287 | 1.450191 | 1.438826 |
|  | 1.5 | 39.000000 | 38.025721 | 37.376202 | 37.159696 | 1.535430 | 1.497073 | 1.719501 | 1.462977 |
|  | 3 | 40.000000 | 38.051443 | 36.752405 | 36.319392 | 1.574800 | 1.498085 | 1.446942 | 1.429894 |
| 40 | 2 | 40.000000 | 38.700962 | 37.834936 | 37.546261 | 1.574800 | 1.523657 | 1.489561 | 1.478196 |
|  | 1.5 | 40.000000 | 39.025821 | 38.376202 | 38.159696 | 1.574800 | 1.536443 | 1.510871 | 1.502347 |
|  | 4.5 | 42.000000 | 39.077164 | 37.128607 | 36.479088 | 1.653540 | 1.538468 | 1.461753 | 1.436182 |
|  | 4 | 42.000000 | 39.401924 | 37.669873 | 37.092523 | 1.653540 | 1.551254 | 1.483063 | 1.460333 |
| 42 | 3 | 42.000000 | 40.051443 | 38.752405 | 38.319392 | 1.653540 | 1.576825 | 1.525682 | 1.508634 |
|  | 2 | 42.000000 | 40.700962 | 39.834936 | 39.546261 | 1.653540 | 1.602397 | 1.568301 | 1.556936 |
|  | 1.5 | 42.000000 | 41.025721 | 40.376202 | 40.159696 | 1.653540 | 1.615183 | 1.589611 | 1.581087 |
|  | 4.5 | 45.000000 | 42.077164 | 40.128607 | 39.479088 | 1.771650 | 1.656578 | 1.579863 | 1.554292 |
|  | 4 | 45.000000 | 42.401924 | 40.669873 | 40.092523 | 1.771650 | 1.669364 | 1.601173 | 1.578443 |
| 45 | 3 | 45.000000 | 43.051443 | 41.752405 | 41.319392 | 1.771650 | 1.694935 | 1.643792 | 1.626744 |
|  | 2 | 45.000000 | 43.700962 | 42.834936 | 42.546261 | 1.771650 | 1.720507 | 1.686411 | 1.675046 |
|  | 1.5 | 45.000000 | 44.025821 | 43.376202 | 43.159696 | 1.771650 | 1.733293 | 1.707721 | 1.699197 |

| Nominal size | Pitch P | Basic major diameter $d$ and $D$, mm | Basic pitch diameter $d_2$ and $D_2$, mm | Basic minor diameter $d_1$ and $D_1$, mm | Minor diameter of external thread $d_{3\max}$, mm | Basic major diameter $d$ and $D$, in | Basic pitch diameter $d_2$ and $D_2$, in | Basic minor diameter $d_1$ and $D_1$, in | Minor diameter of external thread $d_{3\max}$, in |
|---|---|---|---|---|---|---|---|---|---|
|    | 5   | 48.000000 | 44.752405 | 42.587341 | 41.865653 | 1.889760 | 1.761902 | 1.676663 | 1.648251 |
|    | 4   | 48.000000 | 45.401924 | 43.669873 | 43.092523 | 1.889760 | 1.787474 | 1.719283 | 1.696553 |
| 48 | 3   | 48.000000 | 46.051443 | 44.752405 | 44.319392 | 1.889760 | 1.813045 | 1.761902 | 1.744854 |
|    | 2   | 48.000000 | 46.700962 | 45.546261 | 45.834936 | 1.889760 | 1.838617 | 1.793156 | 1.804521 |
|    | 1.5 | 48.000000 | 47.025721 | 46.376202 | 46.159696 | 1.889760 | 1.851403 | 1.825831 | 1.821244 |
|    | 3   | 50.000000 | 48.051443 | 46.752405 | 46.319392 | 1.968500 | 1.891785 | 1.840642 | 1.823594 |
| 50 | 2   | 50.000000 | 48.700962 | 47.834936 | 47.546261 | 1.968500 | 1.917357 | 1.883261 | 1.871896 |
|    | 1.5 | 50.000000 | 49.025721 | 48.376202 | 48.259696 | 1.968500 | 1.930143 | 1.904571 | 1.899984 |
|    | 5   | 52.000000 | 48.752405 | 46.587341 | 45.865653 | 2.047240 | 1.919382 | 1.834144 | 1.805731 |
|    | 4   | 52.000000 | 49.401924 | 47.669873 | 47.092523 | 2.047240 | 1.944954 | 1.876763 | 1.854033 |
| 52 | 3   | 52.000000 | 50.051443 | 48.752405 | 48.319392 | 2.047240 | 1.970525 | 1.919382 | 1.902334 |
|    | 2   | 52.000000 | 50.700962 | 49.834936 | 49.546261 | 2.047240 | 1.996097 | 1.962001 | 1.950636 |
|    | 1.5 | 52.000000 | 51.025721 | 50.376202 | 50.159696 | 2.047240 | 2.008883 | 1.983311 | 1.974787 |
|    | 4   | 55.000000 | 52.401924 | 50.669873 | 50.092523 | 2.165350 | 2.063064 | 1.994873 | 1.972143 |
| 55 | 3   | 55.000000 | 53.051443 | 51.752405 | 51.319392 | 2.165350 | 2.088635 | 2.037492 | 2.020444 |
|    | 2   | 55.000000 | 53.700962 | 52.834936 | 52.546261 | 2.165350 | 2.114207 | 2.080111 | 2.068746 |
|    | 1.5 | 55.000000 | 54.025721 | 53.376202 | 53.159696 | 2.165350 | 2.126993 | 2.101421 | 2.092897 |
|    | 5.5 | 56.000000 | 52.427645 | 50.046075 | 49.252219 | 2.204720 | 2.064076 | 1.970314 | 1.939060 |
|    | 4   | 56.000000 | 53.401924 | 51.669873 | 51.092523 | 2.204720 | 2.102434 | 2.034243 | 2.011513 |
| 56 | 3   | 56.000000 | 54.051443 | 52.752405 | 52.319392 | 2.204720 | 2.128005 | 2.076862 | 2.059814 |
|    | 2   | 56.000000 | 54.700962 | 53.834936 | 53.546261 | 2.204720 | 2.153577 | 2.119481 | 2.108116 |
|    | 1.5 | 56.000000 | 55.025721 | 54.376202 | 54.159696 | 2.204720 | 2.166363 | 2.140791 | 2.132267 |
|    | 4   | 58.000000 | 55.401924 | 53.669873 | 53.092523 | 2.283460 | 2.181174 | 2.112983 | 2.090253 |
| 58 | 3   | 58.000000 | 56.051443 | 54.752405 | 54.319392 | 2.283460 | 2.206745 | 2.155607 | 2.168993 |
|    | 2   | 58.000000 | 56.700962 | 55.834936 | 55.546261 | 2.283460 | 2.232317 | 2.198221 | 2.217294 |
|    | 1.5 | 58.000000 | 57.025721 | 56.376202 | 56.159696 | 2.283460 | 2.245103 | 2.219531 | 2.265596 |
|    | 5.5 | 60.000000 | 56.427645 | 54.046075 | 53.252219 | 2.362200 | 2.221556 | 2.127794 | 2.096540 |
|    | 4   | 60.000000 | 57.401924 | 55.669873 | 55.092523 | 2.362200 | 2.259914 | 2.191723 | 2.168993 |
| 60 | 3   | 60.000000 | 58.051443 | 56.752405 | 56.319392 | 2.362200 | 2.285485 | 2.234342 | 2.217294 |
|    | 2   | 60.000000 | 58.700962 | 57.834936 | 57.546261 | 2.362200 | 2.311057 | 2.276961 | 2.265596 |
|    | 1.5 | 60.000000 | 59.025721 | 58.376202 | 58.159696 | 2.362200 | 2.323843 | 2.298271 | 2.289747 |
|    | 4   | 62.000000 | 59.401924 | 57.669873 | 57.092523 | 2.440940 | 2.338654 | 2.270463 | 2.247733 |
| 62 | 3   | 62.000000 | 60.051443 | 58.752405 | 58.319392 | 2.440940 | 2.364225 | 2.313082 | 2.296034 |
|    | 2   | 62.000000 | 60.700962 | 59.834936 | 59.546261 | 2.440940 | 2.389797 | 2.355701 | 2.344336 |
|    | 1.5 | 62.000000 | 61.025721 | 60.376202 | 60.159696 | 2.440940 | 2.402583 | 2.377011 | 2.368487 |
|    | 6   | 64.000000 | 60.102886 | 57.504809 | 56.638784 | 2.519680 | 2.366251 | 2.263964 | 2.229869 |
|    | 4   | 64.000000 | 61.401924 | 59.669873 | 59.092523 | 2.519680 | 2.417394 | 2.349203 | 2.326473 |
| 64 | 3   | 64.000000 | 62.051443 | 60.752405 | 60.319392 | 2.519680 | 2.442965 | 2.391822 | 2.374774 |
|    | 2   | 64.000000 | 62.700962 | 61.834936 | 61.546261 | 2.519680 | 2.468537 | 2.434441 | 2.423076 |
|    | 1.5 | 64.000000 | 63.025721 | 62.376202 | 62.159696 | 2.519680 | 2.481323 | 2.455751 | 2.447227 |
|    | 4   | 65.000000 | 62.401924 | 60.669873 | 60.092523 | 2.559050 | 2.456764 | 2.388573 | 2.365843 |
| 65 | 3   | 65.000000 | 63.051443 | 61.752405 | 61.319392 | 2.559050 | 2.482335 | 2.431197 | 2.414144 |
|    | 2   | 65.000000 | 63.700962 | 62.834936 | 62.546261 | 2.559050 | 2.507907 | 2.473811 | 2.462446 |
|    | 1.5 | 65.000000 | 64.025721 | 63.376202 | 63.159696 | 2.559050 | 2.520693 | 2.495121 | 2.486597 |

| Nominal size | Pitch $P$ | Basic major diameter $d$ and $D$, mm | Basic pitch diameter $d_2$ and $D_2$, mm | Basic minor diameter $d_1$ and $D_1$, mm | Minor diameter of external thread $d_{3max}$, mm | Basic major diameter $d$ and $D$, in | Basic pitch diameter $d_2$ and $D_2$, in | Basic minor diameter $d_1$ and $D_1$, in | Minor diameter of external thread $d_{3max}$, in |
|---|---|---|---|---|---|---|---|---|---|
|    | 6 | 68.000000 | 64.102886 | 61.504809 | 60.638784 | 2.677160 | 2.523731 | 2.424594 | 2.387349 |
|    | 4 | 68.000000 | 65.401924 | 63.669873 | 63.092523 | 2.677160 | 2.574874 | 2.506683 | 2.483953 |
| 68 | 3 | 68.000000 | 66.051443 | 64.752405 | 64.319392 | 2.677160 | 2.600445 | 2.549307 | 2.532254 |
|    | 2 | 68.000000 | 66.700962 | 65.834936 | 65.546261 | 2.677160 | 2.626016 | 2.560425 | 2.580556 |
|    | 1.5 | 68.000000 | 67.025721 | 66.376202 | 66.159696 | 2.677160 | 2.638803 | 2.613231 | 2.604707 |
|    | 6 | 70.000000 | 66.102886 | 63.504809 | 62.638784 | 2.755900 | 2.602471 | 2.500184 | 2.544829 |
|    | 4 | 70.000000 | 67.401924 | 65.669873 | 65.092523 | 2.755900 | 2.653614 | 2.585423 | 2.641433 |
| 70 | 3 | 70.000000 | 68.051443 | 66.752405 | 66.319392 | 2.755900 | 2.679185 | 2.628042 | 2.689734 |
|    | 2 | 70.000000 | 68.700962 | 67.834936 | 67.546261 | 2.755900 | 2.704757 | 2.670661 | 2.738036 |
|    | 1.5 | 70.000000 | 69.025721 | 68.376202 | 68.159696 | 2.755900 | 2.717543 | 2.691971 | 2.762187 |
|    | 6 | 72.000000 | 68.102886 | 65.504809 | 64.638784 | 2.834640 | 2.681211 | 2.578924 | 2.662939 |
|    | 4 | 72.000000 | 69.401924 | 67.669873 | 67.092523 | 2.834640 | 2.732354 | 2.664163 | 2.759543 |
| 72 | 3 | 72.000000 | 70.051443 | 68.752405 | 68.319392 | 2.834640 | 2.689734 | 2.706782 | 2.807844 |
|    | 2 | 72.000000 | 70.700962 | 69.834936 | 69.546261 | 2.834640 | 2.901607 | 2.749401 | 2.856146 |
|    | 1.5 | 72.000000 | 71.025721 | 70.376202 | 70.159696 | 2.834640 | 2.914393 | 2.770711 | 2.880297 |
|    | 6 | 75.000000 | 71.102886 | 68.504809 | 67.638784 | 2.952750 | 2.799321 | 2.697034 | 2.662939 |
|    | 4 | 75.000000 | 72.401924 | 70.669873 | 70.092523 | 2.952750 | 2.850464 | 2.782273 | 2.759543 |
| 75 | 3 | 75.000000 | 73.051443 | 71.752405 | 71.319392 | 2.952750 | 2.876035 | 2.824892 | 2.807844 |
|    | 2 | 75.000000 | 73.700962 | 72.834936 | 72.546261 | 2.952750 | 2.901607 | 2.867511 | 2.856146 |
|    | 1.5 | 75.000000 | 74.025721 | 73.376202 | 73.159696 | 2.952750 | 2.914393 | 2.888821 | 2.880297 |
|    | 6 | 76.000000 | 72.102886 | 69.504809 | 68.638784 | 2.992120 | 2.838691 | 2.736404 | 2.702309 |
|    | 4 | 76.000000 | 73.401924 | 71.669873 | 71.092523 | 2.992120 | 2.889834 | 2.821643 | 2.798913 |
| 76 | 3 | 76.000000 | 74.051443 | 72.752405 | 72.319392 | 2.992120 | 2.915405 | 2.864262 | 2.847214 |
|    | 2 | 76.000000 | 74.700962 | 73.834936 | 73.546261 | 2.992120 | 2.940977 | 2.906881 | 2.895516 |
|    | 1.5 | 76.000000 | 75.035721 | 74.376202 | 74.159696 | 2.992120 | 2.953763 | 2.928191 | 2.919667 |
| 78 | 2 | 78.000000 | 76.700962 | 75.834936 | 75.546261 | 3.070860 | 3.019717 | 2.985621 | 2.974256 |
|    | 6 | 80.000000 | 76.102886 | 73.504809 | 72.638784 | 3.149600 | 2.996171 | 2.893884 | 2.859789 |
|    | 4 | 80.000000 | 77.401924 | 75.669873 | 75.092523 | 3.149600 | 3.047314 | 2.979123 | 2.956392 |
| 80 | 3 | 80.000000 | 78.051443 | 76.752405 | 76.319392 | 3.149600 | 3.072885 | 3.021742 | 3.004694 |
|    | 2 | 80.000000 | 78.700962 | 77.834936 | 77.546261 | 3.149600 | 3.098457 | 3.064361 | 3.052996 |
|    | 1.5 | 80.000000 | 79.025721 | 78.376202 | 78.159696 | 3.149600 | 3.111243 | 3.085671 | 3.077147 |
| 82 | 2 | 82.000000 | 80.700962 | 79.834936 | 79.546261 | 3.228340 | 3.177197 | 3.143101 | 3.131736 |
|    | 6 | 85.000000 | 81.102886 | 78.504809 | 77.638784 | 3.346450 | 3.193021 | 4.090734 | 3.056639 |
| 85 | 4 | 85.000000 | 82.401924 | 80.669873 | 80.092523 | 3.346450 | 3.244164 | 3.175973 | 3.153243 |
|    | 3 | 85.000000 | 83.051443 | 81.752405 | 81.319392 | 3.346450 | 3.269735 | 3.218592 | 3.201544 |
|    | 2 | 85.000000 | 83.700962 | 82.834936 | 82.546261 | 3.346450 | 3.295307 | 3.261211 | 3.249846 |
|    | 6 | 90.000000 | 86.102886 | 83.504809 | 82.638784 | 3.543300 | 3.389871 | 3.287584 | 3.253489 |
| 90 | 4 | 90.000000 | 87.401924 | 85.669873 | 85.092523 | 3.543300 | 3.441014 | 3.372823 | 3.350093 |
|    | 3 | 90.000000 | 88.051443 | 86.752405 | 86.319392 | 3.543300 | 3.466585 | 3.415442 | 3.398394 |
|    | 2 | 90.000000 | 88.700962 | 87.834936 | 87.546261 | 3.543300 | 3.492157 | 3.458061 | 3.446696 |
|    | 6 | 95.000000 | 91.102886 | 88.504809 | 87.638784 | 3.740150 | 3.586721 | 3.484434 | 3.450339 |
| 95 | 4 | 95.000000 | 92.401924 | 90.669873 | 90.092523 | 3.740150 | 3.637864 | 3.569673 | 3.546943 |
|    | 3 | 95.000000 | 93.051443 | 91.752405 | 91.319392 | 3.740150 | 3.663435 | 3.612292 | 3.595244 |
|    | 2 | 95.000000 | 93.700962 | 92.834936 | 92.546261 | 3.740150 | 3.689007 | 3.654911 | 3.643546 |

| Nominal size | Pitch $P$ | Basic major diameter $d$ and $D$, mm | Basic pitch diameter $d_2$ and $D_2$, mm | Basic minor diameter $d_1$ and $D_1$, mm | Minor diameter of external thread $d_{3max}$, mm | Basic major diameter $d$ and $D$, in | Basic pitch diameter $d_2$ and $D_2$, in | Basic minor diameter $d_1$ and $D_1$, in | Minor diameter of external thread $d_{3max}$, in |
|---|---|---|---|---|---|---|---|---|---|
| 100 | 6 | 100.000000 | 96.102886 | 93.504809 | 92.638784 | 3.937000 | 3.783571 | 3.681284 | 3.647189 |
|  | 4 | 100.000000 | 97.401924 | 95.669873 | 95.092523 | 3.937000 | 3.834714 | 3.766523 | 3.743793 |
|  | 3 | 100.000000 | 98.051443 | 96.752405 | 96.319392 | 3.937000 | 3.860285 | 3.809142 | 3.792094 |
|  | 2 | 100.000000 | 98.700962 | 97.834936 | 97.546261 | 3.937000 | 3.885857 | 3.851761 | 3.840396 |
| 105 | 6 | 105.000000 | 101.102886 | 98.504809 | 97.638784 | 4.133850 | 3.980421 | 3.878134 | 3.844039 |
|  | 4 | 105.000000 | 102.401924 | 100.669873 | 100.092523 | 4.133850 | 4.031564 | 3.963373 | 3.940643 |
|  | 3 | 105.000000 | 103.051443 | 101.752405 | 101.319392 | 4.133850 | 4.057135 | 4.005992 | 3.988944 |
|  | 2 | 105.000000 | 103.700962 | 102.834936 | 102.546261 | 4.133850 | 4.082707 | 4.048611 | 4.037246 |
| 110 | 6 | 110.000000 | 106.102886 | 103.504809 | 102.638784 | 4.330700 | 4.177271 | 4.074984 | 4.040889 |
|  | 4 | 110.000000 | 107.401924 | 105.669873 | 105.092523 | 4.330700 | 4.228414 | 4.160223 | 4.137493 |
|  | 3 | 110.000000 | 108.051443 | 106.752405 | 106.319392 | 4.330700 | 4.253985 | 4.202842 | 4.185794 |
|  | 2 | 110.000000 | 108.700962 | 107.834936 | 107.546261 | 4.330700 | 4.279557 | 4.245461 | 4.234096 |
| 115 | 6 | 115.000000 | 111.102886 | 108.504809 | 107.638784 | 4.527550 | 4.374121 | 4.271834 | 4.237739 |
|  | 4 | 115.000000 | 112.401924 | 110.669873 | 110.092523 | 4.527550 | 4.425264 | 4.357073 | 4.334343 |
|  | 3 | 115.000000 | 113.051443 | 111.752405 | 111.319392 | 4.527550 | 4.450835 | 4.399692 | 4.382644 |
|  | 2 | 115.000000 | 113.700962 | 112.834936 | 112.546261 | 4.527550 | 4.476407 | 4.442311 | 4.430946 |
| 120 | 6 | 120.000000 | 116.102886 | 113.504809 | 112.638784 | 4.724400 | 4.570971 | 4.468684 | 4.434589 |
|  | 4 | 120.000000 | 117.401924 | 115.669873 | 115.092523 | 4.724400 | 4.818964 | 4.553923 | 4.531193 |
|  | 3 | 120.000000 | 118.051443 | 116.752405 | 116.319392 | 4.724400 | 4.844535 | 4.596542 | 4.579494 |
|  | 2 | 120.000000 | 118.700962 | 117.834936 | 117.546261 | 4.724400 | 4.870107 | 4.639161 | 4.627796 |
| 125 | 6 | 125.000000 | 121.102886 | 118.504809 | 117.638784 | 4.921250 | 4.767821 | 4.665534 | 4.631439 |
|  | 4 | 125.000000 | 122.401924 | 120.669873 | 120.092523 | 4.921250 | 4.818964 | 4.750773 | 4.728043 |
|  | 3 | 125.000000 | 123.051443 | 121.752405 | 121.319392 | 4.921250 | 4.844535 | 4.793392 | 4.776344 |
|  | 2 | 125.000000 | 123.700962 | 122.834936 | 122.546261 | 4.921250 | 4.870107 | 4.836011 | 4.824646 |
| 130 | 6 | 130.000000 | 126.102886 | 122.638784 | 122.638784 | 5.118100 | 4.964671 | 4.862384 | 4.828289 |
|  | 4 | 130.000000 | 127.401924 | 125.669873 | 125.092523 | 5.118100 | 5.015814 | 4.947623 | 4.924893 |
|  | 3 | 130.000000 | 128.051443 | 126.752405 | 126.319392 | 5.118100 | 5.041385 | 4.990247 | 4.973194 |
|  | 2 | 130.000000 | 128.700962 | 127.834936 | 127.546261 | 5.118100 | 5.066957 | 5.032861 | 5.021496 |
| 135 | 6 | 135.000000 | 131.102886 | 128.504809 | 127.638784 | 5.314950 | 5.161521 | 5.059234 | 5.025139 |
|  | 4 | 135.000000 | 132.401924 | 130.669873 | 130.092523 | 5.314950 | 5.212664 | 5.144473 | 5.121743 |
|  | 3 | 135.000000 | 133.051443 | 131.752405 | 131.319392 | 5.314950 | 5.238235 | 5.187082 | 5.170044 |
|  | 2 | 135.000000 | 133.700962 | 132.834936 | 132.546261 | 5.314950 | 5.263807 | 5.229711 | 5.218346 |
| 140 | 6 | 140.000000 | 136.102886 | 133.504809 | 132.638784 | 5.511800 | 5.358371 | 5.256084 | 5.221989 |
|  | 4 | 140.000000 | 137.401924 | 135.669873 | 135.092523 | 5.511800 | 5.409514 | 5.341323 | 5.318593 |
|  | 3 | 140.000000 | 138.051443 | 136.752405 | 136.319392 | 5.511800 | 5.435085 | 5.383942 | 5.366894 |
|  | 2 | 140.000000 | 138.700962 | 137.834936 | 137.546261 | 5.511800 | 5.460657 | 5.426561 | 5.415196 |
| 145 | 6 | 145.000000 | 141.102886 | 138.504809 | 137.638784 | 5.708650 | 5.555220 | 5.452934 | 5.418839 |
|  | 4 | 145.000000 | 142.401924 | 140.669873 | 140.092523 | 5.708650 | 5.606364 | 5.538173 | 5.515443 |
|  | 3 | 145.000000 | 143.051443 | 141.752405 | 141.319392 | 5.708650 | 5.631935 | 5.580792 | 5.563744 |
|  | 2 | 145.000000 | 143.700962 | 142.834936 | 142.546261 | 5.708650 | 5.657507 | 5.623411 | 5.612046 |
| 150 | 6 | 150.000000 | 146.102886 | 143.504809 | 142.638784 | 5.905500 | 5.752071 | 5.649784 | 5.615689 |
|  | 4 | 150.000000 | 147.401924 | 145.669873 | 145.092523 | 5.905500 | 5.803214 | 5.735023 | 5.712293 |
|  | 3 | 150.000000 | 148.051443 | 146.752405 | 146.319392 | 5.905500 | 5.828785 | 5.777642 | 5.760594 |
|  | 2 | 150.000000 | 148.700962 | 147.834936 | 147.546261 | 5.905500 | 5.854357 | 5.820261 | 5.808896 |

| Nominal size | Pitch $P$ | Basic major diameter $d$ and $D$, mm | Basic pitch diameter $d_2$ and $D_2$, mm | Basic minor diameter $d_1$ and $D_1$, mm | Minor diameter of external thread $d_{3max}$, mm | Basic major diameter $d$ and $D$, in | Basic pitch diameter $d_2$ and $D_2$, in | Basic minor diameter $d_1$ and $D_1$, in | Minor diameter of external thread $d_{3max}$, in |
|---|---|---|---|---|---|---|---|---|---|
| | 6 | 155.000000 | 151.102886 | 148.504809 | 147.638784 | 6.102350 | 5.948921 | 5.846634 | 5.812539 |
| 155 | 4 | 155.000000 | 152.401924 | 150.669873 | 150.092523 | 6.102350 | 6.000064 | 5.931873 | 5.909143 |
| | 3 | 155.000000 | 153.051443 | 151.752405 | 151.319392 | 6.102350 | 6.025636 | 5.974492 | 5.957444 |
| | 6 | 160.000000 | 156.102886 | 153.504809 | 152.638784 | 6.299200 | 6.145771 | 6.043484 | 6.009389 |
| 160 | 4 | 160.000000 | 157.401924 | 155.669873 | 155.092523 | 6.299200 | 6.196914 | 6.128723 | 6.105993 |
| | 3 | 160.000000 | 158.051443 | 156.752405 | 156.319392 | 6.299200 | 6.222485 | 6.171354 | 6.154294 |
| | 6. | 165.000000 | 161.102886 | 158.504809 | 157.638784 | 6.496050 | 6.342621 | 6.240334 | 6.206239 |
| 165 | 4 | 165.000000 | 162.401924 | 160.669873 | 160.092523 | 6.496050 | 6.393752 | 6.325573 | 6.302843 |
| | 3 | 165.000000 | 163.051443 | 161.752405 | 161.319392 | 6.496050 | 6.379965 | 6.368192 | 6.351144 |
| | 6 | 170.000000 | 166.102886 | 163.504809 | 162.638784 | 6.692900 | 6.539471 | 6.437184 | 6.403089 |
| 170 | 4 | 170.000000 | 167.401924 | 165.669873 | 165.092523 | 6.692900 | 6.590614 | 6.522423 | 6.499693 |
| | 3 | 170.000000 | 168.051443 | 166.752405 | 166.319392 | 6.692900 | 6.616185 | 6.565042 | 6.547994 |
| | 6 | 175.000000 | 171.102886 | 168.504809 | 167.638784 | 6.889750 | 6.736321 | 6.634034 | 6.599939 |
| 175 | 4 | 175.000000 | 172.401924 | 170.669873 | 170.092523 | 6.889750 | 6.787464 | 6.719273 | 6.696543 |
| | 3 | 175.000000 | 173.051443 | 171.752405 | 171.319392 | 6.889750 | 6.813035 | 6.761892 | 6.744844 |
| | 6 | 180.000000 | 176.102886 | 173.504809 | 172.638784 | 7.086600 | 6.933171 | 6.830884 | 6.796789 |
| 180 | 4 | 180.000000 | 177.401924 | 175.669873 | 175.092523 | 7.086600 | 6.984314 | 6.916123 | 6.893393 |
| | 3 | 180.000000 | 178.051443 | 176.752405 | 176.319392 | 7.086600 | 7.009885 | 6.958742 | 6.941694 |
| | 6 | 185.000000 | 181.102886 | 178.504809 | 177.638784 | 7.283450 | 7.130021 | 7.027734 | 6.993639 |
| 185 | 4 | 185.000000 | 182.401924 | 180.669873 | 180.092523 | 7.283450 | 7.181164 | 7.112973 | 7.090243 |
| | 3 | 185.000000 | 183.051443 | 181.752405 | 181.319392 | 7.283450 | 7.206735 | 7.155592 | 7.138544 |
| | 6 | 190.000000 | 186.102886 | 183.504809 | 182.638784 | 7.480300 | 7.326871 | 7.224584 | 7.190489 |
| 190 | 4 | 190.000000 | 187.401924 | 185.669873 | 185.092523 | 7.480300 | 7.378014 | 7.309823 | 7.287093 |
| | 3 | 190.000000 | 188.051443 | 186.752405 | 186.319392 | 7.480300 | 7.403585 | 7.352442 | 7.335394 |
| | 6 | 195.000000 | 191.102886 | 188.504809 | 187.638784 | 7.677150 | 7.523721 | 7.421434 | 7.387339 |
| 195 | 4 | 195.000000 | 192.401924 | 190.669873 | 190.092523 | 7.677150 | 7.574864 | 7.506673 | 7.483943 |
| | 3 | 195.000000 | 193.051443 | 191.752495 | 191.319392 | 7.677150 | 7.600435 | 7.549296 | 7.532244 |
| | 6 | 200.000000 | 196.102886 | 193.504809 | 192.638784 | 7.874000 | 7.720571 | 7.618284 | 7.584189 |
| 200 | 4 | 200.000000 | 197.401924 | 195.669873 | 195.092523 | 7.874000 | 7.771714 | 7.703523 | 7.680793 |
| | 3 | 200.000000 | 198.051443 | 196.752405 | 196.319392 | 7.874000 | 7.797285 | 7.746142 | 7.729094 |
| | 6 | 205.000000 | 201.102886 | 198.504809 | 197.638784 | 8.070850 | 7.917421 | 7.815134 | 7.781039 |
| 205 | 4 | 205.000000 | 202.401924 | 200.669873 | 200.092523 | 8.070850 | 7.968564 | 7.900373 | 7.877643 |
| | 3 | 205.000000 | 203.051443 | 201.752405 | 201.319392 | 8.070850 | 7.994135 | 7.942992 | 7.925944 |
| | 6 | 210.000000 | 206.102886 | 203.504809 | 202.638784 | 8.267700 | 8.114271 | 8.011984 | 7.977889 |
| 210 | 4 | 210.000000 | 207.401924 | 205.669873 | 205.092523 | 8.267700 | 8.165414 | 8.097223 | 8.074493 |
| | 3 | 210.000000 | 208.051443 | 206.752405 | 206.319392 | 8.267700 | 8.190985 | 8.139842 | 8.122794 |
| | 6 | 215.000000 | 211.102886 | 208.504809 | 207.638784 | 8.464550 | 8.311121 | 8.208834 | 8.174739 |
| 215 | 4 | 215.000000 | 212.401924 | 210.669873 | 210.092523 | 8.464550 | 8.362264 | 8.294073 | 8.271343 |
| | 3 | 215.000000 | 213.051443 | 211.752405 | 211.319392 | 8.464550 | 8.387875 | 8.336692 | 8.319644 |
| | 6 | 220.000000 | 216.102886 | 213.504809 | 212.638784 | 8.661400 | 8.507971 | 8.405684 | 8.371589 |
| 220 | 4 | 220.000000 | 217.401924 | 215.669873 | 215.092523 | 8.661400 | 8.559114 | 8.490923 | 8.468193 |
| | 3 | 220.000000 | 218.051443 | 216.752405 | 216.319392 | 8.661400 | 8.584685 | 8.533542 | 8.516494 |
| | 6 | 225.000000 | 221.102886 | 218.504809 | 217.638784 | 8.858250 | 8.704821 | 8.602534 | 8.568439 |
| 225 | 4 | 225.000000 | 222.401924 | 220.669873 | 220.092523 | 8.858250 | 8.755964 | 8.687773 | 8.665043 |
| | 3 | 225.000000 | 223.051443 | 221.752405 | 221.319392 | 8.858250 | 8.781535 | 7.730392 | 8.713344 |

| Nominal size | Pitch $P$ | Basic major diameter $d$ and $D$, mm | Basic pitch diameter $d_2$ and $D_2$, mm | Basic minor diameter $d_1$ and $D_1$, mm | Minor diameter of external thread $d_{3max}$, mm | Basic major diameter $d$ and $D$, in | Basic pitch diameter $d_2$ and $D_2$, in | Basic minor diameter $d_1$ and $D_1$, in | Minor diameter of external thread $d_{3max}$, in |
|---|---|---|---|---|---|---|---|---|---|
| 230 | 6 | 230.000000 | 226.102886 | 223.504809 | 222.638784 | 9.055100 | 8.901671 | 8.799384 | 8.765289 |
|     | 4 | 230.000000 | 227.401924 | 225.669873 | 225.092523 | 9.055100 | 8.952814 | 8.884623 | 8.861893 |
|     | 3 | 230.000000 | 228.051443 | 226.752405 | 226.319892 | 9.055100 | 8.978385 | 8.927242 | 8.910194 |
| 235 | 6 | 235.000000 | 231.102886 | 228.504809 | 227.638784 | 9.251950 | 9.098521 | 8.996234 | 8.962139 |
|     | 4 | 235.000000 | 232.401924 | 230.669873 | 230.092523 | 9.251950 | 9.149664 | 9.081473 | 9.058743 |
|     | 3 | 235.000000 | 233.051443 | 231.752405 | 231.319392 | 9.251950 | 9.175235 | 9.124092 | 9.107044 |
| 240 | 6 | 240.000000 | 236.102886 | 233.504809 | 232.638784 | 9.448800 | 9.295371 | 9.193084 | 9.158989 |
|     | 4 | 240.000000 | 237.401924 | 235.669873 | 235.092523 | 9.448800 | 9.346514 | 9.278323 | 9.255593 |
|     | 3 | 240.000000 | 238.051443 | 236.752405 | 236.319392 | 9.448800 | 9.372085 | 9.320942 | 9.303894 |
| 245 | 6 | 245.000000 | 241.102886 | 238.504809 | 237.638784 | 9.645650 | 9.492221 | 9.389934 | 9.355839 |
|     | 4 | 245.000000 | 242.401924 | 240.669873 | 240.092523 | 9.645650 | 9.543364 | 9.475173 | 9.452443 |
|     | 3 | 245.000000 | 243.051443 | 241.752405 | 241.319392 | 9.645650 | 9.568935 | 9.517792 | 9.500744 |
| 250 | 6 | 250.000000 | 246.102886 | 243.504809 | 242.638784 | 9.842500 | 9.689071 | 9.586784 | 9.552689 |
|     | 4 | 250.000000 | 247.401924 | 245.669873 | 245.092523 | 9.842500 | 9.740214 | 9.672023 | 9.649293 |
|     | 3 | 250.000000 | 248.051443 | 246.752405 | 246.319392 | 9.842500 | 9.765785 | 9.714642 | 9.697594 |
| 255 | 6 | 255.000000 | 251.102886 | 248.504809 | 247.638784 | 10.039350 | 9.885921 | 9.783634 | 9.749539 |
|     | 4 | 255.000000 | 252.401924 | 250.669873 | 250.092523 | 10.039350 | 9.937064 | 9.868873 | 9.846143 |
| 260 | 6 | 260.000000 | 256.102886 | 253.504809 | 252.638784 | 10.236200 | 10.082776 | 9.980484 | 9.946389 |
|     | 4 | 260.000000 | 257.401924 | 255.669873 | 255.092523 | 10.236200 | 10.133914 | 10.065723 | 10.042993 |
| 265 | 6 | 265.000000 | 261.102886 | 258.504809 | 257.638784 | 10.433050 | 10.279621 | 10.177334 | 10.143239 |
|     | 4 | 265.000000 | 262.401924 | 260.669873 | 260.092523 | 10.433050 | 10.330764 | 10.262573 | 10.239843 |
| 270 | 6 | 270.000000 | 266.102886 | 263.504809 | 263.638784 | 10.629900 | 10.476471 | 10.374184 | 10.379459 |
|     | 4 | 270.000000 | 267.401924 | 265.669873 | 265.092523 | 10.629900 | 10.527614 | 10.459423 | 10.436693 |
| 275 | 6 | 275.000000 | 271.102886 | 268.504809 | 267.638784 | 10.826750 | 10.673321 | 10.571034 | 10.536939 |
|     | 4 | 275.000000 | 272.401924 | 270.669873 | 270.092523 | 10.826750 | 10.724464 | 10.656273 | 10.633543 |
| 280 | 6 | 280.000000 | 276.102886 | 273.504809 | 272.638784 | 11.023600 | 10.870171 | 10.767884 | 10.733789 |
|     | 4 | 280.000000 | 277.401924 | 275.669873 | 275.092523 | 11.023600 | 10.921314 | 10.853123 | 10.830393 |
| 285 | 6 | 285.000000 | 281.102886 | 278.504809 | 277.638784 | 11.220450 | 10.067021 | 10.964734 | 10.930639 |
|     | 4 | 285.000000 | 282.401924 | 280.669873 | 280.092523 | 11.220450 | 11.118164 | 11.049973 | 11.027243 |
| 290 | 6 | 290.000000 | 286.102886 | 283.504809 | 282.638784 | 11.417300 | 11.263870 | 11.161584 | 11.127489 |
|     | 4 | 290.000000 | 287.401924 | 285.669873 | 285.092523 | 11.417300 | 11.315014 | 11.246845 | 11.224093 |
| 295 | 6 | 295.000000 | 291.102886 | 288.504809 | 287.638784 | 11.614150 | 11.460721 | 11.358434 | 11.324339 |
|     | 4 | 295.000000 | 292.401924 | 290.669873 | 290.092523 | 11.614150 | 11.511864 | 11.443673 | 11.420943 |
| 300 | 6 | 300.000000 | 296.102886 | 293.504809 | 292.638784 | 11.811000 | 11.657571 | 11.555284 | 11.521189 |
|     | 4 | 300.000000 | 297.401924 | 295.669873 | 295.092523 | 11.81100 | 11.708714 | 11.640523 | 11.617793 |

## Dual (Inch-metric) Dimensioning*

| | | Conversion of tolerances | | |
|---|---|---|---|---|
| **Drawing usage** | | **Reference data** | | |
| | | **Conversion error** | | |
| **Inch** | **Millimeter (roundoff)** | **Millimeter error, inches** | **Percent error** | **Millimeter equivalent** |
| 0.00004 | 0.0010 | −0.00000062 | −1.57 | 0.001016 |
| 0.00005 | 0.0013 | +0.00000118 | +2.36 | 0.00127 |
| 0.0001 | 0.0025 | −0.00000157 | −1.57 | 0.00254 |
| 0.0002 | 0.0051 | +0.00000079 | +0.39 | 0.00508 |
| 0.0003 | 0.0076 | −0.00000079 | −0.26 | 0.00762 |
| 0.0004 | 0.010 | −0.00000629 | −1.57 | 0.01016 |
| 0.0005 | 0.013 | +0.00001181 | +2.36 | 0.01270 |
| 0.0006 | 0.015 | −0.00000945 | −1.57 | 0.01524 |
| 0.0007 | 0.018 | +0.00000866 | +1.24 | 0.01778 |
| 0.0008 | 0.020 | −0.00001260 | −1.57 | 0.02032 |
| 0.0009 | 0.023 | +0.00000551 | +0.62 | 0.02286 |
| 0.0010 | 0.025 | −0.00001575 | −1.57 | 0.0254 |
| 0.0015 | 0.038 | −0.00000394 | −0.26 | 0.0381 |
| 0.0020 | 0.051 | +0.00000787 | +0.39 | 0.0508 |
| 0.0025 | 0.064 | +0.00001968 | +0.79 | 0.0635 |
| 0.0030 | 0.076 | −0.00000787 | −0.26 | 0.0762 |
| 0.004 | 0.10 | −0.00006299 | −1.57 | 0.1016 |
| 0.005 | 0.13 | +0.00011811 | +2.36 | 0.1270 |
| 0.006 | 0.15 | −0.00009449 | −1.57 | 0.1524 |
| 0.007 | 0.18 | +0.00008661 | +1.24 | 0.1778 |
| 0.008 | 0.20 | −0.00012598 | −1.57 | 0.2032 |
| 0.009 | 0.23 | +0.00005511 | +0.62 | 0.2286 |
| 0.010 | 0.25 | −0.00015748 | −1.57 | 0.254 |
| 0.015 | 0.38 | −0.00003937 | −0.26 | 0.381 |
| 0.020 | 0.51 | +0.00007874 | +0.39 | 0.508 |
| 0.030 | 0.76 | −0.00007874 | −0.26 | 0.762 |
| 0.04 | 1.0 | −0.00062993 | −1.57 | 1.016 |
| 0.05 | 1.3 | +0.00118110 | +2.36 | 1.270 |
| 0.06 | 1.5 | −0.00094488 | −1.57 | 1.524 |
| 0.07 | 1.8 | +0.00086614 | +1.24 | 1.778 |
| 0.08 | 2.0 | −0.00125984 | −1.57 | 2.032 |
| 0.09 | 2.3 | +0.00055118 | +0.62 | 2.286 |
| 0.10 | 2.5 | −0.00157480 | −1.57 | 2.54 |
| 0.20 | 5.1 | +0.00078740 | +0.39 | 5.08 |
| 0.30 | 7.6 | −0.00078740 | −0.26 | 7.62 |
| 0.40 | 10.0 | −0.00629922 | −1.57 | 10.16 |

*Adapted from S.A.E. Aero-Space Standards.

# APPENDIX

## Standard parts, sizes, symbols, and abbreviations

**C**

ANSI Standard Unified and American Thread Series, A48

ANSI Standard Square and Hexagon Bolts and Hexagon Cap Screws, A50

ANSI Standard Square and Hexagon Nuts, A51

Sizes of Numbered and Lettered Drills, A52

Acme and Stub Acme Threads, A52

Buttress Threads, A52

American Standard Round-head Bolts, A53

Bolt-length Increments, A54

ANSI Standard Cap Screws—Socket and Slotted Heads, A54

ANSI Standard Machine Screws, A55

ANSI Standard Machine-screw and Stove-bolt Nuts, A56

ANSI Standard Hexagon Socket, Slotted Headless and Square-head Setscrews, A56

ANSI Standard Socket-head Shoulder Screws, A57

ANSI Standard Wood Screws, A57

Small Rivets, A58

Dimensions of Standard Gib-head Keys, Square and Flat, A58

Widths and Heights of Standard Square- and Flat-stock Keys with Corresponding Shaft Diameters, A59

Woodruff-key Dimensions, A59

Woodruff-key-seat Dimensions, A60

Dimensions of Pratt and Whitney Keys, A61

American Standard Plain Washers, A62

American Standard Lock Washers, A63

Dimensions of Taper Pins, A64

ANSI Standard Machine Tapers, Self-holding (Slow) Taper Series, A64

Wire and Sheet-metal Gages, A65

Preferred Basic Sizes for Cylindrical Fits, A66

Tables of Limits for Cylindrical Fits, A67

ANSI Standard Pipe, A69

ANSI Standard 150-lb Malleable-iron Screwed Fittings, A70

ANSI Standard 150-lb Malleable-iron Screwed Fittings: Dimensions of Close-, Medium-, and Open-pattern Return Bends, A71

ANSI Standard Cast-iron Screwed Fittings, A71

Globe, Angle-globe, and Gate Valves, A72

Lengths of Pipe Nipples, A72

Pipe Bushings, A73

ANSI Standard Cast-iron Pipe Flanges and Flanged Fittings, A74

Lengths of Malleable-iron Unions, A75

ANSI Standard Steel Butt-welding Fittings, A75

Threaded Cast-iron Pipe, A76

ANSI Standard Class 125 Cast-iron and 150-lb Steel Valves, A77

Beam Connections, A78

Driving Clearances for Riveting, A78

Gage and Maximum Rivet Size for Angles, A78

Selected Structural Shapes, A79

Standard Jig Bushings, A79

Lock Screws, A79

ANSI Standard Brass Solder-joint Fittings, A82

SAE Standard Cotter Pins, A82

ANSI Standard Graphic Symbols for Piping and Heating, A83

ANSI Standard Plumbing Symbols, A85

ANSI Standard Wiring Symbols, A86

ANSI Standard Graphic Symbols for Electrical Diagrams, A87

Symbols for Materials (Exterior and Section), A92

Weights of Materials, A92

Abbreviations for Use on Drawings, A93

Partial List of Chemical Symbols, A104

## ANSI Standard Unified and American Thread Series[a]

*Threads per inch for coarse, fine, extra-fine, 8-thread, 12-thread, and 16-thread series [b] [tap-drill sizes for approximately 75 per cent depth of thread (not ANSI)]*

| Nominal size (basic major diam.) | Coarse-thd. series UNC and NC[c] in classes 1A, 1B, 2A, 2B, 3A, 3B, 2, 3 | | Fine-thd. series UNF and NF[c] in classes 1A, 1B, 2A, 2B, 3A, 3B, 2, 3 | | Extra-fine thd. series UNEF and NEF[d] in classes 2A, 2B, 2, 3 | | 8-thd. series 8N[c] in classes 2A, 2B, 2, 3 | | 12-thd. series 12UN and 12N[d] in classes 2A, 2B, 2, 3 | | 16-thd. series 16UN and 16N[d] in classes 2A, 2B, 2, 3 | |
|---|---|---|---|---|---|---|---|---|---|---|---|---|
| | Thd./in. | Tap drill | Thd./in. | Tap drill | Thd./in. | Tap drill | Thd./in. | Tap drill | Thd./in. | Tap drill | Thd./in. | Tap drill |
| 0(0.060) | .... | ..... | 80 | $\frac{3}{64}$ | | | | | | | | |
| 1(0.073) | 64 | No. 53 | 72 | No. 53 | | | | | | | | |
| 2(0.086) | 56 | No. 50 | 64 | No. 50 | | | | | | | | |
| 3(0.099) | 48 | No. 47 | 56 | No. 45 | | | | | | | | |
| 4(0.112) | 40 | No. 43 | 48 | No. 42 | | | | | | | | |
| 5(0.125) | 40 | No. 38 | 44 | No. 37 | | | | | | | | |
| 6(0.138) | 32 | No. 36 | 40 | No. 33 | | | | | | | | |
| 8(0.164) | 32 | No. 29 | 36 | No. 29 | | | | | | | | |
| 10(0.190) | 24 | No. 25 | 32 | No. 21 | | | | | | | | |
| 12(0.216) | 24 | No. 16 | 28 | No. 14 | 32 | No. 13 | | | | | | |
| $\frac{1}{4}$ | 20 | No. 7 | 28 | No. 3 | 32 | $\frac{7}{32}$ | | | | | | |
| $\frac{5}{16}$ | 18 | Let. F | 24 | Let. I | 32 | $\frac{9}{32}$ | | | | | | |
| $\frac{3}{8}$ | 16 | $\frac{5}{16}$ | 24 | Let. Q | 32 | $\frac{11}{32}$ | | | | | | |
| $\frac{7}{16}$ | 14 | Let. U | 20 | $\frac{25}{64}$ | 28 | $\frac{13}{32}$ | | | | | | |
| $\frac{1}{2}$ | 13 | $\frac{27}{64}$ | 20 | $\frac{29}{64}$ | 28 | $\frac{15}{32}$ | .. | .... | 12 | $\frac{27}{64}$ | | |
| $\frac{9}{16}$ | 12 | $\frac{31}{64}$ | 18 | $\frac{33}{64}$ | 24 | $\frac{33}{64}$ | .. | .... | 12 | $\frac{31}{64}$ | | |
| $\frac{5}{8}$ | 11 | $\frac{17}{32}$ | 18 | $\frac{37}{64}$ | 24 | $\frac{37}{64}$ | .. | .... | 12 | $\frac{35}{64}$ | | |
| $\frac{11}{16}$ | .... | ..... | .. | ..... | 24 | $\frac{41}{64}$ | .. | .... | 12 | $\frac{39}{64}$ | | |
| $\frac{3}{4}$ | 10 | $\frac{21}{32}$ | 16 | $1\frac{1}{16}$ | 20 | $\frac{45}{64}$ | .. | .... | 12 | $\frac{43}{64}$ | 16 | $1\frac{1}{16}$ |
| $\frac{13}{16}$ | .... | ..... | .. | ..... | 20 | $\frac{49}{64}$ | .. | .... | 12 | $\frac{47}{64}$ | 16 | $\frac{3}{4}$ |
| $\frac{7}{8}$ | 9 | $\frac{49}{64}$ | 14 | $1\frac{3}{16}$ | 20 | $\frac{53}{64}$ | .. | .... | 12 | $\frac{51}{64}$ | 16 | $1\frac{3}{16}$ |
| $\frac{15}{16}$ | .... | ..... | .. | ..... | 20 | $\frac{57}{64}$ | .. | .... | 12 | $\frac{55}{64}$ | 16 | $\frac{7}{8}$ |
| 1 | .... | ..... | 14 | $1\frac{5}{16}$ | .. | ..... | 8 | $\frac{7}{8}$ | | | | |
| 1 | 8 | $\frac{7}{8}$ | 12 | $\frac{59}{64}$ | 20 | $\frac{61}{64}$ | .. | .... | 12 | $\frac{59}{64}$ | 16 | $1\frac{5}{16}$ |
| 1 $\frac{1}{16}$ | .... | ..... | .. | ..... | 18 | 1 | .. | .... | 12 | $\frac{63}{64}$ | 16 | 1 |
| 1 $\frac{1}{8}$ | 7 | $\frac{63}{64}$ | 12 | 1 $\frac{3}{64}$ | 18 | 1 $\frac{5}{64}$ | 8 | 1 | 12 | 1 $\frac{3}{64}$ | 16 | 1 $\frac{1}{16}$ |
| 1 $\frac{3}{16}$ | .... | ..... | .. | ..... | 18 | 1 $\frac{9}{64}$ | .. | .... | 12 | 1 $\frac{7}{64}$ | 16 | 1 $\frac{1}{8}$ |
| 1 $\frac{1}{4}$ | 7 | 1 $\frac{7}{64}$ | 12 | 1 $\frac{11}{64}$ | 18 | 1 $\frac{3}{16}$ | 8 | $1\frac{1}{8}$ | 12 | 1 $\frac{11}{64}$ | 16 | 1 $\frac{3}{16}$ |
| 1 $\frac{5}{16}$ | .... | ..... | .. | ..... | 18 | 1 $\frac{17}{64}$ | .. | .... | 12 | 1 $\frac{15}{64}$ | 16 | 1 $\frac{1}{4}$ |
| 1 $\frac{3}{8}$ | 6 | 1 $\frac{7}{32}$ | 12 | 1 $\frac{19}{64}$ | 18 | 1 $\frac{5}{16}$ | 8 | $1\frac{1}{4}$ | 12 | 1 $\frac{19}{64}$ | 16 | 1 $\frac{5}{16}$ |
| 1 $\frac{7}{16}$ | .... | ..... | .. | ..... | 18 | 1 $\frac{3}{8}$ | .. | .... | 12 | 1 $\frac{23}{64}$ | 16 | 1 $\frac{3}{8}$ |
| 1 $\frac{1}{2}$ | 6 | 1 $\frac{11}{32}$ | 12 | 1 $\frac{27}{64}$ | 18 | 1 $\frac{7}{16}$ | 8 | $1\frac{3}{8}$ | 12 | 1 $\frac{27}{64}$ | 16 | 1 $\frac{7}{16}$ |
| 1 $\frac{9}{16}$ | .... | ..... | .. | ..... | 18 | 1 $\frac{1}{2}$ | .. | .... | .. | ..... | 16 | 1 $\frac{1}{2}$ |
| 1 $\frac{5}{8}$ | .... | ..... | .. | ..... | 18 | 1 $\frac{9}{16}$ | 8 | $1\frac{1}{2}$ | 12 | 1 $\frac{35}{64}$ | 16 | 1 $\frac{9}{16}$ |
| 1 $\frac{11}{16}$ | .... | ..... | .. | ..... | 18 | 1 $\frac{5}{8}$ | .. | .... | .. | ..... | 16 | 1 $\frac{5}{8}$ |

## ANSI Standard Unified and American Thread Series (Cont.)

| Nominal size (basic major diam.) | Coarse-thd. series UNC and NC$^c$ in classes 1A, 1B, 2A, 2B, 3A, 3B, 2, 3 | | Fine-thd. series UNF and NF$^c$ in classes 1A, 1B, 2A, 2B, 3A, 3B, 2, 3 | | Extra-fine thd. series UNEF and NEF$^d$ in classes 2A, 2B, 2, 3 | | 8-thd. series 8N$^c$ in classes 2A, 2B, 2, 3 | | 12-thd. series 12UN and 12N$^d$ in classes 2A, 2B, 2, 3 | | 16-thd. series 16UN and 16N$^d$ in classes 2A, 2B, 2, 3 | |
|---|---|---|---|---|---|---|---|---|---|---|---|---|
| | Thd./in. | Tap drill | Thd./in. | Tap drill | Thd./in. | Tap drill | Thd./in. | Tap drill | Thd./in. | Tap drill | Thd./in. | Tap drill |
| $1\frac{3}{4}$ | 5 | $1\frac{9}{16}$ | .. | ..... | **16** | $1\frac{11}{16}$ | $8^e$ | $1\frac{5}{8}$ | 12 | $1\frac{43}{64}$ | 16 | $1\frac{11}{16}$ |
| $1\frac{13}{16}$ | .... | ..... | .. | ..... | .. | ..... | .. | ... | .. | ... | 16 | $1\frac{3}{4}$ |
| $1\frac{7}{8}$ | .... | ..... | .. | ..... | .. | ..... | 8 | $1\frac{3}{4}$ | 12 | $1\frac{51}{64}$ | 16 | $1\frac{13}{16}$ |
| $1\frac{15}{16}$ | .... | ..... | .. | ..... | .. | ..... | .. | ... | .. | ... | 16 | $1\frac{7}{8}$ |
| 2 | $4\frac{1}{2}$ | $1\frac{25}{32}$ | .. | ..... | **16** | $1\frac{15}{16}$ | $8^e$ | $1\frac{7}{8}$ | 12 | $1\frac{59}{64}$ | 16 | $1\frac{15}{16}$ |
| $2\frac{1}{16}$ | .... | ..... | .. | ..... | .. | ..... | .. | ... | .. | ..... | 16 | 2 |
| $2\frac{1}{8}$ | .... | ..... | .. | ..... | .. | ..... | 8 | 2 | 12 | $2\frac{3}{64}$ | 16 | $2\frac{1}{16}$ |
| $2\frac{3}{16}$ | .... | ..... | .. | ..... | .. | ..... | .. | ... | .. | ..... | 16 | $2\frac{1}{8}$ |
| $2\frac{1}{4}$ | $4\frac{1}{2}$ | $2\frac{1}{32}$ | .. | ..... | .. | ..... | $8^e$ | $2\frac{1}{8}$ | 12 | $2\frac{11}{64}$ | **16** | $2\frac{3}{16}$ |
| $2\frac{5}{16}$ | .... | ..... | .. | ..... | .. | ..... | .. | ... | .. | ..... | 16 | $2\frac{1}{4}$ |
| $2\frac{3}{8}$ | .... | ..... | .. | ..... | .. | ..... | .. | ... | 12 | $2\frac{19}{64}$ | 16 | $2\frac{5}{16}$ |
| $2\frac{7}{16}$ | .... | ..... | .. | ..... | .. | ..... | .. | ... | .. | ... | 16 | $2\frac{3}{8}$ |
| $2\frac{1}{2}$ | 4 | $2\frac{1}{4}$ | .. | ..... | .. | ..... | $8^e$ | $2\frac{3}{8}$ | 12 | $2\frac{27}{64}$ | **16** | $2\frac{7}{16}$ |
| $2\frac{5}{8}$ | .... | ..... | .. | ..... | .. | ..... | .. | ... | 12 | $2\frac{35}{64}$ | 16 | $2\frac{9}{16}$ |
| $2\frac{3}{4}$ | 4 | $2\frac{1}{2}$ | .. | ..... | .. | ..... | $8^e$ | $2\frac{5}{8}$ | 12 | $2\frac{43}{64}$ | 16 | $2\frac{11}{16}$ |
| $2\frac{7}{8}$ | .... | ..... | .. | ..... | .. | ..... | .. | ... | 12 | $2\frac{51}{64}$ | 16 | $2\frac{13}{16}$ |
| 3 | 4 | $2\frac{3}{4}$ | .. | ..... | .. | ..... | $8^e$ | $2\frac{7}{8}$ | 12 | $2\frac{59}{64}$ | **16** | $2\frac{15}{16}$ |
| $3\frac{1}{8}$ | .... | ..... | .. | ..... | .. | ..... | .. | ... | 12 | $3\frac{3}{64}$ | 16 | $3\frac{1}{16}$ |
| $3\frac{1}{4}$ | 4 | 3 | .. | ..... | .. | ..... | $8^e$ | $3\frac{1}{8}$ | 12 | $3\frac{11}{64}$ | **16** | $3\frac{3}{16}$ |
| $3\frac{3}{8}$ | .... | ..... | .. | ..... | .. | ..... | .. | ... | 12 | $3\frac{19}{64}$ | 16 | $3\frac{5}{16}$ |
| $3\frac{1}{2}$ | 4 | $3\frac{1}{4}$ | .. | ..... | .. | ..... | $8^e$ | $3\frac{3}{8}$ | 12 | $3\frac{27}{64}$ | **16** | $3\frac{7}{16}$ |
| $3\frac{5}{8}$ | .... | ..... | .. | ..... | .. | ..... | .. | ... | 12 | $3\frac{35}{64}$ | 16 | $3\frac{9}{16}$ |
| $3\frac{3}{4}$ | 4 | $3\frac{1}{2}$ | .. | ..... | .. | ..... | $8^e$ | $3\frac{5}{8}$ | 12 | $3\frac{43}{64}$ | **16** | $3\frac{11}{16}$ |
| $3\frac{7}{8}$ | .... | ..... | .. | ..... | .. | ..... | .. | ... | 12 | $3\frac{51}{64}$ | 16 | $3\frac{13}{16}$ |
| 4 | 4 | $3\frac{3}{4}$ | .. | ..... | .. | ..... | $8^e$ | $3\frac{7}{8}$ | 12 | $3\frac{59}{64}$ | **16** | $3\frac{15}{16}$ |
| $4\frac{1}{4}$ | .... | ..... | .. | ..... | .. | ..... | $8^e$ | $4\frac{1}{8}$ | 12 | $4\frac{11}{64}$ | 16 | $4\frac{3}{16}$ |
| $4\frac{1}{2}$ | .... | ..... | .. | ..... | .. | ..... | $8^e$ | $4\frac{3}{8}$ | 12 | $4\frac{27}{64}$ | 16 | $4\frac{7}{16}$ |
| $4\frac{3}{4}$ | .... | ..... | .. | ..... | .. | ..... | $8^e$ | $4\frac{5}{8}$ | 12 | $4\frac{43}{64}$ | 16 | $4\frac{11}{16}$ |
| 5 | .... | ..... | .. | ..... | .. | ..... | $8^e$ | $4\frac{7}{8}$ | 12 | $4\frac{59}{64}$ | 16 | $4\frac{15}{16}$ |
| $5\frac{1}{4}$ | .... | ..... | .. | ..... | .. | ..... | $8^e$ | $5\frac{1}{8}$ | 12 | $5\frac{11}{64}$ | 16 | $5\frac{3}{16}$ |
| $5\frac{1}{2}$ | .... | ..... | .. | ..... | .. | ..... | $8^e$ | $5\frac{3}{8}$ | 12 | $5\frac{27}{64}$ | 16 | $5\frac{7}{16}$ |
| $5\frac{3}{4}$ | .... | ..... | .. | ..... | .. | ..... | $8^e$ | $5\frac{5}{8}$ | 12 | $5\frac{43}{64}$ | 16 | $5\frac{11}{16}$ |
| 6 | .... | ..... | .. | ..... | .. | ..... | $8^e$ | $5\frac{7}{8}$ | 12 | $5\frac{59}{64}$ | 16 | $5\frac{15}{16}$ |

$^a$ ANSI B1.1—Dimensions are in inches.   $^b$ Bold type indicates unified combinations.

$^c$ Limits of size for classes are based on a length of engagement equal to the nominal diameter.

$^d$ Limits of size for classes are based on a length of engagement equal to nine times the pitch.

$^e$ These sizes, with specified limits of size, based on a length of engagement of 9 threads in classes 2A and 2B, are designated UN.

*Note.* If a thread is in both the 8-, 12-, or 16-thread series and the coarse, fine, or extra-fine-thread series, the symbols and tolerances of the latter series apply.

## ANSI Standard Square and Hexagon Bolts* and Hexagon Cap Screws†

| Nominal size (basic major diam.) | Regular bolts | | | Heavy bolts | | |
|---|---|---|---|---|---|---|
| | Width across flats $W$, sq.‡ and hex. | Height $H$ | | Width across flats $W$ | Height $H$ | |
| | | Unfaced sq. and hex. | Semifin. hex., fin. hex., and hex. screw | | Unfaced hex. | Semifin. hex. and fin. hex. |
| 1/4 | 3/8(sq.), 7/16(hex.) | 11/64 | 5/32 | | | |
| 5/16 | 1/2 | 7/32 | 13/64 | | | |
| 3/8 | 9/16 | 1/4 | 15/64 | | | |
| 7/16 | 5/8 | 19/64 | 9/32 | | | |
| 1/2 | 3/4 | 11/32 | 5/16 | 7/8 | 7/16 | 13/32 |
| 9/16 | 13/16 | 25/64 | 23/64 | 15/16 | 15/32 | 7/16 |
| 5/8 | 15/16 | 27/64 | 25/64 | 1 1/16 | 17/32 | 1/2 |
| 3/4 | 1 1/8 | 1/2 | 15/32 | 1 1/4 | 5/8 | 19/32 |
| 7/8 | 1 5/16 | 37/64 | 35/64 | 1 7/16 | 23/32 | 11/16 |
| 1 | 1 1/2 | 43/64 | 39/64 | 1 5/8 | 13/16 | 3/4 |
| 1 1/8 | 1 11/16 | 3/4 | 11/16 | 1 13/16 | 29/32 | 27/32 |
| 1 1/4 | 1 7/8 | 27/32 | 25/32 | 2 | 1 | 15/16 |
| 1 3/8 | 2 1/16 | 29/32 | 27/32 | 2 3/16 | 1 3/32 | 1 1/32 |
| 1 1/2 | 2 1/4 | 1 | 15/16 | 2 3/8 | 1 3/16 | 1 1/8 |
| 1 5/8 | 2 7/16 | 1 1/16 | 1 | 2 9/16 | 1 9/32 | 1 7/32 |
| 1 3/4 | 2 5/8 | 1 5/32 | 1 3/32 | 2 3/4 | 1 3/8 | 1 5/16 |
| 1 7/8 | 2 13/16 | 1 7/32 | 1 5/32 | 2 15/16 | 1 15/32 | 1 13/32 |
| 2 | 3 | 1 11/32 | 1 7/32 | 3 1/8 | 1 9/16 | 1 7/16 |
| 2 1/4 | 3 3/8 | 1 1/2 | 1 3/8 | 3 1/2 | 1 3/4 | 1 5/8 |
| 2 1/2 | 3 3/4 | 1 21/32 | 1 17/32 | 3 7/8 | 1 15/16 | 1 13/16 |
| 2 3/4 | 4 1/8 | 1 13/16 | 1 11/16 | 4 1/4 | 2 1/8 | 2 |
| 3 | 4 1/2 | 2 | 1 7/8 | 4 5/8 | 2 5/16 | 2 3/16 |
| 3 1/4 | 4 7/8 | 2 3/16 | 2 | | | |
| 3 1/2 | 5 1/4 | 2 5/16 | 2 1/8 | | | |
| 3 3/4 | 5 5/8 | 2 1/2 | 2 5/16 | | | |
| 4 | 6 | 2 11/16 | 2 1/2 | | | |

* ANSI B18.2—All dimensions in inches.
† ANSI B18.6.2—Hexagon-head cap screw in sizes 1/4 to 1 1/2 only.
‡ Square bolts in (nominal) sizes 1/4 to 1 5/8 only.

For bolt-length increments, see table, page 720. Threads are coarse series, class 2A, except for finished hexagon bolt and hexagon cap screw, which are coarse, fine, or 8-pitch, class 2A. Minimum thread length: $2D + 1/4$ in. for bolts 6 in. or less in length; $2D + 1/2$ in. for bolts over 6 in. in length; bolts too short for formula, thread entire length.

## ANSI Standard Square and Hexagon Nuts *

| Nominal size (basic major diam.) | Reg. nuts — Width across flats W | Reg. nuts Thickness T — Reg. hex. | Reg. hex. jam | Semi-fin. and fin. hex. and hex. slotted | Semi-fin. and fin. hex. jam | Semi-fin. hex. thick, thick slotted,† and castle | Heavy nuts — Width across flats W | Heavy nuts Thickness T — Reg. sq. and hex. | Reg. hex. jam | Semi-fin. hex. and hex. slotted | Semi-fin. hex. jam | Slot Width | Slot Depth | Reg. sq. nuts Width across flats W | Reg. sq. nuts Thickness T |
|---|---|---|---|---|---|---|---|---|---|---|---|---|---|---|---|
| 1/4 | 7/16 | ..... | .... | 7/32 | 5/32 | 9/32 | 1/2 | 1/4 | 3/16 | 15/64 | 11/64 | 5/64 | 3/32 | 7/16 | 7/32 |
| 5/16 | 1/2 | ..... | .... | 17/64 | 3/16 | 21/64 | 9/16 | 5/16 | 7/32 | 19/64 | 13/64 | 3/32 | 3/32 | 9/16 | 17/64 |
| 3/8 | 9/16 | ..... | .... | 21/64 | 7/32 | 13/32 | 11/16 | 3/8 | 1/4 | 23/64 | 15/64 | 1/8 | 1/8 | 5/8 | 21/64 |
| 7/16 | 11/16 | ..... | .... | 3/8 | 1/4 | 29/64 | 3/4 | 7/16 | 9/32 | 27/64 | 17/64 | 1/8 | 5/32 | 3/4 | 3/8 |
| 1/2 | 3/4 | ..... | .... | 7/16 | 5/16 | 9/16 | 7/8 | 1/2 | 5/16 | 31/64 | 19/64 | 5/32 | 5/32 | 13/16 | 7/16 |
| 9/16 | 7/8 | ..... | .... | 31/64 | 5/16 | 39/64 | 15/16 | | | 35/64 | 21/64 | 5/32 | 3/16 | | |
| 5/8 | 15/16 | ..... | .... | 35/64 | 3/8 | 23/32 | 1 1/16 | 5/8 | 3/8 | 39/64 | 23/64 | 3/16 | 7/32 | 1 | 35/64 |
| 3/4 | 1 1/8 | 21/32 | 7/16 | 41/64 | 27/64 | 13/16 | 1 1/4 | 3/4 | 7/16 | 47/64 | 27/64 | 3/16 | 1/4 | 1 1/8 | 21/32 |
| 7/8 | 1 5/16 | 49/64 | 1/2 | 3/4 | 31/64 | 29/32 | 1 7/16 | 7/8 | 1/2 | 55/64 | 31/64 | 3/16 | 1/4 | 1 5/16 | 49/64 |
| 1 | 1 1/2 | 7/8 | 9/16 | 55/64 | 35/64 | 1 | 1 5/8 | 1 | 9/16 | 63/64 | 35/64 | 1/4 | 9/32 | 1 1/2 | 7/8 |
| 1 1/8 | 1 11/16 | 1 | 5/8 | 31/32 | 39/64 | 1 5/32 | 1 13/16 | 1 1/8 | 5/8 | 1 7/64 | 39/64 | 1/4 | 11/32 | 1 11/16 | 1 |
| 1 1/4 | 1 7/8 | 1 3/32 | 3/4 | 1 1/16 | 23/32 | 1 1/4 | 2 | 1 1/4 | 3/4 | 1 7/32 | 23/32 | 5/16 | 3/8 | 1 7/8 | 1 3/32 |
| 1 3/8 | 2 1/16 | 1 13/64 | 13/16 | 1 11/64 | 25/32 | 1 3/8 | 2 3/16 | 1 3/8 | 13/16 | 1 11/32 | 25/32 | 5/16 | 3/8 | 2 1/16 | 1 13/64 |
| 1 1/2 | 2 1/4 | 1 5/16 | 7/8 | 1 9/32 | 27/32 | 1 1/2 | 2 3/8 | 1 1/2 | 7/8 | 1 15/32 | 27/32 | 3/8 | 7/16 | 2 1/4 | 1 5/16 |
| 1 5/8 | 2 7/16 | ..... | ... | 1 25/64 | 29/32 | ..... | 2 9/16 | | .... | 1 19/32 | 29/32 | 3/8 | 7/16 | | |
| 1 3/4 | 2 5/8 | ..... | .... | 1 1/2 | 31/32 | ..... | 2 3/4 | 1 3/4 | 1 | 1 23/32 | 31/32 | 7/16 | 1/2 | | |
| 1 7/8 | 2 13/16 | ..... | .... | 1 39/64 | 1 1/32 | ..... | 2 15/16 | | .... | 1 27/32 | 1 1/32 | 7/16 | 9/16 | | |
| 2 | 3 | ..... | .... | 1 23/32 | 1 3/32 | ..... | 3 1/8 | 2 | 1 1/8 | 1 31/32 | 1 3/32 | 7/16 | 9/16 | | |
| 2 1/4 | 3 3/8 | ..... | .... | 1 59/64 | 1 13/64 | ..... | 3 1/2 | 2 1/4 | 1 1/4 | 2 13/64 | 1 13/64 | 7/16 | 9/16 | | |
| 2 1/2 | 3 3/4 | ..... | .... | 2 9/64 | 1 29/64 | ..... | 3 7/8 | 2 1/2 | 1 1/2 | 2 29/64 | 1 29/64 | 9/16 | 11/16 | | |
| 2 3/4 | 4 1/8 | ..... | .... | 2 23/64 | 1 37/64 | ..... | 4 1/4 | 2 3/4 | 1 5/8 | 2 45/64 | 1 37/64 | 9/16 | 11/16 | | |
| 3 | 4 1/2 | ..... | .... | 2 37/64 | 1 45/64 | ..... | 4 5/8 | 3 | 1 3/4 | 2 61/64 | 1 45/64 | 5/8 | 3/4 | | |
| 3 1/4 | ..... | ..... | .... | ..... | ..... | ..... | 5 | 3 1/4 | 1 7/8 | 3 3/16 | 1 13/16 | 5/8 | 3/4 | | |
| 3 1/2 | ..... | ..... | .... | ..... | ..... | ..... | 5 3/8 | 3 1/2 | 2 | 3 7/16 | 1 15/16 | 5/8 | 3/4 | | |
| 3 3/4 | ..... | ..... | .... | ..... | ..... | ..... | 5 3/4 | 3 3/4 | 2 1/8 | 3 11/16 | 2 1/16 | 5/8 | 3/4 | | |
| 4 | ..... | ..... | .... | ..... | ..... | ..... | 6 1/8 | 4 | 2 1/4 | 3 15/16 | 2 3/16 | 5/8 | 3/4 | | |

* ANSI B18.2—1960.

† Slot dimensions for regular slotted nuts are same as for heavy slotted nuts.

*Thread.* Regular and heavy nuts: coarse series, class 2B; finished and semifinished nuts, regular and heavy (all): coarse, fine, or 8-pitch series, class 2B; thick nuts: coarse or fine series, class 2B.

## Sizes of Numbered and Lettered Drills

| No. | Size | No. | Size | No. | Size | Letter | Size |
|-----|------|-----|------|-----|------|--------|------|
| 80 | 0.0135 | 53 | 0.0595 | 26 | 0.1470 | A | 0.2340 |
| 79 | 0.0145 | 52 | 0.0635 | 25 | 0.1495 | B | 0.2380 |
| 78 | 0.0160 | 51 | 0.0670 | 24 | 0.1520 | C | 0.2420 |
| 77 | 0.0180 | 50 | 0.0700 | 23 | 0.1540 | D | 0.2460 |
| 76 | 0.0200 | 49 | 0.0730 | 22 | 0.1570 | E | 0.2500 |
| 75 | 0.0210 | 48 | 0.0760 | 21 | 0.1590 | F | 0.2570 |
| 74 | 0.0225 | 47 | 0.0785 | 20 | 0.1610 | G | 0.2610 |
| 73 | 0.0240 | 46 | 0.0810 | 19 | 0.1660 | H | 0.2660 |
| 72 | 0.0250 | 45 | 0.0820 | 18 | 0.1695 | I | 0.2720 |
| 71 | 0.0260 | 44 | 0.0860 | 17 | 0.1730 | J | 0.2770 |
| 70 | 0.0280 | 43 | 0.0890 | 16 | 0.1770 | K | 0.2810 |
| 69 | 0.0292 | 42 | 0.0935 | 15 | 0.1800 | L | 0.2900 |
| 68 | 0.0310 | 41 | 0.0960 | 14 | 0.1820 | M | 0.2950 |
| 67 | 0.0320 | 40 | 0.0980 | 13 | 0.1850 | N | 0.3020 |
| 66 | 0.0330 | 39 | 0.0995 | 12 | 0.1890 | O | 0.3160 |
| 65 | 0.0350 | 38 | 0.1015 | 11 | 0.1910 | P | 0.3230 |
| 64 | 0.0360 | 37 | 0.1040 | 10 | 0.1935 | Q | 0.3320 |
| 63 | 0.0370 | 36 | 0.1065 | 9 | 0.1960 | R | 0.3390 |
| 62 | 0.0380 | 35 | 0.1100 | 8 | 0.1990 | S | 0.3480 |
| 61 | 0.0390 | 34 | 0.1110 | 7 | 0.2010 | T | 0.3580 |
| 60 | 0.0400 | 33 | 0.1130 | 6 | 0.2040 | U | 0.3680 |
| 59 | 0.0410 | 32 | 0.1160 | 5 | 0.2055 | V | 0.3770 |
| 58 | 0.0420 | 31 | 0.1200 | 4 | 0.2090 | W | 0.3860 |
| 57 | 0.0430 | 30 | 0.1285 | 3 | 0.2130 | X | 0.3970 |
| 56 | 0.0465 | 29 | 0.1360 | 2 | 0.2210 | Y | 0.4040 |
| 55 | 0.0520 | 28 | 0.1405 | 1 | 0.2280 | Z | 0.4130 |
| 54 | 0.0550 | 27 | 0.1440 | | | | |

## Acme and Stub Acme Threads*

*ANSI-preferred diameter-pitch combinations*

| Nominal (major) diam. | Threads/in. | Nominal (major) diam. | Threads/in. | Nominal (major) diam. | Threads/in. | Nominal (major) diam. | Threads/in. |
|------|------|------|------|------|------|------|------|
| ¼ | 16 | ¾ | 6 | 1½ | 4 | 3 | 2 |
| ⁵⁄₁₆ | 14 | ⅞ | 6 | 1¾ | 4 | 3½ | 2 |
| ⅜ | 12 | 1 | 5 | 2 | 4 | 4 | 2 |
| ⁷⁄₁₆ | 12 | 1⅛ | 5 | 2¼ | 3 | 4½ | 2 |
| ½ | 10 | 1¼ | 5 | 2½ | 3 | 5 | 2 |
| ⅝ | 8 | 1⅜ | 4 | 2¾ | 3 | | |

## Buttress Threads*

| Recommended diam., in. | Assoc. threads/in. |
|------|------|
| ½, ⁹⁄₁₆, ⅝, 1¹⁄₁₆ | 20, 16, 12 |
| ¾, ⅞, 1 | 16, 12, 10 |
| 1⅛, 1¼, 1⅜, 1½ | 16, 12, 10, 8, 6 |
| 1¾, 2, 2¼, 2½, 2¾, 3, 3½, 4 | 16, 12, 10, 8, 6, 5, 4 |
| 4½, 5, 5½, 6 | 12, 10, 8, 6, 5, 4, 3 |
| 7, 8, 9, 10 | 10, 8, 6, 5, 4, 3, 2½, 2 |
| 11, 12, 14, 16 | 10, 8, 6, 5, 4, 3, 2½, 2, 1½, 1¼ |
| 18, 20, 22, 24 | 8, 6, 5, 4, 3, 2½, 2, 1½, 1¼, 1 |

* ANSI B1.5—and B1.9—Diameters in inches.

## ANSI Standard Round-head Bolts[a]

*Proportions for drawing purposes*

| Fastener name | | Nominal diam.[b] (basic major diam.) | A | H | P or M | B | |
|---|---|---|---|---|---|---|---|
| Carriage bolts | Round-head square-neck bolt[c] | No. 10, ¼″ to ½″ by (¹⁄₁₆″), ⅝″ to 1″ by (⅛″) | $2D + \frac{1}{16}$ | $\frac{D}{2}$ | $\frac{D}{2}$ | $D$ | |
| | Round-head ribbed-neck bolt[d] | No. 10, ¼″ to ½″ by (¹⁄₁₆″), ⅝″ and ¾″ | $2D + \frac{1}{16}$ | $\frac{D}{2}$ | $\frac{1}{16}$ | $D + \frac{1}{16}$ | $Q \begin{cases} \frac{3}{16}″ \text{ for } L = \frac{7}{8}″ \text{ or less} \\ \frac{5}{16}″ \text{ for } L = 1″ \text{ and } 1\frac{1}{8}″ \\ \frac{1}{2}″ \text{ for } L = 1\frac{1}{4}″ \text{ or more} \end{cases}$ |
| | Round-head fin.-neck bolt[c] | No. 10, ¼″ to ½″ by (¹⁄₁₆″) | $2D + \frac{1}{16}$ | $\frac{D}{2}$ | $\frac{3}{8}D$ | $1\frac{1}{2}D + \frac{1}{16}$ | |
| | 114° countersunk square-neck bolt[c] | No. 10, ¼″ to ½″ by (¹⁄₁₆″), ⅝″ and ¾″ | $2D + \frac{1}{8}$ | $\frac{1}{32}$ | $D + \frac{1}{32}$ | $D$ | |
| Round-head short square-neck bolt[e] | | ¼″ to ½″ by (¹⁄₁₆″), ⅝″ and ¾″ | $2D + \frac{1}{16}$ | $\frac{D}{2}$ | $\frac{D}{4} + \frac{1}{32}$ | $D$ | |
| T-head bolt[d] | | ¼″ to ½″ by (¹⁄₁₆″), ⅝″ to 1″ by (⅛″) | $2D$ | $\frac{7}{8}D$ | $1\frac{5}{8}D$ | $D$ | |
| Round-head bolt[c] (buttonhead bolt) | | No. 10, ¼″ to ½″ by (¹⁄₁₆″), ⅝″ to 1″ by (⅛″) | $2D + \frac{1}{16}$ | $\frac{D}{2}$ | | | |
| Step bolt[c] | | No. 10, ¼″ to ½″ by (¹⁄₁₆″) | $3D + \frac{1}{16}$ | $\frac{D}{2}$ | $\frac{D}{2}$ | $D$ | |
| Countersunk bolt[c] (may be slotted if so specified) | | ¼″ to ½″ by (¹⁄₁₆″), ⅝″ to 1½″ by (⅛″) | $1.8D$ | Obtain by projection | | | |
| Elevator bolt, flat head, countersunk[c] | | No. 10, ¼″ to ½″ by (¹⁄₁₆″) | $2\frac{1}{2}D + \frac{5}{16}$ | $\frac{D}{3}$ | $\frac{D}{2} + \frac{1}{16}$ | $D$ | Angle $C = 16D + 5°$ (approx.) |
| Elevator bolt, ribbed head[d] (slotted or unslotted as specified) | | ¼″, ⁵⁄₁₆″, ⅜″ | $2D + \frac{1}{16}$ | $\frac{D}{2} - \frac{1}{32}$ | $\frac{D}{2} + \frac{3}{64}$ | $\frac{D}{8} + \frac{1}{16}$ | |

[a] The proportions in this table are in some instances approximate and are intended for drawing purposes only. For exact dimensions see ANSI B18.5—, from which this table was compiled. Dimensions are in inches.

[b] Fractions in parentheses show diameter increments, *e.g.*, ¼ in. to ½ in. by (¹⁄₁₆ in.) includes the diameters ¼ in., ⁵⁄₁₆ in., ⅜ in., ⁷⁄₁₆ in., and ½ in.

Threads are coarse series, class 2A.

Minimum thread length *l*: $2D + \frac{1}{4}$ in. for bolts 6 in. or less in length; $2D + \frac{1}{2}$ in. for bolts over 6 in. in length.

For bolt length increments see table page 720.

[c] Full-size body bolts furnished unless undersize body is specified.

[d] Only full-size body bolts furnished.

[e] Undersize body bolts furnished unless full-size body is specified.

## Bolt-length Increments*

| Bolt diameter | | 1/4 | 5/16 | 3/8 | 7/16 | 1/2 | 5/8 | 3/4 | 7/8 | 1 |
|---|---|---|---|---|---|---|---|---|---|---|
| Length increments | 1/4 | 3/4–3 | 3/4–4 | 3/4–6 | 1–3 | 1–6 | 1–6 | 1–6 | 1–4 1/2 | |
| | 1/2 | 3–4 | 4–5 | 6–9 | 3–6 | 6–13 | 6–10 | 6–15 | 4 1/2–6 | 3–6 |
| | 1 | 4–5 | .... | 9–12 | 6–8 | 13–24 | 10–22 | 15–24 | 6–20 | 6–12 |
| | 2 | .... | .... | ..... | ... | ..... | 22–30 | 24–30 | 20–30 | 12–30 |

\* Compiled from manufacturers' catalogues.

*Example.* 1/4-in. bolt lengths increase by 1/4-in. increments from 3/4- to 3-in. length. 1/2-in. bolt lengths increase by 1/2-in. increments from 6- to 13-in. length. 1-in. bolt lengths increase by 2-in. increments from 12- to 30-in. length.

## ANSI Standard Cap Screws[a]—Socket[b] and Slotted Heads[c]

*For hexagon-head screws, see page* ∎∎

| Nominal diam. | Socket head[d] | | | Flat head[e] | Fillister head[e] | | Round head[e] | |
|---|---|---|---|---|---|---|---|---|
| | A | H | J | A | A | H | A | H |
| 0 | 0.096 | 0.060 | 0.050 | | | | | |
| 1 | 0.118 | 0.073 | 0.050 | | | | | |
| 2 | 0.140 | 0.086 | 1/16 | | | | | |
| 3 | 0.161 | 0.099 | 5/64 | | | | | |
| 4 | 0.183 | 0.112 | 5/64 | | | | | |
| 5 | 0.205 | 0.125 | 3/32 | | | | | |
| 6 | 0.226 | 0.138 | 3/32 | | | | | |
| 8 | 0.270 | 0.164 | 1/8 | | | | | |
| 10 | 5/16 | 0.190 | 5/32 | | | | | |
| 12 | 11/32 | 0.216 | 5/32 | | | | | |
| 1/4 | 3/8 | 1/4 | 3/16 | 1/2 | 3/8 | 11/64 | 7/16 | 3/16 |
| 5/16 | 7/16 | 5/16 | 7/32 | 5/8 | 7/16 | 13/64 | 9/16 | 15/64 |
| 3/8 | 9/16 | 3/8 | 5/16 | 3/4 | 9/16 | 1/4 | 5/8 | 17/64 |
| 7/16 | 5/8 | 7/16 | 5/16 | 13/16 | 5/8 | 19/64 | 3/4 | 5/16 |
| 1/2 | 3/4 | 1/2 | 3/8 | 7/8 | 3/4 | 21/64 | 13/16 | 11/32 |
| 9/16 | 13/16 | 9/16 | 3/8 | 1 | 13/16 | 3/8 | 15/16 | 13/32 |
| 5/8 | 7/8 | 5/8 | 1/2 | 1 1/8 | 7/8 | 27/64 | 1 | 7/16 |
| 3/4 | 1 | 3/4 | 9/16 | 1 3/8 | 1 | 1/2 | 1 1/4 | 17/32 |
| 7/8 | 1 1/8 | 7/8 | 9/16 | 1 5/8 | 1 1/8 | 19/32 | | |
| 1 | 1 5/16 | 1 | 5/8 | 1 7/8 | 1 5/16 | 21/32 | | |
| 1 1/8 | 1 1/2 | 1 1/8 | 3/4 | | | | | |
| 1 1/4 | 1 3/4 | 1 1/4 | 3/4 | | | | | |
| 1 3/8 | 1 7/8 | 1 3/8 | 3/4 | | | | | |
| 1 1/2 | 2 | 1 1/2 | 1 | | | | | |

[a] Dimensions in inches.

[b] ANSI B18.3

[c] ANSI B18.6.2

[d] Thread coarse or fine, class 3A. Thread length $l$: coarse thread, $2D + 1/2$ in.; fine thread, $1\frac{1}{2}D + 1/2$ in.

[e] Thread coarse, fine, or 8-pitch, class 2A. Thread length $l$: $2D + 1/4$ in.

Slot proportions vary with size of screw; draw to look well. All body-length increments for screw lengths 1/4 in. to 1 in. = 1/8 in., for screw lengths 1 in. to 4 in. = 1/4 in., for screw lengths 4 in. to 6 in. = 1/2 in.

# ANSI Standard Machine Screws *

*Heads may be slotted or recessed*

Round Head     Flat Head     Fillister Head     Oval Head     Truss Head

| Nominal diam. | Round head | | Flat head | Fillister head | | | Oval head | | Truss head | |
|---|---|---|---|---|---|---|---|---|---|---|
| | A | H | A | A | H | O | A | C | A | H |
| 0 | 0.113 | 0.053 | 0.119 | 0.096 | 0.045 | 0.059 | 0.119 | 0.021 | | |
| 1 | 0.138 | 0.061 | 0.146 | 0.118 | 0.053 | 0.071 | 0.146 | 0.025 | 0.194 | 0.053 |
| 2 | 0.162 | 0.069 | 0.172 | 0.140 | 0.062 | 0.083 | 0.172 | 0.029 | 0.226 | 0.061 |
| 3 | 0.187 | 0.078 | 0.199 | 0.161 | 0.070 | 0.095 | 0.199 | 0.033 | 0.257 | 0.069 |
| 4 | 0.211 | 0.086 | 0.225 | 0.183 | 0.079 | 0.107 | 0.225 | 0.037 | 0.289 | 0.078 |
| 5 | 0.236 | 0.095 | 0.252 | 0.205 | 0.088 | 0.120 | 0.252 | 0.041 | 0.321 | 0.086 |
| 6 | 0.260 | 0.103 | 0.279 | 0.226 | 0.096 | 0.132 | 0.279 | 0.045 | 0.352 | 0.094 |
| 8 | 0.309 | 0.120 | 0.332 | 0.270 | 0.113 | 0.156 | 0.332 | 0.052 | 0.384 | 0.102 |
| 10 | 0.359 | 0.137 | 0.385 | 0.313 | 0.130 | 0.180 | 0.385 | 0.060 | 0.448 | 0.118 |
| 12 | 0.408 | 0.153 | 0.438 | 0.357 | 0.148 | 0.205 | 0.438 | 0.068 | 0.511 | 0.134 |
| $\frac{1}{4}$ | 0.472 | 0.175 | 0.507 | 0.414 | 0.170 | 0.237 | 0.507 | 0.079 | 0.573 | 0.150 |
| $\frac{5}{16}$ | 0.590 | 0.216 | 0.635 | 0.518 | 0.211 | 0.295 | 0.635 | 0.099 | 0.698 | 0.183 |
| $\frac{3}{8}$ | 0.708 | 0.256 | 0.762 | 0.622 | 0.253 | 0.355 | 0.762 | 0.117 | 0.823 | 0.215 |
| $\frac{7}{16}$ | 0.750 | 0.328 | 0.812 | 0.625 | 0.265 | 0.368 | 0.812 | 0.122 | 0.948 | 0.248 |
| $\frac{1}{2}$ | 0.813 | 0.355 | 0.875 | 0.750 | 0.297 | 0.412 | 0.875 | 0.131 | 1.073 | 0.280 |
| $\frac{9}{16}$ | 0.938 | 0.410 | 1.000 | 0.812 | 0.336 | 0.466 | 1.000 | 0.150 | 1.198 | 0.312 |
| $\frac{5}{8}$ | 1.000 | 0.438 | 1.125 | 0.875 | 0.375 | 0.521 | 1.125 | 0.169 | 1.323 | 0.345 |
| $\frac{3}{4}$ | 1.250 | 0.547 | 1.375 | 1.000 | 0.441 | 0.612 | 1.375 | 0.206 | 1.573 | 0.410 |

Binding Head     Pan Head     Pan Head (recessed)     Hexagon Head     100° Flat Head

| Nominal diam. | Binding head | | | | Pan head | | | Hexagon head | | 100° flat head |
|---|---|---|---|---|---|---|---|---|---|---|
| | A | O | F | U | A | H | O | A | H | A |
| 2 | 0.181 | 0.046 | 0.018 | 0.141 | 0.167 | 0.053 | 0.062 | 0.125 | 0.050 | |
| 3 | 0.208 | 0.054 | 0.022 | 0.162 | 0.193 | 0.060 | 0.071 | 0.187 | 0.055 | |
| 4 | 0.235 | 0.063 | 0.025 | 0.184 | 0.219 | 0.068 | 0.080 | 0.187 | 0.060 | 0.225 |
| 5 | 0.263 | 0.071 | 0.029 | 0.205 | 0.245 | 0.075 | 0.089 | 0.187 | 0.070 | |
| 6 | 0.290 | 0.080 | 0.032 | 0.226 | 0.270 | 0.082 | 0.097 | 0.250 | 0.080 | 0.279 |
| 8 | 0.344 | 0.097 | 0.039 | 0.269 | 0.322 | 0.096 | 0.115 | 0.250 | 0.110 | 0.332 |
| 10 | 0.399 | 0.114 | 0.045 | 0.312 | 0.373 | 0.110 | 0.133 | 0.312 | 0.120 | 0.385 |
| 12 | 0.454 | 0.130 | 0.052 | 0.354 | 0.425 | 0.125 | 0.151 | 0.312 | 0.155 | |
| $\frac{1}{4}$ | 0.513 | 0.153 | 0.061 | 0.410 | 0.492 | 0.144 | 0.175 | 0.375 | 0.190 | 0.507 |
| $\frac{5}{16}$ | 0.641 | 0.193 | 0.077 | 0.513 | 0.615 | 0.178 | 0.218 | 0.500 | 0.230 | 0.635 |
| $\frac{3}{8}$ | 0.769 | 0.234 | 0.094 | 0.615 | 0.740 | 0.212 | 0.261 | 0.562 | 0.295 | 0.762 |

\* ANSI B18.6.3—1962. Dimensions given are maximum values, all in inches.

Thread length: screws 2 in. long or less, thread entire length; screws over 2 in. long, thread length $l = 1\frac{3}{4}$ in. Threads are coarse or fine series, class 2. Heads may be slotted or recessed as specified, excepting hexagon form, which is plain or may be slotted if so specified. Slot and recess proportions vary with size of fastener; draw to look well.

## ANSI Standard Machine-screw * and Stove-bolt† Nuts‡

| Nominal size | 0 | 1 | 2 | 3 | 4 | 5 | 6 | 8 | 10 | 12 | 1/4 | 5/16 | 3/8 |
|---|---|---|---|---|---|---|---|---|---|---|---|---|---|
| "W" | 5/32 | 5/32 | 3/16 | 3/16 | 1/4 | 5/16 | 5/16 | 11/32 | 3/8 | 7/16 | 7/16 | 9/16 | 5/8 |
| "T" | 3/64 | 3/64 | 1/16 | 1/16 | 3/32 | 7/64 | 7/64 | 1/8 | 1/8 | 5/32 | 3/16 | 7/32 | 1/4 |

* Machine-screw nuts are hexagonal and square.
† Stove-bolt nuts are square.
‡ ANSI B18.6.3—Dimensions are in inches.
Thread is coarse series for square nuts and coarse or fine series for hexagon nuts; class 2B.

## ANSI Standard Hexagon Socket, * Slotted headless,† and Square-head‡ Setscrews

Cup Point     Flat Point

Oval Point     Cone Point

Full Dog Point     Half Dog Point

*(All six point types are available in all three head types)*

| Diam. $D$ | Cup and flat-point diam. $C$ | Oval-point radius $R$ | Cone-point angle $Y$ 118° for these lengths and shorter | Cone-point angle $Y$ 90° for these lengths and longer | Full and half dog points Diam. $P$ | Full and half dog points Length Full $Q$ | Full and half dog points Length Half $q$ | Socket width $J$ |
|---|---|---|---|---|---|---|---|---|
| 5 | 1/16 | 3/32 | 1/8 | 3/16 | 0.083 | 0.06 | 0.03 | 1/16 |
| 6 | 0.069 | 7/64 | 1/8 | 3/16 | 0.092 | 0.07 | 0.03 | 1/16 |
| 8 | 5/64 | 1/8 | 3/16 | 1/4 | 0.109 | 0.08 | 0.04 | 5/64 |
| 10 | 3/32 | 9/64 | 3/16 | 1/4 | 0.127 | 0.09 | 0.04 | 3/32 |
| 12 | 7/64 | 5/32 | 3/16 | 1/4 | 0.144 | 0.11 | 0.06 | 3/32 |
| 1/4 | 1/8 | 3/16 | 1/4 | 5/16 | 5/32 | 1/8 | 1/16 | 1/8 |
| 5/16 | 11/64 | 15/64 | 5/16 | 3/8 | 13/64 | 5/32 | 5/64 | 5/32 |
| 3/8 | 13/64 | 9/32 | 3/8 | 7/16 | 1/4 | 3/16 | 3/32 | 3/16 |
| 7/16 | 15/64 | 21/64 | 7/16 | 1/2 | 19/64 | 7/32 | 7/64 | 7/32 |
| 1/2 | 9/32 | 3/8 | 1/2 | 9/16 | 11/32 | 1/4 | 1/8 | 1/4 |
| 9/16 | 5/16 | 27/64 | 9/16 | 5/8 | 25/64 | 9/32 | 9/64 | 1/4 |
| 5/8 | 23/64 | 15/32 | 5/8 | 3/4 | 15/32 | 5/16 | 5/32 | 5/16 |
| 3/4 | 7/16 | 9/16 | 3/4 | 7/8 | 9/16 | 3/8 | 3/16 | 3/8 |
| 7/8 | 33/64 | 21/32 | 7/8 | 1 | 21/32 | 7/16 | 7/32 | 1/2 |
| 1 | 19/32 | 3/4 | 1 | 1 1/8 | 3/4 | 1/2 | 1/4 | 9/16 |
| 1 1/8 | 43/64 | 27/32 | 1 1/8 | 1 1/4 | 27/32 | 9/16 | 9/32 | 9/16 |
| 1 1/4 | 3/4 | 15/16 | 1 1/4 | 1 1/2 | 15/16 | 5/8 | 5/16 | 5/8 |
| 1 3/8 | 53/64 | 1 1/32 | 1 3/8 | 1 5/8 | 1 1/32 | 11/16 | 11/32 | 5/8 |
| 1 1/2 | 29/32 | 1 1/8 | 1 1/2 | 1 3/4 | 1 1/8 | 3/4 | 3/8 | 3/4 |
| 1 3/4 | 1 1/16 | 1 5/16 | 1 3/4 | 2 | 1 5/16 | 7/8 | 7/16 | 1 |
| 2 | 1 7/32 | 1 1/2 | 2 | 2 1/4 | 1 1/2 | 1 | 1/2 | 1 |

* ANSI B18.3—Dimensions are in inches. Threads coarse or fine, class 3A. Length increments: 1/4 in. to 5/8 in. by (1/16 in.); 5/8 in. to 1 in. by (1/8 in.); 1 in. to 4 in. by (1/4 in.); 4 in. to 6 in. by (1/2 in.). Fractions in parentheses show length increments; for example, 5/8 in. to 1 in. by (1/8 in.) includes the lengths 5/8 in., 3/4 in., 7/8 in., and 1 in.
† ANSI B18.6.2—Threads coarse or fine, class 2A. Slotted headless screws standardized in sizes No. 5 to 3/4 in. only. Slot proportions vary with diameter. Draw to look well.
‡ ANSI B18.6.2—Threads coarse, fine, or 8-pitch, class 2A. Square-head setscrews standardized in sizes No. 10 to 1 1/2 in. only.

## ANSI Standard Socket-head Shoulder Screws[a]

| Shoulder diameter $D$ | | | Head[b] | | | Thread | | Shoulder lengths[d] |
|---|---|---|---|---|---|---|---|---|
| Nominal | Max. | Min. | Diam. $A$ | Height $H$ | Hexagon[c] $J$ | Specification $E$ | Length $I$ | |
| 1/4 | 0.2480 | 0.2460 | 3/8 | 3/16 | 1/8 | 10-24NC-3 | 3/8 | 3/4–2 1/2 |
| 5/16 | 0.3105 | 0.3085 | 7/16 | 7/32 | 5/32 | 1/4-20NC-3 | 7/16 | 1  –3 |
| 3/8 | 0.3730 | 0.3710 | 9/16 | 1/4 | 3/16 | 5/16-18NC-3 | 1/2 | 1  –4 |
| 1/2 | 0.4980 | 0.4960 | 3/4 | 5/16 | 1/4 | 3/8-16NC-3 | 5/8 | 1 1/4–5 |
| 5/8 | 0.6230 | 0.6210 | 7/8 | 3/8 | 5/16 | 1/2-13NC-3 | 3/4 | 1 1/2–6 |
| 3/4 | 0.7480 | 0.7460 | 1 | 1/2 | 3/8 | 5/8-11NC-3 | 7/8 | 1 1/2–8 |
| 1 | 0.9980 | 0.9960 | 1 5/16 | 5/8 | 1/2 | 3/4-10NC-3 | 1 | 1 1/2–8 |
| 1 1/4 | 1.2480 | 1.2460 | 1 3/4 | 3/4 | 5/8 | 7/8-9NC-3 | 1 1/8 | 1 1/2–8 |

[a] ANSI B18.3—Dimensions are in inches.
[b] Head chamfer is 30°.
[c] Socket depth = $\frac{3}{4}H$.
[d] Shoulder-length increments: shoulder lengths from 3/4 in. to 1 in., 1/8-in. intervals; shoulder lengths from 1 in. to 5 in., 1/4-in. intervals; shoulder lengths from 5 in. to 7 in., 1/2-in. intervals; shoulder lengths from 7 in. to 8 in., 1-in. intervals. Shoulder-length tolerance ±0.005.

## ANSI Standard Wood Screws *

| Nominal size | Basic diam. of screw $D$ | No. of threads per in.† | Slot width ‡ $J$ (all heads) | Round head | | Flat head | Oval head | |
|---|---|---|---|---|---|---|---|---|
| | | | | $A$ | $H$ | $A$ | $A$ | $C$ |
| 0 | 0.060 | 32 | 0.023 | 0.113 | 0.053 | 0.119 | 0.119 | 0.021 |
| 1 | 0.073 | 28 | 0.026 | 0.138 | 0.061 | 0.146 | 0.146 | 0.025 |
| 2 | 0.086 | 26 | 0.031 | 0.162 | 0.069 | 0.172 | 0.172 | 0.029 |
| 3 | 0.099 | 24 | 0.035 | 0.187 | 0.078 | 0.199 | 0.199 | 0.033 |
| 4 | 0.112 | 22 | 0.039 | 0.211 | 0.086 | 0.225 | 0.225 | 0.037 |
| 5 | 0.125 | 20 | 0.043 | 0.236 | 0.095 | 0.252 | 0.252 | 0.041 |
| 6 | 0.138 | 18 | 0.048 | 0.260 | 0.103 | 0.279 | 0.279 | 0.045 |
| 7 | 0.151 | 16 | 0.048 | 0.285 | 0.111 | 0.305 | 0.305 | 0.049 |
| 8 | 0.164 | 15 | 0.054 | 0.309 | 0.120 | 0.332 | 0.332 | 0.052 |
| 9 | 0.177 | 14 | 0.054 | 0.334 | 0.128 | 0.358 | 0.358 | 0.056 |
| 10 | 0.190 | 13 | 0.060 | 0.359 | 0.137 | 0.385 | 0.385 | 0.060 |
| 12 | 0.216 | 11 | 0.067 | 0.408 | 0.153 | 0.438 | 0.438 | 0.068 |
| 14 | 0.242 | 10 | 0.075 | 0.457 | 0.170 | 0.491 | 0.491 | 0.076 |
| 16 | 0.268 | 9 | 0.075 | 0.506 | 0.187 | 0.544 | 0.544 | 0.084 |
| 18 | 0.294 | 8 | 0.084 | 0.555 | 0.204 | 0.597 | 0.597 | 0.092 |
| 20 | 0.320 | 8 | 0.084 | 0.604 | 0.220 | 0.650 | 0.650 | 0.100 |
| 24 | 0.372 | 7 | 0.094 | 0.702 | 0.254 | 0.756 | 0.756 | 0.116 |

*Round Head*
*Flat Head*
*Oval Head*

* ANSI B18.6.1—Dimensions given are maximum values, all in inches. Heads may be slotted or recessed as specified.
† Thread length $l = \frac{2}{3}L$.
‡ Slot depths and recesses vary with type and size of screw; draw to look well.

## Small Rivets*

TINNERS'

COOPERS'

BELT

### Tinner's

| Size No.† | D Diam. body | L Length |
|---|---|---|
| 8 oz | 0.089 | 0.16 |
| 12 | 0.105 | 0.19 |
| 1 lb | 0.111 | 0.20 |
| 1½ | 0.130 | 0.23 |
| 2 | 0.144 | 0.27 |
| 2½ | 0.148 | 0.28 |
| 3 | 0.160 | 0.31 |
| 4 | 0.176 | 0.34 |
| 6 | 0.203 | 0.39 |
| 8 | 0.224 | 0.44 |
| 10 | 0.238 | 0.47 |
| 12 | 0.259 | 0.50 |
| 14 | 0.284 | 0.52 |
| 16 | 0.300 | 0.53 |

Approx. proportions:
$A = 2.25 \times D$
$H = 0.30 \times D$

### Cooper's

| Size No. | D Diam. body | L Length |
|---|---|---|
| 1 lb | 0.109 | 0.219 |
| 1½ | 0.127 | 0.256 |
| 2 | 0.141 | 0.292 |
| 2½ | 0.148 | 0.325 |
| 3 | 0.156 | 0.358 |
| 4 | 0.165 | 0.392 |
| 6 | 0.203 | 0.466 |
| 8 | 0.238 | 0.571 |
| 10 | 0.250 | 0.606 |
| 12 | 0.259 | 0.608 |
| 14 | 0.271 | 0.643 |
| 16 | 0.281 | 0.677 |

Approx. proportions:
$A = 2.25 \times D$, $d = 0.90 \times D$
$E = 0.40 \times D$, $H = 0.30 \times D$
Included $\angle \theta = 144°$
$\angle \alpha = 18°$

### Belt

| Size No.‡ | D Diam. body | L Length |
|---|---|---|
| 7 | 0.180 | From 3/8 to 3/4 by 1/8" increments |
| 8 | 0.165 | |
| 9 | 0.148 | |
| 10 | 0.134 | |
| 11 | 0.120 | |
| 12 | 0.109 | |
| 13 | 0.095 | |

Approx. proportions:
$A = 2.8 \times D$, $d = 0.9 \times D$
$E = 0.4 \times D$, $H = 0.3 \times D$
Tolerances on the nominal diameter:
$+0.002$
$-0.004$
Finished rivets shall be free from injurious defects.

* ANSI B18.1—1955. All dimensions given in inches.
† Size numbers refer to the Trade Name or weight of 1,000 rivets.
‡ Size number refers to the Stubs iron-wire gage number of the stock used in the body of the rivet.

## Dimensions of Standard Gib-head Keys, Square and Flat

L = 4 W (Min.)
L = 16 W (Max.)
Taper 1/8" in 12" (1:96)

| Diameters of shafts | Square type Key | | Gib head | | | Flat type Key | | Gib head | | |
|---|---|---|---|---|---|---|---|---|---|---|
| | W | H | C | D | E | W | H | C | D | E |
| ½ – 9/16 | 1/8 | 1/8 | 1/4 | 7/32 | 5/32 | 1/8 | 3/32 | 3/16 | 1/8 | 1/8 |
| 5/8 – 7/8 | 3/16 | 3/16 | 5/16 | 9/32 | 7/32 | 3/16 | 1/8 | 1/4 | 3/16 | 5/32 |
| 15/16–1¼ | 1/4 | 1/4 | 7/16 | 11/32 | 11/32 | 1/4 | 3/16 | 5/16 | 1/4 | 3/16 |
| 1 5/16–1 3/8 | 5/16 | 5/16 | 9/16 | 13/32 | 13/32 | 5/16 | 1/4 | 3/8 | 5/16 | 1/4 |
| 1 7/16–1¾ | 3/8 | 3/8 | 11/16 | 15/32 | 15/32 | 3/8 | 1/4 | 7/16 | 3/8 | 5/16 |
| 1 13/16–2¼ | 1/2 | 1/2 | 7/8 | 19/32 | 5/8 | 1/2 | 3/8 | 5/8 | 1/2 | 7/16 |
| 2 5/16–2¾ | 5/8 | 5/8 | 1 1/16 | 23/32 | 3/4 | 5/8 | 7/16 | 3/4 | 5/8 | 1/2 |
| 2 7/8 –3¼ | 3/4 | 3/4 | 1 1/4 | 7/8 | 7/8 | 3/4 | 1/2 | 7/8 | 3/4 | 5/8 |
| 3 3/8 –3¾ | 7/8 | 7/8 | 1 1/2 | 1 | 1 | 7/8 | 5/8 | 1 1/16 | 7/8 | 3/4 |
| 3 7/8 –4½ | 1 | 1 | 1 3/4 | 1 3/16 | 1 3/16 | 1 | 3/4 | 1 1/4 | 1 | 1 13/16 |
| 4 ¾ –5½ | 1¼ | 1¼ | 2 | 1 7/16 | 1 7/16 | 1¼ | 7/8 | 1 1/2 | 1¼ | 1 |
| 5 ¾ –6 | 1½ | 1½ | 2 ½ | 1 ¾ | 1 ¾ | 1½ | 1 | 1¾ | 1½ | 1 ¼ |

* ANSI B17.1—Dimensions in inches.

## Widths and Heights of Standard Square-and Flat-stock Keys with Corresponding Shaft Diameters*

| Shaft diam. $d$ (inclusive) | Square-stock keys $W$ | Flat-stock keys, $W \times H$ | Shaft diam. $d$ (inclusive) | Square-stock keys $W$ | Flat-stock keys, $W \times H$ |
|---|---|---|---|---|---|
| ½ – 9⁄16 | ⅛ | ⅛ × 3⁄32 | 2⅞–3¼ | ¾ | ¾ × ½ |
| ⅝ – ⅞ | 3⁄16 | 3⁄16 × ⅛ | 3⅜–3¾ | ⅞ | ⅞ × ⅝ |
| 15⁄16–1¼ | ¼ | ¼ × 3⁄16 | 3⅞–4½ | 1 | 1 × ¾ |
| 1 5⁄16–1⅜ | 5⁄16 | 5⁄16 × ¼ | 4¾–5½ | 1¼ | 1¼ × ⅞ |
| 1 7⁄16–1¾ | ⅜ | ⅜ × ¼ | 5¾–6 | 1½ | 1½ × 1 |
| 1 13⁄16–2¼ | ½ | ½ × ⅜ | | | |
| 2 5⁄16–2¾ | ⅝ | ⅝ × 7⁄16 | | | |

* Compiled from manufacturers' catalogues.

## Woodruff-key Dimensions

| Key* No. | Nominal size $A \times B$ | Max. width of key $A$ | Max. diam. of key $B$ | Max. height of key | | Distance below center $E$ |
|---|---|---|---|---|---|---|
| | | | | $C$ | $D$ | |
| 204 | 1⁄16 × ½ | 0.0635 | 0.500 | 0.203 | 0.194 | 3⁄64 |
| 304 | 3⁄32 × ½ | 0.0948 | 0.500 | 0.203 | 0.194 | 3⁄64 |
| 305 | 3⁄32 × ⅝ | 0.0948 | 0.625 | 0.250 | 0.240 | 1⁄16 |
| 404 | ⅛ × ½ | 0.1260 | 0.500 | 0.203 | 0.194 | 3⁄64 |
| 405 | ⅛ × ⅝ | 0.1260 | 0.625 | 0.250 | 0.240 | 1⁄16 |
| 406 | ⅛ × ¾ | 0.1260 | 0.750 | 0.313 | 0.303 | 1⁄16 |
| 505 | 5⁄32 × ⅝ | 0.1573 | 0.625 | 0.250 | 0.240 | 1⁄16 |
| 506 | 5⁄32 × ¾ | 0.1573 | 0.750 | 0.313 | 0.303 | 1⁄16 |
| 507 | 5⁄32 × ⅞ | 0.1573 | 0.875 | 0.375 | 0.365 | 1⁄16 |
| 606 | 3⁄16 × ¾ | 0.1885 | 0.750 | 0.313 | 0.303 | 1⁄16 |
| 607 | 3⁄16 × ⅞ | 0.1885 | 0.875 | 0.375 | 0.365 | 1⁄16 |
| 608 | 3⁄16 × 1 | 0.1885 | 1.000 | 0.438 | 0.428 | 1⁄16 |
| 609 | 3⁄16 × 1⅛ | 0.1885 | 1.125 | 0.484 | 0.475 | 5⁄64 |
| 807 | ¼ × ⅞ | 0.2510 | 0.875 | 0.375 | 0.365 | 1⁄16 |
| 808 | ¼ × 1 | 0.2510 | 1.000 | 0.438 | 0.428 | 1⁄16 |
| 809 | ¼ × 1⅛ | 0.2510 | 1.125 | 0.484 | 0.475 | 5⁄64 |
| 810 | ¼ × 1¼ | 0.2510 | 1.250 | 0.547 | 0.537 | 5⁄64 |
| 811 | ¼ × 1⅜ | 0.2510 | 1.375 | 0.594 | 0.584 | 3⁄32 |
| 812 | ¼ × 1½ | 0.2510 | 1.500 | 0.641 | 0.631 | 7⁄64 |
| 1008 | 5⁄16 × 1 | 0.3135 | 1.000 | 0.438 | 0.428 | 1⁄16 |
| 1009 | 5⁄16 × 1⅛ | 0.3135 | 1.125 | 0.484 | 0.475 | 5⁄64 |
| 1010 | 5⁄16 × 1¼ | 0.3135 | 1.250 | 0.547 | 0.537 | 5⁄64 |
| 1011 | 5⁄16 × 1⅜ | 0.3135 | 1.375 | 0.594 | 0.584 | 3⁄32 |
| 1012 | 5⁄16 × 1½ | 0.3135 | 1.500 | 0.641 | 0.631 | 7⁄64 |
| 1210 | ⅜ × 1¼ | 0.3760 | 1.250 | 0.547 | 0.537 | 5⁄64 |
| 1211 | ⅜ × 1⅜ | 0.3760 | 1.375 | 0.594 | 0.584 | 3⁄32 |
| 1212 | ⅜ × 1½ | 0.3760 | 1.500 | 0.641 | 0.631 | 7⁄64 |

* Dimensions in inches. Key numbers indicate the nominal key dimensions. The last two digits give the nominal diameter $B$ in eighths of an inch, and the digits preceding the last two give the nominal width $A$ in thirty-seconds of an inch. Thus 204 indicates a key 2⁄32 by ⅛, or 1⁄16 by ½ in.

## Woodruff-key-seat Dimensions

| Key* No. | Nominal size | Key slot | | | |
|---|---|---|---|---|---|
| | | Width $W$ | | Depth $H$ | |
| | | Max. | Min. | Max. | Min. |
| 204 | $\frac{1}{16} \times \frac{1}{2}$ | 0.0630 | 0.0615 | 0.1718 | 0.1668 |
| 304 | $\frac{3}{32} \times \frac{1}{2}$ | 0.0943 | 0.0928 | 0.1561 | 0.1511 |
| 305 | $\frac{3}{32} \times \frac{5}{8}$ | 0.0943 | 0 0928 | 0.2031 | 0.1981 |
| 404 | $\frac{1}{8} \times \frac{1}{2}$ | 0.1255 | 0.1240 | 0.1405 | 0.1355 |
| 405 | $\frac{1}{8} \times \frac{5}{8}$ | 0.1255 | 0.1240 | 0.1875 | 0.1825 |
| 406 | $\frac{1}{8} \times \frac{3}{4}$ | 0.1255 | 0.1240 | 0.2505 | 0.2455 |
| 505 | $\frac{5}{32} \times \frac{5}{8}$ | 0.1568 | 0.1553 | 0.1719 | 0.1669 |
| 506 | $\frac{5}{32} \times \frac{3}{4}$ | 0.1568 | 0.1553 | 0.2349 | 0.2299 |
| 507 | $\frac{5}{32} \times \frac{7}{8}$ | 0.1568 | 0.1553 | 0.2969 | 0.2919 |
| 606 | $\frac{3}{16} \times \frac{3}{4}$ | 0.1880 | 0.1863 | 0.2193 | 0.2143 |
| 607 | $\frac{3}{16} \times \frac{7}{8}$ | 0.1880 | 0.1863 | 0.2813 | 0.2763 |
| 608 | $\frac{3}{16} \times 1$ | 0.1880 | 0.1863 | 0.3443 | 0.3393 |
| 609 | $\frac{3}{16} \times 1\frac{1}{8}$ | 0.1880 | 0.1863 | 0.3903 | 0.3853 |
| 807 | $\frac{1}{4} \times \frac{7}{8}$ | 0.2505 | 0.2487 | 0.2500 | 0.2450 |
| 808 | $\frac{1}{4} \times 1$ | 0.2505 | 0.2487 | 0.3130 | 0.3080 |
| 809 | $\frac{1}{4} \times 1\frac{1}{8}$ | 0.2505 | 0.2487 | 0.3590 | 0.3540 |
| 810 | $\frac{1}{4} \times 1\frac{1}{4}$ | 0.2505 | 0.2487 | 0.4220 | 0.4170 |
| 811 | $\frac{1}{4} \times 1\frac{3}{8}$ | 0.2505 | 0.2487 | 0.4690 | 0.4640 |
| 812 | $\frac{1}{4} \times 1\frac{1}{2}$ | 0.2505 | 0.2487 | 0.5160 | 0.5110 |
| 1008 | $\frac{5}{16} \times 1$ | 0.3130 | 0.3111 | 0.2818 | 0.2768 |
| 1009 | $\frac{5}{16} \times 1\frac{1}{8}$ | 0.3130 | 0.3111 | 0.3278 | 0.3228 |
| 1010 | $\frac{5}{16} \times 1\frac{1}{4}$ | 0.3130 | 0.3111 | 0.3908 | 0.3858 |
| 1011 | $\frac{5}{16} \times 1\frac{3}{8}$ | 0.3130 | 0.3111 | 0.4378 | 0.4328 |
| 1012 | $\frac{5}{16} \times 1\frac{1}{2}$ | 0.3130 | 0.3111 | 0.4848 | 0.4798 |
| 1210 | $\frac{3}{8} \times 1\frac{1}{4}$ | 0.3755 | 0.3735 | 0.3595 | 0.3545 |
| 1211 | $\frac{3}{8} \times 1\frac{3}{8}$ | 0.3755 | 0.3735 | 0.4060 | 0.4015 |
| 1212 | $\frac{3}{8} \times 1\frac{1}{2}$ | 0.3755 | 0.3735 | 0.4535 | 0.4485 |

\* Dimensions in inches. Key numbers indicate the nominal key dimensions. The last two digits give the nominal diameter $B$ in eighths of an inch, and the digits preceding the last two give the nominal width $A$ in thirty-seconds of an inch. Thus 204 indicates a key $\frac{2}{32}$ by $\frac{4}{8}$, or $\frac{1}{16}$ by $\frac{1}{2}$ in.

## Dimensions of Pratt and Whitney Keys

| Key No. | $L$ | $W$ | $H$ | $D$ | Key No. | $L$ | $W$ | $H$ | $D$ |
|---|---|---|---|---|---|---|---|---|---|
| 1 | 1/2 | 1/16 | 3/32 | 1/16 | 22 | 1 3/8 | 1/4 | 3/8 | 1/4 |
| 2 | 1/2 | 3/32 | 9/64 | 3/32 | 23 | 1 3/8 | 5/16 | 15/32 | 5/16 |
| 3 | 1/2 | 1/8 | 3/16 | 1/8 | F | 1 3/8 | 3/8 | 9/16 | 3/8 |
| 4 | 5/8 | 3/32 | 9/64 | 3/32 | 24 | 1 1/2 | 1/4 | 3/8 | 1/4 |
| 5 | 5/8 | 1/8 | 3/16 | 1/8 | 25 | 1 1/2 | 5/16 | 15/32 | 5/16 |
| 6 | 5/8 | 5/32 | 15/64 | 5/32 | G | 1 1/2 | 3/8 | 9/16 | 3/8 |
| 7 | 3/4 | 1/8 | 3/16 | 1/8 | 51 | 1 3/4 | 1/4 | 3/8 | 1/4 |
| 8 | 3/4 | 5/32 | 15/64 | 5/32 | 52 | 1 3/4 | 5/16 | 15/32 | 5/16 |
| 9 | 3/4 | 3/16 | 9/32 | 3/16 | 53 | 1 3/4 | 3/8 | 9/16 | 3/8 |
| 10 | 7/8 | 5/32 | 15/64 | 5/32 | 26 | 2 | 3/16 | 9/32 | 3/16 |
| 11 | 7/8 | 3/16 | 9/32 | 3/16 | 27 | 2 | 1/4 | 3/8 | 1/4 |
| 12 | 7/8 | 7/32 | 21/64 | 7/32 | 28 | 2 | 5/16 | 15/32 | 5/16 |
| A | 7/8 | 1/4 | 3/8 | 1/4 | 29 | 2 | 3/8 | 9/16 | 3/8 |
| 13 | 1 | 3/16 | 9/32 | 3/16 | 54 | 2 1/4 | 1/4 | 3/8 | 1/4 |
| 14 | 1 | 7/32 | 21/64 | 7/32 | 55 | 2 1/4 | 5/16 | 15/16 | 5/16 |
| 15 | 1 | 1/4 | 3/8 | 1/4 | 56 | 2 1/4 | 3/8 | 9/16 | 3/8 |
| B | 1 | 5/16 | 15/32 | 5/16 | 57 | 2 1/4 | 7/16 | 21/32 | 7/16 |
| 16 | 1 1/8 | 3/16 | 9/32 | 3/16 | 58 | 2 1/2 | 5/16 | 15/32 | 5/16 |
| 17 | 1 1/8 | 7/32 | 21/64 | 7/32 | 59 | 2 1/2 | 3/8 | 9/16 | 3/8 |
| 18 | 1 1/8 | 1/4 | 3/8 | 1/4 | 60 | 2 1/2 | 7/16 | 21/32 | 7/16 |
| C | 1 1/8 | 5/16 | 15/32 | 5/16 | 61 | 2 1/2 | 1/2 | 3/4 | 1/2 |
| 19 | 1 1/4 | 3/16 | 9/32 | 3/16 | 30 | 3 | 3/8 | 9/16 | 3/8 |
| 20 | 1 1/4 | 7/32 | 21/64 | 7/32 | 31 | 3 | 7/16 | 21/32 | 7/16 |
| 21 | 1 1/4 | 1/4 | 3/8 | 1/4 | 32 | 3 | 1/2 | 3/4 | 1/2 |
| D | 1 1/4 | 5/16 | 15/32 | 5/16 | 33 | 3 | 9/16 | 27/32 | 9/16 |
| E | 1 1/4 | 3/8 | 9/16 | 3/8 | 34 | 3 | 5/8 | 15/16 | 5/8 |

Dimensions in inches. Key is 2/3 in shaft; 1/3 in hub. Keys are 0.001 in. oversize in width to ensure proper fitting in keyway. Keyway size: width = $W$; depth = $H - D$. Length $L$ should never be less than $2W$.

## ANSI Standard Plain Washers *

| Size | Light | | | Medium | | | Heavy | | | Extra heavy | | |
|---|---|---|---|---|---|---|---|---|---|---|---|---|
| | ID | OD | Thickness | ID | OD | Thickness | ID | OD | Thickness | ID | OD | Thickness |
| 0 | 5/64 | 3/16 | 0.020 | | | | | | | | | |
| 1 | 3/32 | 7/32 | 0.020 | | | | | | | | | |
| 2 | 3/32 | 1/4 | 0.020 | | | | | | | | | |
| 3 | 1/8 | 1/4 | 0.022 | | | | | | | | | |
| 4 | 1/8 | 1/4 | 0.022 | 1/8 | 5/16 | 0.032 | | | | | | |
| 5 | 5/32 | 5/16 | 0.035 | 5/32 | 3/8 | 0.049 | | | | | | |
| 6 | 5/32 | 5/16 | 0.035 | 5/32 | 3/8 | 0.049 | | | | | | |
| 7 | 11/64 | 13/32 | 0.049 | 3/16 | 3/8 | 0.049 | | | | | | |
| 8 | 3/16 | 3/8 | 0.049 | 3/16 | 7/16 | 0.049 | | | | | | |
| 9 | 13/64 | 15/32 | 0.049 | 7/32 | 1/2 | 0.049 | | | | | | |
| 3/16 | 7/32 | 7/16 | 0.049 | 7/32 | 1/2 | 0.049 | 1/4 | 9/16 | 0.049 | | | |
| 10 | 7/32 | 7/16 | 0.049 | 1/4 | 9/16 | 0.049 | 1/4 | 9/16 | 0.065 | | | |
| 11 | 15/64 | 17/32 | 0.049 | 1/4 | 9/16 | 0.049 | 1/4 | 9/16 | 0.065 | | | |
| 12 | 1/4 | 1/2 | 0.049 | 1/4 | 9/16 | 0.049 | 1/4 | 9/16 | 0.065 | | | |
| 14 | 17/64 | 5/8 | 0.049 | 5/16 | 3/4 | 0.065 | 5/16 | 7/8 | 0.065 | | | |
| 1/4 | 9/32 | 5/8 | 0.065 | 5/16 | 3/4 | 0.065 | 5/16 | 3/4 | 0.065 | 5/16 | 7/8 | 0.065 |
| 16 | 9/32 | 5/8 | 0.065 | 5/16 | 3/4 | 0.065 | 5/16 | 7/8 | 0.065 | 5/16 | 7/8 | 0.065 |
| 18 | 5/16 | 3/4 | 0.065 | 3/8 | 3/4 | 0.065 | 3/8 | 7/8 | 0.083 | 3/8 | 1 1/8 | 0.065 |
| 5/16 | 11/32 | 11/16 | 0.065 | 3/8 | 3/4 | 0.065 | 3/8 | 7/8 | 0.083 | 3/8 | 1 1/8 | 0.065 |
| 20 | 11/32 | 11/16 | 0.065 | 3/8 | 3/4 | 0.065 | 3/8 | 7/8 | 0.083 | 3/8 | 1 1/8 | 0.065 |
| 24 | 13/32 | 13/16 | 0.065 | 7/16 | 7/8 | 0.083 | 7/16 | 1 | 0.083 | 7/16 | 1 3/8 | 0.083 |
| 3/8 | 13/32 | 13/16 | 0.065 | 7/16 | 7/8 | 0.083 | 7/16 | 1 | 0.083 | 7/16 | 1 3/8 | 0.083 |
| 7/16 | 15/32 | 59/64 | 0.065 | 1/2 | 1 1/8 | 0.083 | 1/2 | 1 1/4 | 0.083 | 1/2 | 1 5/8 | 0.083 |
| 1/2 | 17/32 | 1 1/16 | 0.095 | 9/16 | 1 1/4 | 0.109 | 9/16 | 1 3/8 | 0.109 | 9/16 | 1 7/8 | 0.109 |
| 9/16 | 19/32 | 1 3/16 | 0.095 | 5/8 | 1 3/8 | 0.109 | 5/8 | 1 1/2 | 0.109 | 5/8 | 2 1/8 | 0.134 |
| 5/8 | 21/32 | 1 5/16 | 0.095 | 11/16 | 1 1/2 | 0.134 | 11/16 | 1 3/4 | 0.134 | 11/16 | 2 3/8 | 0.165 |
| 3/4 | 13/16 | 1 1/2 | 0.134 | 13/16 | 1 3/4 | 0.148 | 13/16 | 2 | 0.148 | 13/16 | 2 7/8 | 0.165 |
| 7/8 | 15/16 | 1 3/4 | 0.134 | 15/16 | 2 | 0.165 | 15/16 | 2 1/4 | 0.165 | 15/16 | 3 3/8 | 0.180 |
| 1 | 1 1/16 | 2 | 0.134 | 1 1/16 | 2 1/4 | 0.165 | 1 1/16 | 2 1/2 | 0.165 | 1 1/16 | 3 3/8 | 0.238 |
| 1 1/8 | ..... | ..... | ..... | 1 3/16 | 2 1/2 | 0.165 | 1 1/4 | 2 3/4 | 0.165 | | | |
| 1 1/4 | ..... | ..... | ..... | 1 5/16 | 2 3/4 | 0.165 | 1 3/8 | 3 | 0.165 | | | |
| 1 3/8 | ..... | ..... | ..... | 1 7/16 | 3 | 0.180 | 1 1/2 | 3 1/4 | 0.180 | | | |
| 1 1/2 | ..... | ..... | ..... | 1 9/16 | 3 1/4 | 0.180 | 1 5/8 | 3 1/2 | 0.180 | | | |
| 1 5/8 | ..... | ..... | ..... | 1 11/16 | 3 1/2 | 0.180 | 1 3/4 | 3 3/4 | 0.180 | | | |
| 1 3/4 | ..... | ..... | ..... | 1 13/16 | 3 3/4 | 0.180 | 1 7/8 | 4 | 0.180 | | | |
| 1 7/8 | ..... | ..... | ..... | 1 15/16 | 4 | 0.180 | 2 | 4 1/4 | 0.180 | | | |
| 2 | ..... | ..... | ..... | 2 1/16 | 4 1/4 | 0.180 | 2 1/8 | 4 1/2 | 0.180 | | | |
| 2 1/4 | ..... | ..... | ..... | ..... | ..... | ..... | 2 3/8 | 4 3/4 | 0.220 | | | |
| 2 1/2 | ..... | ..... | ..... | ..... | ..... | ..... | 2 5/8 | 5 | 0.238 | | | |
| 2 3/4 | ..... | ..... | ..... | ..... | ..... | ..... | 2 7/8 | 5 1/4 | 0.259 | | | |
| 3 | ..... | ..... | ..... | ..... | ..... | ..... | 3 1/8 | 5 1/2 | 0.284 | | | |

* ANSI B27.2—All dimensions in inches.

## ANSI Standard Lock Washers *

| Nominal size | Inside diam., min. | Light | | Medium | | Heavy | | Extra heavy | |
|---|---|---|---|---|---|---|---|---|---|
| | | Min. thickness | Outside diam., max. | Min. thickness | Outside diam., max. | Min. thickness | Outside diam., max. | Min. thickness | Outside diam., max. |
| 0.086 (No.  2) | 0.088 | 0.015 | 0.165 | 0.020 | 0.175 | 0.025 | 0.185 | 0.027 | 0.211 |
| 0.099 (No.  3) | 0.102 | 0.020 | 0.188 | 0.025 | 0.198 | 0.031 | 0.212 | 0.034 | 0.242 |
| 0.112 (No.  4) | 0.115 | 0.020 | 0.202 | 0.025 | 0.212 | 0.031 | 0.226 | 0.034 | 0.256 |
| 0.125 (No.  5) | 0.128 | 0.025 | 0.225 | 0.031 | 0.239 | 0.040 | 0.255 | 0.045 | 0.303 |
| 0.138 (No.  6) | 0.141 | 0.025 | 0.239 | 0.031 | 0.253 | 0.040 | 0.269 | 0.045 | 0.317 |
| 0.164 (No.  8) | 0.168 | 0.031 | 0.280 | 0.040 | 0.296 | 0.047 | 0.310 | 0.057 | 0.378 |
| 0.190 (No. 10) | 0.194 | 0.040 | 0.323 | 0.047 | 0.337 | 0.056 | 0.353 | 0.068 | 0.437 |
| 0.216 (No. 12) | 0.221 | 0.047 | 0.364 | 0.056 | 0.380 | 0.063 | 0.394 | 0.080 | 0.500 |
| $\frac{1}{4}$ | 0.255 | 0.047 | 0.489 | 0.062 | 0.493 | 0.077 | 0.495 | 0.084 | 0.539 |
| $\frac{5}{16}$ | 0.319 | 0.056 | 0.575 | 0.078 | 0.591 | 0.097 | 0.601 | 0.108 | 0.627 |
| $\frac{3}{8}$ | 0.382 | 0.070 | 0.678 | 0.094 | 0.688 | 0.115 | 0.696 | 0.123 | 0.746 |
| $\frac{7}{16}$ | 0.446 | 0.085 | 0.780 | 0.109 | 0.784 | 0.133 | 0.792 | 0.143 | 0.844 |
| $\frac{1}{2}$ | 0.509 | 0.099 | 0.877 | 0.125 | 0.879 | 0.151 | 0.889 | 0.162 | 0.945 |
| $\frac{9}{16}$ | 0.573 | 0.113 | 0.975 | 0.141 | 0.979 | 0.170 | 0.989 | 0.182 | 1.049 |
| $\frac{5}{8}$ | 0.636 | 0.126 | 1.082 | 0.156 | 1.086 | 0.189 | 1.100 | 0.202 | 1.164 |
| $\frac{11}{16}$ | 0.700 | 0.138 | 1.178 | 0.172 | 1.184 | 0.207 | 1.200 | 0.221 | 1.266 |
| $\frac{3}{4}$ | 0.763 | 0.153 | 1.277 | 0.188 | 1.279 | 0.226 | 1.299 | 0.241 | 1.369 |
| $\frac{13}{16}$ | 0.827 | 0.168 | 1.375 | 0.203 | 1.377 | 0.246 | 1.401 | 0.261 | 1.473 |
| $\frac{7}{8}$ | 0.890 | 0.179 | 1.470 | 0.219 | 1.474 | 0.266 | 1.504 | 0.285 | 1.586 |
| $\frac{15}{16}$ | 0.954 | 0.191 | 1.562 | 0.234 | 1.570 | 0.284 | 1.604 | 0.308 | 1.698 |
| 1 | 1.017 | 0.202 | 1.656 | 0.250 | 1.672 | 0.306 | 1.716 | 0.330 | 1.810 |
| 1 $\frac{1}{16}$ | 1.081 | 0.213 | 1.746 | 0.266 | 1.768 | 0.326 | 1.820 | 0.352 | 1.922 |
| 1 $\frac{1}{8}$ | 1.144 | 0.224 | 1.837 | 0.281 | 1.865 | 0.345 | 1.921 | 0.375 | 2.031 |
| 1 $\frac{3}{16}$ | 1.208 | 0.234 | 1.923 | 0.297 | 1.963 | 0.364 | 2.021 | 0.396 | 2.137 |
| 1 $\frac{1}{4}$ | 1.271 | 0.244 | 2.012 | 0.312 | 2.058 | 0.384 | 2.126 | 0.417 | 2.244 |
| 1 $\frac{5}{16}$ | 1.335 | 0.254 | 2.098 | 0.328 | 2.156 | 0.403 | 2.226 | 0.438 | 2.350 |
| 1 $\frac{3}{8}$ | 1.398 | 0.264 | 2.183 | 0.344 | 2.253 | 0.422 | 2.325 | 0.458 | 2.453 |
| 1 $\frac{7}{16}$ | 1.462 | 0.273 | 2.269 | 0.359 | 2.349 | 0.440 | 2.421 | 0.478 | 2.555 |
| 1 $\frac{1}{2}$ | 1.525 | 0.282 | 2.352 | 0.375 | 2.446 | 0.458 | 2.518 | 0.496 | 2.654 |

* ANSI B27.1—1958. All dimensions in inches.

**Tapers.** Taper means the difference in diameter or width in 1 ft of length; see figure below. *Taper pins*, much used for fastening cylindrical parts and for doweling, have a standard taper of ¼ in. per ft.

**Machine tapers.** The ANSI Standard for self-holding (slow) machine tapers is designed to replace the various former standards. The table below shows its derivation. Detailed dimensions and tolerances for taper tool shanks and taper sockets will be found in ANSI B5.10—1963.

## Dimensions of Taper Pins

*Taper ¼ in. per ft*

| Size No. | Diam., large end | Drill size for reamer | Max. length |
|---|---|---|---|
| 000000 | 0.072 | 53 | ⅝ |
| 00000 | 0.092 | 47 | ⅝ |
| 0000 | 0.108 | 42 | ¾ |
| 000 | 0.125 | 37 | ¾ |
| 00 | 0.147 | 31 | 1 |
| 0 | 0.156 | 28 | 1 |
| 1 | 0.172 | 25 | 1¼ |
| 2 | 0.193 | 19 | 1½ |
| 3 | 0.219 | 12 | 1¾ |
| 4 | 0.250 | 3 | 2 |
| 5 | 0.289 | ¼ | 2¼ |
| 6 | 0.341 | ⁹⁄₃₂ | 3¼ |
| 7 | 0.409 | 1¹⁄₃₂ | 3¾ |
| 8 | 0.492 | 1³⁄₃₂ | 4½ |
| 9 | 0.591 | ³¹⁄₆₄ | 5¼ |
| 10 | 0.706 | 1⁹⁄₃₂ | 6 |
| 11 | 0.857 | 2³⁄₃₂ | 7¼ |
| 12 | 1.013 | 5⁵⁄₆₄ | 8¾ |
| 13 | 1.233 | 1 ¹⁄₆₄ | 10¾ |

All dimensions in inches.

## ANSI Standard Machine Tapers,* Self-holding (Slow) Taper Series

*Basic dimensions*

| Origin of series | No. of taper | Taper /ft | Diam. at gage line | Means of driving and holding | | | |
|---|---|---|---|---|---|---|---|
| Brown and Sharpe taper series | | 0.239 | 0.500 | 0.239 | Tongue drive with shank held in by friction | | | |
| | | 0.299 | 0.500 | 0.299 | | | | |
| | | 0.375 | 0.500 | 0.375 | | | | |
| Morse taper series | 1 | 0.600 | 0.475 | | Tongue drive with shank held in by key | | |
| | 2 | 0.600 | 0.700 | | | | |
| | 3 | 0.602 | 0.938 | | | | |
| | 4 | 0.623 | 1.231 | | | | |
| | 4½ | 0.623 | 1.500 | | | | |
| | 5 | 0.630 | 1.748 | | | | |
| ¾-in./ft taper series | 200 | 0.750 | 2.000 | | | Key drive with shank held in by key | Key drive with shank held in by drawbolt |
| | 250 | 0.750 | 2.500 | | | | |
| | 300 | 0.750 | 3.000 | | | | |
| | 350 | 0.750 | 3.500 | | | | |
| | 400 | 0.750 | 4.000 | | | | |
| | 500 | 0.750 | 5.000 | | | | |
| | 600 | 0.750 | 6.000 | | | | |
| | 800 | 0.750 | 8.000 | | | | |
| | 1,000 | 0.750 | 10.000 | | | | |
| | 1,200 | 0.750 | 12.000 | | | | |

* ANSI B5.10—All dimensions in inches.

## Wire and Sheet-metal Gages

*Dimensions in decimal parts of an inch*

| No. of gage | American or Brown and Sharpe[a] | Washburn & Moen or American Steel & Wire Co.[b] | Birmingham or Stubs iron wire[c] | Music wire[d] | Imperial wire gage[e] | U.S. Std. for plate[f] |
|---|---|---|---|---|---|---|
| 0000000 | . . . . | 0.4900 | . . . . | . . . . | 0.5000 | 0.5000 |
| 000000 | 0.5800 | 0.4615 | . . . . | 0.004 | 0.4640 | 0.4688 |
| 00000 | 0.5165 | 0.4305 | 0.500 | 0.005 | 0.4320 | 0.4375 |
| 0000 | 0.4600 | 0.3938 | 0.454 | 0.006 | 0.4000 | 0.4063 |
| 000 | 0.4096 | 0.3625 | 0.425 | 0.007 | 0.3720 | 0.3750 |
| 00 | 0.3648 | 0.3310 | 0.380 | 0.008 | 0.3480 | 0.3438 |
| 0 | 0.3249 | 0.3065 | 0.340 | 0.009 | 0.3240 | 0.3125 |
| 1 | 0.2893 | 0.2830 | 0.300 | 0.010 | 0.3000 | 0.2813 |
| 2 | 0.2576 | 0.2625 | 0.284 | 0.011 | 0.2760 | 0.2656 |
| 3 | 0.2294 | 0.2437 | 0.259 | 0.012 | 0.2520 | 0.2500 |
| 4 | 0.2043 | 0.2253 | 0.238 | 0.013 | 0.2320 | 0.2344 |
| 5 | 0.1819 | 0.2070 | 0.220 | 0.014 | 0.2120 | 0.2188 |
| 6 | 0.1620 | 0.1920 | 0.203 | 0.016 | 0.1920 | 0.2031 |
| 7 | 0.1443 | 0.1770 | 0.180 | 0.018 | 0.1760 | 0.1875 |
| 8 | 0.1285 | 0.1620 | 0.165 | 0.020 | 0.1600 | 0.1719 |
| 9 | 0.1144 | 0.1483 | 0.148 | 0.022 | 0.1440 | 0.1563 |
| 10 | 0.1019 | 0.1350 | 0.134 | 0.024 | 0.1280 | 0.1406 |
| 11 | 0.0907 | 0.1205 | 0.120 | 0.026 | 0.1160 | 0.1250 |
| 12 | 0.0808 | 0.1055 | 0.109 | 0.029 | 0.1040 | 0.1094 |
| 13 | 0.0720 | 0.0915 | 0.095 | 0.031 | 0.0920 | 0.0938 |
| 14 | 0.0641 | 0.0800 | 0.083 | 0.033 | 0.0800 | 0.0781 |
| 15 | 0.0571 | 0.0720 | 0.072 | 0.035 | 0.0720 | 0.0703 |
| 16 | 0.0508 | 0.0625 | 0.065 | 0.037 | 0.0640 | 0.0625 |
| 17 | 0.0453 | 0.0540 | 0.058 | 0.039 | 0.0560 | 0.0563 |
| 18 | 0.0403 | 0.0475 | 0.049 | 0.041 | 0.0480 | 0.0500 |
| 19 | 0.0359 | 0.0410 | 0.042 | 0.043 | 0.0400 | 0.0438 |
| 20 | 0.0320 | 0.0348 | 0.035 | 0.045 | 0.0360 | 0.0375 |
| 21 | 0.0285 | 0.0317 | 0.032 | 0.047 | 0.0320 | 0.0344 |
| 22 | 0.0253 | 0.0286 | 0.028 | 0.049 | 0.0280 | 0.0313 |
| 23 | 0.0226 | 0.0258 | 0.025 | 0.051 | 0.0240 | 0.0281 |
| 24 | 0.0201 | 0.0230 | 0.022 | 0.055 | 0.0220 | 0.0250 |
| 25 | 0.0179 | 0.0204 | 0.020 | 0.059 | 0.0200 | 0.0219 |
| 26 | 0.0159 | 0.0181 | 0.018 | 0.063 | 0.0180 | 0.0188 |
| 27 | 0.0142 | 0.0173 | 0.016 | 0.067 | 0.0164 | 0.0172 |
| 28 | 0.0126 | 0.0162 | 0.014 | 0.071 | 0.0148 | 0.0156 |
| 29 | 0.0113 | 0.0150 | 0.013 | 0.075 | 0.0136 | 0.0141 |
| 30 | 0.0100 | 0.0140 | 0.012 | 0.080 | 0.0124 | 0.0125 |
| 31 | 0.0089 | 0.0132 | 0.010 | 0.085 | 0.0116 | 0.0109 |
| 32 | 0.0080 | 0.0128 | 0.009 | 0.090 | 0.0108 | 0.0102 |
| 33 | 0.0071 | 0.0118 | 0.008 | 0.095 | 0.0100 | 0.0094 |
| 34 | 0.0063 | 0.0104 | 0.007 | 0.100 | 0.0092 | 0.0086 |
| 35 | 0.0056 | 0.0095 | 0.005 | 0.106 | 0.0084 | 0.0078 |
| 36 | 0.0050 | 0.0090 | 0.004 | 0.112 | 0.0076 | 0.0070 |
| 37 | 0.0045 | 0.0085 | . . . . | 0.118 | 0.0068 | 0.0066 |
| 38 | 0.0040 | 0.0080 | . . . . | 0.124 | 0.0060 | 0.0063 |
| 39 | 0.0035 | 0.0075 | . . . . | 0.130 | 0.0052 | |
| 40 | 0.0031 | 0.0070 | . . . . | 0.138 | 0.0048 | |

[a] Recognized standard in the United States for wire and sheet metal of copper and other metals except steel and iron.

[b] Recognized standard for steel and iron wire. Called the "U.S. steel wire gage."

[c] Formerly much used, now nearly obsolete.

[d] American Steel & Wire Co.'s music wire gage. Recommended by U.S. Bureau of Standards.

[e] Official British Standard.

[f] Legalized U.S. Standard for iron and steel plate, although plate is now always specified by its thickness in decimals of an inch.

Preferred thicknesses for uncoated thin flat metals (under 0.250 in.): ANSI B32.1—1959 gives recommended sizes for sheets.

## Preferred Basic Sizes for Cylindrical Fits*

| | | | | | | | | |
|---|---|---|---|---|---|---|---|---|
| ... | 0.0100 | ... | ... | ... | 0.80 | 5½ | 5.5000 | 5.5 |
| ... | 0.0125 | ... | ⅞ | 0.8750 | ... | 5¾ | 5.7500 | 5.75 |
| 1/64 | 0.015625 | ... | ... | ... | 0.90 | 6 | 6.0000 | 6.0 |
| ... | 0.0200 | ... | 1 | 1.0000 | 1.0 | 6½ | 6.5000 | 6.5 |
| ... | 0.0250 | ... | ... | ... | 1.1 | 7 | 7.0000 | 7.0 |
| 1/32 | 0.03125 | ... | 1⅛ | 1.1250 | ... | 7½ | 7.5000 | 7.5 |
| ... | 0.0400 | 0.04 | ... | ... | ... | 8 | 8.0000 | 8.0 |
| ... | 0.0500 | ... | 1¼ | 1.2500 | 1.25 | 8½ | 8.5000 | 8.5 |
| ... | ... | 0.06 | 1⅜ | 1.3750 | ... | 9 | 9.0000 | 9.0 |
| 1/16 | 0.0625 | ... | ... | ... | 1.40 | 9½ | 9.5000 | 9.5 |
| ... | 0.0800 | ... | 1½ | 1.5000 | 1.50 | 10 | 10.0000 | 10.0 |
| 3/32 | 0.09375 | ... | 1⅝ | 1.6250 | ... | 10½ | 10.5000 | 10.5 |
| ... | 0.1000 | 0.10 | 1¾ | 1.7500 | 1.75 | 11 | 11.0000 | 11.0 |
| 1/8 | 0.1250 | ... | 1⅞ | 1.8750 | ... | 11½ | 11.5000 | 11.5 |
| ... | ... | 0.15 | 2 | 2.0000 | 2.0 | 12 | 12.0000 | 12.0 |
| 5/32 | 0.15625 | ... | 2⅛ | 2.1250 | ... | 12½ | 12.5000 | 12.5 |
| 3/16 | 0.1875 | ... | 2¼ | 2.2500 | 2.25 | 13 | 13.0000 | 13.0 |
| ... | ... | 0.20 | 2⅜ | 2.3750 | ... | 13½ | 13.5000 | 13.5 |
| 1/4 | 0.2500 | 0.25 | 2½ | 2.5000 | 2.5 | 14 | 14.0000 | 14.0 |
| ... | ... | 0.30 | 2⅝ | 2.6250 | ... | 14½ | 14.5000 | 14.5 |
| 5/16 | 0.3125 | ... | 2¾ | 2.7500 | 2.75 | 15 | 15.0000 | 15.0 |
| ... | ... | 0.35 | 2⅞ | 2.8750 | ... | 15½ | 15.5000 | 15.5 |
| 3/8 | 0.3750 | ... | 3 | 3.0000 | 3.0 | 16 | 16.0000 | 16.0 |
| ... | ... | 0.40 | 3¼ | 3.2500 | 3.25 | 16½ | 16.5000 | 16.5 |
| 7/16 | 0.4375 | ... | 3½ | 3.5000 | 3.5 | 17 | 17.0000 | 17.0 |
| 1/2 | 0.5000 | 0.50 | 3¾ | 3.7500 | 3.75 | 17½ | 17.5000 | 17.5 |
| 9/16 | 0.5625 | ... | 4 | 4.0000 | 4.0 | 18 | 18.0000 | 18.0 |
| ... | ... | 0.60 | 4¼ | 4.2500 | 4.25 | 18½ | 18.5000 | 18.5 |
| 5/8 | 0.6250 | ... | 4½ | 4.5000 | 4.5 | 19 | 19.0000 | 19.0 |
| 11/16 | 0.6875 | ... | 4¾ | 4.7500 | 4.75 | 19½ | 19.5000 | 19.5 |
| ... | ... | 0.70 | 5 | 5.0000 | 5.0 | 20 | 20.0000 | 20.0 |
| 3/4 | 0.7500 | 0.75 | 5¼ | 5.2500 | 5.25 | 20½ | 20.5000 | ... |
| | | | | | | 21 | 21.0000 | ... |

* Adapted from ANSI B4.1—1955

# TABLES OF LIMITS FOR CYLINDRICAL FITS*
## Table 1.  Running and Sliding Fits

| Nominal size, in. | Class RC 1 Clear- ance | Std. lim. Hole | Std. lim. Shaft | Class RC 2 Clear- ance | Std. lim. Hole | Std. lim. Shaft | Class RC 3 Clear- ance | Std. lim. Hole | Std. lim. Shaft | Class RC 4 Clear- ance | Std. lim. Hole | Std. lim. Shaft |
|---|---|---|---|---|---|---|---|---|---|---|---|---|
| 0.04–0.12 | 0.1 | +0.2 | −0.1 | 0.1 | +0.25 | −0.1 | 0.3 | +0.25 | −0.3 | 0.3 | +0.4 | −0.3 |
|  | 0.45 | 0 | −0.25 | 0.55 | 0 | −0.3 | 0.8 | 0 | −0.55 | 1.1 | 0 | −0.7 |
| 0.12–0.24 | 0.15 | +0.2 | −0.15 | 0.15 | +0.3 | −0.15 | 0.4 | +0.3 | −0.4 | 0.4 | +0.5 | −0.4 |
|  | 0.5 | 0 | −0.3 | 0.65 | 0 | −0.35 | 1.0 | 0 | −0.7 | 1.4 | 0 | −0.9 |
| 0.24–0.40 | 0.2 | +0.25 | −0.2 | 0.2 | +0.4 | −0.2 | 0.5 | +0.4 | −0.5 | 0.5 | +0.6 | −0.5 |
|  | 0.6 | 0 | −0.35 | 0.85 | 0 | −0.45 | 1.3 | 0 | −0.9 | 1.7 | 0 | −1.1 |
| 0.40–0.71 | 0.25 | +0.3 | −0.25 | 0.25 | +0.4 | −0.25 | 0.6 | +0.4 | −0.6 | 0.6 | +0.7 | −0.6 |
|  | 0.75 | 0 | −0.45 | 0.95 | 0 | −0.55 | 1.4 | 0 | −1.0 | 2.0 | 0 | −1.3 |
| 0.71–1.19 | 0.3 | +0.4 | −0.3 | 0.3 | +0.5 | −0.3 | 0.8 | +0.5 | −0.8 | 0.8 | +0.8 | −0.8 |
|  | 0.95 | 0 | −0.55 | 1.2 | 0 | −0.7 | 1.8 | 0 | −1.3 | 2.4 | 0 | −1.6 |
| 1.19–1.97 | 0.4 | +0.4 | −0.4 | 0.4 | +0.6 | −0.4 | 1.0 | +0.6 | −1.0 | 1.0 | +1.0 | −1.0 |
|  | 1.1 | 0 | −0.7 | 1.4 | 0 | −0.8 | 2.2 | 0 | −1.6 | 3.0 | 0 | −2.0 |
| 1.97–3.15 | 0.4 | +0.5 | −0.4 | 0.4 | +0.7 | −0.4 | 1.2 | +0.7 | −1.2 | 1.2 | +1.2 | −1.2 |
|  | 1.2 | 0 | −0.7 | 1.6 | 0 | −0.9 | 2.6 | 0 | −1.9 | 3.6 | 0 | −2.4 |
| 3.15–4.73 | 0.5 | +0.6 | −0.5 | 0.5 | +0.9 | −0.5 | 1.4 | +0.9 | −1.4 | 1.4 | +1.4 | −1.4 |
|  | 1.5 | 0 | −0.9 | 2.0 | 0 | −1.1 | 3.2 | 0 | −2.3 | 4.2 | 0 | −2.8 |
| 4.73–7.09 | 0.6 | +0.7 | −0.6 | 0.6 | +1.0 | −0.6 | 1.6 | +1.0 | −1.6 | 1.6 | +1.6 | −1.6 |
|  | 1.8 | 0 | −1.1 | 2.3 | 0 | −1.3 | 3.6 | 0 | −2.6 | 4.8 | 0 | −3.2 |

| Nominal size, in. | Class RC 5 | Hole | Shaft | Class RC 6 | Hole | Shaft | Class RC 7 | Hole | Shaft | Class RC 8 | Hole | Shaft | Class RC 9 | Hole | Shaft |
|---|---|---|---|---|---|---|---|---|---|---|---|---|---|---|---|
| 0.04–0.12 | 0.6 | +0.4 | −0.6 | 0.6 | +0.6 | −0.6 | 1.0 | +1.0 | −1.0 | 2.5 | +1.6 | −2.5 | 4.0 | +2.5 | −4.0 |
|  | 1.4 | 0 | −1.0 | 1.8 | 0 | −1.2 | 2.6 | 0 | −1.6 | 5.1 | 0 | −3.5 | 8.1 | 0 | −5.6 |
| 0.12–0.24 | 0.8 | +0.5 | −0.8 | 0.8 | +0.7 | −0.8 | 1.2 | +1.2 | −1.2 | 2.8 | +1.8 | −2.8 | 4.5 | +3.0 | −4.5 |
|  | 1.8 | 0 | −1.3 | 2.2 | 0 | −1.5 | 3.1 | 0 | −1.9 | 5.8 | 0 | −4.0 | 9.0 | 0 | −6.0 |
| 0.24–0.40 | 1.0 | +0.6 | −1.0 | 1.0 | +0.9 | −1.0 | 1.6 | +1.4 | −1.6 | 3.0 | +2.2 | −3.0 | 5.0 | +3.5 | −5.0 |
|  | 2.2 | 0 | −1.6 | 2.8 | 0 | −1.9 | 3.9 | 0 | −2.5 | 6.6 | 0 | −4.4 | 10.7 | 0 | −7.2 |
| 0.40–0.71 | 1.2 | +0.7 | −1.2 | 1.2 | +1.0 | −1.2 | 2.0 | +1.6 | −2.0 | 3.5 | +2.8 | −3.5 | 6.0 | +4.0 | −6.0 |
|  | 2.6 | 0 | −1.9 | 3.2 | 0 | −2.2 | 4.6 | 0 | −3.0 | 7.9 | 0 | −5.1 | 12.8 | 0 | −8.8 |
| 0.71–1.19 | 1.6 | +0.8 | −1.6 | 1.6 | +1.2 | −1.6 | 2.5 | +2.0 | −2.5 | 4.5 | +3.5 | −4.5 | 7.0 | +5.0 | −7.0 |
|  | 3.2 | 0 | −2.4 | 4.0 | 0 | −2.8 | 5.7 | 0 | −3.7 | 10.0 | 0 | −6.5 | 15.5 | 0 | −10.5 |
| 1.19–1.97 | 2.0 | +1.0 | −2.0 | 2.0 | +1.6 | −2.0 | 3.0 | +2.5 | −3.0 | 5.0 | +4.0 | −5.0 | 8.0 | +6.0 | −8.0 |
|  | 4.0 | 0 | −3.0 | 5.2 | 0 | −3.6 | 7.1 | 0 | −4.6 | 11.5 | 0 | −7.5 | 18.0 | 0 | −12.0 |
| 1.97–3.15 | 2.5 | +1.2 | −2.5 | 2.5 | +1.8 | −2.5 | 4.0 | +3.0 | −4.0 | 6.0 | +4.5 | −6.0 | 9.0 | +7.0 | −9.0 |
|  | 4.9 | 0 | −3.7 | 6.1 | 0 | −4.3 | 8.8 | 0 | −5.8 | 13.5 | 0 | −9.0 | 20.5 | 0 | −13.5 |
| 3.15–4.73 | 3.0 | +1.4 | −3.0 | 3.0 | +2.2 | −3.0 | 5.0 | +3.5 | −5.0 | 7.0 | +5.0 | −7.0 | 10.0 | +9.0 | −10.0 |
|  | 5.8 | 0 | −4.4 | 7.4 | 0 | −5.2 | 10.7 | 0 | −7.2 | 15.5 | 0 | −10.5 | 24.0 | 0 | −15.0 |
| 4.73–7.09 | 3.5 | +1.6 | −3.5 | 3.5 | +2.5 | −3.5 | 6.0 | +4.0 | −6.0 | 8.0 | +6.0 | −8.0 | 12.0 | +10.0 | −12.0 |
|  | 6.7 | 0 | −5.1 | 8.5 | 0 | −6.0 | 12.5 | 0 | −8.5 | 18.0 | 0 | −12.0 | 28.0 | 0 | −18.0 |

## Table 2.  Clearance Locational Fits

| Nominal size, in. | Class LC 1 | Hole | Shaft | Class LC 2 | Hole | Shaft | Class LC 3 | Hole | Shaft | Class LC 4 | Hole | Shaft | Class LC 5 | Hole | Shaft |
|---|---|---|---|---|---|---|---|---|---|---|---|---|---|---|---|
| 0.04–0.12 | 0 | +0.25 | +0 | 0 | +0.4 | +0 | 0 | +0.6 | +0 | 0 | +1.0 | +0 | 0.1 | +0.4 | −0.1 |
|  | 0.45 | −0 | −0.2 | 0.65 | −0 | −0.25 | 1 | −0 | −0.4 | 2.0 | −0 | −1.0 | 0.75 | −0 | −0.35 |
| 0.12–0.24 | 0 | +0.3 | +0 | 0 | +0.5 | +0 | 0 | +0.7 | +0 | 0 | +1.2 | +0 | 0.15 | +0.5 | −0.15 |
|  | 0.5 | −0 | −0.2 | 0.8 | −0 | −0.3 | 1.2 | −0 | −0.5 | 2.4 | −0 | −1.2 | 0.95 | −0 | −0.45 |
| 0.24–0.40 | 0 | +0.4 | +0 | 0 | +0.6 | +0 | 0 | +0.9 | +0 | 0 | +1.4 | +0 | 0.2 | +0.6 | −0.2 |
|  | 0.65 | −0 | −0.25 | 1.0 | −0 | −0.4 | 1.5 | −0 | −0.6 | 2.8 | −0 | −1.4 | 1.2 | −0 | −0.6 |
| 0.40–0.71 | 0 | +0.4 | +0 | 0 | +0.7 | +0 | 0 | +1.0 | +0 | 0 | +1.6 | +0 | 0.25 | +0.7 | −0.25 |
|  | 0.7 | −0 | −0.3 | 1.1 | −0 | −0.4 | 1.7 | −0 | −0.7 | 3.2 | −0 | −1.6 | 1.35 | −0 | −0.65 |
| 0.71–1.19 | 0 | +0.5 | +0 | 0 | +0.8 | +0 | 0 | +1.2 | +0 | 0 | +2.0 | +0 | 0.3 | +0.8 | −0.3 |
|  | 0.9 | −0 | −0.4 | 1.3 | −0 | −0.5 | 2 | −0 | −0.8 | 4 | −0 | −2.0 | 1.6 | −0 | −0.8 |
| 1.19–1.97 | 0 | +0.6 | +0 | 0 | +1.0 | +0 | 0 | +1.6 | +0 | 0 | +2.5 | +0 | 0.4 | +1.0 | −0.4 |
|  | 1.0 | −0 | −0.4 | 1.6 | −0 | −0.6 | 2.6 | −0 | −1 | 5 | −0 | −2.5 | 2.0 | −0 | −1.0 |
| 1.97–3.15 | 0 | +0.7 | +0 | 0 | +1.2 | +0 | 0 | +1.8 | +0 | 0 | +3 | +0 | 0.4 | +1.2 | −0.4 |
|  | 1.2 | −0 | −0.5 | 1.9 | −0 | −0.7 | 3 | −0 | −1.2 | 6 | −0 | −3 | 2.3 | −0 | −1.1 |
| 3.15–4.73 | 0 | +0.9 | +0 | 0 | +1.4 | +0 | 0 | +2.2 | +0 | 0 | +3.5 | +0 | 0.5 | +1.4 | −0.5 |
|  | 1.5 | −0 | −0.6 | 2.3 | −0 | −0.9 | 3.6 | −0 | −1.4 | 7 | −0 | −3.5 | 2.8 | −0 | −1.4 |
| 4.73–7.09 | 0 | +1.0 | +0 | 0 | +1.6 | +0 | 0 | +2.5 | +0 | 0 | +4 | +0 | 0.6 | +1.6 | −0.6 |
|  | 1.7 | −0 | −0.7 | 2.6 | −0 | −1.0 | 4.1 | −0 | −1.6 | 8 | −0 | −4 | 3.2 | −0 | −1.6 |

| Nominal size, in. | Class LC 6 | Hole | Shaft | Class LC 7 | Hole | Shaft | Class LC 8 | Hole | Shaft | Class LC 9 | Hole | Shaft | Class LC 10 | Hole | Shaft | Class LC 11 | Hole | Shaft |
|---|---|---|---|---|---|---|---|---|---|---|---|---|---|---|---|---|---|---|
| 0.04–0.12 | 0.3 | +0.6 | −0.3 | 0.6 | +1.0 | −0.6 | 1.0 | +1.6 | −1.0 | 2.5 | +2.5 | −2.5 | 4 | +4 | −4 | 5 | +6 | −5 |
|  | 1.5 | −0 | −0.9 | 2.6 | −0 | −1.6 | 3.6 | −0 | −2.0 | 7.5 | −0 | −5.0 | 12 | −0 | −8 | 17 | −0 | −11 |
| 0.12–0.24 | 0.4 | +0.7 | −0.4 | 0.8 | +1.2 | −0.8 | 1.2 | +1.8 | −1.2 | 2.8 | +3.0 | −2.8 | 4.5 | +5 | −4.5 | 6 | +7 | −6 |
|  | 1.8 | −0 | −1.1 | 3.2 | −0 | −2.0 | 4.2 | −0 | −2.4 | 8.8 | −0 | −5.8 | 14.5 | −0 | −9.5 | 20 | −0 | −13 |
| 0.24–0.40 | 0.5 | +0.9 | −0.5 | 1.0 | +1.4 | −1.0 | 1.6 | +2.2 | −1.6 | 3.0 | +3.5 | −3.0 | 5 | +6 | −5 | 7 | +9 | −7 |
|  | 2.3 | −0 | −1.4 | 3.8 | −0 | −2.4 | 5.2 | −0 | −3.0 | 10.0 | −0 | −6.5 | 17 | −0 | −11 | 25 | −0 | −16 |
| 0.40–0.71 | 0.6 | +1.0 | −0.6 | 1.2 | +1.6 | −1.2 | 2.0 | +2.8 | −2.0 | 3.5 | +4.0 | −3.5 | 6 | +7 | −6 | 8 | +10 | −8 |
|  | 2.6 | −0 | −1.6 | 4.4 | −0 | −2.8 | 6.4 | −0 | −3.6 | 11.5 | −0 | −7.5 | 20 | −0 | −13 | 28 | −0 | −18 |
| 0.71–1.19 | 0.8 | +1.2 | −0.8 | 1.6 | +2.0 | −1.6 | 2.5 | +3.5 | −2.5 | 4.5 | +5.0 | −4.5 | 7 | +8 | −7 | 10 | +12 | −10 |
|  | 3.2 | −0 | −2.0 | 5.6 | −0 | −3.6 | 8.0 | −0 | −4.5 | 14.5 | −0 | −9.5 | 23 | −0 | −15 | 34 | −0 | −22 |
| 1.19–1.97 | 1.0 | +1.6 | −1.0 | 2.0 | +2.5 | −2.0 | 3.0 | +4.0 | −3.0 | 5 | +6 | −5 | 8 | +10 | −8 | 12 | +16 | −12 |
|  | 4.2 | −0 | −2.6 | 7.0 | −0 | −4.5 | 9.5 | −0 | −5.5 | 17 | −0 | −11 | 28 | −0 | −18 | 44 | −0 | −28 |
| 1.97–3.15 | 1.2 | +1.8 | −1.2 | 2.5 | +3.0 | −2.5 | 4.0 | +4.5 | −4.0 | 6 | +7 | −6 | 10 | +12 | −10 | 14 | +18 | −14 |
|  | 4.8 | −0 | −3.0 | 8.5 | −0 | −5.5 | 11.5 | −0 | −7.0 | 20 | −0 | −13 | 34 | −0 | −22 | 50 | −0 | −32 |
| 3.15–4.73 | 1.4 | +2.2 | −1.4 | 3.0 | +3.5 | −3.0 | 5.0 | +5.0 | −5.0 | 7 | +9 | −7 | 11 | +14 | −11 | 16 | +22 | −16 |
|  | 5.8 | −0 | −3.6 | 10.0 | −0 | −6.5 | 13.5 | −0 | −8.5 | 25 | −0 | −16 | 39 | −0 | −25 | 60 | −0 | −38 |
| 4.73–7.09 | 1.6 | +2.5 | −1.6 | 3.5 | +4.0 | −3.5 | 6 | +6 | −6 | 8 | +10 | −8 | 12 | +16 | −12 | 18 | +25 | −18 |
|  | 6.6 | −0 | −4.1 | 11.5 | −0 | −7.5 | 16 | −0 | −10 | 28 | −0 | −18 | 44 | −0 | −28 | 68 | −0 | −43 |

*Adapted from ANSI B4.1—Limits for hole and shaft are applied algebraically to basic size to obtain limits of size for parts. Data in boldface in accordance with ABC agreements.

## Table 3.  Transition Locational Fits

| Nominal size, in. | Class LT 1 Fit† | LT 1 Hole | LT 1 Shaft | Class LT 2 Fit | LT 2 Hole | LT 2 Shaft | Class LT 3 Fit | LT 3 Hole | LT 3 Shaft | Class LT 4 Fit | LT 4 Hole | LT 4 Shaft | Class LT 6 Fit | LT 6 Hole | LT 6 Shaft | Class LT 7 Fit | LT 7 Hole | LT 7 Shaft |
|---|---|---|---|---|---|---|---|---|---|---|---|---|---|---|---|---|---|---|
| 0.04–0.12 | −0.15 | +0.4 | +0.15 | −0.3 | +0.6 | +0.3 |  |  |  |  |  |  | −0.55 | +0.6 | +0.55 | −0.5 | +0.4 | +0.5 |
|  | +0.5 | −0 | −0.1 | +0.7 | −0 | −0.1 |  |  |  |  |  |  | +0.45 | −0 | +0.15 | +0.15 | −0 | +0.25 |
| 0.12–0.24 | −0.2 | +0.5 | +0.2 | −0.4 | +0.7 | +0.4 |  |  |  |  |  |  | −0.7 | +0.7 | +0.7 | −0.6 | +0.5 | +0.6 |
|  | +0.6 | −0 | −0.1 | +0.8 | −0 | −0.1 |  |  |  |  |  |  | +0.5 | −0 | +0.2 | +0.2 | −0 | +0.3 |
| 0.24–0.40 | −0.3 | +0.6 | +0.3 | −0.4 | +0.9 | +0.4 | −0.5 | +0.6 | +0.5 | −0.7 | +0.9 | +0.7 | −0.8 | +0.9 | +0.8 | −0.8 | +0.6 | +0.8 |
|  | +0.7 | −0 | −0.1 | +1.1 | −0 | −0.2 | +0.5 | −0 | +0.1 | +0.8 | −0 | +0.1 | +0.7 | −0 | +0.2 | +0.2 | −0 | +0.4 |
| 0 40–0.71 | −0.3 | +0.7 | +0.3 | −0.5 | +1.0 | +0.5 | −0.5 | +0.7 | +0.5 | −0.8 | +1.0 | +0.8 | −1.0 | +1.0 | +1.0 | −0.9 | +0.7 | +0.9 |
|  | +0.8 | −0 | −0.1 | +1.2 | −0 | −0.2 | +0.6 | −0 | +0.1 | +0.9 | −0 | +0.1 | +0.7 | −0 | +0.3 | +0.2 | −0 | +0.5 |
| 0.71–1.19 | −0.3 | +0.8 | +0.3 | −0.5 | +1.2 | +0.5 | −0.6 | +0.8 | +0.6 | −0.9 | +1.2 | +0.9 | −1.1 | +1.2 | +1.1 | −1.1 | +0.8 | +1.1 |
|  | +1.0 | −0 | −0.2 | +1.5 | −0 | −0.3 | +0.7 | −0 | +0.1 | +1.1 | −0 | +0.1 | +0.9 | −0 | +0.3 | +0.2 | −0 | +0.6 |
| 1.19–1.97 | −0.4 | +1.0 | +0.4 | −0.6 | +1.6 | +0.6 | −0.7 | +1.0 | +0.7 | −1.1 | +1.6 | +1.1 | −1.4 | +1.6 | +1.4 | −1.3 | +1.0 | +1.3 |
|  | +1.2 | −0 | −0.2 | +2.0 | −0 | −0.4 | +0.9 | −0 | +0.1 | +1.5 | −0 | +0.1 | +1.2 | −0 | +0.4 | +0.3 | −0 | +0.7 |
| 1.97–3.15 | −0.4 | +1.2 | +0.4 | −0.7 | +1.8 | +0.7 | −0.8 | +1.2 | +0.8 | −1.3 | +1.8 | +1.3 | −1.7 | +1.8 | +1.7 | −1.5 | +1.2 | +1.5 |
|  | +1.5 | −0 | −0.3 | +2.3 | −0 | −0.5 | +1.1 | −0 | +0.1 | +1.7 | −0 | +0.1 | +1.3 | −0 | +0.5 | +0.4 | −0 | +0.8 |
| 3.15–4.73 | −0.5 | +1.4 | +0.5 | −0.8 | +2.2 | +0.8 | −1.0 | +1.4 | +1.0 | −1.5 | +2.2 | +1.5 | −1.9 | +2.2 | +1.9 | −1.9 | +1.4 | +1.9 |
|  | +1.8 | −0 | −0.4 | +2.8 | −0 | −0.6 | +1.3 | −0 | +0.1 | +2.1 | −0 | +0.1 | +1.7 | −0 | +0.5 | +0.4 | −0 | +1.0 |
| 4.73–7.09 | −0.6 | +1.6 | +0.6 | −0.9 | +2.5 | +0.9 | −1.1 | +1.6 | +1.1 | −1.7 | +2.5 | +1.7 | −2.2 | +2.5 | +2.2 | −2.2 | +1.6 | +2.2 |
|  | +2.0 | −0 | −0.4 | +3.2 | −0 | −0.7 | +1.5 | −0 | +0.1 | +2.4 | −0 | +0.1 | +1.9 | −0 | +0.6 | +0.4 | −0 | +1.2 |

† "Fit" represents the maximum interference (minus values) and the maximum clearance (plus values).

## Table 4.  Interference Locational Fits

| Nominal size, in. | Class LN 2 Interference | LN 2 Hole | LN 2 Shaft | Class LN 3 Interference | LN 3 Hole | LN 3 Shaft |
|---|---|---|---|---|---|---|
| 0.04–0.12 | 0 | +0.4 | +0.65 | 0.1 | +0.4 | +0.75 |
|  | 0.65 | −0 | +0.4 | 0.75 | −0 | +0.5 |
| 0.12–0.24 | 0 | +0.5 | +0.8 | 0.1 | +0.5 | +0.9 |
|  | 0.8 | −0 | +0.5 | 0.9 | −0 | +0.6 |
| 0.24–0.40 | 0 | +0.6 | +1.0 | 0.2 | +0.6 | +1.2 |
|  | 1.0 | −0 | +0.6 | 1.2 | −0 | +0.8 |
| 0.40–0.71 | 0 | +0.7 | +1.1 | 0.3 | +0.7 | +1.4 |
|  | 1.1 | −0 | +0.7 | 1.4 | −0 | +1.0 |
| 0.71–1.19 | 0 | +0.8 | +1.3 | 0.4 | +0.8 | +1.7 |
|  | 1.3 | −0 | +0.8 | 1.7 | −0 | +1.2 |
| 1.19–1.97 | 0 | +1.0 | +1.6 | 0.4 | +1.0 | +2.0 |
|  | 1.6 | −0 | +1.0 | 2.0 | −0 | +1.4 |
| 1.97–3.15 | 0.2 | +1.2 | +2.1 | 0.4 | +1.2 | +2.3 |
|  | 2.1 | −0 | +1.4 | 2.3 | −0 | +1.6 |
| 3.15–4.73 | 0.2 | +1.4 | +2.5 | 0.6 | +1.4 | +2.9 |
|  | 2.5 | −0 | +1.6 | 2.9 | −0 | +2.0 |
| 4.73–7.09 | 0.2 | +1.6 | +2.8 | 0.9 | +1.6 | +3.5 |
|  | 2.8 | −0 | +1.8 | 3.5 | −0 | +2.5 |

## Table 5.  Force and Shrink Fits

| Nominal size | Class FN 1 | FN 1 Hole | FN 1 Shaft | Class FN 2 | FN 2 Hole | FN 2 Shaft | Class FN 3 | FN 3 Hole | FN 3 Shaft | Class FN 4 | FN 4 Hole | FN 4 Shaft | Class FN 5 | FN 5 Hole | FN 5 Shaft |
|---|---|---|---|---|---|---|---|---|---|---|---|---|---|---|---|
| 0.04–0.12 | 0.05 | +0.25 | +0.5 | 0.2 | +0.4 | +0.85 |  |  |  | 0.3 | +0.4 | +0.95 | 0.5 | +0.4 | +1.3 |
|  | 0.5 | −0 | +0.3 | 0.85 | −0 | +0.6 |  |  |  | 0.95 | −0 | +0.7 | 1.3 | −0 | +0.9 |
| 0.12–0.24 | 0.1 | +0.3 | +0.6 | 0.2 | +0.5 | +1.0 |  |  |  | 0.4 | +0.5 | +1.2 | 0.7 | +0.5 | +1.7 |
|  | 0.6 | −0 | +0.4 | 1.0 | −0 | +0.7 |  |  |  | 1.2 | −0 | +0.9 | 1.7 | −0 | +1.2 |
| 0.24–0.40 | 0.1 | +0.4 | +0.75 | 0.4 | +0.6 | +1.4 |  |  |  | 0.6 | +0.6 | +1.6 | 0.8 | +0.6 | +2.0 |
|  | 0.75 | −0 | +0.5 | 1.4 | −0 | +1.0 |  |  |  | 1.6 | −0 | +1.2 | 2.0 | −0 | +1.4 |
| 0.40–0.56 | 0.1 | +0.4 | +0.8 | 0.5 | +0.7 | +1.6 |  |  |  | 0.7 | +0.7 | +1.8 | 0.9 | +0.7 | +2.3 |
|  | 0.8 | −0 | +0.5 | 1.6 | −0 | +1.2 |  |  |  | 1.8 | −0 | +1.4 | 2.3 | −0 | +1.6 |
| 0.56–0.71 | 0.2 | +0.4 | +0.9 | 0.5 | +0.7 | +1.6 |  |  |  | 0.7 | +0.7 | +1.8 | 1.1 | +0.7 | +2.5 |
|  | 0.9 | −0 | +0.6 | 1.6 | −0 | +1.2 |  |  |  | 1.8 | −0 | +1.4 | 2.5 | −0 | +1.8 |
| 0.71–0.95 | 0.2 | +0.5 | +1.1 | 0.6 | +0.8 | +1.9 |  |  |  | 0.8 | +0.8 | +2.1 | 1.4 | +0.8 | +3.0 |
|  | 1.1 | −0 | +0.7 | 1.9 | −0 | +1.4 |  |  |  | 2.1 | −0 | +1.6 | 3.0 | −0 | +2.2 |
| 0.95–1.19 | 0.3 | +0.5 | +1.2 | 0.6 | +0.8 | +1.9 | 0.8 | +0.8 | +2.1 | 1.0 | +0.8 | +2.3 | 1.7 | +0.8 | +3.3 |
|  | 1.2 | −0 | +0.8 | 1.9 | −0 | +1.4 | 2.1 | −0 | +1.6 | 2.3 | −0 | +1.8 | 3.3 | −0 | +2.5 |
| 1.19–1.58 | 0.3 | +0.6 | +1.3 | 0.8 | +1.0 | +2.4 | 1.0 | +1.0 | +2.6 | 1.5 | +1.0 | +3.1 | 2.0 | +1.0 | +4.0 |
|  | 1.3 | −0 | +0.9 | 2.4 | −0 | +1.8 | 2.6 | −0 | +2.0 | 3.1 | −0 | +2.5 | 4.0 | −0 | +3.0 |
| 1.58–1.97 | 0.4 | +0.6 | +1.4 | 0.8 | +1.0 | +2.4 | 1.2 | +1.0 | +2.8 | 1.8 | +1.0 | +3.4 | 3.0 | +1.0 | +5.0 |
|  | 1.4 | −0 | +1.0 | 2.4 | −0 | +1.8 | 2.8 | −0 | +2.2 | 3.4 | −0 | +2.8 | 5.0 | −0 | +4.0 |
| 1.97–2.56 | 0.6 | +0.7 | +1.8 | 0.8 | +1.2 | +2.7 | 1.3 | +1.2 | +3.2 | 2.3 | +1.2 | +4.2 | 3.8 | +1.2 | +6.2 |
|  | 1.8 | −0 | +1.3 | 2.7 | −0 | +2.0 | 3.2 | −0 | +2.5 | 4.2 | −0 | +3.5 | 6.2 | −0 | +5.0 |
| 2.56–3.15 | 0.7 | +0.7 | +1.9 | 1.0 | +1.2 | +2.9 | 1.8 | +1.2 | +3.7 | 2.8 | +1.2 | +4.7 | 4.8 | +1.2 | +7.2 |
|  | 1.9 | −0 | +1.4 | 2.9 | −0 | +2.2 | 3.7 | −0 | +3.0 | 4.7 | −0 | +4.0 | 7.2 | −0 | +6.0 |
| 3.15–3.94 | 0.9 | +0.9 | +2.4 | 1.4 | +1.4 | +3.7 | 2.1 | +1.4 | +4.4 | 3.6 | +1.4 | +5.9 | 5.6 | +1.4 | +8.4 |
|  | 2.4 | −0 | +1.8 | 3.7 | −0 | +2.8 | 4.4 | −0 | +3.5 | 5.9 | −0 | +5.0 | 8.4 | −0 | +7.0 |
| 3.94–4.73 | 1.1 | +0.9 | +2.6 | 1.6 | +1.4 | +3.9 | 2.6 | +1.4 | +4.9 | 4.6 | +1.4 | +6.9 | 6.6 | +1.4 | +9.4 |
|  | 2.6 | −0 | +2.0 | 3.9 | −0 | +3.0 | 4.9 | −0 | +4.0 | 6.9 | −0 | +6.0 | 9.4 | −0 | +8.0 |
| 4.73–5.52 | 1.2 | +1.0 | +2.9 | 1.9 | +1.6 | +4.5 | 3.4 | +1.6 | +6.0 | 5.4 | +1.6 | +8.0 | 8.4 | +1.6 | +11.6 |
|  | 2.9 | −0 | +2.2 | 4.5 | −0 | +3.5 | 6.0 | −0 | +5.0 | 8.0 | −0 | +7.0 | 11.6 | −0 | +10.0 |
| 5.52–6.30 | 1.5 | +1.0 | +3.2 | 2.4 | +1.6 | +5.0 | 3.4 | +1.6 | +6.0 | 5.4 | +1.6 | +8.0 | 10.4 | +1.6 | +13.6 |
|  | 3.2 | −0 | +2.5 | 5.0 | −0 | +4.0 | 6.0 | −0 | +5.0 | 8.0 | −0 | +7.0 | 13.6 | −0 | +12.0 |
| 6.30–7.09 | 1.8 | +1.0 | +3.5 | 2.9 | +1.6 | +5.5 | 4.4 | +1.6 | +7.0 | 6.4 | +1.6 | +9.0 | 10.4 | +1.6 | +13.6 |
|  | 3.5 | −0 | +2.8 | 5.5 | −0 | +4.5 | 7.0 | −0 | +6.0 | 9.0 | −0 | +8.0 | 13.6 | −0 | +12.0 |

## ANSI Standard Pipe[a,b]

*Welded wrought iron*

| Nominal pipe size | Actual outside diam. $D$ | Tap-drill size $S$ | Thd. /in. $T$ | Distance pipe enters fittings $L$ | Wall thickness $X$ | | | Weight, lb/ft[f] | | |
|---|---|---|---|---|---|---|---|---|---|---|
| | | | | | Stand-ard 40[c] | Extra strong 80[d] | Double extra strong[e] | Stand-ard 40[c] | Extra strong 80[d] | Double extra strong[e] |
| 1/8 | 0.405 | 11/32 | 27 | 3/16 | 0.070 | 0.098 | .... | 0.25 | 0.32 | |
| 1/4 | 0.540 | 7/16 | 18 | 9/32 | 0.090 | 0.122 | .... | 0.43 | 0.54 | |
| 3/8 | 0.675 | 37/64 | 18 | 19/64 | 0.093 | 0.129 | .... | 0.57 | 0.74 | |
| 1/2 | 0.840 | 23/32 | 14 | 3/8 | 0.111 | 0.151 | 0.307 | 0.86 | 1.09 | 1.714 |
| 3/4 | 1.050 | 59/64 | 14 | 13/32 | 0.115 | 0.157 | 0.318 | 1.14 | 1.48 | 2.440 |
| 1 | 1.315 | 1 5/32 | 11 1/2 | 1/2 | 0.136 | 0.183 | 0.369 | 1.68 | 2.18 | 3.659 |
| 1 1/4 | 1.660 | 1 1/2 | 11 1/2 | 35/64 | 0.143 | 0.195 | 0.393 | 2.28 | 3.00 | 5.214 |
| 1 1/2 | 1.900 | 1 47/64 | 11 1/2 | 9/16 | 0.148 | 0.204 | 0.411 | 2.72 | 3.64 | 6.408 |
| 2 | 2.375 | 2 7/32 | 11 1/2 | 37/64 | 0.158 | 0.223 | 0.447 | 3.66 | 5.03 | 9.029 |
| 2 1/2 | 2.875 | 2 5/8 | 8 | 7/8 | 0.208 | 0.282 | 0.565 | 5.80 | 7.67 | 13.695 |
| 3 | 3.5 | 3 1/4 | 8 | 15/16 | 0.221 | 0.306 | 0.615 | 7.58 | 10.3 | 18.583 |
| 3 1/2 | 4.0 | 3 3/4 | 8 | 1 | 0.231 | 0.325 | .... | 9.11 | 12.5 | |
| 4 | 4.5 | 4 1/4 | 8 | 1 1/16 | 0.242 | 0.344 | 0.690 | 10.8 | 15.0 | 27.451 |
| 5 | 5.563 | 5 5/16 | 8 | 1 5/32 | 0.263 | 0.383 | 0.768 | 14.7 | 20.8 | 38.552 |
| 6 | 6.625 | 6 5/16 | 8 | 1 1/4 | 0.286 | 0.441 | 0.884 | 19.0 | 28.6 | 53.160 |
| 8 | 8.625 | ..... | 8 | 1 15/32 | 0.329 | 0.510 | 0.895 | 28.6 | 43.4 | 72.424 |
| 10 | 10.75 | ..... | 8 | 1 43/64 | 0.372 | 0.606 | .... | 40.5 | 64.4 | |
| 12 | 12.75 | ..... | 8 | 1 7/8 | 0.414 | 0.702 | .... | 53.6 | 88.6 | |
| 14 OD | 14.0 | ..... | 8 | 2 | 0.437 | 0.750 | .... | 62.2 | 104. | |
| 16 OD | 16.0 | ..... | 8 | 2 13/64 | 0.500 | .... | .... | 82.2 | | |
| 18 OD | 18.0 | ..... | 8 | 2 13/32 | 0.562 | .... | .... | 103. | | |
| 20 OD | 20.0 | ..... | 8 | 2 19/32 | 0.562 | .... | .... | 115. | | |
| 24 OD | 24.0 | ..... | 8 | 3 | | | | | | |

[a] For welded and seamless steel pipe. See ANSI B36.10—Dimensions in inches.

[b] A pipe size may be designated by giving the nominal pipe size and wall thickness or by giving the nominal pipe size and weight per linear foot.

[c] Refers to ANSI Standard schedule numbers, approximate values for the expression $1,000 \times P/S$. Schedule 40—standard weight.

[d] Schedule 80—extra strong.

[e] Not ANSI Standard, but commercially available in both wrought iron and steel.

[f] Plain ends.

## ANSI Standard 150-lb Malleable-iron Screwed Fittings[a]

90° ELBOW

TEE    CROSS

45° ELL    Y BRANCH

PLUG

| Nominal pipe size | A | B | C | E | F | G | H | J | K | L | M |
|---|---|---|---|---|---|---|---|---|---|---|---|
| 1/8 | 0.69 | 0.25 | .... | 0.200 | 0.405 | 0.090 | 0.693 | 1.00 b | .... | 0.264 | |
| 1/4 | 0.81 | 0.32 | 0.73 | 0.215 | 0.540 | 0.095 | 0.844 | 1.19 | 0.94 | 0.402 | 1.00 |
| 3/8 | 0.95 | 0.36 | 0.80 | 0.230 | 0.675 | 0.100 | 1.015 | 1.44 | 1.03 | 0.408 | 1.13 |
| 1/2 | 1.12 | 0.43 | 0.88 | 0.249 | 0.840 | 0.105 | 1.197 | 1.63 | 1.15 | 0.534 | 1.25 |
| 3/4 | 1.31 | 0.50 | 0.98 | 0.273 | 1.050 | 0.120 | 1.458 | 1.89 | 1.29 | 0.546 | 1.44 |
| 1 | 1.50 | 0.58 | 1.12 | 0.302 | 1.315 | 0.134 | 1.771 | 2.14 | 1.47 | 0.683 | 1.69 |
| 1 1/4 | 1.75 | 0.67 | 1.29 | 0.341 | 1.660 | 0.145 | 2.153 | 2.45 | 1.71 | 0.707 | 2.06 |
| 1 1/2 | 1.94 | 0.70 | 1.43 | 0.368 | 1.900 | 0.155 | 2.427 | 2.69 | 1.88 | 0.724 | 2.31 |
| 2 | 2.25 | 0.75 | 1.68 | 0.422 | 2.375 | 0.173 | 2.963 | 3.26 | 2.22 | 0.757 | 2.81 |
| 2 1/2 | 2.70 | 0.92 | 1.95 | 0.478 | 2.875 | 0.210 | 3.589 | 3.86 | 2.57 | 1.138 | 3.25 |
| 3 | 3.08 | 0.98 | 2.17 | 0.548 | 3.500 | 0.231 | 4.285 | 4.51 | 3.00 | 1.200 | 3.69 |
| 3 1/2 | 3.42 | 1.03 | 2.39 | 0.604 | 4.000 | 0.248 | 4.843 | 5.09 b | .... | 1.250 | 4.00 |
| 4 | 3.79 | 1.08 | 2.61 | 0.661 | 4.500 | 0.265 | 5.401 | 5.69 | 3.70 | 1.300 | 4.38 |
| 5 | 4.50 | 1.18 | 3.05 | 0.780 | 5.563 | 0.300 | 6.583 | 6.86 b | .... | 1.406 | 5.12 |
| 6 | 5.13 | 1.28 | 3.46 | 0.900 | 6.625 | 0.336 | 7.767 | 8.03 b | .... | 1.513 | 5.86 |

| Nominal pipe size | N | P | T | U | V | W | X | Y | Z[c] | O[d] | Thickness of ribs on caps, couplings |
|---|---|---|---|---|---|---|---|---|---|---|---|
| 1/8 | 0.20 | .... | .... | .... | .... | 0.96 | 0.37 | 0.24 | 9/32 | .... | 0.090 |
| 1/4 | 0.26 | .... | .... | .... | .... | 1.06 | 0.44 | 0.28 | 3/8 | .... | 0.095 |
| 3/8 | 0.37 | .... | 0.50 | 1.43 | 1.93 | 1.16 | 0.48 | 0.31 | 7/16 | .... | 0.100 |
| 1/2 | 0.51 | 0.87 | 0.61 | 1.71 | 2.32 | 1.34 | 0.56 | 0.38 | 9/16 | 0.16 | 0.105 |
| 3/4 | 0.69 | 0.97 | 0.72 | 2.05 | 2.77 | 1.52 | 0.63 | 0.44 | 5/8 | 0.18 | 0.120 |
| 1 | 0.91 | 1.16 | 0.85 | 2.43 | 3.28 | 1.67 | 0.75 | 0.50 | 13/16 | 0.20 | 0.134 |
| 1 1/4 | 1.19 | 1.28 | 1.02 | 2.92 | 3.94 | 1.93 | 0.80 | 0.56 | 15/16 | 0.22 | 0.145 |
| 1 1/2 | 1.39 | 1.33 | 1.10 | 3.28 | 4.38 | 2.15 | 0.83 | 0.62 | 1 1/8 | 0.24 | 0.155 |
| 2 | 1.79 | 1.45 | 1.24 | 3.93 | 5.17 | 2.53 | 0.88 | 0.68 | 1 5/16 | 0.26 | 0.173 |
| 2 1/2 | 2.20 | 1.70 | 1.52 | 4.73 | 6.25 | 2.88 | 1.07 | 0.74 | 1 1/2 | 0.29 | 0.210 |
| 3 | 2.78 | 1.80 | 1.71 | 5.55 | 7.26 | 3.18 | 1.13 | 0.80 | 1 11/16 | 0.31 | 0.231 |
| 3 1/2 | 3.24 | 1.90 | .... | .... | .... | 3.43 | 1.18 | 0.86 | 1 7/8 | 0.34 | 0.248 |
| 4 | 3.70 | 2.08 | 2.01 | 6.97 | 8.98 | 3.69 | 1.22 | 1.00 | 2 1/8 | 0.37 | 0.265 |
| 5 | 4.69 | 2.32 | .... | .... | .... | .... | 1.31 | 1.00 | 2 5/16 | 0.46 | 0.300 |
| 6 | 5.67 | 2.55 | .... | .... | .... | .... | 1.40 | 1.25 | 2 1/2 | 0.52 | 0.336 |

[a] ANSI B16.3—Street tee not made in 1/8-in. size. Dimensions in inches. Left-hand couplings have four or more ribs. Right-hand couplings have two ribs.   [b] Street ell only.

[e] These dimensions are the nominal size of wrench (ANSI B18.2—). Square-head plugs are designed to fit these wrenches.

[d] Solid plugs are provided in sizes 1/8 to 3 1/2 in. inclusive; cored plugs 1/2 to 3 1/2 in. inclusive. Cored plugs have min. metal thickness at all points equal to dimension O except at the end of thread.

90° STREET ELBOW    45° STREET ELBOW    STREET TEE

COUPLING

REDUCING COUPLING

CAP

## ANSI Standard 150-lb Malleable-iron Screwed Fittings: Dimensions of Close-, Medium-, and Open-pattern Return Bends*†

CLOSE

MEDIUM

| Nominal pipe size | Length of thread, min. $B$ | Width of band, min. $E$ | Inside diam. of fittings $F$ Max. | Inside diam. of fittings $F$ Min. | Metal thickness ‡ $G$ | Outside diam. of band, min. $H$ | Center to center (close pattern) $R_1$ | Center to center (medium pattern) $R_2$ | Center to center (open pattern) $R_3$ |
|---|---|---|---|---|---|---|---|---|---|
| ½ | 0.43 | 0.249 | 0.840 | 0.897 | 0.116 | 1.197 | 1.000 | 1.250 | 1.50 |
| ¾ | 0.50 | 0.273 | 1.050 | 1.107 | 0.133 | 1.458 | 1.250 | 1.500 | 2.00 |
| 1 | 0.58 | 0.302 | 1.315 | 1.385 | 0.150 | 1.771 | 1.500 | 1.875 | 2.50 |
| 1¼ | 0.67 | 0.341 | 1.660 | 1.730 | 0.165 | 2.153 | 1.750 | 2.250 | 3.00 |
| 1½ | 0.70 | 0.368 | 1.900 | 1.970 | 0.178 | 2.427 | 2.188 | 2.500 | 3.50 |
| 2 | 0.75 | 0.422 | 2.375 | 2.445 | 0.201 | 2.963 | 2.625 | 3.000 | 4.00 |
| 2½ | 0.92 | 0.478 | 2.875 | 2.975 | 0.244 | 3.589 | .... | .... | 4.50 |
| 3 | 0.98 | 0.548 | 3.500 | 3.600 | 0.272 | 4.285 | .... | .... | 5.00 |
| 4 | 1.08 | 0.661 | 4.500 | 4.600 | 0.315 | 5.401 | .... | .... | 6.00 |

All dimensions given in inches.

* Adapted from ANSI B16.3—.

† It is permissible to furnish close-pattern return bends not banded. Close-pattern return bends will not make up parallel coils, as the distance center to center of two adjacent bends is greater than the center to center of openings of a single bend.

‡ Patterns shall be designed to produce castings of metal thicknesses given in the table. Metal thickness at no point shall be less than 90 per cent of the thickness given in the table.

OPEN

## ANSI Standard Cast-iron Screwed Fittings *

*For maximum working saturated steam pressure of 125 and 250 lb per sq in.*

| Nominal pipe size | $A$ | $B$ min. | $C$ | $E$ min. | $F$ Min. | $F$ Max. | $G$ min. | $H$ min. |
|---|---|---|---|---|---|---|---|---|
| ¼ | 0.81 | 0.32 | 0.73 | 0.38 | 0.540 | 0.584 | 0.110 | 0.93 |
| ⅜ | 0.95 | 0.36 | 0.80 | 0.44 | 0.675 | 0.719 | 0.120 | 1.12 |
| ½ | 1.12 | 0.43 | 0.88 | 0.50 | 0.840 | 0.897 | 0.130 | 1.34 |
| ¾ | 1.31 | 0.50 | 0.98 | 0.56 | 1.050 | 1.107 | 0.155 | 1.63 |
| 1 | 1.50 | 0.58 | 1.12 | 0.62 | 1.315 | 1.385 | 0.170 | 1.95 |
| 1¼ | 1.75 | 0.67 | 1.29 | 0.69 | 1.660 | 1.730 | 0.185 | 2.39 |
| 1½ | 1.94 | 0.70 | 1.43 | 0.75 | 1.900 | 1.970 | 0.200 | 2.68 |
| 2 | 2.25 | 0.75 | 1.68 | 0.84 | 2.375 | 2.445 | 0.220 | 3.28 |
| 2½ | 2.70 | 0.92 | 1.95 | 0.94 | 2.875 | 2.975 | 0.240 | 3.86 |
| 3 | 3.08 | 0.98 | 2.17 | 1.00 | 3.500 | 3.600 | 0.260 | 4.62 |
| 3½ | 3.42 | 1.03 | 2.39 | 1.06 | 4.000 | 4.100 | 0.280 | 5.20 |
| 4 | 3.79 | 1.08 | 2.61 | 1.12 | 4.500 | 4.600 | 0.310 | 5.79 |
| 5 | 4.50 | 1.18 | 3.05 | 1.18 | 5.563 | 5.663 | 0.380 | 7.05 |
| 6 | 5.13 | 1.28 | 3.46 | 1.28 | 6.625 | 6.725 | 0.430 | 8.28 |
| 8 | 6.56 | 1.47 | 4.28 | 1.47 | 8.625 | 8.725 | 0.550 | 10.63 |
| 10 | 8.08 | 1.68 | 5.16 | 1.68 | 10.750 | 10.850 | 0.690 | 13.12 |
| 12 | 9.50 | 1.88 | 5.97 | 1.88 | 12.750 | 12.850 | 0.800 | 15.47 |

ELBOW          TEE

CROSS

45° ELBOW

* ANSI B16.4—Dimensions in inches.

## Globe, Angle-globe, and Gate Valves*

| Size | A (globe only) | B (open) | C | D (angle only) | E | F (open) | G |
|---|---|---|---|---|---|---|---|
| 1/8 | 2 | 4 | 1 3/4 | 1 | | | |
| 1/4 | 2 | 4 | 1 3/4 | 1 | 1 7/8 | 5 1/8 | 1 3/4 |
| 3/8 | 2 1/4 | 4 1/2 | 2 | 1 1/8 | 2 | 5 1/8 | 1 3/4 |
| 1/2 | 2 3/4 | 5 1/4 | 2 1/2 | 1 1/4 | 2 1/8 | 5 1/2 | 2 |
| 3/4 | 3 3/16 | 6 | 2 3/4 | 1 1/2 | 2 3/8 | 6 5/8 | 2 1/2 |
| 1 | 3 3/4 | 6 3/4 | 3 | 1 3/4 | 2 7/8 | 7 7/8 | 2 3/4 |
| 1 1/4 | 4 1/4 | 7 1/4 | 3 5/8 | 2 | 3 1/4 | 9 1/2 | 3 |
| 1 1/2 | 4 3/4 | 8 1/4 | 4 | 2 1/4 | 3 1/2 | 10 7/8 | 3 5/8 |
| 2 | 5 3/4 | 9 1/2 | 4 3/4 | 2 3/4 | 3 7/8 | 13 1/8 | 4 |
| 2 1/2 | 6 3/4 | 11 | 6 | 3 1/4 | 4 1/2 | 15 3/8 | 4 3/4 |
| 3 | 8 | 12 1/4 | 7 | 3 3/4 | 5 | 17 7/8 | 5 3/8 |

\* Dimensions in inches and compiled from manufacturers' catalogues for drawing purposes.

## Lengths of Pipe Nipples*

| Size | Length | | Size | Length | | Size | Length | | Size | Length | |
|---|---|---|---|---|---|---|---|---|---|---|---|
| | Close | Short | | Close | Short | | Close | Short | | Close | Short |
| 1/8 | 3/4 | 1 1/2 | 1/2 | 1 1/8 | 1 1/2 | 1 1/4 | 1 5/8 | 2 1/2 | 2 1/2 | 2 1/2 | 3 |
| 1/4 | 7/8 | 1 1/2 | 3/4 | 1 3/8 | 2 | 1 1/2 | 1 3/4 | 2 1/2 | 3 | 2 5/8 | 3 |
| 3/8 | 1 | 1 1/2 | 1 | 1 1/2 | 2 | 2 | 2 | 2 1/2 | | | |

\* Compiled from manufacturers' catalogues. Dimensions in inches.

Long-nipple lengths: from short-nipple lengths to 6 in. in 1/2 in. increments; from 6 in. nipple lengths to 12 in. in 1 in. increments; from 12 in. nipple lengths to 24 in. in 2 in. increments.

## Pipe Bushings*

*Dimensions of outside-head, inside-head, and face bushings in inches*

FACE BUSHING        OUTSIDE HEAD        INSIDE HEAD

| Size | Length of external thread,† min., A | Height of head, min., D | Width of head,‡ min., C — Outside | Width of head,‡ min., C — Inside | Size | Length of external thread,‡ min., A | Height of head, min., D | Width of head,† min., C — Outside | Width of head,† min., C — Inside |
|---|---|---|---|---|---|---|---|---|---|
| ¼ × ⅛ | 0.44 | 0.14 | 0.64 | | 1½ × ¼ | 0.83 | 0.37 | .... | 1.12 |
| ⅜ × ¼ | 0.48 | 0.16 | 0.68 | | 2 × 1½ | 0.88 | 0.34 | 2.48 | |
| ⅜ × ⅛ | 0.48 | 0.16 | 0.68 | | 2 × 1¼ | 0.88 | 0.34 | 2.48 | |
| ½ × ⅜ | 0.56 | 0.19 | 0.87 | | 2 × 1 | 0.88 | 0.41 | .... | 1.95 |
| ½ × ¼ | 0.56 | 0.19 | 0.87 | | 2 × ¾ | 0.88 | 0.41 | .... | 1.63 |
| ½ × ⅛ | 0.56 | 0.19 | 0.87 | | 2 × ½ | 0.88 | 0.41 | .... | 1.34 |
| ¾ × ½ | 0.63 | 0.22 | 1.15 | | 2 × ⅜ | 0.88 | 0.41 | .... | 1.12 |
| ¾ × ⅜ | 0.63 | 0.22 | 1.15 | | 2 × ¼ | 0.88 | 0.41 | .... | 1.12 |
| ¾ × ¼ | 0.63 | 0.22 | 1.15 | | 2½ × 2 | 1.07 | 0.37 | 2.98 | |
| ¾ × ⅛ | 0.63 | 0.22 | 1.15 | | 2½ × 1½ | 1.07 | 0.44 | 2.68 | |
| 1 × ¾ | 0.75 | 0.25 | 1.42 | | 2½ × 1¼ | 1.07 | 0.44 | .... | 2.39 |
| 1 × ½ | 0.75 | 0.25 | 1.42 | | 2½ × 1 | 1.07 | 0.44 | .... | 1.95 |
| 1 × ⅜ | 0.75 | 0.30 | .... | 1.12 | 2½ × ¾ | 1.07 | 0.44 | .... | 1.63 |
| 1 × ¼ | 0.75 | 0.30 | .... | 1.12 | 2½ × ½ | 1.07 | 0.44 | .... | 1.34 |
| 1 × ⅛ | 0.75 | 0.30 | .... | 1.12 | 3 × 2½ | 1.13 | 0.40 | 3.86 | |
| 1¼ × 1 | 0.80 | 0.28 | 1.76 | | 3 × 2 | 1.13 | 0.48 | 3.28 | |
| 1¼ × ¾ | 0.80 | 0.28 | 1.76 | | 3 × 1½ | 1.13 | 0.48 | .... | 2.68 |
| 1¼ × ½ | 0.80 | 0.34 | .... | 1.34 | 3 × 1¼ | 1.13 | 0.48 | .... | 2.39 |
| 1¼ × ⅜ | 0.80 | 0.34 | .... | 1.12 | 3 × 1 | 1.13 | 0.48 | .... | 1.95 |
| 1¼ × ¼ | 0.80 | 0.34 | .... | 1.12 | 3 × ¾ | 1.13 | 0.48 | .... | 1.63 |
| 1½ × 1¼ | 0.83 | 0.31 | 2.00 | | 3 × ½ | 1.13 | 0.48 | .... | 1.34 |
| 1½ × 1 | 0.83 | 0.31 | 2.00 | | | | | | |
| 1½ × ¾ | 0.83 | 0.37 | .... | 1.63 | | | | | |
| 1½ × ½ | 0.83 | 0.37 | .... | 1.34 | | | | | |
| 1½ × ⅜ | 0.83 | 0.37 | .... | 1.12 | | | | | |

* ANSI B16.14.

† In the case of outside-head bushings, length A includes provisions for imperfect threads.

‡ Heads of bushings shall be hexagonal or octagonal, except that on the larger sizes of outside-head bushings the heads may be made round with lugs instead of hexagonal or octagonal.

## ANSI Standard Cast-iron Pipe Flanges and Flanged Fittings *

*For maximum working saturated steam pressure of 125 lb per sq in. (gage)*

90°ELL   LONG RAD ELL   45°ELL   REDUCING ELL   SIDE OUTLET ELL   TRUE "Y"   TEE

SIDE OUTLET TEE   CROSS   LATERAL   REDUCER   ECCENTRIC REDUCER   FLANGES

| Nominal pipe size N | A | B | C | D | E | F | G | H | K min. |
|---|---|---|---|---|---|---|---|---|---|
| 1 | 3½ | 5 | 1¾ | 7½ | 5¾ | 1¾ | .... | 4¼ | 7/16 |
| 1¼ | 3¾ | 5½ | 2 | 8 | 6¼ | 1¾ | .... | 4⅝ | ½ |
| 1½ | 4 | 6 | 2¼ | 9 | 7 | 2 | .... | 5 | 9/16 |
| 2 | 4½ | 6½ | 2½ | 10½ | 8 | 2½ | 5 | 6 | ⅝ |
| 2½ | 5 | 7 | 3 | 12 | 9½ | 2½ | 5½ | 7 | 11/16 |
| 3 | 5½ | 7¾ | 3 | 13 | 10 | 3 | 6 | 7½ | ¾ |
| 3½ | 6 | 8½ | 3½ | 14½ | 11½ | 3 | 6½ | 8½ | 13/16 |
| 4 | 6½ | 9 | 4 | 15 | 12 | 3 | 7 | 9 | 15/16 |
| 5 | 7½ | 10¼ | 4½ | 17 | 13½ | 3½ | 8 | 10 | 15/16 |
| 6 | 8 | 11½ | 5 | 18 | 14½ | 3½ | 9 | 11 | 1 |
| 8 | 9 | 14 | 5½ | 22 | 17½ | 4½ | 11 | 13½ | 1 ⅛ |
| 10 | 11 | 16½ | 6½ | 25½ | 20½ | 5 | 12 | 16 | 1 3/16 |
| 12 | 12 | 19 | 7½ | 30 | 24½ | 5½ | 14 | 19 | 1 ¼ |

| Nominal pipe size N | L | M | No. of bolts | Diam. of bolts | Length of bolts | X min. | Y min. | Wall thickness | V |
|---|---|---|---|---|---|---|---|---|---|
| 1 | 3⅛ | ⅝ | 4 | ½ | 1¾ | 1 15/16 | 1 1/16 | 5/16 | ⅜ |
| 1¼ | 3½ | ⅝ | 4 | ½ | 2 | 2 5/16 | 1 3/16 | 5/16 | 7/16 |
| 1½ | 3⅞ | ⅝ | 4 | ½ | 2 | 2 9/16 | ⅞ | 5/16 | ½ |
| 2 | 4¾ | ¾ | 4 | ⅝ | 2¼ | 3 1/16 | 1 | 5/16 | 9/16 |
| 2½ | 5½ | ¾ | 4 | ⅝ | 2½ | 3 9/16 | 1 ⅛ | 5/16 | ⅝ |
| 3 | 6 | ¾ | 4 | ⅝ | 2½ | 4 ¼ | 1 3/16 | ⅜ | 11/16 |
| 3½ | 7 | ¾ | 8 | ⅝ | 2¾ | 4 13/16 | 1 ¼ | 7/16 | ¾ |
| 4 | 7½ | ¾ | 8 | ⅝ | 3 | 5 5/16 | 1 5/16 | ½ | ⅞ |
| 5 | 8½ | ⅞ | 8 | ¾ | 3 | 6 7/16 | 1 7/16 | ½ | ⅞ |
| 6 | 9½ | ⅞ | 8 | ¾ | 3¼ | 7 9/16 | 1 9/16 | 9/16 | 1 5/16 |
| 8 | 11¾ | ⅞ | 8 | ¾ | 3½ | 9 11/16 | 1 ¾ | ⅝ | 1 1/16 |
| 10 | 14¼ | 1 | 12 | ⅞ | 3¾ | 11 15/16 | 1 15/16 | ¾ | 1 ⅛ |
| 12 | 17 | 1 | 12 | ⅞ | 3¾ | 14 1/16 | 2 3/16 | 13/16 | |

* ANSI B16.1—Dimensions in inches.

## Lengths of Malleable-iron Unions*

*Ground joint*

| Nominal size . . . . . . . . | $\frac{1}{8}$ | $\frac{1}{4}$ | $\frac{3}{8}$ | $\frac{1}{2}$ | $\frac{3}{4}$ | 1 | $1\frac{1}{4}$ | $1\frac{1}{2}$ | 2 | $2\frac{1}{2}$ | 3 |
|---|---|---|---|---|---|---|---|---|---|---|---|
| End to end . . . . . . . . | $1\frac{1}{2}$ | $1\frac{9}{16}$ | $1\frac{5}{8}$ | $1\frac{13}{16}$ | $2\frac{1}{16}$ | $2\frac{1}{4}$ | $2\frac{1}{2}$ | $2\frac{5}{8}$ | 3 | $3\frac{9}{16}$ | $3\frac{15}{16}$ |

* Compiled from manufacturers' catalogues. Dimensions in inches.

## ANSI Standard Steel Butt-welding Fittings*·†

*Elbows, tees, caps, and stub ends*

90° ELL    45° ELL    TEE

LAP JOINT STUB END    CAP

| Nominal pipe size | Outside diam. at bevel | Center to end | | | Welding caps E | Lapped-joint stub ends | | |
|---|---|---|---|---|---|---|---|---|
| | | 90° welding elbow A | 45° welding elbow B | Of run, welding tee C | | Lengths F | Radius of fillet R | Diameter of lap G |
| 1 | 1.310 | $1\frac{1}{2}$ | $\frac{7}{8}$ | $1\frac{1}{2}$ | $1\frac{1}{2}$ | 4 | $\frac{1}{8}$ | 2 |
| $1\frac{1}{4}$ | 1.660 | $1\frac{7}{8}$ | 1 | $1\frac{7}{8}$ | $1\frac{1}{2}$ | 4 | $\frac{3}{16}$ | $2\frac{1}{2}$ |
| $1\frac{1}{2}$ | 1.900 | $2\frac{1}{4}$ | $1\frac{1}{8}$ | $2\frac{1}{4}$ | $1\frac{1}{2}$ | 4 | $\frac{1}{4}$ | $2\frac{7}{8}$ |
| 2 | 2.375 | 3 | $1\frac{3}{8}$ | $2\frac{1}{2}$ | $1\frac{1}{2}$ | 6 | $\frac{5}{16}$ | $3\frac{5}{8}$ |
| $2\frac{1}{2}$ | 2.875 | $3\frac{3}{4}$ | $1\frac{3}{4}$ | 3 | $1\frac{1}{2}$ | 6 | $\frac{5}{16}$ | $4\frac{1}{8}$ |
| 3 | 3.500 | $4\frac{1}{2}$ | 2 | $3\frac{3}{8}$ | 2 | 6 | $\frac{3}{8}$ | 5 |
| $3\frac{1}{2}$ | 4.000 | $5\frac{1}{4}$ | $2\frac{1}{4}$ | $3\frac{3}{4}$ | $2\frac{1}{2}$ | 6 | $\frac{3}{8}$ | $5\frac{1}{2}$ |
| 4 | 4.500 | 6 | $2\frac{1}{2}$ | $4\frac{1}{8}$ | $2\frac{1}{2}$ | 6 | $\frac{7}{16}$ | $6\frac{3}{16}$ |

*Butt-welding reducers*

CONCENTRIC REDUCER

ECCENTRIC REDUCER

| Nominal pipe size | Outside diam. at bevel | | End to end H | Nominal pipe size | Outside diam. at bevel | | End to end H |
|---|---|---|---|---|---|---|---|
| | Large end | Small end | | | Large end | Small end | |
| 1 × $\frac{3}{4}$ | 1.315 | 1.050 | 2 | 3 × $2\frac{1}{2}$ | 3.500 | 2.875 | $3\frac{1}{2}$ |
| 1 × $\frac{1}{2}$ | | 0.840 | | 3 × 2 | | 2.375 | |
| 1 × $\frac{3}{8}$ | | 0.675 | | 3 × $1\frac{1}{2}$ | | 1.900 | |
| | | | | 3 × $1\frac{1}{4}$ | | 1.660 | |
| $1\frac{1}{4}$ × 1 | 1.660 | 1.315 | 2 | $3\frac{1}{2}$ × 3 | 4.000 | 3.500 | 4 |
| $1\frac{1}{4}$ × $\frac{3}{4}$ | | 1.050 | | $3\frac{1}{2}$ × $2\frac{1}{2}$ | | 2.875 | |
| $1\frac{1}{4}$ × $\frac{1}{2}$ | | 0.840 | | $3\frac{1}{2}$ × 2 | | 2.375 | |
| | | | | $3\frac{1}{2}$ × $1\frac{1}{2}$ | | 1.900 | |
| $1\frac{1}{2}$ × $1\frac{1}{4}$ | 1.900 | 1.660 | $2\frac{1}{2}$ | $3\frac{1}{2}$ × $1\frac{1}{4}$ | | 1.660 | |
| $1\frac{1}{2}$ × 1 | | 1.315 | | 4 × $3\frac{1}{2}$ | 4.500 | 4.000 | 4 |
| $1\frac{1}{2}$ × $\frac{3}{4}$ | | 1.050 | | 4 × 3 | | 3.500 | |
| $1\frac{1}{2}$ × $\frac{1}{2}$ | | 0.840 | | 4 × $2\frac{1}{2}$ | | 2.875 | |
| 2 × $1\frac{1}{2}$ | 2.375 | 1.900 | 3 | 4 × 2 | | 2.375 | |
| 2 × $1\frac{1}{4}$ | | 1.660 | | 4 × $1\frac{1}{2}$ | | 1.900 | |
| 2 × 1 | | 1.315 | | 5 × 4 | 5.563 | 4.500 | 5 |
| 2 × $\frac{3}{4}$ | | 1.050 | | 5 × $3\frac{1}{2}$ | | 4.000 | |
| $2\frac{1}{2}$ × 2 | 2.875 | 2.375 | $3\frac{1}{2}$ | 5 × 3 | | 3.500 | |
| $2\frac{1}{2}$ × $1\frac{1}{2}$ | | 1.900 | | 5 × $2\frac{1}{2}$ | | 2.875 | |
| $2\frac{1}{2}$ × $1\frac{1}{4}$ | | 1.660 | | | | | |
| $2\frac{1}{2}$ × 1 | | 1.315 | | | | | |

* For larger sizes, see ANSI B16.3—.
† ANSI B16.9—Dimensions in inches.

## Threaded Cast-iron Pipe*

*Dimension of pipe and drainage hubs*

| Pipe size | Pipe | | | Drainage hubs | | | | | Nominal weights | |
| --- | --- | --- | --- | --- | --- | --- | --- | --- | --- | --- |
| | Nominal diam. | | Wall thick-ness, min., | Thread length* | Diam. of groove, max., | End to shoulder † | Min. band | | Type *A* and barrel of type *B* per foot | Addi-tional weight of hubs for type *B* |
| | | | | | | | Diam. | Length | | |
| | Outside D | Inside A | G | B | K | C | H | E | | |
| 1¼ | 1.66 | 1.23 | 0.187 | 0.42 | 1.73 | 0.71 | 2.39 | 0.71 | 3.033 | 0.60 |
| 1½ | 1.90 | 1.45 | 0.195 | 0.42 | 1.97 | 0.72 | 2.68 | 0.72 | 3.666 | 0.90 |
| 2 | 2.38 | 1.89 | 0.211 | 0.43 | 2.44 | 0.76 | 3.28 | 0.76 | 5.041 | 1.00 |
| 2½ | 2.88 | 2.32 | 0.241 | 0.68 | 2.97 | 1.14 | 3.86 | 1.14 | 7.032 | 1.35 |
| 3 | 3.50 | 2.90 | 0.263 | 0.76 | 3.60 | 1.20 | 4.62 | 1.20 | 9.410 | 2.80 |
| 4 | 4.50 | 3.83 | 0.294 | 0.84 | 4.60 | 1.30 | 5.79 | 1.30 | 13.751 | 3.48 |
| 5 | 5.56 | 4.81 | 0.328 | 0.93 | 5.66 | 1.41 | 7.05 | 1.41 | 19.069 | 5.00 |
| 6 | 6.63 | 5.76 | 0.378 | 0.95 | 6.72 | 1.51 | 8.28 | 1.51 | 26.223 | 6.60 |
| 8 | 8.63 | 7.63 | 0.438 | 1.06 | 8.72 | 1.71 | 10.63 | 1.71 | 39.820 | 10.00 |
| 10 | 10.75 | 9.75 | 0.438 | 1.21 | 10.85 | 1.92 | 13.12 | 1.93 | 50.234 | |
| 12 | 12.75 | 11.75 | 0.438 | 1.36 | 12.85 | 2.12 | 15.47 | 2.13 | 60.036 | |

* ANSI A40.5—All dimensions are given in inches, except where otherwise stated. Type *A* has external threads both ends. Type *B* as shown.

† The length of thread *B* and the end to shoulder *C* shall not vary from the dimensions shown by more than plus or minus the equivalent of the pitch of one thread.

## ANSI Standard Class-125 Cast-iron and 150-lb Steel Valves: Valves with Class-125 or 150-lb End Flanges, or with Welding Ends—Face-to-face and End-to-end Dimensions[a]

| Nominal valve size | Class-125 cast iron — Flanged end—plain face | | | | | | | | 150-lb steel — Flanged end (1/16 in. raised face) and welding end | | | | | | | |
|---|---|---|---|---|---|---|---|---|---|---|---|---|---|---|---|---|
| | Gate | | Plug | | | Globe and lift check | Swing check[b] | Control | Gate | | | | Plug | | | |
| | Solid wedge | Double disc | Short pattern | Regular | Venturi | | | | Solid wedge | Double disc | Solid wedge | Double disc | Short pattern | Regular | Venturi | Round port full bore |
| | A | A | A | A | A | A | A | A | A | A | B | B | A | A | A | A |
| ¼ | ... | ... | ... | ... | ... | ... | ... | ... | 4 | 4 | 4 | 4 | ... | ... | ... | ... |
| ⅜ | ... | ... | ... | ... | ... | ... | ... | ... | 4 | 4 | 4 | 4 | ... | ... | ... | ... |
| ½ | ... | ... | ... | ... | ... | ... | ... | ... | 4¼ | 4¼ | 4¼ | 4¼ | ... | ... | ... | ... |
| ¾ | ... | ... | ... | ... | ... | ... | ... | ... | 4⅝ | 4⅝ | 4⅝ | 4⅝ | ... | ... | ... | ... |
| 1 | ... | ... | 5½ | 5½ | ... | ... | ... | 7¼ | 5 | 5 | 5 | 5 | 5½ | ... | ... | 7 |
| 1¼ | ... | ... | ... | 6½ | ... | ... | ... | ... | 5½ | 5½ | 5½ | 5½ | ... | ... | ... | ... |
| 1½ | ... | ... | 6½ | 6½ | ... | ... | ... | 8¾ | 6½ | 6½ | 6½ | 6½ | 6½ | ... | ... | 8¾ |
| 2 | 7 | 7 | 7 | 7½ | ... | 8 | 8 | 10 | 7 | 7 | 8½ | 8½ | 7 | ... | ... | 10½ |
| 2½ | 7½ | 7½ | 7½ | 8¼ | ... | 8½ | 8½ | 10⅞ | 7½ | 7½ | 9½ | 9½ | 7½ | ... | ... | 11¾ |
| 3 | 8 | 8 | 8 | 9 | ... | 9½ | 9½ | 11¾ | 8 | 8 | 11⅛ | 11⅛ | 8 | ... | ... | 13½ |
| 3½ | 8½ | 8½ | ... | ... | ... | ... | ... | ... | ... | ... | ... | ... | ... | ... | ... | ... |
| 4 | 9 | 9 | 9 | 9 | ... | 11½ | 11½ | 13⅞ | 9 | 9 | 12 | 12 | 9 | ... | ... | 17 |
| 5 | 10 | 10 | 10 | 14 | ... | 13 | 13 | ... | 10 | 10 | 15 | 15 | ... | ... | ... | ... |
| 6 | 10½ | 10½ | 10½ | 15½ | 15½ | 14 | 14 | 17¾ | 10½ | 10½ | 15⅞ | 15⅞ | 10½ | 15½ | ... | 21 |
| 8 | 11½ | 11½ | 11½ | 18 | 18 | 19½ | 19½ | 21⅜ | 11½ | 11½ | 16½ | 16½ | 11½ | 18 | ... | 25 |

### 150-lb steel

| Nominal valve size | Flanged end (1/16 in. raised face) and welding end | | | | Flanged end (ring joint) | | | | | | | |
|---|---|---|---|---|---|---|---|---|---|---|---|---|
| | Globe and lift check | Swing check | "Y" pattern globe | Control | Gate | | Plug | | | Round port full bore | Globe and lift check | Swing check |
| | | | | | Solid wedge | Double disc | Short pattern | Regular | Venturi | | | |
| | A&B | A&B | A&B | A | C | C | C | C | C | C | C | C |
| ¼ | 4 | 4 | ... | ... | ... | ... | ... | ... | ... | ... | ... | ... |
| ⅜ | 4 | 4 | ... | ... | ... | ... | ... | ... | ... | ... | ... | ... |
| ½ | 4¼ | 4¼ | 5½ | ... | 4¹¹⁄₁₆ | 4¹¹⁄₁₆ | ... | ... | ... | ... | 4¹¹⁄₁₆ | 4¹¹⁄₁₆ |
| ¾ | 4⅝ | 4⅝ | 6 | ... | 5⅛ | 5⅛ | ... | ... | ... | ... | 5⅛ | 5⅛ |
| 1 | 5 | 5 | 6½ | 7¼ | 5½ | 5½ | 6 | ... | ... | 7½ | 5½ | 5½ |
| 1¼ | 5½ | 5½ | 7¼ | ... | 6 | 6 | ... | ... | ... | ... | 6 | 6 |
| 1½ | 6½ | 6½ | 8 | 8¾ | 7 | 7 | 7 | ... | ... | 9¼ | 7 | 7 |
| 2 | 8 | 8 | 9 | 10 | 7½ | 7½ | 7½ | ... | ... | 11 | 8½ | 8½ |
| 2½ | 8½ | 8½ | 11 | 10⅞ | 8 | 8 | 8 | ... | ... | 12¼ | 9 | 9 |
| 3 | 9½ | 9½ | 12½ | 11¾ | 8½ | 8½ | 8½ | ... | ... | 14 | ... | 10 |
| 3½ | ... | ... | ... | ... | ... | ... | ... | ... | ... | ... | ... | ... |
| 4 | 11½ | 11½ | 14½ | 13⅞ | 9½ | 9½ | 9½ | ... | ... | 17½ | 12 | 12 |
| 5 | 14 | 13 | ... | ... | 10½ | 10½ | ... | ... | ... | ... | 14½ | 13½ |
| 6 | 16 | 14 | 18½ | 17¾ | 11 | 11 | 11 | 16 | ... | 21½ | 16½ | 14½ |
| 8 | 19½ | 19½ | 23½ | 21⅜ | 12 | 12 | 12 | 18½ | ... | 25½ | 20 | 20 |

[a] Adapted from ANSI B16.10—API Standards 6D and 600 conform to the dimensions shown for corresponding sizes, valve type and flange class or welding end.

[b] These dimensions are not intended to cover the type of check valve having the seat angle at approximately 45 deg to the run of the valve or the "Underwriter Pattern" or other patterns where large clearances are required.

## Beam Connections

## Driving Clearances for Riveting

|   | Diam. of rivet | | | | | | | | |
|---|---|---|---|---|---|---|---|---|---|
|   | $\frac{1}{2}$ | $\frac{5}{8}$ | $\frac{3}{4}$ | $\frac{7}{8}$ | 1 | $1\frac{1}{8}$ | $1\frac{1}{4}$ | $1\frac{3}{8}$ | $1\frac{1}{2}$ |
| $D$ | $1\frac{3}{4}$ | 2 | $2\frac{1}{4}$ | $2\frac{1}{2}$ | $2\frac{3}{4}$ | 3 | $3\frac{1}{4}$ | $3\frac{1}{2}$ | $3\frac{3}{4}$ |
| $C$ | 1 | $1\frac{1}{8}$ | $1\frac{1}{4}$ | $1\frac{3}{8}$ | $1\frac{1}{2}$ | $1\frac{5}{8}$ | $1\frac{3}{4}$ | $1\frac{7}{8}$ | 2 |

Dimensions in inches.

## Gage and Maximum Rivet Size for Angles

| Leg | 8 | 7 | 6 | 5 | 4 | $3\frac{1}{2}$ | 3 | $2\frac{1}{2}$ | 2 | $1\frac{3}{4}$ | $1\frac{1}{2}$ | $1\frac{3}{8}$ | $1\frac{1}{4}$ | 1 |
|---|---|---|---|---|---|---|---|---|---|---|---|---|---|---|
| $g$ | $4\frac{1}{2}$ | 4 | $3\frac{1}{2}$ | 3 | $2\frac{1}{2}$ | 2 | $1\frac{3}{4}$ | $1\frac{3}{8}$ | $1\frac{1}{8}$ | 1 | $\frac{7}{8}$ | $\frac{7}{8}$ | $\frac{3}{4}$ | $\frac{5}{8}$ |
| $g_1$ | 3 | $2\frac{1}{2}$ | $2\frac{1}{4}$ | 2 | | | | | | | | | | |
| $g_2$ | 3 | 3 | $2\frac{1}{2}$ | $1\frac{3}{4}$ | | | | | | | | | | |
| Max. rivet | $1\frac{1}{8}$ | $1\frac{1}{8}$ | 1 | 1 | $\frac{7}{8}$ | $\frac{7}{8}$ | $\frac{7}{8}$ | $\frac{3}{4}$ | $\frac{5}{8}$ | $\frac{1}{2}$ | $\frac{3}{8}$ | $\frac{3}{8}$ | $\frac{3}{8}$ | $\frac{1}{4}$ |

Dimensions in inches.

## Selected Structural Shapes*

*Dimensions for detailing*

| Name | Depth of section, in. | Weight per foot, lb | Flange Width, in. | Flange Mean thickness, in. | Web Thickness, in. | Web Half thickness, in. | $T$, in. | $k$, in. | $g_1$, in. | $c$, in. | Grip, in. | Max. flange rivet, in. | Usual gage $g$, in. | Clearance $m$ |
|---|---|---|---|---|---|---|---|---|---|---|---|---|---|---|
| Channels | 18 | 58.0 | 4¼ | ⅝ | 11/16 | ⅜ | 15⅜ | 1 5/16 | 2¾ | ¾ | ⅝ | 1 | 2½ | |
| | 15 | 40.0 | 3½ | ⅝ | 9/16 | ¼ | 12⅜ | 1 5/16 | 2¾ | ⅝ | ⅝ | 1 | 2 | |
| | 12 | 30.0 | 3⅛ | ½ | ½ | ¼ | 9⅞ | 1 1/16 | 2½ | 9/16 | ½ | ⅞ | 1¾ | |
| | 10 | 15.3 | 2⅝ | 7/16 | ¼ | ⅛ | 8⅛ | 15/16 | 2½ | 5/16 | 7/16 | ¾ | 1½ | |
| | 9 | 13.4 | 2⅜ | 7/16 | ¼ | ⅛ | 7¼ | ⅞ | 2½ | 5/16 | ⅜ | ¾ | 1⅜ | |
| | 8 | 18.75 | 2½ | ⅜ | ½ | ¼ | 6⅜ | 13/16 | 2¼ | 9/16 | ⅜ | ¾ | 1½ | |
| | 7 | 12.25 | 2¼ | ⅜ | 5/16 | 3/16 | 5⅜ | 13/16 | 2 | ⅜ | ⅜ | ⅝ | 1¼ | |
| | 6 | 10.5 | 2 | ⅜ | 5/16 | 3/16 | 4½ | ¾ | 2 | ⅜ | ⅜ | ⅝ | 1⅛ | |
| | 5 | 9.0 | 1⅞ | 5/16 | 5/16 | 3/16 | 3⅝ | 11/16 | 2 | ⅜ | 5/16 | ½ | 1⅛ | |
| | 4 | 7.25 | 1¾ | 5/16 | 5/16 | 3/16 | 2¾ | ⅝ | 2 | ⅜ | 5/16 | ½ | 1 | |
| | 3 | 6.0 | 1⅝ | ¼ | ⅜ | 3/16 | 1¾ | ⅝ | ... | 7/16 | 5/16 | ½ | ⅞ | |
| WF shapes | 21†(21¼) | 127 | 13 | 1 | 9/16 | 5/16 | 17¾ | 1 ¾ | 3 | ⅜ | ..... | | 5½ | 25 |
| | 16 (16⅜) | 96 | 11½ | ⅞ | 9/16 | 5/16 | 13⅛ | 1 ⅝ | 2¾ | ⅜ | ..... | | 5½ | 20 |
| | 14 (14⅛) | 84 | 12 | ¾ | 7/16 | ¼ | 11⅜ | 1 ⅜ | 2¾ | 5/16 | ..... | | 5½ | 18⅝ |
| | 14 (13¾) | 48 | 8 | 9/16 | ⅜ | 3/16 | 11⅜ | 1 3/16 | 2½ | ¼ | ..... | | 5½ | 16 |
| | 12 (12¼) | 50 | 8⅛ | ⅝ | ⅜ | 3/16 | 9¾ | 1 ¼ | 2½ | ¼ | ..... | | 5½ | 14⅝ |
| | 10 (10) | 49 | 10 | 9/16 | ⅜ | 3/16 | 7⅞ | 1 1/16 | 2½ | ¼ | ..... | | 5½ | 14⅛ |
| | 10 (9¾) | 33 | 8 | 7/16 | 5/16 | 3/16 | 7⅞ | 15/16 | 2¼ | ¼ | ..... | | 5½ | 12⅝ |
| | 8 (8) | 28 | 6½ | 7/16 | 5/16 | ⅛ | 6⅜ | 13/16 | 2¼ | 3/16 | ..... | | 3½ | 10½ |
| Beams | 24 | 120.0 | 8 | 1 ⅛ | 13/16 | 7/16 | 20⅛ | 1 15/16 | 3¼ | ½ | 1 ⅛ | 1 | 4 | |
| | 20 | 85.0 | 7 | 15/16 | 11/16 | 5/16 | 16½ | 1 ¾ | 3¼ | ⅜ | ⅞ | 1 | 4 | |
| | 18 | 70.0 | 6¼ | 11/16 | ¾ | ⅜ | 15¼ | 1 ⅜ | 2¾ | 7/16 | 11/16 | ⅞ | 3½ | |
| | 15 | 50.0 | 5⅝ | ⅝ | 9/16 | 5/16 | 12½ | 1 ¼ | 2¾ | ⅜ | 9/16 | ¾ | 3½ | |
| | 12 | 31.8 | 5 | 9/16 | ⅜ | 3/16 | 9¾ | 1 ⅛ | 2½ | ¼ | ½ | ¾ | 3 | |
| | 10 | 35.0 | 5 | ½ | ⅝ | 5/16 | 8 | 1 | 2½ | ⅜ | ½ | ¾ | 2¾ | |
| | 8 | 23.0 | 4⅛ | 7/16 | 7/16 | ¼ | 6¼ | ⅞ | 2¼ | 5/16 | 7/16 | ¾ | 2¼ | |
| | 7 | 20.0 | 3⅞ | ⅜ | 7/16 | ¼ | 5⅜ | 13/16 | 2 | 5/16 | ⅜ | ⅝ | 2¼ | |
| | 6 | 17.25 | 3⅝ | ⅜ | ½ | ¼ | 4½ | ¾ | 2 | 5/16 | ⅜ | ⅝ | 2 | |
| | 5 | 10.0 | 3 | 5/16 | ¼ | ⅛ | 3⅝ | 11/16 | 2 | 3/16 | 5/16 | ½ | 1¾ | |
| | 4 | 9.5 | 2¾ | 5/16 | 5/16 | 3/16 | 2¾ | ⅝ | 2 | ¼ | 5/16 | ½ | 1½ | |
| | 3 | 7.5 | 2½ | ¼ | ⅜ | 3/16 | 1⅞ | 9/16 | ... | ¼ | ¼ | ⅜ | 1½ | |

CHANNEL

WF SHAPE

BEAM

* From Steel Construction Handbook.
† Nominal depth; ( ) indicates actual depth.

## Standard Jig Bushings*

*RENEWABLE SLIP*

*RENEWABLE FIXED*

*LINER
(used with Renewable
Type bushings)*

| Renewable Slip Type and Renewable Fixed Type | | | | | | | | | | | | | Liners | | |
|---|---|---|---|---|---|---|---|---|---|---|---|---|---|---|---|
| | | | | | Head dimensions | | | | | | | | | | |
| Slip type Hole size A | Fixed type Hole size A | Tolerance on hole | Body diameter limits B | Lengths available C | F | G Slip type | G Fixed type | H | J | L | R | Lock screw no. | Hole limits A | OD limits B | Lengths |
| 0.052 to 0.089 | 0.055 to 0.089 | +0.0004 +0.0001 | 0.3125 / 0.3123 | 5/16, 1/2, 3/4, 1 | 35/64 | 3/8 | 1/4 | 1/8 | 11/64 | 65° | 1/2 | | 0.3126 / 0.3129 | 0.5017 / 0.5014 | Same as the bushing |
| 0.0935 to 0.1562 | 0.0935 to 0.1562 | +0.0004 +0.0001 | 0.3125 / 0.3123 | 5/16, 1/2, 3/4, 1 | 35/64 | 3/8 | 1/4 | 1/8 | 11/64 | 65° | 1/2 | | 0.3126 / 0.3129 | 0.5017 / 0.5014 | |
| 0.1406 to 0.3437 | 0.1570 to 0.3125 | Incl 1/4 +0.0004 +0.0001 Over 1/4 +0.0005 +0.0001 | 0.5000 / 0.4998 | 5/16, 1/2, 3/4, 1, 1 3/8, 1 3/4 | 51/64 | 7/16 | 1/4 | 1/8 | 19/64 | 65° | 5/8 | 1 | 0.5002 / 0.5005 | 0.7518 / 0.7515 | |
| 0.2812 to 0.5312 | 0.3160 to 0.5000 | +0.0005 +0.0001 | 0.7500 / 0.7498 | 1/2, 3/8, 1, 1 3/8, 1 3/4, 2 1/8 | 1 3/64 | 7/16 | 1/4 | 1/8 | 27/64 | 50° | 3/4 | | 0.7503 / 0.7506 | 1.0015 / 1.0018 | |
| 0.4687 to 0.7812 | 0.5156 to 0.750 | +0.0005 +0.0001 | 1.0000 / 0.9998 | 3/4, 1, 1 3/8, 1 3/4, 2 1/8, 2 1/2 | 1 27/64 | 7/16 | 3/8 | 3/16 | 19/32 | 35° | 59/64 | 2 | 1.0004 / 1.0007 | 1.3772 / 1.3768 | |
| 0.7817 to 1.0312 | 0.7656 to 1.0000 | +0.0006 +0.0002 | 1.3750 / 1.3747 | 3/4, 1, 1 3/8, 1 3/4, 2 1/8, 2 1/2 | 1 5/64 | 7/16 | 3/8 | 3/16 | 25/32 | 30° | 1 7/64 | 2 | 1.3756 / 1.3760 | 1.7523 / 1.7519 | |
| 0.9687 to 1.4062 | 1.0156 to 1.3750 | +0.0006 +0.0002 | 1.7500 / 1.7497 | 1, 1 3/8, 1 3/4, 2 1/8, 2 1/2, 3 | 2 19/64 | 5/8 | 3/8 | 3/16 | 1 | 30° | 1 25/64 | 3 | 1.7508 / 1.7512 | 2.2521 / 2.2525 | |
| 1.3437 to 1.7812 | 1.3906 to 1.750 | Incl 1 1/2 +0.0006 +0.0002 Over 1 1/2 +0.0007 +0.0003 | 2.2500 / 2.2496 | 1, 1 3/8, 1 3/4, 2 1/8, 2 1/2, 3 | 2 5/64 | 5/8 | 3/8 | 3/16 | 1 1/4 | 25° | 1 41/64 | 3 | 2.2510 / 2.2515 | 2.7526 / 2.7522 | |

Dimensions in inches.
*ANSI B5.6—.
Head design in accordance with manufacturer's practice; slip type usually knurled.

## Lock Screws

| Screw No. | A | B | C | D | E | F | ASA thd |
|---|---|---|---|---|---|---|---|
| 1 | 3/8 | 3/8 | 5/8 | 1/16 | 1/4 | .138 / .132 | 5/16–18 |
| 2 | 7/8 | 3/8 | 5/8 | 3/32 | 3/8 | .200 / .194 | 5/16–18 |
| 3 | 1 | 7/16 | 3/4 | 1/8 | 3/8 | .200 / .194 | 3/8–16 |

## Standard Jig Bushings (Cont.)

| Press-fit Headless and Press-fit Head Types | | | | | |
|---|---|---|---|---|---|
| Hole size *A* | Tolerance on hole | OD limits *B* | Lengths available *C* | Head type dimension *F* | *G* |
| 0.055 to 0.0995 | +0.0004 +0.0001 | $\frac{0.2046}{0.2043}$ | �5⁄16, ½ | 1⁹⁄64 | ³⁄32 |
| 0.1015 to 0.1360 | | $\frac{0.2516}{0.2513}$ | ⁵⁄16, ½ | 2³⁄64 | ³⁄32 |
| 0.1405 to 0.1875 | | $\frac{0.3141}{0.3138}$ | ⁵⁄16, ½ <br> ¾, 1 | 2⁷⁄64 | ⅛ |
| 0.1890 to 0.2500 | | $\frac{0.4078}{0.4075}$ | ⁵⁄16, ½ | ½ | ⁵⁄32 |
| 0.2570 to 0.3125 | +0.0005 +0.0001 | $\frac{0.5017}{0.5014}$ | ¾, 1 <br> 1⅜, 1¾ | 3⁹⁄64 | ⁷⁄32 |
| 0.316 to 0.4219 | | $\frac{0.6267}{0.6264}$ | ½, ¾ | 5¹⁄64 | ⁷⁄32 |
| 0.4375 to 0.500 | | $\frac{0.7518}{0.7515}$ | 1, 1⅜ <br> 1¾, 2⅛ | 5⁹⁄64 | ⁷⁄32 |
| 0.5156 to 0.625 | | $\frac{0.8768}{0.8765}$ | ¾, 1 | 1⁷⁄64 | ¼ |
| 0.6406 to 0.7500 | | $\frac{1.0018}{1.0015}$ | 1⅜, 1¾ | 1¹⁵⁄64 | ⁵⁄16 |
| 0.7656 to 1.0000 | +0.0006 +0.0002 | $\frac{1.3772}{1.3768}$ | 2⅛, 2½ | 1³⁹⁄64 | ⅜ |
| 1.0156 to 1.3750 | | $\frac{1.7523}{1.7519}$ | | 1⁶³⁄64 | ⅜ |
| 1.3906 to 1.7500 | Incl 1½ +0.0006 +0.0002 <br> Over 1½ +0.0007 +0.0003 | $\frac{2.2525}{2.2521}$ | 1, 1⅜ <br> 1¾, 2⅛ <br> 2½, 3 | 2³¹⁄64 | ⅜ |

*PRESS FIT HEADLESS TYPE*

*PRESS FIT HEAD TYPE*

## ANSI Standard Brass Solder-joint Fittings*

90° ELL

45° ELL

CROSS

COUPLING     ECCENTRIC COUPLING

TEE

| Tubing size | A | B | C | D | Reducing size | F | E | G |
|---|---|---|---|---|---|---|---|---|
| 3/8 | 5/16 | 1 1/16 | 3/16 | 1/8 | 3/4 × 1/2 | 5/8 | 1 | 1 3/16 |
| 1/2 | 7/16 | 1 3/16 | 3/16 | 1/8 | 1 × 3/4 | 1 1/16 | 1 1/16 | 1 |
| 3/4 | 9/16 | 1 | 1/4 | 1/8 | 1 × 1/2 | 1/2 | 1 1/16 | 1 3/16 |
| 1 | 3/4 | 1 1/16 | 5/16 | 1/8 | 2 × 1 1/2 | 1 1/8 | 1 3/8 | 1 3/16 |
| 1 1/2 | 1 | 1 3/16 | 1/2 | 1/8 | 2 1/2 × 2 | 1 3/16 | 1 5/8 | 1 3/8 |
| 2 | 1 1/4 | 1 3/8 | 9/16 | 3/16 | 2 1/2 × 1 1/2 | 1 5/16 | 1 5/8 | 1 3/16 |
| 2 1/2 | 1 1/2 | 1 5/8 | 5/8 | 3/16 | 3 × 2 1/2 | 1 1/4 | 1 7/8 | 1 5/8 |
| 3 | 1 3/4 | 1 7/8 | 3/4 | 3/16 | 3 1/2 × 3 | 1 1/8 | 2 1/16 | 1 7/8 |
| 3 1/2 | 2 | 2 1/16 | 7/8 | 3/16 | 3 1/2 × 2 1/2 | 1 5/16 | 2 1/16 | 1 5/8 |
| 4 | 2 1/4 | 2 1/4 | 15/16 | 1/4 | 4 × 3 1/2 | 1 3/16 | 2 1/4 | 2 1/16 |

* Adapted from ANSI B16.18—.

## SAE STANDARD COTTER PINS

STANDARD

MITER   EXTENDED MITER   DOUBLE MITER

SQUARE CUT   BEVEL POINT   HAMMER LOCK

| | Pin diameter A | | | Eye diameter, min. | | Recommended hole diam., drill size |
|---|---|---|---|---|---|---|
| | Nominal | Max. | Min. | Inside B | Outside C | |
| | 1/32 (0.031) | 0.032 | 0.028 | 1/32 | 1/16 | 3/64 (0.0469) |
| | 3/64 (0.047) | 0.048 | 0.044 | 3/64 | 3/32 | 1/16 (0.0625) |
| | 1/16 (0.062) | 0.060 | 0.056 | 1/16 | 1/8 | 5/64 (0.0781) |
| | 5/64 (0.078) | 0.076 | 0.072 | 5/64 | 5/32 | 3/32 (0.0937) |
| | 3/32 (0.094) | 0.090 | 0.086 | 3/32 | 3/16 | 7/64 (0.1094) |
| | 1/8 (0.125) | 0.120 | 0.116 | 1/8 | 1/4 | 9/64 (0.1406) |
| | 5/32 (0.156) | 0.150 | 0.146 | 5/32 | 5/16 | 11/64 (0.1719) |
| | 3/16 (0.188) | 0.176 | 0.172 | 3/16 | 3/8 | 13/64 (0.2031) |
| | 7/32 (0.219) | 0.207 | 0.202 | 7/32 | 7/16 | 15/64 (0.2344) |
| | 1/4 (0.250) | 0.225 | 0.220 | 1/4 | 1/2 | 17/64 (0.2656) |
| | 5/16 (0.312) | 0.280 | 0.275 | 5/16 | 5/8 | 5/16 (0.3125) |
| | 3/8 (0.375) | 0.335 | 0.329 | 3/8 | 3/4 | 3/8 (0.3750) |
| | 1/2 (0.500) | 0.473 | 0.467 | 1/2 | 1 | 1/2 (0.5000) |

# ANSI Standard Graphic Symbols for Piping and Heating *

| PIPING | | |
|---|---|---|
| Piping, General | (Lettered with name of material conveyed) | |
| Non-intersecting Pipes | | |
| (To differentiate lines of piping on a drawing the following symbols may be used.) | | |

| Air | Cold Water | Steam |
|---|---|---|
| Gas | Hot Water | Condensate |
| Oil | Vacuum | Refrigerant |

| PIPE FITTINGS AND VALVES | | | | | |
|---|---|---|---|---|---|
|  | Flanged | Screwed | Bell and Spigot | Welded | Soldered |
| Joint | | | | | |
| Elbow—90 deg | | | | | |
| Elbow—45 deg | | | | | |
| Elbow—Turned Up | | | | | |
| Elbow—Turned Down | | | | | |
| Elbow—Long Radius | | | | | |
| Side Outlet Elbow Outlet Down | | | | | |
| Side Outlet Elbow Outlet Up | | | | | |
| Base Elbow | | | | | |
| Double Branch Elbow | | | | | |
| Reducing Elbow | | | | | |
| Reducer | | | | | |
| Eccentric Reducer | | | | | |
| Tee-Outlet Up | | | | | |
| Tee-Outlet Down | | | | | |
| Tee | | | | | |
| Side Outlet Tee Outlet Up | | | | | |
| Side Outlet Tee Outlet Down | | | | | |
| Single Sweep Tee | | | | | |
| Double Sweep Tee | | | | | |
| Cross | | | | | |
| Lateral | | | | | |
| Gate Valve | | | | | |

* ANSI Y32.2.3—.

# Symbols for Piping and Heating (Cont.)

## PIPING

| | Flanged | Screwed | Bell and Spigot | Welded | Soldered |
|---|---|---|---|---|---|
| Globe Valve | | | | | |
| Angle Globe Valve | | | | | |
| Angle Gate Valve | | | | | |
| Check Valve | | | | | |
| Angle Check Valve | | | | | |
| Stop Cock | | | | | |
| Safety Valve | | | | | |
| Quick Opening Valve | | | | | |
| Float Operating Valve | | | | | |
| Motor Operated Gate Valve | | | | | |
| Motor Operated Globe Valve | | | | | |
| Expansion Joint Flanged | | | | | |
| Reducing Flange | | | | | |
| Union | (See Joint) | | | | |
| Sleeve | | | | | |
| Bushing | | | | | |

## HEATING AND VENTILATING

| | | | | | |
|---|---|---|---|---|---|
| Lock and Shield Valve | | Tube Radiator (Plan) (Elev.) | | Exhaust Duct, Section | |
| Reducing Valve | | Wall Radiator (Plan) (Elev.) | | Butterfly Damper (Plan or Elev.) (Elev. or Plan) | |
| Diaphragm Valve | | Pipe Coil (Plan) (Elev.) | | Deflecting Damper Rectangular Pipe | |
| Thermostat | (T) | Indirect Radiator (Plan) (Elev.) | | Vanes | |
| Radiator Trap (Plan) (Elev.) | | Supply Duct, Section | | Air Supply Outlet | |
| | | | | Exhaust Inlet | |

## HEAT-POWER APPARATUS

| | | | | | |
|---|---|---|---|---|---|
| Flue Gas Reheater (Intermediate Superheater) | | Steam Turbine | | Automatic By-pass Valve | |
| Steam Generator (Boiler) | | Condensing Turbine | | Automatic Valve Operated by Governor | |
| Live Steam Superheater | | Open Tank | | Pumps Air Service Boiler Feed Condensate Circulating Water Reciprocating | |
| Feed Heater With Air Outlet | | Closed Tank | | | |
| Surface Condenser | | Automatic Reducing Valve | | Dynamic Pump (Air Ejector) | |

## ANSI Standard Plumbing Symbols *

| | | | |
|---|---|---|---|
| Corner Bath | Recessed Bath | Roll Rim Bath | S B<br>Sitz Bath |
| F B<br>Foot Bath | B<br>Bidet | Shower Stall | (Plan)  (Elev.)<br>Shower Head |
| (Plan)<br>(Elev.)<br>Overhead Gang Shower | M L<br>Manicure Lavatory<br>Medical Lavatory | LAV<br>Corner Lavatory | W L<br>Wall Lavatory |
| DENTAL LAV<br>Dental Lavatory | S<br>Plain Kitchen Sink | Kitchen Sink<br>R&L Drain Board | P L<br>Pedestal Lavatory |
| Kitchen Sink<br>L.H. Drain Board | Combination Sink<br>and Dishwasher | S& T<br>Combination Sink<br>and Laundry Tray | S S<br>Service Sink |
| L T<br>Laundry Tray | Wash Sink<br>(Wall Type) | Water Closet<br>(No Tank) | Wash Sink |
| Water Closet (Low Tank) | Urinal<br>(Pedestal Type) | Urinal<br>(Corner Type) | Urinal<br>(Wall Type) |
| DF<br>Drinking Fountain<br>(Pedestal Type) | DF<br>Drinking Fountain<br>(Wall Type) | Urinal<br>(Stall Type) | T U<br>Urinal<br>(Trough Type) |
| DF<br>Drinking Fountain<br>(Trough Type) | HW T<br>Hot Water Tank | WH<br>Water Heater | M<br>Meter | HR<br>Hose Rack |
| Vacuum Outlet | HB<br>Hose Bib | G<br>Gas Outlet | D<br>Drain | G<br>Grease Separator |
| O<br>Oil Separator | C O<br>Cleanout | Garage Drain | Floor Drain With<br>Backwater Valve | Roof Sump |

* ANSI Y32.4—.

## Wiring Symbols for Architecture

Ceiling Outlet

" " for Extensions

" Lamp Receptacle, Specifications to describe type, as Key, Keyless or Pull Chain

Ceiling Fan Outlet

Pull Switch

Drop Cord

Wall Bracket

" Outlet for Extensions

" Lamp Receptacle, as specified

" Fan Outlet

Single Convenience Outlet

Double " " "

Junction Box

Special Purpose Outlets
Lighting, Heating and Power
as described in specifications

Exit Light

Floor Outlet

Floor Elbow $O^E$, Floor Tee $O^T$

Local Switch, Single Pole $S^1$
Double Pole $S^2$, 3-Way $S^3$, 4-Way $S^4$

Automatic Door Switch $S^D$

Key Push Button Switch $S^K$

Electrolier Switch $S^E$

Push Button Switch and Pilot $S^P$

Remote Control Push Button Switch $S^R$

Tank Switch — T.S.

Motor, Motor Controller — M.C.

Lighting Panel

Power Panel

Heating Panel

Pull Box

Cable Supporting Box

Meter

Transformer

Push Button

Pole Line

Buzzer, Bell

Annunciator

Branch Circuit, Run Exposed -----

Run Concealed Under Floor — — —

" " " Floor Above — · — · —

Feeder Run Exposed ══════

Run Concealed Under Floor ───────

" " " Floor Above —··—··—

Telephone, Interior, Public

Clock, Secondary, Master

Time Stamp

Electric Door Opener

Local Fire Alarm Gong

City Fire Alarm Station

Local " " "

Fire Alarm Central Station

Speaking Tube

Nurse's Signal Plug — N

Maid's Plug — M

Horn Outlet

District Messenger Call

Watchman Station — W

Watchman Central Station Detector — W

Public Telephone - P.B.X. Switchboard — PBX

Interior Telephone Central Switchboard — IX

Interconnection Cabinet

Telephone Cabinet

Telegraph "

Special Outlet for Signal System as Specified

Battery —|||||||||||—

Signal Wires in Conduit Under Floor ————

" " " " " Floor Above ————

This Character Marked on Tap Circuits Indicates

2 No.14 Conductors in $\frac{1}{2}$" Conduit          ||

3 " 14    "    " $\frac{1}{2}$" "          |||

4 " 14    "    " $\frac{3}{4}$" ( Unless Marked $\frac{1}{2}$" )   ||||

5 " 14    "    " $\frac{3}{4}$" "          |||||

6 " 14    "    " 1" ( Unless Marked $\frac{3}{4}$" )   ||||||

7 " 14    "    " 1" "          |||||||

8 " 14    "    " 1" "          ||||||||

(Radio Outlet )

(Public Speaker Outlet )

# ANSI Standard Graphic Symbols for Electrical Diagrams *

*Single-line symbols are shown at the left, complete symbols at the right, and symbols for both purposes are centered in each column.*
*Listing is alphabetical.*

## ADJUSTABLE
## CONTINUOUSLY ADJUSTABLE (Variable)
*The shaft of the arrow is drawn at about 45 degrees across the body of the symbol.*

## AMPLIFIER
See also MACHINE, ROTATING

General
The triangle is pointed in the direction of transmission.

Amplifier type may be indicated in the triangle by words, standard abbreviations, or a letter combination from the following list.

| | | | |
|---|---|---|---|
| BDG | Bridging | MON | Monitoring |
| BST | Booster | PGM | Program |
| CMP | Compression | PRE | Preliminary |
| DC | Direct Current | PWR | Power |
| EXP | Expansion | TRQ | Torque |
| LIM | Limiting | | |

Applications

Booster amplifier with two inputs

Monitoring amplifier with two outputs

Amplifier with associated power supply

## ANTENNA

General
Types or functions may be indicated by words or abbreviations adjacent to the symbol.

Dipole

## ARRESTER (Electric Surge, Lightning, etc.)
## GAP

General

Carbon block
*The sides of the rectangle are to be approximately in the ratio of 1 to 2 and the space between rectangles shall be approximately equal to the width of a rectangle.*

Electrolytic or aluminum cell
*This symbol is not composed of arrowheads.*

Protective gap
*These arrowheads shall not be filled.*

Sphere gap

Multigap, general

## ATTENUATOR
See also PAD

General

Balanced, general

Unbalanced, general

## BATTERY
The long line is always positive, but polarity may be indicated in addition.
Example:

Generalized direct-current source

## BREAKER, CIRCUIT
If it is desired to show the condition causing the breaker to trip, the relay-protective-function symbols      may be used alongside the breaker symbol.

General
Note 1—Use appropriate number of single-line diagram symbols.

SEE NOTE I

Air or, if distinction is needed, for alternating-current circuit breaker rated at 1,500 volts or less and for direct-current circuit breaker.

SEE NOTE I

## CAPACITOR
See also TERMINATION

General
If it is necessary to identify the capacitor electrodes, the curved element shall represent the outside electrode in fixed paper-dielectric and ceramic-dielectric capacitors, the negative electrode in electrolytic capacitors, the moving element in adjustable and variable capacitors, and the low-potential element in feed-through capacitors.

Application: shielded capacitor

Application: adjustable or variable capacitor

If it is necessary to identify trimmer capacitors, the letter T should appear adjacent to the symbol.

Application: adjustable or variable capacitors with mechanical linkage of units

Shunt capacitor

Feed-through capacitor (with terminals shown on feed-through element)
Commonly used for bypassing high-frequency currents to chassis.

Application: feed-through capacitor between 2 inductors with third lead connected to chassis

## CELL, PHOTOSENSITIVE (Semiconductor)

$\lambda$ indicates that the primary characteristic of the element within the circle is designed to vary under the influence of light.

Asymmetrical photoconductive transducer (resistive)
*This arrowhead shall be solid.*

Symmetrical photoconductive transducer; selenium cell

## CHASSIS
## FRAME
See also GROUND
The chassis or frame is not necessarily at ground potential.

## COIL, BLOWOUT

## COIL, OPERATING
See also INDUCTOR; WINDING
Note 3—The asterisk is not a part of the symbol. Always replace the asterisk by a device designation.

* SEE NOTE 3

## CONNECTION, MECHANICAL —
## MECHANICAL INTERLOCK
The preferred location of the mechanical connection is as shown in the various applications, but other locations may be equally acceptable.

Mechanical connection (*short dashes*)

- - - - -

---

* Adapted from ANSI Y32.2—.

## Symbols for Electrical Diagrams (Cont.)

Mechanical connection or interlock with fulcrum (*short dashes*)

Mechanical interlock, other

INDICATE BY A NOTE

**CONNECTOR**
**DISCONNECTING DEVICE**
*The connector symbol is not an arrowhead. It is larger and the lines are drawn at a 90-degree angle.*

Female contact

Male contact

Connector assembly, movable or stationary portion; jack, plug, or receptacle
Note 4—Use appropriate number of contact symbols.

SEE NOTE 4

Commonly used for a jack or receptacle (usually stationary)

SEE NOTE 4  OR

Commonly used for a plug (usually movable)

SEE NOTE 4  OR

Separable connectors (engaged)

SEE NOTE 4  OR

Application: engaged 4-conductor connectors; the plug has 1 male and 3 female contacts

OR

Communication switchboard-type connector

2-conductor (jack)

2-conductor (plug)

Jacks with circuit normalled through one way

Jacks with circuit normalled through both ways

Jacks in multiple, one set with circuit normalled through both ways

Connectors of the type commonly used for power-supply purposes (convenience outlets and mating connectors)

Female contact

Male contact

2-conductor nonpolarized connector with female contacts

2-conductor nonpolarized connector with male contacts

2-conductor polarized connector with female contacts

2-conductor polarized connector with male contacts

3-conductor polarized connector with female contacts

3-conductor polarized connector with male contacts

4-conductor polarized connector with female contacts

4-conductor polarized connector with male contacts

Test blocks

Female portion with short-circuiting bar (with terminals shown)

Male portion (with terminals shown)

**CONTACT, ELECTRIC**
For build-ups or forms using electric contacts, see applications under CONNECTOR

Fixed contact

Fixed contact for jack, key, relay, etc.

OR          OR

Fixed contact for switch

OR

Fixed contact for momentary switch
See SWITCH

Sleeve

OR          OR

Moving contact

Adjustable or sliding contact for resistor, inductor, etc.

OR

Locking

Nonlocking

Closed contact (break)

OR

Open contact (make)

OR

Transfer

OR

Make-before-break

Application: open contact with time closing (TC or TDC) feature

TC  OR  TDC

Application: closed contact with time opening (TO or TDO) feature

TO  OR  TDO

Time sequential closing

OR

**COUPLER, DIRECTIONAL**
Commonly used in coaxial and waveguide diagrams.
The arrows indicate the direction of power flow.
Number of coupling paths, type of coupling, and transmission loss may be indicated.

General

# Symbols for Electrical Diagrams (Cont.)

**Applications**

*E*-plane aperture coupling, 30-decibel transmission loss

Loop coupling, 30-decibel transmission loss

Probe coupling, 30-decibel transmission loss

Resistance coupling, 30-decibel transmission loss

## DIRECTION OF FLOW OF POWER, SIGNAL, OR INFORMATION

One-way

Note 10—The lower symbol is used if it is necessary to conserve space. *The arrowhead in the lower symbol shall be filled.*

OR

SEE NOTE 10

Both ways

OR

SEE NOTE 10

## DISCONTINUITY

A component that exhibits throughout the frequency range of interest the properties of the type of circuit element indicated by the symbol within the triangle.

Commonly used for coaxial and waveguide transmission.

Equivalent series element, general

Capacitive reactance

## ELEMENT, CIRCUIT (General)

Note 12—The asterisk is not a part of the symbol. Always indicate the type of apparatus by appropriate words or letters in the rectangle.

＊ SEE NOTE 12

Accepted abbreviations in the latest edition of American Standard Z32.13 may be used in the rectangle.

The following letter combinations may be used in the rectangle.

| | | | |
|---|---|---|---|
| CB | Circuit breaker | NET | Network |
| DIAL | Telephone dial | PS | Power supply |
| EQ | Equalizer | RU | Reproducing |
| FAX | Facsimile set | | unit |
| FL | Filter | RG | Recording unit |
| FL-BE | Filter, band elimination | TEL | Telephone station |
| FL-BP | Filter, band pass | TPR | Teleprinter |
| FL-HP | Filter, high pass | TTY | Teletypewriter |
| FL-LP | Filter, low pass | | |

## ELEMENT, THERMAL

Thermomechanical transducer

Note 13—Use appropriate number of single-line diagram symbols.

SEE NOTE 13

## FUSE

SEE NOTE 14

Note 14—Use appropriate number of single-line diagram symbols.

Fusible element

SEE NOTE 14

## GROUND

See also CHASSIS; FRAME

## INDUCTOR WINDING

General

Either symbol may be used in the following subparagraphs.

OR

If it is desired especially to distinguish magnetic-core inductors

## LAMP

Ballast lamp; ballast tube

The primary characteristic of the element within the circle is designed to vary nonlinearly with the temperature of the element.

Fluorescent lamp

2-terminal

4-terminal

Glow lamp; cold-cathode lamp; neon lamp

Alternating-current type

Direct-current type
See also TUBE, ELECTRON

Incandescent-filament illuminating lamp

## MACHINE, ROTATING

Basic

Generator, general

Motor, general

Motor, multispeed

USE BASIC MOTOR SYMBOL AND NOTE SPEEDS

Rotating armature with commutator and brushes

Wound rotor

**Field, generator or motor**

Compensating or commutating

Series

Shunt, or separately excited

Permanent magnet

Winding symbols

Motor and generator winding symbols may be shown in the basic circle using the following representations.

1-phase

2-phase

3-phase wye (ungrounded)

3-phase wye (grounded)

3-phase delta

6-phase diametrical

6-phase double-delta

## MOTION, MECHANICAL

Translation, one direction

Translation, both directions

Rotation, one direction

Rotation, both directions

# Symbols for Electrical Diagrams (Cont.)

## NETWORK

General

NET

## OSCILLATOR
## GENERALIZED ALTERNATING-CURRENT SOURCE

## PAD

See also ATTENUATOR

General

Balanced, general

Unbalanced, general

## PATH, TRANSMISSION

Air or space path

Dielectric path other than air

Commonly used for coaxial and waveguide transmission.

DIEL

Crossing of paths or conductors not connected
*The crossing is not necessarily at a 90-degree angle.*

Junction of paths or conductors

Junction (if desired)

Application: junction of different-size cables

Junction of connected paths, conductors, or wires

OR

OR ONLY IF REQUIRED
BY SPACE LIMITATION

## RELAY
See also CONTACT

Fundamental symbols for contacts, mechanical connections, coils, etc., are the basis of relay symbols and should be used to represent relays on complete diagrams.

Basic

Relay coil

Note 21—The asterisk is not a part of the symbol. Always replace the asterisk by a device designation.

* SEE NOTE 21

Application: 2-pole double-make

## RESISTOR

Note 22—The asterisk is not a part of the symbol. Always add identification within or adjacent to the rectangle.

General

* SEE NOTE 22

Tapped resistor

* SEE NOTE 22

Application: with adjustable contact

* SEE NOTE 22

Application: adjustable or continuously adjustable (variable) resistor

* SEE NOTE 22

Heating resistor

* SEE NOTE 22

## SWITCH

Fundamental symbols for contacts, mechanical connections, etc., may be used for switch symbols.

The standard method of showing switches is in a position with no operating force applied. For switches that may be in any one of two or more positions with no operating force applied and for switches actuated by some mechanical device (as in air-pressure, liquid-level, rate-of-flow, etc., switches), a clarifying note may be necessary to explain the point at which the switch functions.

Single throw, general

Double throw, general

Application: 2-pole double-throw switch with terminals shown

Knife switch, general

Switch, nonlocking; momentary or spring return

The symbols to the left are commonly used for spring buildups in key switches, relays, and jacks.
The symbols to the right are commonly used for toggle switches.

Circuit closing (make)

OR

Circuit opening (break)

OR

Switch, locking

The symbols to the left are commonly used for spring buildups in key switches, relays, and jacks.
The symbols to the right are commonly used for toggle switches.

Circuit closing (make)

OR

Circuit opening (break)

OR

Transfer, 2-position

 OR

## TERMINAL, CIRCUIT

Terminal board or terminal strip with 4 terminals shown; group of 4 terminals

Number and arrangement as convenient.

## TRANSFORMER

General

Either winding symbol may be used.

Additional windings may be shown or indicated by a note.

For power transformers, use polarity marking $H_1$-$X_1$, etc., from American Standard C6.1.

In coaxial and waveguide circuits, this symbol will represent a taper or step transformer without mode change.

 OR

If it is desired especially to distinguish a magnetic-core transformer

## TUBE, ELECTRON

Tube-component symbols are shown first. These are followed by typical applications showing the use of these specific symbols in the various classes of devices such as thermionic, cold-cathode, and photoemissive tubes of varying structures and combinations of elements (triodes, pentodes, cathode-ray tubes, magnetrons, etc.).

Lines outside of the envelope are not part of the symbol but are electrical connections thereto.

*Connections between the external circuit and electron tube symbols within the envelope may be located as required to simplify the diagram.*

Emitting electrode

Directly heated (filamentary) cathode
Note—Leads may be connected in any convenient manner to ends of the ∧ provided the identity of the ∧ is retained.

Indirectly heated cathode
Lead may be connected to either extreme end of the ⌐ or, if required, to both ends, in any convenient manner.

Cold cathode (including ionically heated cathode)

Photocathode

Pool cathode

Ionically heated cathode with provision for supplementary heating

8

## Symbols for Electrical Diagrams (Cont.)

**Controlling electrode**

Grid (including beam-confining or beam-forming electrodes)

Deflecting electrodes (used in pairs); reflecting or repelling electrode (used in velocity-modulated tube)

Ignitor (in pool tubes) (should extend into pool)
Starter (in gas tubes)

Exciter (contactor type)

**Collecting electrode**

Anode or plate (including collecting electrode and fluorescent target)

Target or X-ray anode
*Drawn at about a 45-degree angle.*

**Collecting and emitting electrode**

Dynode

Alternately collecting and emitting

Composite anode-photocathode

Composite anode-cold cathode

Composite anode-ionically heated cathode with provision for supplementary heating

Heater

---

**Envelope (shell)**

The general envelope symbol identifies the envelope or enclosure regardless of evacuation or pressure. When used with electron-tube component symbols, the general envelope symbol indicates a vacuum enclosure unless otherwise specified. A gas-filled electron device may be indicated by a dot within the envelope symbol.

General

 OR

Split envelope
If necessary, envelope may be split.

Gas-filled
*The dot may be located as convenient.*

Shield

This is understood to shield against electric fields unless otherwise noted.

Any shield against electric fields that is within the envelope and that is connected to an independent terminal

Outside envelope of X-ray tube

Coupling by loop (electromagnetic type)
Coupling loop may be shown inside or outside envelope as desired, but if inside it should be shown grounded.

---

**Resonators (cavity type)**

Single-cavity envelope and grid-type associated electrodes

Double-cavity envelope and grid-type associated electrodes

Multicavity magnetron anode and envelope

Associated parts of a circuit, such as focusing coils, deflecting coils, field coils, etc., are not a part of the tube symbol but may be added to the circuit in the form of standard symbols. For example, resonant-type magnetron with permanent magnet may be shown:

External and internal shields, whether integral parts of tubes or not, shall be omitted from the circuit diagram unless the circuit diagram requires their inclusion.

In line with standard drafting practice, straight-line crossovers are recommended.

Typical applications

Triode with directly heated filamentary cathode and envelope connection to base terminal

Equipotential-cathode pentode showing use of elongated envelope

---

Typical wiring figure
This figure illustrates how tube symbols may be placed in any convenient position in a circuit.

For tubes with keyed bases
Explanatory word and arrow are not a part of the symbol shown.

KEY

For tubes with bayonets, bosses, and other reference points

Base terminals
Explanatory words and arrows are not a part of the symbol.

SMALL PIN

LARGE PIN

Envelope terminals
Explanatory words and arrows are not a part of the symbol.

RIGID TERMINAL

FLEXIBLE LEAD

Applications

Triode with indirectly heated cathode and envelope connected to base terminal

Triode-heptode with rigid envelope connection

---

## Transistor Symbols

PNP transistor     NPN transistor

## Symbols for Materials (Exterior)

Brick

Stone

Transparent Material
Glass. Celluloid. Etc.

Wood

## Symbols for Materials (Section)

Cast Iron

Steel

Bronze, Brass, Copper
and Composition

White Metal. Zinc,
Lead, Babbitt & Alloys

Earth

Rock

Aluminum

(Show solid for narrow sections)

Electric Insulation, Mica,
Fibre, Vulcanite, Bakelite, Etc.

Sound or Heat Insulation
Cork, Asbestos, Packing, Etc.

Flexible Material
Fabric, Rubber, Etc.

Sand

Water & Other Liquids

Fire Brick and
Refractory Material

Concrete

Brick or Stone
Masonry

Marble, Slate, Glass,
Porcelain, Etc.

## Weights of Materials

| Metals | lb/cu in. | Wood | lb/cu in. |
|---|---|---|---|
| Aluminum alloy, cast | 0.099 | Ash | 0.024 |
| Aluminum, cast | 0.094 | Balsa | 0.0058 |
| Aluminum, wrought | 0.097 | Cedar | 0.017 |
| Babbitt metal | 0.267 | Cork | 0.009 |
| Brass, cast or rolled | 0.303–0.313 | Hickory | 0.0295 |
| Brass, drawn | 0.323 | Maple | 0.025 |
| Bronze, aluminum cast | 0.277 | Oak (white) | 0.028 |
| Bronze, phosphor | 0.315–0.321 | Pine (white) | 0.015 |
| Chromium | 0.256 | Pine (yellow) | 0.025 |
| Copper, cast | 0.311 | Poplar | 0.018 |
| Copper, rolled, drawn or wire | 0.322 | Walnut (black) | 0.023 |
| Dowmetal *A* | 0.065 | **Miscellaneous materials** | **lb/ cu ft** |
| Duralumin | 0.101 | Asbestos | 175 |
| Gold | 0.697 | Bakelite | 79.5 |
| Iron, cast | 0.260 | Brick, common | 112 |
| Iron, wrought | 0.283 | Brick, fire | 144 |
| Lead | 0.411 | Celluloid | 86.4 |
| Magnesium | 0.063 | Earth, packed | 100 |
| Mercury | 0.491 | Fiber | 89.9 |
| Monel metal | 0.323 | Glass | 163 |
| Silver | 0.379 | Gravel | 109 |
| Steel, cast or rolled | 0.274–0.281 | Limestone | 163 |
| Steel, tool | 0.272 | Plexiglass | 74.3 |
| Tin | 0.263 | Sandstone | 144 |
| Zinc | 0.258 | Water | 62.4 |

## Abbreviations for Use on Drawings

Abbreviations are shortened forms of words or expressions and their use on drawings is entirely for the purpose of conserving space and drafting time. Since they must be interpreted by shopmen, assemblers, construction men, and others, abbreviations should not be used where the meaning may not be clear. In case of doubt, spell out. Only words that are abbreviated on drawings are included in this list, which is selected from the more complete list given in the SAE Aerospace-Automotive Drawing Standards (Section Z.1)

The same abbreviation is used for all forms of a given word. Periods are used only to avoid misinterpretation of an abbreviation. Spaces between letters are for clarity only; the abbreviation of word combinations or phrases is sometimes spaced. The use of hyphens and slant bars has been avoided where practicable. Upper-case letters are used for abbreviations. Subscripts are not used in abbreviations. Whenever a special abbreviation is used (one not appearing in the SAE Aerospace-Automotive standard), it should be explained in a table on the drawing. For metals, chemicals, formulas, and equations, the abbreviations, symbols, and rules in general use and established by long acceptance are employed.

Abbreviations for colors and a partial list of chemical symbols are given after the general abbreviations.

| Word | Abbreviation | Word | Abbreviation |
|------|--------------|------|--------------|
| Abbreviate | ABBR | American Wire Gage | AWG |
| Absolute | ABS | Ammeter | AM |
| Accelerate | ACCEL | Amount | AMT |
| Acceleration due to gravity | G | Ampere | AMP |
| Access panel | AP | Ampere hour | AMP HR |
| Accessory | ACCESS. | Amplifier | AMPL |
| Actual | ACT. | Anneal | ANL |
| Adapter | ADPT | Antenna | ANT. |
| Addendum | ADD. | Apparatus | APP |
| Adjust | ADJ | Approved | APPD |
| Advance | ADV | Approximate | APPROX |
| After | AFT. | Arc weld | ARC/W |
| Aggregate | AGGR | Area | A |
| Aileron | AIL | Armature | ARM. |
| Air-break switch | ABS | Arrange | ARR. |
| Air-circuit breaker | ACB | Arrester | ARR. |
| Aircraft | ACFT | Asbestos | ASB |
| Airplane | APL | Assemble | ASSEM |
| Airtight | AT | Assembly | ASSY |
| Alarm | ALM | Atomic | AT |
| Allowance | ALLOW | Attach | ATT |
| Alloy | ALY | Audio-frequency | AF |
| Alteration | ALT | Automatic | AUTO |
| Alternate | ALT | Auto-transformer | AUTO TR |
| Alternating current | AC | Auxiliary | AUX |
| Alternator | ALT | Average | AVG |
| Altitude | ALT | | |
| Aluminum | AL | Babbitt | BAB |
| American Standard | AMER STD | Back to back | B to B |

| Word | Abbreviation | Word | Abbreviation |
|---|---|---|---|
| Baffle | BAF | Cast (used with other materials) | C |
| Balance | BAL | Cast iron | CI |
| Ball bearing | BB | Cast-iron pipe | CIP |
| Base line | BL | Cast steel | CS |
| Base plate | BP | Casting | CSTG |
| Battery | BAT. | Castle nut | CAS NUT |
| Bearing | BRG | Cement | CEM |
| Bent | BT | Center | CTR |
| Between | BET. | Center line | CL |
| Between centers | BC | Center to center | C to C |
| Between perpendiculars | BP | Centering | CTR |
| Bevel | BEV | Centigrade | C |
| Bill of material | B/M | Centigram | CG |
| Birmingham Wire Gage | BWG | Centiliter | CL |
| Blank | BLK | Centimeter | CM |
| Block | BLK | Centrifugal | CENT. |
| Blueprint | BP | Centrifugal force | CF |
| Bolt circle | BC | Ceramic | CER |
| Bottom | BOT | Chain | CH |
| Bottom chord | BC | Chamfer | CHAM |
| Brake | BK | Change | CHG |
| Brass | BRS | Change notice | CN |
| Brazing | BRZG | Change order | CO |
| Break | BRK | Channel | CHAN |
| Breaker | BKR | Check | CHK |
| Brinnell hardness | BH | Check valve | CV |
| British Standard | BR STD | Chemical | CHEM |
| British thermal units | BTU | Chord | CHD |
| Broach | BRO | Chrome molybdenum | CR MOLY |
| Bronze | BRZ | Chromium plate | CR PL |
| Brown & Sharp | B&S | Chrome vanadium | CR VAN |
| Brush | BR | Circle | CIR |
| Burnish | BNH | Circuit | CKT |
| Bushing | BUSH. | Circular | CIR |
| Bypass | BYP | Circular pitch | CP |
|  |  | Circulate | CIRC |
| Cadmium plate | CD PL | Circumference | CIRC |
| Calculate | CALC | Clamp | CLP |
| Calibrate | CAL | Class | CL |
| Calking | CLKG | Clear | CLR |
| Capacitor | CAP | Clearance | CL |
| Capacity | CAP | Clockwise | CW |
| Cap screw | CAP. SCR | Closing | CL |
| Carburize | CARB | Clutch | CL |
| Caseharden | CH | Coated | CTD |
| Casing | CSG | Coaxial | COAX |

| Word | Abbreviation | Word | Abbreviation |
|---|---|---|---|
| Coefficient | COEF | Counterbore | CBORE |
| Cold drawn | CD | Counterdrill | CDRILL |
| Cold-drawn steel | CDS | Counterpunch | CPUNCH |
| Cold rolled | CR | Countersink | CSK |
| Cold-rolled steel | CRS | Countersink other side | CSK-O |
| Column | COL | Coupling | CPLG |
| Combination | COMB. | Cover | COV |
| Combustion | COMB | Crank | CRK |
| Communication | COMM | Cross connection | XCONN |
| Commutator | COMM | Cross section | XSECT |
| Complete | COMPL | Cubic | CU |
| Composite | CX | Current | CUR |
| Composition | COMP | Cyanide | CYN |
| Compressor | COMPR | Cycle | CY |
| Concentric | CONC | Cycles per minute | CPM |
| Concrete | CONC | Cycles per second | CPS |
| Condition | COND | Cylinder | CYL |
| Conduct | COND | | |
| Conductor | COND | Decibel | DB |
| Conduit | CND | Decimal | DEC |
| Connect | CONN | Dedendum | DED |
| Constant | CONST | Deep drawn | DD |
| Contact | CONT | Deflect | DEFL |
| Container | CNTR | Degree | (°) DEG |
| Continue | CONT | Density | D |
| Continuous wave | CW | Describe | DESCR |
| Contract | CONT | Design | DSGN |
| Contractor | CONTR | Designation | DESIG |
| Control | CONT | Detail | DET |
| Control relay | CR | Detector | DET |
| Control switch | CS | Detonator | DET |
| Controller | CONT | Develop | DEV |
| Convert | CONV | Diagonal | DIAG |
| Conveyor | CNVR | Diagram | DIAG |
| Cooled | CLD | Diameter | DIA |
| Copper oxide | CUO | Diametral pitch | DP |
| Copper plate | COP. PL | Diaphragm | DIAPH |
| Cord | CD | Differential | DIFF |
| Correct | CORR | Dimension | DIM. |
| Corrosion resistant | CRE | Diode | DIO |
| Corrosion-resistant steel | CRES | Direct current | DC |
| Corrugate | CORR | Directional | DIR |
| Cotter | COT | Discharge | DISCH |
| Counter | CTR | Disconnect | DISC. |
| Counterclockwise | CCW | Distance | DIST |
| Counterbalance | CBAL | Distribute | DISTR |

| Word | Abbreviation | Word | Abbreviation |
|---|---|---|---|
| Ditto | DO. | Fahrenheit | F |
| Double | DBL | Fairing | FAIR. |
| Dovetail | DVTL | Farad | F |
| Dowel | DWL | Far side | FS |
| Down | DN | Feed | FD |
| Drafting | DFTG | Feeder | FDR |
| Draftsman | DFTSMN | Feet | (') FT |
| Drain | DR | Feet per minute | FPM |
| Drawing | DWG | Feet per second | FPS |
| Drawing list | DL | Female | FEM |
| Drill | DR | Fiber | FBR |
| Drill rod | DR | Field | FLD |
| Drive | DR | Figure | FIG. |
| Drive fit | DF | Filament | FIL |
| Drop | D | Fillet | FIL |
| Drop forge | DF | Filling | FILL. |
| Duplex | DX | Fillister | FIL |
| Duplicate | DUP | Filter | FLT |
| Dynamic | DYN | Finish | FIN. |
| Dynamo | DYN | Finish all over | FAO |
|  |  | Fireproof | FPRF |
| Each | EA | Fitting | FTG |
| Eccentric | ECC | Fixture | FIX. |
| Effective | EFF | Flange | FLG |
| Electric | ELEC | Flashing | FL |
| Elevation | EL | Flat | F |
| Enclose | ENCL | Flat head | FH |
| End to end | E to E | Flexible | FLEX. |
| Envelope | ENV | Float | FLT |
| Equal | EQ | Floor | FL |
| Equation | EQ | Fluid | FL |
| Equipment | EQUIP. | Fluorescent | FLUOR |
| Equivalent | EQUIV | Flush | FL |
| Estimate | EST | Focus | FOC |
| Evaporate | EVAP | Foot | (') FT |
| Excavate | EXC | Force | F |
| Exhaust | EXH | Forging | FORG |
| Expand | EXP | Forward | FWD |
| Exterior | EXT | Foundation | FDN |
| External | EXT | Foundry | FDRY |
| Extra heavy | X HVY | Fractional | FRAC |
| Extra strong | X STR | Frame | FR |
| Extrude | EXTR | Freezing point | FP |
|  |  | Frequency | FREQ |
| Fabricate | FAB | Frequency, high | HF |
| Face to face | F to F | Frequency, low | LF |

| Word | Abbreviation | Word | Abbreviation |
|---|---|---|---|
| Frequency, medium | MF | Harden | HDN |
| Frequency modulation | FM | Hardware | HDW |
| Frequency, super high | SHF | Head | HD |
| Frequency, ultra high | UHF | Headless | HDLS |
| Frequency, very high | VHF | Heat | HT |
| Frequency, very low | VLF | Heat treat | HT TR |
| Friction horsepower | FHP | Heater | HTR |
| From below | FR BEL | Heavy | HVY |
| Front | FR | Height | HGT |
| Fuel | F | Henry | H |
| Furnish | FURN | Hexagon | HEX |
| Fusible | FSBL | High | H |
| Fusion point | FNP | High frequency | HF |
| | | High point | H PT |
| Gage or Gauge | GA | High pressure | HP |
| Gallon | GAL | High speed | HS |
| Galvanize | GALV | High-speed steel | HSS |
| Galvanized iron | GI | High tension | HT |
| Galvanized steel | GS | High voltage | HV |
| Galvanized steel wire rope | GSWR | Highway | HWY |
| Gas | G | Holder | HLR |
| Gasket | GSKT | Hollow | HOL |
| Gasoline | GASO | Horizontal | HOR |
| General | GEN | Horsepower | HP |
| Glaze | GL | Hot rolled | HR |
| Government | GOVT | Hot-rolled steel | HRS |
| Government furnished equipment | GFE | Hour | HR |
| Governor | GOV | Hydraulic | HYD |
| Grade | GR | | |
| Graduation | GRAD | Identify | IDENT |
| Gram | G | Ignition | IGN |
| Graphic | GRAPH. | Illuminate | ILLUM |
| Graphite | GPH | Illustrate | ILLUS |
| Grating | GRTG | Impact | IMP |
| Gravity | G | Impedance | IMP. |
| Grid | G | Inch | ($''$) IN. |
| Grind | GRD | Inches per second | IPS |
| Groove | GRV | Include | INCL |
| Ground | GRD | Increase | INCR |
| | | Indicate | IND |
| Half hard | ½H | Inductance or induction | IND |
| Half round | ½RD | Industrial | IND |
| Handle | HDL | Information | INFO |
| Hanger | HGR | Injection | INJ |
| Hard | H | Inlet | IN |
| Hard-drawn | HD | Inspect | INSP |

| Word | Abbreviation | Word | Abbreviation |
|---|---|---|---|
| Install | INSTL | Lacquer | LAQ |
| Instantaneous | INST | Laminate | LAM |
| Instruct | INST | Lateral | LAT |
| Instrument | INST | Lead-coated metal | LCM |
| Insulate | INS | Lead covered | LC |
| Interchangeable | INTCHG | Leading edge | LE |
| Interior | INT | Left | L |
| Interlock | INTLK | Left hand | LH |
| Intermediate | INTER | Length | LG |
| Intermittent | INTMT | Length over all | LOA |
| Internal | INT | Letter | LTR |
| Interrupt | INTER | Light | LT |
| Interrupted continuous wave | ICW | Limit | LIM |
| Interruptions per minute | IPM | Line | L |
| Interruptions per second | IPS | Linear | LIN |
| Intersect | INT | Link | LK |
| Inverse | INV | Liquid | LIQ |
| Invert | INV | Liter | L |
| Iron | I | Locate | LOC |
| Iron-pipe size | IPS | Long | LG |
| Irregular | IRREG | Longitude | LONG. |
| Issue | ISS | Low explosive | LE |
|  |  | Low frequency | LF |
| Jack | J | Low pressure | LP |
| Job order | JO | Low tension | LT |
| Joint | JT | Low voltage | LV |
| Junction | JCT | Low speed | LS |
|  |  | Low torque | LT |
| Kelvin | K | Lubricate | LUB |
| Key | K | Lubricating oil | LO |
| Keyseat | KST | Lumen | L |
| Keyway | KWY | Lumens per watt | LPW |
| Kilo | K |  |  |
| Kilocycle | KC | Machine | MACH |
| Kilocycles per second | KC | Magnet | MAG |
| Kilogram | KG | Main | MN |
| Kiloliter | KL | Male and female | M&F |
| Kilometer | KM | Malleable | MALL |
| Kilovolt | KV | Malleable iron | MI |
| Kilovolt-ampere | KVA | Manual | MAN. |
| Kilovolt-ampere hour | KVAH | Manufacture | MFR |
| Kilowatt | KW | Manufactured | MFD |
| Kilowatt-hour | KWH | Manufacturing | MFG |
| Kip (1000 lb) | K | Material | MATL |
| Knots | KN | Material list | ML |
|  |  | Maximum | MAX |
| Laboratory | LAB |  |  |

| Word | Abbreviation | Word | Abbreviation |
|---|---|---|---|
| Maximum working pressure | MWP | Mounted | MTD |
| Mean effective pressure | MEP | Mounting | MTG |
| Mechanical | MECH | Multiple | MULT |
| Mechanism | MECH | Multiple contact | MC |
| Medium | MED | | |
| Mega | M | National | NATL |
| Megacycles | MC | Natural | NAT |
| Megawatt | MW | Near face | NF |
| Megohm | MEG | Near side | NS |
| Melting point | MP | Negative | NEG |
| Metal | MET. | Network | NET |
| Meter (instrument or measure | | Neutral | NEUT |
| of length) | M | Nickel-silver | NI-SIL |
| Micro | $\mu$ or U | Nipple | NIP. |
| Microampere | $\mu$A or UA | Nominal | NOM |
| Microfarad | $\mu$F or UF | Normal | NOR |
| Microhenry | $\mu$H or UH | Normally closed | NC |
| Micro-inch-root-mean square | $\mu$-IN-RMS or | Normally open | NO |
| | U-IN-RMS | Not to scale | NTS |
| Micrometer | MIC | Number | NO. |
| Micron | $\mu$ or U | | |
| Microvolt | $\mu$V or UV | Obsolete | OBS |
| Microwatt | $\mu$W or UW | Octagon | OCT |
| Miles | MI | Ohm | $\Omega$ |
| Miles per gallon | MPG | Oil-circuit breaker | OCB |
| Miles per hour | MPH | Oil insulated | OI |
| Milli | M | Oil switch | OS |
| Milliampere | MA. | On center | OC |
| Milligram | MG | One pole | 1 P |
| Millihenry | MH | Opening | OPNG |
| Millimeter | MM | Operate | OPR |
| Milliseconds | MS | Opposite | OPP |
| Millivolt | MV | Optical | OPT |
| Milliwatt | MW | Ordnance | ORD |
| Minimum | MIN | Orifice | ORF |
| Minute | (') MIN | Original | ORIG |
| Miscellaneous | MISC | Oscillate | OSC |
| Mixture | MIX. | Ounce | OZ |
| Model | MOD | Out to Out | O to O |
| Modify | MOD | Outlet | OUT. |
| Modulated continuous wave | MCW | Output | OUT. |
| Modulator | MOD | Outside diameter | OD |
| Molecular weight | MOL WT | Outside face | OF |
| Monument | MON | Outside radius | OR |
| Morse taper | MOR T | Over-all | OA |
| Motor | MOT | Overhead | OVHD |

| Word | Abbreviation | Word | Abbreviation |
|---|---|---|---|
| Overload | OVLD | Pounds per cubic foot | PCF |
| Overvoltage | OVV | Pounds per square foot | PSF |
| Oxidized | OXD | Pounds per square inch | PSI |
| | | Pounds per square inch absolute | PSIA |
| Pack | PK | Power | PWR |
| Packing | PKG | Power amplifier | PA |
| Painted | PTD | Power directional relay | PDR |
| Pair | PR | Power factor | PF |
| Panel | PNL | Preamplifier | PREAMP |
| Parallel | PAR. | Precast | PRCST |
| Part | PT | Prefabricated | PREFAB |
| Pattern | PATT | Preferred | PFD |
| Perforate | PERF | Premolded | PRMLD |
| Permanent | PERM | Prepare | PREP |
| Permanent magnet | PM | Press | PRS |
| Perpendicular | PERP | Pressure | PRESS. |
| Phase | PH | Pressure angle | PA. |
| Phosphor bronze | PH BRZ | Primary | PRI |
| Photograph | PHOTO | Process | PROC |
| Physical | PHYS | Production | PROD |
| Piece | PC | Profile | PF |
| Piece mark | PC MK | Project | PROJ |
| Pierce | PRC | Punch | PCH |
| Pipe Tap | PT | Purchase | PUR |
| Pitch | P | Push-pull | P-P |
| Pitch circle | PC | | |
| Pitch diameter | PD | Quadrant | QUAD |
| Plastic | PLSTC | Quality | QUAL |
| Plate | PL | Quantity | QTY |
| Plotting | PLOT. | Quart | QT |
| Pneumatic | PNEU | Quarter | QTR |
| Point | PT | Quarter hard | ¼ H |
| Point of compound curve | PCC | Quarter round | ¼ RD |
| Point of curve | PC | Quartz | QTZ |
| Point of intersection | PI | | |
| Point of reverse curve | PRC | Radial | RAD |
| Point of switch | PS | Radio frequency | RF |
| Point of tangent | PT | Radius | R |
| Polar | POL | Reactive | REAC |
| Pole | P | Reactive kilovolt-ampere | KVAR |
| Polish | POL | Reactive volt-ampere | VAR |
| Port | P | Reactive voltmeter | RVM |
| Position | POS | Reactor | REAC |
| Positive | POS | Ream | RM |
| Potential | POT. | Reassemble | REASM |
| Pound | LB | Received | RECD |

| Word | Abbreviation | Word | Abbreviation |
|------|-------------|------|-------------|
| Receiver | REC | Saddle | SDL |
| Receptacle | RECP | Safe working pressure | SWP |
| Recriprocate | RECIP | Safety | SAF |
| Recirculate | RECIRC | Sand blast | SD BL |
| Reclosing | RECL | Saturate | SAT. |
| Record | REC | Schedule | SCH |
| Rectangle | RECT | Schematic | SCHEM |
| Rectifier | RECT | Scleroscope hardness | SH |
| Reduce | RED. | Screen | SCRN |
| Reference | REF | Screw | SCR |
| Reference line | REF L | Second | SEC |
| Regulator | REG | Section | SECT |
| Reinforce | REINF | Segment | SEG |
| Relay | REL | Select | SEL |
| Release | REL | Semifinished | SF |
| Relief | REL | Semifixed | SFXD |
| Remove | REM | Semisteel | SS |
| Repair | REP | Separate | SEP |
| Replace | REPL | Sequence | SEQ |
| Reproduce | REPRO | Serial | SER |
| Require | REQ | Series | SER |
| Required | REQD | Serrate | SERR |
| Resistance | RES | Service | SERV |
| Resistor | RES | Set screw | SS |
| Retainer | RET. | Shaft | SFT |
| Retard | RET. | Shield | SHLD |
| Return | RET. | Shipment | SHPT |
| Reverse | REV | Shop order | SO |
| Revise | REV | Short wave | SW |
| Revolution | REV | Shunt | SH |
| Revolutions per minute | RPM | Side | S |
| Revolutions per second | RPS | Signal | SIG |
| Rheostat | RHEO | Sink | SK |
| Right | R | Sketch | SK |
| Right hand | RH | Sleeve | SLV |
| Ring | R | Slide | SL |
| Rivet | RIV | Slotted | SLOT. |
| Rockwell hardness | RH | Small | SM |
| Roller bearing | RB | Smoke | SMK |
| Root diameter | RD | Smokeless | SMKLS |
| Root mean square | RMS | Socket | SOC |
| Rotary | ROT. | Soft | S |
| Rotate | ROT. | Solder | SLD |
| Rough | RGH | Solenoid | SOL |
| Round | RD | Sound | SND |
| Rubber | RUB. | South | S |

| Word | Abbreviation | Word | Abbreviation |
|------|-------------|------|-------------|
| Space | SP | Tangent | TAN. |
| Spare | SP | Taper | TPR |
| Speaker | SPKR | Technical | TECH |
| Special | SPL | Tee | T |
| Specific | SP | Teeth per inch | TPI |
| Specific gravity | SP GR | Television | TV |
| Specific heat | SP HT | Temperature | TEMP |
| Specification | SPEC | Template | TEMP |
| Speed | SP | Tensile strength | TS |
| Spherical | SPHER | Tension | TENS. |
| Spindle | SPDL | Terminal | TERM. |
| Split phase | SP PH | Terminal board | TB |
| Spot-faced | SF | That is | IE |
| Spring | SPG | Theoretical | THEO |
| Square | SQ | Thermal | THRM |
| Stabilize | STAB | Thermostat | THERMO |
| Stainless | STN | Thick | THK |
| Standard | STD | Thousand | M |
| Static pressure | SP | Thread | THD |
| Station | STA | Throttle | THROT |
| Stationary | STA | Through | THRU |
| Steel | STL | Time | T |
| Stiffener | STIFF. | Time delay | TD |
| Stock | STK | Time-delay closing | TDC |
| Storage | STG | Time-delay opening | TDO |
| Straight | STR | Tinned | TD |
| Strip | STR | Tobin bronze | TOB BRZ |
| Structural | STR | Toggle | TGL |
| Substitute | SUB | Tolerance | TOL |
| Suction | SUCT | Tongue and groove | T&G |
| Summary | SUM. | Tool steel | TS |
| Supervise | SUPV | Tooth | T |
| Supply | SUP | Total | TOT |
| Surface | SUR | Total indicator reading | TIR |
| Survey | SURV | Trace | TR |
| Switch | SW | Tracer | TCR |
| Symbol | SYM | Transfer | TRANS |
| Symmetrical | SYM | Transformer | TRANS |
| Synchronous | SYN | Transmission | XMSN |
| Synthetic | SYN | Transmitter | XMTR |
| System | SYS | Transmitting | XMTG |
| | | Transportation | TRANS |
| Tabulate | TAB. | Transverse | TRANSV |
| Tachometer | TACH | Trimmer | TRIM. |
| Tandem | TDM | Triode | TRI |

| Word | Abbreviation | Word | Abbreviation |
|---|---|---|---|
| True air speed | TAS | Voltmeter | VM |
| Truss | T | Volts per mil | VPM |
| Tubing | TUB | Volume | VOL |
| Tuned radio frequency | TRF | | |
| Turbine | TURB | Washer | WASH |
| Typical | TYP | Water | W |
| | | Water line | WL |
| Ultimate | ULT | Watertight | WT |
| Ultra-high frequency | UHF | Watt | W |
| Under voltage | UV | Watt-hour | WHR |
| Unit | U | Watt-hour meter | WHM |
| United States Gage | USG | Wattmeter | WM |
| United States Standard | USS | Weight | WT |
| Universal | UNIV | West | W |
| | | Wet bulb | WB |
| Vacuum | VAC | Width | W |
| Vacuum tube | VT | Wind | WD |
| Valve | V | Winding | WDG |
| Vapor proof | VAP PRF | Wire | W |
| Variable | VAR | With | W/ |
| Variable-frequency oscillator | VFO | With equipment and spare parts | W/E&SP |
| Velocity | V | Without | W/O |
| Ventilate | VENT. | Without equipment and spare parts | W/O E&SP |
| Versed sine | VERS | | |
| Versus | VS | Wood | WD |
| Vertical | VERT | Woodruff | WDF |
| Very-high frequency | VHF | Working point | WP |
| Very-low frequency | VLF | Working pressure | WP |
| Video-frequency | VDF | Wrought | WRT |
| Vibrate | VIB | Wrought iron | WI |
| Viscosity | VISC | | |
| Vitreous | VIT | Yard | YD |
| Voice frequency | VF | Year | YR |
| Volt | V | Yield point | YP |
| Volt-ampere | VA | Yield strength | YS |

## Abbreviations for Colors

| | | | |
|---|---|---|---|
| Amber | AMB | Green | GRN |
| Black | BLK | Orange | ORN |
| Blue | BLU | White | WHT |
| Brown | BRN | Yellow | YEL |

## Partial List of Chemical Symbols

| Word | Abbreviation | Word | Abbreviation |
|---|---|---|---|
| Aluminum | Al | Molybdenum | Mo |
| Antimony (stibium) | Sb | Neon | Ne |
| Barium | Ba | Nickel | Ni |
| Beryllium | Be | Nitrogen | N |
| Bismuth | Bi | Oxygen | O |
| Boron | B | Phosphorus | P |
| Bromine | Br | Platinum | Pt |
| Cadmium | Cd | Potassium (kalium) | K |
| Calcium | Ca | Radium | Ra |
| Carbon | C | Rhodium | Rh |
| Chlorine | Cl | Ruthenium | Ru |
| Chromium | Cr | Selenium | Se |
| Cobalt | Co | Silicon | Si |
| Copper | Cu | Silver (argentum) | Ag |
| Fluorine | F | Sodium (natrium) | Na |
| Gold (aurium) | Au | Strontium | Sr |
| Helium | He | Sulfur | S |
| Hydrogen | H | Tantalum | Ta |
| Indium | In | Tellurium | Te |
| Iodine | I | Thallium | Tl |
| Iridium | Ir | Tin (stannum) | Sn |
| Iron (ferrum) | Fe | Titanium | Ti |
| Lead (plumbum) | Pb | Tungsten (wolframium) | W |
| Lithium | Li | Uranium | U |
| Magnesium | Mg | Vanadium | V |
| Manganese | Mn | Zinc | Zn |
| Mercury (hydrargyrum) | Hg | Zirconium | Zr |

# APPENDIX

## Supply catalogues and technical manuals

**SOUTHCO**

## Blind riveting is easier when you follow these simple
## TIPS ON INSTALLATION

### BEFORE YOU START THE JOB

Measure your total material thickness. Select the correct grip length rivet from the tables on the next three pages by comparing your sheet thickness with the min. and max. grips shown in the column headed GRIP LENGTH.

**IN METAL**     **IN WOOD**

GRIP LENGTH = Total thickness of sheets to be fastened

USE "L" DIMENSION (length under head) instead of grip length. L = M (thickness of metal) + D (hole depth in wood.)

### DRILL A HOLE

Select drill size according to your drilling method (hand or jig drilling)

| RIVET DIA. | JIG DRILLING | | | | | HAND DRILLING | | | | |
|---|---|---|---|---|---|---|---|---|---|---|
| | $\frac{1}{8}$ | $\frac{5}{32}$ | $\frac{3}{16}$ | $\frac{1}{4}$ | $\frac{5}{16}$ | $\frac{1}{8}$ | $\frac{5}{32}$ | $\frac{3}{16}$ | $\frac{1}{4}$ | $\frac{5}{16}$ |
| DRILL SIZE | 30 | 20 | 11 | F | P | $\frac{1}{8}$ | $\frac{5}{32}$ | $\frac{3}{16}$ | $\frac{1}{4}$ | $\frac{5}{16}$ |
| HOLE DIA. | .128 | .161 | .191 | .257 | .323 | | | | | |

### INSERT A RIVET

seat head firmly against outer surface

### HIT THE PIN

Drive pin flush with rivet head

Expanding prongs clinch sheets tightly, eliminating gaps.

Metal and wood pulled tightly together. Nothing protrudes through wood.

The following compilation shows representative pages from suppliers' and manufacturers' catalogues and technical manuals. These publications offer varied technical information, engineering specifications, and design data in many fields, material which is of great value both to students and to practicing engineers, designers, draftsmen, architects, and builders. In addition to their first-line role as sources of technical information and data, these publications exemplify interesting solutions to the graphic design problems involved in presenting large amounts of detail in readable form.

The student should take every opportunity to start and maintain a collection of such catalogues and manuals. He will find it a valuable addition to his technical library.

**SOUTHCO** # STEEL DRIVE RIVETS

SEE PAGE 3 FOR INSTALLATION DATA

USE THESE COLUMNS FIRST TO LOCATE YOUR CORRECT

# GRIP LENGTH

GRIP = TOTAL THICKNESS OF ALL SHEETS FASTENED TOGETHER

## 1/8" DIA. PART NUMBERS

## 3/16" DIA. PART NUMBERS

| length under head L | UNIVERSAL HEAD | 78° CSK. HEAD | length under head L | UNIVERSAL HEAD | 78° CSK. HEAD | FULL BRAZIER HEAD | MINIMUM GRIP | NOMINAL GRIP | MAXIMUM GRIP |
|---|---|---|---|---|---|---|---|---|---|
| 5/32 | 38-404-01-93* | | | | | | .015 | 1/32 | .046 |
| 5/32 | 38-104-02-93 | | 7/32 | | | 38-206-02-91 | .046 | 1/16 | .078 |
| 3/16 | 38-104-03-93 | 38-604-03-93‡ | 7/32 | 38-106-03-91 | | 38-206-03-91 | .078 | 3/32 | .109 |
| 7/32 | 38-104-04-93 | 38-604-04-93 | 1/4 | 38-106-04-91 | | 38-206-04-91 | .109 | 1/8 | .140 |
| 1/4 | 38-104-05-93 | 38-604-05-93 | 9/32 | 38-106-05-91 | | 38-206-05-91 | .140 | 5/32 | .171 |
| 9/32 | 38-104-06-93 | 38-604-06-93 | 5/16 | 38-106-06-91 | 38-606-06-91 | 38-206-06-91 | .171 | 3/16 | .203 |
| 5/16 | 38-104-07-93 | 38-604-07-93 | 11/32 | 38-106-07-91 | 38-606-07-91 | 38-206-07-91 | .203 | 7/32 | .234 |
| 11/32 | 38-104-08-93 | 38-604-08-93 | 3/8 | 38-106-08-91 | 38-606-08-91 | 38-206-08-91 | .234 | 1/4 | .265 |
| 3/8 | 38-104-09-93 | 38-604-09-93 | 13/32 | 38-106-09-91 | 38-606-09-91 | 38-206-09-91 | .265 | 9/32 | .296 |
| 13/32 | 38-104-10-93 | 38-604-10-93 | 7/16 | 38-106-10-91 | 38-606-10-91 | 38-206-10-91 | .296 | 5/16 | .328 |
| 7/16 | 38-104-11-93 | 38-604-11-93 | 15/32 | 38-106-11-91 | 38-606-11-91 | | .328 | 11/32 | .359 |
| 15/32 | 38-104-12-93 | 38-604-12-93 | 1/2 | 38-106-12-91 | 38-606-12-91 | | .359 | 3/8 | .390 |
| 1/2 | 38-104-13-93 | 38-604-13-93 | 17/32 | 38-106-13-91 | 38-606-13-91 | 38-206-13-91 | .390 | 13/32 | .421 |
| | | | 9/16 | 38-106-14-91 | 38-606-14-91 | | .421 | 7/16 | .453 |
| | * SPECIAL CONE HEAD | | 19/32 | 38-106-15-91 | 38-606-15-91 | | .453 | 15/32 | .484 |
| | | | 5/8 | 38-106-16-91 | 38-606-16-91 | 38-206-16-91 | .484 | 1/2 | .515 |
| | ‡ L = 7/32" | | 21/32 | 38-106-17-91 | 38-606-17-91 | | .515 | 17/32 | .546 |
| | | | 11/16 | 38-106-18-91 | 38-606-18-91 | | .546 | 9/16 | .578 |
| | | | 23/32 | 38-106-19-91 | 38-606-19-91 | | .578 | 19/32 | .609 |
| | | | 3/4 | 38-106-20-91 | 38-606-20-91 | 38-206-20-91 | .609 | 5/8 | .640 |

## New: handy kit for experimental & short run jobs.

### DRIVE RIVET SHOP ASSORTMENT

Contains approximately 1,000 Aluminum Drive Rivets, ³⁄₁₆" & ¼" diameter, with Full Brazier Heads. Grip lengths are carefully selected to handle the fastening of almost every common sheet thickness.

The package is a permanent metal-edge kit with refillable inner compartments. Complete instructions for drive riveting are a part of the box itself.

Part No. 38-88-101-10

## MATERIAL

Steel, cadmium plated. ⅛ dia. rivets have stainless steel pins. 3/16 & 1/4 dia. rivets have carbon steel pins*, cad. plated

*Stainless steel pins may be substituted at Manufacturer's option.

universal head .500 .107

78° csk. head 78° .452 .125

## 1/4" DIA.

### PART NUMBERS

| length under head L | UNIVERSAL HEAD | 78° CSK. HEAD |
|---|---|---|
| | | |
| 7/32 | 38-108-03-91 | |
| 1/4 | 38-108-04-91 | |
| 9/32 | 38-108-05-91 | |
| 5/16 | 38-108-06-91 | 38-608-06-91 |
| 11/32 | 38-108-07-91 | 38-608-07-91 |
| 3/8 | 38-108-08-91 | 38-608-08-91 |
| 13/32 | 38-108-09-91 | 38-608-09-91 |
| 7/16 | 38-108-10-91 | 38-608-10-91 |
| 15/32 | 38-108-11-91 | 38-608-11-91 |
| 1/2 | 38-108-12-91 | 38-608-12-91 |
| 17/32 | 38-108-13-91 | 38-608-13-91 |
| 9/16 | 38-108-14-91 | 38-608-14-91 |
| 19/32 | 38-108-15-91 | 38-608-15-91 |
| 5/8 | 38-108-16-91 | 38-608-16-91 |
| 21/32 | 38-108-17-91 | 38-608-17-91 |
| 11/16 | 38-108-18-91 | 38-608-18-91 |
| 23/32 | 38-108-19-91 | 38-608-19-91 |
| 3/4 | 38-108-20-91 | 38-608-20-91 |

## New... Larger diameter Steel Drive Rivets

SEE PAGE 3 FOR INSTALLATION DATA

USE THESE COLUMNS FIRST TO LOCATE YOUR CORRECT

### GRIP LENGTH

GRIP = TOTAL THICKNESS OF ALL SHEETS FASTENED TOGETHER

| MINIMUM GRIP | NOMINAL GRIP | MAXIMUM GRIP |
|---|---|---|
| .094 | 1/8 | .156 |
| .157 | 3/16 | .218 |
| .219 | 1/4 | .281 |
| .282 | 5/16 | .343 |
| .344 | 3/8 | .406 |
| .407 | 7/16 | .468 |
| .469 | 1/2 | .531 |
| .532 | 9/16 | .593 |
| .594 | 5/8 | .656 |
| .657 | 11/16 | .718 |
| .719 | 3/4 | .781 |
| .782 | 13/16 | .843 |
| .844 | 7/8 | .906 |
| .907 | 15/16 | .968 |
| .969 | 1 | 1.031 |

universal head .625 .138

5/16

## 5/16" DIA.

### PART NUMBER

| length under head L | UNIVERSAL HEAD |
|---|---|
| 5/16 | 38-110-04-91 |
| 3/8 | 38-110-06-91 |
| 7/16 | 38-110-08-91 |
| 1/2 | 38-110-10-91 |
| 9/16 | 38-110-12-91 |
| 5/8 | 38-110-14-91 |
| 11/16 | 38-110-16-91 |
| 3/4 | 38-110-18-91 |
| 13/16 | 38-110-20-91 |
| 7/8 | 38-110-22-91 |
| 15/16 | 38-110-24-91 |
| 1" | 38-110-26-91 |
| 1-1/16 | 38-110-28-91 |
| 1-1/8 | 38-110-30-91 |
| 1-3/16 | 38-110-32-91 |

MATERIAL—RIVET and PIN
Steel, cadmium plated

## Repair Truck Bodies in Minutes with RIVETPATCH™

Each Kit contains an aircraft type, Alclad Aluminum patch with formed edges, rounded corners, pilot holes. Sealant, Drive Rivets and instructions included. Only a drill and hammer needed to do a professional job in ten minutes!

4 sizes available: 3 x 5 (Part 38-99-337-11), 5 x 8 (Part 38-99-338-11), 8 x 11⅝ (Part 38-99-339-11), and 11⅝ x 24 (Part 38-99-353-11).

### OPTIONAL TOOLS AND ACCESSORIES

| For Rivet Dia. | PART NUMBERS | | |
|---|---|---|---|
| | Rivet Set Tools | 100° Countersinks | Southco Rivet Selectors |
| 1/8 | 29-1-504-10 | 29-13-101-11 | 29-12-101-23 |
| 5/32 | 29-1-505-10 | | |
| 3/16 | 29-1-506-10 | 29-13-102-11 | |
| 1/4 | 29-1-508-10 | | |

# SOUTHCO

# ALUMINUM

| universal head | 100° csk. head | full brazier head |
|---|---|---|
| .250 / .054 / ⅛ | .216 / 100° / .042 / ⅛ | .312 / .062 / ⅛ |

| universal head | 100° csk. head |
|---|---|
| .312 / .067 / 5/32 | .278 / 100° / .055 / 5/32 |

USE THESE COLUMNS
FIRST TO LOCATE
YOUR CORRECT

## GRIP LENGTH

GRIP =
TOTAL THICKNESS OF ALL
SHEETS FASTENED TOGETHER

## 1/8" DIA.
### PART NUMBERS

## 5/32" DIA.
### PART NUMBERS

| length under head L | UNIVERSAL HEAD | 100° CSK. HEAD | FULL BRAZIER HEAD | length under head L | UNIVERSAL HEAD | 100° CSK. HEAD | MINIMUM GRIP | NOMINAL GRIP | MAXIMUM GRIP |
|---|---|---|---|---|---|---|---|---|---|
| 5/32 | 38-404-01-13* | | | 3/16 | 38-105-01-13 | | .015 | 1/32 | .046 |
| 5/32 | 38-104-02-13 | | | 3/16 | 38-105-02-13 | | .046 | 1/16 | .078 |
| 3/16 | 38-104-03-13 | 38-504-03-13‡ | 38-204-03-13 | 7/32 | 38-105-03-13 | | .078 | 3/32 | .109 |
| 7/32 | 38-104-04-13 | 38-504-04-13 | 38-204-04-13 | 1/4 | 38-105-04-13 | 38-505-04-13 | .109 | 1/8 | .140 |
| 1/4 | 38-104-05-13 | 38-504-05-13 | 38-204-05-13 | 9/32 | 38-105-05-13 | 38-505-05-13 | .140 | 5/32 | .171 |
| 9/32 | 38-104-06-13 | 38-504-06-13 | 38-204-06-13 | 5/16 | 38-105-06-13 | 38-505-06-13 | .171 | 3/16 | .203 |
| 5/16 | 38-104-07-13 | 38-504-07-13 | 38-204-07-13 | 11/32 | 38-105-07-13 | 38-505-07-13 | .203 | 7/32 | .234 |
| 11/32 | 38-104-08-13 | 38-504-08-13 | | 3/8 | 38-105-08-13 | 38-505-08-13 | .234 | 1/4 | .265 |
| 3/8 | 38-104-09-13 | 38-504-09-13 | | 13/32 | 38-105-09-13 | 38-505-09-13 | .265 | 9/32 | .296 |
| 13/32 | 38-104-10-13 | 38-504-10-13 | | 7/16 | 38-105-10-13 | 38-505-10-13 | .296 | 5/16 | .328 |
| 7/16 | 38-104-11-13 | 38-504-11-13 | | 15/32 | 38-105-11-13 | 38-505-11-13 | .328 | 11/32 | .359 |
| 15/32 | 38-104-12-13 | 38-504-12-13 | 38-204-12-13 | 1/2 | 38-105-12-13 | 38-505-12-13 | .359 | 3/8 | .390 |
| 1/2 | 38-104-13-13 | 38-504-13-13 | | 17/32 | 38-105-13-13 | 38-505-13-13 | .390 | 13/32 | .421 |
| 17/32 | 38-104-14-13 | | | 9/16 | 38-105-14-13 | 38-505-14-13 | .421 | 7/16 | .453 |
| | | | | 19/32 | 38-105-15-13 | 38-505-15-13 | .453 | 15/32 | .484 |
| | * SPECIAL CONE HEAD | | ‡ L = 7/32" | 5/8 | 38-105-16-13 | 38-505-16-13 | .484 | 1/2 | .515 |
| | | | | 21/32 | 38-105-17-13 | 38-505-17-13 | .515 | 17/32 | .546 |
| | | | | 11/16 | 38-105-18-13 | 38-505-18-13 | .546 | 9/16 | .578 |
| | | | | 23/32 | 38-105-19-13 | 38-505-19-13 | .578 | 19/32 | .609 |
| 23/32 | 38-104-20-13 | | | 3/4 | 38-105-20-13 | 38-505-20-13 | .609 | 5/8 | .640 |
| | | | | | | | .671 | 11/16 | .703 |
| | | | | | | | .734 | 3/4 | .765 |
| | | | | | | | .796 | 13/16 | .828 |
| | | | | | | | .859 | 7/8 | .890 |
| | | | | | | | .921 | 15/16 | .953 |
| | | | | | | | .984 | 1 | 1.015 |

## ALUMINUM
## EXTRA LONG

RIVETS FOR METAL OR MASONRY
(Head dimensions same as on page 5)

Subtract ⅛" from "L" to determine grip

| length under head L | 3/16" DIA. FULL BRAZIER HEAD | ¼" DIA. FULL BRAZIER HEAD | ¼" DIA. 100° CSK. HEAD |
|---|---|---|---|
| 1⅛ | 38-206-32-13 | 38-208-32-13 | 38-508-32-13 |
| 1¼ | | 38-208-36-13 | |
| 1⁵⁄₁₆ | | 38-208-38-13 | |
| 1⅜ | | 38-208-40-13 | |
| 1⁷⁄₁₆ | | 38-208-42-13 | |
| 1½ | | 38-208-44-13 | 38-508-44-13 |
| 1⁹⁄₁₆ | | 38-208-46-13 | |
| 1⅝ | | 38-208-48-13 | |
| 1¾ | | 38-208-52-13 | 38-508-52-13 |

# DRIVE RIVETS

**2117-T4 ALUMINUM RIVET BODY WITH 300 SERIES STAINLESS STEEL PINS.**

SEE PAGE 3 FOR INSTALLATION DATA

(Aluminum rivets of other alloys, and pins of aluminum available on special request)

## 3/16" DIA.
### PART NUMBERS

## 1/4" DIA.
### PART NUMBERS

| length under head L * | UNIVERSAL HEAD | 100° CSK. HEAD | FULL BRAZIER HEAD | ROUND PLY-HEAD® | FLAT PLY-HEAD® | All Purpose LINER HEAD | length under head L * | UNIVERSAL HEAD | 100° CSK. HEAD | FULL BRAZIER HEAD |
|---|---|---|---|---|---|---|---|---|---|---|
| 7/32 | 38-106-02-13 | | 38-206-02-13 | | | | 7/32 | 38-108-02-13 | | |
| 7/32 | 38-106-03-13 | | 38-206-03-13 | | | 38-99-269-03 | 7/32 | 38-108-03-13 | | 38-208-03-13 |
| 1/4 | 38-106-04-13 | | 38-206-04-13 | | | 38-99-269-04 | 1/4 | 38-108-04-13 | | 38-208-04-13 |
| 9/32 | 38-106-05-13 | | 38-206-05-13 | | | 38-99-269-05 | 9/32 | 38-108-05-13 | | 38-208-05-13 |
| 5/16 | 38-106-06-13 | 38-506-06-13 | 38-206-06-13 | | | 38-99-269-06 | 5/16 | 38-108-06-13 | 38-508-06-13 | 38-208-06-13 |
| 11/32 | 38-106-07-13 | 38-506-07-13 | 38-206-07-13 | | | 38-99-269-07 | 11/32 | 38-108-07-13 | 38-508-07-13 | 38-208-07-13 |
| 3/8 | 38-106-08-13 | 38-506-08-13 | 38-206-08-13 | 38-906-08-13 | | 38-99-269-08 | 3/8 | 38-108-08-13 | 38-508-08-13 | 38-208-08-13 |
| 13/32 | 38-106-09-13 | 38-506-09-13 | 38-206-09-13 | 38-906-09-13 | | 38-99-269-09 | 13/32 | 38-108-09-13 | 38-508-09-13 | 38-208-09-13 |
| 7/16 | 38-106-10-13 | 38-506-10-13 | 38-206-10-13 | 38-906-10-13 | 38-806-10-13 | 38-99-269-10 | 7/16 | 38-108-10-13 | 38-508-10-13 | 38-208-10-13 |
| 15/32 | 38-106-11-13 | 38-506-11-13 | 38-206-11-13 | 38-906-11-13 | | 38-99-269-11 | 15/32 | 38-108-11-13 | 38-508-11-13 | 38-208-11-13 |
| 1/2 | 38-106-12-13 | 38-506-12-13 | 38-206-12-13 | 38-906-12-13 | 38-806-12-13 | 38-99-269-12 | 1/2 | 38-108-12-13 | 38-508-12-13 | 38-208-12-13 |
| 17/32 | 38-106-13-13 | 38-506-13-13 | 38-206-13-13 | | 38-806-13-13 | 38-99-269-13 | 17/32 | 38-108-13-13 | 38-508-13-13 | 38-208-13-13 |
| 9/16 | 38-106-14-13 | 38-506-14-13 | 38-206-14-13 | 38-906-14-13 | 38-806-14-13 | 38-99-269-14 | 9/16 | 38-108-14-13 | 38-508-14-13 | 38-208-14-13 |
| 19/32 | 38-106-15-13 | 38-506-15-13 | 38-206-15-13 | | | | 19/32 | 38-108-15-13 | 38-508-15-13 | 38-208-15-13 |
| 5/8 | 38-106-16-13 | 38-506-16-13 | 38-206-16-13 | 38-906-16-13 | 38-806-16-13 | 38-99-269-16 | 5/8 | 38-108-16-13 | 38-508-16-13 | 38-208-16-13 |
| 21/32 | 38-106-17-13 | 38-506-17-13 | 38-206-17-13 | | | | 21/32 | 38-108-17-13 | 38-508-17-13 | 38-208-17-13 |
| 11/16 | 38-106-18-13 | 38-506-18-13 | 38-206-18-13 | 38-906-18-13 | 38-806-18-13 | 38-99-269-18 | 11/16 | 38-108-18-13 | 38-508-18-13 | 38-208-18-13 |
| 23/32 | 38-106-19-13 | 38-506-19-13 | 38-206-19-13 | | | | 23/32 | 38-108-19-13 | 38-508-19-13 | 38-208-19-13 |
| 3/4 | 38-106-20-13 | 38-506-20-13 | 38-206-20-13 | 38-906-20-13 | 38-806-20-13 | 38-99-269-20 | 3/4 | 38-108-20-13 | 38-508-20-13 | 38-208-20-13 |
| 13/16 | 38-106-22-13 | | 38-206-22-13 | | | | 13/16 | 38-108-22-13 | 38-508-22-13 | 38-208-22-13 |
| | | | | | | | 27/32 | | | 38-208-23-13 |
| 7/8 | 38-106-24-13 | | 38-206-24-13 | | | 38-99-269-24 | 7/8 | 38-108-24-13 | 38-508-24-13 | 38-208-24-13 |
| 15/16 | 38-106-26-13 | | 38-206-26-13 | | | | 15/16 | 38-108-26-13 | 38-508-26-13 | 38-208-26-13 |
| 1 | 38-106-28-13 | 38-506-28-13 | 38-206-28-13 | | | 38-99-269-28 | 1 | 38-108-28-13 | 38-508-28-13 | 38-208-28-13 |
| 1-1/16 | 38-106-30-13 | | 38-206-30-13 | | | | 1-1/16 | 38-108-30-13 | 38-508-30-13 | 38-208-30-13 |
| 1-1/8 | 38-106-32-13 | | 38-206-32-13 | | | | 1-1/8 | 38-108-32-13 | 38-508-32-13 | 38-208-32-13 |

**\* NOTE:** The L dimension (length under head) is of importance only as a means of dimensional description. This dimension is undergoing modification in some of our rivets. In cases where actual shank length differs from that given in the L column above, a note both inside the rivet package and on its label gives corrected information. The grip length of the rivet is not affected by shank length modification.

# PALNUT® Self-Threading Nuts

## SPECIFIC ADVANTAGES OF OTHER TYPES

### Regular Type

This is the lowest-cost PALNUT Self-threading Nut. Especially adapted to assemblies where available space is at a premium. Uses shorter studs. Competitive with push types, yet assembles fast and assures better, tighter assemblies. Shape of nut makes it possible to assemble with internal wrench. (See wrench illustration below).

### Acorn Type

All the cost and assembly advantages of the PALNUT Self-threading principle are gained in this decorative, crowned nut. It covers up the ends of studs for pleasing appearance and protection against scratching.

**PALNUT Magnetic Socket Wrench**          **PALNUT Internal Hex Wrench**

## EASY, FAST ASSEMBLY

PALNUT Self-threading Nuts are formed to standard hex shape. They may be fastened with any standard tools, manual or power. *However, by using the PALNUT Magnetic Socket Wrench, which fits all standard tools, greater convenience and higher-speed assembly are obtained.*

### Power Tool Assembly

Air or electric tools, equipped with the PALNUT Magnetic Socket Wrench, permit starting, running on and tightening in one simple, fast operation.

### Manual Assembly

PALNUT Self-threading Nuts may be applied with hand tools, but power tools are recommended for production runs. The PALNUT Magnetic Socket Wrench facilitates manual assembly.

**Write for Bulletin WR-516 giving full details on PALNUT Wrenches and Tools**

## STUD INFORMATION

Studs of any malleable material may be used, including zinc or aluminum die cast, steel, brass or aluminum wire and high tensile plastic materials.

Tolerance on stud diameter ±0.003″

Permissible tolerance on stud diameter is plus or minus 0.003″. Tapered studs should be within this tolerance at the point where the nut grips the stud at completion of assembly. (See illustration).

PALNUT Self-threading Nuts will perform satisfactorily on blunt end studs or wire, but a 1/32″ x 45° chamfered end is preferred for easier starting. Full data on stud specifications is available on request.

## TORQUE-TENSILE DATA

Recommended assembly torque and resulting stud tension for each Self-Threading Nut is shown on page 7. The nominal torque figures given can be used for preliminary adjustment of power tools.

Power tools with ratcheting (slip) type clutch or the "one shot" quick-breaking type clutch may be used. For Self-Threaders with Sealer we specifically recommend power tools with the ratcheting type clutch.

Where assembly conditions deviate from the normal, a downward or upward revision of the recommended assembly torque setting on the power tool clutch may be necessary to meet the current conditions. Examples given below.

1. Dry, flat, non-slippery surfaces and stud diameters on the high side of the tolerance will require higher assembly torque.
2. Irregular, slippery surfaces and stud diameters on the low side of the tolerance will require lower assembly torque.

# PALNUT® Self-Threading Nuts

## DIMENSIONS AND TORQUE-TENSILE DATA

| WASHER TYPE | WASHER TYPE with SEALER (See Sealer Information on Page 5) | WASHER TYPE STYLE SG | REGULAR TYPE | ACORN TYPE |
|---|---|---|---|---|

| Type | Stud Size | PALNUT Part No. | Hex Width "W" | PALNUT Height "H" | Washer Diameter "D" | Teeth Depth "T" | Net Wgt. Lbs./M Pieces | TORQUE-TENSILE DATA (See page 6 for details) | | | |
|---|---|---|---|---|---|---|---|---|---|---|---|
| | | | | | | | | ON ZINC DIE CAST STUDS | | ON STEEL STUDS | |
| | | | | | | | | Recommended Torque (Inch Lbs.) | Stud Tension Lbs. | Recommended Torque (Inch Lbs.) | Stud Tension (Lbs.) |
| WASHER TYPE | 1/8" | SD125007 | 5/16" | .186" | 7/16" | | 1.2 | 15 | 90 | 26 | 130 |
| | 1/8 | SD125085 | 5/16 | .194 | 17/32 | | 1.5 | 20 | 80 | 30 | 110 |
| | 5/32 | SD156009 | 11/32 | .212 | 9/16 | | 1.9 | 30 | 100 | 45 | 240 |
| | 3/16 | SD188008 | 3/8 | .217 | 1/2 | | 1.7 | 38 | 210 | 50 | 330 |
| | 3/16 | SD188009 | 3/8 | .225 | 9/16 | | 2.0 | 42 | 180 | 55 | 300 |
| | 3/16 | SD188010 | 3/8 | .235 | 5/8 | | 2.3 | 46 | 160 | 60 | 280 |
| | 1/4 | SD250095 | 7/16 | .232 | 19/32 | | 2.5 | 65 | 260 | 95 | 370 |
| | 1/4 | SD250011 | 7/16 | .247 | 11/16 | | 3.1 | 75 | 240 | 90 | 340 |
| WASHER TYPE STYLE SG | 1/8 | SG125085 | 5/16 | .194 | 17/32 | .013" | 1.5 | 20 | 80 | 32 | 110 |
| | 3/16 | SG188008 | 3/8 | .207 | 1/2 | .013 | 1.6 | 38 | 210 | 55 | 330 |
| | 3/16 | SG188010 | 3/8 | .235 | 5/8 | .013 | 2.3 | 46 | 160 | 60 | 280 |
| | 1/4 | SG250095 | 7/16 | .219 | 19/32 | .015 | 2.4 | 65 | 260 | 95 | 370 |
| REGULAR TYPE | 3/32 | SR094005 | 5/16 | .100 | | | 0.7 | 5 | 40 | 8 | 80 |
| | 1/8 | SR125004 | 1/4 | .088 | | | 0.4 | 8 | 50 | 10 | 110 |
| | 1/8 | SR125 | 5/16 | .100 | | | 0.7 | 11 | 65 | 16 | 130 |
| | 1/8 | SR125009 | 9/16 | .114 | | | 2.0 | 18 | 90 | 30 | 190 |
| | 5/32 | SR156 | 11/32 | .110 | | | 0.9 | 18 | 100 | 30 | 250 |
| | 3/16 | SR188006 | 3/8 | .116 | | | 1.1 | 26 | 140 | 32 | 280 |
| | 3/16 | SR188 | 1/2 | .129 | | | 2.1 | 30 | 140 | 38 | 250 |
| TYPE ACORN | 1/8 | SC125 | 5/16 | .265 | | | 1.6 | 11 | 65 | 18 | 130 |
| | 1/8 | SC125006 | 3/8 | .324 | | | 2.2 | 14 | 80 | 25 | 160 |
| | 1/8 | SC125007 | 7/16 | .372 | | | 3.2 | 16 | 90 | 28 | 210 |
| | 3/16 | SC188 | 7/16 | .380 | | | 3.7 | 26 | 140 | 45 | 320 |
| | 3/16 | SC188008 | 1/2 | .437 | | | | 30 | 140 | 55 | 350 |
| | 3/16 | SC188009 | 9/16 | .474 | | | 6.3 | 32 | 140 | 65 | 400 |
| | 1/4 | SC250009 | 9/16 | .484 | | | | 53 | 180 | 77 | 500 |

Standard Finish: Cadmium.*  Material: Carbon Spring Steel. Optional Finishes: Phosphate, and Mechanical Zinc plus Chromate. (Except Acorn Type)
*Cadmium Plated PALNUT Self-Threading Nuts are usually colored yellow to differentiate them from PALNUT Lock Nuts. An exception is the Acorn Type which is plated with bright cadmium for appearance.

## WRITE FOR FREE SAMPLES, STATING TYPE, SIZE AND APPLICATION

Detailed information can be supplied on recommended assembly torques, stud tension, stud dimensions and types of sealers.

# PUSHNUT® FASTENERS—FLAT ROUND TYPE

## USER ADVANTAGES

Flat Round PUSHNUT Fasteners provide the advantages common to push-on type fasteners generally, which are made by several manufacturers. These include ● Exceptionally fast assembly ● Elimination of more expensive screws and nuts, washers or retaining rings ● Elimination of costly threading, hole drilling or annular grooving.

Compared with push-on nuts generally, Flat Round PUSHNUT Fasteners offer these *specific advantages*

**SUPERIOR GRIP** Means less slippage   **GREATER HOLDING POWER** Means higher axial load strength
The number, shape, length and angularity of teeth in each part represent a carefully-designed balance between relative ease of assembly and ultimate strength. Steel thickness and relation of outside diameter to stud size are also part of this "balanced design".

**LOWER COST** Production in large quantities on ultra-modern equipment, plus advanced finishing techniques, assure a per-thousand cost that is invariably below competitive designs.

## STUD RECOMMENDATIONS

**DIAMETER TOLERANCE:** $+.002''$ $-.003''$, INCLUDING PLATING. Chamfered (45°) stud ends are preferred, but if cut or sheared from longer rod or wire, square ends should be free of burrs or mushrooming.

**STUD MATERIALS:** May be mild steel, aluminum, die cast zinc, or other malleable materials.

Subject to verification by your own tests, Style PV parts, and Styles PS and PD parts in the thinner gauges, are likely to perform satisfactorily on plastics having good tensile strength and toughness.

**STUD FINISHES:** Steel studs—any commercial finish is satisfactory, except nickel or chrome plating. Die cast zinc—any finish is satisfactory, except nickel-chrome plating should not exceed .003'' maximum thickness.

**STUD HARDNESS:** See comments under PERFORMANCE DATA.

## PERFORMANCE DATA

Push-on Force and Removal Resistance values are "averages" obtained when parts are applied to low carbon, cold drawn steel rod of hardness not exceeding 78 on Rockwell 30T scale (this is corrected reading allowing for curvature error on round rod).

| PUSHNUT PART NO. | PUSH-ON FORCE LBS. | REMOV. RESIST. LBS. | PUSHNUT PART NO. | PUSH-ON FORCE LBS. | REMOV. RESIST. LBS. | PUSHNUT PART NO. | PUSH-ON FORCE LBS. | REMOV. RESIST. LBS |
|---|---|---|---|---|---|---|---|---|
| PS062032 | 20 | 100 | PZ001514 | 130 | 600 | PR375010 | 50 | 400 |
| PS094032 | 25 | 150 | PS240085 | 30 | 600 | PS375312 | 50 | 650 |
| PS125306 | 10 | 100 | PS250385 | 25 | 400 | PS375612 | 70 | 800 |
| PS125006 | 20 | 250 | PS250085 | 45 | 650 | PS375012 | 85 | 1000 |
| PD156307 | 10 | 250 | PV250015 | 10 | 140 | PV375015 | 15 | 450 |
| PD156007 | 20 | 300 | PR312075 | 40 | 300 | PS438014 | 70 | 1400 |
| PS188307 | 10 | 250 | PS312310 | 30 | 650 | PV438014 | 20 | 600 |
| PS188007 | 25 | 420 | PS312010 | 50 | 900 | PS500016 | 160 | 2000 |
| PD219385 | 20 | 300 | PV312015 | 15 | 300 | | | |

**NOTE:** Application of PUSHNUT Fasteners to studs or rods exceeding specified size tolerances, or maximum hardness, can result in considerable variation in REMOVAL RESISTANCE values. Nevertheless, careful pre-testing may show the resulting strength is still more than adequate, depending on degree of performance needed.

# PUSHNUT® FASTENERS—FLAT ROUND TYPE

## AVAILABLE SIZES AND STYLES

**DESIGN 1 STYLE PS**

**DESIGN 2 STYLE PS STYLE PD AND PART PZ001514**

**DESIGN 3 STYLE PS**

**STYLE PV**

**PART PR312075**

**PART PR375010**

| STUD DIAM. | PUSHNUT PART NO. | DES # | STEEL THICK. | TOTAL HEIGHT H | WASH. DIAM. INSIDE F | WASH. DIAM. OUTSIDE D | WGT. LBS/M. |
|---|---|---|---|---|---|---|---|
| .062" (1/16") | PS062032 | 3 | .009" | .038" | .142" | .195" | 0.09 |
| .094" (3/32") | PS094032 | 3 | .009" | .038" | .142" | .194" | 0.09 |
| .125" (1/8") | PS125306 | 1 | .009" | .045" | .228" | 3/8" | 0.3 |
| .125" (1/8") | PS125006 | 1 | .013" | .052" | .228" | 3/8" | 0.4 |
| .156" (5/32") | PD156307 | 2 | .010" | .047" | .320" | 7/16" | 0.4 |
| .156" (5/32") | PD156007 | 2 | .013" | .058" | .320" | 7/16" | 0.5 |
| .188" (3/16") | PS188307 | 2 | .010" | .056" | .320" | 7/16" | 0.4 |
| .188" (3/16") | PS188007 | 2 | .015" | .064" | .320" | 7/16" | 0.5 |
| .219" (7/32") | PD219385 | 2 | .012" | .067" | .388" | 17/32" | 0.7 |
| .237" | PZ001514 | 2 | .021" | .050" | .420" | 3/4" | 2.5 |
| .240" | PS240085 | 2 | .017" | .069" | .388" | 17/32" | 0.9 |
| .250" (1/4") | PS250385 | 2 | .012" | .057" | .388" | 17/32" | 0.6 |
| .250" (1/4") | PS250085 | 2 | .017" | .066" | .388" | 17/32" | 1.0 |
| .250" (1/4") | PV250015 | | .015" | .083" | .750" | 15/16" | 2.5 |
| .312" (5/16") | PR312075 | | .013" | .040" | .385" | 15/32" | 0.35 |
| .312" (5/16") | PS312310 | 2 | .015" | .059" | .456" | 5/8" | 1.1 |
| .312" (5/16") | PS312010 | 2 | .021" | .070" | .456" | 5/8" | 1.5 |
| .312" (5/16") | PV312015 | | .015" | .097" | .750" | 15/16" | 2.7 |
| .375" (3/8") | PR375010 | | .015" | .056" | .500" | 5/8" | 0.87 |
| .375" (3/8") | PS375312 | 2 | .017" | .061" | .546" | 3/4" | 1.7 |
| .375" (3/8") | PS375612 | 2 | .021" | .068" | .546" | 3/4" | 2.1 |
| .375" (3/8") | PS375012 | 2 | .027" | .081" | .546" | 3/4" | 2.8 |
| .375" (3/8") | PV375015 | | .015" | .093" | .750" | 15/16" | 2.5 |
| .438" (7/16") | PS438014 | 2 | .030" | .097" | .638" | 7/8" | 4.1 |
| .438" (7/16") | PV438015 | | .015" | .075" | .750" | 15/16" | 2.5 |
| .500" (1/2") | PS500016 | 2 | .035" | .112" | .730" | 1" | 6.4 |

**MATERIALS:** Spring Steel (some parts available in stainless steel on special order).

**FINISHES:** Mechanical Zinc, Cadmium or Electro-Zinc (see note), Phosphate and Oil, Plain. Other finishes on special order.

**NOTE:** Style PS and PD parts, and part PZ001514, are not available with Cadmium, Electro-Zinc or any other electroplated finish.

For aluminum, and some plastic studs, or wooden dowels, Style PV, Styles PS and PD in the thinner gauges, or Style PR parts are recommended.

# RIVNUT ENGINEERING DATA
# PREPARATION PROCEDURES

### FOR FLAT HEAD INSTALLATION

Flat head Rivnuts require only a punched or drilled hole of the proper size for their installation. Use a lead drill if desired and follow with a sharp finish drill held at right angles to the work. Always see that dirt or metal particles are removed from between metal sheets and burrs removed wherever possible. Sheets should be clamped or pressed into contact to reduce air gap to a minimum.

### FOR MACHINE COUNTERSINK INSTALLATION

Machine countersinking can be used only in metal thicker than the head thickness of the Rivnut. A precision hole and countersink can best be obtained by following these simple steps:

### FOR DIMPLE COUNTERSINK INSTALLATION

Metal thinner than Rivnut head thickness requires a dimple countersink.

The ideal bulge on any Rivnut application will always be formed against a flat under surface. The bell-mouth that results from ordinary dimpling will not permit proper formation of the bulge. Rivnuts upset against the sharp edge will form a weak bulge, a spread shank, or possibly shear.

To provide a flat surface in the dimpling operation, a ledge at the bottom of the dimpling die can be used. The "flat" on the dimple will save costly deburring before dimpling and enables the Rivnut bulge to form normally, providing maximum strength.

Dimple or press countersink hole. Note ledge at bottom of dimpling die

1. Drill undersized hole with lead drill

2. Countersink

3. Drill correct diameter hole with sharp finish drill

4. If keyed Rivnut is to be used cut keyway with key cutter tool or by the use of a guided drill

5. Install Rivnut

### FOR KEYED RIVNUTS

Keyways can be cut with a B.F.Goodrich Key Cutter tool. Standard tool will cut 3/32″ aluminum, 1/16″ mild steel, 1/32″ stainless steel. To cut keyways in metal too thick for this tool, use a small, round file or guided drill.

### TORQUE WITHOUT KEYS

A hexagonal hole with a keyless Rivnut is an excellent substitute for keyed Rivnuts and eliminates the need to match the key to the keyway. This type installation offers torque resistance approximating that of keyed Rivnuts.

### HOW TO REMOVE RIVNUTS

Should it be necessary to remove an installed Rivnut, it can be drilled out by using the same size drill as used for the original hole. Drill through the head of the Rivnut and then punch out the shank. The counterbore will act as a guide for the drill. The same size Rivnut can then be installed in the same hole.

# ADJUSTING ANVIL TO SUIT RIVNUT LENGTH

Rivnut threads may be deformed or stripped if the pull-up stud does not engage all threads in the Rivnut. Speed Headers and all power tools have pull-up studs or anvils which can be adjusted easily to suit the Rivnut blank length. The following sketches demonstrate proper relation between face of anvil and end of pull-up stud.

### OPEN END RIVNUT

Point of the pull-up stud should extend just beyond the end of the Rivnut, as in Fig. 1. Fig. 2 shows Rivnut head tight against anvil, ready for upset.

### CLOSED END RIVNUT

Thread Rivnut on pull-up stud all the way to bottom of threads. Back Rivnut off one complete turn, then adjust anvil so it contacts Rivnut head, as in Fig. 3.

### FILLISTER HEAD RIVNUT

Adjust stud to same reference points as conventional types.

# FINDING GRIP RANGE

### GRIP RANGE

### MEASURING "GRIP"

Maximum grip (above) represents the greatest material thickness in which a specific Rivnut should be properly installed. Minimum grip represents material of least thickness in which a specific Rivnut can be properly installed.

"Grip Range" is that area between maximum and minimum —the zone of thickness best suited to the installation of a specific Rivnut.

The grip ranges for various Rivnuts can be found on pages 8 and 9.

IMPORTANT: When material thickness (grip) approaches minimum or maximum for a given size Rivnut, a trial installation should be made.

In order to select the correct Rivnut, physical measurements must be made. When installing flat head Rivnuts in a surface installation or countersunk types in machine countersunk holes, "grip" is the same as metal thickness (left above).

For dimpled or press countersunk holes, "grip" is the measurement from the metal surface to the underside of the dimpled hole (right above).

IMPORTANT: Measurements should include air gaps, paint and any burrs which cannot be removed.

# RIVNUT ENGINEERING DATA—FLAT HEAD

| First No. of Type No. | Thread Size* | B ±.015 | C Nom. | D +.000 −.004 | E Max. | F +.005 −.000 | G Max. | Install Drill Size (Ref.) | Install Hole Size Min. | Max. | Keyway Dimensions P+.003 −.000 | H |
|---|---|---|---|---|---|---|---|---|---|---|---|---|
| 4 | ♯ 4-40 UNC-3B | .270 | .025 | .155 | .198 | .054 | .023 | ⁵⁄₃₂ | .155 | .157 | .062 | .046-.048 |
| 6 | ♯ 6-32 UNC-3B | .325 | .032 | .189 | .240 | .054 | .023 | ♯12 | .189 | .193 | .062 | .056-.058 |
| 8 | ♯ 8-32 UNC-3B | .357 | .032 | .221 | .271 | .054 | .023 | ♯2 | .221 | .226 | .062 | .056-.058 |
| 10 | ♯10-32 UNF-3B | .406 | .038 | .250 | .302 | .054 | .023 | E | .250 | .256 | .062 | .056-.058 |
| 25 | ¼-20 UNC-3B | .475 | .058 | .332 | .382 | .054 | .035 | Q | .332 | .338 | .062 | .056-.058 |
| 31 | ⁵⁄₁₆-18 UNC-3B | .665 | .062 | .413 | .505 | .120 | .040 | Z | .413 | .423 | .128 | .097-.102 |
| 37 | ⅜-16 UNC-3B | .781 | .088 | .490 | .597 | .120 | .040 | 12.5 MM | .490 | .500 | .128 | .110-.115 |
| 50 | ½-13 UNC-3B | 1.000 | .125 | .640 | .772 | .151 | .040 | 4¹⁄₆₄ | .640 | .650 | .159 | .135-.140 |

*Both UNC and UNF threads are available in No. 10 and larger thread sizes.

CODE: Diameter and grip range as tabulated. First letter of type number indicates material and finish: "A" for aluminum alloy, "S" for C-1108 or C-1110 steel, "SS" for Type 430 corrosion resistant steel, "CH" for heat treated C-4037 steel and "BR" for brass. Letter between dash numbers indicate type "—" for keyless open end, "K" for keyed open end, "B" for keyless closed end, "KB" for keyed closed end.

EXAMPLES: A25K80 = Aluminum alloy keyed open end ¼-20 internal thread .020 to .080 grip range.    SS6KB200 = Corrosion resistant steel keyed closed end No. 6-32 internal thread .160 to .200 grip range.

WEIGHTS — For Brass Rivnuts multiply weight of aluminum Rivnuts by 3.31. Weights for "CH" Rivnuts (C-4037 steel) and "SS" Rivnuts (Type 430 corrosion resistant steel ) same as for "S" Rivnuts.

| TYPE NUMBER | GRIP RANGE | IDENT. MARK | OPEN END KEYED AND KEYLESS A ±.015 | M REF. | Wt. (LBS./1000) ALUM. | STEEL | CLOSED END KEYLESS L ±.015 | J REF. | K REF. | WT. (LBS./1000) ALUM. | STEEL | CLOSED END KEYED L ±.015 | J REF. | K REF. | WT. (LBS./1000) ALUM.- | STEEL |
|---|---|---|---|---|---|---|---|---|---|---|---|---|---|---|---|---|
| 4-60 | .010-.060 | Blank | .345 | .230 | .4 | 1.3 | .500 | .385 | .230 | .6 | 1.9 | .500 | .385 | .230 | .6 | 1.9 |
| 4-85 | .060-.085 | 1-Rad. | .370 | .230 | .4 | 1.4 | .525 | .385 | .230 | .7 | 2.0 | .525 | .385 | .230 | .7 | 2.0 |
| 4-110 | .085-.110 | 2-Rad. | .400 | .230 | .5 | 1.4 | .555 | .390 | .230 | .7 | 2.0 | .555 | .390 | .230 | .7 | 2.0 |
| 4-135 | .110-.135 | 3-Rad. | .425 | .230 | .5 | 1.5 | .580 | .385 | .230 | .7 | 2.1 | .580 | .385 | .230 | .7 | 2.1 |
| 4-160 | .135-.160 | 4-Rad. | .450 | .230 | .5 | 1.5 | .605 | .385 | .230 | .7 | 2.1 | .605 | .385 | .230 | .7 | 2.1 |
| 4-185 | .160-.185 | 5-Rad. | .480 | .230 | .5 | 1.6 | .635 | .385 | .230 | .7 | 2.2 | .635 | .385 | .230 | .7 | 2.2 |
| 6-75 | .010-.075 | 1-Rad. | .438 | .300 | .8 | 2.4 | .625 | .490 | .305 | 1.2 | 3.5 | .750 | .615 | .405 | 1.4 | 4.1 |
| 6-120 | .075-.120 | 3-Rad. | .500 | .315 | .9 | 2.6 | .625 | .440 | .255 | 1.1 | 3.4 | .750 | .565 | .355 | 1.3 | 4.0 |
| 6-160 | .120-.160 | 5-Rad. | .500 | .270 | .9 | 2.6 | .750 | .520 | .260 | 1.3 | 4.0 | .750 | .520 | .310 | 1.3 | 4.0 |
| 6-200 | .160-.200 | 1-Circ. | .562 | .290 | .9 | 2.8 | .750 | .480 | .260 | 1.3 | 3.9 | .750 | .480 | .260 | 1.3 | 3.9 |
| 6-240 | .200-.240 | 2-Circ. | .625 | .310 | 1.0 | 3.0 | .750 | .435 | .260 | 1.3 | 3.8 | .750 | .435 | .260 | 1.3 | 3.8 |
| 6-280 | .240-.280 | 3-Circ. | .687 | .330 | 1.1 | 3.3 | .812 | .455 | .265 | 1.3 | 4.1 | .812 | .455 | .265 | 1.3 | 4.1 |
| 8-75 | .010-.075 | 1-Rad. | .438 | .300 | 1.0 | 3.0 | .625 | .490 | .305 | 1.5 | 4.5 | .750 | .615 | .405 | 1.7 | 5.3 |
| 8-120 | .075-.120 | 3-Rad. | .500 | .315 | 1.1 | 3.3 | .625 | .440 | .255 | 1.4 | 4.4 | .750 | .565 | .355 | 1.7 | 5.2 |
| 8-160 | .120-.160 | 5-Rad. | .500 | .270 | 1.1 | 3.2 | .750 | .520 | .260 | 1.7 | 5.1 | .750 | .520 | .310 | 1.7 | 5.1 |
| 8-200 | .160-.200 | 1-Circ. | .625 | .350 | 1.3 | 3.9 | .750 | .475 | .265 | 1.6 | 5.0 | .750 | .475 | .265 | 1.6 | 5.0 |
| 8-240 | .200-.240 | 2-Circ. | .625 | .305 | 1.2 | 3.8 | .875 | .555 | .310 | 1.9 | 5.6 | .875 | .555 | .310 | 1.9 | 5.6 |
| 8-280 | .240-.280 | 3-Circ. | .687 | .340 | 1.3 | 4.1 | .875 | .530 | .265 | 1.8 | 5.6 | .875 | .530 | .290 | 1.8 | 5.6 |
| 10-80 | .010-.080 | Blank | .531 | .380 | 1.5 | 4.5 | .781 | .630 | .380 | 2.3 | 6.8 | .781 | .630 | .380 | 2.3 | 6.8 |
| 10-130 | .080-.130 | 1-Rad. | .594 | .390 | 1.6 | 4.9 | .843 | .640 | .390 | 2.4 | 7.2 | .843 | .640 | .390 | 2.4 | 7.2 |
| 10-180 | .130-.180 | 2-Rad. | .641 | .390 | 1.7 | 5.1 | .891 | .640 | .390 | 2.4 | 7.4 | .891 | .640 | .390 | 2.4 | 7.4 |
| 10-230 | .180-.230 | 3-Rad. | .703 | .395 | 1.8 | 5.4 | .953 | .645 | .395 | 2.6 | 7.8 | .953 | .645 | .395 | 2.6 | 7.8 |
| 10-280 | .230-.280 | 4-Rad. | .750 | .395 | 1.9 | 5.7 | 1.000 | .645 | .395 | 2.6 | 8.0 | 1.000 | .645 | .395 | 2.6 | 8.0 |
| 10-330 | .280-.330 | 5-Rad. | .797 | .385 | 1.9 | 5.9 | 1.047 | .630 | .385 | 2.7 | 8.2 | 1.047 | .630 | .385 | 2.7 | 8.2 |
| 25-80 | .020-.080 | Blank | .625 | .450 | 3.2 | 9.7 | .937 | .760 | .440 | 4.9 | 15.1 | .937 | .760 | .440 | 5.0 | 15.1 |
| 25-140 | .080-.140 | 1-Rad. | .687 | .450 | 3.4 | 10.3 | 1.000 | .760 | .440 | 5.1 | 15.7 | 1.000 | .760 | .440 | 5.1 | 15.7 |
| 25-200 | .140-.200 | 2-Rad. | .750 | .450 | 3.6 | 10.9 | 1.062 | .760 | .440 | 5.3 | 16.3 | 1.062 | .760 | .440 | 5.3 | 16.3 |
| 25-260 | .200-.260 | 3-Rad. | .812 | .445 | 3.8 | 11.5 | 1.125 | .755 | .445 | 5.5 | 16.8 | 1.125 | .755 | .445 | 5.5 | 16.8 |
| 25-320 | .260-.320 | 4-Rad. | .875 | .445 | 4.0 | 12.0 | 1.187 | .755 | .445 | 5.7 | 17.4 | 1.187 | .755 | .445 | 5.7 | 17.4 |
| 25-380 | .320-.380 | 5-Rad. | .937 | .445 | 4.1 | 12.6 | 1.250 | .755 | .445 | 5.9 | 18.0 | 1.250 | .755 | .445 | 5.9 | 18.0 |
| 31-125 | .030-.125 | Blank | .750 | .505 | 6.0 | 18.2 | 1.187 | .940 | .550 | 9.6 | 29.1 | 1.187 | .940 | .550 | 9.6 | 29.2 |
| 31-200 | .125-.200 | 1-Rad. | .875 | .555 | 6.7 | 20.3 | 1.281 | .960 | .555 | 10.1 | 30.6 | 1.281 | .960 | .555 | 10.1 | 30.7 |
| 31-275 | .200-.275 | 2-Rad. | .937 | .540 | 6.9 | 21.1 | 1.343 | .950 | .560 | 10.3 | 31.4 | 1.343 | .950 | .560 | 10.3 | 31.5 |
| 31-350 | .275-.350 | 3-Rad. | 1.032 | .560 | 7.4 | 22.6 | 1.437 | .965 | .570 | 10.8 | 32.9 | 1.437 | .965 | .570 | 10.8 | 32.9 |
| 31-425 | .350-.425 | 4-Rad. | 1.125 | .580 | 7.9 | 24.0 | 1.531 | .985 | .575 | 11.3 | 34.3 | 1.531 | .985 | .575 | 11.3 | 34.4 |
| 31-500 | .425-.500 | 5-Rad. | 1.187 | .565 | 8.2 | 24.9 | 1.593 | .975 | .580 | 11.5 | 35.1 | 1.593 | .975 | .580 | 11.6 | 35.2 |
| 37-115 | .030-.115 | Blank | .844 | .585 | 9.7 | 29.7 | 1.281 | 1.020 | .660 | 14.8 | 45.0 | 1.281 | 1.020 | .660 | 14.8 | 45.1 |
| 37-200 | .115-.200 | 1-Rad. | .938 | .595 | 10.3 | 31.4 | 1.375 | 1.030 | .670 | 15.4 | 46.8 | 1.375 | 1.030 | .670 | 15.4 | 46.9 |
| 37-285 | .200-.285 | 2-Rad. | 1.031 | .605 | 10.9 | 33.2 | 1.468 | 1.040 | .680 | 15.9 | 48.5 | 1.468 | 1.040 | .680 | 16.0 | 48.6 |
| 37-370 | .285-.370 | 3-Rad. | 1.125 | .615 | 11.5 | 34.9 | 1.562 | 1.050 | .690 | 16.5 | 50.3 | 1.562 | 1.050 | .690 | 16.5 | 50.4 |
| 37-455 | .370-.455 | 4-Rad. | 1.218 | .630 | 12.0 | 36.7 | 1.656 | 1.065 | .710 | 17.1 | 52.1 | 1.656 | 1.065 | .710 | 17.1 | 52.2 |
| 37-540 | .455-.540 | 5-Rad. | 1.312 | .635 | 12.6 | 38.5 | 1.750 | 1.075 | .715 | 17.7 | 53.8 | 1.750 | 1.075 | .715 | 17.7 | 53.9 |
| 50-145 | .025-.145 | Blank | 1.062 | .730 | 20.8 | 63.5 | 1.656 | 1.325 | .855 | 31.9 | 97.3 | 1.656 | 1.325 | .855 | 32.0 | 97.5 |
| 50-265 | .145-.265 | 1-Rad. | 1.188 | .735 | 22.2 | 67.5 | 1.781 | 1.330 | .865 | 33.2 | 101.3 | 1.781 | 1.330 | .865 | 33.3 | 101.4 |
| 50-385 | .265-.385 | 2-Rad. | 1.312 | .740 | 23.4 | 71.4 | 1.906 | 1.335 | .875 | 34.5 | 105.1 | 1.906 | 1.335 | .875 | 34.5 | 105.2 |
| 50-505 | .385-.505 | 3-Rad. | 1.453 | .765 | 24.9 | 76.0 | 2.156 | 1.465 | .895 | 37.8 | 115.1 | 2.156 | 1.465 | .895 | 37.8 | 115.3 |
| 50-625 | .505-.625 | 4-Rad. | 1.578 | .770 | 26.2 | 80.0 | 2.297 | 1.485 | .905 | 39.3 | 119.9 | 2.297 | 1.485 | .905 | 39.4 | 120.0 |
| 50-745 | .625-.745 | 5-Rad. | 1.719 | .790 | 27.8 | 84.6 | 2.438 | 1.510 | .920 | 40.9 | 124.5 | 2.438 | 1.510 | .920 | 40.9 | 124.6 |

# • RIVNUT ENGINEERING DATA—COUNTERSUNK HEAD

| First No. of Type No. | Thread Size* | B Ref. | C Max. | D +.000 −.004 | E Max. | F +.005 −.000 | Install Drill Size (Ref.) | Install Hole Size Min. | Install Hole Size Max. | Keyway Dimensions P +.003 −.000 | Keyway Dimensions H |
|---|---|---|---|---|---|---|---|---|---|---|---|
| 4 | # 4-40 UNC-3B | .263 | .051 | .155 | .198 | .054 | 5/32 | .155 | .157 | .062 | .046-.048 |
| 6 | # 6-32 UNC-3B | .323 | .063 | .189 | .240 | .054 | #12 | .189 | .193 | .062 | .056-.058 |
| 8 | # 8-32 UNC-3B | .355 | .063 | .221 | .271 | .054 | # 2 | .221 | .226 | .062 | .056-.058 |
| 10 | #10-32 UNF-3B | .391 | .065 | .250 | .302 | .054 | E | .250 | .256 | .062 | .056-.058 |
| 25 | ¼-20 UNC-3B | .529 | .089 | .332 | .382 | .054 | Q | .332 | .338 | .062 | .056-.058 |
| 31 | 5/16-18 UNC-3B | .656 | .104 | .413 | .505 | .120 | Z | .413 | .423 | .128 | .097-.102 |
| 37 | ⅜-16 UNC-3B | .770 | .124 | .490 | .597 | .120 | 12.5 MM | .490 | .500 | .128 | .110-.115 |
| 50 | ½-13 UNC-3B | .990 | .154 | .640 | .772 | .151 | 41/64 | .640 | .650 | .159 | .135-.140 |

*Both UNC and UNF threads are available in No. 10 and larger thread sizes.

CODE: Diameter and grip range as tabulated. First letter of type number indicates material and finish: "A" for aluminum alloy, "S" for C-1108 or C-1110 steel, "SS" for Type 430 corrosion resistant steel, "CH" for heat treated C-4037 steel and "BR" for brass. Letter between dash numbers indicate type "—" for keyless open end, "K" for keyed open end, "B" for keyless closed end, "KB" for keyed closed end.

EXAMPLES: A25K151 = Aluminum alloy keyed open end ¼-20   SS6KB241 = Corrosion resistant steel keyed closed end No. internal thread .089 to .151 grip range.   6-32 internal thread .201 to .241 grip range.

WEIGHTS — For Brass Rivnuts multiply weight of aluminum Rivnuts by 3.31. Weights for "CH" Rivnuts (C-4037 steel) and "SS" Rivnuts (Type 430 corrosion resistant steel) same as for "S" Rivnuts.

| TYPE NUMBER | GRIP RANGE | IDENT. MARK | OPEN END KEYED AND KEYLESS A ±.015 | OPEN END KEYED AND KEYLESS M REF. | OPEN END WT. (LBS./1000) ALUM. | OPEN END WT. (LBS./1000) STEEL | CLOSED END KEYLESS L ±.015 | CLOSED END KEYLESS J REF. | CLOSED END KEYLESS K REF. | CLOSED END KEYLESS WT. (LBS./1000) ALUM. | CLOSED END KEYLESS WT. (LBS./1000) STEEL | CLOSED END KEYED L ±.015 | CLOSED END KEYED J REF. | CLOSED END KEYED K REF. | CLOSED END KEYED WT. (LBS./1000) ALUM. | CLOSED END KEYED WT. (LBS./1000) STEEL |
|---|---|---|---|---|---|---|---|---|---|---|---|---|---|---|---|---|
| 4-81 | .050-.081 | Blank | .370 | .235 | .4 | 1.3 | .525 | .390 | .235 | .6 | 1.9 | .525 | .390 | .235 | .6 | 1.9 |
| 4-106 | .081-.106 | 1-Rad. | .395 | .235 | .4 | 1.3 | .550 | .390 | .235 | .6 | 1.9 | .550 | .390 | .235 | .6 | 1.9 |
| 4-131 | .106-.131 | 2-Rad. | .420 | .235 | .4 | 1.4 | .575 | .390 | .235 | .7 | 2.0 | .575 | .390 | .235 | .7 | 2.0 |
| 4-156 | .131-.156 | 3-Rad. | .450 | .235 | .5 | 1.4 | .600 | .390 | .235 | .7 | 2.0 | .600 | .390 | .235 | .7 | 2.0 |
| 4-181 | .156-.181 | 4-Rad. | .475 | .235 | .5 | 1.5 | .625 | .390 | .235 | .7 | 2.1 | .625 | .390 | .235 | .7 | 2.1 |
| 4-206 | .181-.206 | 5-Rad. | .500 | .235 | .5 | 1.5 | .650 | .390 | .235 | .7 | 2.1 | .650 | .390 | .235 | .7 | 2.1 |
| 6-106 | .065-.106 | Blank | .500 | .325 | .8 | 2.5 | .687 | .510 | .325 | 1.2 | 3.6 | .812 | .635 | .425 | 1.4 | 4.2 |
| 6-161 | .106-.161 | 2-Rad. | .500 | .280 | .8 | 2.4 | .687 | .465 | .280 | 1.2 | 3.5 | .812 | .590 | .380 | 1.3 | 4.1 |
| 6-201 | .161-.201 | 4-Rad. | .562 | .295 | .9 | 2.6 | .687 | .420 | .260 | 1.1 | 3.4 | .812 | .545 | .335 | 1.3 | 4.0 |
| 6-241 | .201-.241 | 1-Circ. | .625 | .315 | .9 | 2.9 | .812 | .505 | .295 | 1.3 | 4.0 | .812 | .505 | .295 | 1.3 | 4.0 |
| 6-281 | .241-.281 | 2-Circ. | .625 | .270 | .9 | 2.8 | .812 | .465 | .265 | 1.3 | 3.9 | .812 | .465 | .265 | 1.3 | 3.9 |
| 6-321 | .281-.321 | 3-Circ. | .687 | .290 | 1.0 | 3.0 | .844 | .455 | .265 | 1.3 | 4.0 | .844 | .455 | .265 | 1.3 | 4.0 |
| 8-106 | .065-.106 | Blank | .500 | .325 | 1.0 | 3.1 | .687 | .510 | .325 | 1.5 | 4.6 | .812 | .635 | .425 | 1.8 | 5.4 |
| 8-161 | .106-.161 | 2-Rad. | .500 | .280 | 1.0 | 3.0 | .687 | .465 | .280 | 1.5 | 4.5 | .812 | .590 | .380 | 1.7 | 5.3 |
| 8-201 | .161-.201 | 4-Rad. | .562 | .290 | 1.1 | 3.3 | .687 | .415 | .255 | 1.4 | 4.4 | .812 | .540 | .330 | 1.7 | 5.2 |
| 8-241 | .201-.241 | 1-Circ. | .625 | .310 | 1.2 | 3.6 | .875 | .560 | .290 | 1.8 | 5.5 | .875 | .560 | .290 | 1.8 | 5.5 |
| 8-281 | .241-.281 | 2-Circ. | .687 | .325 | 1.1 | 3.2 | .875 | .515 | .290 | 1.8 | 5.4 | .875 | .515 | .290 | 1.8 | 5.4 |
| 8-321 | .281-.321 | 3-Circ. | .687 | .295 | 1.2 | 3.8 | .875 | .485 | .300 | 1.7 | 5.2 | .875 | .485 | .300 | 1.7 | 5.2 |
| 10-116 | .065-.116 | Blank | .578 | .395 | 1.4 | 4.3 | .828 | .645 | .395 | 2.2 | 6.7 | .828 | .645 | .395 | 2.2 | 6.7 |
| 10-166 | .116-.166 | 1-Rad. | .625 | .385 | 1.5 | 4.6 | .875 | .635 | .385 | 2.3 | 6.9 | .875 | .635 | .385 | 2.3 | 6.9 |
| 10-216 | .166-.216 | 2-Rad. | .687 | .400 | 1.6 | 4.9 | .938 | .650 | .400 | 2.4 | 7.2 | .938 | .650 | .400 | 2.4 | 7.2 |
| 10-266 | .216-.266 | 3-Rad. | .734 | .390 | 1.7 | 5.1 | .984 | .640 | .390 | 2.5 | 7.5 | .984 | .640 | .390 | 2.5 | 7.5 |
| 10-316 | .266-.316 | 4-Rad. | .781 | .385 | 1.8 | 5.4 | 1.031 | .635 | .385 | 2.5 | 7.7 | 1.031 | .635 | .385 | 2.5 | 7.7 |
| 10-366 | .316-.366 | 5-Rad. | .844 | .400 | 1.9 | 5.7 | 1.094 | .650 | .400 | 2.6 | 8.0 | 1.094 | .650 | .400 | 2.6 | 8.0 |
| 25-151 | .089-.151 | Blank | .687 | .440 | 3.2 | 9.8 | 1.000 | .750 | .435 | 5.0 | 15.1 | 1.000 | .750 | .435 | 5.0 | 15.1 |
| 25-211 | .151-.211 | 1-Rad. | .750 | .440 | 3.4 | 10.3 | 1.062 | .750 | .435 | 5.2 | 15.7 | 1.062 | .750 | .435 | 5.2 | 15.7 |
| 25-271 | .211-.271 | 2-Rad. | .812 | .440 | 3.6 | 10.9 | 1.125 | .750 | .435 | 5.4 | 16.3 | 1.125 | .750 | .435 | 5.4 | 16.3 |
| 25-331 | .271-.331 | 3-Rad. | .875 | .435 | 3.8 | 11.5 | 1.187 | .750 | .435 | 5.7 | 16.9 | 1.187 | .750 | .435 | 5.7 | 16.9 |
| 25-391 | .331-.391 | 4-Rad. | .937 | .435 | 4.0 | 12.1 | 1.250 | .750 | .435 | 5.7 | 17.5 | 1.250 | .750 | .435 | 5.7 | 17.5 |
| 25-451 | .391-.451 | 5-Rad. | 1.000 | .445 | 4.2 | 12.7 | 1.312 | .760 | .445 | 5.9 | 18.1 | 1.312 | .760 | .445 | 5.9 | 18.1 |
| 31-181 | .106-.181 | Blank | .844 | .540 | 5.9 | 17.9 | 1.218 | .915 | .540 | 9.0 | 27.5 | 1.218 | .915 | .540 | 9.0 | 27.5 |
| 31-256 | .181-.256 | 1-Rad. | .937 | .560 | 6.3 | 19.3 | 1.312 | .935 | .560 | 9.5 | 28.9 | 1.312 | .935 | .560 | 9.5 | 29.0 |
| 31-331 | .256-.331 | 2-Rad. | 1.000 | .550 | 6.6 | 20.1 | 1.406 | .955 | .550 | 10.0 | 30.4 | 1.406 | .955 | .550 | 10.0 | 30.5 |
| 31-406 | .331-.406 | 3-Rad. | 1.093 | .565 | 7.1 | 21.5 | 1.468 | .940 | .565 | 10.2 | 31.1 | 1.468 | .940 | .565 | 10.2 | 31.2 |
| 31-481 | .406-.481 | 4-Rad. | 1.156 | .555 | 7.3 | 22.3 | 1.562 | .960 | .555 | 10.7 | 32.6 | 1.562 | .960 | .555 | 10.8 | 32.7 |
| 31-556 | .481-.556 | 5-Rad. | 1.250 | .575 | 7.8 | 23.7 | 1.625 | .950 | .575 | 10.9 | 33.3 | 1.625 | .950 | .575 | 11.0 | 33.4 |
| 37-211 | .125-.211 | Blank | .938 | .580 | 8.9 | 27.0 | 1.375 | 1.020 | .655 | 13.9 | 42.3 | 1.375 | 1.020 | .655 | 13.9 | 42.4 |
| 37-296 | .211-.296 | 1-Rad. | 1.031 | .590 | 9.4 | 28.7 | 1.468 | 1.030 | .655 | 14.5 | 44.1 | 1.468 | 1.030 | .655 | 14.5 | 44.1 |
| 37-381 | .296-.381 | 2-Rad. | 1.125 | .600 | 10.0 | 30.5 | 1.562 | 1.040 | .675 | 15.0 | 45.8 | 1.562 | 1.040 | .675 | 15.1 | 45.9 |
| 37-466 | .381-.466 | 3-Rad. | 1.219 | .615 | 10.6 | 32.3 | 1.656 | 1.050 | .690 | 15.6 | 47.6 | 1.656 | 1.050 | .690 | 15.7 | 47.7 |
| 37-551 | .466-.551 | 4-Rad. | 1.312 | .625 | 11.2 | 34.0 | 1.750 | 1.065 | .705 | 16.2 | 49.4 | 1.750 | 1.065 | .705 | 16.3 | 49.5 |
| 37-636 | .551-.636 | 5-Rad. | 1.422 | .650 | 11.9 | 36.2 | 1.859 | 1.090 | .715 | 16.9 | 51.6 | 1.859 | 1.090 | .715 | 17.0 | 51.7 |
| 50-276 | .156-.276 | Blank | 1.188 | .725 | 18.4 | 56.1 | 1.781 | 1.320 | .850 | 29.5 | 89.9 | 1.781 | 1.320 | .850 | 29.5 | 90.0 |
| 50-396 | .276-.396 | 1-Rad. | 1.312 | .730 | 19.7 | 60.0 | 1.906 | 1.325 | .865 | 30.8 | 93.7 | 1.906 | 1.325 | .865 | 30.8 | 93.9 |
| 50-516 | .396-.516 | 2-Rad. | 1.438 | .735 | 21.0 | 63.9 | 2.031 | 1.330 | .880 | 32.0 | 97.6 | 2.031 | 1.330 | .880 | 32.1 | 97.7 |
| 50-636 | .516-.636 | 3-Rad. | 1.578 | .750 | 22.5 | 68.5 | 2.172 | 1.350 | .890 | 33.6 | 102.3 | 2.172 | 1.350 | .890 | 33.6 | 102.4 |
| 50-756 | .636-.756 | 4-Rad. | 1.718 | .765 | 24.0 | 73.1 | 2.312 | 1.360 | .900 | 35.1 | 106.9 | 2.312 | 1.360 | .900 | 35.1 | 107.0 |
| 50-876 | .756-.876 | 5-Rad. | 1.843 | .780 | 25.3 | 77.1 | 2.437 | 1.375 | .930 | 36.3 | 110.8 | 2.437 | 1.375 | .930 | 36.4 | 110.9 |

# GENERAL DESIGN

## for Utilizing the unique,

**DOUBLE SHEAR STRENGTH** will be the determining factor in specifying the Rollpin size for a given stress in numerous applications. These values, for carbon steel and corrosion resistant steel Rollpins, are given in the table on Pages 14 and 16.

In soft metals ultimate failure depends more upon the strength of the hole material than upon shear strength of the Rollpin. Under increased load, hole edges are deformed by compressive forces. At ultimate loads the soft material may be expected to fail before the Rollpin.

In many cases, particularly with smaller mechanisms, the choice of Rollpin sizes will be influenced by space considerations and by general proportions of related mem-

bers. Recommended practices are given in the accompanying tables.

As with other types of pins, good Rollpin design practice is to avoid conditions where the direction of vibration parallels the axis of the pin. For most applications the Rollpin's capacity to withstand the loosening effects of vibration can be consistently relied upon; however, unusual conditions can result in an axial vibration component that should be carefully evaluated. In such instances accelerated vibration tests should be made of prototypes of the proposed design. ESNA will be glad to provide sample Rollpins for such tests.

### EFFECT OF GAP ORIENTATION ON ROLLPIN SHEAR STRENGTH

Data obtained in tests made to determine the effect of gap orientation on Rollpin shear strength revealed that random slot positioning is satisfactory for all but the most extreme Rollpin applications. For special applications maximum shear strength is assured by inserting the Rollpin so the gap is in line with the direction of the load.

LOAD — 90°

**SHEAR VALUE — 100%**     **SHEAR VALUE — 106%**

Double shear strength figures listed on Page 14 are based on tests made in accordance with Spec. MIL-P-10971 with pins in 90° position fastened in hardened steel fixtures.

LOAD — 180°

### RECOMMENDED ROLLPIN SIZES FOR VARIOUS SHAFT DIAMETERS

**When used as a Transverse Pin**

| D Shaft Diameter Inches | d Rollpin Nominal Diameter, Inches | Rollpin Size Number |
|---|---|---|
| 3/16 | 1/16 | 062 |
| 7/32 | 5/64 | 078 |
| 1/4 | 3/32 | 094 |
| 5/16-3/8 | 1/8 | 125 |
| 7/16-1/2 | 5/32 | 156 |
| 9/16 & 5/8 | 3/16 | 187 |
| 11/16 & 3/4 | 7/32 | 219 |
| 13/16 | 1/4 | 250 |
| 7/8 | 5/16 | 312 |
| 1 | 3/8 | 375 |
| 1-1/4 | 7/16 | 437 |
| 1-1/2 | 1/2 | 500 |

**When used as a Longitudinal Key**

| D Shaft Diameter Inches | d Rollpin Nominal Diameter, Inches | Rollpin Size Number |
|---|---|---|
| 1/4 | 1/16 | 062 |
| 5/16 | 5/64 | 078 |
| 3/8 | 3/32 | 094 |
| 7/16-1/2 | 1/8 | 125 |
| 9/16-5/8 | 5/32 | 156 |
| 11/16-3/4 | 3/16 | 187 |
| 7/8 | 7/32 | 219 |
| 15/16-1 | 1/4 | 250 |
| 1-1/4 | 5/16 | 312 |
| 1-3/8 | 3/8 | 375 |
| 1-1/2 - 1-5/8 | 7/16 | 437 |
| 2 | 1/2 | 500 |

**When used as a Replacement for Taper Pins**

## EQUIVALENT TAPER PINS

| Taper Pin Number | Equivalent Rollpin | |
|---|---|---|
| | Nominal Diameter, Inches | Rollpin Size Number |
| 7/0 | 1/16 | 062 |
| 6/0 | 5/64 | 078 |
| 5/0 | 3/32 | 094 |
| 4/0 | 1/8 | 125 |
| 3/0 | 1/8 | 125 |
| 2/0 | 5/32 | 156 |
| 0 | 5/32 | 156 |
| 1 | 3/16 | 187 |
| 2 | 3/16 | 187 |
| 3 | 7/32 | 219 |
| 4 | 1/4 | 250 |
| 5 | 5/16 | 312 |
| 6 | 3/8 | 375 |
| 7 | 7/16 | 437 |
| 8 | 1/2 | 500 |

# PRINCIPLES

## inherent ROLLPIN advantages

Fig. 1

Because of the inherent flexibility of the Rollpin it has found wide acceptance as a clevis joint pin. It is recommended, for greater bearing area, that the Rollpin be held by the outer members of the clevis (Fig. 1). However, the design may require that the inner member of the clevis (Fig. 2) be used to retain the pin. This alternate technique will also provide satisfactory performance results. The table below gives the average Rollpin spring back data relative to the span length.

Fig. 2

**BLACK — Minimum Hole**  **Fig. 1**  **RED — Maximum Hole**

| NOMINAL ROLLPIN DIAMETER | (A) HOLE DIAMETER | C — Rollpin Diameter at Center of Clevis Span —Thousandths / B — Span Length — Inches ||||||||||||||
|---|---|---|---|---|---|---|---|---|---|---|---|---|---|---|---|
| | | ⅛ | ¼ | ⅜ | ½ | ⅝ | ¾ | ⅞ | 1 | 1¼ | 1½ | 1¾ | 2 | 2½ | 3 |
| .062 | .062 | .063 | .065 | .065 | .065 | .065 | | | | | | | | | |
| | .065 | .066 | .067 | .067 | .067 | .067 | | | | | | | | | |
| .078 | .078 | .079 | .079 | .079 | .079 | .079 | .079 | | | | | | | | |
| | .081 | .083 | .084 | .084 | .084 | .084 | .084 | | | | | | | | |
| .094 | .094 | .096 | .097 | .098 | .098 | .098 | .098 | | | | | | | | |
| | .097 | .097 | .099 | .099 | .099 | .099 | .099 | | | | | | | | |
| .125 | .125 | .127 | .129 | .129 | .130 | .130 | .130 | .131 | .131 | | | | | | |
| | .129 | .129 | .130 | .130 | .130 | .131 | .131 | .131 | .131 | | | | | | |
| .156 | .156 | .158 | .159 | .160 | .161 | .162 | .162 | .163 | .163 | .163 | .163 | | | | |
| | .160 | .160 | .161 | .161 | .162 | .162 | .162 | .163 | .163 | .163 | .163 | | | | |
| .187 | .187 | .188 | .189 | .190 | .191 | .192 | .193 | .194 | .195 | .195 | .195 | .195 | .195 | | |
| | .192 | .192 | .193 | .193 | .194 | .194 | .194 | .195 | .195 | .195 | .195 | .195 | .195 | | |
| .219 | .219 | .220 | .221 | .222 | .223 | .223 | .224 | .225 | .225 | .226 | .226 | .226 | .226 | | |
| | .224 | .224 | .225 | .225 | .225 | .225 | .226 | .226 | .226 | .226 | .226 | .226 | .226 | | |
| .250 | .250 | .252 | .254 | .255 | .256 | .256 | .257 | .258 | .258 | .259 | .260 | .260 | .260 | | |
| | .256 | .257 | .258 | .258 | .259 | .259 | .260 | .260 | .260 | .260 | .260 | .260 | .260 | | |
| .312 | .312 | .313 | .313 | .313 | .314 | .315 | .315 | .316 | .317 | .317 | .318 | .319 | .320 | .321 | .321 |
| | .318 | .319 | .319 | .319 | .319 | .319 | .319 | .320 | .320 | .320 | .320 | .320 | .321 | .321 | .321 |
| .375 | .375 | .376 | .376 | .376 | .376 | .377 | .378 | .379 | .380 | .382 | .383 | .384 | .384 | .385 | .385 |
| | .382 | .383 | .383 | .383 | .383 | .383 | .383 | .384 | .384 | .384 | .384 | .385 | .385 | .385 | .385 |
| .500 | .500 | .501 | .502 | .502 | .503 | .504 | .504 | .505 | .506 | .507 | .508 | .510 | .512 | .514 | .514 |
| | .510 | .511 | .511 | .511 | .512 | .512 | .512 | .512 | .513 | .513 | .513 | .514 | .514 | .514 | .514 |

**Black-unsized end**  **Fig. 2**  **Red-Sized end**

(Each cell shows two values: black-unsized end (left) / red-sized end (right). Two rows per nominal diameter correspond to hole diameter min/max.)

| NOMINAL | HOLE DIA. | ⅛ | ¼ | ⅜ | ½ | ⅝ | ¾ | ⅞ | 1 | 1¼ | 1½ | 2 |
|---|---|---|---|---|---|---|---|---|---|---|---|---|
| .062 | .062 | .066 .063 | .068 .064 | .068 .064 | .068 .064 | .068 .064 | .068 .064 | | | | | |
| | .065 | .067 .066 | .067 .067 | .067 .067 | .067 .067 | .067 .067 | .067 .067 | | | | | |
| .078 | .078 | .081 .080 | .083 .080 | .083 .080 | .083 .080 | .083 .080 | .083 .080 | | | | | |
| | .081 | .083 .083 | .084 .083 | .084 .083 | .084 .083 | .084 .083 | .084 .083 | | | | | |
| .094 | .094 | .098 .095 | .100 .097 | .101 .098 | .101 .098 | .101 .098 | .101 .098 | | | | | |
| | .097 | .099 .098 | .101 .099 | .101 .099 | .101 .099 | .101 .099 | .101 .099 | | | | | |
| .125 | .125 | .128 .126 | .130 .127 | .131 .127 | .131 .128 | .132 .128 | .132 .128 | | | | | |
| | .129 | .130 .130 | .131 .131 | .131 .131 | .132 .132 | .132 .132 | .132 .132 | | | | | |
| .156 | .156 | .160 .159 | .161 .159 | .161 .160 | .161 .160 | .162 .160 | .162 .161 | .163 .161 | .163 .161 | | | |
| | .160 | .162 .162 | .163 .163 | .163 .163 | .163 .163 | .163 .163 | .163 .163 | .163 .163 | .163 .163 | | | |
| .187 | .187 | .190 .188 | .192 .189 | .194 .190 | .195 .191 | .195 .191 | .195 .191 | .195 .191 | | | | |
| | .192 | .194 .193 | .195 .194 | .195 .194 | .195 .195 | .195 .195 | .195 .195 | .195 .195 | | | | |
| .219 | .219 | .223 .221 | .225 .222 | .226 .224 | .227 .225 | .227 .225 | .228 .226 | .228 .226 | .228 .226 | | | |
| | .224 | .225 .225 | .226 .226 | .227 .226 | .227 .227 | .227 .227 | .227 .227 | .227 .227 | .227 .227 | | | |
| .250 | .250 | .252 .251 | .253 .252 | .255 .253 | .257 .254 | .258 .255 | .258 .255 | .258 .255 | .259 .256 | .259 .256 | .259 .256 | .259 |
| | .256 | .257 .257 | .258 .258 | .259 .259 | .259 .259 | .260 .260 | .260 .260 | .260 .260 | .260 .260 | .260 .260 | .260 .260 | .260 |
| .312 | .312 | .316 .314 | .319 .316 | .320 .318 | .322 .319 | .323 .320 | .324 .321 | .325 .321 | .325 .321 | .325 .321 | .325 .321 | .325 |
| | .318 | .320 .319 | .322 .320 | .323 .321 | .324 .321 | .324 .322 | .325 .322 | .325 .322 | .325 .322 | .325 .322 | .325 .322 | .325 |
| .375 | .375 | .378 .376 | .380 .378 | .384 .380 | .385 .381 | .386 .383 | .387 .384 | .388 .385 | .388 .385 | .388 .385 | .389 .385 | .385 |
| | .382 | .384 .383 | .385 .384 | .386 .385 | .386 .385 | .387 .386 | .387 .386 | .388 .387 | .388 .387 | .389 .387 | .389 .387 | .387 |
| .500 | .500 | .502 .502 | .503 .502 | .505 .504 | .506 .505 | .508 .506 | .509 .507 | .511 .508 | .512 .509 | .513 .510 | .514 .510 | .515 .511 |
| | .510 | .511 .511 | .512 .512 | .513 .513 | .514 .514 | .514 .514 | .514 .514 | .515 .515 | .515 .515 | .515 .515 | .515 .515 | .515 .515 |

(ESNA) ─ FASTENER DIVISION ─

SHAPE OF CHAMFER OPTIONAL

\* APPROX

TOLERANCE ON SPECIFIED
LENGTH "L"
0.187 TO 1.000 ±.015
1.001 TO 2.000 ±.020
2.001 TO 3.000 ±.025
3.001 TO 4.000 ±.030
4.001 & ABOVE ±.035

| A | | | B | C | | STOCK THICKNESS | RECOMMENDED HOLE SIZE | | MINIMUM DOUBLE SHEAR STRENGTH POUNDS |
|---|---|---|---|---|---|---|---|---|---|
| NOMINAL | MAXIMUM (GO RING GAGE) | MINIMUM 1/3(D₁+D₂+D₃) | MAX | MIN | MAX | | MIN | MAX | |
| .062 | .069 | .066 | .059 | .007 | .028 | .012 | .062 | .065 | 425 |
| .078 | .086 | .083 | .075 | .008 | .032 | .018 | .078 | .081 | 650 |
| .094 | .103 | .099 | .091 | .008 | .038 | .022 | .094 | .097 | 1,000 |
| .125 | .135 | .131 | .122 | .008 | .044 | .028 | .125 | .129 | 2,100 |
| .140 | .149 | .145 | .137 | .008 | .044 | .028 | .140 | .144 | 2,200 |
| .156 | .167 | .162 | .151 | .010 | .048 | .032 | .156 | .160 | 3,000 |
| .187 | .199 | .194 | .182 | .011 | .055 | .040 | .187 | .192 | 4,400 |
| .219 | .232 | .226 | .214 | .011 | .065 | .048 | .219 | .224 | 5,700 |
| .250 | .264 | .258 | .245 | .012 | .065 | .048 | .250 | .256 | 7,700 |
| .312 | .328 | .321 | .306 | .014 | .080 | .062 | .312 | .318 | 11,500 |
| .375 | .392 | .385 | .368 | .016 | .095 | .077 | .375 | .382 | 17,600 |
| .437 | .456 | .448 | .430 | .017 | .095 | .077 | .437 | .445 | 20,000 |
| .500 | .521 | .513 | .485 | .025 | .110 | .094 | .500 | .510 | 25,800 |

CODE: PARTS ARE IDENTIFIED BY A SERIES OF DASH NUMBERS IN ACCORDANCE WITH THE EXAMPLE SHOWN BELOW.

EXAMPLE: 5 9 - 028 - 125 - 0750 = ROLLPIN, .125 DIA., .028 WALL, 3/4 LONG, PLAIN CARBON STEEL.

THIRD DASH NUMBER INDICATES LENGTH IN THOUSANDTHS OF AN INCH. PREFIX "O" IF LESS THAN ONE INCH.

SECOND DASH NUMBER INDICATES NOMINAL DIAMETER IN THOUSANDTHS OF AN INCH AND IS DETERMINED FROM "A-NOMINAL" COLUMN IN TABULATION ABOVE.

FIRST DASH NUMBER INDICATES WALL THICKNESS IN THOUSANDTHS OF AN INCH AND IS DETERMINED FROM "STOCK THICKNESS" COLUMN IN TABULATION ABOVE.

SECOND DIGIT INDICATES FINISH:
9 = PLAIN.
⑦ 3 = PHOSPHATE COATING, MIL SPEC MIL-C-16232(N Ord) TYPE Z.
2 = CADMIUM PLATE, FEDERAL SPEC QQ-P-416, TYPE 1, CLASS 3.
1 = ZINC PLATE, FEDERAL SPEC QQ-Z-325, TYPE 1, CLASS 3.

FIRST DIGIT INDICATES MATERIAL:
5 = HIGH CARBON STEEL.

FINISH: CARBON STEEL PINS ARE NORMALLY FURNISHED WITH A BLACK OILED FINISH.

HARDNESS: ROCKWELL "C" 46-53 OR EQUIVALENT.

ISSUED: 20 APR 50 REVISED: ⑦ 9 JAN 62

| PERFORMANCE SPECIFICATION | ROLLPIN | RP-A |
|---|---|---|
| ESNA SPEC 401 (SEE NOTE: 1) | HIGH CARBON STEEL | PAGE 1 OF 2 |

—FASTENER DIVISION——————————————

NOTE 1. CARBON STEEL ROLLPINS MEET THE REQUIREMENTS OF THE FOLLOWING:
 SPECIFICATIONS: MIL-P-10971, AMS 7205.
 STANDARDS: ARMY ORDNANCE DRAWINGS BFSX2 AND BFSX2.1
  NAVY DRAWINGS NAVORD OSTD 600, PAGES 5-6.11 TO 5-6.13
  MS9047(ASG), MS9048(ASG), MSI6562, NAS561

**PLAIN CARBON STEEL ROLLPIN AVAILABILITY**

NOMINAL DIAMETER

| LENGTH | .062 | .078 | .094 | .125 | .140 | .156 | .187 | .219 | .250 | .312 | .375 | .437 | .500 |
|---|---|---|---|---|---|---|---|---|---|---|---|---|---|
| 0.125" | * | | | | | | | | | | | | |
| 0.187 | * | * | * | | | | | | | | | | |
| 0.250 | * | * | * | | Δ | * | | | | | | | |
| 0.312 | * | * | * | * | Δ | * | * | | | | | | |
| 0.375 | * | * | * | * | Δ | * | * | * | * | * | | | |
| 0.437 | * | * | * | * | Δ | * | * | * | * | * | | | |
| 0.500 | * | * | * | * | Δ | * | * | * | * | * | | | |
| 0.562 | * | * | * | * | Δ | * | * | * | * | * | | | |
| 0.625 | * | * | * | * | Δ | * | * | * | * | Δ | * | | |
| 0.687 | * | * | * | * | Δ | * | * | * | * | Δ | Δ | | |
| 0.750 | * | * | * | * | * | * | * | * | * | * | * | Δ | * |
| 0.812 | * | * | * | * | * | * | * | * | * | * | Δ | Δ | Δ |
| 0.875 | * | * | * | * | * | * | * | * | * | * | * | Δ | * |
| 0.937 | * | * | * | * | * | * | * | * | * | * | * | Δ | Δ |
| 1.000 | * | * | * | * | * | * | * | * | * | * | * | * | * |
| 1.125 | * | * | * | Δ | * | * | * | * | * | * | * | Δ | * |
| 1.250 | Δ | * | * | * | * | * | * | * | * | * | * | * | * |
| 1.375 | * | * | * | * | * | * | * | * | * | * | * | * | * |
| 1.500 | | * | * | * | Δ | * | * | * | * | * | * | * | * |
| 1.625 | | | Δ | Δ | * | * | * | * | * | * | * | Δ | * |
| 1.750 | | | Δ | Δ | * | * | * | * | * | * | * | * | * |
| 1.875 | | | Δ | Δ | * | * | * | * | * | * | * | * | * |
| 2.000 | | | Δ | Δ | * | * | * | * | * | * | * | * | * |
| 2.250 | | | | * | Δ | * | * | * | * | * | * | * | * |
| 2.500 | | | | | Δ | * | * | * | * | * | * | * | * |
| 2.750 | | | | | | * | Δ | Δ | * | * | * | * | * |
| 3.000 | | | | | | | Δ | * | * | * | * | * | * |
| 3.250 | | | | | | | Δ | * | * | * | * | Δ | * |
| 3.500 | | | | | | | Δ | * | * | * | * | Δ | * |
| 3.750 | | | | | | | Δ | Δ | * | * | * | Δ | * |
| 4.000 | | | | | | | Δ | Δ | * | * | * | Δ | * |
| 4.250 | | | | | | | Δ | Δ | Δ | Δ | Δ | Δ | Δ |
| 4.500 | | | | | | | Δ | Δ | Δ | Δ | Δ | * | * |
| 4.750 | | | | | | | | Δ | Δ | Δ | * | Δ | * |
| 5.000 | | | | | | | | Δ | Δ | Δ | * | Δ | * |
| 5.250 | | | | | | | | | | | Δ | Δ | Δ |
| 5.500 | | | | | | | | | | | Δ | Δ | Δ |

ESNA AVAILABILITY CODE:
 * - STANDARD PARTS NORMALLY CARRIED IN STOCK.
 Δ - STANDARD PARTS NOT ALWAYS CARRIED IN INVENTORY AND ON WHICH IT MAY BE NECESSARY TO
  REQUIRE MINIMUM ECONOMICAL RUNS.

IN ADDITION TO THE STOCK LENGTHS APPEARING IN THIS TABLE, ROLLPINS LESS THAN 1" LONG CAN BE MADE
AVAILABLE IN LENGTH INCREMENTS OF 1/32". ROLLPINS LONGER THAN 1" CAN BE MADE AVAILABLE IN LENGTH
INCREMENTS OF 1/16", BUT IN BOTH CASES SUCH PARTS ARE NON-STOCK ITEMS AND MUST BE MADE TO ORDER.
ROLLPINS OF SPECIAL DECIMAL LENGTHS ARE SUBJECT TO SPECIAL ORDER REQUIREMENTS.

ISSUED: 20 APR 50  REVISED: ⑦9 JAN 62

| PERFORMANCE SPECIFICATION | ROLLPIN | RP-A |
|---|---|---|
| ESNA SPEC 401 (SEE NOTE: 1) | HIGH CARBON STEEL | PAGE 2 OF 2 |

ELASTIC STOP NUT CORPORATION OF AMERICA, 2330 VAUXHALL ROAD, UNION, NEW JERSEY.

BASIC *internal series* **N5000**

**WALDES TRUARC RETAINING RINGS**

U.S. Pat. No. 2,861,824

See Fig. 2, Page 27

section I-I

- Sizes identified by this symbol are available in tape-wrapped **Rol-Pak®** cartridges.

| HOUSING DIA. | | | MIL-R-21248 MS 16625 NAS 669 INTERNAL SERIES **N5000** | TRUARC RING DIMENSIONS | | | | Approx. weight per 1000 pieces | GROOVE DIMENSIONS | | | | Nominal groove depth | APPLICATION DATA | | | |
|---|---|---|---|---|---|---|---|---|---|---|---|---|---|---|---|---|---|
| | | | | FREE DIA. | | THICKNESS | | | DIAMETER | | WIDTH | | | CLEARANCE DIAMETER | | ALLOW. THRUST LOAD (lbs.) Sharp Corner Abutment | |
| Dec. equiv. inch | Approx. fract. equiv. inch | Approx. mm. | size—no. | D | tol. | t | tol. | lbs. | G | tol. | W | tol. | d | When sprung into housing | When sprung into groove | RINGS Safety factor = 4 Important! See Page 19 | GROOVES Safety factor = 2 Important! See Page 19 |
| **S** | **S** | **S** | | | | | | | **G** | | **W** | | | **C₁** | **C₂** | **Pᵣ** | **Pᵍ** |
| .250 | ¼ | 6.4 | • N5000-25 | .280 | | .015 | | .08 | .268 | ±.001 .0015 | .018 | +.002 −.000 | .009 | .115 | .133 | 420 | 190 |
| .312 | ⁵⁄₁₆ | 7.9 | • N5000-31 | .346 | | .015 | | .11 | .330 | ±T.I.R. | .018 | | .009 | .173 | .191 | 530 | 240 |
| .375 | ⅜ | 9.5 | • N5000-37 | .415 | | .025 | | .25 | .397 | ±.002 | .029 | | .011 | .204 | .226 | 1050 | 350 |
| .438 | ⁷⁄₁₆ | 11.1 | • N5000-43 | .482 | | .025 | | .37 | .461 | .002 | .029 | | .012 | .23 | .254 | 1220 | 440 |
| .453 | ²⁹⁄₆₄ | 11.5 | • N5000-45 | .498 | | .025 | | .43 | .477 | T.I.R. | .029 | | .012 | .25 | .274 | 1280 | 460 |
| .500 | ½ | 12.7 | • N5000-50 | .548 | +.010 −.005 | .035 | | .70 | .530 | | .039 | | .015 | .26 | .29 | 1980 | 510 |
| .512 | — | 13.0 | • N5000-51 | .560 | | .035 | | .77 | .542 | ±.002 .004 T.I.R. | .039 | | .015 | .27 | .30 | 2030 | 520 |
| .562 | ⁹⁄₁₆ | 14.3 | • N5000-56 | .620 | | .035 | | .86 | .596 | | .039 | | .017 | .275 | .305 | 2220 | 710 |
| .625 | ⅝ | 15.9 | • N5000-62 | .694 | | .035 | | 1.0 | .665 | | .039 | | .020 | .34 | .38 | 2470 | 1050 |
| .688 | ¹¹⁄₁₆ | 17.5 | • N5000-68 | .763 | | .035 | | 1.2 | .732 | | .039 | +.003 −.000 | .022 | .40 | .44 | 2700 | 1280 |
| .750 | ¾ | 19.0 | • N5000-75 | .831 | | .035 | | 1.3 | .796 | | .039 | | .023 | .45 | .49 | 3000 | 1460 |
| .777 | — | 19.7 | • N5000-77 | .859 | | .042 | | 1.7 | .825 | | .046 | | .024 | .475 | .52 | 4550 | 1580 |
| .812 | ¹³⁄₁₆ | 20.6 | • N5000-81 | .901 | | .042 | | 1.9 | .862 | | .046 | | .025 | .49 | .54 | 4800 | 1710 |
| .866 | — | 22.0 | • N5000-86 | .961 | | .042 | | 2.0 | .920 | | .046 | | .027 | .54 | .59 | 5100 | 1980 |
| .875 | ⅞ | 22.2 | • N5000-87 | .971 | +.015 −.010 | .042 | | 2.1 | .931 | ±.003 .004 T.I.R. | .046 | | .028 | .545 | .60 | 5150 | 2080 |
| .901 | — | 22.9 | • N5000-90 | 1.000 | | .042 | ±.002 | 2.2 | .959 | | .046 | | .029 | .565 | .62 | 5350 | 2200 |
| .938 | ¹⁵⁄₁₆ | 23.8 | • N5000-93 | 1.041 | | .042 | | 2.4 | 1.000 | | .046 | | .031 | .61 | .67 | 5600 | 2450 |
| 1.000 | 1 | 25.4 | • N5000-100 | 1.111 | | .042 | | 2.7 | 1.066 | | .046 | | .033 | .665 | .73 | 5950 | 2800 |
| 1.023 | — | 26.0 | • N5000-102 | 1.136 | | .042 | | 2.8 | 1.091 | | .046 | | .034 | .69 | .755 | 6050 | 3000 |
| 1.062 | 1¹⁄₁₆ | 27.0 | • N5000-106 | 1.180 | | .050 | | 3.7 | 1.130 | | .056 | | .034 | .685 | .75 | 7450 | 3050 |
| 1.125 | 1⅛ | 28.6 | • N5000-112 | 1.249 | | .050 | | 4.0 | 1.197 | | .056 | | .036 | .745 | .815 | 7900 | 3400 |
| 1.181 | — | 30.0 | • N5000-118 | 1.319 | | .050 | | 4.3 | 1.255 | | .056 | | .037 | .79 | .86 | 8400 | 3700 |
| 1.188 | 1³⁄₁₆ | 30.2 | • N5000-118 | 1.319 | | .050 | | 4.3 | 1.262 | | .056 | | .037 | .80 | .87 | 8400 | 3700 |
| 1.250 | 1¼ | 31.7 | • N5000-125 | 1.388 | | .050 | | 4.8 | 1.330 | | .056 | | .040 | .875 | .955 | 8800 | 4250 |
| 1.259 | — | 32.0 | • N5000-125 | 1.388 | +.025 −.020 | .050 | | 4.8 | 1.339 | | .056 | | .040 | .885 | .965 | 8800 | 4250 |
| 1.312 | 1⁵⁄₁₆ | 33.3 | • N5000-131 | 1.456 | | .050 | | 5.0 | 1.396 | | .056 | | .042 | .93 | 1.01 | 9300 | 4700 |
| 1.375 | 1⅜ | 34.9 | • N5000-137 | 1.526 | | .050 | | 5.1 | 1.461 | ±.004 .005 T.I.R. | .056 | +.004 −.000 | .043 | .99 | 1.07 | 9700 | 5050 |
| 1.378 | — | 35.0 | • N5000-137 | 1.526 | | .050 | | 5.1 | 1.464 | | .056 | | .043 | .99 | 1.07 | 9700 | 5050 |
| 1.438 | 1⁷⁄₁₆ | 36.5 | • N5000-143 | 1.596 | | .050 | | 5.8 | 1.528 | | .056 | | .045 | 1.06 | 1.15 | 10200 | 5500 |
| 1.456 | — | 37.0 | • N5000-145 | 1.616 | | .050 | | 6.4 | 1.548 | | .056 | | .046 | 1.08 | 1.17 | 10300 | 5700 |
| 1.500 | 1½ | 38.1 | • N5000-150 | 1.660 | | .050 | | 6.5 | 1.594 | | .056 | | .047 | 1.12 | 1.21 | 10550 | 6000 |
| 1.562 | 1⁹⁄₁₆ | 39.7 | • N5000-156 | 1.734 | | .062 | | 8.1 | 1.658 | | .068 | | .048 | 1.14 | 1.23 | 13700 | 6350 |
| 1.575 | — | 40.0 | • N5000-156 | 1.734 | | .062 | ±.003 | 8.1 | 1.671 | ±.005 .005 T.I.R. | .068 | | .048 | 1.15 | 1.24 | 13700 | 6350 |
| 1.625 | 1⅝ | 41.3 | • N5000-162 | 1.804 | +.035 −.025 | .062 | | 10.0 | 1.725 | | .068 | | .050 | 1.15 | 1.25 | 14200 | 6900 |
| 1.653 | — | 42.0 | • N5000-165 | 1.835 | | .062 | | 10.4 | 1.755 | | .068 | | .051 | 1.17 | 1.27 | 14500 | 7200 |
| 1.688 | 1¹¹⁄₁₆ | 42.9 | • N5000-168 | 1.874 | | .062 | | 10.8 | 1.792 | | .068 | | .052 | 1.21 | 1.31 | 14800 | 7450 |

Thickness t applies only to un-plated rings. For plated and stainless steel (Type H) rings, add .002″ to the listed maximum thickness. Maximum ring thickness will be at least .0002″ less than the listed minimum groove width (**W**).

T.I.R. (total indicator reading) is the maximum allowable deviation of concentricity between groove and housing.

**Additional sizes appear on Pages 28 and 30**

WALDES
TRUARC
RETAINING
RINGS

BASIC
*internal series*
**N5000**

**FIG. 1:**
**MAXIMUM ALLOWABLE**
**CORNER RADIUS (R_max.)**
**AND CHAMFER (Ch_max.)**

**FIG. 2:**
**ENLARGED DETAIL**
**OF GROOVE PROFILE**
**AND EDGE MARGIN (Z)**

| MAXIMUM BOTTOM RADII | |
|---|---|
| Ring Size | R |
| -25 thru -100 | .005 |
| -102 thru -1000 | .010 |

**FIG. 3:**
**SUPPLEMENTARY**
**RING DIMENSIONS**

**FIG. 4:**
**MINIMUM GAP WIDTH**
(Ring installed in groove)

| | SUPPLEMENTARY APPLICATION DATA | | | | SUPPLEMENTARY RING DIMENSIONS | | | | | | | | |
|---|---|---|---|---|---|---|---|---|---|---|---|---|---|
| INTERNAL SERIES **N5000** | Maximum allowable corner radii and chamfers of retained parts (Fig. 1). See Page 21. | | Allowable assembly load with R_max. or Ch_max. | Edge margin (Fig. 2) See Page 19 | LUG | | LARGE SECTION | | SMALL SECTION | | HOLE DIAMETER | | MIN. GAP WIDTH (Fig. 4) Ring installed in groove |
| size—no. | R_max. | Ch_max. | P'_r(lbs.) | Z | B | tol. | E | tol. | J | tol. | P | tol. | A |
| • N5000-25 | .011 | .0085 | 190 | .027 | .065 | | .025 | | .015 | | .031 | | .047 |
| • N5000-31 | .016 | .013 | 190 | .027 | .066 | | .033 | ±.002 | .018 | ±.002 | .031 | | .055 |
| • N5000-37 | .023 | .018 | 530 | .033 | .082 | | .040 | | .028 | | .041 | | .063 |
| • N5000-43 | .027 | .021 | 530 | .036 | .098 | ±.003 | .049 | ±.003 | .029 | ±.003 | .041 | | .063 |
| • N5000-45 | .027 | .021 | 530 | .036 | .098 | | .050 | | .030 | | .047 | | .071 |
| • N5000-50 | .027 | .021 | 1100 | .045 | .114 | | .053 | | .035 | | .047 | | .090 |
| • N5000-51 | .027 | .021 | 1100 | .045 | .114 | | .053 | | .035 | | .047 | | .092 |
| • N5000-56 | .027 | .021 | 1100 | .051 | .132 | | .053 | | .035 | | .047 | | .095 |
| • N5000-62 | .027 | .021 | 1100 | .060 | .132 | | .060 | ±.004 | .035 | ±.004 | .062 | | .104 |
| • N5000-68 | .027 | .021 | 1100 | .066 | .132 | | .063 | | .036 | | .062 | +.010 −.002 | .118 |
| • N5000-75 | .032 | .025 | 1100 | .069 | .142 | | .070 | | .040 | | .062 | | .143 |
| • N5000-77 | .035 | .028 | 1650 | .072 | .146 | | .074 | | .044 | | .062 | | .145 |
| • N5000-81 | .035 | .028 | 1650 | .075 | .155 | | .077 | | .044 | | .062 | | .153 |
| • N5000-86 | .035 | .028 | 1650 | .081 | .155 | | .081 | | .045 | | .062 | | .172 |
| • N5000-87 | .035 | .028 | 1650 | .084 | .155 | | .084 | ±.005 | .045 | ±.005 | .062 | | .179 |
| • N5000-90 | .038 | .030 | 1650 | .087 | .155 | | .087 | | .047 | | .062 | | .188 |
| • N5000-93 | .038 | .030 | 1650 | .093 | .155 | | .091 | | .050 | | .062 | | .200 |
| • N5000-100 | .042 | .034 | 2400 | .099 | .155 | | .104 | | .052 | | .062 | | .212 |
| • N5000-102 | .042 | .034 | 2400 | .102 | .155 | | .106 | | .054 | | .062 | | .220 |
| • N5000-106 | .044 | .035 | 2400 | .102 | .180 | | .110 | | .055 | | .078 | | .213 |
| • N5000-112 | .047 | .036 | 2400 | .108 | .180 | ±.005 | .116 | | .057 | | .078 | | .232 |
| (S=1.181) • N5000-118 | .047 | .036 | 2400 | .111 | .180 | | .120 | | .058 | | .078 | | .226 |
| (S=1.188) • N5000-118 | .047 | .036 | 2400 | .111 | .180 | | .120 | | .058 | | .078 | | .245 |
| (S=1.250) • N5000-125 | .048 | .038 | 2400 | .120 | .180 | | .124 | | .062 | | .078 | | .265 |
| (S=1.259) • N5000-125 | .048 | .038 | 2400 | .120 | .180 | | .124 | | .062 | | .078 | | .290 |
| • N5000-131 | .048 | .038 | 2400 | .126 | .180 | | .130 | ±.006 | .062 | ±.006 | .078 | | .284 |
| (S=1.375) • N5000-137 | .048 | .038 | 2400 | .129 | .180 | | .130 | | .063 | | .078 | | .297 |
| (S=1.378) • N5000-137 | .048 | .038 | 2400 | .129 | .180 | | .130 | | .063 | | .078 | | .305 |
| • N5000-143 | .048 | .038 | 2400 | .135 | .180 | | .133 | | .065 | | .078 | | .313 |
| • N5000-145 | .048 | .038 | 2400 | .138 | .180 | | .133 | | .065 | | .078 | +.015 −.002 | .320 |
| • N5000-150 | .048 | .038 | 2400 | .141 | .180 | | .133 | | .066 | | .078 | | .340 |
| (S=1.562) • N5000-156 | .064 | .050 | 3900 | .144 | .202 | | .157 | | .078 | | .078 | | .338 |
| (S=1.575) • N5000-156 | .064 | .050 | 3900 | .144 | .202 | | .157 | | .078 | | .078 | | .374 |
| • N5000-162 | .064 | .050 | 3900 | .150 | .227 | | .164 | ±.007 | .082 | ±.007 | .078 | | .339 |
| • N5000-165 | .064 | .050 | 3900 | .153 | .227 | | .167 | | .083 | | .078 | | .348 |
| • N5000-168 | .064 | .050 | 3900 | .156 | .227 | | .170 | | .085 | | .078 | | .357 |

**BASIC** *internal series* **N5000** (Continued)

**WALDES TRUARC RETAINING RINGS**

U. S. Pat. No. 2,861,824

See Fig. 2, Page 29

section I-I

• Sizes identified by this symbol are available in tape-wrapped **Rol-Pak®** cartridges.

| HOUSING DIA. | | | MIL-R-21248 MS 16625 NAS 669 INTERNAL SERIES **N5000** | TRUARC RING DIMENSIONS | | | | | GROOVE DIMENSIONS | | | | | APPLICATION DATA | | | |
|---|---|---|---|---|---|---|---|---|---|---|---|---|---|---|---|---|---|
| | | | | Thickness t applies only to unplated rings. For plated and stainless steel (Type H) rings, add .002″ to the listed maximum thickness. Maximum ring thickness will be at least .0002″ less than the listed minimum groove width (W). | | | | Approx. weight per 1000 pieces | T.I.R. (total indicator reading) is the maximum allowable deviation of concentricity between groove and housing. | | | | | CLEARANCE DIAMETER | | ALLOW. THRUST LOAD (lbs.) Sharp Corner Abutment | |
| | | | | FREE DIA. | | THICKNESS | | | DIAMETER | | WIDTH | | Nominal groove depth | When sprung into housing | When sprung into groove | RINGS Safety factor = 4 Important! See Page 19 | GROOVES Safety factor = 2 Important! See Page 19 |
| Dec. equiv. inch | Approx. fract. equiv. inch | Approx. mm. | size—no. | | | | | | | | | | | | | | |
| S | S | S | size—no. | D | tol. | t | tol. | lbs. | G | tol. | W | tol. | d | C₁ | C₂ | Pr | Pg |
| 1.750 | 1¾ | 44.4 | • N5000-175 | 1.942 | | .062 | | 10.3 | 1.858 | | .068 | | .054 | 1.26 | 1.36 | 15350 | 8050 |
| 1.812 | 1 13/16 | 46.0 | • N5000-181 | 2.012 | | .062 | | 11.5 | 1.922 | | .068 | | .055 | 1.32 | 1.43 | 15900 | 8450 |
| 1.850 | — | 47.0 | • N5000-185 | 2.054 | +.035 | .062 | | 12.8 | 1.962 | ± .005 | .068 | +.004 | .056 | 1.36 | 1.47 | 16200 | 8750 |
| 1.875 | 1⅞ | 47.6 | • N5000-187 | 2.072 | −.025 | .062 | | 12.8 | 1.989 | .005 T.I.R. | .068 | −.000 | .057 | 1.39 | 1.50 | 16450 | 9050 |
| 1.938 | 1 15/16 | 49.2 | • N5000-193 | 2.141 | | .062 | | 13.3 | 2.056 | | .068 | | .059 | 1.45 | 1.56 | 17000 | 9700 |
| 2.000 | 2 | 50.8 | • N5000-200 | 2.210 | | .062 | | 14.0 | 2.122 | | .068 | | .061 | 1.50 | 1.62 | 17500 | 10300 |
| 2.047 | — | 52.0 | N5000-206 | 2.280 | | .078 | | 18.0 | 2.171 | | .086 | | .062 | 1.52 | 1.64 | 22750 | 10850 |
| 2.062 | 2 1/16 | 52.4 | N5000-206 | 2.280 | | .078 | | 18.0 | 2.186 | | .086 | | .062 | 1.54 | 1.66 | 22750 | 10850 |
| 2.125 | 2⅛ | 54.0 | N5000-212 | 2.350 | | .078 | | 19.4 | 2.251 | | .086 | | .063 | 1.58 | 1.70 | 23400 | 11350 |
| 2.165 | — | 55.0 | N5000-218 | 2.415 | | .078 | | 19.6 | 2.295 | | .086 | | .065 | 1.61 | 1.74 | 24100 | 12050 |
| 2.188 | 2 3/16 | 55.6 | N5000-218 | 2.415 | | .078 | | 19.6 | 2.318 | | .086 | | .065 | 1.64 | 1.77 | 24100 | 12050 |
| 2.250 | 2¼ | 57.1 | N5000-225 | 2.490 | | .078 | | 21.8 | 2.382 | | .086 | | .066 | 1.69 | 1.82 | 24850 | 12600 |
| 2.312 | 2 5/16 | 58.7 | N5000-231 | 2.560 | | .078 | | 22.6 | 2.450 | | .086 | | .069 | 1.75 | 1.88 | 25450 | 13550 |
| 2.375 | 2⅜ | 60.3 | N5000-237 | 2.630 | | .078 | | 23.2 | 2.517 | | .086 | | .071 | 1.81 | 1.95 | 26150 | 14300 |
| 2.440 | 2 7/16 | 62.0 | N5000-244 | 2.702 | | .078 | | 25.4 | 2.584 | | .086 | | .072 | 1.86 | 2.00 | 26900 | 14900 |
| 2.500 | 2½ | 63.5 | N5000-250 | 2.775 | +.040 | .078 | | 25.5 | 2.648 | | .086 | | .074 | 1.91 | 2.05 | 27600 | 15650 |
| 2.531 | 2 17/32 | 64.3 | N5000-250 | 2.775 | −.030 | .078 | | 25.5 | 2.681 | | .086 | | .075 | 1.94 | 2.09 | 27600 | 15650 |
| 2.562 | 2 9/16 | 65.1 | N5000-256 | 2.844 | | .093 | | 34.0 | 2.714 | | .103 | | .076 | 1.95 | 2.10 | 33700 | 16500 |
| 2.625 | 2⅝ | 66.7 | N5000-262 | 2.910 | | .093 | ±.003 | 34.5 | 2.781 | | .103 | | .078 | 2.02 | 2.17 | 34550 | 17350 |
| 2.677 | — | 68.0 | N5000-268 | 2.980 | | .093 | | 35.0 | 2.837 | | .103 | | .080 | 2.05 | 2.21 | 35400 | 18250 |
| 2.688 | 2 11/16 | 68.3 | N5000-268 | 2.980 | | .093 | | 35.0 | 2.848 | ± .006 | .103 | +.005 | .080 | 2.06 | 2.22 | 35400 | 18250 |
| 2.750 | 2¾ | 69.8 | N5000-275 | 3.050 | | .093 | | 35.5 | 2.914 | .006 T.I.R. | .103 | −.000 | .082 | 2.12 | 2.28 | 36100 | 19200 |
| 2.812 | 2 13/16 | 71.4 | N5000-281 | 3.121 | | .093 | | 36.0 | 2.980 | | .103 | | .084 | 2.18 | 2.34 | 36950 | 20050 |
| 2.835 | — | 72.0 | N5000-281 | 3.121 | | .093 | | 36.0 | 3.006 | | .103 | | .085 | 2.21 | 2.38 | 36950 | 20050 |
| 2.875 | 2⅞ | 73.0 | N5000-287 | 3.191 | | .093 | | 41.0 | 3.051 | | .103 | | .088 | 2.22 | 2.39 | 37800 | 21500 |
| 2.953 | — | 75.0 | N5000-300 | 3.325 | | .093 | | 42.5 | 3.135 | | .103 | | .091 | 2.30 | 2.48 | 39500 | 23150 |
| 3.000 | 3 | 76.2 | N5000-300 | 3.325 | | .093 | | 42.5 | 3.182 | | .103 | | .091 | 2.35 | 2.53 | 39500 | 23150 |
| 3.062 | 3 1/16 | 77.8 | N5000-306 | 3.418 | | .109 | | 53.0 | 3.248 | | .120 | | .093 | 2.41 | 2.59 | 47100 | 24100 |
| 3.125 | 3⅛ | 79.4 | N5000-312 | 3.488 | | .109 | | 56.0 | 3.315 | | .120 | | .095 | 2.47 | 2.66 | 48100 | 25200 |
| 3.149 | — | 80.0 | N5000-315 | 3.523 | | .109 | | 57.0 | 3.341 | | .120 | | .096 | 2.49 | 2.68 | 48600 | 25700 |
| 3.156 | 3 5/32 | 80.2 | N5000-315 | 3.523 | | .109 | | 57.0 | 3.348 | | .120 | | .096 | 2.50 | 2.69 | 48600 | 25700 |
| 3.250 | 3¼ | 82.5 | N5000-325 | 3.623 | ±.055 | .109 | | 60.0 | 3.446 | | .120 | | .098 | 2.54 | 2.73 | 50000 | 27000 |
| 3.346 | 3 11/32 | 85.0 | N5000-334 | 3.734 | | .109 | | 65.0 | 3.546 | | .120 | | .100 | 2.63 | 2.83 | 51600 | 28300 |
| 3.469 | 3 15/32 | 88.1 | N5000-347 | 3.857 | | .109 | | 69.0 | 3.675 | | .120 | | .103 | 2.76 | 2.96 | 53400 | 30200 |
| 3.500 | 3½ | 88.9 | N5000-350 | 3.890 | | .109 | | 71.0 | 3.710 | | .120 | | .105 | 2.79 | 3.00 | 53900 | 31200 |
| 3.543 | — | 90.0 | N5000-354 | 3.936 | | .109 | | 72.0 | 3.755 | | .120 | | .106 | 2.83 | 3.04 | 54600 | 31800 |
| 3.562 | 3 9/16 | 90.5 | N5000-354 | 3.936 | | .109 | | 72.0 | 3.776 | | .120 | | .107 | 2.85 | 3.06 | 54600 | 31800 |

Additional sizes appear on Pages 26 and 30

WALDES **TRUARC** RETAINING RINGS

BASIC
*internal series*
**N5000**
(Continued)

| FIG. 1: MAXIMUM ALLOWABLE CORNER RADIUS ($R_{max.}$) AND CHAMFER ($Ch_{max.}$) | FIG. 2: ENLARGED DETAIL OF GROOVE PROFILE AND EDGE MARGIN (Z) | FIG. 3: SUPPLEMENTARY RING DIMENSIONS | FIG. 4: MINIMUM GAP WIDTH (Ring installed in groove) | FIG. 5: LUG DESIGN: SIZES -206 THRU -275 AND -300 THRU -1000 |

| MAXIMUM BOTTOM RADII | |
|---|---|
| **Ring Size** | **R** |
| -25 thru -100 | .005 |
| -102 thru -1000 | .010 |

## SUPPLEMENTARY APPLICATION DATA / SUPPLEMENTARY RING DIMENSIONS

| INTERNAL SERIES **N5000** size—no. | Maximum allowable corner radii and chamfers of retained parts (Fig. 1). See Page 21. $R_{max.}$ | $Ch_{max.}$ | Allowable assembly load with $R_{max.}$ or $Ch_{max.}$ $P_r'$(lbs.) | Edge margin (Fig. 2) See Page 19 Z | LUG B | tol. | LARGE SECTION E | tol. | SMALL SECTION J | tol. | HOLE DIAMETER P | tol. | MIN. GAP WIDTH (Fig. 4) Ring installed in groove A |
|---|---|---|---|---|---|---|---|---|---|---|---|---|---|
| • N5000-175 | .064 | .050 | 3900 | .162 | .234 | | .171 | | .083 | | .078 | | .372 |
| • N5000-181 | .064 | .050 | 3900 | .165 | .234 | | .170 | | .084 | | .093 | | .382 |
| • N5000-185 | .064 | .050 | 3900 | .168 | .234 | | .170 | | .085 | | .093 | | .392 |
| • N5000-187 | .064 | .050 | 3900 | .171 | .234 | | .170 | | .085 | | .093 | | .420 |
| • N5000-193 | .064 | .050 | 3900 | .177 | .234 | | .170 | | .085 | | .093 | | .438 |
| • N5000-200 | .064 | .050 | 3900 | .183 | .240 | | .170 | | .085 | | .093 | | .453 |
| (S=2.047) N5000-206 | .076 | .061 | 6200 | .186 | .250 | | .186 | | .091 | | .093 | | .428 |
| (S=2.062) N5000-206 | .078 | .062 | 6200 | .186 | .250 | | .186 | | .091 | | .093 | | .468 |
| N5000-212 | .078 | .062 | 6200 | .189 | .260 | | .195 | | .096 | | .093 | | .460 |
| (S=2.165) N5000-218 | .078 | .062 | 6200 | .195 | .264 | | .199 | | .098 | | .093 | | .439 |
| (S=2.188) N5000-218 | .078 | .062 | 6200 | .195 | .264 | | .199 | | .098 | | .093 | | .489 |
| N5000-225 | .078 | .062 | 6200 | .198 | .270 | | .203 | | .099 | | .093 | | .478 |
| N5000-231 | .078 | .062 | 6200 | .207 | .270 | | .206 | | .100 | | .093 | | .486 |
| N5000-237 | .078 | .062 | 6200 | .213 | .270 | ±.005 | .207 | ±.007 | .102 | ±.007 | .093 | | .504 |
| N5000-244 | .078 | .062 | 6200 | .216 | .280 | | .209 | | .103 | | .110 | | .518 |
| (S=2.500) N5000-250 | .078 | .062 | 6200 | .222 | .280 | | .210 | | .103 | | .110 | | .532 |
| (S=2.531) N5000-250 | .078 | .062 | 6200 | .225 | .280 | | .210 | | .103 | | .110 | | .597 |
| N5000-256 | .088 | .070 | 9000 | .228 | .290 | | .222 | | .109 | | .110 | | .540 |
| N5000-262 | .088 | .070 | 9000 | .234 | .290 | | .226 | | .111 | | .110 | +.015 −.002 | .558 |
| (S=2.677) N5000-268 | .090 | .072 | 9000 | .240 | .300 | | .230 | | .113 | | .110 | | .539 |
| (S=2.688) N5000-268 | .090 | .072 | 9000 | .240 | .300 | | .230 | | .113 | | .110 | | .568 |
| N5000-275 | .092 | .074 | 9000 | .246 | .300 | | .234 | | .115 | | .110 | | .590 |
| (S=2.812) N5000-281 | .088 | .070 | 9000 | .252 | .300 | | .230 | | .115 | | .110 | | .615 |
| (S=2.835) N5000-281 | .088 | .070 | 9000 | .255 | .300 | | .230 | | .115 | | .110 | | .676 |
| N5000-287 | .092 | .074 | 9000 | .264 | .310 | | .240 | | .120 | | .110 | | .626 |
| (S=2.953) N5000-300 | .092 | .074 | 9000 | .273 | .310 | | .250 | | .122 | | .110 | | .619 |
| (S=3.000) N5000-300 | .092 | .074 | 9000 | .273 | .310 | | .250 | | .122 | | .110 | | .738 |
| N5000-306 | .097 | .078 | 12000 | .279 | .310 | | .254 | | .126 | | .125 | | .651 |
| N5000-312 | .099 | .079 | 12000 | .285 | .310 | | .259 | | .129 | | .125 | | .655 |
| (S=3.149) N5000-315 | .100 | .080 | 12000 | .288 | .310 | | .262 | | .129 | | .125 | | .650 |
| (S=3.156) N5000-315 | .100 | .080 | 12000 | .288 | .310 | | .262 | | .129 | | .125 | | .669 |
| N5000-325 | .104 | .083 | 12000 | .294 | .342 | | .269 | | .135 | | .125 | | .698 |
| N5000-334 | .108 | .086 | 12000 | .300 | .342 | ±.008 | .276 | ±.008 | .140 | ±.008 | .125 | | .705 |
| N5000-347 | .108 | .086 | 12000 | .309 | .342 | | .286 | | .144 | | .125 | | .763 |
| N5000-350 | .110 | .088 | 12000 | .315 | .342 | | .289 | | .142 | | .125 | | .774 |
| (S=3.543) N5000-354 | .110 | .088 | 12000 | .318 | .342 | | .292 | | .142 | | .125 | | .788 |
| (S=3.562) N5000-354 | .110 | .088 | 12000 | .321 | .342 | | .292 | | .142 | | .125 | | .842 |

See Fig. 2, Page 33

**BASIC** *external series* **5100** — WALDES **TRUARC** RETAINING RINGS

• Sizes identified by this symbol are available in tape-wrapped Rol-Pak® cartridges

| SHAFT DIAMETER | | | MIL-R-21248 MS 16624 NAS 670 EXTERNAL SERIES 5100 | TRUARC RING DIMENSIONS | | | | | GROOVE DIMENSIONS | | | | | APPLICATION DATA | | | |
|---|---|---|---|---|---|---|---|---|---|---|---|---|---|---|---|---|---|
| Dec. equiv. inch | Approx fract. equiv. inch | Approx. mm | size — no. | FREE DIA. | tol. | THICKNESS | tol. | Approx. weight per 1000 pieces | DIAMETER | tol. | WIDTH | tol. | Nominal groove depth | CLEARANCE DIAMETER When sprung over shaft | When sprung into groove | RINGS | GROOVES |
| S | S | S | | D | tol. | t | tol. | lbs. | G | tol. | W | tol. | d | C₁ | C₂ | Pr | Pg |
| .125 | 1/8 | 3.2 | 5100-12 | .112 | | .010 | ±.001 | .018 | .117 | | .012 | | .004 | .222 | .214 | 110 | 35 |
| .156 | 5/32 | 4.0 | 5100-15 | .142 | | .010 | | .037 | .146 | | .012 | | .005 | .270 | .260 | 130 | 55 |
| .188 | 3/16 | 4.8 | 5100-18 | .168 | +.002 -.004 | .015 | | .059 | .175 | ±.0015 .0015 T.I.R. | .018 | +.002 -.000 | .006 | .298 | .286 | 240 | 80 |
| .197 | — | 5.0 | 5100-19 | .179 | | .015 | | .063 | .185 | | .018 | | .006 | .319 | .307 | 250 | 85 |
| .219 | 7/32 | 5.6 | 5100-21 | .196 | | .015 | | .074 | .205 | | .018 | | .007 | .338 | .324 | 280 | 110 |
| .236 | 15/64 | 6.0 | 5100-23 | .215 | | .015 | | .086 | .222 | | .018 | | .007 | .355 | .341 | 310 | 120 |
| .250 | 1/4 | 6.4 | • 5100-25 | .225 | | .025 | | .21 | .230 | | .029 | | .010 | .45 | .43 | 590 | 175 |
| .276 | — | 7.0 | 5100-27 | .250 | | .025 | | .23 | .255 | | .029 | | .010 | .48 | .46 | 650 | 195 |
| .281 | 9/32 | 7.1 | • 5100-28 | .256 | | .025 | | .24 | .261 | | .029 | | .010 | .49 | .47 | 660 | 200 |
| .312 | 5/16 | 7.9 | • 5100-31 | .281 | | .025 | | .27 | .290 | | .029 | | .011 | .54 | .52 | 740 | 240 |
| .344 | 11/32 | 8.7 | 5100-34 | .309 | +.002 -.005 | .025 | | .31 | .321 | ±.002 .002 T.I.R. | .029 | | .011 | .57 | .55 | 800 | 265 |
| .354 | — | 9.0 | 5100-35 | .320 | | .025 | | .35 | .330 | | .029 | | .012 | .59 | .57 | 820 | 300 |
| .375 | 3/8 | 9.5 | • 5100-37 | .338 | | .025 | | .39 | .352 | | .029 | | .012 | .61 | .59 | 870 | 320 |
| .394 | — | 10.0 | 5100-39 | .354 | | .025 | | .42 | .369 | | .029 | | .012 | .62 | .60 | 940 | 335 |
| .406 | 13/32 | 10.3 | 5100-40 | .366 | | .025 | | .43 | .382 | | .029 | | .012 | .63 | .61 | 950 | 350 |
| .438 | 7/16 | 11.1 | • 5100-43 | .395 | | .025 | | .50 | .412 | | .029 | | .013 | .66 | .64 | 1020 | 400 |
| .469 | 15/32 | 11.9 | 5100-46 | .428 | | .025 | | .54 | .443 | | .029 | | .013 | .68 | .66 | 1100 | 450 |
| .500 | 1/2 | 12.7 | • 5100-50 | .461 | | .035 | | .91 | .468 | ±.002 .004 T.I.R. | .039 | | .016 | .77 | .74 | 1650 | 550 |
| .551 | — | 14.0 | 5100-55 | .509 | | .035 | | .90 | .519 | | .039 | | .016 | .81 | .78 | 1800 | 600 |
| .562 | 9/16 | 14.3 | • 5100-56 | .521 | | .035 | | 1.1 | .530 | | .039 | +.003 -.000 | .016 | .82 | .79 | 1850 | 650 |
| .594 | 19/32 | 15.1 | 5100-59 | .550 | | .035 | ±.002 | 1.2 | .559 | | .039 | | .017 | .86 | .83 | 1950 | 750 |
| .625 | 5/8 | 15.9 | • 5100-62 | .579 | | .035 | | 1.3 | .588 | | .039 | | .018 | .90 | .87 | 2060 | 800 |
| .669 | — | 17.0 | 5100-66 | .621 | | .035 | | 1.4 | .629 | | .039 | | .020 | .93 | .89 | 2200 | 950 |
| .672 | 43/64 | 17.1 | 5100-66 | .621 | | .035 | | 1.4 | .631 | | .039 | | .020 | .93 | .89 | 2200 | 950 |
| .688 | 11/16 | 17.5 | • 5100-68 | .635 | +.005 -.010 | .042 | | 1.8 | .646 | ±.003 .004 T.I.R. | .046 | | .021 | 1.01 | .97 | 3400 | 1000 |
| .750 | 3/4 | 19.0 | • 5100-75 | .693 | | .042 | | 2.1 | .704 | | .046 | | .023 | 1.09 | 1.05 | 3700 | 1200 |
| .781 | 25/32 | 19.8 | 5100-78 | .722 | | .042 | | 2.2 | .733 | | .046 | | .024 | 1.12 | 1.08 | 3900 | 1300 |
| .812 | 13/16 | 20.6 | 5100-81 | .751 | | .042 | | 2.5 | .762 | | .046 | | .025 | 1.15 | 1.10 | 4000 | 1450 |
| .875 | 7/8 | 22.2 | 5100-87 | .810 | | .042 | | 2.8 | .821 | | .046 | | .027 | 1.21 | 1.16 | 4300 | 1650 |
| .938 | 15/16 | 23.8 | 5100-93 | .867 | | .042 | | 3.1 | .882 | | .046 | | .028 | 1.34 | 1.29 | 4650 | 1850 |
| .984 | 63/64 | 25.0 | 5100-98 | .910 | | .042 | | 3.5 | .926 | | .046 | | .029 | 1.39 | 1.34 | 4850 | 2000 |
| 1.000 | 1 | 25.4 | 5100-100 | .925 | | .042 | | 3.6 | .940 | | .046 | | .030 | 1.41 | 1.35 | 4950 | 2100 |
| 1.023 | — | 26.0 | 5100-102 | .946 | | .042 | | 3.9 | .961 | | .046 | | .031 | 1.43 | 1.37 | 5050 | 2250 |
| 1.062 | 1 1/16 | 27.0 | 5100-106 | .982 | | .050 | | 4.8 | .998 | | .056 | | .032 | 1.50 | 1.44 | 6200 | 2400 |
| 1.125 | 1 1/8 | 28.6 | 5100-112 | 1.041 | | .050 | | 5.1 | 1.059 | | .056 | | .033 | 1.55 | 1.49 | 6600 | 2600 |
| 1.188 | 1 3/16 | 30.2 | 5100-118 | 1.098 | | .050 | | 5.6 | 1.118 | ±.004 .005 T.I.R | .056 | | .035 | 1.61 | 1.54 | 7000 | 2950 |
| 1.250 | 1 1/4 | 31.7 | 5100-125 | 1.156 | +.010 -.015 | .050 | | 5.9 | 1.176 | | .056 | | .037 | 1.69 | 1.62 | 7350 | 3250 |
| 1.312 | 1 5/16 | 33.3 | 5100-131 | 1.214 | | .050 | | 6.8 | 1.232 | | .056 | | .040 | 1.75 | 1.67 | 7750 | 3700 |
| 1.375 | 1 3/8 | 34.9 | 5100-137 | 1.272 | | .050 | | 7.2 | 1.291 | | .056 | | .042 | 1.80 | 1.72 | 8100 | 4100 |
| 1.438 | 1 7/16 | 36.5 | 5100-143 | 1.333 | | .050 | | 8.1 | 1.350 | | .056 | | .044 | 1.87 | 1.79 | 8500 | 4500 |
| 1.500 | 1 1/2 | 38.1 | 5100-150 | 1.387 | | .050 | | 9.0 | 1.406 | | .056 | +.004 -.000 | .047 | 1.99 | 1.90 | 8800 | 5000 |
| 1.562 | 1 9/16 | 39.7 | 5100-156 | 1.446 | | .062 | | 12.4 | 1.468 | | .068 | | .047 | 2.10 | 2.01 | 11400 | 5200 |
| 1.625 | 1 5/8 | 41.3 | 5100-162 | 1.503 | | .062 | | 13.2 | 1.529 | | .068 | | .048 | 2.17 | 2.08 | 11850 | 5500 |
| 1.688 | 1 11/16 | 42.9 | 5100-168 | 1.560 | | .062 | | 14.8 | 1.589 | | .068 | | .049 | 2.24 | 2.15 | 12350 | 5850 |
| 1.750 | 1 3/4 | 44.4 | 5100-175 | 1.618 | | .062 | ±.003 | 15.3 | 1.650 | ±.005 .005 T.I.R. | .068 | | .050 | 2.31 | 2.21 | 12800 | 6200 |
| 1.772 | — | 45.0 | 5100-177 | 1.637 | +.013 -.020 | .062 | | 15.4 | 1.669 | | .068 | | .051 | 2.33 | 2.23 | 12950 | 6400 |
| 1.812 | 1 13/16 | 46.0 | 5100-181 | 1.675 | | .062 | | 16.2 | 1.708 | | .068 | | .052 | 2.38 | 2.28 | 13250 | 6650 |
| 1.875 | 1 7/8 | 47.6 | 5100-187 | 1.735 | | .062 | | 17.3 | 1.769 | | .068 | | .053 | 2.44 | 2.34 | 13700 | 7000 |
| 1.969 | 1 31/32 | 50.0 | 5100-196 | 1.819 | | .062 | | 18.0 | 1.857 | | .068 | | .056 | 2.54 | 2.43 | 14350 | 7800 |
| 2.000 | 2 | 50.8 | 5100-200 | 1.850 | | .062 | | 19.0 | 1.886 | | .068 | | .057 | 2.55 | 2.44 | 14600 | 8050 |

Thickness **t** applies only to unplated rings. For plated and stainless steel (Type H) rings, add .002″ to the listed maximum thickness. Maximum ring thickness will be at least .0002″ less than the listed minimum groove width (**W**).

T.I.R. (total indicator reading) is the maximum allowable deviation of concentricity between groove and shaft.

ALLOW. THRUST LOAD (lbs.) Sharp Corner Abutment — RINGS Safety factor = 4 Important! See Page 19 — GROOVES Safety factor = 2 Important! See Page 19

(For continuation of sizes, see Page 34)

*TRUARC RETAINING RINGS DIVISION*

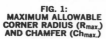

BASIC
*external series*
5100

**FIG. 1:**
MAXIMUM ALLOWABLE CORNER RADIUS (R_max.) AND CHAMFER (Ch_max.)

**FIG. 2:**
ENLARGED DETAIL OF GROOVE PROFILE AND EDGE MARGIN (Z)

| MAXIMUM BOTTOM RADII | |
|---|---|
| Ring size | R |
| -12 thru -23 | Sharp corners |
| -25 thru -35 | .003 |
| -37 thru -100 | .005 |
| -102 thru -200 | .010 |

**FIG. 3:**
SUPPLEMENTARY RING DIMENSIONS

**FIG. 4:**
MAXIMUM GAGING DIAMETER (Ring installed in groove)

**FIG. 5:**
LUG DESIGN Sizes -12 thru -23

| EXTERNAL SERIES 5100 size — no. | Max. allow. corner radii and chamfers of retained parts. (Fig. 1). See Pg. 21 R_max. | Ch_max. | Allow. assembly load with R_max. or Ch_max. P′ (lbs.) | Edge margin (Fig. 2) See Page 19 Z | Safe rotational speeds rpm | LUG B | tol. | LARGE SECTION E | tol. | SMALL SECTION J | tol. | HOLE DIAMETER P | tol. | MAX. GAGING DIA. Ring installed in groove (Fig. 4) K |
|---|---|---|---|---|---|---|---|---|---|---|---|---|---|---|
| 5100-12 | .010 | .006 | 45 | .012 | 80000 | .046 | | .018 | ±.0015 | .011 | ±.0015 | .026 | | .148 |
| 5100-15 | .015 | .009 | 45 | .015 | 80000 | .054 | | .026 | | .016 | | .026 | | .189 |
| 5100-18 | .014 | .0085 | 105 | .018 | 80000 | .050 | ±.002 | .025 | | .016 | | .025 | | .218 |
| 5100-19 | .0145 | .009 | 105 | .018 | 80000 | .056 | | .026 | ±.002 | .016 | ±.002 | .026 | | .229 |
| 5100-21 | .015 | .009 | 105 | .021 | 80000 | .056 | | .028 | | .017 | | .026 | | .252 |
| 5100-23 | .0165 | .010 | 105 | .021 | 80000 | .056 | | .030 | | .019 | | .026 | | .272 |
| • 5100-25 | .018 | .011 | 470 | .030 | 80000 | .080 | | .035 | | .025 | | .041 | | .290 |
| 5100-27 | .0175 | .0105 | 470 | .031 | 76000 | .081 | | .035 | | .024 | | .041 | | .315 |
| • 5100-28 | .020 | .012 | 470 | .030 | 74000 | .080 | | .038 | | .0255 | | .041 | | .326 |
| • 5100-31 | .020 | .012 | 470 | .033 | 70000 | .087 | | .040 | | .026 | | .041 | | .357 |
| 5100-34 | .021 | .0125 | 470 | .033 | 64000 | .087 | | .042 | | .0265 | | .041 | | .390 |
| 5100-35 | .023 | .014 | 470 | .036 | 62000 | .087 | | .046 | ±.003 | .029 | ±.003 | .041 | | .405 |
| • 5100-37 | .026 | .0155 | 470 | .036 | 60000 | .088 | | .050 | | .0305 | | .041 | | .433 |
| 5100-39 | .027 | .016 | 470 | .037 | 56500 | .087 | | .052 | | .031 | | .041 | | .452 |
| 5100-40 | .0285 | .017 | 470 | .036 | 55000 | .087 | ±.003 | .054 | | .033 | | .041 | +.010 −.002 | .468 |
| • 5100-43 | .029 | .0175 | 470 | .039 | 50000 | .088 | | .055 | | .033 | | .041 | | .501 |
| 5100-46 | .031 | .018 | 470 | .039 | 42000 | .088 | | .060 | | .035 | | .041 | | .540 |
| • 5100-50 | .034 | .020 | 910 | .048 | 40000 | .108 | | .065 | | .040 | | .047 | | .574 |
| 5100-55 | .027 | .0165 | 910 | .048 | 36000 | .108 | | .053 | | .036 | | .047 | | .611 |
| • 5100-56 | .038 | .023 | 910 | .048 | 35000 | .108 | | .072 | | .041 | | .047 | | .644 |
| 5100-59 | .0395 | .0235 | 910 | .052 | 32000 | .109 | | .076 | ±.004 | .043 | ±.004 | .047 | | .680 |
| • 5100-62 | .0415 | .025 | 910 | .055 | 30000 | .110 | | .080 | | .045 | | .047 | | .715 |
| (S=.669) 5100-66 | .040 | .024 | 910 | .060 | 29000 | .110 | | .082 | | .043 | | .047 | | .756 |
| (S=.672) 5100-66 | .040 | .024 | 910 | .060 | 29000 | .110 | | .082 | | .043 | | .047 | | .758 |
| • 5100-68 | .042 | .025 | 1340 | .063 | 28000 | .136 | | .084 | | .048 | | .052 | | .779 |
| • 5100-75 | .046 | .0275 | 1340 | .069 | 26500 | .136 | | .092 | | .051 | | .052 | | .850 |
| 5100-78 | .047 | .028 | 1340 | .072 | 25500 | .136 | | .094 | | .052 | | .052 | | .883 |
| 5100-81 | .047 | .028 | 1340 | .075 | 24500 | .136 | | .096 | | .054 | | .052 | | .914 |
| 5100-87 | .051 | .0305 | 1340 | .081 | 23000 | .137 | | .104 | ±.005 | .057 | ±.005 | .052 | | .987 |
| 5100-93 | .055 | .033 | 1340 | .084 | 21500 | .166 | | .110 | | .063 | | .078 | | 1.054 |
| 5100-98 | .056 | .0335 | 1340 | .087 | 20500 | .167 | | .114 | | .0645 | | .078 | | 1.106 |
| 5100-100 | .057 | .034 | 1340 | .090 | 20000 | .167 | | .116 | | .065 | | .078 | | 1.122 |
| 5100-102 | .058 | .035 | 1340 | .093 | 19500 | .168 | | .118 | | .066 | | .078 | | 1.147 |
| 5100-106 | .060 | .036 | 1950 | .096 | 19000 | .181 | | .122 | | .069 | | .078 | | 1.192 |
| 5100-112 | .063 | .038 | 1950 | .099 | 18800 | .182 | ±.004 | .128 | | .071 | | .078 | | 1.261 |
| 5100-118 | .064 | .0385 | 1950 | .105 | 18000 | .182 | | .132 | | .072 | | .078 | | 1.325 |
| 5100-125 | .068 | .041 | 1950 | .111 | 17000 | .183 | | .140 | | .076 | | .078 | | 1.396 |
| 5100-131 | .068 | .041 | 1950 | .120 | 16500 | .183 | | .146 | | .0765 | | .078 | | 1.458 |
| 5100-137 | .072 | .043 | 1950 | .126 | 16000 | .184 | | .152 | | .082 | | .078 | | 1.529 |
| 5100-143 | .076 | .045 | 1950 | .132 | 15000 | .184 | | .160 | | .086 | | .078 | +.015 −.002 | 1.600 |
| 5100-150 | .079 | .047 | 1950 | .141 | 14800 | .214 | | .168 | | .091 | | .120 | | 1.668 |
| 5100-156 | .082 | .049 | 3000 | .141 | 14000 | .235 | | .172 | ±.006 | .093 | ±.006 | .125 | | 1.740 |
| 5100-162 | .087 | .052 | 3000 | .144 | 13200 | .235 | | .180 | | .097 | | .125 | | 1.812 |
| 5100-168 | .090 | .054 | 3000 | .148 | 13000 | .235 | | .184 | | .099 | | .125 | | 1.877 |
| 5100-175 | .091 | .054 | 3000 | .150 | 12200 | .237 | | .188 | | .101 | | .125 | | 1.945 |
| 5100-177 | .092 | .055 | 3000 | .154 | 11700 | .237 | | .190 | | .102 | | .125 | | 1.967 |
| 5100-181 | .092 | .055 | 3000 | .156 | 11500 | .238 | | .192 | | .102 | | .125 | | 2.010 |
| 5100-187 | .094 | .056 | 3000 | .159 | 11000 | .239 | | .196 | | .104 | | .125 | | 2.076 |
| 5100-196 | .094 | .056 | 3000 | .168 | 10500 | .245 | | .200 | | .106 | | .125 | | 2.170 |
| 5100-200 | .096 | .057 | 3000 | .171 | 10500 | .239 | | .204 | | .108 | | .125 | | 2.205 |

*WALDES KOHINOOR, INC. • LONG ISLAND CITY 1, N.Y.*

E-RING external series 5133 — X5133 • Y5133 — WALDES TRUARC RETAINING RINGS

section I-I

See Fig. 2, Page 47

• Sizes identified by this symbol are available in tape-wrapped **Rol-Pak®** cartridges for use with Truarc Applicators, Dispensers and other assembly tools

| SHAFT DIA. | | | MIL-R-21248 MS 16633 EXTERNAL SERIES 5133 X5133 Y5133 | TRUARC RING DIMENSIONS | | | | | GROOVE DIMENSIONS | | | | | APPLICATION DATA | | | |
|---|---|---|---|---|---|---|---|---|---|---|---|---|---|---|---|---|---|
| | | | | FREE DIA. | | THICKNESS | | Approx. weight per 1000 pieces | DIAMETER | | WIDTH | | Nominal groove depth | CLEARANCE DIAMETER | | ALLOW. THRUST LOAD (lbs.) Sharp Corner Abutment | |
| | | | | | | | | | | | | | | Free outside dia. (REF.) | When sprung into groove G | RINGS Safety factor=3 | GROOVES Safety factor=2 |
| Dec. equiv. inch | Approx. fract. equiv. inch | Approx. mm. | size—no. | D | tol. | t | tol. | lbs. | G | tol. | W | tol. | d | Y | C₂ | Pᵣ | P_g |
| S | S | S | size—no. | D | tol. | t | tol. | lbs. | G | tol. | W | tol. | d | Y | C₂ | $P_r$ | $P_g$ |
| .040 | — | 1.0 | X5133-4 | .025 | | .010 | ±.001 | .009 | .026 | | .012 | | .007 | .079 | .090 | 13 | 6 |
| .062 | 1/16 | 1.6 | X5133-6 | .051 | +.001 | .010 | ±.001 | .028 | .052 | | .012 | | .005 | .140 | .150 | 20 | 7 |
| .062 | 1/16 | 1.6 | Y5133-6 | .051 | −.003 | .010 | ±.002 | .094 | .052 | | .012 | | .005 | .187 | .200 | 40 | 7 |
| .062 | 1/16 | 1.6 | 5133-6 | .051 | | .010 | ±.001 | .030 | .052 | +.002 −.000 .0015 T.I.R. | .012 | +.002 −.000 | .005 | .156 | .165 | 20 | 7 |
| .094 | 3/32 | 2.4 | X5133-9 | .069 | +.002 −.003 | .015 | | .10 | .074 | | .018 | | .010 | .230 | .245 | 45 | 20 |
| .094 | 3/32 | 2.4 | 5133-9 | .073 | | .015 | | .058 | .074 | | .018 | | .010 | .187 | .200 | 45 | 20 |
| .110 | 7/64 | 2.8 | X5133-11 | .076 | | .015 | | .31 | .079 | | .018 | | .015 | .375 | .390 | 60 | 40 |
| .125 | 1/8 | 3.2 | 5133-12 | .094 | | .015 | | .087 | .095 | | .018 | | .015 | .230 | .240 | 65 | 45 |
| .140 | 9/64 | 3.6 | X5133-14 | .100 | | .015 | | .060 | .102 | | .018 | | .019 | .203 | .215 | 75 | 60 |
| .140 | 9/64 | 3.6 | Y5133-14 | .108 | | .015 | | .10 | .110 | | .018 | | .015 | .250 | .265 | 75 | 45 |
| .140 | 9/64 | 3.6 | 5133-14 | .102 | +.001 −.003 | .025 | | .21 | .105 | +.002 −.000 .002 T.I.R. | .029 | | .017 | .270 | .285 | 170 | 60 |
| .156 | 5/32 | 4.0 | 5133-15 | .114 | | .025 | | .21 | .116 | | .029 | | .020 | .282 | .295 | 175 | 75 |
| .172 | 11/64 | 4.4 | X5133-17 | .125 | | .025 | | .24 | .127 | | .029 | | .022 | .312 | .325 | 180 | 90 |
| .188 | 3/16 | 4.8 | X5133-18 | .122 | | .025 | | .45 | .125 | | .029 | | .031 | .375 | .39 | 200 | 135 |
| .188 | 3/16 | 4.8 | 5133-18 | .145 | | .025 | | .29 | .147 | | .029 | | .020 | .335 | .35 | 190 | 90 |
| .219 | 7/32 | 5.6 | X5133-21 | .185 | | .025 | | .47 | .188 | | .029 | | .015 | .437 | .45 | 225 | 75 |
| .250 | 1/4 | 6.3 | 5133-25 | .207 | | .025 | ±.002 | .76 | .210 | | .029 | | .020 | .527 | .54 | 255 | 115 |
| .312 | 5/16 | 7.9 | X5133-31 | .243 | | .025 | | .57 | .250 | | .029 | +.003 −.000 | .031 | .500 | .52 | 325 | 225 |
| .375 | 3/8 | 9.5 | 5133-37 | .300 | | .035 | | 1.5 | .303 | | .039 | | .036 | .660 | .68 | 690 | 315 |
| .438 | 7/16 | 11.1 | 5133-43 | .337 | +.002 −.004 | .035 | | 1.5 | .343 | | .039 | | .047 | .687 | .71 | 830 | 480 |
| .438 | 7/16 | 11.1 | X5133-43 | .375 | | .035 | | 1.0 | .380 | +.003 −.000 .004 T.I.R. | .039 | | .029 | .600 | .62 | 800 | 280 |
| .500 | 1/2 | 12.7 | 5133-50 | .392 | | .042 | | 2.5 | .396 | | .046 | | .052 | .800 | .82 | 1110 | 600 |
| .625 | 5/8 | 15.9 | 5133-62 | .480 | | .042 | | 3.2 | .485 | | .046 | | .070 | .940 | .96 | 1420 | 1050 |
| .744 | — | 18.9 | X5133-74 | .616 | | .050 | | 4.3 | .625 | | .056 | | .059 | 1.000 | 1.02 | 1900 | 1050 |
| .750 | 3/4 | 19.0 | X5133-74 | .616 | +.003 −.005 | .050 | | 4.3 | .625 | | .056 | | .062 | 1.000 | 1.02 | 1950 | 1100 |
| .750 | 3/4 | 19.0 | 5133-75 | .574 | | .050 | | 5.8 | .580 | | .056 | | .085 | 1.120 | 1.14 | 2000 | 1500 |
| .875 | 7/8 | 22.2 | 5133-87 | .668 | | .050 | | 7.6 | .675 | | .056 | | .100 | 1.300 | 1.32 | 2350 | 2050 |
| .984 | 63/64 | 25.0 | X5133-98 | .822 | | .050 | | 9.2 | .835 | | .056 | | .074 | 1.500 | 1.53 | 2600 | 1750 |
| 1.000 | 1 | 25.4 | X5133-98 | .822 | | .050 | | 9.2 | .835 | | .056 | | .082 | 1.500 | 1.53 | 2650 | 1900 |
| 1.188 | 1 3/16 | 30.2 | X5133-118 | 1.066 | +.006 −.010 | .062 | ±.003 | 11.3 | 1.079 | +.005 −.000 T.I.R. | .068 | +.004 −.000 | .054 | 1.626 | 1.67 | 3450 | 1500 |
| 1.375 | 1 3/8 | 34.9 | X5133-137 | 1.213 | | .062 | | 15.4 | 1.230 | | .068 | | .072 | 1.875 | 1.92 | 4100 | 2350 |

Thickness t applies only to un-plated rings. For **plated** and **stainless steel** (Type H) rings, add .002″ to the listed maximum thickness. Maximum ring thickness will be at least .0002″ less than the listed minimum groove width (**W**).

T.I.R. (total indicator reading) is the maximum allowable deviation of concentricity between groove and shaft.

Important! See Page 19   Important! See Page 19

**FIG. 1:**
**MAXIMUM ALLOWABLE**
**CORNER RADIUS (R_max.)**
**AND CHAMFER (Ch_max.)**

**FIG. 2:**
**ENLARGED DETAIL**
**OF GROOVE PROFILE**
**AND EDGE MARGIN (Z)**

### MAXIMUM BOTTOM RADII

| Ring size | R |
|---|---|
| X-4 thru -6 | Sharp Corners |
| X-9 thru -25 | .005 |
| X-31 thru X-43 | .010 |
| -50 thru X-137 | .015 |

## SUPPLEMENTARY APPLICATION DATA

| EXTERNAL SERIES 5133 X5133 Y5133 size—no. | Maximum allowable corner radii and chamfers of retained parts (Fig. 1). See Page 21. $R_{max.}$ | $Ch_{max.}$ | Allow. assembly load with $R_{max.}$ or $Ch_{max.}$ $P'_{r(lbs.)}$ | Edge margin (Fig. 2) See Page 19 Z | Safe rotational speeds rpm |
|---|---|---|---|---|---|
| X5133-4 | .015 | .010 | 13 | .014 | 40000 |
| • X5133-6 | .030 | .020 | 20 | .010 | 40000 |
| • Y5133-6 | .035 | .025 | 40 | .010 | 40000 |
| • 5133-6 | .030 | .020 | 20 | .010 | 40000 |
| • X5133-9 | .053 | .040 | 45 | .020 | 36000 |
| • 5133-9 | .040 | .030 | 45 | .020 | 36000 |
| • X5133-11 | .080 | .060 | 60 | .030 | 35000 |
| • 5133-12 | .040 | .030 | 65 | .030 | 35000 |
| • X5133-14 | .029 | .022 | 75 | .038 | 32000 |
| • Y5133-14 | .040 | .030 | 75 | .030 | 32000 |
| • 5133-14 | .060 | .045 | 170 | .034 | 32000 |
| • 5133-15 | .060 | .045 | 175 | .040 | 31000 |
| • X5133-17 | .060 | .045 | 180 | .044 | 30000 |
| • X5133-18 | .060 | .045 | 200 | .062 | 30000 |
| • 5133-18 | .060 | .045 | 190 | .040 | 30000 |
| • X5133-21 | .060 | .045 | 225 | .030 | 26000 |
| • 5133-25 | .060 | .045 | 255 | .040 | 25000 |
| • X5133-31 | .060 | .045 | 325 | .062 | 22000 |
| • 5133-37 | .065 | .050 | 690 | .072 | 20000 |
| • 5133-43 | .065 | .050 | 830 | .094 | 16500 |
| • X5133-43 | .050 | .035 | 800 | .058 | 16500 |
| • 5133-50 | .080 | .060 | 1110 | .104 | 14000 |
| • 5133-62 | .080 | .060 | 1420 | .140 | 12000 |
| X5133-74 | .060 | .045 | 1900 | .118 | 11000 |
| X5133-74 | .057 | .042 | 1900 | .124 | 11000 |
| • 5133-75 | .085 | .065 | 2000 | .170 | 10500 |
| 5133-87 | .085 | .065 | 2350 | .200 | 9000 |
| X5133-98 | .085 | .065 | 2700 | .148 | 6500 |
| X5133-98 | .077 | .057 | 2700 | .164 | 6500 |
| X5133-118 | .090 | .070 | 3450 | .108 | 5500 |
| X5133-137 | .090 | .070 | 4100 | .144 | 4000 |

# BIBLIOGRAPHY OF ALLIED SUBJECTS

The following classified list is given to supplement this book, whose scope as a general treatise on engineering drawing permits only the mention or brief explanation of some subjects.

## Aeronautical Drafting and Design

Anderson, Newton H.: "Aircraft Layout and Detail Design," McGraw-Hill, New York.

Apalategui, J. J., and J. J. Adams: "Aircraft Analytic Geometry," McGraw-Hill, New York.

Davis, D. J., and C. H. Goen: "Aircraft Mechanical Drawing," McGraw-Hill, New York.

Katz, H. H.: "Aircraft Drafting," Macmillan, New York.

Leaveli, S., and S. Bungay: "Aircraft Production Standards," McGraw-Hill, New York.

LeMaster, C. A.: "Aircraft Sheet Metal Work," American Technological Society, Chicago.

Liming, Roy: "Practical Analytic Geometry with Applications of Aircraft," Macmillan, New York.

Liston, Joseph: "Powerplants for Aircraft," McGraw-Hill, New York.

Meadowcroft, Norman: "Aircraft Detailing Drafting," McGraw-Hill, New York.

"SAE Aeronautical Drafting Manual," Society of Automotive Engineers, Inc., New York.

Svensen, C. L.: "A Manual of Aircraft Drafting," Van Nostrand, Princeton, N.J.

Tharratt, George: "Aircraft Production Illustration," McGraw-Hill, New York.

## Architectural Drawing

Crane, T.: "Architectural Construction," Wiley, New York.

Field, W. B.: "Architectural Drawing," McGraw-Hill, New York.

Kenney, Joseph E., and John P. McGrail: "Architectural Drawing for the Building Trades," McGraw-Hill, New York.

Martin, C. L.: "Architectural Graphics," Macmillan, New York.

Morgan, Sherley W.: "Architectural Drawing," McGraw-Hill, New York.

Ramsey, C. G., and H. R. Sleeper: "Architectural Graphic Standards," Wiley, New York.

Saylor, H. H.: "Dictionary of Architecture," Wiley, New York.

Sleeper, H. R.: "Architectural Specifications," Wiley, New York.

## Careers in Engineering

Beakley, G. C., and H. W. Leach: "Careers in Engineering and Technology," Macmillan, New York.

Carlisle, N. D.: "Your Career in Engineering," Dutton, New York.

Grinter, L. E., and H. C. Spencer: "Engineering Preview," Macmillan, New York.

McGuire, J. G., and H. W. Barlow: "An Introduction to the Engineering Profession," Addison-Wesley, Reading, Mass.

Smith, R. V.: "Engineering as a Career," McGraw-Hill, New York.

Williams, C. C.: "Building an Engineering Career," McGraw-Hill, New York.

## Computers and Programming

Arms, W. Y., J. E. Baker, and R. M. Pengally: "A Practical Approach to Computing," Wiley, New York.

Banks, P. M., and P. M. Doupnik: "Introduction to Computer Science," Wiley, New York.

Bartee, Thomas: "Introduction to Computer Science," McGraw-Hill, New York.

Crowley, T. H.: "Understanding Computers," McGraw-Hill, New York.

Gear, C. W.: "Computer Organization and Programming," McGraw-Hill, New York.

Hare, V.: "Introduction to Programming," Harcourt, Brace, New York.

Jackson, P. C.: "Introduction to Artificial Intelligence," Petrocelli-Charter, New York.

Jacquez, J. A.: "A First Course in Computing and Numerical Methods," Addison-Wesley, Reading, Mass.

Kernighan, B. W., and P. J. Plauger: "Software Tools," Addison-Wesley, Reading, Mass.

Kindred, A. R.: "Introduction to Computers, "Prentice-Hall, Englewood Cliffs, N.J.

Laver, M.: "An Introduction to the Uses of Computers," Cambridge University Press, New York.

Maurer, W. D.: "Programming: An Introduction to Computer Techniques," Holden-Day, San Francisco.

Montgomery, G.: "Digital Calculating Machines and their Application to Scientific and Engineering Work," Blackie, London.

Nicholls, J. E.: "The Structure and Design of Programming Languages," Addison-Wesley, Reading, Mass.

Ralston, Anthony: "Introduction to Programming and Computer Science," McGraw-Hill, New York.

Randall, Brian: "The Origins of Digital Computers," Springer-Verlag, New York.

Tucker, A. B.: "Programming Languages," McGraw-Hill, New York.

Walker, T. M.: "Introduction to Computer Science," Allyn and Bacon, Boston.

Wilde, D. U.: "An Introduction to Computing," Prentice-Hall, Englewood Cliffs, N.J.

Wright, E. W.: "Structural Design by Computer," Van Nostrand, New York.

## Computer Graphics

Anderson, R. H.: "Multi-access Computing Research, "RAND Corporation, Santa Monica, Calif.

Anderson, R. H.: "A Selective Bibliography of Computer Graphics," RAND Corporation, Santa Monica, Calif.

Auerbach: "Digital Plotters and Image Digitizers," Princeton, N.J.

Bell, T.: "Modeling the Video Graphics System: Procedure and Model Description," RAND Corporation, Santa Monica, Calif.

Bell, T.: "GAPPS (Graphical Analysis Procedures for System Simulation)," RAND Corporation, Santa Monica, Calif.

Boehm, Barry W.: "POGO: Programmer Oriented Graphics Operations," RAND Corporation, Santa Monica, Calif.

Browne, L. V. I. (ed.),: "Curved Surfaces in Engineering: Computer Methods for Design and Manufacture," Science and Technology Press.

Fariman, M., and V. Nievergelt (eds.): "Pertinent Concepts in Computer Graphics," University of Illinois Press, Urbana.

Fetter, W. A.: "Computer Graphics in Communication," McGraw-Hill, New York.

Forrest, A.: "Curves and Surfaces for Computer Aided Design," University of Cambridge Press, New York.

Franke, H. W.: "Computer Graphics; Computer Art," Phaidon, New York.

Hyman, Anthony: "The Computer in Design," Studio Vista, London.

Lee, Theodore, M.: "Three Dimensional Curves and Surfaces for Rapid Computer Display," Management Information Series.

Newman, W. M.: "Principles of Interactive Computer Graphics," McGraw-Hill, New York.

Parslow, R. D., and R. Elliott Green: "Advanced Computer Graphics: Economics, Techniques and Applications," Plenum, New York.

Prince, M. David: "Interactive Graphics for Computer Aided Design," Addison-Wesley, Reading, Mass.

Priver, Arthur S.: "An Interactive Graphic System for Curve Fitting and Editing," RAND Corporation, Santa Monica, Calif.

Rogers, David F., and J. Alan Adams: "Mathematical Elements for Computer Graphics," McGraw-Hill, New York.

Seacrest, Don, and Jing Wievergelt (eds.). "Emerging Concepts in Computer Graphics," Benjamin, New York.

## Descriptive Geometry

Bradley, H. C., and E. H. Uhler: "Descriptive Geometry for Engineers," International Textbook, Scranton, Pa.

Grant, Hiram E.: "Practical Descriptive Geometry," McGraw-Hill, New York.

Higbee, F. G.: "Drawing-board Geometry," Wiley, New York.

Hood, George J., and Albert S. Palmerlee: "Geometry of Engineering Drawing," McGraw-Hill, New York.

Howe, H. B.: "Descriptive Geometry," Ronald, New York.

Johnson, L. O., and I. Wladaver: "Elements of Descriptive Geometry," Prentice-Hall, Englewood Cliffs, N.J.

Paré, E. G., R. O. Loving, and I. L. Hill: "Descriptive Geometry," Macmillan, New York.

Rowe, C. E., and James Dorr McFarland: "Engineering Descriptive Geometry," Van Nostrand, Princeton, N.J.

Shupe, Hollie W., and Paul E. Machovina: "A Manual of Engineering Geometry and Graphics for Students and Draftsmen," McGraw-Hill, New York.

Slaby, S. M.: "Engineering Descriptive Geometry," Barnes & Noble," New York.

Street, W. E.: "Technical Descriptive Geometry," Van Nostrand, Princeton, N.J.

## Drawing-instrument Catalogues

Theo. Alteneder and Sons, Philadelphia.
Eugene Dietzgen Co., Chicago.
Gramercy Guild Group, Inc., Denver, Colo.
Keuffel & Esser Co., Hoboken, N.J.
The Frederick Post Co., Chicago.
V and E Manufacturing Co., Pasadena, Calif.

## Electrical Drafting

Baer, C. J.: "Electrical and Electronics Drawing," McGraw-Hill, New York.
Bishop, Calvin C., C. T. Gilliam, and Associates: "Electrical Drafting and Design," McGraw-Hill, New York.
Carini, L. F. B.: "Drafting for Electronics," McGraw-Hill, New York.
Kocher, S. E.: "Electrical Drafting," International Textbook, Scranton, Pa.
Shiers, G.: "Electronic Drafting," McGraw-Hill, New York.
Van Gieson, D. Walter: "Electrical Drafting," McGraw-Hill, New York.

## Engineering Design

Edel, D. H.: "Introduction to Creative Design," Prentice-Hall, Englewood Cliffs, N.J.
Beakley, G. C., and H. W. Leach: "Engineering—An Introduction to a Creative Profession," Macmillan, New York.
Hill, P. H.: "The Science of Engineering Design," Holt, New York.
Krick, E. V.: "An Introduction to Engineering and Engineering Design," Wiley, New York.
Spotts, M. F.: "Design Engineering Projects," Prentice-Hall, Englewood Cliffs, N.J.

## Engineering Drawing

French, Thomas E., and Charles J. Vierck: "Fundamentals of Engineering Drawing," McGraw-Hill, New York.
French, Thomas E., and Charles J. Vierck: "Graphic Science and Design," McGraw-Hill, New York.
French, Thomas E., and Charles J. Vierck: "A Manual of Engineering Drawing for Students and Draftsmen," McGraw-Hill, New York.
Giesecke, F. E., A. Mitchell, and H. C. Spencer: "Technical Drawing," Macmillan, New York.
Healy, W. L., and A. H. Rau: "Simplified Drafting Practice," Wiley, New York.
Hoelscher, R. P., and C. H. Springer: "Engineering Drawing and Geometry," Wiley, New York.

Katz, H. H.: "Handbook of Layout and Dimensioning for Production," Macmillan, New York.
Lent, Deane: "Machine Drawing," Prentice-Hall, Englewood Cliffs, N.J.
Luzadder, W. J.: "Graphics for Engineers," Prentice-Hall, Englewood Cliffs, N.J., 1957.
Spencer, H. C., and J. T. Dygdon: "Basic Technical Drawing," Macmillan, New York.
Zozzora, Frank: "Engineering Drawing," McGraw-Hill, New York.

## Engineering Drawing Problems

Baer, C. J.: "Electrical and Electronic Drawing," McGraw-Hill, New York.
Bishop, C. C.: "Electrical Drafting and Design," McGraw-Hill, New York.
Cooper, C. D., and P. E. Machovina: "Engineering Drawing Problems," McGraw-Hill, New York.
Dygdon, J. T., R. O. Loving, and J. E. Halicki: "Basic Problems in Engineering Drawing," Holt, New York.
Giesecke, F. E., A. Mitchell, H. C. Spencer, and I. L. Hill: "Technical Problems," Macmillan, New York.
Higbee, F. G., and J. M. Russ: "Engineering Drawing and Problems," Wiley, New York.
Johnson, L. O., and I. Wladaver: "Engineering Drawing Problems," Prentice-Hall, Englewood Cliffs, N.J.
Levens, S. A., and A. E. Edstrom: "Problems in Engineering Drawing," McGraw-Hill, New York.
Luzadder, W. J., et al.: "Problems in Engineering Drawing," Prentice-Hall, Englewood Cliffs, N.J.
McNeary, M., E. R. Weidhas, and E. A. Kelso: "Creative Problems for Basic Engineering Drawing," McGraw-Hill, New York.
Paré, E. G., and F. M. Hrachovsky: "Graphic Representation," Macmillan, New York.
Paré, E. G., and E. F. Tozer: "Engineering Drawing Problems," Van Nostrand, Princeton, N.J.
Spencer, H. C., and I. L. Hill: "Technical Drawing Problems," series 2 and 3, Macmillan, New York.
Turner, W. W., C. P. Buck, and H. P. Ackert: "Basic Engineering Drawing," Ronald, New York.
Vierck, C. J., C. D. Cooper, and P. E. Machovina: "Engineering Drawing Problems," McGraw-Hill, New York.
Vierck, C. J., and R. I. Hang: "Fundamental Engineering Drawing Problems," McGraw-Hill, New York.
Vierck, C. J., and R. I. Hang: "Engineering Drawing Problems," McGraw-Hill, New York.
Vierck, C. J., and R. I. Hang: "Graphic Science Problems," McGraw-Hill, New York.

Zozzora, F.: "Engineering Drawing Problems," McGraw-Hill, New York.

## Engineering Graphics

Arnod, J. N.: "Introductory Graphics," McGraw-Hill, New York.

Earle, J. H.: "Engineering Design Graphics," Addison-Wesley, Reading, Mass.

French, T. E., and C. J. Vierck: "Graphic Science and Design," McGraw-Hill, New York.

Giesecke, F. E., A. Mitchell, H. C. Spencer, I. L. Hill, and R. O. Loving: "Engineering Graphics," Macmillan, New York.

Hammand, R., et al.: "Engineering Graphics for Design & Analysis," Ronald, New York.

Hoelscher, R. P., and C. H. Springer: "Engineering Design and Geometry," Wiley, New York.

Levens, A. S.: "Graphics in Engineering and Science," Wiley, New York.

Luzadder, W. J.: "Basic Graphics," Prentice-Hall, Englewood Cliffs, N.J.

Paré, E. G.: "Engineering Drawing," Holt, New York.

Rising, J. S., M. W. Almfeldt, and P. S. DeJong: "Engineering Graphics," W. C. Brown, Dubuque, Iowa.

Rule, J. T., and E. F. Watts: "Engineering Graphics," McGraw-Hill, New York.

Shupe, J. T., and E. F. Watts: "Engineering Geometry and Graphics," McGraw-Hill, New York.

Svenson, C. L., and W. E. Street: "Engineering Graphics," Van Nostrand, New York.

Wellman, B. L.: "Introduction to Graphical Analysis and Design," McGraw-Hill, New York.

## Handbooks

A great many handbooks, with tables, formulas, and information, are published for the different branches of the engineering profession, and are useful for ready reference. Handbook formulas and figures should of course be used only with an understanding in the principles upon which they are based. Among the best-known handbooks are the following, alphabetized by title:

"ASHRAE Guide and Data Book," American Society of Heating, Refrigerating and Air-Conditioning, New York.

"ASME Handbook," McGraw-Hill, New York.

"American Machinists Handbook," F. H. Colvin and F. A. Stanley (eds.), McGraw-Hill, New York.

"Architects and Builders Handbook," A. E. Knowlton (ed.), McGraw-Hill, New York.

"Building Construction," W. C. Huntington (ed.), Wiley, New York.

"Chambers Technical Dictionary," C. F. Tweney, and L. E. C. Hughes (eds.), Macmillan, New York.

"Chemical Engineers Handbook," R. H. Perry (ed.), McGraw-Hill, New York.

"Civil Engineers Handbook," L. C. Urquhart (ed.), McGraw-Hill, New York.

"Cutting of Metals," American Society of Mechanical Engineers, New York.

"Definitions of Occupational Specialties in Engineering," American Society of Mechanical Engineers, New York.

"Design Data and Methods—Applied Mechanics," American Society of Mechanical Engineers, New York.

"Dynamics of Automatic Controls," American Society of Mechanical Engineers, New York.

"Electronic Drafting Handbook," N. M. Raskhodoff (ed.), Macmillan, New York.

"Engineering Tables," American Society of Mechanical Engineers, New York.

"Frequency Response," American Society of Mechanical Engineers, New York.

"Gear Handbook," D. W. Dudley (ed.), McGraw-Hill, New York.

"General Engineering Handbook," E. O. O'Rourke (ed.), McGraw-Hill, New York.

"Glossary of Terms in Nuclear Science and Technology," American Society of Mechanical Engineers, New York.

"Machinery's Handbook," The Industrial Press, New York.

"Manual on Cutting Metals," American Society of Mechanical Engineers, New York.

"Manual of Standard Practice for Detailing Reinforced Concrete Structures," American Concrete Institute, Detroit, Mich.

"Mechanical Engineers' Handbook," Lionel S. Marks (ed.), McGraw-Hill, New York.

"Mechanical Engineers' Handbook," William Kent (ed.), Wiley, New York.

"Metals Engineering—Design," American Society of Mechanical Engineers, New York.

"Metals Properties," American Society of Mechanical Engineers, New York.

"New American Machinists' Handbook," Rupert Le Grand (ed.), McGraw-Hill, New York.

"Operation and Flow Process Charts," American Society of Mechanical Engineers, New York.

"Piping Handbook," Sabin Crocker, McGraw-Hill, New York.

A134

"Plant Layout Templates and Models," American Society of Mechanical Engineers, New York.

"Riveted Joints," American Society of Mechanical Engineers, New York.

"SAE Automotive Drafting Standards," Society of Automotive Engineers, New York.

"SAE Handbook," Society of Automotive Engineers, New York.

"Shock and Vibration Instrumentation," American Society of Mechanical Engineers, New York.

"Smoley's New Combined Tables," Smoley and Sons, Pennsylvania.

"Standard Handbook of Electrical Engineers," A. E. Knowlton (ed.), McGraw-Hill, New York.

"Standard Handbook for Electrical Engineers," D. G. Fink and J. M. Carroll (eds.), McGraw-Hill, New York.

"Standard Handbook for Mechanical Engineers," T. Baumeister and L. S. Marks, (eds.), McGraw-Hill, New York.

"Steel Castings Handbook," Steel Founders Society of America, Cleveland, Ohio.

"Steel Construction," American Institute of Steel Construction, New York.

"Structural Shop Drafting," American Institute of Steel Construction, New York.

"Tool Engineers' Handbook," Frank W. Wilson (ed.), McGraw-Hill, New York.

## Illustration

Hoelscher, Randolph Philip, Clifford Harry Springer, and Richard F. Pohle: "Industrial Production Illustration," McGraw-Hill, New York.

Tharratt, George: "Aircraft Production Illustration," McGraw-Hill, New York.

Treacy, John: "Production Illustration," Wiley, New York.

## Kinematics and Machine Design

Albert, C. D.: "Machine Design Drawing Room Problems," Wiley, New York.

Black, Paul H.: "Machine Design," McGraw-Hill, New York.

Faires, V. M.: "Design of Machine Elements," Macmillan, New York.

Guillet, G. L.: "Kinematics of Machines," Wiley, New York.

Ham, C. W., E. J. Crane, and W. L. Rogers: "Mechanics of Machinery," McGraw-Hill, New York.

Hyland, P. H., and J. B. Kommers: "Machine Design," McGraw-Hill, New York.

Keown, Robert McArdle, and Virgil Moring Faires: "Mechanism," McGraw-Hill, New York.

Prageman, I. H.: "Mechanism," International Textbook, Scranton, Pa.

Sahag, L. M.: "Kinematics of Machines," Ronald, New York.

Schwamb, Peter, and others: "Elements of Mechanism," Wiley, New York.

Sloane, A.: "Engineering Kinematics," Macmillan, New York.

Doughtie, Venton L., and Alex Vallance: "Design of Machine Members," McGraw-Hill, New York.

## Lettering

French, T. E., and R. Meiklejohn: "Essentials of Lettering," McGraw-Hill, New York.

French, T. E., and W. D. Turnbull: "Lessons in Lettering," McGraw-Hill, New York.

George, R. F.: "Modern Lettering for Pen and Brush Poster Design," Hunt Pen Co., Camden, New Jersey.

Hornung, C. P.: "Lettering form A to Z," Ziff-Davis, New York.

Ogg, Oscar: "An Alphabet Source Book," Harper, New York.

Svenson, C. L.: "The Art of Lettering," Van Nostrand, Princeton, N.J.

## Map and Topographic Drawing

Deetz, C. H.: "Elements of Map Projection," U.S. Government Printing Office, Washington, D.C.

Greitzer, S. L.: "Elementary Topography and Map Reading," McGraw-Hill, New York.

Hinks, A. R.: "Maps and Surveys," Macmillan, New York.

"Manual of Surveying Instructions for the Survey of the Public Lands of the United States," U.S. Government Printing Office, Washington, D.C.

Robinson, A. H.: "Elements of Calligraphy, Wiley, New York.

Sloane, R. C., and J. M. Montz: "Elements of Topographic Drawing," McGraw-Hill, New York.

Whitmore, G. D.: "Advanced Surveying and Mapping," International Textbook, Scranton, Pa.

Whitmore, G. D.: "Elements of Photogrammetry," International Textbook, Scranton, Pa.

## Patent Drawing

Guide for Patent Draftsmen, U.S. Government Printing Office, Washington, D.C.

Radzinsky, H.: "Making Patent Drawings," Macmillan, New York.

## Perspective

Freese, E. I.: "Perspective Projection," Reinhold, New York.

Lawson, P. J.: "Practical Perspective Drawing," McGraw-Hill, New York.

Lubchez, B.: "Perspective," Van Nostrand, Princeton, N.J.

Morehead, J. C., and J. C. Morehead: "A Handbook of Perspective Drawing," Princeton, N.J.

Norling, E.: "Perspective Made Easy," Macmillan, New York.

Turner, W. W.: "Simplified Perspective," Ronald, New York.

## Piping

Babbit, H. E.: "Plumbing," McGraw-Hill, New York.

"Catalog of Valves, Fittings and Piping," Crane Co.

"Catalogue," Walworth Co., New York.

Crocker, S.: "Piping Handbook," McGraw-Hill, New York.

Day, L. J.: "Standard Plumbing Details," Wiley, New York.

"Handbook of Cast Iron Pipe," Cast Iron Pipe Research Association, Oak Brook, Ill.

Littleton, C. T.: "Industrial Piping," McGraw-Hill, New York.

NAVCO Piping Catalog," National Valve and Manufacturing Co.

Nielson, L. S.: "Standard Plumbing Engineering Design," McGraw-Hill, New York.

"Piping Design and Engineering," Grinnell Co.

Plum, S.: "Plumbing Practice and Design," Wiley, New York.

"Primer in Power Plant Piping," Grinnell Co.

Rase, H. F.: "Piping Design for Process Plants," Wiley, New York.

Thompson, C. H.: "Fundamentals of Pipe Drafting," Wiley, New York.

## Shop Practice and Tools

Begeman, M. L.: "Manufacturing Processes," Wiley, New York.

Benedict, Otis J., Jr.: "Manual of Foundry and Pattern Shop Practice," McGraw-Hill, New York.

Boston, O. W.: "Metal Processing," Wiley, New York.

Burghardt, Henry D., Aaron Axelrod, and James Anderson: "Machine Tool Operation," McGraw-Hill, New York.

Campbell, H. L.: "Metal Castings," Wiley, New York.

Campbell, James S., Jr.: "Casting and Forming Processes in Manufacturing," McGraw-Hill, New York.

Clapp, W. H.: "Metal Castings," Wiley, New York.

Colvin, Fred H., and Lucian L. Haas: "Jigs and Fixtures," McGraw-Hill, New York.

Doe, E. W.: "Foundry Work," Wiley, New York.

Dubois, J. H.: "Plastics," American Technical Society, Illinois.

Hesse, H. C.: "Engineering Tools and Processes," Van Nostrand, Princeton, N.J.

Hine, C. R.: "Machine Tools for Engineers," McGraw-Hill, New York.

Hinman, C. W.: "Die Engineering Layout and Formulas," McGraw-Hill, New York.

Johnson, C. G.: "Forging Practice," American Technical Society, Illinois.

Schaller, G. S.: "Engineering Manufacturing Methods," McGraw-Hill, New York.

Wendt, R. E.: "Foundry Work," McGraw-Hill, New York.

Young, J. F.: "Materials and Processes," Wiley, New York.

## Structural Drawing and Design

See Handbooks, as well as the following:

Bishop, C. T.: "Structural Drafting," Wiley, New York.

Shedd and Vawter: "Theory of Simple Structures," Wiley, New York.

Urquhart, L. C., and C. E. O'Rourke: "Design of Steel Structures," McGraw-Hill, New York.

Winter, George, and Leonard C. Urquhart, C. E. O'Rourke, and A. H. Nelson: "Design of Concrete Structures," McGraw-Hill, New York.

## American National Standards

The American National Standards Institute, 1430 Broadway, New York, N.Y. 10018.

For a complete listing of standards, see the ANSI price list and index

## Bolts and Screws

Hexagon Head Cap Screws, Slotted Head Cap Screws, Square Head Set Screws and Slotted Headless Screws, B18.6.2—1956

Plow Bolts, B18.9—1958

Round Head Bolts, B18.9—1952 (R 1959)

Slotted and Recessed Head Machine Screws and Machine Screw Nuts, B18.6.3—1962

Slotted and Recessed Head Wood Screws, B18.6.1—1961

Socket Cap, Shoulder and Set Screws, B18.3—1969

Square and Hex Bolts and Screws, B18.2.1—1965

Square and Hex Nuts, B19.2.2—1965

Track Bolts and Nuts, B18.10—1963

### Charts and Graphs

Illustrations for Publication and Projection, Y15.1—1959
Time-Series Charts, Y15.2—1960

### Dimensioning and Surface Finish

Preferred Limits and Fits for Cylindrical Parts, B4.1—1967
Rules for Rounding Off Numerical Values, Z25.1—1940 (R 1961)
Scales to Use with Decimal—Inch Dimensioning, Z75.1—1955
Surface Texture, B46.1—1962

### Drafting Manual

Sect. 1 Size and Format, Y14.1—1957
Sect. 2 Line Conventions and Lettering, Y14.2—1973
Sect. 3 Projections, Y14.3—1957
Sect. 4 Pictorial Drawing, Y14.4—1957
Sect. 5 Dimensioning and Tolerancing for Engineering Drawings, Y14.5—1966
Sect. 6 Screw Threads, Y14.6—1957
Sect. 7 Gears, Splines and Serrations, Y14.7—1958
Sect. 8 Castings, Y14.8—1976
Sect. 9 Forging Y14.9—1958
Sect. 10 Metal Stampings, Y14.10—1959
Sect. 11 Plastics, Y14.11—1958
Sect. 12 Die Castings, Y14.12—1976
Sect. 13 Springs, Helical and Flat, Y14.3—1976
Sect. 14 Mechanical Assemblies, Y14.14—1966
Sect. 15 Electrical Diagrams Y14.15—1966
Sect. 16 Tools, Dies and Gages, Y14.16—1966
Sect. 17 Fluid Power Diagrams, Y14.17—1966

### Gears

System for Straight Bevel Gears, B6.13—1965
Tooth Proportions for Coarse-Pitch Involute Spur Gears, B6.1—1968
Tooth Proportions for Fine-Pitch Involute Spur and Helical Gears, B6.7—1967

### Graphic Symbols

Graphic Electrical Wiring Symbols for Architectural and Electrical Layout Drawings, Y32.9—1962
Graphic Symbols for Electric and Electronics Diagrams, Y32.2—1970
Graphic Symbols for Heat-Power Apparatus, Z32.2.6—1950 (R 1956)
Graphic Symbols for Heating, Ventilating and Air-Conditioning, Z32.2.4—1949

Graphic Symbols for Pipe Fittings, Valves and Piping, Z32.2.3—1949 (R 1953)
Graphic Symbols for Plumbing, Y32.4—1955
Graphic Symbols for Use on Railroad Maps and Profiles, Y32.7—1957
Graphic Symbols for Welding, Y32.3—1969
Symbols for Engineering Mathematics, Y10.17—1961

### Keys and Pins

Machine Pins, B5.20—1969
Woodruff Keys and Keyseats, B17.2—1961

### Piping

Cast-Iron Pipe Centrifugally Cast in Sand-Lined Molds, A21.8—1962
Cast-Iron Pipe Flanges and Flanged Fittings, 25, 125, 250 and 800 lb, B16.1—1967
Cast-Iron Screwed Fittings, 125 and 250 lb., B16.4—1963
Ferrous Plugs, Bushings and Locknuts with Pipe Threads, B16.14—1965
Malleable-Iron Screwed Fittings, 150 and 300 lb., B16.5—1963
Wrought-Steel and Wrought-Iron Pipe, B36.10—1959

### Rivets

Large Rivets (½ Inch Nominal Dia. and Larger), B18.4—1960
Small Solid Rivets, B18.1—1965

### Small Tools and Machine Tool Elements

Jig Bushings, B5.6—1962
Machine Tapers, B5.10—1963
Milling Cutters and End Mills, B94.19—1958
Reamers, B94.2—1964
T-Slots—Their Bolts, Nuts, Tongues and Cutters, B5.1—1949
Taps, Cut and Ground Threads, B94.9—1967
Twist Drills, Straight Shank and Taper Shank, B94.11—1967

### Threads

Dry Seal Pipe Threads, B2.2—1968
Nomenclature, Definitions and Letter Symbols for Screw Threads, B1.7—1965
Pipe Threads, B2.1—1968
Unified and American National Screw Threads, B1.1—1960

## Washers

Lock Washers, B27.1—1965
Plain Washers, B27.2—1965

## Miscellaneous

Knurling, B94.6—1966
Preferred Thicknesses for Uncoated Thin Flat Medals, B32.1—1952 (R 1968)

## Publications of National Societies

The national engineering organizations publish a wide variety of manuals, standards, handbooks, and pamphlets. Information concerning these publications is available directly from the societies. The following is a selected list of American organizations:

Aerospace Industries Assn. of America, 1725 Desales St. NW, Washington, D.C. 20036.

Air Conditioning and Refrigeration Institute, 1815 North Fort Myer Drive, Arlington, Va. 22209

American Association for the Advancement of Science (AAAS), 1515 Massachusetts Ave. NW, Washington, D.C. 20035

American Association of Petroleum Geologists, Inc. (AAPG), 1444 S. Boulder, Tulsa, Okla. 74101

American Association of University Professors, 1 Dupont Circle, NW, Massachusetts Ave. NW, Washington, D.C. 20036

American Ceramic Society (ACerS), 65 Ceramic Drive, Columbus, Ohio 43214

American Chemical Society (ACS), ACS Bldg., 1155 16th St. NW, Washington, D.C. 20036

American Concrete Institute (ACI), 22400 W. 7 Mi. Redford Station, Detroit, Mich. 48219

American Gas Association (AGA), 1515 Wilson Blvd., Arlington, Va. 22209

American Inst. of Aeronautics and Astronautics (AIAA), 1290 Avenue of the Americas, New York, N.Y. 10019

American Institute of Chemical Engineers (AIChE), 345 East 47th St., New York, N.Y. 10017

American Institute of Mining and Metallurgical, and Petroleum Engineers, Inc. (AIME), 345 East 47th St., New York, N.Y. 10017

American Institute of Steel Construction, Inc. (AISC), 1221 Avenue of the Americas, New York, N.Y. 10020

American Mining Congress, Ring Bldg., Washington, D.C. 20036

American Petroleum Institute (API), 2101 L St. NW, Washington, D.C.

American Society for Engineering Education (ASEE), 1 Dupont Circle, Washington, D.C. 20036

American Society for Metals (ASM), Metals Park, Ohio 44073

American Society for Quality Control (ASQC), 161 West Wisconsin Ave., Milwaukee, Wis. 53203

American Society for Testing and Materials (ASTM), 1916 Race St., Philadelphia, Pa. 19103

American Society of Agricultural Engineers (ASAE), 2950 Niles Rd. St. Joseph, Mich. 44085

American Society of Civil Engineers (ASCE), 345 East 47th St., New York, N.Y. 10017

American Society of Heating, Refrigerating, and Air-conditioning Engineers (ASHRAE), 345 East 47th St., New York, N.Y. 10017

American Society of Lubrication Engineers, 838 Busse Highway, Park Ridge, Ill. 60068

American Society of Mechanical Engineers (ASME), 345 East 47th St., New York, N.Y. 10017

American Society of Photogrammetry, 105 N. Virginia Ave., Falls Church, Va. 22046

American Society of Safety Engineers, 850 Busse Highway, Park Ridge, Ill. 60068

Asphalt Institute, Asphalt Institute Bldg., College Park, Md. 20740

Association of Iron and Steel Engineers (AISE), 3 Gateway Ctr., Suite 2350, Pittsburgh, Pa. 15222

Federation of American Scientists (FAS), 307 Massachusetts Ave. NW, 20002

Illuminating Engineering Society, 345 East 47th St., New York, N.Y. 10017

Industrial Research Institute, Inc., 100 Park Ave., New York 17, N.Y.

Institute of Electrical and Electronics Engineers (IEEE), 345 East 47th St., New York, N.Y. 10017

Mining and Metallurgical Society of America (MMSA), 345 East 47th St., New York, N.Y. 10017

National Aeronautic Association (NAA), 8211 15th St., NW, Washington, D.C. 20036

National Society of Professional Engineers (NSPE), 2029 K St. NW, Washington, D.C. 20036

National Standards Institute, 1430 Broadway, New York, N.Y. 10018

Society of Automotive Engineers (SAE), 400 Commonwealth Drive, Warrendale, Pa. 15096

Society of Manufacturing Engineers (SME), 20501 Ford Road, Dearborn, Mich. 48128

Standards Engineers Society (SES), 6700 Pennsylvania Avenue S., Minneapolis, Minn. 55423

# INDEX

Abbreviations, 349
  on charts, 408
  list, A93—A104
Accordion folding, 491
Accuracy, 490
Acme threads, table, A52
Across-corners construction of hexagon, 83
Across-flats construction of hexagon, 83
Acute angle, 103
Adjacent areas, 171
Adjacent-part lines, 59, 60
Aligned dimensioning, 346
Aligned sections, 249
Aligned views, 250
Allowance, 371
Alphabet of lines:
  ink, 59
  pencil, 59
Alphabets:
  inclined, 119
  vertical, 116
Alterations, 497
Alternate-position lines, 59, 60, 330
Alternate-position views, 137
Altitude, 428
American Standard thread series, table, A48—A49
Ames lettering instrument, 116
Ampersand, 118, 119
Analog plotter, 468
Angles, 103
  to bisect, 82
  dimensional tolerance, 380
  dimensioning, 361
  to divide, 82
  drawing, 47
  of elevation, 285
  gage for, A78
  in isometric, 270
  lateral, 284
  to lay out, 80, 81
  for oblique drawing, 275
  for perspective drawing, 284
  positioning by, in dimensioning, 351, 352
  rivet size for, A78
  trigonometric functions for, table, A26—A31
Animation, 468
Annealing, 339

Applied geometry, 5, 71—103
  problems, 104—112
Archimedes, spiral of, 98
Architect's scales, 34—35, 50
Arcs:
  dashed-line, 143
  dimensioning, 363
  in isometric, 273
  to lay off length on line, 87
  length of chord, table, A32
  in oblique, 278
  tangent: to arc, 58
    to straight line, 58
    and circle, 77
    to two lines, 76
Areas:
  adjacent, 171
  hidden, 171
  meaning, 169
  reading, 171—172
Arms in section, 246
Arrowheads, 343
Arrowless dimensioning, 504
Artgum eraser, 32
Assembly:
  of parts, 339
  selective, 373
Assembly drawings, 484—487
  bill of material for, 489
  number for, 495
  problems, 515—518
  tabular, 488
Assembly sections, 241
Assembly sketches, 497
Auxiliary-adjacent auxiliary view, 223
Auxiliary elevation, 222
Auxiliary sections, 240
Auxiliary surface, 140, 171
Auxiliary views, 204—223
  basic concepts, 204—205
  defined, 204
  dimensioning, 358
  procedure, 211—215
  reference plane, 207
  sectional, 240
  terms for, 222
  types of, 222
  (*See also* Edge views; Normal views)
Axes, pictorial, sketching, 295
Axis:
  of surfaces, 428
  of vision, 284

Axonometric projection, 6, 266—267
  from orthographic views, 8, 279
Axonometric sketching, 294, 302
  problems, 306

Bar charts, 415
Barch-Payzant pen, 126
Base line for lettering, 119
Base-line dimensioning, 378
  simplified, 504
Basic-hole system, 372, 373
Basic-shaft system, 372, 373
Batters, dimensioning, 363
Beam compasses, 40, 736
Beam connections, A78
Beams, structural, table, A79
Bearing, symbol for, 254
Belt rivets, table, A58
Bevel scale, 35
Bibliography, A130—A137
Bilateral tolerance, 373, 374
Bill of material, 489
Binding-head screws, table, A55
Bisecting:
  angle, 82
  line, 52, 78
Blueprints, 31, 501
  of charts, 408
Boards, drawing, 30, 43
Boldface lettering, 114
Bolt circles, 367
Bolt-length increments, table, A54
Bolts:
  diameters, table, A54
  lengths, table, A54
  round-head, table, A53
  in sectional views, 246
  specification, 369
  square- and hexagon-head, table, A50
Border pens, 42, 736
Borders on drawing sheet, 52
Boring, 333
Bottleholder, 33
Bottom view, 137
Bow instrument, 37, 38, 55
Box construction:
  for isometric drawings, 269
  in sketching, 297
Box method for transferring polygon, 83
Braddock-Rowe triangle, 114

1

Brass fittings, table, A82
Breaks, conventional, 252
Broaches, 337
Broaching, 337
  dimensioning, 375
Broken-out sections, 238
Broken views, 239, 253
Bushings:
  pipe, table, A73
  tables, A80−A81
Butt-welding pipe fittings, table, A75
Buttress threads, table, A52
BW prints, 501

Cabinet drawing, 278
  problems, 314
  sketching, 294
Cable, symbol for, 254
Calculator, hand, A9−A21
    problems, A20−A21
Calculus, graphic, 413
Calipers, 500
Cams, 24
Cap line for lettering, 119
Cap screws:
  hexagon-head, table, A50
  socket- and slotted-head, table, A54
Capital lettering, 116−119
Captions for scales, 408−409
Carburizing, 339
Carriage bolts, table, A53
Casehardening, 339
Cast-iron fittings, tables, A71, A74
Cast-iron pipe, threaded, table, A76
Cast-iron valves, table, A77
Casting drawings, 391
Castings:
  die, 330
  dimensioning, 390−391
  sand (see Sand castings)
Catalogues, A105−A129
Cathode-ray tube display, 468−469
Cavalier projection, 275
  sketching, 294
Center of vision, 284
Center lines, 59, 60, 143, 492
  in dimensioning, 350, 351, 357, 380
Centers:
  of arc, 85
  of circle, 86
  location in sketching, 298
Chain, symbol, 254
Chamfers, dimensioning, 361
Change-record blocks, 495, 497
Changes in drawings, 497
Channels, steel, table, A79
Charts, 11, 406−421
  ANSI Standards for, 417
  bar, 415
  captions, 408−409

Charts:
  choice of, 410−411
  classification, 413
  classification of, 406
  computer graphics, 474
  curves for, 406, 408
  data for, 411−413
  for display, 417
  to draw, 407−409
  of equations, 411−413
  flow sheets, 413
  keys for, 408
  lettering, 409
  line widths, 409
  logarithmic, 409
  multiple-bar, 415
  nomograms, 413
  notation for, 407
  pictorial, 416
  pie, 415
  polar, 409
  popular, 414
  problems, 422
  ratio, 409
  rectilinear, 406
  for reproduction, 416
  route, 413
  scale values, 408
  scales, 408, 409
  semilogarithmic, 409
  strata, 416
  titles, 407, 408
  trilinear, 411
  volume, 416
Checking of drawings, 496
  order of, 496
Chemical symbols, A104
Chordal method for laying out angle, 80
Chords, table of length for circle arcs,
    A32
Circle arcs (see Arcs)
Circle method for ellipse, 91
Circles:
  center, 86
  concentric, 55
  as conic section, 88
  to draw, 53−56
  freehand, 163
  to inscribe: in polygon, 87
    in square, 87
    in triangle, 86
  instruments for drawing, 36−44
  in isometric, 272−273, 299
  in oblique, 278
  in perspective, 289
  in pictorial, sketched, 298
  size of, 51
  sketching, 163
  tangent to line, 74
  tangents to, 58, 75, 101, 102
  through three points, 85

Circular features, drawing, 6
Circumscribed figures, 83
Civil engineer's scale, 34
Classification charts, 413
Clay modeling, 174
Clearance fits, 372
  table, A67
Cloth (see Tracing cloth)
Coarse-thread series, table, A48−A49
Colors, abbreviations for, A103
Combination square, 500
Commercial Gothic lettering (see
      Gothic lettering)
Commercial practices, 26, 502−508
  overlays, 506
  simplified drawing (see Simplified
      drawing)
  templates, 506, 507
Common-fraction dimensioning, 346
Compasses, 37−38, 53−56
Compressed lettering, 114, 122
Computation sketches, 497
Computer graphics, 466−480
  interactive, 476
  numerical control, 477−480
  programming, 471
  tools and machines for, 416−471, 476
  uses of, 473−477
Concentric-circle method for ellipse, 92
Concentric circles, 55
Condensed lettering, 114, 122
Cone-point setscrews, A56
Cones, 103
  defined, 428
  development, 453−455
  dimensioning, 349−350
  intersection: with cylinder, 440
    with line, 437
    with plane, 438, 439
  of revolution, 88
Conic helix, 99
Conic sections, 88
Conjugate diameters, 89
  for projection of ellipse, 154
Connections, beam, A78
Connectors for sheet metal, 458
Conoid, 103
Contour pen, 42
Conventional breaks, 252
Conventional practices, 245, 250−254
  problems, 254−261
Conventional symbols, 253, 254
Conversion tables:
  English to metric, A23
  fractions to decimals, A23
  weights and measures, A33−A44
Cooper's rivets, table, A58
Coordinate paper for sketching,
    160−162, 499
Coordinate rulings for charts, 408
Copying methods, 502

Core box, 328
Corners:
    identification, 154
    reading, 173
Corrections of drawings, 497
Cosine values, table, A26—A31
Cotangent values, table, A26—A31
Cotter pins, table, A82
Counterbore, 337
Counterbored holes, dimensioning, 366
Countersink, 337
Countersunk bolts, A53
Countersunk holes, dimensioning, 365
Couplings, pipe, table, A70
Cross-section paper, 160—161, 499
Crossed-belt tangents, 101
Crosshatching (see Section-lining)
Cubic prism, 103
Cumulative errors, 50
Cumulative tolerances, 378
Cup-point setscrews, A56
Curved features, drawing, 6
Curved lines:
    defined, 85
    division, 52
    double-, 85
    geometry, 85—100
    normal to, 85
    single-, 85
    tangent to, 85
Curved rulers, 35
Curved surfaces:
    dimensioning, 368
    intersection, 436
Curves:
    on charts, 407—409
    cycloidal, 97
    dimensioning, 364
    in isometric, 271
    noncircular, to draw, 93
    in perspective, 290
    plane, 88
    projection of, 154
    sketching, 298
Cutting-plane lines, 59, 60
Cutting planes, 237
    for alignment of holes, 249
    offset, 238
    symbol for, 242
Cycloids, 97
Cylinder, 103
    defined, 428
    development, 450
    dimensioning, 349, 355—356, 367
    intersection, 439
        with cone, 440
        with line, 436
        with plane, 437
    measurement, 499
Cylindrical fits, table: of limits, A67—A68
    of preferred sizes, A66

Cylindrical helix, 99
Cylindroid, 103

Dashed lines, 142
Dashes in dimensioning, 349
Data, analysis of, 411
Datum dimensioning, 354, 364
Decagon, 103
Decimal dimensioning, 347, 352
Decimal equivalents of fractions, table,
    A25
Decimal places, 374
Decimal points, 349
Decimals, rounding off, 348
Depth, defined, 348
Depth gage, 500
Depth measurements, transfer, 152
Design drawing, 484
Design sketches, 497
Detail drawings, 484, 487—488
    inking order, 492
    penciling order, 492
    problems, 510—515, 519—526
Detail sketches, 497
Developed views, 250
Developments, 10, 448—453
    cone, 453—455
    cylinder, 450
    defined, 448
    double-curved surfaces, 457
    elbow, 452
    gore method, 458
    octagonal dome, 452
    prism, 448
    problems, 459—462
    pyramids, 449
    sphere, 457
    theory, 448
    transition pieces, 455
    by triangulation, 453
    warped surfaces, 457
    zone method, 458
Diagonals:
    in sketching, 298
    uses of, 83
Diagram drawings, 487
Diagrammatic sketches, 497
Die castings, 330
    dimensioning, 375
Dies, 330
Dihedral angle, 448
Dimension lines, 59, 60
    placement, 355
    spacing, 355
Dimension values, placement, 345
Dimensioning, 342—397
    abbreviations for, 349
    aligned system, 346
    by angle, 351
    angles, 361, 380

Dimensioning:
    arcs, 363
    arrowheads for, 343
    arrowless, 504
    auxiliary views, 358
    basic, 354
    batters, 363
    bolt circles, 367
    broaching, 375
    castings, 390
    center lines, 350, 356—358, 380
    chamfers, 361
    common-fraction, 346
    cones, 349
    continuous, 356
    contour, 355
    correlation, 352
    counterbore, 366
    countersink, 366
    curves, 364, 368
    cylinders, 349, 356, 367
    from datum, 354, 364
    decimal, 347
    die castings, 375
    dimension lines, 59, 60, 358
    distance, 345
    distances, selection of, 342
    drilling, 375
    drop forgings, 392
    dual, 394—396, 536—559
    extension lines for, 343
    fasteners, in feet and inches, 345
    figures for, 345
    finish marks, 348
    fit (see Fit)
    foreign, 346
    forgings, 375, 392
    fractions for, 345, 346
    grades, 363
    grinding, 375
    half section, 356
    hole circles, 367
    holes, 365—366
    keys, 369
    lathe, 375
    leaders, 344
    limit (see Limit dimensioning)
    in limited space, 358
    line weights for, 342
    mating parts, 352
    maximum and minimum, 355
    measuring for, 499
    method, 342
    metric, 368, 394—396, 536—559
    milling, 375
    normal views, 358
    notes for (see Notes, for dimension-
        ing)
    by offset, 351, 364
    order of, 358
    of orthographic drawings, 20

Dimensioning:
   parts machined from stock, 389
   pictorial drawings, 20, 360—361
   of pipe drawings, 609
   of pipe fittings, 605
   of pipe threads, 606
   placement, 342, 355—361
   planing, 375
   plastic molding, 375
   position, 348, 365
     selection of, 350
   practice, 342
   prisms, 350, 355
   problems, 397—398
   production methods, 385
   pyramids, 350
   radii, 364
   reading direction for, 346
   reaming, 375
   reference, 354
   revision, 348
   roughness symbols, 384
   rounded-end shapes, 365
   sand castings, 375
   scale, 348
   sectional views, 358
   selection, 350
     principles, 387
     procedure, 387
   shaping, 375
   sheet-metal parts, 392
   simplified, 506
   size, 349
     maximum and minimum, 355
     section of, 350
   slopes, 366
   spheres, 350, 367
   spot-faced holes, 366
   staggered, 357
   standard features, 342, 369
   superfluous, 352
   surface quality symbols, 384
   tapers, 361
   theory, 349
   threads, 375
   tolerances (see Tolerances)
   true-position, 382
   unnecessary, 352
   values for, 345, 347
     placement, 356, 357
   views, 355, 357
   warped surfaces, 350
   working sketches, 497
Dimensioning systems, 166
Dimensions, 20, 371
   used by machinist, 333
Dimetric drawings, 7, 266, 274
   from orthographic views, 8, 279—281
   problems, 313
   sketching, 295
Directo prints, 501

Directrix, 93
Display charts, 417
Displays, 468—471
Distance dimensioning, 345
Distances in sketching, 299
Dividers, 37, 38, 52, 152
   use of, 52
Dodecagon, 103
Dodecahedron, 103
Dotted lines, 142n.
Double-curved line, 85, 99
Double-curved surfaces, 103, 428
   development, 457
Drafting pencils, 31—32
Drafting tape, 31
Drawing:
   angles, 48
   isometric (see Isometric drawings)
   layout, 52, 151
   lines, 46
   oblique (see Oblique drawings)
   order of, 155, 492
   orthographic (see Orthographic free-
      hand drawing; Orthographic pro-
      jection)
   perspective, 281—293
   pictorial (see Pictorial drawings)
   preparation for, 43
   procedure for, 155
   and shop, 326—339
   speed in, 490
Drawing board, 30
   height, 43
   inclination, 43
Drawing ink, 33
Drawing number, 495
Drawing paper, 30
   placement, 45
   texture, 157—159
Drawing pencils, 31
Drawing sheet, layout, 52
Drawings:
   alteration, 497
   assembly, 484—487
   auxiliary, 9
   axonometric, 8
   chart, 407
   checking, 496
   design, 484
   detail, 484, 487
     penciling, 492
   dimensioning (see Dimensioning)
   dimetric, 7
   distribution, 326
   duplication, 20
   electrical (see Electrical drafting)
   filing, 491, 502
   folding, 491
   forging, 330
   freehand, 6, 157
   illustrated, 725—732

Drawings:
   inking, 492—494
   instrument, 30, 36—43
   isometric, 7
   layout, 149
   manufacturing method and, 326
   multiple system, 327
   oblique, 8
   one-view, 7
   orthographic, 132—179
   pattern, 328
   penciling, 492
   pictorial, 6
   piping, 24
   printing, 492
   prints of, 501—502
   reading (see Reading of drawings)
   reproduction, 501
   scale, 49—52, 348
   set, 489
   and shop, 326
   sizes of, 149, 491—492
   standard, 488
   tabular, 488
   three-view, 7
   title blocks for, 702
   trimetric, 8
   two-view, 7, 144
   welding, 23
   working (see Working drawings)
Drill, 337
Drill press, 333
Drilling, 333
   dimensioning, 375
Drills:
   specification, 369
   table of sizes, A52
Drop forgings, 326, 330
   dimensioning, 392
Drop line for lettering, 119
Drum line plotter, 467—468
Dual dimensioning, 394—396
   problems, 536—559

Edco pen, 126
Edge views, 141, 205
   of skew surface, 217—218
Edges:
   projection, 140
   reading, 173
   true length, 140
Eight-centered ellipse, 93, 94
Elbows:
   development, 452
   drawing, 215, 216
   pipe, 604
     table, A70
Electrical diagrams, table of graphic
     symbols for, A87—A91

Electrical drafting, 24
Elevation auxiliary, 222
Elevation view, 134
Elevator bolts, table, A53
Ellipses:
  approximate, 94
  as conic section, 88
  construction, 90−92
  defined, 88
  diameters, 88−89
  equation, 89
  in isometric, 272
  projection, 153
  in sketching, 294
  tangents to, 92
Ellipsoids, 103, 428
Empirical equations, charts for, 411
End elevation, 135
End view of line, 217
Engineering drawing, 4
Engineering graphics:
  defined, 4
  importance of, 4
Engineering orders, 497
Enlarged drawings, 51
Enlarged view, 358
Enlarging geometric shape, 83
Epicycloid, 97
Equiangular triangle, 103
Equilateral hyperbola, 96
Equilateral triangle, 103
  to locate center, 82
Erasers, 32
Erasing, 62
  ink, 492
Erasing machine, 63
Erasing shield, 32
Executive sketches, 497
Exploded drawings, 527−532
Exponential equations, plots, 413
Extended lettering, 114, 122
Extension lines, 59, 60, 343
Exterior materials, symbols, A92
Extra-fine thread series:
  table, A48−A49

Facing, 332, 333
Fasteners, 22, A105−A121
  Palnuts, A110−A111
  Pushnuts, A112−A113
  Rivets, A105−A109
  Rivnuts, A114−A117
  Rollpins, A118−A121
Filing of drawings, 491, 502
Fillet-and-round gage, 500
Fillets, 251
  dimensioning, 349, 354
Fillister-head screws, tables, A54, A55
Fine-thread series, table, A48−A49
Finish marks, 348

Finish turning, dimensioning, 375
Fit, 372, 376
  ANSI, 376
    examples, 377
  classes, 372
    cylindrical, 376
    limits, tables, A67−A68
    preferred sizes, table, A66
  description, 376
  of mating parts, 370
Fittings, tables: brass, A82
  butt-welding, A75
  cast-iron, A74
  flanged, A74
  malleable-iron, A70−A71
  screwed, A70−A71
Fixtures, 339
Flanged fittings, table, A74
Flash, 330
Flat-head screws, tables, A54, A55, A57
Flat keys, tables, A58, A59
Flat-point setscrews, A56
Flatbed plotter, 467
Flow sheets, 414
Focus, 88, 95
Fold lines, 451
Folding of drawings, 491
Foot marks, 345
Force fits, table, A68
Forgings, 330
  dimensioning, 375
  drop, 330
    dimensioning, 392
Form tolerances, 383
Formed surface, 332
Foundry, 329
Four-centered ellipse, 92
Fractions:
  lettering, 118
  table of decimal equivalents, A23
Freehand drawing (see Sketching)
French curves, 35
  use of, 61−62
Front-adjacent auxiliary, 222
Front auxiliary, 222
Front elevation view, 134
Front view, 134
Frontal line, 141
  end view, 217
Frontal plane, 134
Frustum, 103
  of pyramid, 428
Full-divided scales, 34
Full-dog-point setscrews, A56
Full section, 237
  isometric, 274

Gages:
  for angles, A78
  measuring, 499

Gang milling, 336
Gate valves, table, A72
Gears, 23
  rack, 68
Generated surface, 332
Generatrix, 428
Geometric shapes, 103
  breakdown, 175
Geometry:
  applied, 5, 72−103
  problems, 104−109
Gib-head keys, table, A58
Girth line, 448
Globe valves, table, A72
Glossary, A1−A8
Gore method, 458
Gothic lettering, 5, 114−121
Grades, dimensioning, 363
Graphic calculus, 413
Graphic language, 4−5
Graphic symbols (see Symbols)
Graphics (see Computer graphics)
Graphics program, 471
Graphs (see Charts)
Graticulation, 290
Grinder, 336
Grinding, dimensioning, 375
Ground plane, 284
Guide lines:
  for fractions, 119
  for lettering, 114, 115
  for slope, 120

Half-dog-point setscrews, A56
Half section, 238
  dimensioning, 360
  isometric, 274
Half views, 250
Hand calculator, A9−A21
  problems, A20−A21
Hard-copy devices, 470−471
Hardening, 339
Heat-power symbols, table A84
Heat-treatment, 339
Heating symbols, table, A84
Heel radius, 452
Height, defined, 138
Helicoid, 103
Helix, 99, 100
Hems for sheet metal, 458
Henry tank pen, 127
Heptagon, 103
Hexagon, 103
  to construct, 83
Hexagon-head bolts, table, A50
Hexagon-head cap screws, table, A50
Hexagon-head screws, tables, A50, A55,
  A56
Hexagon nuts, table, A51
Hexahedron, 103

Hidden features, 142
  in sectioning, 242
Hidden lines, 59, 60, 142
  in isometric, 269
Hole circles, 367
Holes:
  dimensioning, 365
  in isometric, 272
  measurement, 499
  positioning, 367
Honing, 337
Horizon, 284
Horizon plane, 284
Horizontal lettering strokes, 116
Horizontal lines, 140
  drawing, 45
  end view, 217
  sketching, 161 – 162
Horizontal plane, 134 – 135
Horizontal projection, 151
Hyperbola, 96
  as conic section, 88
Hyperbolic paraboloid, 103
Hyperboloid, 103, 428
Hypocycloid, 97

Icosahedron, 103
Idea sketches, 497
Illustration, 303
Inch fractions, table of decimal equiva-
  lents, A23
Inch marks, 345
Inclined lettering, 119 – 121
Inclined lines:
  intersection with plane, 429
  in perspective, 292
Inclined planes in perspective, 293
Inclined surface, 9, 206
  normal view, 206 – 210
    on practical objects, 210 – 213
    problems, 223 – 231
  (See also Auxiliary surface)
Ink cartridge, 33
Ink lines, 60, 61
Ink shading, 305
Inkholder for pens, 127
Inking, 56 – 58
  drawings, order of, 492 – 493
  order of, 68, 155 – 157, 492 – 493
  pens for, 33
Inks, 33
  erasing, 492
  printing of, 492
Inscribed figures, 84
Inspection of parts, 339
Instrument drawing, 6
Instruments, 30 – 63
  Ames lettering, 116
  beam compasses, 736
  bow, 37 – 40, 53 – 56

Instruments:
  Braddock-Rowe, 115
  compasses, 37 – 40, 53 – 56
  dividers, 37 – 40, 52
  french curve, 35, 61
  lettering, 41
  pencil pointer, 32
  pencils, 31 – 32, 44 – 45
  pens, 33, 36 – 37, 56 – 58
  scales, 34 – 35, 49 – 51
  selection, 30
  sets, 42, 43
  T square, 33, 45 – 48
  triangles, 33, 46 – 49
  use of, 5, 43
    exercises in, 63 – 69
    rules for, 63
Interactive graphics, 476
Intercept, 288
Interference fits, 373
  table, A68
Intersections, 10, 141, 428
  of cone: with cylinder, 439
    with line, 437
  conventional, 250, 252, 428
  of curved surfaces, 436
  of cylinders, 439
    with cone, 439
    with line, 437
  defined, 428
  of lines (see Lines)
  of planes (see Planes)
  plotted, 428
  prisms, 434
    and pyramid, 436
  problems, 459 – 463
  procedure, 430
  of pyramids, 435
    and prism, 436
Involute, 98
Irregular curves, 36
Isometric drawings, 7, 268 – 281
  angles, 270 – 271
  arcs, 273
  axes, 267, 268
  box construction, 270
  circles, 272
  curves, 271 – 272
  ellipses, 272
  to make, 268 – 269
  measurements, 269
  nonisometric lines in, 269
    offset method, 271
  from orthographic views, 7, 281
  problems, 306 – 308
  reversed, 273 – 274
  scale for, 268
  sections, 274
  sketching, 294 – 297
  spheres, 273
Isometric line, 268

Isometric planes, 268
Isometric projection, 267 – 268
Isosceles triangle, 103

Jig bushings, table, A80
Jigs, 23, 339
Joints, for sheet metal, 458

Key for chart, 408
Key seats, tables, A59 – A61
Keys, 22
  specification, 369
  tables, A58 – A59, A61
Keyways, tables, A59 – A61
Knurl, conventional symbol, 254

Lap joints, 458
Lapping, 337
Lathe, 332
  turret, 335
Lathe turning, dimensioning, 375
Lay, symbols for, 387
Layout drawings, 60
Leaders, 344
Left auxiliary, 222
Left-handed draftsman, 47, 49
  lettering for, 122 – 123
Leroy pen, 126
Lettering, 5, 114 – 127
  alphabets, 118 – 123
  boldface, 114
  capital, 117 – 121
  for charts, 409
  composition, 123 – 124
  condensed, 114, 122
  extended, 114, 122
  fractions, 119
  Gothic, 116 – 119
  guide lines for, 114, 119
  horizontal strokes for, 116
  inclined, 119 – 120, 122
  instruments for, 115, 116
  for left-hander, 122 – 123
  lightface, 114
  lower-case, 119, 121
  numerals, 118 – 119
  pencil, 115 – 116
  pens, 127
  problems, 128 – 129
  proportions, 114, 117
  Reinhardt, 121
  single-stroke, 114 – 124
    inclined, 119 – 121
    vertical, 116 – 119
  slope lines, 120
  spacing, 124 – 125
  stability rule for, 114
  strokes, 117 – 119, 122, 123

Lettering:
  thickness, 116
  titles, 124, 495−496
  vertical, 116−119
  vertical strokes for, 117
  words, 124
Lettering instruments, 41
Lettering pens, 126
Lettering strokes, 116, 122
  for left-handers, 122
Light and shade, 303
Light pen, 476
Lightface lettering, 114
Limit dimensioning, 369−372
  allowance, 370
  angles, 380
  basic-hole and basic-shaft, 372,
    373
  for cylindrical parts, 376
  fit (see Fit)
  for manufacturing, 374−375
  nomenclature, 371−372
  problems, 397−398
  surface quality, 383
    symbols, 386
  tolerances (see Tolerances)
  values, 373
  (See also Dimensioning)
Line connections, 73−78
Line gage, 61
Line plotter, 617
Line practice, 61
Line relationships, 73−78
Line widths, 58−61
  for curves on charts, 409
  for dimensioning, 342
Linear equations, plots of, 411
Lines, 5
  alphabets, 59−60
  to bisect, 53, 78
  center, 143
  classification, 140−141
  curved (see Curved lines)
    to divide, 52, 79, 80
  double-curved, 99
  drawing, 46
  end view, 217
  fold, 448
  freehand, 157−158
  frontal, 141
  gage for, 61
  hidden, 142
  horizontal, 46, 141
  inclined, 292, 429
  ink, 60, 61
  intersection: with cone, 437
    with cylinder, 436
    with plane, 429−431
    with skew surface, 431
  to lay off on circle, 88
  long-break, 59, 60, 254

Lines:
  meaning, 169
  measuring, 286, 451
  normal view, 215−217
  oblique, 140
  parallel, 49, 72, 73, 100
  pencil, 59, 61
  perpendicular, 49, 73, 100
  positions, 140, 141
  precedence, 143
  problems, 104
  profile, 141
  representation of, 141
  sketched, 157−162
  skew, 140, 215, 430
  tangents to, 74, 76, 101
  transverse, 296
  to trisect, 79
  true length of, 140, 215
  through two points, 72
  vertical, 47
    sketching, 296
  widths, 60
Lists, computer graphics, 473−474
Lithoprinting, 502
Locational fits, table, A67−A68
Lock screws, table, A80
Lock washers, table, A63
Logarithmic charts, 409
Logarithmic functions of angles, table,
    A26−A31
Logarithmic paper, 409
Logarithmic plots for data, 412
Logarithmic scales, 409
Logarithms of numbers, table,
    A24−A25
Long-break lines, 59, 60, 254
Lower-case lettering, 119−121
Lugs in section, 247
Lunes, 457

Machine elements, 21
Machine parts:
  method of production, 326
  sheet-metal, 326, 337
  from stock, 326, 335
  welded, 327, 337
Machine screws:
  nuts for, table, A56
  table, A55
Machine shop, 330
Machine tapers, table, A64
Machining, 331, 334, 335
Machining drawings, 488
Malleable-iron fittings, table, A70
Malleable-iron unions, table, A75
Manufacturing methods, drawings for,
    327
Maps, 25
  computer graphics, 474−475

Masking tape, 31
Materials:
  symbols, A92
  weights, A92
Mathematical models, 471−472
Mathematical tables, A22−A46
  problems, 112
Mathematical values, table, A22
Mating parts:
  dimensioning, 352, 354
  fits (see Fits)
Mating surfaces, 370
Matrix plotter, 468
Measurements, 50, 499
  conversion tables, A23, A35−A46
  in isometric, 268
  making, 499
  metric, 11−19, 394−396
  reading, 52
  scale, 51
  SI units, 11
  for sketches, 299
  transfer, 152
Measures, conversion table, A33−A76
Measuring lines, 451
  in perspective, 286−288
Measuring-point method for perspec-
    tives, 291
Mechanical engineer's scales, 34, 50
Meridian sections, 457
Metric dimensioning, 368, 394−396
  problems, 536−559
Metric equivalents, table, A23
Metric measurement:
  dual dimensioning, 394−396,
    536−559
  SI units, 11−19
Micrometer caliper, 501
Milling, 336
  dimensioning, 375
Milling cutters, specification, 369
Milling machine, 335−336
Missing-line exercises, 179
Miter projection, 151
Modeling, 175
Molds, 330
Multiple-bar charts, 415
Multiple-drawing system, 327
Multiple-view drawings, 492
Mylar, 491

Nominal size, 371
Nomograms, 413
Nonagon, 103
Noncircular curves to draw, 93
Nonisometric line, 269
Normal to tangent line, 85
Normal views, 9, 204
  for constructing other views, 215
  dimensioning, 357, 358

Normal views:
  drawing procedure, 211–213
  inclined surface, 206–210
    on practical object, 210–213
    problems, 223–231
  line, 215–217
  purposes, 213–215
  skew surfaces, 219–222
    on practical objects, 221–222
    problems, 223–231
Normalizing, 339
Normals to ellipse, 92
Notation for charts, 407
Notes:
  for dimensioning, 343, 351, 357, 361
  general, 361
  reading direction for, 346
  specific, 361
  in title blocks, 126
  wording, 368
Numbers, logarithms of, table,
    A24–A25
Numerals, lettering, 118–119
Numerical control:
  programming, 479–480
  tools and machines for, 477–480
  uses of, 480
Nuts:
  Palnuts, A110–A111
  Pushnuts, A112–A113
  Rivnuts, A114–A117
  tables: hexagon, A51
    machine-screw, A56
    square, A51
    stove-bolt, A56

Object orientation, 144
Oblate ellipsoid, 103
Oblique cone, 103
Oblique cylinder, 103
Oblique drawings, 8, 266, 276–278
  cabinet, 278
  circles, 278
  to make, 276
  object orientation, 276
  offset method. 278
  from orthographic views, 281–282
  problems, 310–314
  rules for, 276
  sketching, 294, 295, 302
    problems, 315
  starting plane, 278
Oblique hexagonal pyramid, 103
Oblique line, 141
Oblique pentagonal prism, 103
Oblique pentagonal pyramid, 103
Oblique projection, 6, 8, 275
  from orthographic views, 8
  rays of sight for, 132
Oblique pyramid, 428

Oblique rectangular prism, 103
Oblique surface, 140, 171, 205, 428
Oblique view, 222
Obtuse angle, 103
Octagon, 103
  to inscribe in square, 84
Octagonal dome, development, 452
Octahedron, 103
Offset method:
  for isometric, 269–274
  for parabola, 88, 95
  for transferring polygon, 83
Offsets, positioning by, in dimension-
    ing, 351, 364
Ogee curve, to draw, 77
One-view drawings, 7
Open-belt tangents, 102
Open-divided scales, 35, 50, 52
Orthogonal projection (see Ortho-
    graphic projection)
Orthographic drawing, 132–179
  (See also Orthographic projection)
Orthographic freehand drawing,
    157–166
  (See also Sketching)
Orthographic projection, 132–155
  defined, 132
  freehand, 157, 165
  introduction to, 6
  planes for, 134–135
  rays of sight, 132, 133
  reading exercises, 145, 146, 167–179
  in sketching, 157–166
  theory, 132
  types of, 7
  views, 6–9, 132–137
    projection from: axonometric, 279
    isometric, 281
    oblique, 283
Orthographic reading (see Reading of
    drawings)
Orthographic views, 475–476
  (See also Orthographic projection,
    views)
Outline assembly drawing, 486
Outline lines, 59, 60
Outline sectioning, 245
Outline sketches, 497
Oval-head screws, tables, A55, A57
Oval-point setscrews, tables, A56
Overlays, 506–507
Ozalid prints, 502

Pan-head screws, table, A55
Paper:
  sizes, 491
  for sketching, 160–161
  texture, 158–159
  tracing, 31, 490
  (See also Drawing paper)

Parabola:
  as conic section, 88
  construction, 95
  defined, 93
  equation, 94
  method for: parabolic-envelope,
    95
    parallelogram, 95
Paraboloid, 103, 428
Parallel lines to draw, 49, 73, 100
Parallel perspective, 288
Parallelepipeds, 103
Parallelogram method:
  for ellipse, 92
  for parabola, 95
Part machined from stock, dimension-
    ing, 389
Part number, 495
Partial views, 215
  auxiliary section, 240
Parts list, 489
Parts production, 388–389
Pattern drawing, 328, 488
Pattern shop, 328
Pen shading, 305
Pen strokes for lettering, 126
Pencil cloth, 490
Pencil lines, 59, 61
Pencil pointer, 32
Pencil shading, 304
Penciling:
  order of, 492
  stages in, 156
Pencils, 31–32
  choice, 44–45
  for lettering, 115–116
  to sharpen, 44
  sketched lines with, 157
  use of, 45
Penholders, 33
Pens, 33, 37, 42, 56
  for lettering, 126
  to sharpen, 56
  use of, 57
  (See also specific types of pens)
Pentagon, 103
  construction, 84
  to inscribe in circle, 84
Penwiper, 33
Perpendicular lines, 49, 73, 101
Perspective, 9
Perspective drawings, 266, 282–288
  angles for, 284–285
  circles, 289–290
  cruves, 290
  definitions, 284
  elevation view angle, 285
  inclined lines, 292
  inclined planes, 293
  lateral view angle, 284
  to make, 285, 288

Perspective drawings:
  measuring lines, 286
  measuring points, 291
  parallel, 288
  problems, 306–307
  sketching, 294–297, 302
    problems, 313
  terms used, 284
  theory, 282
  vanishing points, 286–287
Perspective projection, 6, 9
  rays of sight for, 132, 133
Photostat prints, 502
Pictorial charts, 416
Pictorial drawings, 6, 266–305
  axonometric, 266
  dimensioning, 20, 360–361
  dimetric, 266, 274
  isometric, 268–274
  methods, 268
  oblique, 276–278
  perspective, 189, 282–293
  problems, 313
  trimetric, 266, 275
  (See also Pictorial sketching)
Pictorial illustration, 303
Pictorial projection, three main divisions, 6
Pictorial sketching, 174, 293–304
  axes, 295
  axonometric, 294, 302
  box construction in, 299
  circles, 298
  classification, 294
  curves, 298
  diagonals, 298
  dimetric, 295
  distances in, 299
  illustration, 303
  isometric, 295
  light and shade, 303–304
  materials for, 294
  measurements, 299–301
  oblique, 294, 296, 302
  perspective, 294, 302
  principal lines, 296–297
  problems, 307–312
  proportioning, 299
  shading, 304
  stages in, 174
  steps in, 301
  for symmetry, 297
  trimetric, 296
  type of selection, 296
  view direction, 294
  working drawings, 497–501
Pictorial views, 132
Pictorials (see Pictorial drawings)
Picture plane in perspective, 284
Pictures, computer graphics, 475
Pie charts, 415

Pin and string construction of ellipse, 90
Pipe:
  cast-iron, table, A76
  conventional break for, 252
  table, A69
  (See also Piping)
Pipe bushings, table, A73
Pipe fittings, tables, A70–A77
Pipe flanges, table, A74
Pipe nipples, table, A72
Piping, table, A69
Piping drawings, 24
Piping symbols, table, A83–A84
Plan view, 134
Plane curves, 88
  problems, 107
Plane figures, problems, 104
Plane surfaces (see Planes)
Planer, 334
Planes:
  generation, 428
  inclined, in perspective, 293
  intersection, 429
    with cone, 438
    with cylinder, 437
    with lines, 429–433
    with planes, 433
  of projection, 134–135, 139
  receding, 432
  relationship, 139
  skew, intersection, 434
Planing, dimensioning, 375
Plasma panel display, 469–470
Plastic molding, dimensioning, 375
Plastics, 337, 491
Plinths, 103
Plotters, 467–469
Plumbing symbols, table, A85
Points:
  measuring, in perspective, 291
  of tangency, 74
  vanishing, 286–288
Polar charts, 409
Polygons, 103
  construction, 85
    to inscribe circle on, 87
    to transfer to new base, 83
Popular charts, 414
Position dimensions, 350
  selection, 350–352
Positional tolerances, 382
Power equations, plots of, 412
Pratt and Whitney keys, table, A61
Precision, 369
Primary auxiliary (see Normal views)
Printing of drawings, 492
Prints of drawings, 501–502
Prisms, 103
  defined, 428
  development, 449

Prisms:
  dimensioning, 355, 361
  intersection, 434–435
    with pyramid, 436
Profile line, 141
  end view, 217
Profile plane, 133
Programs:
  computer graphics, 471–473
  numerical control, 479–480
Projection:
  auxiliary, 206–210
  axonometric, 6, 266
    from orthographic views, 279
  cabinet, 278
  cavalier, 275
  dimetric, 274
  isometric, 267
  methods, 151–155
  oblique, 6, 8, 276–278
  orthographic (see Orthographic projection)
  perspective, 8, 132
  pictorial (see Pictorial sketching)
  reversed, 273–274
  trimetric, 275
  from views, 151
Projection studies, 145, 146
Projective geometry, 132
Projective theory, 132
Projectors, 133
Prolate ellipsoid, 103
Protractor, 43
Pyramids, 103
  defined, 428
  development, 449
  dimensioning, 350
  intersection, 435
    with prism, 436

Quadrilaterals, 103
Quarter scale, 51
Quarter size, 51

Radii:
  dimensioning, 364
  throat and heel, 452
Railroad pens, 41
Raster cathode-ray tube, 469
Rate-of-change chart, 409
Ratio chart, 409
Reading of drawings, 167–168
  areas, 169–171
  calculation of volume as aid, 175
  corners, 173
  defined, 167
  edges, 173
  exercises in, 177–179
  lines, 169, 171

Reading of drawings:
  by modeling, 174
  by pictorials, 174
  procedure, 168
  by sketching, 173—174
  surfaces, 169—172
Reamers, 338
  specifications, 369
Reaming, dimensioning, 375
Rear auxiliary, 222
Rear view, 137
  alternate position for, 137
Record strip, 496
Rectangle, 103
  locating center, 83
Rectangular plots for data, 411
Rectilinear charts, 406
Rectilinear coordinates, 407, 411
Reducing geometric shape, 83
Reference dimensions, 354
Reference line, 150
Reference plane for auxiliary view, 207
Refresh cathode-ray tube, 469
Regular polygons, 103
Regular solids, 103
Reinhardt alphabet, 121
Removed section, 239
Repeat lines, 59, 60
Reproduction of drawings, 495
Retaining rings, A122—A129
Return bends, pipe, table, A71
Reverse curve, to draw, 78
Reversed isometric, 273
Revolved section, 239
Rhomboid, 103
Rhombus, 103
Ribs:
  dimensioning, 354
  in section, 247
Right angle, 103
Right auxiliary, 222
Right circular cone, 88
Right cone, 103, 349
Right cylinder, 103
Right hexagonal prism, 103
Right prism, defined, 428
Right pyramid, 103, 350, 428
Right-side view, alternate position for, 138
Right surfaces, 428
Right triangular prism, 103
Ring, 428
Riveting:
  connections, A78
  driving clearances, table, A78
Rivets, 23, A105—A109
  specification, 369
  table, A58
Rolled shapes, conventional symbol, 254
Rope, conventional symbol, 254

Rotated section, 239
Rotation, difference between auxiliary projection and, 267
Roughness symbols, 386—388
Round-head bolts, table, A53
Round-head screws, tables, A54, A55, A57
Round section, conventional break for, 254
Rounds, 251
  dimensioning, 349, 354
Route charts, 413
Ruby eraser, 32
Ruled surfaces, 428
Rulers (see Scales)
Ruling pens, 34—35, 42, 56
  to sharpen, 56
  use of, 57
Running fits, table, A67
Runouts, 252

Sand castings, 327
  dimensioning, 391
  drawings of, 329
Sandpaper pad, 44
Scale, 495
  of drawings, 49—51, 348
  reading, 51—52
  statement on drawing, 51
  unnecessary, 504
Scalene triangles, 103
Scales:
  for charts, 408
  reading, 51
  types of, 34—35
  use of, 49—51
Screw-pitch gage, 500
Screw threads, 22
  Acme, table, A52
  American National, table, A48—A49
  buttress, table, A52
  coarse, table, A48—A49
  dimensioning, 375
  external, 337
  extra-fine, table, A48—A49
  fine table, A48—A49
  measurement, 500
  stub Acme, table, A52
  tables, A48—A52
  unified, table, A48—A49
Screwed pipe fittings, tables, A70—A71
Screws:
  cap, table, A54
  lock, table, A80
  machine, table, A55
  set, table, A56
  shoulder, table, A57
  specification, 369
  tables, A50, A54—A57
  wood, table, A57

Seams, sheet-metal, 448—458
  types, 458
Secondary auxiliary (see Normal views)
Section-lining, 59, 60, 236, 242
  alternate, 247
  outline, 245
  simplified, 504
Sectional views, 10, 236—248
  defined, 236
  detail in, 242
  dimensioning, 358
  drawing practices, 242—245
  (See also sections)
Sectioning (see Section-lining)
Sections, 236—250
  aligned, 249
  alignment of features, 248—250
  arms, 246
  assembly, 241
  broken-out, 239
  code for, 245
  cutting plane, 237, 242
  defined, 236
  full, 237
  half, 238
  hidden details, 242
  holes, 249—250
  identification, 236
  in isometric, 274
  lugs, 247
  materials code, 245
  parts not sectioned, 245
  problems, 254—261
  removed, 239
  ribs, 247
  rotated, 239
  spokes, 246, 247
  types of, 236
Selective assembly, 373
Semilogarithmic charts, 409
Semilogarithmic plots, 413
Serpentine, 103
Setscrews:
  points for, A56
  table, A56
Shade lines, 304
Shading, 304, 305
Shape breakdown, 176
Shape description, 132
  methods, 6—10
Shape representation, 132
Shaper, 334
Shaping, dimensioning, 375
Sharpening pencils, 44
Sheet-metal designs, 448
Sheet-metal gages, specification, 369
  table, A65
Sheet-metal parts, 326
  dimensioning, 392—394
Shop practices, 23, 326
Shop processes, 166

Short-break lines, 59, 60
Shoulder screws, table, A57
Shrink fits, table, A82
Shrink rule, 328
SI units, 11—19
  dual dimensioning, 394—396
  problems, 536—559
Side-adjacent auxiliary, 222
Side elevation, 135
Side view, 135
  alternate position for, 138
Significant figures, 374
Simplified drawing, 503—506
  abbreviations, 506
  cautions for, 506
  dimensioning in, 504—505
  freehand, 506
  scale, 503—504
  sketching, 504
  structural, 504, 505
  views, 503
Sine values, table, A26—A31
Single-curved lines, 85
Single-curved surfaces, 428
Single-stroke lettering, 114
  defined, 114
Size, 370
Size description, method, 10
Size dimensioning, 349
  selection, 350—351
Sizes for drawings, 149
Sketches:
  checking, 501
  computation, 497
  on coordinate paper, 499
  types of, 497
  working, 497—501
Sketching:
  circles, 163
  on coordinate paper, 160—162,
    499
  learning to read by, 173—174
  lines, 163—164
    horizontal, 163
    inclined, 164
    quality, 157
    straight, 162
    vertical, 163
  orthographic, 157—166, 497
    order of, 164—165
    problems, 180—187
    projection in, 165
  paper, 158—161
  pencils for, 157
  pictorial (see Pictorial sketching)
  problems, 180—202
  reading by, 173—174
  simplified, 504
  steps in, 165—166
  technique, 161—165
  working (see Working sketches)

Skew lines, 140, 215
  intersection: with plane, 430—431
    with skew surface, 431—433
Skew surfaces, 9, 205, 217
  edge views, 217—219
  intersection: with lines, 431
    with skew plane, 433
  normal views, 219—222
  on practical objects, 221—222
  problems, 228—231
  (See also Oblique surface)
Sliding fits, table, A67
Slopes, dimensioning, 363
Slotted headless setscrews, table, A56
Socket-head cap screws, table, A54
Socket-head screw, tables, A54, A56,
    A57
Solids, geometric, 103
Space dimensions, 138
Spacing of lettering, 124, 125
Specifications, 484
Speed in drawing, 490
Speedball pen, 126
Spheres, 103, 428
  development, 457
  dimensioning, 350, 367
  in isometric, 273
Spiral, 98
  of Archimedes, 98
Spokes in section, 246
Spot-faced holes, dimensioning, 366
Spot-facing tool, 337
Springs, 22
Square, 103
  to inscribe in circle, 87
Square-head bolts, table, A50
Square-head setscrews, table, A56
Square keys, tables, A58, A59
Square nuts, table, A51
Square prism, 103
Stability rule, 114
Standard drawings, 488
Standard parts, 488
Standard sheet, parts from, 337
Standard stock, parts from, 326, 334
Standing seams, 448
Station point in perspective, 284
Steel structural shapes, table, A79
Steel valves, table, A77
Step bolts, table, A53
Stock machining, 326, 334
Storage cathode-ray tube, 619
Storing of drawings, 502
Stove bolts, nuts for, table, A56
Straight-line figures, 78—85
Straight lines, to draw, 73
Strata charts, 416
Stretchout line, 448
Structural drawing, 26
  simplified, 503
Structural-steel shapes, table, A79

Stub Acme thread, table, A52
Subassembly drawing, 486
Superfinishing, 337
Surface limit, 141
Surface positions, classification,
    140—141
Surface quality, 389—390
  symbols, 391—393
Surface roughness, 391
  values, 394
Surfaces:
  auxiliary, 171
  classification, 140, 205, 428
  curved: dimensioning, 368
    intersection, 436
  double-curved, 428, 457
    development, 457
  formed, 338
  generated, 338
  inclined, 7, 206—210
    drawing, 210—213
    normal view, 206—213
    problems, 223—228
  oblique, 171
  plane (see Planes)
  projection, 140—141, 152
  reading, 169—172
  of revolution, 428
  ruled, 428
  single-curved, 428
  skew, 9, 217
    edge view, 217—219
    intersection, 431—433
    normal view, 219—222
  true shape, 141
  warped, 350, 428
    development, 457
Symbols:
  conventional, 253—254
  for electrical diagrams, table,
    A87—A91
  graphical, table, A83—A92
  for points on chart, 408
  table, A83—A92

T square, 33
  testing, 46
  use of, 45—46
Tabular drawings, 696
Tacks, 31
Tangency point, 58
Tangent method for laying out angle, 80
Tangent points, 74
Tangent values, table, A26—A31
Tangents, 58
  of circle: and line, 75
    passing through two points, 75
  to circle, 101,102
  crossed-belt, 102
  to curved line, 74, 85
  to draw, 74, 75

Tangents:
  to ellipse, 92
  to hyperbola, 96
  inking, 60
  normal to, 85
  open-belt, 102
  problems, 104—107
  to two circles, 76, 77
  to two lines, 75
Tape, 31
Taper pins, table, A64
Tapers:
  dimensioning, 349 361
  table, A64
Taps, 338
Technical manuals, A105—A129
Tees, pipe, table, A70
Tempering, 339
Templates, 506—507
Tetrahedron, 103
Texture of drawing surfaces, 157—159
Threading, 332
Threads per inch (see Screw threads)
Three-view drawings, 7
Throat radius, 452
Thumbtacks, 31
Tinner's rivets, table, A58
Title blocks, 494
Titles, 125—126
  for charts, 407, 408
  form, 495
  position, 492
Tolerances:
  for angles, 380
  bilateral, 372—374
  between centers, 379
  of concentricity, 379
  cumulative, 378
  decimal places for, 374
  defined, 371
  expression of, 373
  fits (see Fit)
  form, 383
  for manufacturing, 374—375
  maximum material condition, 382
  positional, 382
  for screw threads, 515
  selection, 375
    fundamentals, 374
  types of, 372
  unilateral, 372—374
  zone, 383
Tools:
  machine, 372
  for mass production, 339
Top-adjacent auxiliary, 222
Top view, 135
Topographic drawing, 25, 676
Torus, 103, 428

Tracing, order of, 156
Tracing cloth, 31, 490, 492
  pencil, 490
  sizes, 491—492
  use of, 156—157
Tracing paper, 31, 490
  coordinate, 160—161
  use of, 156—157
Tracings, 492—493
  duplicating, 502
Trammel construction of ellipse, 90
Transistors, symbols for, A91
Transition fits, 372
  table, A68
Transition pieces, development, 455
Transparent box, 135
Transverse lines, 296
Trapezium, 103
Trapezoid, 103
Triangles, (instrument), 33
  testing, 49
  use of, 46—49
Triangles (shape), 103
  center, to locate, 82
  to construct, 82
  equilateral: to inscribe circle in, 86
    to locate center, 82
Triangular scales, 35, 51
Triangulation:
  for development, 453—454
    of warped surfaces, 457
  for transferring polygon, 83
Triaxial diagram, 411
Trigonometric functions, table,
    A26—A31
Trilinear charts, 411
Trimetric drawings, 6, 266, 275
  from orthographic views, 8, 279—281
  sketching, 294
Trisection of line, 79
True length of line, 140
True shape of surface, 141
Truncated cone, 103
Truncated pyramid, 103
Truncated surfaces, defined, 428
Truncated triangular prism, 103
Truss-head screws, table, A55
Tubing, conventional break for, 254
Turret lathe, 335
Twist drill, 337
Two-view drawings, 7, 147

Unidirectional dimensioning, 346
Unified threads, table, A48—A49
Unilateral tolerance, 373, 374
Unions, pipe, table, A75
Unit assembly drawings, 486
Units, conversion table, A33—A46

Valves, pipe, tables, A72, A77
Vandyke paper, 501
Vanishing points, 286
Ventilating symbols, table, A84
Vertex, defined, 428
Vertical lettering, 116—119
Vertical lettering strokes, 116
Vertical lines:
  drawing, 46—47
  sketching, 161, 296
Vertical projection, 151
View directions, 139
Views:
  aligned, 250
  alternate-position, 137
  arrangement, 149
  auxiliary (see Auxiliary views)
  combination of, 137
  developed, 250
  dimensioning, 355—361
  edge, 205, 217—219
  enlarged, 358
  half, 250
  normal, 7, 206—217
  orthographic (see Orthographic
    views)
  partial, 215
  principal, 134
  projection of, 151
  relative position, 136
  sectional, 10, 236—247
  selection, 144—149
  spacing, 149
  unnecessary, 711
Volume calculations, 175
  problems, 201
Volume diagrams, 416

Waist line for lettering, 119
Warped surfaces, 103
  development, 457
  dimensioning, 350
Washers:
  lock, table, A63
  plain, table, A62
Weight calculations, 175
  problems, 201
Weights, conversion table, A33—A46
Welded parts, 327, 337
WF shapes, structural, table, A79
Width, defined, 138
Wire-frame pictures, 475
Wire gages:
  specification, 369
  table, A65
Wiring symbols, table, A86
Wood screws, table, A57
Woodruff key seats, table, A60

Woodruff keys, table, A59
Word spacing, 125
Words, lettering, 124
Working drawings, 484—501
    architectural: assembly, 484—487
        outline, 486
        unit, 486
    checking, 496
    defined, 484
    design, 484

Working drawings:
    detail, 484, 486—488
    making, 489—490
    materials for, 490—491
    pictorial, problems, 322—323
    problems, 509—535
    sketching (*see* Working sketches)
    (*See also* Drawings)
Working sketches, 497—501
    checking, 501

Working sketches:
    on coordinate paper, 499
    dimensioning, 499—501
    making, 497—499
Wrought-iron pipe, table, A69


Zone method, 458
Zone tolerances, 383
Zoning of drawings, 496